Bioactive Compounds

The potential of bioactive compounds can be unlocked through an in-depth examination of their properties in the book titled ***Bioactive Compounds: Identification and Characterization of their Food and Pharmacological Potential***. This comprehensive volume provides an in-depth examination of the complexities of bioactive compounds, offering a meticulous exploration of their identification, characterization, and multifaceted roles in food and pharmaceutical applications.

Through comprehensive analyses and illustrative case studies, this book elucidates the intricacies of isolating and understanding bioactive compounds, highlighting their multifaceted biological activities and therapeutic potential. Readers will gain insights into the latest techniques for extracting, purifying, and analyzing these compounds, crucial for identifying them in diverse natural sources.

Each chapter elucidates the scientific principles that underpin the biological effects of bioactive compounds, also addressing practical considerations for their application in functional foods, nutraceuticals, and pharmaceutical formulations. The scope of these effects encompasses antioxidants and antimicrobials, anti-inflammatory agents, and other biological activities. This scholarly work bridges the gap between theory and application, making it an indispensable resource for researchers, academics, and professionals in the fields of food science, pharmacology, and biomedical research.

Key Features

- **Comprehensive Coverage:** Explores bioactive compounds in depth, covering identification, characterization, and applications in food and pharmaceutical industries
- **Cutting-Edge Research:** Integrates the latest methods for extracting, purifying, and analyzing bioactive compounds, ensuring relevance and timeliness
- **Practical Applications:** Provides insights into developing functional foods, nutraceuticals, and pharmaceuticals with bioactive compounds, bridging theory and practice
- **Multidisciplinary Approach:** Addresses the roles of bioactive compounds from nutritional and therapeutic perspectives, appealing to researchers and professionals across fields
- **Authoritative Resource:** Offers a valuable reference for understanding and utilizing bioactive compounds in diverse scientific and industrial settings

Food Analysis & Properties
Series Editor
Leo M.L. Nollet
University College Ghent, Belgium

This CRC series **Food Analysis and Properties** is designed to provide a state-of-art coverage on topics to the understanding of physical, chemical and functional properties of foods: including (1) recent analysis techniques of a choice of food components; (2) developments and evolutions in analysis techniques related to food; (3) recent trends in analysis techniques of specific food components and/or a group of related food components.

Sequencing Technologies in Microbial Food Safety and Quality
Edited by Devarajan Thangadurai, Leo M.L. Nollet, Saher Islam, and Jeyabalan Sangeetha

Nanoemulsions in Food Technology: Development, Characterization, and Applications
Edited by Javed Ahmad and Leo M.L. Nollet

Mass Spectrometry in Food Analysis
Edited by Leo M.L. Nollet and Robert Winkler

Bioactive Peptides from Food: Sources, Analysis, and Functions
Edited by Leo M.L. Nollet and Semih Ötleş

Nutriomics: Well-being through Nutrition
Edited by Devarajan Thangadurai, Saher Islam, Leo M.L. Nollet, and Juliana Bunmi Adetunji

Analysis of Naturally Occurring Food Toxins of Plant Origin
Edited by Leo M.L. Nollet and Javed Ahmad

Analysis of Food Spices: Identification and Authentication
Edited by Leo M.L. Nollet, Javed Ahmad, and Javed Ahamad

Flavoromics: An Integrated Approach to Flavor and Sensory Assessment
Edited by Leo M.L. Nollet and Matteo Bordiga

Bioactive Compounds from Food: Benefits and Analysis
Edited by Leo M.L. Nollet and Javed Ahmad

Bioactive Compounds: Identification and Characterization of their Food and Pharmacological Potential
Edited by Mozaniel Santana de Oliveira and Leo M.L. Nollet

Handbook of Analysis and Extraction Methods of Anthocyanins
Edited by Leo M.L. Nollet and Semih Otles

For more information, please visit the Series Page: https://www.crcpress.com/Food-Analysis-Properties/book-series/CRCFOODANPRO

Bioactive Compounds

Identification and Characterization of their Food and Pharmacological Potential

Edited by
Mozaniel Santana de Oliveira
and Leo M.L. Nollet

CRC Press
Taylor & Francis Group
Boca Raton London New York

CRC Press is an imprint of the
Taylor & Francis Group, an informa business

Designed cover image: Shutterstock

First edition published 2025

by CRC Press
2385 Executive Center Drive, Suite 320, Boca Raton, FL 33431

and by CRC Press
4 Park Square, Milton Park, Abingdon, Oxon, OX14 4RN

CRC Press is an imprint of Taylor & Francis Group, LLC

ISBN: 978-1-032-59899-4 (hbk)
ISBN: 978-1-032-61267-6 (pbk)
ISBN: 978-1-003-46280-4 (ebk)

DOI: 10.1201/9781003462804

Typeset in Times New Roman
by Apex CoVantage, LLC

I dedicate this book to my parents, Maria de Oliveira and Manoel de Oliveira; my seven brothers; my wife, Joyce Fonte; and my beautiful daughter, Isabela de Oliveira.

Contents

SECTION 3 Essential Oils

SECTION 4 Fixed Oils

SECTION 5 Oil Resins

SECTION 6 Polar Compounds and Functional Foods

SECTION 7 Phenolic Compounds

About the Editors

Dr. Mozaniel Santana de Oliveira holds a bachelor's degree in chemistry from the Federal University of Pará (UFPA), obtained in 2014. He also holds a master's degree and a PhD in food science and technology from the same institution, which he obtained in 2016 and 2018, respectively. With an extensive academic record, Oliveira has authored 95 international articles, 8 international books, 45 chapters in international books, and 62 conference abstracts. His expertise spans several fields, including natural products chemistry, food science and technology, methods for prospecting bioactive molecules, natural products biotechnology, and allelopathy. Dr. Oliveira has an H-index in Scopus of 21, Web of Science h-index of 21, and Google Scholar h-index of 26.

Currently, Dr. Oliveira is a professor in the graduate program in biological sciences, with an emphasis in tropical botany, at the Federal Rural University of the Amazon/Museu Paraense Emílio Goeldi (UFRA/MPEG). He is also an associate professor in the graduate programs of food science and technology and pharmaceutical sciences at UFPA. Previously, he served as an ambassador for Bentham Science Publishers and as an academic mentor for Web of Science.

Dr. Oliveira also makes significant contributions as a peer reviewer, having reviewed over 413 articles for 70 international journals. As an editorial board member, he has edited more than 250 scientific articles for journals such as *Future Integrative Medicine*, *PLoS One*, *The Natural Products Journal*, *International Journal of Food Science*, *Plants-Basel*, *Advances in Pharmacological and Pharmaceutical Sciences*, *Open Chemistry*, *Frontiers in Pharmacology*, *Frontiers in Plant Science*, *Frontiers in Nutrition*, *Molecules*, *Journal of Food Quality*, and *Journal of Herbal Medicine*. Currently, Oliveira is a postdoctoral researcher in the graduate program in biological sciences, specializing in tropical botany, at the Federal Rural University of the Amazon/Museu Paraense Emílio Goeldi (UFRA/MPEG).

Leo M.L. Nollet earned an MS (1973) and a PhD (1978) in biology from the Katholieke Universiteit Leuven, Belgium. He is an editor and associate editor of numerous books. He edited for M. Dekker, New York—now CRC Press of the Taylor & Francis Publishing Group—the first, second, and third editions of *Food Analysis by HPLC* and *Handbook of Food Analysis*. The last edition is a two-volume book. Dr. Nollet also edited the *Handbook of Water Analysis* (first, second, and third editions) and *Chromatographic Analysis of the Environment*, third and fourth editions (CRC Press). With F. Toldrá, he coedited two books published in 2006, 2007, and 2017: *Advanced Technologies for Meat Processing* (CRC Press) and *Advances in Food Diagnostics*. With M. Poschl, he coedited the book *Radionuclide Concentrations in Foods and the Environment*, also published in 2006 (CRC Press). Dr. Nollet has also coedited with Y. H. Hui and other colleagues on several books: *Handbook of Food Product Manufacturing* (2007); *Handbook of Food Science, Technology, and Engineering* (CRC Press, 2005); *Food Biochemistry and Food Processing* (first and second editions, 2006 and 2012); and the *Handbook of Fruits and Vegetable Flavors* (2010). In addition, he edited the *Handbook of Meat, Poultry, and Seafood Quality*, first and second editions (2007 and 2012). From 2008 to 2011, he published five volumes on animal product–related books with F. Toldrá: *Handbook of Muscle Foods Analysis*, *Handbook of Processed Meats and Poultry Analysis*, *Handbook of Seafood and Seafood*

Products Analysis, *Handbook of Dairy* Foods Analysis (second edition, 2021), and *Handbook of Analysis of Edible Animal By-Products*. Also, in 2011, with F. Toldrá, he coedited two volumes for CRC Press: *Safety Analysis of Foods of Animal Origin* and *Sensory Analysis of Foods of Animal Origin*. In 2012, they published the *Handbook of Analysis of Active Compounds in Functional Foods*. In a coedition with Hamir Rathore, *Handbook of Pesticides: Methods of Pesticides Residues Analysis* was marketed in 2009; *Pesticides: Evaluation of Environmental Pollution* in 2012; *Biopesticides Handbook* in 2015; and *Green Pesticides Handbook: Essential Oils for Pest Control* in 2017. Other finished book projects include *Food Allergens: Analysis, Instrumentation, and Methods* (with A. van Hengel; CRC Press, 2011) and *Analysis of Endocrine Compounds in Food* (2011). Dr. Nollet's recent projects include *Proteomics in Foods* with F. Toldrá (2013) and *Transformation Products of Emerging Contaminants in the Environment: Analysis, Processes, Occurrence, Effects, and Risks* with D. Lambropoulou (2014). In the series Food Analysis & Properties, he edited (with C. Ruiz-Capillas) *Flow Injection Analysis of Food Additives* (CRC Press, 2015) and *Marine Microorganisms: Extraction and Analysis of Bioactive Compounds* (CRC Press, 2016). With A. S. Franca, he coedited *Spectroscopic Methods in Food Analysis* (CRC Press, 2017), and with Horacio Heinzen and Amadeo R. Fernandez-Alba, he coedited *Multiresidue Methods for the Analysis of Pesticide Residues in Food* (CRC Press, 2017). Further volumes in the series Food Analysis & Properties are *Phenolic Compounds in Food: Characterization and Analysis* (with Janet Alejandra Gutierrez-Uribe, 2018), *Testing and Analysis of GMO-containing Foods and Feed* (with Salah E. O. Mahgoub, 2018), *Fingerprinting Techniques in Food Authentication and Traceability* (with K. S. Siddiqi, 2018), *Hyperspectral Imaging Analysis and Applications for Food Quality* (with N. C. Basantia, Leo M.L. Nollet, and Mohammed Kamruzzaman, 2018), *Ambient Mass Spectroscopy Techniques in Food and the Environment* (with Basil K. Munjanja, 2019), *Food Aroma Evolution: During Food Processing, Cooking, and Aging* (with M. Bordiga, 2019), *Mass Spectrometry Imaging in Food Analysis* (2020), *Proteomics in Food Authentication* (with S. Ötleş, 2020), *Analysis of Nanoplastics and Microplastics in Food* (with K. S. Siddiqi, 2020), *Chiral Organic Pollutants, Monitoring and Characterization in Food and the Environment* (with Edmond Sanganyado and Basil K. Munjanja, 2020), *Sequencing Technologies in Microbial Food Safety and Quality* (with Devarajan Thangardurai, Saher Islam, Jeyabalan Sangeetha, 2021), *Nanoemulsions in Food Technology: Development, Characterization, and Applications* (with Javed Ahmad, 2021), *Mass Spectrometry in Food Analysis* (with Robert Winkler, 2022), *Bioactive Peptides from Food: Sources, Analysis, and Functions* (with Semih Ötles, 2022), and *Nutriomics: Well-being through Nutrition* (with Devarajan Thangadurai, Saher Islam, Juliana Bunmi Adetunji, 2022). In 2023, he published with Javed Ahmad *Analysis of Naturally Occurring Food Toxins of Plant Origin*.

The last published books are *Biopesticides Handbook*, 2nd ed. (with S. Mir), *Handbook of Seafood and Seafood Products Analysis* (with F. Toldrá), *Flavoromics, an Integrated to Flavor and Sensory Assessment* (with M. Bordiga), *Analysis of Food Spices* (with J. Ahmad and J. Ahamad), *Global Regulations of Medicinal, Pharmaceutical, and Food Products* (with F. Ali), and *Bioactive Compounds from Food Benefits and Analysis* (with J. Ahmad).

Contributors

Javed Ahamad
Department of Pharmacy
Tishk International University
Erbil, Iraq

Javed Ahmad
College of Pharmacy
Najran University
Najran, Saudi Arabia

Showket Ahmad Dar
Faculty of Forestry
Ganderbal-SKUAST-K
Srinagar, India

Lorena C. Albernaz
Laboratório de Farmacognosia
Universidade de Brasília
Brasilia, Brazil

Ammar Al-Farga
College of Science
University of Jeddah
Jeddah, KSA

Zaryab Ali
Charoen Pokphand Pakistan Pvt. Ltd.
Lahore, Pakistan

Ayah T. Abdullah Al-Kilani
Faculty of Pharmacy
Petra University
Amman, Jordan

Ivone M. C. Almeida
Departamento de Ciências Química
Universidade de Porto
Porto, Portugal

Mohmmad Al-Salem
Faculty of Agriculture
Jordan University of Science and Technology
Irbid, Jordan

Rita C. Alves
Departamento de Ciências Química
Universidade de Porto
Porto, Portugal

Eloisa Helena Aguiar Andrade
Adolpho Ducke Laboratory
Museu Paraense Emilio Goelde
Belém, Brazil

Mohammad Aslam
Faculty of Pharmacy
Al Hawash Private University
Al-Mouzeina, Syria

Dev Karan Bairwa
SKN College of Agriculture
Jobner, India

João Paulo Barreto Sousa
Laboratório de Farmacognosia
Universidade de Brasília
Brasília, Brazil

Leonardo Oliveira Bittencourt
Institute of Biological Sciences
Federal University of Pará
Belém, Brazil

Fábio José Bonfim Cardoso
Multidisciplinary Chemistry Laboratory
Federal University of Pará
Belém, Brazil

Hina Bukhari
University of Agriculture
Faisalabad, Pakistan

Anna Rafaela Cavalcante Braga
Nutrition and Food Research Center
Universidade Federal de São Paulo
São Paulo, Brazil

Kangjam Bumpy
Pandit Deen Dayal Upadhyay Institute of Agriculture Sciences
Yarou Meitram, India

Monize Bürck
Nutrition and Food Research Center
Universidade Federal de São Paulo
São Paulo, Brazil

Renan Campos Chisté
Institute of Technology
Federal University of Pará
Belém, Brazil

Carlos Ariel Cardona Alzate
Departamento de Ingeniería Química
Universidad Nacional de Colombia
Bogotá, Colombia

Xu Chen
School of Pharmacy
Linyi University
Linyi, P. R. China

Muhammad Farhan Jahangir Chughtai
Khwaja Fareed University of Engineering and Information Technology
Rahim Yar Khan, Pakistan

Márlia Coelho-Ferreira
Laboratory of Ethnobotany
Museu Paraense Emílio Goeldi
Belém, Brazil

Julia Helena da Silva Martins
Federal University of Pará
Faculty of Electrical Engineering (FEE)
Belém, Brazil

Luiza Helena Da Silva Martins
Federal Rural University of Amazonia
Belém, Brazil

Joyce Kelly da Silva
Programa de Pós-Graduação em Biotecnologia
Federal University of Pará
Belém, Brazil

Rafaela da Trindade
Programa de Pós-Graduação em Biotecnologia
Federal University of Pará
Belém, Brazil

Rogério M. Dallago
Department of Food Engineering
URI Erechim
Fátima, Brazil

Georgeos Deeb
Faculty of Science
Damascus University
Damascus, Syria

Cristina Delerue-Matos
Instituto Superior de Engenharia do Porto
Instituto Politécnico do Porto
Porto, Portugal

Marielba de Los Angeles Rodriguez Salazar
Institute of Technology
Federal University of Pará
Belém, Brazil

Carolina E. Demaman Oro
Department of Food Engineering
URI Erechim
Fátima, Brazil

Jonilson de Melo e Silva
Instituto Federal do Pará
Belém, Brazil

Paula Maria Correa de Oliveira
Laboratório de Farmacognosia
Universidade de Brasília
Brasília, Brazil

Anderson de Santana Botelho
Adolpho Ducke Laboratory
Museu Paraense Emilio Goelde
Belém, Brazil

Karyme do Socorro de Souza Vilhena
Adolpho Ducke Laboratory
Museu Paraense Emilio Goelde
Belém, Brazil

Yendrembam K. Devi
Department of Entomology
Lovely Professional University
Phagwara, India

Maicon Sérgio N. dos Santos
Laboratory of Agroindustrial Processes Engineering
Federal University of Santa Maria
Santa Maria, Brazil

Akanksha Dwivedi
Acropolis Institute of Pharmaceutical Education and Research
Indore, India

Hany S. Abd El-Raouf
Faculty of Agriculture
Al-Azhar University
Cairo, Egypt

Laila Salmen Espindola
Laboratório de Farmacognosia
Universidade de Brasília
Brasília, Brazil

Ahmed M. El-Taher
Faculty of Agriculture
Al-Azhar University
Cairo, Egypt

Oberdan Oliveira Ferreira
Adolpho Ducke Laboratory
Museu Paraense Emilio Goelde
Belém, Brazil

Ellem Suane Ferreira Alves
Universidade Federal do Pará
Belém, Brazil

Ingrid Ferreira de Souza
Nutrition and Food Research Center
Universidade Federal de São Paulo
São Paulo, Brazil

Fernanda Wariss Figueiredo Bezerra
Postgraduate Program in Food Science and Technology
Federal University of Pará
Belém, Brazil

Ebtesam A. Gashash
Department of Chemistry,
Faculty of Arts and Science in Balijurashi
Al-Baha University
Baha, Saudi Arabia

Magally Araújo Gonçalves
Instituto de Saúde e Produção Animal
Universidade Federal Rural da Amazônia
Belém, Brazil

Saulo Edgar Gomes de Souza
Institute of Technology
Federal University of Pará
Belém, Brazil

Letícia Guerreiro Trindade
Department of Chemical Engineering
Universidade Federal de São Paulo
São Paulo, Brazil

Shaofen Guo
School of Pharmacy
Linyi University
Linyi, P. R. China

Wajid Hasan
Bihar Agriculture University
Bhagalpur, India

Kounser Javeed
Faculty of Agriculture, Wadura
FOA-Wadura
Sopore, India

Bruna Kaori Feitosa Tanaka
Department of Biosciences
Universidade Federal de São Paulo
São Paulo, Brazil

Bruno Douglas Kerber
Laboratory of Agroindustrial Processes Engineering
Federal University of Santa Maria
Santa Maria, Brazil

Rakhi Khabiya
Acropolis Institute of Pharmaceutical Education and Research
Indore, India

Andrea Komesu
Universidade de Campinas (UNICAMP)
Campinas, Brazil

Kátia Lamarão-Vieira
Institute of Biological Sciences
Federal University of Pará
Belém, Brazil

Yanli Li
School of Pharmacy
Linyi University
Linyi, P. R. China

Rafael Rodrigues Lima
Institute of Biological Sciences
Federal University of Pará
Belém, Brazil

I. Yimjenjang Longkumer
Department of Entomology
Lovely Professional University
Phagwara, India

Mônica Masako Nakamoto
Nutrition and Food Research Center
Universidade Federal de São Paulo
São Paulo, Brazil

Mohd. Amir Mirza
Department of Pharmaceutics
School of Pharmaceutical Education and Research
Jamia Hamdard
New Delhi, India

Naveena Nazim
GDC–Khanabal
Anantnag, India

Vinicius Ruan Neves dos Santos
Institute of Biological Sciences
Federal University of Pará
Belém, Brazil

Renata Nolasco Braga Souto
Nutrition and Food Research Center
Universidade Federal de São Paulo
São Paulo, Brazil

Felipe Oliveira Nunes
Department of Chemistry
Federal University of Mato Grosso do Sul
Campo Grande, Brazil

Elivaldo Nunes Modesto Junior
Institute of Technology
Federal University of Pará
Belém, Brazil

M. Beatriz P. P. Oliveira
Departamento de Ciências Química
Universidade de Porto
Porto, Portugal

M. Teresa Oliva-Teles
Porto Higher Institute of Engineering R
Dr. António Bernardino de Almeida 431
Porto, Portugal

Magdy A. Omar
Faculty of Agriculture
Al-Azhar University
Cairo, Egypt

Mariana Ortiz-Sánchez
Departamento de Ingeniería Química
Universidad Nacional de Colombia
Bogotá, Colombia

Marcilene Paiva da Silva
Adolpho Ducke Laboratory
Museu Paraense Emilio Goelde
Belém, Brazil

Imran Pasha
University of Agriculture
Faisalabad, Pakistan

Sara Piedrahita-Rodríguez
Departamento de Ingeniería Química
Universidad Nacional de Colombia
Bogotá, Colombia

Roberta S. Pinho
Departamento de Biología
Universidade Federal Rural de Pernambuco
Recife, Brazil

Lissara Polano Ody
Laboratory of Agroindustrial Processes Engineering
Federal University of Santa Maria
Santa Maria, Brazil

Devyani Rajput
Department of Pharmaceutical Sciences
Dr. Harisingh Gour Vishwavidyalaya
Sagar, India

Rizwan Rashid
Division of Vegetable Science
Ganderbal-SKUAST-K
Srinagar, India

Daissy Lorena Restrepo Serna
Departamento de Ingeniería Química
Universidad Nacional de Colombia
Bogotá, Colombia

Johnatt Allan Rocha de Oliveira
Universidade Federal do Pará
Belém, Brazil

Ingryd Rodrigues Martins
Instituto Federal do Pará
Belém, Brazil

Hervé Rogez
College of Biotechnology
Federal University of Pará
Belém, Brazil

Kanza Saeed
Khwaja Fareed University of Engineering and Information Technology
Rahim Yar Khan, Pakistan

Joana Santos
Departamento de Ciências Química
Universidade de Porto
Porto, Portugal

William N. Setzer
Department of Chemistry
University of Alabama
Huntsville, Alabama

Suzene I. Silva
Departamento de Biología
Universidade Federal Rural de Pernambuco
Recife, Brazil

Cristian Kallahan Silva Chagas
Institute of Biological Sciences
Federal University of Pará
Belém, Brazil

Norton M. Silveira
Laboratory of Agroindustrial Processes Engineering
Federal University of Santa Maria
Santa Maria, Brazil

Priyanka Soni
Faculty of Pharmacy
B. R. Nahata College of Pharmacy
Bhunyakhedi, India

Vishal Soni
Faculty of Pharmacy
B. R. Nahata College of Pharmacy
Bhunyakhedi, India

Leonardo Souza da Costa
Adolpho Ducke Laboratory
Museu Paraense Emilio Goelde
Belém, Brazil

Deiweson Souza-Monteiro
Institute of Biological Sciences
Federal University of Pará
Belém, Brazil

Zhen Wang
School of Pharmacy
Linyi University
Linyi, P. R. China

Wallacy Watson Pereira Melo
Institute of Biological Sciences
Federal University of Pará
Belém, Brazil

Wenqiang Yang
School of Pharmacy
Linyi University
Linyi, P. R. China

Xiuling Yu
School of Pharmacy
Linyi University
Linyi, P. R. China

Giovani L. Zabot
Laboratory of Agroindustrial Processes Engineering
Federal University of Santa Maria
Santa Maria, Brazil

Syed Naseem Zaffar
Faculty of Forestry
Ganderbal-SKUAST-K
Srinagar, India

Muhammad Zuhair
University of Agriculture
Faisalabad, Pakistan

Series Preface

There will always be a need to analyze food compounds and their properties. Current trends in analyzing methods include automation, increasing the speed of analyses, and miniaturization. Over the years, the unit of detection has evolved from micrograms to pictograms.

A classical pathway of analysis is sampling, sample preparation, cleanup, derivatization, separation, and detection. At every step, researchers are working and developing new methodologies. A large number of papers are published every year on all facets of analysis, so there is a need for books that gather information on one kind of analysis technique or on the analysis methods for a specific group of food components.

The scope of the CRC Series on Food Analysis & Properties aims to present a range of books edited by distinguished scientists and researchers who have significant experience in scientific pursuits and critical analysis. This series is designed to provide state-of-the-art coverage on topics such as:

1. Recent analysis techniques on a range of food components
2. Developments and evolution in analysis techniques related to food
3. Recent trends in analysis techniques for specific food components and/or a group of related food components
4. The understanding of physical, chemical, and functional properties of foods

The book *Bioactive Compounds: Identification and Characterization of their Food and Pharmacological Potential* is volume number 25 of this series.

I am happy to be a series editor of such books, for the following reasons:

- I am able to pass on my experience in editing high-quality books related to food.
- I get to know colleagues from all over the world more personally.
- I continue to learn about interesting developments in food analysis.

Much work is involved in the preparation of a book. I have been assisted and supported by a number of people, all of whom I would like to thank. I would especially like to thank the team at CRC Press/Taylor & Francis, with a special word of thanks to Steve Zollo, senior editor.

Many, many thanks to all the editors and authors of this volume and future volumes. I very much appreciate all their effort, time, and willingness to do a great job.

I dedicate this series to:

- My wife, for her patience with me (and all the time I spend on my computer)
- All patients suffering from prostate cancer

Knowing what this means, I am hoping they will have some relief.

Dr. Leo M.L. Nollet (Retired)
University College Ghent
Ghent, Belgium

Preface

Because of their broad range of uses and significant potential for the amelioration of human health and well-being, bioactive compounds caught a growing interest from the science community as well as industry during the previous years. These compounds, extensively from natural sources, have drawn much attention owing to their prospective roles in nutrition and interventions for the prevention of diseases. The demand for natural bioactive compounds in the global market is developing rapidly, so several requirements have emerged: an increasing need for an updated and extensive reference tome to help students, researchers, or any professional working in this vibrant field.

The book *Bioactive Compounds: Identification and Characterization of Their Food and Pharmacological Potential*, edited by Mozaniel Santana de Oliveira and Leo M.L. Nollet, provides a contemporary account of the methods in this field as well as expresses an exciting vision for future developments. Featuring the research of leading experts in plant and marine bioactives, it starts with an introductory overview before going on to cover fruits, vegetables, and algae for their functionality (e.g., skin health) and formulation methods, including nanoencapsulation, which are particularly useful for increasing solubility while maintaining stability throughout extraction processes.

The book is divided into seven sections, each covering a diverse array of topics, including natural compounds, methods of extraction, and their applications in the food industry and pharmacology, with 22 chapters in total. Each chapter is thoughtfully designed to give a thorough overview of its specific subject. This structure ensures that readers gain a strong understanding of both the fundamental principles and the latest research developments related to bioactive compounds.

SECTION 1 NATURAL COMPOUNDS

This section offers an in-depth exploration of natural bioactive compounds, highlighting their sources, compositions, and various applications. It emphasizes the significance of plants and microalgae as abundant sources of these compounds and delves into their potential uses in the food and pharmaceutical industries.

SECTION 2 SUPERCRITICAL FLUID EXTRACTION

This section concentrates on the cutting-edge technique of supercritical fluid extraction, celebrated for its efficiency and environmental friendliness. The chapters explore the use of this method in extracting bioactive compounds from agri-food products and by-products. They highlight recent advancements and prefeasibility analyses for various metabolites, such as polyphenols.

SECTION 3 ESSENTIAL OILS

This section delves into the various applications of essential oils, particularly emphasizing their role in the food industry. It discusses the challenges and innovations related to using essential oils for food preservation, controlling plant pathogens, and the pharmacological properties obtained from medicinal and food plants.

SECTION 4 FIXED OILS

This part is all about solidified oils, and it describes the potential health benefits and medicinal values of carotenoid-rich vegetable oils found in the Amazon, as well as those from Brazil's semi-arid areas. Additionally, it gives a full account of andiroba oil, which is obtained from the Amazon, including detailed description and examination of its pharmaceutical properties.

SECTION 5 OIL RESINS

The chapters in this section provide a review of the phytochemistry and pharmacology of *Copaifera* species and other oil resins. The authors discuss the characterization, chemical compositions, food applications, and biological activities of these resins, emphasizing their significance and potential uses.

SECTION 6 POLAR COMPOUNDS AND FUNCTIONAL FOODS

This section addresses the pharmacological activities of various polar compounds, including terpenoids, flavonoids, and alkaloids. This section highlights the advancements, trends, applications, and safety of these compounds, with a particular focus on the functional food potential of açaí, which is not only a traditional food in the Amazon but also a promising raw material for pharmaceutical innovation.

SECTION 7 PHENOLIC COMPOUNDS

In the last section of this study, the actor phenolic compounds in plants, its biosynthetic pathway, and its pharmacological activities are discussed. The text then moves on to look at the antibacterial role of resin that bees use, and concludes with talks about lignin uses in drug/gene delivery and tissue engineering.

This book is illustrated with rich empirical data, case studies, and practical insights that make it an invaluable reference for both academia and as well as industry. This volume presents a collection of diverse and interdisciplinary chapters in the field of bioactive compounds research. This type of strategy underscores the broad potential uses for these agents and stresses that a united front will be required in moving forward with human utilization of natural bioactive substances.

This book is anticipated to act as a sore point for the field of bioactive compounds to advance further than before, becoming the trigger for untrodden new directions in research and applications which will be of benefit to society at large, as this chemical composition has applications in various industries and agriculture. The book has a cornucopia of information and ideas that help researchers, scientists, and professionals in the food and pharmaceutical industries excel even if they are new. It helps attain mastery of the discoverable bioactive compounds.

Dr. Mozaniel Santana de Oliveira and Dr. Leo M.L. Nollet
Editors, Bioactive Compounds: Identification and Characterization of their Food and Pharmacological Potential

Acknowledgments

We would like to thank all the contributors to this project, each chapter author, and other contributors.

I would like to thank CRC Press for their support during the execution of the project.

Editor Mozaniel Santana de Oliveira thanks PDPG-POSDOC—Programa de Desenvolvimento da Pós-Graduação (PDPG) Pós-Doutorado Estratégico—as well as CAPES for the scholarship (process number: (88887.852405/2023-00).

Section 1

Natural Compounds

1 Natural Compounds
Sources, Composition, and Applications

Javed Ahamad

1.1 INTRODUCTION

Natural compounds from ancient times are used for the prevention and treatment of human ailments. They are an important part of human civilization, used as foods, health supplements, nutraceuticals, and in the treatment of various human diseases (Nobili et al., 2009; Rai et al., 2021; Kussmann and Cunha, 2022). Bioactive compounds are obtained from various natural sources, such as plants, animals, and microbes. Minerals and marine products are also considered natural sources of bioactive compounds (Azimova, 2013). Mankind has been exploring natural resources and using them as cures for illnesses since prehistoric times. Traditional medical practices, including homeopathy, Chinese medicine, Unani, and Ayurveda, are utilized to treat illnesses. Traditional systems medicine (complementary alternative medicine) cures human ailments by utilizing different natural materials, including trees, herbs, minerals, and animal products (Houghton, 1995; Arnason et al., 1981; Pieroni, 2017).

Historically, most of the new drugs have been discovered from natural products and compounds derived from natural products. During the past 15 years, pharmaceutical industry research into natural products has declined, in part because of an emphasis on high-throughput screening of synthetic libraries (Lahlou, 2007). Currently, there is a substantial decline in new drug approvals and an impending loss of patent protection for important medicines. However, untapped biological resources, "smart screening" methods, robotic separation with structural analysis, metabolic engineering, and synthetic biology offer exciting technologies for new natural product drug discovery. Advances in rapid genetic sequencing, coupled with manipulation of biosynthetic pathways, may provide a vast resource for the future discovery of pharmaceutical agents (Atanasov et al., 2021).

Natural products and their structural analogs have historically made a major contribution to pharmacotherapy, especially for cancer and infectious diseases. Nevertheless, natural products also present challenges for drug discovery, such as technical barriers to screening, isolation, characterization, and optimization, which contributed to a decline in their pursuit by the pharmaceutical industry from the 1990s onward (Clark, 1996; Li et al., 2009). In recent years, several technological and scientific developments, including improved analytical tools, genome mining and engineering strategies, and microbial culturing advances, have addressed such challenges and opened new opportunities. Consequently, interest in natural products as drug leads is being revitalized, particularly for tackling antimicrobial resistance (Newman et al., 2003). Considering the focus of the current book, we comprehensively included plants and animals used as foods as sources of bioactive natural compounds in the present chapter.

1.2 FOOD AS A SOURCE OF NATURAL BIOACTIVE COMPOUNDS

Plants are the major source of food for mankind (Figure 1.1). Plants contain simple molecules synthesized *via* photosynthesis and are known as primary plant metabolites, for example,

DOI: 10.1201/9781003462804-2

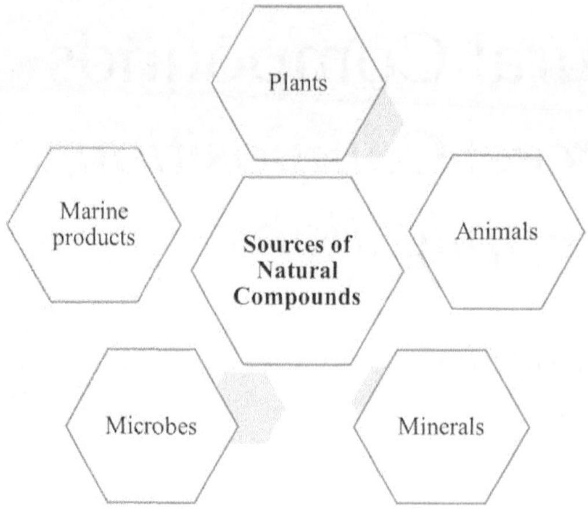

FIGURE 1.1 Different sources of natural products.

simple sugars, lipids, proteins, and peptides (Rates, 2001; Zaynab et al., 2019; Ahamad et al., 2023). These primary plant metabolites contain high-energy bonds that provide energy and make structural components of the body. These primary plant metabolites provide energy (carbohydrates, lipids, fats), make components of cell membranes (lipids), help in the development of the human body (proteins and peptides), etc. These primary plant metabolites further lead to the biogenesis of complex plant compounds known as secondary plant metabolites, and these compounds are used as medicine, nutraceuticals, and health supplements (Shakya, 2016; Erb and Kliebenstein, 2020).

A healthy food diet is essential in maintaining human health. Bioactive compounds present in plants, besides providing nutrition, also play an important role in disease prevention through antioxidants, anti-inflammation, neuroprotection, and immunomodulation (de Ridder et al., 2017; Cena and Calder, 2020). Different categories of secondary plant metabolites are present in food products, such as anthocyanins, stilbenes, polyphenols, flavonoids, phytosterols, cucurbitacins, proteins, peptides, lipids, and terpenes (González Mera et al., 2019). These classes of natural compounds are known for their health benefits. The phytochemicals isolated from food plants are used for the prevention and treatment of various human diseases, including chronic diseases, such as cancer, diabetes, infections, and cardiovascular complications (Poiroux-Gonord et al., 2010; Badyal et al., 2020). Table 1.1 comprises a list of bioactive compounds in different plants consumed as food.

1.3 PLANTS AS A SOURCE OF NATURAL BIOACTIVE COMPOUNDS

Plants are considered major sources of bioactive natural compounds, and these are utilized as food, spice, and remedies for the treatment of human diseases (Verpoorte, 2000). Plants consist of a vast variety of secondary plant metabolites belonging to different categories, such as alkaloids, glycosides, tannins, lipids, terpenes, polyphenols, flavonoids, etc. (Figure 1.2) (Ahamad et al., 2021). These bioactive compounds play important roles in the interaction of plants with their environment, particularly with their biotic environment, where they work as a defense against natural enemies and attractants for pollinators. Thousands of bioactive compounds are isolated and characterized from natural sources; for example, 22,000 terpenes, 21,000 alkaloids, and 5,000 flavonoids and tannins are known (Kroymann, 2011).

TABLE 1.1
Bioactive Compounds in Food Plants, Their Sources, Chemical Class, and Pharmacological Properties

Category	Bioactive Compounds	Plants	Pharmacological Actions	Ref.
Flavonoids	Quercetin, kaempferol, myricetin, catechin, epicatechin	Fruits and berries (e.g., grape, apple, banana, olive, apricate, strawberry, peach, and orange) Vegetables, legumes, and nuts (e.g., tomatoes, spinach, sugar beet, turnips, peanuts, soybeans) Spices and herbs (e.g., tea, coffee, thyme, sage, basil, oregano)	Flavonoids are potent antioxidants and are used for the prevention and treatment of metabolic syndrome and various forms of cancers.	Terahara, 2015; Yao et al., 2004; Guven et al., 2019; Gupta et al., 2023
Polyphenols	Oleuropein, hydroxytyrosol, epigallocatechin gallate, and chlorogenic acid	Olive, green tea, coffee	Antioxidants, metabolic disorders, neuroprotective, and cardioprotective.	Ahamad et al., 2019; Zhang et al., 2021; Habauzit and Morand, 2012; Silva and Pogačnik, 2020
Anthocyanins	Malvidin, delphinidin, cyanidin, petunidin, pelargonidin, peonidin	Apples, onion, grapes, soybeans, blackberries, cranberries, blueberries	Anthocyanins are potent antioxidant molecules used for the treatment of various cancers, metabolic disorders, and CVD.	Shrikhande and Francis, 1976; Mannino et al., 2021; Wang and Stoner, 2008; Jiang et al., 2019
Stilbenes	Resveratrol, pterostilbene, piceatannol, piceid, pinosylvin, and viniferin	Grapes, blackberries, cranberries, blueberries, peanuts, sprouts	Stilbenes are used for the treatment of chronic human ailments, such as cancer, diabetes, and CVD.	Goufo et al., 2020; Baur and Sinclair, 2006; Frémont, 2000; Szkudelska and Szkudelski, 2010
Cucurbitacins	Cucurbitacins A, B, C, D, K, E, etc.	Bitter gourd, pumpkin, cucumber, melon, etc.	Anticancer, antidiabetic, anti-inflammatory, and for CVD.	Hussain et al., 2019; Ahamad et al., 2017; Jayaprakasam et al., 2003
Carotenoids	α-carotene, β-carotene, crocin, crocetin, lutein, and α-cryptoxanthin	Carrots, saffron, papaya, mandarin orange, tomatoes	Carotenoids are used to treat cancer, coronary heart disease, skin and eye diseases, and to improve immunity. Source of vitamin A.	Britton et al., 2009; Saini et al., 2015; Johnson, 2002; Nishino et al., 2002
Phytosterols	*β*-sitosterol, stigmasterol, campesterol, sitostanol, and campestanol	Bitter gourd, corn, rice, peanut, linseed, and mustard	Antidiabetic, anticancer, hypolipidemic, and cardioprotective.	Woyengo et al., 2009; Ostlund, 2002; Lagarda et al., 2006; Ogbe et al., 2015; Marangoni and Poli, 2010

(*Continued*)

TABLE 1.1 *(Continued)*
Bioactive Compounds in Food Plants, Their Sources, Chemical Class, and Pharmacological Properties

Category	Bioactive Compounds	Plants	Pharmacological Actions	Ref.
Resins	Gingerol, shogaol, and zingerone; capsaicin and capsanthin	Ginger and capsicum	As spice and for the treatment of cough; antiseptic, antibacterial. Capsicum is used as a counterirritant.	Abascal and Yarnell, 2009; Dissanayake et al., 2020
Polysaccharides	Pectin, starch	*Citrus* sp., apples, berries; potato, rice, wheat, and corn.	Nutrition, and pectins are used as probiotic, anticancer, and antihyperlipidemic.	Kamal et al., 2021; Blanco-Pérez et al., 2021; Zhang et al., 2015; Carvalho, 2008

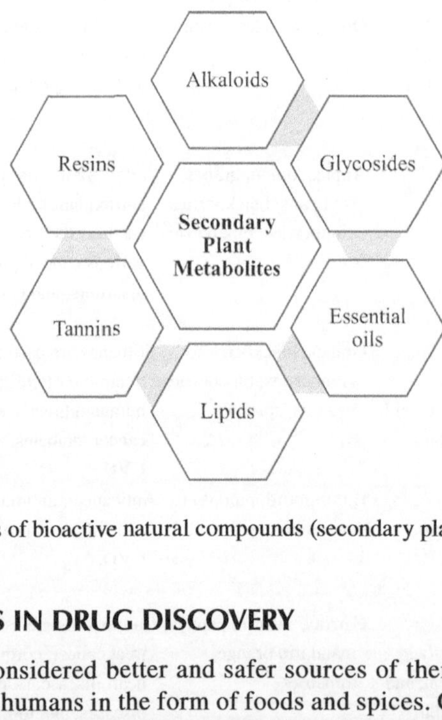

FIGURE 1.2 Various classes of bioactive natural compounds (secondary plant metabolites).

1.4 ROLE OF PLANTS IN DRUG DISCOVERY

Natural compounds are considered better and safer sources of therapeutic agents due to their longtime association with humans in the form of foods and spices. Chemical compounds found in natural sources have great chemical diversity and have many biological actions. The chemical diversity of natural compounds plays an important role in drug discovery and development (Farnsworth et al., 1985; Rates, 2001). The drug derived from natural products offers options for prevention and treatment of diseases still uncurable, such as different forms of cancer, metabolic disorders such as diabetes, obesity, cardiovascular diseases, and infections. In general, natural compounds derived from plants are safer than synthetic medications. Additionally, plant-derived secondary metabolites are thought to possess a greater chemical diversity and to be superior to synthetic combinatorial chemicals (Wolfender and Queiroz, 2012; Singh and Pelaez, 2008). Several natural compounds, including vincristine, vinblastine, paclitaxel, docetaxel, and others, have recently demonstrated cytotoxic and antitumor properties in clinical trials and have been effectively developed into anticancer medications (Dubey et al., 2004). Today, finding drugs from plants is a costly and time-consuming procedure. To secure patent protection, phytochemicals derived from medicinal plants are utilized as lead compounds in the drug discovery process and, subsequently, in the production of synthetic or semisynthetic drugs (Thomford et al., 2018). As an alternative to conventional medical care, traditional medicines have gained popularity in recent years as therapies for chronic illnesses and lifestyle disorders, like diabetes, obesity, cancer, tuberculosis, malaria, and cardiovascular problems. A total number of 51% new drugs approved by the FDA from 1981 to 2014 only came from natural products–based drugs (Newman and Cragg, 2016; Ahmad and Ahamad, 2020). For example, a list of some important drugs that are developed from natural sources is presented in Table 1.2.

1.5 ANIMAL-DERIVED NATURAL BIOACTIVE COMPOUNDS

Animals produce a vast variety of natural bioactive compounds with diverse pharmacological activities. The major bioactive compounds derived from animals include polyunsaturated fatty acids (e.g., omega-3 fatty acids, including α-linolenic acid, eicosapentaenoic acid (EPA), and docosahexaenoic acid (DHA), and omega-6 fatty acids, like linoleic acid, γ-linolenic acid, etc.), polysaccharides and amino-polysaccharides (e.g., chitin and chitosan), and proteins and peptides from animal milk. These bioactive compounds have several health benefits in humans, such as cardioprotective,

TABLE 1.2
Bioactive Phytochemicals Derived from Medicinal Plants

S. No.	Plant Source	Natural Compounds	Pharmacological Uses
1.	*Adonis vernalis* L.	Adoniside	Cardiotonic
2.	*Aesculus hippocastanum* L.	Aescin	Anti-inflammatory
3.	*Agrimonia eupatoria* L.	Agrimophol	Anthelmintic
4.	*Ammi majus* L.	Xanthototoxin	Leukoderma, vitiligo treatment
5.	*Ammi visnaga* L.	Khellin	Bronchodilator
6.	*Anamirtacocculus* L.	Picrotoxin	Analeptic
7.	*Ananascomosus* L.	Bromelain	Anti-inflammatory
8.	*Andrographis paniculata* N.	Andrographolide	Bacillary dysentery
9.	*Anisodus tanguticus* Maxim.	Anisodamine, anisodine	Anticholinergic
10.	*Ardisia japonica* Bl.	Bergenin	Antitussive
11.	*Areca catechu* L.	Arecoline	Anthelmintic
12.	*Artemisia maritima* L.	Santonin	Ascaricide
13.	*Artemisia annua* L.	Artemisinin	Antimalarial
14.	*Atropa belladonna* L.	Atropine	Anticholinergic
15.	*Berberis vulgaris* L.	Berberine	Bacillary dysentery
16.	*Brassica nigra* L.	Allyl isothiocyanate	Rubefacient
17.	*Camptotheca acuminata* D	Camptothecin	Anticancer
18.	*Camellia sinensis* L.	Caffeine	CNS stimulant, diuretic
19.	*Carica papaya* L.	Papain, chymopapain	Proteolytic, mucolytic
20.	*Catharanthus roseus* L.	Vincristine, vinblastine	Anticancer

(Continued)

TABLE 1.2 *(Continued)*
Bioactive Phytochemicals Derived from Medicinal Plants

S. No.	Plant Source	Natural Compounds	Pharmacological Uses
21.	*Cassia* spp.	Sennosides A and B	Laxative
22.	*Centella asiatica* L.	Asiaticoside	Vulnerary, nervine tonic
23.	*Cephaelis ipecacuanha*	Emetine	Amebicide, emetic
24.	*Chondodendron tomentosum*	Tubocurarine	Skeletal muscle relaxant
25.	*Cinchona* spp.	Quinine, quinidine	Antimalarial, antiarrhythmic
26.	*Colchium autumnale* L.	Colchicine	Antitumor agent, antigout
27.	*Convallaria sessiliflora* L.	Convoallotoxin	Cardiotonic
28.	*Crotolaria sessiliflora* L.	Monocrotaline	Antitumor agent
29.	*Curcuma longa* L.	Curcumin	Choleretic
30.	*Cynara scolyumus* L.	Cynarin	Choleretic
31.	*Datura metel* L.	Scopolamine	Sedative
32.	*Digenea simplex* Wulf.	Kainic acid	Ascaricide
33.	*Digitalis lanata* Ehrh.	Lanatosides, digoxin	Cardiotonic
34.	*Digitalis purpurea* L.	Digoxin, digitoxin, gitalin	Cardiotonic
35.	*Ephedra sinica* Stapf.	Ephedrine, pseudoephedrine	Sympathomimetic
36	*Erythroxylum coca* Lamk.	Cocaine	Local anesthetic
37.	*Fraxinus rhynchophylla* H.	Aesculetin	Antidysentery
38.	*Glycyrrhiza globra* L.	Glycyrrhizin	Sweetener
39.	*Gossypium* spp.	Gossypol	Male contraceptive
40.	*Hydrangea macrophylla*	Phyllodulcin	Sweetener
41.	*Hydrastis canadensis* L.	Hydrastine	Hemostatic, astringent
42.	*Hyoscamus niger* L.	Hyoscamine	Anticholinergic
43.	*Lobelia inflata* L.	Lobeline	Smoking deterrent, respiratory stimulant
44.	*Papaver somniferum* L.	Morphine, codeine, noscapine	Analgesic, antitussive
45.	*Pausinystalia yohimbe*	Yohimbine	Aphrodisiac
46.	*Physostigma venenosum* Balf.	Physostigmine	Cholinesterase inhibitor
47	*Pilocarpus jaborandi* Holmes	Pilocarpine	Parasympathomimetic
48.	*Podophyllum peltatum* L.	Etoposide, teniposide podophyllotoxin	Antitumor agent
49.	*Potentilla fragaroides* L.	(+)-Catechin	Hemostatic
50.	*Rauvolfia serpentina* L.	Ajmalicine	Circulatory disorders
51.	*Rauvolfia serpentina* L.	Reserpine	Antihypertensive, tranquilizer
52.	*Salix alba* L.	Salicin	Analgesic
53.	*Silybum marianum* L.	Silymarin	Antihepatotoxic
54.	*Simarouba glauca* DC.	Glaucaroubin	Amebicide
55.	*Stephania sinica* Diels	Rotundine	Analgesic; sedative
56.	*Stevia rebaudiana*	Stevioside	Sweetener
57.	*Strophanthus gratus* Baill.	Ouabain	Cardiotonic
58.	*Strychnos nux-vomica* L.	Strychnine	CNS stimulant
59.	*Taxus brevifolia* Nutt.	Taxol	Anticancer
60.	*Theobroma cacao* L.	Theobromine	Diuretic, bronchodilator
61.	*Thymus vulgaris* L.	Trichosanthin	Abortifacient
62.	*Urginea matritima* L.	Scillarin A	Cardiotonic
63.	*Valeriana officinalis* L.	Valepotriates	Sedative
64.	*Veratrum album* L.	Protoveratrins A and B	Antihypertensive

TABLE 1.3
Animal-Derived Bioactive Natural Compounds

Animal-Derived Bioactives	Source	Health Benefits	Ref.
Polyunsaturated fatty acids (e.g., EPA, DHA, DPA, linoleic acid, linolenic acids, etc.)	Fish oils (cod, salmon, herring, etc.), meat, and dairy products	Cardioprotective, anticancer, antidiabetic, etc.	Kapoor et al., 2021; Lunn and Theobald, 2006; Siriwardhana et al., 2012
Chitin and chitosan	Crab and shrimp shells	Antimicrobial, anticancer, hypolipidemic, and preservative	Jeon et al., 2000; Kumar, 2000
Milk peptides (e.g., lactoferrins, phosphopeptides, lactoferricin, immunopeptides, etc.)	Milk of cow, buffalo, and sheep	Opioid antagonist, antimicrobial, immunostimulant, anticarcinogenic, etc.	Meisel, 2005

anti-inflammatory, antidiabetic, anticancer, immunomodulatory, etc. (Jeon et al., 2000; Kapoor et al., 2021; Lunn and Theobald, 2006). See Table 1.3.

1.6 CONCLUSION

Natural products have served as a major source of drugs that are used for the prevention and treatment of various human conditions, including chronic conditions, such as cancer, diabetes, hypertension, cardiovascular diseases, arthritis, etc. Various kinds of food, like fruits, vegetables, berries, cereals, legumes, etc., contain many bioactive compounds belonging to different classes that are responsible for their therapeutic benefits. Several bioactive compounds are isolated and characterized from these food materials. Still, there is a need to discover more natural product–based drugs for the prevention and treatment of human ailments, like cancer, metabolic syndrome, COVID-19, etc., because natural products provide more chemical diversity and are safer compared to synthetic drugs.

REFERENCES

Abascal K, Yarnell E. Clinical uses of *Zingiber officinale* (ginger). *Alternative and Complimentary Therapies*. 2009; 15(5): 231–7.

Ahamad J, Ahmad J, Nollet LM. Food spices: An overview. *Analysis of Food Spices*. 2023: 3–18.

Ahamad J, Amin S, Mir SR. *Momordica charantia* Linn. (Cucurbitaceae): Review on phytochemistry and pharmacology. *Phytochemistry*. 2017; 11(2): 53–65.

Ahamad J, Toufeeq I, Khan MA, Ameen MS, Anwer ET, Uthirapathy S, Mir SR, Ahmad J. Oleuropein: A natural antioxidant molecule in the treatment of metabolic syndrome. *Phytotherapy Research*. 2019; 33(12): 3112–28.

Ahmad J, Ahamad J, editors. *Bioactive Phytochemicals: Drug Discovery to Product Development*. Bentham Science Publishers; 2020.

Arnason T, Hebda RJ, Johns T. Use of plants for food and medicine by Native Peoples of eastern Canada. *Canadian Journal of Botany*. 1981; 59(11): 2189–325.

Atanasov AG, Zotchev SB, Dirsch VM, The International Natural Product Sciences Taskforce, Supuran CT. Natural products in drug discovery: Advances and opportunities. *Nature Reviews Drug Discovery*. 2021; 20: 200–16.

Azimova SS and Yunusov MS, editors. *Natural Compounds Alkaloids: Plant Sources, Structure and Properties*. Springer; 2013.

Badyal S, Singh H, Yadav AK, Sharma S, Bhushan I. Plant secondary metabolites and their uses. *Plant Archives*. 2020; 20(2): 3336–40.

Baur JA, Sinclair DA. Therapeutic potential of resveratrol: The in vivo evidence. *Nature Reviews Drug Discovery*. 2006; 5(6): 493–506.

Blanco-Pérez F, Steigerwald H, Schülke S, Vieths S, Toda M, Scheurer S. The dietary fiber pectin: Health benefits and potential for the treatment of allergies by modulation of gut microbiota. *Current Allergy and Asthma Reports*. 2021; 21: 1–9.

Britton G, Khachik F. Carotenoids in food. *Carotenoids: Volume 5: Nutrition and Health*. Basel: Birkhäuser Basel; 2009 Dec 29: 45–66.

Carvalho AJ. Starch: Major sources, properties and applications as thermoplastic materials. *Monomers, Polymers and Composites from Renewable Resources*. 2008: 321–42.

Cena H, Calder PC. Defining a healthy diet: Evidence for the role of contemporary dietary patterns in health and disease. *Nutrients*. 2020; 12(2): 334.

Clark AM. Natural products as a resource for new drugs. *Pharmaceutical Research*. 1996; 13: 1133–41.

de Ridder D, Kroese F, Evers C, Adriaanse M, Gillebaart M. Healthy diet: Health impact, prevalence, correlates, and interventions. *Psychology & Health*. 2017; 32(8): 907–41.

Dissanayake KG, Waliwita WA, Liyanage RP. A review on medicinal uses of *Zingiber officinale* (ginger). *International Journal of Health Sciences and Research*. 2020; 10(6): 142–8.

Dubey NK, Kumar R, Tripathi P. Global promotion of herbal medicine: India's opportunity. *Current Science*. 2004; 86(1): 37–41.

Erb M, Kliebenstein DJ. Plant secondary metabolites as defenses, regulators, and primary metabolites: The blurred functional trichotomy. *Plant Physiology*. 2020; 184(1): 39–52.

Farnsworth NR, Akerele O, Bingel AS, Soejarto DD, Guo Z. Medicinal plants in therapy. *Bulletin of the World Health Organization*. 1985; 63: 965–81.

Frémont L. Biological effects of resveratrol. *Life Sciences*. 2000; 66(8): 663–73.

González Mera IF, González Falconí DE, Morera Córdova V. Secondary metabolites in plants: Main classes, phytochemical analysis and pharmacological activities. *Bionatura*. 2019; 4(4): 1000–9.

Goufo P, Singh RK, Cortez I. A reference list of phenolic compounds (including stilbenes) in grapevine (*Vitis vinifera* L.) roots, woods, canes, stems, and leaves. *Antioxidants*. 2020; 9(5): 398.

Gupta M, Ahmad J, Ahamad J, Kundu S, Goel A, Mishra A. Flavonoids as promising anticancer therapeutics: Contemporary research, nanoantioxidant potential, and future scope. *Phytotherapy Research*. 2023; 37(11): 5159–92.

Guven H, Arici A, Simsek O. Flavonoids in our foods: A short review. *Journal of Basic and Clinical Health Sciences*. 2019; 3(2): 96–106.

Habauzit V, Morand C. Evidence for a protective effect of polyphenols-containing foods on cardiovascular health: An update for clinicians. *Therapeutic Advances in Chronic Disease*. 2012; 3(2): 87–106.

Houghton PJ. The role of plants in traditional medicine and current therapy. *The Journal of Alternative and Complementary Medicine*. 1995; 1(2): 131–43.

Hussain H, Green IR, Saleem M, Khattak KF, Irshad M, Ali M. Cucurbitacins as anticancer agents: A patent review. *Recent Patents on Anti-Cancer Drug Discovery*. 2019; 14(2): 133–43.

Jayaprakasam B, Seeram NP, Nair MG. Anticancer and antiinflammatory activities of cucurbitacins from *Cucurbita andreana*. *Cancer Letters*. 2003; 189(1): 11–6.

Jeon YJ, Shahidi F, Kim SK. Preparation of chitin and chitosan oligomers and their applications in physiological functional foods. *Food Reviews International*. 2000; 16(2): 159–76.

Jiang X, Li X, Zhu C, Sun J, Tian L, Chen W, Bai W. The target cells of anthocyanins in metabolic syndrome. *Critical Reviews in Food Science and Nutrition*. 2019; 59(6): 921–46.

Johnson EJ. The role of carotenoids in human health. *Nutrition in Clinical Care*. 2002; 5(2): 56–65.

Kamal MM, Kumar J, Mamun MA, Ahmed MN, Shishir MR, Mondal SC. Extraction and characterization of pectin from *Citrus sinensis* peel. *Journal of Biosystems Engineering*. 2021; 46: 16–25.

Kapoor B, Kapoor D, Gautam S, Singh R, Bhardwaj S. Dietary polyunsaturated fatty acids (PUFAs): Uses and potential health benefits. *Current Nutrition Reports*. 2021; 10: 232–42.

Kroymann J. Natural diversity and adaptation in plant secondary metabolism. *Current Opinion in Plant Biology*. 2011; 14(3): 246–51.

Kumar MN. A review of chitin and chitosan applications. *Reactive and Functional Polymers*. 2000; 46(1): 1–27.

Kussmann M, Cunha DH. Nature has the answers: Discovering and validating natural bioactives for human health. *Bioactive Compounds in Health and Disease*. 2022; 5(11): 222–35.

Lagarda MJ, García-Llatas G, Farré R. Analysis of phytosterols in foods. *Journal of Pharmaceutical and Biomedical Analysis*. 2006; 41(5): 1486–96.

Lahlou M. Screening of natural products for drug discovery. *Expert Opinion on Drug Discovery*. 2007; 2(5): 697–705.

Li JW, Vederas JC. Drug discovery and natural products: End of an era or an endless frontier? *Science*. 2009; 325(5937): 161–5.

Lunn J, Theobald HE. The health effects of dietary unsaturated fatty acids. *Nutrition Bulletin*. 2006; 31(3): 178–224.

Mannino G, Gentile C, Ertani A, Serio G, Bertea CM. Anthocyanins: Biosynthesis, distribution, ecological role, and use of biostimulants to increase their content in plant foods—a review. *Agriculture*. 2021; 11(3): 212.

Marangoni F, Poli A. Phytosterols and cardiovascular health. *Pharmacological Research*. 2010; 61(3): 193–9.

Meisel H. Biochemical properties of peptides encrypted in bovine milk proteins. *Current Medicinal Chemistry*. 2005; 12(16): 1905–19.

Newman DJ, Cragg GM. Natural products as sources of new drugs from 1981 to 2014. *Journal of Natural Products*. 2016; 79(3): 629–61.

Newman DJ, Cragg GM, Snader KM. Natural products in therapeutic. *Journal of Natural Products*. 2003; 66: 1022.

Nishino H, Murakoshi M, Ii T, Takemura M, Kuchide M, Kanazawa M, Yang Mou X, Wada S, Masuda M, Ohsaka Y, Yogosawa S. Carotenoids in cancer chemoprevention. *Cancer and Metastasis Reviews*. 2002; 21: 257–64.

Nobili S, Lippi D, Witort E, Donnini M, Bausi L, Mini E, Capaccioli S. Natural compounds for cancer treatment and prevention. *Pharmacological Research*. 2009; 59(6): 365–78.

Ogbe RJ, Ochalefu DO, Mafulul SG, Olaniru OB. A review on dietary phytosterols: Their occurrence, metabolism and health benefits. *Asian Journal of Plant Science & Research*. 2015; 5(4):10–21.

Ostlund Jr RE. Phytosterols in human nutrition. *Annual Review of Nutrition*. 2002; 22(1): 533–49.

Pieroni A. Traditional uses of wild food plants, medicinal plants, and domestic remedies in Albanian, Aromanian and Macedonian villages in South-Eastern Albania. *Journal of Herbal Medicine*. 2017; 9: 81–90.

Poiroux-Gonord F, Bidel LP, Fanciullino AL, Gautier H, Lauri-Lopez F, Urban L. Health benefits of vitamins and secondary metabolites of fruits and vegetables and prospects to increase their concentrations by agronomic approaches. *Journal of Agricultural and Food Chemistry*. 2010; 58(23): 12065–82.

Rai N, Kumari Keshri P, Verma A, Kamble SC, Mishra P, Barik S, Kumar Singh S, Gautam V. Plant associated fungal endophytes as a source of natural bioactive compounds. *Mycology*. 2021; 12(3): 139–59.

Rates SM. Plants as source of drugs. *Toxicon*. 2001; 39(5): 603–13.

Saini RK, Nile SH, Park SW. Carotenoids from fruits and vegetables: Chemistry, analysis, occurrence, bioavailability and biological activities. *Food Research International*. 2015; 76: 735–50.

Shakya AK. Medicinal plants: Future source of new drugs. *International Journal of Herbal Medicine*. 2016; 4(4): 59–64.

Shrikhande AJ, Francis FJ. Anthocyanins in foods. *Critical Reviews in Food Science & Nutrition*. 1976; 7(3): 193–218.

Silva RF, Pogačnik L. Polyphenols from food and natural products: Neuroprotection and safety. *Antioxidants*. 2020; 9(1): 61.

Singh SB, Pelaez F. Biodiversity, chemical diversity and drug discovery. *Natural Compounds as Drugs Volume I*. 2008: 141–74.

Siriwardhana N, Kalupahana NS, Moustaid-Moussa N. Health benefits of n-3 polyunsaturated fatty acids: Eicosapentaenoic acid and docosahexaenoic acid. *Advances in Food and Nutrition Research*. 2012; 65: 211–22.

Szkudelska K, Szkudelski T. Resveratrol, obesity and diabetes. *European Journal of Pharmacology*. 2010; 635(1–3): 1–8.

Terahara N. Flavonoids in foods: A review. *Natural Product Communications*. 2015; 10(3): 1934578X1501000334.

Thomford NE, Senthebane DA, Rowe A, Munro D, Seele P, Maroyi A, Dzobo K. Natural products for drug discovery in the 21st century: Innovations for novel drug discovery. *International Journal of Molecular Sciences*. 2018; 19(6): 1578.

Verpoorte R. Pharmacognosy in the new millennium: Lead finding and biotechnology. *Journal of Pharmacy and Pharmacology*. 2000; 52: 253–62.

Wang LS, Stoner GD. Anthocyanins and their role in cancer prevention. *Cancer Letters*. 2008; 269(2): 281–90.

Wolfender JL, Queiroz EF. Chemical diversity of natural resources and the bioactivity of their constituents. *Chimia*. 2012; 66(5): 324.

Woyengo TA, Ramprasath VR, Jones PJ. Anticancer effects of phytosterols. *European Journal of Clinical Nutrition*. 2009; 63(7): 813–20.

Yao LH, Jiang YM, Shi J, Tomas-Barberan FA, Datta N, Singanusong R, Chen SS. Flavonoids in food and their health benefits. *Plant Foods for Human Nutrition*. 2004; 59: 113–22.

Zaynab M, Fatima M, Sharif Y, Zafar MH, Ali H, Khan KA. Role of primary metabolites in plant defense against pathogens. *Microbial Pathogenesis*. 2019; 137: 103728.

Zhang L, Han Z, Granato D. Polyphenols in foods: Classification, methods of identification, and nutritional aspects in human health. *Advances in Food and Nutrition Research*. 2021; 98: 1–33.

Zhang W, Xu P, Zhang H. Pectin in cancer therapy: A review. *Trends in Food Science & Technology*. 2015; 44(2): 258–71.

2 Plants as a Source of Bioactive Compounds

Javed Ahamad

2.1 INTRODUCTION

Higher plants are an important part of human necessities, like food, clothes, and shelter. Besides providing fulfilling human needs, they are also important sources of therapeutics, nutritional supplements, various vitamins, and minerals. Plants are an important source of organic compounds that are synthesized through photosynthesis (Evans, 2009). Plants synthesize primary plant metabolites, which further leads to the biogenesis of various kinds of complex chemical compounds known as secondary plant metabolites (Sitarek et al., 2020; Nunes et al., 2020). Most of the secondary plant metabolites are biologically active and used for the prevention and treatment of human ailments (Ahamad, 2021).

The most important classes of secondary plant metabolites include alkaloids (e.g., vincristine, vinblastine, quinine, morphine, cocaine, lobeline, hyoscine, reserpine, etc.), glycosides (e.g., digitoxin, panaxoside, charantin, glycyrrhizin, etc.), terpenes (e.g., citral, camphor, menthol, carvone, geraniol, eugenol, fenchone, etc.), resins (e.g., etoposide, gingerol, capsaicin, etc.), tannins (e.g., arjunolic acid, hematoxylin, catechin, mangiferin, etc.), lipids (e.g., linoleic acid, linolenic acid, oleic acid, EPA, DHA, etc.), flavonoids (e.g., rutin, quercitrin, quercetin, kaempferol, apigenin, luteolin, etc.), etc. (Pinheiro et al., 2018; Masoro, 1977).

The bioactive compounds of plants, since ancient times, have been a major source of therapeutic agents. Due to the increase in side effects and adverse effects of synthetic drugs, natural products are gaining popularity in the treatment of chronic disorders, such as cancer (e.g., vincristine, vinblastine, paclitaxel, topotecan, etc.), malaria (e.g., quinine, artemisinin and its derivatives), etc. (Ahmad and Ahamad, 2020; Esmeeta et al., 2022). The present chapter highlights the sources, chemistry, and therapeutic applications of important bioactive compounds of alkaloids, glycosides, resins, tannins, lipids, terpenes, and flavonoids.

2.2 SOURCE, CHEMISTRY, AND MEDICINAL USES OF BIOACTIVE COMPOUNDS

In recent years, advances in extraction, isolation, and characterization techniques have significantly increased the number of new lead molecules from natural products, especially from plants. The important classes of secondary plant metabolites that have more lead molecules or bioactive compounds include alkaloids, glycosides, terpenes, tannins, resins, flavonoids, polyphenols, lipids, etc. These classes of bioactive compounds are discussed in detail in what follows:

2.2.1 Alkaloids

Higher plants contain many bioactive secondary metabolites. *Alkaloids* are considered as the most potent secondary plant metabolites and are abundantly found in plants. Alkaloids are "alkali-like" complex compounds present in plants and a few animals. These are derived from amino acids, contain nitrogen atoms in their heterocycle, and have marked pharmacological actions (Pelletier, 1983).

DOI: 10.1201/9781003462804-3

Alkaloids are subdivided into true, proto-, and pseudo-alkaloids, depending on their biogenesis. Those alkaloids derived from amino acids and that have a nitrogen atom in their heterocycle are known as true alkaloids, those derived from amino acids and do not contain a nitrogen atom in the cycle are known as proto-alkaloids, and the third ones, which are not derived from amino acids, are called pseudo-alkaloids (Cordell, 2008). *Alkaloids* are potent bioactive compounds and are a source of several potential natural products used for the treatment of many human ailments, such as cancer (e.g., vincristine, vinblastine), malaria (e.g., quinine), pain (as analgesic, for example, morphine, and as anesthetic, for example, cocaine), asthma (e.g., lobeline), motion sickness (e.g., hyoscine), dysentery (e.g., ergotamine), hypertension (e.g., reserpine), etc. (Street et al., 2017; Gutiérrez-Grijalva et al., 2020; Thibane and Mudau, 2022). Table 2.1 and Figure 2.1 comprise important bioactive alkaloids present in higher plants.

2.2.2 Terpenes and Terpenoids

Essential oils (volatile oils) are odorous, volatile principles of plants and animals. They evaporate when exposed to air at ordinary temperature, and these are chemically derived from terpenes (*terpenes* are hydrocarbons derived from two isoprene units, C_5H_{10}) and their oxygenated derivatives, known as terpenoids. Essential oils usually consist of monoterpene ($C_{10}H_{16}$), sesquiterpene

TABLE 2.1
Important Bioactive Alkaloids' Occurrence in Higher Plants

Bioactive Alkaloids	Common Name	Botanical Source	Therapeutic Uses
Vincristine and vinblastine	Vinca	*Catharanthus roseus* (Apocynaceae)	Anticancer
Morphine and codeine	Opium	*Papaver somniferum* (Papaveraceae)	Analgesic and antitussive
Colchicine	Colchicum	*Colchicum autumnale* (Liliaceae)	Treatment of gout and cancer
Ephedrine	Ephedra	*Ephedra sinica* (Ephedraceae)	Treatment of asthma
Piperine	Black pepper	*Piper nigrum* (Piperaceae)	Treatment of gonorrhea and chronic bronchitis
Cocaine	Coca	*Erythroxylum coca* (Erythroxylaceae)	Local anesthetics
l-Hyoscyamine, hyoscine, atropine	Datura, belladona, and Hyoscyamus	*Datura stramonium, D. metal, Atropa belladonna,* and *Hyoscyamus niger* (Solanaceae)	Anticholinergic, CNS depressant effects, motion sickness, and ophthalmic uses
Emetine	Ipecac	*Cephaelis acuminata* (Rubiaceae)	Expectorant, emetic, and in the treatment of amoebic dysentery
Reserpine	Rauwolfia	*Rauwolfia serpentina* (Apocynaceae)	Treatment of high blood pressure
Quinine	Cinchona	*Cinchona officinalis* (Rubiaceae)	Antimalarial
Caffeine	Tea and coffee	*Thea sinensis* (Ternstroemiaceae) and *Coffea arabica* (Rubiaceae)	CNS stimulant and diuretic

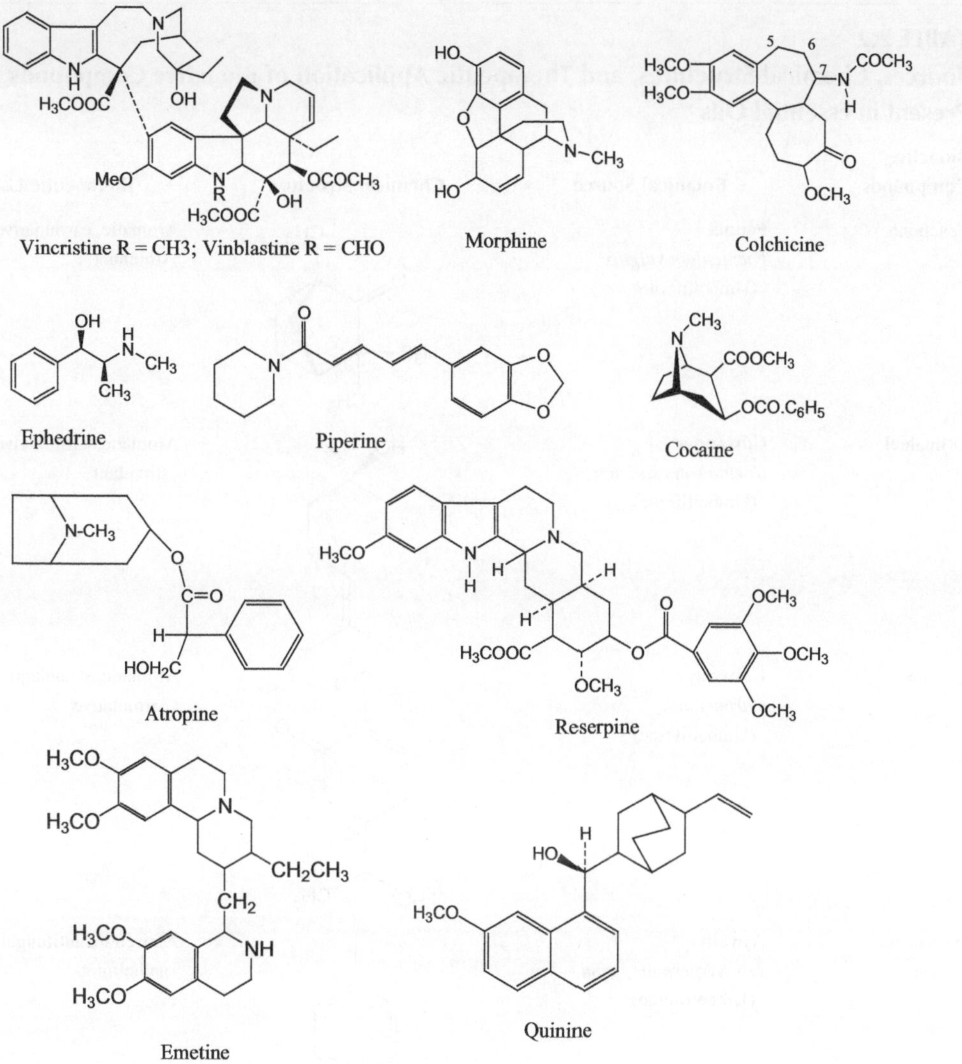

FIGURE 2.1 Chemical structures of important alkaloids.

($C_{15}H_{24}$), and diterpenes ($C_{20}H_{32}$) (Ahamad et al., 2023). Essential oils are mainly found in plant families, such as Labiatae (e.g., basil, mentha), Zingiberaceae (e.g., cardamom, ginger), Umbelliferae (e.g., coriander, fennel, caraway, cumin), Myrtaceae (e.g., clove, eucalyptus), Lauraceae (e.g., cinnamon), and Graminae (e.g., lemongrass) (Ahamad et al., 2023; Zuzarte and Salgueiro, 2015).

Essential oils have great importance in perfumery, cosmetics, food, and the pharmaceutical industry. Therapeutically, these are used as antiseptics, stimulants, carminatives, diuretics, anthelmintics, analgesics, antirheumatics, aromatics, and counter-irritants. Sometimes these are also used as repellant and deodorant. These are also used as important medicinal agents for therapeutic purposes, like carminatives (umbelliferous fruits), anthelmintics (*Chenopodium* oil), diuretics (juniper), antiseptics (eucalyptus), counter-irritants (oil of wintergreen), local anesthetics (clove), sedatives (jatamansi), and as a source of vitamin A (lemongrass) (Başer and Demirci, 2007; Jugreet et al., 2020). Table 2.2 presents various bioactive compounds of essential oils, their sources, and their therapeutic uses.

TABLE 2.2
Sources, Chemical Structures, and Therapeutic Application of Bioactive Compounds Present in Essential Oils

Bioactive Compounds	Botanical Source	Chemical Structure	Therapeutic Uses
Fenchone	**Fennel** *Foeniculum vulgare* (Umbelliferae)	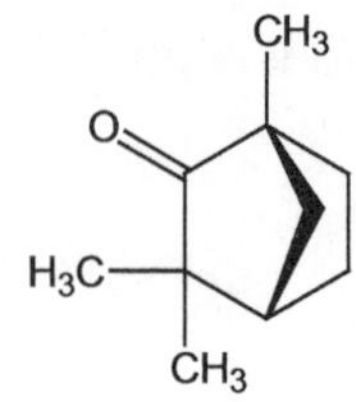	Aromatic, carminative, stimulant
D-linalool	**Coriander** *Coriandrum sativum* (Umbelliferae)		Aromatic, carminative, stimulant
Carvone	**Caraway** *Carum carvi* (Umbelliferae)	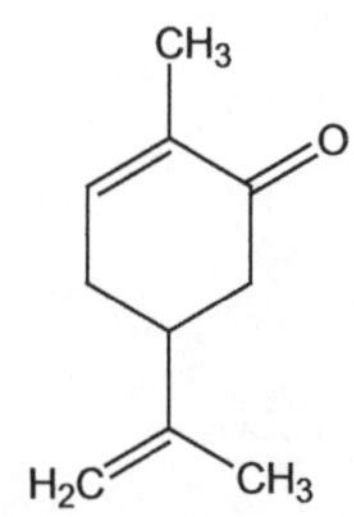	Aromatic, stimulant, carminative
p-Cymene	**Ajwain** *Trachyspermum ammi* (Umbelliferae)	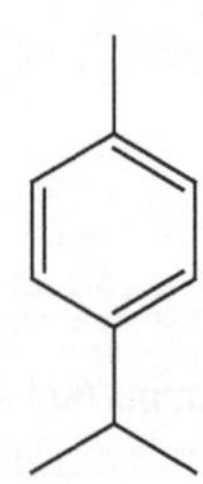	Antiseptic, antifungal, anthelmintic
Menthol	**Peppermint** *Mentha piperita* (Labiatae)	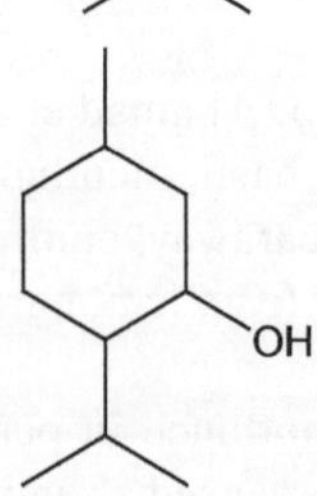	Carminative, flavoring agents
Borneol	**Rosemary** *Rosmarinus officinalis* (Labiatae)	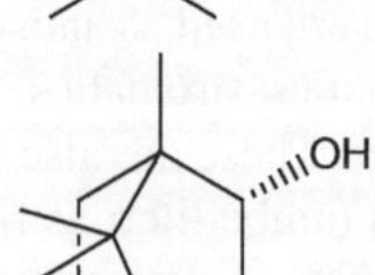	Rubefacient

TABLE 2.2 ***(Continued)***
Sources, Chemical Structures, and Therapeutic Application of Bioactive Compounds Present in Essential Oils

Bioactive Compounds	Botanical Source	Chemical Structure	Therapeutic Uses
Thymol	**Thyme** *Thymus vulgaris* (Labiatae)	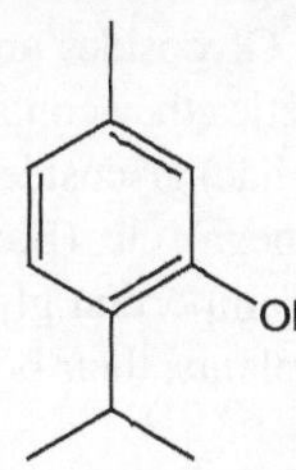	Expectorant, antispasmodic, carminative
Linalyl acetate	**Lavender** *Lavandula angustifolia* (Labiatae)	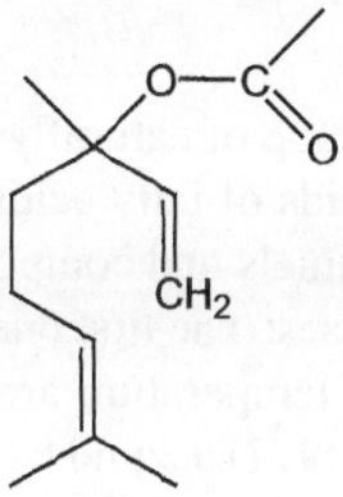	Aromatic, carminative, flavoring agent
Cinnamaldehyde	**Cinnamon** *Cinnamomum zeylanicum* (Lauraceae)	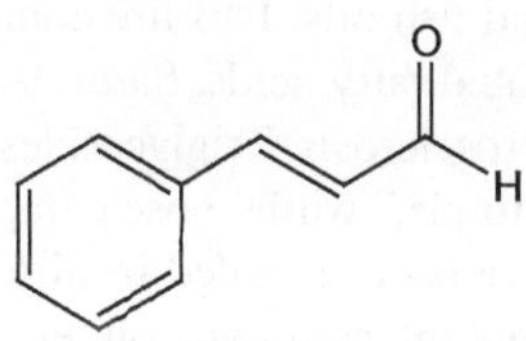	Spice, carminative
Eugenol	**Clove** *Eugenia caryophyllus* (Myrtaceae)	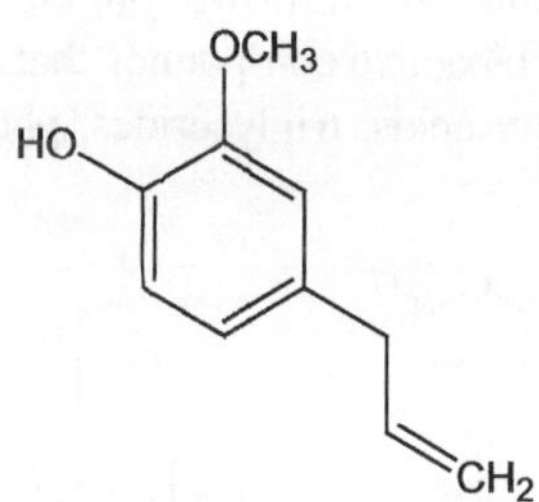	Dental analgesic, antiseptic
1,8-cineol	**Eucalyptus** *Eucalyptus globulus* (Myrtaceae)	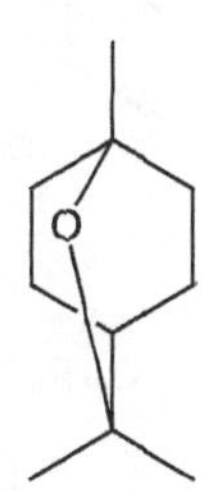	Counter-irritant, antiseptic, expectorant
Citral	**Lemongrass** *Cymbopogon flexuosus* (Graminae)	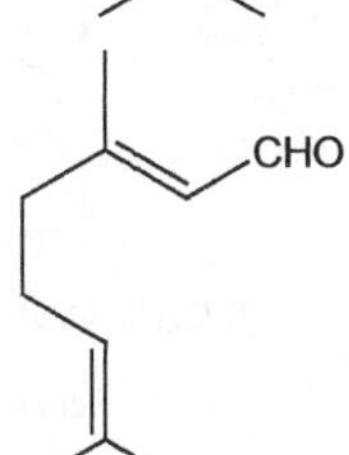	Flavoring agent, source of vitamin A

2.2.3 Glycosides

Glycosides are a diverse naturally secondary plant metabolites, and they contain sugar (glycone) bonded with non-sugar (aglycone or genin) parts. The non-sugar part of a glycoside is responsible for pharmacological actions, and the sugar helps in the absorption and distribution of aglycone (Prassas and Diamandis, 2008). The most common sugar found in glycoside is *β-D*-glucose, and sometimes they also contain diverse types of sugar molecules, such as digitoxose, rhamnose, galactose, mannose, etc. (Kren and Martínková, 2001). Glycosides are categorized into several subclasses, depending on the different types of aglycone moieties they contain, such as anthraquinone glycosides (e.g., aloe, rhubarb, cascara, senna), steroidal (cardiac) glycosides (e.g., digitalis, onion, strophanthus), saponin glycosides (e.g., ginseng, glycyrrhiza, senega), etc. (Bartnik and Facey, 2024; Levy et al., 1995). Figure 2.2 presents the chemical structures of important glycosides (aglycone), and Table 2.3 comprises important bioactive glycosides in higher plants, their botanical sources, and their medicinal applications.

2.2.4 Lipids

Lipids are a large and diverse group of naturally occurring organic compounds, and these are esters of fatty acids with glycerol. Lipids of fatty acids with glycerol, commonly known as triglycerides, are found in both plants and animals and compose one of the major food groups of our diet. Lipids are categorized into two subclasses: The first one is fixed oils, and the second one is fats. Lipids that are solid or semi-solid at room temperature are classified as fats and are found predominantly in animals (Evans, 2009; Masoro, 1977). Liquid triglycerides are called oils and originate/are obtained mainly from vegetable and fish oils. Fats are composed of mainly saturated fatty acids, while oils are composed of unsaturated fatty acids. Saturated and *trans*-fatty acid glycerides in the diet have been associated with atherosclerosis. Triglycerides having three identical acyl chains, like tristearin and triolein, are called "simple," while those composed of different acyl chains are called "mixed." Several vegetable oils have been included in different pharmacopeias, like almond oil, castor oil, olive oil, sesame oil, peanut oil, sunflower oil, etc. (Hoffmann, 2013).

Lipids contain several bioactive compounds that are categorized into different classes, such as saturated and unsaturated fatty acids, triglycerides, phospholipids, polyphenols, flavonoids, phytosterols,

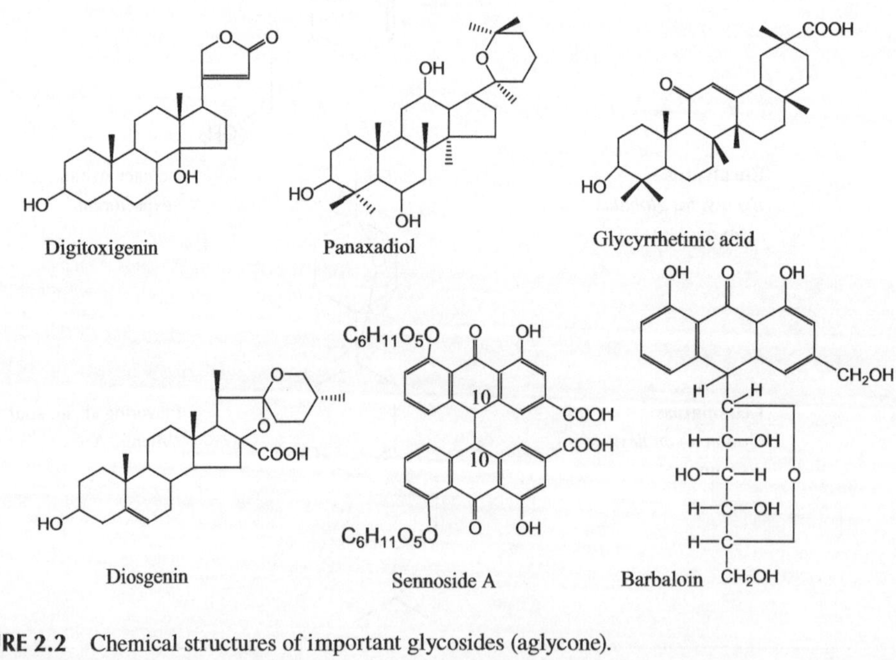

FIGURE 2.2 Chemical structures of important glycosides (aglycone).

TABLE 2.3
Important Bioactive Glycoside Compounds' Occurrence in Higher Plants

Bioactive Glycosides	Common Name	Botanical Source	Medicinal Application
Barbaloin and aloesin	Aloe	*Aloe barbadensis, A. perryi, A. ferox* (Liliaceae)	Cosmetic uses, purgative, anti-inflammatory
Sennosides	Senna	*Cassia acutifolia, C. angustifolia* (Leguminosae)	Purgative
Rhein	Rhubarb	*Rheum emodi, R. palmatum* (Polygonaceae)	Purgative
Digitoxin	Digitalis	*Digitalis purpurea* (Scrophulariaceae)	Cardiotonic
Scillaren	Onion	*Urginea indica* (Liliaceae)	Cardiotonic, expectorant
Bacoposides	Brahmi	*Bacopa moniera* (Scrophulariaceae)	Nervine tonic
Diosgenin	Dioscorea	*Dioscorea deltoidei* (Dioscoreaceae)	Synthesis of steroidal hormones and treatment of rheumatic arthritis
Panaxosides	Ginseng	*Panax ginseng* (Araliaceae)	Immunostimulant and aphrodisiac
Charantin	Bitter gourd	*Momordica charantia* (Cucurbitaceae)	Antidiabetic
Glycyrrhizin	Glycyrrhiza	*Glycyrrhiza glabra* (Leguminosae)	Peptic ulcer treatment, demulcent

TABLE 2.4
Important Bioactive Compounds of Lipids in Higher Plants and Animals

Bioactive Compounds	Common Name	Botanical Source	Pharmaceutical Application
Ricinoleic acid	Castor	*Ricinus communis* (Euphorbiaceae)	Cathartic
Oleic acid and linoleic acid	Sesame	*Sesamum indicum* (Pedaliaceae)	Nutritive and laxative
EPA, DHA, and vitamins D and E	Cod-liver oil	*Gadus callarias* (Gadidae)	Treatment of rickets, rheumatic pains
Oleic acid	Olive	*Olea europaea* (Oleaceae)	Emollient, antidiabetic, and neuroprotective
Chaulmoogric acid	Chaulmoogra	*Hydnocarpus anthelmintic* (Flacourtriaceae)	Treatment of tuberculosis, leprosy, psoriasis, and rheumatism

carotenoids, vitamins, enzymes, minerals, etc. (Dorni et al., 2018). Monounsaturated fatty acids (MUFAs) are a class of chemical compounds in fatty acids that contain one double bond. MUFAs are known for their health benefits and are abundantly found in olive oil, avocadoes, and nuts (Yaqoob, 2002). Oleic acid is considered a major MUFA and is abundantly found in olive oil. Oleic acid is known for its beneficial effects on metabolic disorders and cancer (Gillingham et al., 2011; Kris-Etherton, 1999). Polyunsaturated fatty acids (PUFAs) found in edible oils like olive, fish, soybean, sunflower, linseed, etc. are known for their cardioprotective, immunomodulatory, and anticancer potential. These beneficial activities of polyunsaturated fatty acids are due to the presence of many health-promoting constituents in these edible oils, such as omega-3 fatty acids (e.g., α-linolenic acid, eicosapentaenoic acid (EPA), and docosahexaenoic acid (DHA) and omega-6 fatty acids (e.g., linoleic acid, γ-linolenic acid, and arachidonic acid) (Kapoor et al., 2021; Saini and Keum, 2018). Table 2.4 comprises the important bioactive

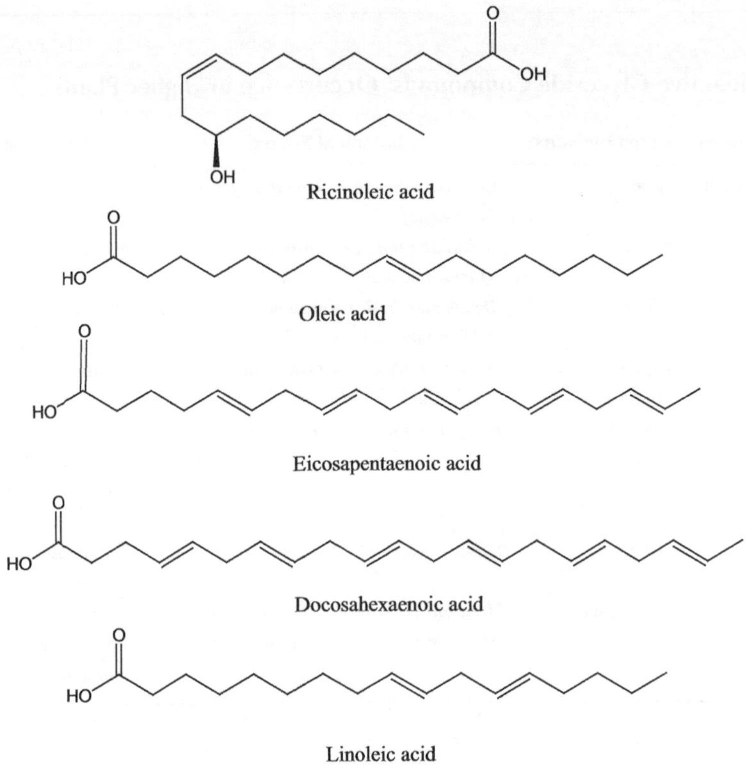

FIGURE 2.3 Chemical structures of important bioactive compounds present in lipids.

compounds of lipids in higher plants and animals, and Figure 2.3 shows the chemical structures of important bioactive compounds present in lipids.

2.2.5 Resins

Resins are an amorphous mixture of essential oils, oxygenated products of terpenes, and carboxylic acids. Resins are obtained as exudates from plants and considered as the end product of metabolism. These are solid or semi-solid amorphous products of a complex chemical nature usually insoluble in water (Parimal et al., 2011; Dilworth et al., 2024). Generally, resins are complex mixtures of many compounds; however, isoprene (C_5H_8) units are the fundamental building blocks of all true resins. Resins are usually found in homogenous combinations with other plant metabolites and, hence, collectively known as resin combinations. The exact mechanism of formation is unknown, but it seems to be the product of the oxidation of terpenes, followed by polymerization. They are usually formed in schizogenous and schizolysigenous cavities or ducts (Hernández-Vázquez et al., 2010). Table 2.5 and Figure 2.4 represent the botanical source, therapeutic application, and chemical structures of important bioactive resins.

2.2.6 Tannins

Tannins are a mixture of complex organic, non-nitrogenous substances that are derivatives of polyhydroxy benzoic acid (polyphenols) with the ability to precipitate proteins and have high molecular weight. Tannins are one of the most widely occurring groups of natural compounds in different plant families. These secondary metabolites are present in solution form in the cell sap and distinct

TABLE 2.5
Important Bioactive Resins' Occurrence in Higher Plants

Bioactive Resins	Common Name	Botanical Source	Medicinal Uses
Δ^9-THC and cannabinoids	Cannabis	*Cannabis sativa* (Cannabaceae)	Hypnotic, anticonvulsant, analgesic, and anti-anxiety
Etoposide	Podophyllum	*Podophyllum peltatum* (Podophyllaceae)	Anticancer
Gingerol	Ginger	*Zingiber officinale* (Zingiberaceae)	Cough remedy
Capsaicin	Red chili	*Capsicum annum* (Solanaceae)	Counter-irritant and carminative
Guggulosterols	Guggul	*Commiphora weightii* (Burseraceae)	Anti-inflammatory, anti-rheumatic, hypolipidemic

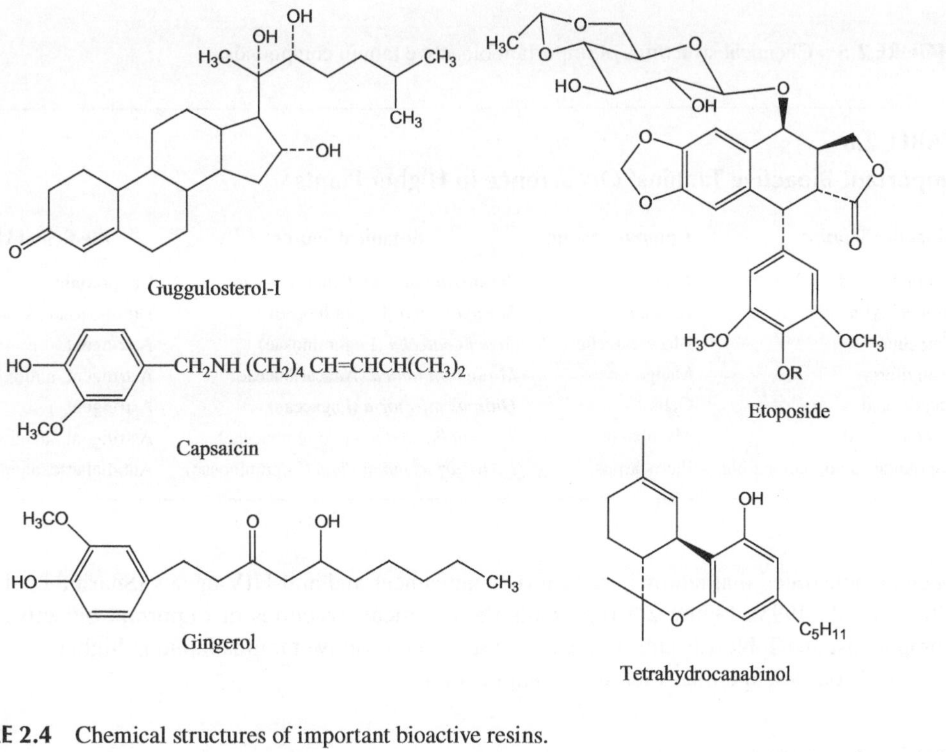

FIGURE 2.4 Chemical structures of important bioactive resins.

vacuoles. They have been known for a long time as astringent substances, having the capacity to combine with tissue proteins and precipitate them (Sieniawska and Baj, 2017; Koleckar et al., 2008). The basis of its use therapeutically depends on its astringent properties. Tannins are used as a mild antiseptic in the treatment of diarrhea, GIT problems, and burns; in the leather industry; and for the formation of inks. It precipitates proteins and is traditionally used as a styptic agent and internally for the protection of the surface of the mouth and throat. These also act as antidiarrheal agents and antidotes in the poisoning of heavy metals, alkaloids, and glycosides. The use of tannins is reduced nowadays because tannic acid causes necrosis of the liver. Recently, it has

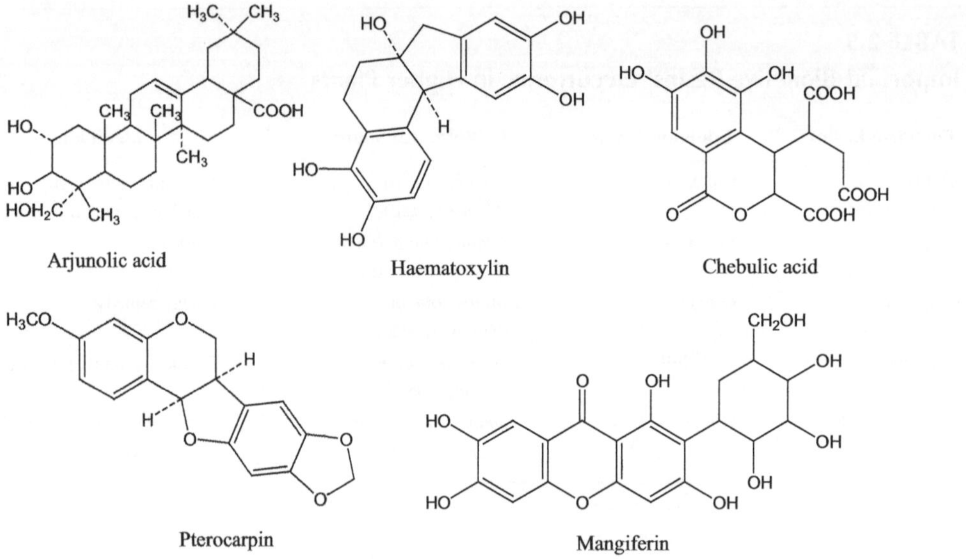

FIGURE 2.5 Chemical structures of important bioactive tannin compounds.

TABLE 2.6
Important Bioactive Tannins' Occurrence in Higher Plants

Bioactive Tannins	Common Name	Botanical Source	Medicinal Uses
Arjunolic acid	Arjuna	*Terminalia arjuna* (Combretaceae)	Cardiotonic
Hematoxylin	Ashoka	*Saraca indica* (Leguminosae)	Uterine tonic, sedative
Catechin	Black catechu	*Acacia catechu* (Leguminosae)	Astringent
Mangiferin	Mango	*Mangifera indica* (Anacardiaceae)	Astringent, antioxidant
Tannic acid	Galls	*Quircus infectoria* (Fagaceae)	Astringent
Chebulic acid	Myrobalan	*Terminalia chebula* (Combretaceae)	Astringent, laxative, tonic
Kinotannic acid, pterocarpin	Pterocarpus	*Pterocarpus marsupium* (Leguminosae)	Antidiabetic, antidiarrheal

been demonstrated that tannins are useful as anticancer and anti-HIV agents (Sharma et al., 2021; Chung et al., 1998). Figure 2.5 represents the chemical structures of important bioactive tannin compounds, and Table 2.6 summarizes the important bioactive tannins found in higher plants, their botanical sources, and their therapeutic applications.

2.2.7 Flavonoids

Flavonoids are polyphenolic compounds abundantly found in plants. About 6,000 different flavonoid compounds are isolated and characterized in different plant species. Flavonoids can be categorized into three classes: isoflavonoids, neoflavonoids, and bioflavonoids (Figure 2.6). Isoflavonoids possess a 3-phenyl chroman skeleton and are derived from the 2-phenylchroman skeleton of flavonoids. Genistein, daidzein, and glycitin are major isoflavonoids found in soybeans (*Glycine max*) (Miadoková, 2009; Naquvi et al., 2023). Neoflavonoids possess a 4-phenylchroman skeleton with no OH group substitution at position 2. Bioflavonoids (flavonoids) in which the B-ring is attached to position 2 of the C-ring are called bioflavonoids. These bioflavonoids are further subdivided into many subclasses, depending on the structural features of the C-ring, and they include flavonols, flavones, flavanones, anthocyanidins, chalcones, flavan-3-ol/catechins, and prenyl-flavonoids. The important bioflavonoids include flavonols

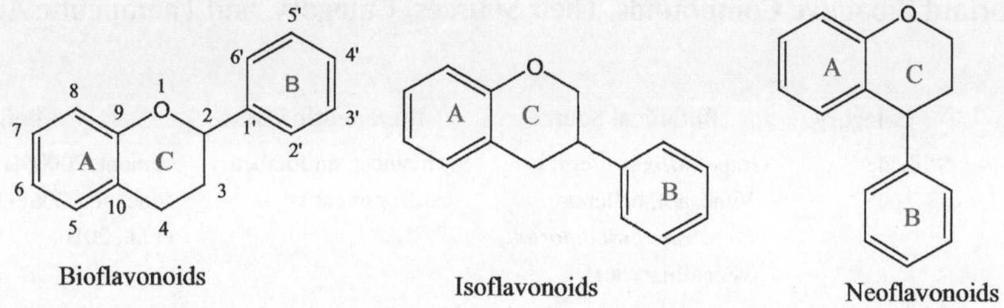

FIGURE 2.6 Different categories of flavonoids.

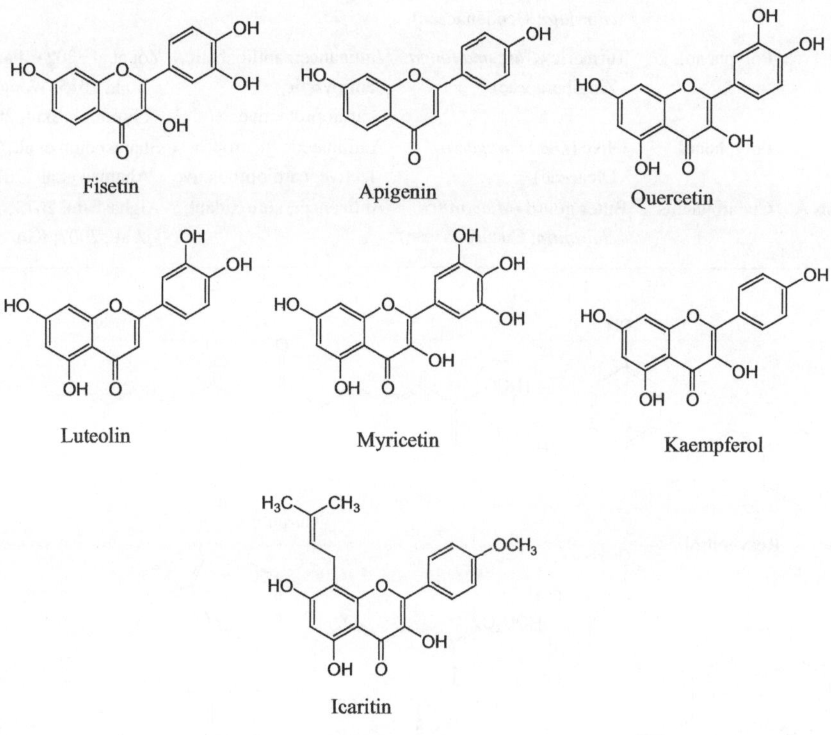

FIGURE 2.7 Chemical structure of some important flavonoids.

(e.g., quercetin, kaempferol, fisetin, and myricetin), flavones (e.g., apigenin and luteolin), flavanones (e.g., hesperidin), and prenyl-flavonoids (e.g., icaritin) (Ramesh et al., 2021; Li et al., 2017).

Flavonoids are known for their potent antioxidant activity, and these flavonoids are used for the prevention of various types of cancer (Romagnolo and Selmin, 2012; Gupta et al., 2023), diabetes (Ahamad et al., 2019; Manavi et al., 2021), cardiovascular diseases (Ciumărnean et al., 2020; Gross, 2004), COVID-19 (Russo et al., 2020), etc. Figure 2.7 depicts the chemical structure of important flavonoid compounds.

2.2.8 Other Bioactive Compounds in Plants

There are several bioactive compounds present in plants which we discussed earlier. Plants contain various vitamins, minerals, polyphenols and phenolic acids, proteins and peptides, polysaccharides, iridoids, stilbenes, anthocyanins, etc. These bioactive compounds are summarized in Table 2.7 and Figure 2.8.

TABLE 2.7
Some Important Bioactive Compounds, Their Sources, Category, and Therapeutic Applications

Bioactive Compound	Category	Botanical Source	Therapeutic Uses	Ref.
Resveratrol	Stilbene	Grape (*Vitis vinifera*; Vitaceae), hellebore (*Veratrum grandiflorum*; Melanthiaceae)	Anticancer, antidiabetic, cardioprotective	Frémont, 2000; Baur and Sinclair, 2006; Diaz-Gerevini et al., 2016
Malvidin	Anthocyanin	Berries (*Vaccinium* sp.; Ericaceae)	Anticancer, antidiabetic, antioxidant	Fritz et al., 2006; Baba et al., 2017; Merecz-Sadowska et al., 2023
Swertiamarin	Secoiridoids	Chirata (*Enicostemma littorale* and *Swertia chiryiata*; Gentianaceae)	Antidiabetic, bitter tonic, hepatoprotective	Ahamad et al., 2016; Leong et al., 2016; Ahamad, 2022
Curcumin	Polyphenol	Turmeric (*Curcuma longa*; Zingiberaceae)	Anticancer, antidiabetic, antipyretic, cardioprotective	Zoi et al., 2021; Fazel Nabavi et al., 2015; Wongcharoen and Phrommintikul, 2009
Oleuropein	Polyphenol	Olive (*Olea europaea*; Oleaceae)	Antidiabetic, neuroprotective, cardioprotective	Shamshoum et al., 2017; Ahamad et al., 2019
Cucurbitacins A	Cucurbitacins	Bitter gourd (*Momordica charantia*; Cucurbitaceae)	Anticancer, antioxidant,	Alghasham, 2013; Tannin-Spitz et al., 2007; Kim et al., 2018

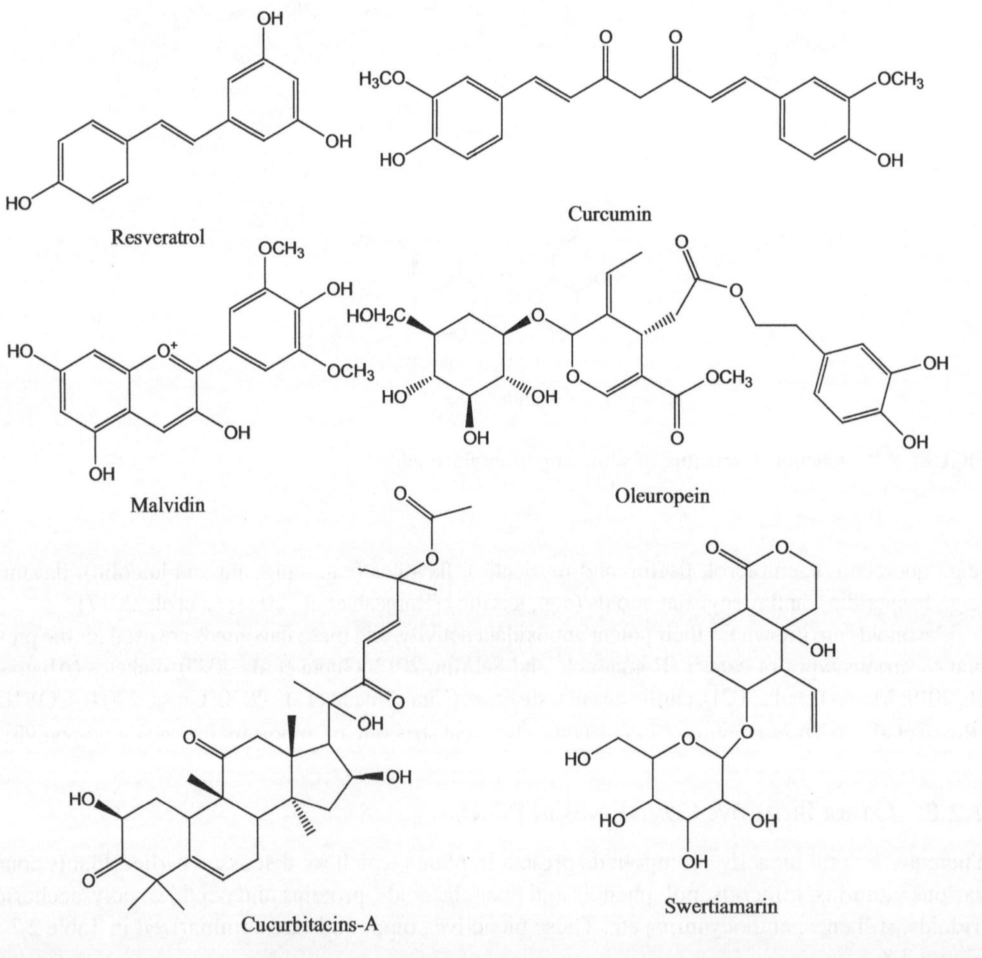

FIGURE 2.8 Chemical structures of some important bioactive compounds of different classes.

2.3 CONCLUSION

Due to an exponential increase in infectious diseases and resistance to existing antibiotics, there is an increasing demand of new drugs derived from natural products. Plants are considered a major source of new drugs because they offer greater chemical diversity and a unique mechanism of action. Bioactive compounds of plants play an important role in the discovery of new drugs. Novel drug candidates discovered from natural products will assist in sustaining human health and fighting against disease and illness. The current chapter provides readers with an overview of different bioactive compounds isolated and characterized from plants. These bioactive compounds may be, in the future, approved as new drug candidates and used for the prevention and treatment of human ailments.

REFERENCES

Ahamad J. Swertiamarin for the treatment of metabolic syndrome. *Frontiers in Natural Product Chemistry*. 2022; 9: 75–94.

Ahamad J. *Text Book of Advanced Pharmacognosy and Natural Products*. 1st Edition. Ahmedabad, India: Mahi Publication; 2021.

Ahamad J, Ahmad J, Nollet LM. Food spices: An overview. *Analysis of Food Spices*. 2023: 3–18.

Ahamad J, Hassan N, Amin S, Mir SR. Swertiamarin contributes to glucose homeostasis via inhibition of carbohydrate metabolizing enzymes. *Journal of Natural Remedies*. 2016: 125–30.

Ahamad J, Toufeeq I, Khan MA, Ameen MS, Anwer ET, Uthirapathy S, Mir SR, Ahmad J. Oleuropein: A natural antioxidant molecule in the treatment of metabolic syndrome. *Phytotherapy Research*. 2019; 33(12): 3112–28.

Ahmad J, Ahamad J, editors. *Bioactive Phytochemicals: Drug Discovery to Product Development*. Bentham Science Publishers; 2020.

Alghasham AA. Cucurbitacins—a promising target for cancer therapy. *International Journal of Health Sciences*. 2013; 7(1): 77.

Baba AB, Nivetha R, Chattopadhyay I, Nagini S. Blueberry and malvidin inhibit cell cycle progression and induce mitochondrial-mediated apoptosis by abrogating the JAK/STAT-3 signalling pathway. *Food and Chemical Toxicology*. 2017; 109: 534–43.

Bartnik M, Facey P. Glycosides. In *Pharmacognosy*. Academic Press; 2024: 103–65.

Başer KH, Demirci F. Chemistry of essential oils. In *Flavours and Fragrances: Chemistry, Bioprocessing and Sustainability*, edited by Berger RG. New York: Springer; 2007: 43–86.

Baur JA, Sinclair DA. Therapeutic potential of resveratrol: The in vivo evidence. *Nature Reviews Drug Discovery*. 2006; 5(6): 493–506.

Chung KT, Wong TY, Wei CI, Huang YW, Lin Y. Tannins and human health: A review. *Critical Reviews in Food Science and Nutrition*. 1998; 38(6): 421–64.

Ciumărnean L, Milaciu MV, Runcan O, Vesa ŞC, Răchişan AL, Negrean V, Perné MG, Donca VI, Alexescu TG, Para I, Dogaru G. The effects of flavonoids in cardiovascular diseases. *Molecules*. 2020; 25(18): 4320.

Cordell GA. *The Alkaloids: Chemistry and Biology*. Academic Press; 2008.

Diaz-Gerevini GT, Repossi G, Dain A, Tarres MC, Das UN, Eynard AR. Beneficial action of resveratrol: How and why? *Nutrition*. 2016; 32(2): 174–8.

Dilworth LL, Riley CK, Stennett DK. Plant constituents: Carbohydrates, oils, resins, balsams, and plant hormones. In *Pharmacognosy*. Academic Press; 2024: 49–74.

Dorni C, Sharma P, Saikia G, Longvah T. Fatty acid profile of edible oils and fats consumed in India. *Food Chemistry*. 2018; 238: 9–15.

Esmeeta A, Adhikary S, Dharshnaa V, Swarnamughi P, Ummul Maqsummiya Z, Banerjee A, Pathak S, Duttaroy AK. Plant-derived bioactive compounds in colon cancer treatment: An updated review. *Biomedicine & Pharmacotherapy*. 2022; 153: 113384.

Evans WC. *Trease and Evans' Pharmacognosy*. Elsevier Health Sciences; 2009.

Fazel Nabavi S, Thiagarajan R, Rastrelli L, Daglia M, Sobarzo-Sanchez E, Alinezhad H, Mohammad Nabavi S. Curcumin: A natural product for diabetes and its complications. *Current Topics in Medicinal Chemistry*. 2015; 15(23): 2445–55.

Frémont L. Biological effects of resveratrol. *Life Sciences*. 2000; 66(8): 663–73.

Fritz J, Kern M, Pahlke G, Vatter S, Marko D. Biological activities of malvidin, a red wine anthocyanidin. *Molecular Nutrition & Food Research*. 2006; 50(4–5): 390–5.

Gillingham LG, Harris-Janz S, Jones PJ. Dietary monounsaturated fatty acids are protective against metabolic syndrome and cardiovascular disease risk factors. *Lipids*. 2011; 46: 209–28.

Gross M. Flavonoids and cardiovascular disease. *Pharmaceutical Biology*. 2004; 42(sup1): 21–35.

Gupta M, Ahmad J, Ahamad J, Kundu S, Goel A, Mishra A. Flavonoids as promising anticancer therapeutics: Contemporary research, nanoantioxidant potential, and future scope. *Phytotherapy Research*. 2023; 37(11): 5159–92.

Gutiérrez-Grijalva EP, López-Martínez LX, Contreras-Angulo LA, Elizalde-Romero CA, Heredia JB. Plant alkaloids: Structures and bioactive properties. *Plant-Derived Bioactives: Chemistry and Mode of Action*. 2020: 85–117.

Hernández-Vázquez L, Mangas S, Palazón J, Navarro-Ocaña A. Valuable medicinal plants and resins: Commercial phytochemicals with bioactive properties. *Industrial Crops and Products*. 2010; 31(3): 476–80.

Hoffmann G. *The Chemistry and Technology of Edible Oils and Fats and Their High Fat Products*. Academic Press; 2013.

Jugreet BS, Suroowan S, Rengasamy RK, Mahomoodally MF. Chemistry, bioactivities, mode of action and industrial applications of essential oils. *Trends in Food Science & Technology*. 2020; 101: 89–105.

Kapoor B, Kapoor D, Gautam S, Singh R, Bhardwaj S. Dietary polyunsaturated fatty acids (PUFAs): Uses and potential health benefits. *Current Nutrition Reports*. 2021; 10: 232–42.

Kim KH, Lee IS, Park JY, Kim Y, An EJ, Jang HJ. Cucurbitacin B induces hypoglycemic effect in diabetic mice by regulation of AMP-activated protein kinase alpha and glucagon-like peptide-1 via bitter taste receptor signaling. *Frontiers in Pharmacology*. 2018; 9: 1071.

Koleckar V, Kubikova K, Rehakova Z, Kuca K, Jun D, Jahodar L, Opletal L. Condensed and hydrolysable tannins as antioxidants influencing the health. *Mini Reviews in Medicinal Chemistry*. 2008; 8(5): 436–47.

Kren V, Martínková L. Glycosides in medicine: "The role of glycosidic residue in biological activity". *Current Medicinal Chemistry*. 2001; 8(11): 1303–28.

Kris-Etherton PM. Monounsaturated fatty acids and risk of cardiovascular disease. *Circulation*. 1999; 100(11): 1253–8.

Leong XY, Thanikachalam PV, Pandey M, Ramamurthy S. A systematic review of the protective role of swertiamarin in cardiac and metabolic diseases. *Biomedicine & Pharmacotherapy*. 2016; 84: 1051–60.

Levy DE, Tang C, editors. *The Chemistry of C-glycosides*. Elsevier; 1995.

Li F, Du BW, Lu DF, Wu WX, Wongkrajang K, Wang L, Pu WC, Liu CL, Liu HW, Wang MK, Wang F. Flavonoid glycosides isolated from *Epimedium brevicornum* and their estrogen biosynthesis-promoting effects. *Scientific Reports*. 2017; 7: 7760.

Manavi SP, Amiri T, Mozafaryan MJ. Role of flavonoids in diabetes. *Journal of Reviews in Medical Sciences*. 2021; 1(3): 149–61.

Masoro EJ. Lipids and lipid metabolism. *Annual Review of Physiology*. 1977; 39(1): 301–21.

Merecz-Sadowska A, Sitarek P, Kowalczyk T, Zajdel K, Jęcek M, Nowak P, Zajdel R. Food anthocyanins: Malvidin and its glycosides as promising antioxidant and anti-inflammatory agents with potential health benefits. *Nutrients*. 2023; 15(13): 3016.

Miadoková E. Isoflavonoids—an overview of their biological activities and potential health benefits. *Interdisciplinary Toxicology*. 2009; 2(4): 211–8.

Naquvi KJ, Ahamad A, Salma A, Tariq E, Ahamad J. Qualitative diagnostic features of polyphenols and flavonoids containing food spices. In *Analysis of Food Spices*. CRC Press; 2023: 151–62.

Nunes CDR, Barreto Arantes M, Menezes de Faria Pereira S, Leandro da Cruz L, de Souza Passos M, Pereira de Moraes L, Vieira IJC, Barros de Oliveira D. Plants as sources of anti-inflammatory agents. *Molecules*. 2020; 25(16): 3726.

Parimal K, Khale A, Pramod K. Resins from herbal origin and a focus on their applications. *International Journal of Pharmaceutical Sciences and Research*. 2011; 2(5): 1077–85.

Pelletier SW, editor. *Alkaloids: Chemical and Biological Perspectives*. New York: Wiley; 1983.

Pinheiro GKLO, Araújo Filho I, Araújo Neto I, Rêgo ACM, Azevedo EP, Pinheiro FI, Lima Filho AAS. Nature as a source of drugs for ophthalmology. *Arquivos Brasileiros de Oftalmologia*. 2018; 81(5): 443–54.

Prassas I, Diamandis EP. Novel therapeutic applications of cardiac glycosides. *Nature Reviews Drug Discovery*. 2008; 7(11): 926–35.

Ramesh P, Jagadeesan R, Sekaran S, Dhanasekaran A, Vimalraj S. Flavonoids: Classification, function, and molecular mechanisms involved in bone remodelling. *Frontiers in Endocrinology*. 2021; 12: 779638. doi: 10.3389/fendo.2021.779638.

Romagnolo DF, Selmin OI. Flavonoids and cancer prevention: A review of the evidence. *Journal of Nutrition in Gerontology and Geriatrics*. 2012; 31(3): 206–38.

Russo M, Moccia S, Spagnuolo C, Tedesco I, Russo GL. Roles of flavonoids against coronavirus infection. *Chemico-Biological Interactions*. 2020; 328: 109211.

Saini RK, Keum YS. Omega-3 and omega-6 polyunsaturated fatty acids: Dietary sources, metabolism, and significance—a review. *Life Sciences*. 2018; 203: 255–67.

Shamshoum H, Vlavcheski F, Tsiani E. Anticancer effects of oleuropein. *Biofactors*. 2017; 43(4): 517–28.

Sharma K, Kumar V, Kaur J, Tanwar B, Goyal A, Sharma R, Gat Y, Kumar A. Health effects, sources, utilization and safety of tannins: A critical review. *Toxin Reviews*. 2021; 40(4): 432–44.

Sieniawska E, Baj T. Tannins. In *Pharmacognosy*. Academic Press; 2017: 199–232.

Sitarek P, Kowalczyk T, Wieczfinska J, Merecz-Sadowska A, Górski K, Śliwiński T, Skała E. Plant extracts as a natural source of bioactive compounds and potential remedy for the treatment of certain skin diseases. *Current Pharmaceutical Design*. 2020; 26(24): 2859–75.

Street RA, Prinsloo G, McGaw LJ. Alkaloids potential health benefits and toxicity. In *Utilisation of Bioactive Compounds from Agricultural and Food Production Waste*. CRC Press; 2017: 60–85.

Tannin-Spitz T, Bergman M, Grossman S. Cucurbitacin glucosides: Antioxidant and free-radical scavenging activities. *Biochemical and Biophysical Research Communications*. 2007; 364(1): 181–6.

Thibane VS, Mudau FN. Herbal alkaloids in healthcare. In *Herbal Biomolecules in Healthcare Applications*. Academic Press; 2022: 283–94.

Wongcharoen W, Phrommintikul A. The protective role of curcumin in cardiovascular diseases. *International Journal of Cardiology*. 2009; 133(2): 145–51.

Yaqoob P. Monounsaturated fatty acids and immune function. *European Journal of Clinical Nutrition*. 2002; 56(3): S9–13.

Zoi V, Galani V, Lianos GD, Voulgaris S, Kyritsis AP, Alexiou GA. The role of curcumin in cancer treatment. *Biomedicines*. 2021; 9(9): 1086.

Zuzarte M, Salgueiro L. Essential oils chemistry. *Bioactive Essential Oils and Cancer*. 2015: 19–61.

3 Microalgae as a Source of Bioactive Compounds

Monize Bürck, Mônica Masako Nakamoto, Letícia Guerreiro Trindade, Renata Nolasco Braga Souto, Bruna Kaori Feitosa Tanaka, Ingrid Ferreira de Souza, and Anna Rafaela Cavalcante Braga

3.1 INTRODUCTION

Algae are a broad group of organisms that can be subclassified based on their size, that being macroalgae, also known as seaweeds, and microalgae (Martínez-Ruiz et al., 2022; Stark & O'Gara, 2012). *Microalgae* refer to microorganisms that are unicellular and, straightforward, multicellular (Stark & O'Gara, 2012), which can be prokaryotic and eukaryotic (M. G. de Morais et al., 2015). This subgroup is broad and diverse, composed of thousands of species. Studies estimate that about 30,000 species have been identified and analyzed (Zhou et al., 2023). *Arthrospira* (also known as *Spirulina*), *Chlorella*, and *Tetraselmis* are the most studied species, mainly used for human nutrition (Ramos-Romero et al., 2021).

In general, microalgae measure about 3–10 µm, changing in diverse shapes (M. G. de Morais et al., 2015), and thanks to their photosynthetic activity, these microorganisms have a significant role in all ecosystems. It is estimated that microalgae are responsible for 40% of global photosynthesis and are the primary producers of organic compounds and biomass in marine systems (de Vera et al., 2018). Several species can be categorized by their colors and the kind of pigments they have, including green algae, red algae, and diatoms, among others (Li et al., 2008). Three main types of photosynthetic pigments (chlorophylls, carotenoids, and phycobilin) are of importance due to their use as a natural dye and for their biological potential (Begum et al., 2016).

Furthermore, microalgae are also known for their rich nutritional composition, mainly of proteins, lipids, carbohydrates, and nucleic acids (Encarnação et al., 2015). A significant amount of proteins is reported by species like *Arthrospira* (50–70% of dry weight) and *Chlorella* (next to 58% of dry weight) (Hoseini et al., 2013). Additionally, with this considerable level of macronutrients, microalgae also contain high-value biological active compounds, including polysaccharides, polyphenols, carotenoids, phycobiliproteins, vitamins, enzymes, sterols, and particularly, polyunsaturated fatty acids (PUFAs) (Ampofo & Abbey, 2022; Dolganyuk et al., 2020; Ramos-Romero et al., 2021). Several new research studies have been conducted on the biopotential of these compounds, leading to 15,000 different compounds characterized nowadays (Cardozo et al., 2007).

Microalgae have a metabolic plasticity, able to grow in different environments and extreme conditions, such as temperature, pH, light intensity, CO_2 accessibility, salinity, and nutrients (Ma et al., 2020; Ramos-Romero et al., 2021; Saide et al., 2021). It is well-known that several growing conditions and different species are directly related to the proportion of each compound available in the biomass (Dolganyuk et al., 2020; Encarnação et al., 2015). Still, their cultivation is relatively simple, having a related fast growth (Saide et al., 2021), and contrary to heterotrophic organisms, those that grow require different organic compounds; these photosynthetic microorganisms are capable of oxidizing inorganic substances and mineral elements, producing their biomass (Dolganyuk et al., 2020).

DOI: 10.1201/9781003462804-4

Likewise, this growth system uses a small quantity of water and consumes carbon dioxide while generating oxygen, not polluting the atmosphere (Dolganyuk et al., 2020). Industrial-scale microalgae cultivation is increasing globally due to sustainable and eco-friendly cultivating, a social concern discussed (Saide et al., 2021). In parallel, large-scale cultivation is directly associated with the presence and types of bioactive compounds and the rich application possibilities in many biotechnology sectors (Bilal et al., 2017).

The presence of multiple bioactive compounds in microalgae can lead to many biological effects, especially antioxidant activity (Chen et al., 2010; Chester et al., 2019), immunomodulatory (Riccio et al., 2020), antibacterial (Sukhikh et al., 2022), anticancer, and antidiabetes (Lauritano et al., 2016), anti-inflammatory (De Jesus Raposo et al., 2016), and neuroprotection (Parameswari & Lakshmi, 2022). Therefore, the application of whole microalgae biomass, or just their biopotential compounds, has significantly increased worldwide. Both are demanded and already are being applied in many biotechnological sectors, like in health (Eze et al., 2023), pharmaceuticals (Saeed et al., 2022), food (Sidari & Tofalo, 2019), the energy industry, and biomaterials (Saide et al., 2021).

3.2 METABOLITE PROFILING OF MICROALGAE BIOMASS AND ITS INDUSTRIAL AND BIOLOGICAL APPLICATIONS

Microalgae are unicellular prokaryotic or eukaryotic organisms and can be cultivated with little use of land in systems that aim to treat wastewater and produce biomass with metabolites of biological and industrial value for different applications (Álvarez-González et al., 2023; Singh et al., 2023). Cultivation methods and their physicochemical compositions are decisive for the development and composition of microalgae biomass, the most common being photoautotrophic methods, heterotrophic modes, and mixotrophic methods (combined cultivation of the previous modes) (Cheirsilp et al., 2023).

The primary metabolites produced by microalgae, depending on the species, are proteins, bioactive peptides, essential fatty acids, polysaccharides, carotenoids, chlorophylls, and phycocyanins, which have interesting biotechnological functions for the food, pharmaceutical, cosmetics, feed, biofuels, and biofertilizers industries (Wu et al., 2021).

Therefore, to obtain compounds for industrial use through the cultivation of microalgae, which contributes to sustainable development, the following texts will show the primary metabolites of these organisms and their applications.

3.2.1 Proteins, Peptides, and Amino Acids

Within the context of the circular economy and growing food in the field, microalgae biomasses cultivated in wastewater can be used as biofertilizers instead of chemical fertilizers (Goyal et al., 2023), with their proteins being a source of nitrogen (N) for plants to grow correctly and develop secondary metabolites, such as phenolic compounds (Braun & Colla, 2023). An example is the study that compared wheat plants fertilized with microalgae biomass (grown in wastewater) and plants fertilized with mineral fertilizer. The researchers reached the main conclusion that the use of microalgae is suitable as an organic fertilizer for wheat; however, when using this biomass, maintaining a combined regime with mineral fertilizer should be evaluated to have increased agronomic efficiency (Mückschel et al., 2023).

Still, in food, microalgae are alternative sources of protein to include in the diet as a substitute for proteins of animal origin (Williamson et al., 2023). They produce biomasses rich in proteins, bioactive peptides, and essential amino acids for human and animal nutrition (Verni et al., 2023). Most of the microalgae already explored have a high protein content (30–80%), which can be nutritionally improved through the enzymatic hydrolysis process to increase their solubility, digestibility, and bioactivity (Eilam et al., 2023). When microalgae biomasses are added to formulations such as bread, dairy products, and snacks, the protein content tends to increase, as well as being sources of protein and other nutrients for addition to fish, pig, and poultry feed (Amorim et al., 2021; Bhatnagar et al., 2023).

Furthermore, microalgae proteins are also beneficial for the pharmaceutical industry, as they are precursors of antibodies, cytokines, and enzymes (Shi et al., 2021) and can be used in the development of protein-based drugs, such as vaccines, antimicrobial peptides, immunotoxins, and hormones at lower costs, and with more excellent safety than when using some traditional organisms (Goshtasbi et al., 2023; Xie et al., 2023). They are also being used to obtain peptides with anti-inflammatory, antihypertensive, antioxidant, antimicrobial, and anticollagenase activities for use in pharmaceuticals, cosmetics, and food (Apone et al., 2019; Lin et al., 2018; Montalvo et al., 2019; Suttisuwan et al., 2019).

3.2.2 Lipids and Fatty Acids

Lipids from microalgae stand out for their application in nutrition and health, mainly due to the presence of essential fatty acids, such as omega-3, in the forms of docosahexaenoic acid (DHA) and eicosapentaenoic acid (EPA), which have anti-inflammatory and antitumor action and can prevent and reduce neurodegenerative and cardiovascular diseases, in addition to participating in the regulation of glucose and lipid metabolism (Li et al., 2019; T.-T. Zhang et al., 2019).

However, one of the primary sources for obtaining DHA and EPA is the consumption of these compounds after their extraction from fish and crustacean oil or the direct consumption of fish oil, which contains a high amount of saturated fatty acids; in addition, both forms come from a non-ecological and unsustainable practice over time (Peltomaa et al., 2019; Sherratt et al., 2020). For these reasons, the cultivation of microalgae to produce DHA and EPA, such as *Nannochloropsis oceanica*, is justified, and metabolic engineering techniques and physicochemical and microbiological modifications can be applied to the cultivation medium to increase the productivity and stability of these compounds (Jesionowska et al., 2023; Kumari et al., 2023).

From the point of view of the aesthetic area, lipids from microalgae, due to their technological properties, can be used in cosmetics, acting as structuring, emulsifying, and emollient agents, also providing antioxidant, anti-inflammatory, and antimicrobial activities (De Luca et al., 2021). Regarding the structuring property, researchers have found that lipsticks containing common waxes for cosmetics and alkenone wax, a lipid structure from microalgae, have good stability and texture results, with emphasis on the formulation with the highest amount of alkenone, which consumers prefer (Huynh et al., 2020). Studies like this, which focus on microalgae products, highlight the importance of looking for sustainable compounds of natural origin and without animal origin for the formulation of cosmetics (McIntosh et al., 2018; Morocho-Jácome et al., 2022).

Finally, the most common application of microalgae oil is in producing third-generation biofuels. This is relevant for the environment due to the carbon dioxide fixation capacity of microalgae, their cultivation in agro-industry wastewater, and the possibility of producing biodiesel in hybrid refineries to achieve maximum use, also considering other microalgae products (Hoang et al., 2023; S. Zhang et al., 2022). For biodiesel synthesis to be efficient, it is also recommended to use microalgae with more significant potential for lipid accumulation, for example, *Chlorella vulgaris* (49–52%), *Dunaliella salina* (6–25%), and *Neochloris oleoabundans* (35–54%). Appropriate choice of cultivation method and harvesting and lipid extraction processes with low loss rates must be carried out, followed by the lipid transesterification process (Demirbas & Fatih Demirbas, 2011; Goh et al., 2019; Ramesh Kumar et al., 2019; Sajjadi et al., 2018).

3.2.3 Polysaccharides

The sustainable cultivation of microalgae from the bioremediation of wastewater from agro-industries can be a means for the production of carbohydrates, specifically biocompatible, biodegradable, stable, and versatile polysaccharides (Franco-Morgado et al., 2023). The main polysaccharides in microalgae cell walls are cellulose, hemicellulose, and pectin (Sirohi et al., 2021).

These biopolymers benefit different areas, for example, the generation of bioethanol, a third-generation biofuel that contributes to carbon neutralization through its production (Condor et al., 2022; Maity & Mallick, 2022). But they also impact human health, as they can have anti-inflammatory, antioxidant, antitumor, and antiviral activity (Srimongkol et al., 2023; Tounsi et al., 2022).

For use in the food and feed manufacturing chain, microalgae polysaccharides can act as thickeners, gelling agents, probiotics, and compounds for bioactive biofilms in packaging (Tounsi et al., 2022). One of the main reasons polysaccharides can be used as food additives is, when there are sulfate groups, which make the polysaccharide hydrophilic to retain water molecules and become more viscous (Caetano et al., 2022).

When microalgal polysaccharides are applied as additives or microalgae are used entirely in food or feed processing, heat and mechanical agitation may be necessary and are essential for the rheological properties of the products. Heat treatment and high-speed shear homogenization increase the viscosity and bioaccessibility of bioactive compounds from *Synechococcus microalgae* sp. PCC 7002 and *Spirulina plantensis* due to the dissolution of polysaccharides attached to their cell walls (Wang et al., 2022).

3.2.4 Pigments

Microalgae can synthesize pigments with greenish, bluish, yellowish, and orange tones, which can be classified into the following three large groups: chlorophylls, carotenoids, and phycobilins (Cao et al., 2023). In addition to the diversity of colors for application in different products, these pigments are also capable of benefiting health by providing, for example, antioxidant, anti-inflammatory, and anticancer activities and protective effects against neurodegenerative diseases (Mapelli-Brahm et al., 2023; Pentón-Rol et al., 2021).

Between 2009 and 2019, the main microalgae that emerged on the market for the production of chlorophyll pigments, phycocyanin, astaxanthin, and β-carotene, were *Chlorella vulgaris*, *Spirulina platensis*, *Haematococcus pluvialis*, and *Dunaliella salina*, respectively (Silva et al., 2020). These pigments are of interest to the food, pharmaceutical, and cosmetic industries that are looking to use natural dyes to increase the visual attractiveness of their products, increase their bioactive properties, and meet the population's demand for more natural products, or limiting legislation on synthetic additives (Chini Zittelli et al., 2023; Novoveská et al., 2019).

Microalgae, depending on their species, can produce biomasses with great diversity and quantity of metabolites, grouped, as exemplified here, in proteins, lipids, polysaccharides, and pigments, for application in the development of food, feed, medicines, vaccines, cosmetics, biofuels, and biofertilizers, promoting biotechnological benefits and benefits for human and animal health. These metabolites also stand out in these applications because they come from natural sources, contributing to environmental preservation and sustainable development.

3.3 MICROALGAL PIGMENTS: BIOTECHNOLOGICAL ASPECTS FOR INDUSTRIAL APPLICATIONS

Microalgae play an essential role in the world regarding sustainability due to their wide range of biotechnological applications, such as wastewater treatment, biofuels, animal feed production, cosmetics, dyes, pharmaceuticals, and food fields (Tounsi et al., 2023). They can produce various pigments that play fundamental roles in photosynthesis (Tounsi et al., 2023), and chlorophyll is one of the most important and widely distributed of all the existing natural pigments (Fernandes et al., 2020).

Chlorophyll is a fat-soluble pigment (Jeevanandam et al., 2020), classified into five groups (a, b, c, d, and f) (Fernandes et al., 2020), with the basic structural unit represented by a hydrophobic phytol tail and a tetrapyrrole ring consisting of four pyrrole rings, called porphyrins, with the central magnesium atom linked to the ring's nitrogen molecules (A. Srivastava et al., 2022). Chlorophylls a and b have structural differences, resulting in different colors and the absorption of different wavelengths. Chlorophyll a is green and blue, with maximum optical absorption at 660–665 nm, while chlorophyll b is yellow and green, with maximum absorption in the 642–652 nm range (Jeevanandam et al., 2020). Chlorophylls b, c, d, and f are accessory pigments extending the light absorption range, capturing more solar energy (López-Hernández et al., 2022).

Like any other compound, it is unstable due to different temperatures, oxygen, and pH changes. Some structural modifications involving the replacement of the central magnesium with another atom, such as copper or zinc, improve this pigment's chemical stability and enhance its color. The more expressive greenish hue and their excellent stability make these chlorophyll derivatives potential colorants to be used worldwide. The presence of central zinc, copper, and magnesium ions in the structure of chlorophyll is closely related to the inhibition of free radicals. In addition, they have several other relevant properties, such as antimicrobial, antimutagenic, cicatrizing, and anti-inflammatory activity, which allow them to be used in various biotechnological areas (Kholssi et al., 2021).

Chlorophyll as a food additive has been an alternative to artificial preservatives, especially with the possibility of developing toxic effects that can trigger allergic reactions. In addition, its use as a colorant is closely linked to the health benefits it provides (Fernandes et al., 2020). *Spirulina platensis* is a microalga that is widely used in this regard because it contains large quantities of this pigment, used in both the cosmetics and food areas in the development of products such as butter cookies, pasta, and chewing gums made from the microalgae *Chlorella vulgaris*, *Arthrospira platensis*, and *Isochrysis*, respectively (Rai et al., 2021; A. K. Srivastava et al., 2018). Chlorophyll derivatives can also be used as dyes in the textile industry and exhibit antimutagenic properties mainly related to reducing the risk of developing colorectal cancer (Jeevanandam et al., 2020).

Other pigments with relevant biological activity are in microalgae, including carotenoids and phycobiliproteins, especially astaxanthin, β -carotene, phycocyanin, and phycoerythrin (A. Srivastava et al., 2022). *Carotenoids* are lipophilic compounds varying in color from yellow to red (Rajesh et al., 2017), with an optical absorption range of 400–550 nm (López-Hernández et al., 2022). They can be divided into two classes: hydrocarbon-based, known as carotenes, corresponding to non-oxygen molecules (lycopene, α-carotene, β -carotene), and alcohol-based, known as xanthophylls, which have oxygen molecules in their composition (lutein, xanthophylls, astaxanthin, and canthaxanthin) (Honda, 2020).

They act as accessory light-capturing pigments, structural components, and reactive oxygen species scavengers in photosynthetic organisms. Due to their antioxidant power, the most commercialized carotenoids are astaxanthin, β -carotene, zeaxanthin, lycopene, and lutein. They are widely used as additives for animal feed, cosmetics, food coloring, provitamins, antioxidants, and therapeutic applications, such as immune stimulators, antiaging, and anticancer agents (Feng et al., 2021; Jeevanandam et al., 2020).

Carotenoids are used as additives and color enhancers (A. Srivastava et al., 2022), with the microalgae species *H. pluvialis*, *D. salina*, *S. platensis*, *B. braunii*, *Chlorella* spp., *Scenedesmus* spp., and *Chlorococcum* spp. being commonly used, showing several health benefits due to their action against degenerative diseases, cardiovascular diseases, cancer, and diabetes (Ambati et al., 2019). β -carotene from the microalga *Dunaliella salina* can be used as a natural colorant approved by the Food and Drug Administration and has potential application in the animal feed and food supplement sector and is currently gaining visibility in the cosmetics industry with the production of antiaging creams, sunscreens, and nail polishes. It also has an important antioxidant function in protecting brain cells from deterioration, reducing memory loss in diseases such as Alzheimer's, and reducing cancer development. The use of *Haematococcus* sp. microalgae extract containing astaxanthin has been shown in studies to have anti-inflammatory activity in eye treatment in rats and antimicrobial activity, reducing the growth of *H. pylori* in mice infected with gastric inflammation, also acting to prevent cancer and degenerative diseases and improve brain function and skin structure (Jeevanandam et al., 2020).

Another pigment that is gaining visibility for its biological potential is phycobiliprotein. Phycobiliproteins are water-soluble and organized into phycobilisome complexes. Phycoerythrin, allophycocyanin, and phycocyanin are the main phycobiliproteins in microalgae and function as accessory pigments that capture light during photosynthesis, with wavelength absorption in the 500–650 nm range (Jeevanandam et al., 2020; López-Hernández et al., 2022). The use of phycobiliproteins as fluorescent markers for cells and macromolecules in gel electrophoresis, gel exclusion chromatography, and isoelectric focusing is one of the main applications of this pigment, precisely because of its high molar absorptivity in the visible wavelength, favoring this application (Jeevanandam et al., 2020; Kuddus et al., 2013). In addition, they have antioxidant, hepatoprotective, and neuroprotective properties and are also commonly used as food coloring (Jeevanandam et al., 2020).

Among the phycobiliproteins, phycocyanin is the main constituent present in phycobilisome (Kuddus et al., 2013). It has a bluish color, with an optical absorbance of 615–640 nm (Jeevanandam et al., 2020) and critical biological properties, such as anti-inflammatory, antioxidant (Fratelli et al., 2021, 2022), hepatoprotective (Kuddus et al., 2013), neuroprotective (Jeevanandam et al., 2020), and antitumor activity (M. G. de Morais et al., 2018; Salgado et al., 2022), which allow it to be widely applied in pharmaceuticals, cosmetics, and diagnostic tests as a fluorescent marker (A. Srivastava et al., 2022); in the development of colorimetric indicators for food monitoring (Moreira et al., 2018; Terra et al., 2021); as a natural colorant for the development of ice cream, dairy products, jams, chewing gums, popsicles, and soft drinks (Muñoz-Miranda & Iñiguez-Moreno, 2023); and as a food supplement (M. G. de Morais et al., 2018).

Regarding its biological effects and potential health benefits, studies on rats have shown that phycocyanin significantly reduced cell death in the hippocampus and decreased liver inflammation in these animals (Jeevanandam et al., 2020). It has anticancer properties, acting against cancer cells in the colon and rectum, liver, and mammary glands, especially the highly purified phycocyanin extracted from the cyanobacterium *Thermosynechococcus elongatus* (López-Hernández et al., 2022).

Interest in microalgae has been growing in recent years precisely because of the population's growing concern and search for natural sources that promote health and the environment, and especially for the biological potential that their pigments exhibit, enabling a vast number of applications in different areas, aimed at promoting health, sustainability, and encouraging the circular economy. Therefore, studies have been carried out to improve biotechnological applications in developing new products and synthesizing pigments.

3.4 ANTIBACTERIAL POTENTIALS OF MICROALGAE

Microalgae are an attractive alternative to be used as bactericides as they have a broad spectrum of biological activities (Ribeiro et al., 2022). Bioactive compounds synthesized by this seawater algae may possess antibacterial, antialgal, antioxidant, antiviral, and antitumor properties (Saeed et al., 2022). These properties, together with the possibility of their application in high-added-value products, arouse the interest of many industries, such as pharmaceuticals, cosmetics, and refineries (Pina-Pérez et al., 2017).

Pharmaceutical industries are particularly interested in antimicrobial compounds derived from microalgae due to their wide range of applications, such as in food supplements, antibiotics, and wastewater treatment (Borowitzka, 2013; Leng et al., 2020; Senhorinho et al., 2015). The lack of new antibiotics and the increase in bacterial resistance to existing antibiotics have increased the search for new bactericidal compounds due to the threat to public health that microbial infections and/or reinfections represent (Senhorinho et al., 2015).

Among the different bioactive molecules that contribute to microalgae antimicrobial activity are polyunsaturated fatty acids (PUFAs), terpene derivatives, indoles, acetogenins, volatile halogenated hydrocarbons, and antioxidants (polyphenols, flavonoids, and carotenoids) (Hussein et al., 2020; Maadane et al., 2017).

To assess whether different microalgae species have bactericidal activity, they must extract their bioactive molecules with solvents of different polarities (Stirk & van Staden, 2022). Different methods can be used for cell disruptions and extraction, such as stirring, sonication, bead milling, hominization, Soxhlet extraction, pressurized liquid extraction, and supercritical CO_2 extraction (Falaise et al., 2016). After extraction, the extracts can be tested against a scope of microorganisms encompassing Gram-positive and Gram-negative bacteria in an in vitro antimicrobial test. To consider a microalgae species antibacterial, it must be active for at least one microorganism. The antibacterial activity of microalgae extracts depends on two factors: (1) the chosen solvent, since the solvent must be capable of extracting certain bioactive compounds, and (2) the sensitivity of bacteria to these bioactives (Pina-Pérez et al., 2017).

One of the microalgae species that has the most bactericidal activity is *Chlorella*. This microalga is widely studied because it is approved as a "generally recognized as safe" (GRAS) categorized single-cell protein nutritional supplement (Kose et al., 2017). In addition to having bactericidal properties by itself, microalgae can be used in the green synthesis of nanoparticles (NPs) for pharmaceutical applications due to numerous advantages, such as high biomass productivity, production of bioactive metabolites, low toxicity, and biodegradability (Stirk & van Staden, 2022).

Although several microalgae species are used in the NPs' synthesis, there are still few works in which these NPs have their bactericidal activity tested (Shankar et al., 2016). Among these research, silver nanoparticles (AgNPs) with antibacterial properties are most frequently synthesized (Muthusamy et al., 2017; Patel et al., 2015). Alone or in NP synthesis, the microalgae can generate potent bactericidal biocomposites for Gram-positive and Gram-negative bacteria. These biocomposites with antimicrobial power can be used to develop textile products, food packaging, surgical equipment, and other industrial and health applications (da Silva Ferreira et al., 2017). Another way of using these biocomposites is through their coupling in nanofibers or nanotubes, thus enhancing the bactericidal action (Karaduman et al., 2022).

Nanofibers with encapsulated bioproducts can help prevent reinfection by microorganisms as they can be incorporated for commercialization in food packaging, wound bandages, and surgical equipment. However, although microalgae have a promising future as a potent bactericidal agent alone or incorporated into nanoparticles, their diversity has yet to be fully explored, as not more than 30,000 have been studied (Sathasivam et al., 2019). Table 3.1 presents a list of microalgae whose compounds were extracted or used to synthesize NPs to test their antibacterial potential.

3.5 MICROALGAE FOR INDUSTRIAL EXTRACTION USING BIOREFINERY

Climate change is a fundamental concern in a society that depends on nonrenewable fossil fuels (i.e., natural gases, coal, and petroleum), which are highly associated with greenhouse gas emissions. Thus, addressing the demands for renewable resources for energy generation is mandatory. The concept of "waste" must be eliminated from current design frameworks, and instead focus should be on material and energy flows. Biorefineries combine biogenic sources with technologies intermediated by industries that process final products (Kamm & Kamm, 2004), reimagining the chemical value chain using non-fossil feedstocks (Zimmerman et al., 2020). Microalgal biorefinery has the potential for carbon net-zero emission through carbon dioxide sequestration (Wang et al., 2023).

The attractiveness of the microalgae biomass for the industrial sector begins with wastewater treatment and less land use, which is related to even more sustainability of the process, also generating biochar and bio-oil and pyrolytic gas (E. G. de Morais et al., 2023) through the pyrolysis method. Additionally, biodiesel from replacing traditional diesel or as a mixture and other valuable compounds were obtained from *Spirulina* (*Arthrospira*), with lipid yield of 19.8% and 47.9% for transesterification in situ or conventionally, respectively (E. G. de Morais et al., 2018), while biofixed $1.07 g.L^{-1}$ of carbon dioxide.

TABLE 3.1
Antibacterial Activity from Microalgae

Microalgae	Solvent	Compound	Form	Bacteria	Inhibition Zone (mm) and/or Minimum Inhibition Concentration (MIC/mg/mL)	Reference
Chlorella sp. (UKM8)	Methanol	Phenol, hexadecanoic acid, phytol, 9,12-octadecadienoic acid, and bicyclo[3.1.1] heptane	Microalgae extract	*P. aeruginosa*, *S. aureus*, *E. coli*, MRSA, *K. pneumoniae*	10.9/5 11/1.25 9/2.5 13.8/0.39 8.6/5	(Sathasivam et al., 2019)
I. galbana	Ethanol	Unsaturated fatty acids and carotenoids	Microalgae extract	*S. aureus*	20/-	(Alsenani et al., 2020)
Chlorella sp. FN1	Methanol/water	Unsaturated fatty acids and carotenoids	Microalgae extract	*S. aureus*	16.33/-	(Alsenani et al., 2020)
Scenedesmus sp. NT8c	Methanol/water	Unsaturated fatty acids and carotenoids	Microalgae extract	*S. aureus*	17.67/-	(Alsenani et al., 2020)
Coelastrella sp. BGV	Chloroform/methanol	Fatty acids	Microalgae extract	*E. coli* UPEC	25/0.3	(Toshkova-Yotova et al., 2022)
Chlorococcum humicola	Benzene	Carbohydrates, fatty acids, amino acids, saponins, carotenoids, flavonoids, alkaloids	Microalgae extract	*S. aureus*, *E. coli*	14/-, 12/-	(Bhagavathy et al., 2011)
Coccomyxa onubensis	Chloroform	Fatty acids	Microalgae extract	*E. coli*	17.2/-	(Navarro et al., 2017)
Spirulina plantensis	Ethanol	Alkaloids, saponins, phenols, and quinones	Microalgae suspension	*E. coli*	-/1,000	(Pratita & Fathurohman, 2020)
Spirulina	Chloroform	Phenolic compounds	Nanofibers/food packaging	*S. aureus* ATCC 25923	7.5/-	(Kuntzler et al., 2020)
Spirulina sp. LEB 18	Methanol	Phenolic compounds	Nanofibers	*S. aureus* ATCC 25923 *E. coli* ATCC 25972	6.4 /- 5.5/-	(Kuntzler et al., 2018)
Dunaliella salina/gold NPs		Algal suspension	NPs	*S. aureus*, *E. coli*	15/- 11 /-	(Basiratnia et al., 2021)
Chlorella vulgaris/AgCl (NPs)			NPs	*S. aureus*, *K. pneumoniae*	-/0.01	(Bhagavathy et al., 2011)

A study conducted by Farrokheh et al. (2022) aimed to process biodiesel at an industrial level from *Chlorella vulgaris* and *Spirulina platensis* by lipid extraction using electrocatalysis with nano-magnetic KOH/CaO-Fe_3O_4 and KF/KOH-Fe_3O_4. *Chlorella* was superior to *Spirulina* in terms of lipid content, and the results showed that 1.5% KF/KOH-Fe_3O_4 with microwave and reflux decreases the transesterification time reaction and improves the final product efficiency. The process yield was 96.8%, and the biodiesel followed the international standard ASTM D975 of fuels.

Recently, an improvement in the lipid content of microalgae *Tetraselmis elliptica* by optimizing the growth medium varying freshwater percentage, different nitrogen, and carbon sources, and trace metals such as $FeSO_4$, Mn SO_4, and Cu SO_4, showed the potential of its microalgae as biodiesel feedstock (Marey et al., 2022). In addition, the study has achieved the highest carbon sources destined for bioethanol production, once again contributing to a circular economy.

Additionally, lower-cost methods are of great interest. Thus, the harvest and exploitation of harmful algal blooms with toxicity that usually occurs in effluents are an arrangement for cost reduction. In this context, a recent approach demonstrated the use of microwave-assisted alkali-freeze pretreatment of algal blooms suitable as cheaper growing mediums and microwave-assisted hydrolysate for biomass and lipid accumulation in *Chlorella minutissima*, which showed twice more lipid content than the control medium (Kumar et al., 2022). It is worth mentioning that the lipid profile and the quality of the biodiesel properties of the obtained product were acceptable by international ASTM and EU standards.

Along these lines, biorefinery has been academically suggested from fruits rich in lipids and marine organisms such as algae (Amarante & Braga, 2021) and in running forward the circular economy since these matrices also provide bioactive compounds, pigments, and proteins for food and feed, pharmaceutical, and technological applications (Nakamoto et al., 2023; Fratelli et al., 2021). Commercializing biorefinery final products includes a call for energy savings (Wang et al., 2023) and cost reduction, such as low-cost growing medium, as studied by Kumar et al. (2022), for the accumulation of marine organisms and microalgae rich in lipids. These challenges must be overcome for industrial microalgal biorefinery in daily life.

Applying just one product from microalgae leads to less profitability and restricts the feasibility of utilizing such a rich matrix (Lopes et al., 2022). Thereby, using microalgae biomass for obtaining several bioproducts for various industry segments is a strategy for better handling with natural and renewable resources that must drive the industry's attention in an integrated framework, an output being an input for another processing.

Through literature search, microalgae are a promising approach for simultaneous biodiesel and bioethanol production concomitantly with reduced greenhouse gas emissions and carbon dioxide fixation. The methods evolved ultrasound and microwave-assisted extraction of lipids, hydrolysis, and pyrolysis, and other valuable compounds such as proteins and bioactive compounds with significant positive biological effects can be accessed, increasing the biorefinery range and cost-effectiveness.

3.6 MICROALGAE: CURRENT APPLICATIONS IN DAILY LIFE AND PROSPECTS FOR THE FUTURE

Microalgae have significantly impacted human life, especially in coastal cultures. In recent decades, there has been a remarkable increase in the interest and utilization of microalgae across a wide range of everyday applications. Due to their substantial nutritional value, microalgae are experiencing a notable resurgence in the global culinary landscape. As presented in this chapter, they are packed with vitamins, minerals, dietary fiber, and protein and are a rich source of omega-3 fatty acids known for their health benefits. As a result, products derived from microalgae, such as seaweed salads, snacks, condiments, and even algae-based burgers, are gaining popularity. These food options cater to the growing demand for vegetarian and vegan diets and are champions of a health-conscious and sustainable nutritional regimen (E. G. de Morais et al., 2023).

The modern era of biotechnological advancements in microalgae and cyanobacteria dates back to the early 1940s, gaining significant momentum with the algal mass culture symposium held at Stanford University in 1952. Since then, various technologies and methodologies for large-scale microalgae cultivation, including open-air ponds and specialized photobioreactors, have been introduced to enable commercial exploitation and sustainable harvesting. Microalgae find applications in diverse industries, including food processing, wastewater treatment, and pharmaceuticals.

Chinese dietary practices integrated *Nostoc* microalgae species approximately two millennia ago, followed by the recognition of the nutritional value of *Chlorella* and *Spirulina* species in regions like Taiwan, Japan, and Mexico (Hachicha et al., 2022). In recent decades, these four specific microalgae (*Spirulina*, *Chlorella*, *Dunaliella salina*, and *Haematococcus pluvialis*) have received significant attention for their biotechnological and nutritional applications, with *Chlorella* and *Spirulina* dominating the global microalgae market due to their nutritional profiles (Becker, 2013; Koyande et al., 2019).

Spirulina was recognized as "the best food for the future" at the 1974 United Nations World Food Conference. The World Health Organization (WHO) endorsed *Spirulina* for its abundance of iron and protein and suitability for children. The International Institution for the Use of *Spirulina* against Malnutrition (IIMSAM) and the Food and Agriculture Organization of the United Nations (FAO) highlighted the role of *Spirulina* in fighting hunger and malnutrition and for sustainable societal development (FDA, 2001).

Apart from the food sector, microalgae have found a substantial niche in the cosmetics and skin care industry. Their moisturizing, antioxidant, and anti-inflammatory properties have made them popular ingredients in creams, lotions, facial masks, and antiaging formulations. Numerous studies have documented bioactive compounds derived from microalgae, particularly *Spirulina*, showcasing beneficial effects on skin health and promoting wound healing, though the underlying mechanisms are still being studied (da Costa et al., 2023).

As research and development delve deeper into the potential of microalgae, we find ourselves on the brink of a revolution in harnessing these marine organisms. With the rising demand for sustainable and environmentally friendly alternatives, microalgae are poised to become essential for meeting future needs. The versatility of microalgae and their ability to address diverse needs in our modern world make them remarkable agents of change. Continued research and innovation in microalgae promise a healthier, more eco-friendly, and prosperous society. We are just scratching the surface of what microalgae can offer, and the potential is genuinely fascinating.

REFERENCES

Álvarez-González, A., Greque de Morais, E., Planas-Carbonell, A., & Uggetti, E. (2023). Enhancing sustainability through microalgae cultivation in urban wastewater for biostimulant production and nutrient recovery. *Science of the Total Environment*, *904*, 166878. https://doi.org/10.1016/j.scitotenv.2023.166878

Amarante, M. C. A. De, & Braga, A. R. C. (2021). Biofuels and oils from Amazon crops: Challenges and opportunities for the sustainable use of biodiversity resources. *17*(4), 204–213. https://doi.org/10.1089/IND.2021.0005. https://Home.Liebertpub.Com/Ind

Ambati, R. R., Gogisetty, D., Aswathanarayana, R. G., Ravi, S., Bikkina, P. N., Bo, L., & Yuepeng, S. (2019). Industrial potential of carotenoid pigments from microalgae: Current trends and future prospects. *Critical Reviews in Food Science and Nutrition*, *59*(12), 1880–1902. https://doi.org/10.1080/10408398.2018.1432561

Amorim, M. L., Soares, J., Coimbra, J. S. dos R., Leite, M. de O., Albino, L. F. T., & Martins, M. A. (2021). Microalgae proteins: Production, separation, isolation, quantification, and application in food and feed. *Critical Reviews in Food Science and Nutrition*, *61*(12), 1976–2002. https://doi.org/10.1080/10408398.2020.1768046

Ampofo, J., & Abbey, L. (2022). Microalgae: Bioactive composition, health benefits, safety and prospects as potential high-value ingredients for the functional food industry. *Foods (Basel, Switzerland)*, *11*(12). https://doi.org/10.3390/FOODS11121744

Apone, F., Barbulova, A., & Colucci, M. G. (2019). Plant and microalgae derived peptides are advantageously employed as bioactive compounds in cosmetics. *Frontiers in Plant Science*, *10*. https://doi.org/10.3389/fpls.2019.00756

Basiratnia, E., Einali, A., Azizian-Shermeh, O., Mollashahi, E., & Ghasemi, A. (2021). Biological synthesis of gold nanoparticles from suspensions of green microalga Dunaliella salina and their antibacterial potential. *BioNanoScience*, *11*(4), 977–988.

Becker, E. W. (2013). Microalgae for human and animal nutrition. In *Handbook of microalgal culture* (pp. 461–503). Wiley. https://doi.org/10.1002/9781118567166.ch25

Begum, H., Yusoff, F. M. D., Banerjee, S., Khatoon, H., & Shariff, M. (2016). Availability and utilization of pigments from microalgae. *Critical Reviews in Food Science and Nutrition*, *56*(13), 2209–2222. https://doi.org/10.1080/10408398.2013.764841

Bhagavathy, S., Sumathi, P., & Bell, I. J. S. (2011). Green algae Chlorococcum humicola-a new source of bioactive compounds with antimicrobial activity. *Asian Pacific Journal of Tropical Biomedicine*, *1*(1), S1–S7.

Bhatnagar, P., Gururani, P., Singh, N., Gautam, P., Vlaskin, M. S., & Kumar, V. (2023). Review on microalgae protein and its current and future utilisation in the food industry. *International Journal of Food Science & Technology*. https://doi.org/10.1111/ijfs.16586

Bilal, M., Rasheed, T., Ahmed, I., & Iqbal, H. M. N. (2017). High-value compounds from microalgae with industrial exploitability – a review. *Frontiers in Bioscience (Scholar Edition)*, *9*(3), 319–342. https://doi.org/10.2741/S490

Borowitzka, M. A. (2013). High-value products from microalgae-their development and commercialisation. *Journal of Applied Phycology*, *25*(3), 743–756. https://doi.org/10.1007/s10811-013-9983-9

Braun, J. C. A., & Colla, L. M. (2023). Use of microalgae for the development of biofertilizers and biostimulants. *BioEnergy Research*, *16*(1), 289–310. https://doi.org/10.1007/s12155-022-10456-8

Caetano, P. A., do Nascimento, T. C., Fernandes, A. S., Nass, P. P., Vieira, K. R., Maróstica Junior, M. R., Jacob-Lopes, E., & Zepka, L. Q. (2022). Microalgae-based polysaccharides: Insights on production, applications, analysis, and future challenges. *Biocatalysis and Agricultural Biotechnology*, *45*, 102491. https://doi.org/10.1016/j.bcab.2022.102491

Cao, K., Cui, Y., Sun, F., Zhang, H., Fan, J., Ge, B., Cao, Y., Wang, X., Zhu, X., Wei, Z., Yao, Q., Ma, J., Wang, Y., Meng, C., & Gao, Z. (2023). Metabolic engineering and synthetic biology strategies for producing high-value natural pigments in Microalgae. *Biotechnology Advances*, *68*, 108236. https://doi.org/10.1016/j.biotechadv.2023.108236

Cardozo, K. H. M., Guaratini, T., Barros, M. P., Falcão, V. R., Tonon, A. P., Lopes, N. P., Campos, S., Torres, M. A., Souza, A. O., Colepicolo, P., & Pinto, E. (2007). Metabolites from algae with economical impact. *Comparative Biochemistry and Physiology Part C: Toxicology & Pharmacology*, *146*(1–2), 60–78. https://doi.org/10.1016/J.CBPC.2006.05.007

Cheirsilp, B., Maneechote, W., Srinuanpan, S., & Angelidaki, I. (2023). Microalgae as tools for bio-circular-green economy: Zero-waste approaches for sustainable production and biorefineries of microalgal biomass. *Bioresource Technology*, *387*, 129620. https://doi.org/10.1016/j.biortech.2023.129620

Chen, B., You, W., Huang, J., Yu, Y., & Chen, W. (2010). Isolation and antioxidant property of the extracellular polysaccharide from *Rhodella reticulata*. *World Journal of Microbiology and Biotechnology*, *26*(5), 833–840. https://doi.org/10.1007/S11274-009-0240-Y

Chester, K., Zahiruddin, S., Ahmad, A., Khan, W., Paliwal, S., & Ahmad, S. (2019). Bioautography-based identification of antioxidant metabolites of *Solanum nigrum* L. and exploration its hepatoprotective potential against D-galactosamine-induced hepatic fibrosis in rats. *Pharmacognosy Magazine*, *2019*, 104–110.

Chini Zittelli, G., Lauceri, R., Faraloni, C., Silva Benavides, A. M., & Torzillo, G. (2023). Valuable pigments from microalgae: Phycobiliproteins, primary carotenoids, and fucoxanthin. *Photochemical & Photobiological Sciences*, *22*(8), 1733–1789. https://doi.org/10.1007/s43630-023-00407-3

Condor, B. E., de Luna, M. D. G., Abarca, R. R. M., Chang, Y., Leong, Y. K., Chen, C., Chen, P., Lee, D., & Chang, J. (2022). Optimization and modeling of carbohydrate production in microalgae for use as feedstock in bioethanol fermentation. *International Journal of Energy Research*, *46*(13), 19300–19312. https://doi.org/10.1002/er.7709

da Costa, S. F. F., dos Passos Ramos, S., Bürck, M., & Braga, A. R. C. (2023). Development of eco-friendly solid shampoo containing natural pigments: Physical-chemical, microbiological characterization and analysis of antioxidant activity. *Industrial Biotechnology*, *19*(6), 337–346. https://doi.org/10.1089/ind.2023.0011

da Silva Ferreira, V., ConzFerreira, M. E., Lima, L. M. T. R., Frasés, S., de Souza, W., & Sant'Anna, C. (2017). Green production of microalgae-based silver chloride nanoparticles with antimicrobial activity against pathogenic bacteria. *Enzyme and Microbial Technology*, *97*, 114–121. https://doi.org/10.1016/j.enzmictec.2016.10.018

De Jesus Raposo, M. F., de Morais, A. M. M. B., & de Morais, R. M. S. C. (2016). Emergent sources of prebiotics: Seaweeds and microalgae. *Marine Drugs*, *14*(2). https://doi.org/10.3390/MD14020027

De Luca, M., Pappalardo, I., Limongi, A. R., Viviano, E., Radice, R. P., Todisco, S., Martelli, G., Infantino, V., & Vassallo, A. (2021). Lipids from microalgae for cosmetic applications. *Cosmetics*, *8*(2), 52. https://doi.org/10.3390/cosmetics8020052

de Morais, E. G., Cassuriag, A. P. A., Callejas, N., Martinez, N., Vieitez, I., Jachmanián, I., Santos, L. O., de Morais, M. G., & Costa, J. A. V. (2018). Evaluation of CO2 biofixation and biodiesel production by Spirulina (Arthospira) cultivated in air-lift photobioreactor. *Brazilian Archives of Biology and Technology*, *61*, e18161339. https://doi.org/10.1590/1678-4324-2018161339

de Morais, E. G., da Silveira, J. T., Schüler, L. M., de Freitas, B. C. B., Costa, J. A. V., de Morais, M. G., Ferrer, I., & Barreira, L. (2023). Biomass valorization via pyrolysis in microalgae-based wastewater treatment: Challenges and opportunities for a circular bioeconomy. *Journal of Applied Phycology*, *1*, 1–20. https://doi.org/10.1007/S10811-023-03104-X

de Morais, M. G., da Fontoura Prates, D., Moreira, J. B., Duarte, J. H., & Costa, J. A. V. (2018). Phycocyanin from microalgae: Properties, extraction and purification, with some recent applications. *Industrial Biotechnology*, *14*(1), 30–37. https://doi.org/10.1089/ind.2017.0009

de Morais, M. G., Vaz, B. D. S., de Morais, E. G., & Costa, J. A. V. (2015). Biologically active metabolites synthesized by microalgae. *BioMed Research International*, *2015*. https://doi.org/10.1155/2015/835761

de Vera, C. R., Crespín, G. D., Daranas, A. H., Looga, S. M., Lillsunde, K. E., Tammela, P., Perälä, M., Hongisto, V., Virtanen, J., Rischer, H., Muller, C. D., Norte, M., Fernández, J. J., & Souto, M. L. (2018). Marine microalgae: Promising source for new bioactive compounds. *Marine Drugs*, *16*(9). https://doi.org/10.3390/MD16090317

Demirbas, A., & Fatih Demirbas, M. (2011). Importance of algae oil as a source of biodiesel. *Energy Conversion and Management*, *52*(1), 163–170. https://doi.org/10.1016/j.enconman.2010.06.055

Dolganyuk, V., Belova, D., Babich, O., Prosekov, A., Ivanova, S., Katserov, D., Patyukov, N., & Sukhikh, S. (2020). Microalgae: A promising source of valuable bioproducts. *Biomolecules*, *10*(8), 1–24. https://doi.org/10.3390/BIOM10081153

Eilam, Y., Khattib, H., Pintel, N., & Avni, D. (2023). Microalgae—sustainable source for alternative proteins and functional ingredients promoting gut and liver health. *Global Challenges*, *7*(5). https://doi.org/10.1002/gch2.202200177

Encarnação, T., Pais, A. A. C. C., Campos, M. G., & Burrows, H. D. (2015). Cyanobacteria and microalgae: A renewable source of bioactive compounds and other chemicals. *Science Progress*, *98*(Pt 2), 145–168. https://doi.org/10.3184/003685015X14298590596266

Eze, C. N., Onyejiaka, C. K., Ihim, S. A., Ayoka, T. O., Aduba, C. C., Ndukwe, J. K., Nwaiwu, O., & Onyeaka, H. (2023). Bioactive compounds by microalgae and potentials for the management of some human disease conditions. *AIMS Microbiology*, *9*(1), 55–74. https://doi.org/10.3934/MICROBIOL.2023004

Falaise, C., François, C., Travers, M.-A., Morga, B., Haure, J., Tremblay, R., Turcotte, F., Pasetto, P., Gastineau, R., Hardivillier, Y., Leignel, V., & Mouget, J.-L. (2016). Antimicrobial compounds from eukaryotic microalgae against human pathogens and diseases in aquaculture. *Marine Drugs*, *14*(9), 159. https://doi.org/10.3390/md14090159

Farrokheh, A., Tahvildari, K., & Nozari, M. (2022). Biodiesel production from the *Chlorella vulgaris* and *Spirulina platensis* microalgae by electrolysis using CaO/KOH-Fe3O4 and KF/KOH-Fe3O4 as magnetic nanocatalysts. *Biomass Conversion and Biorefinery*, *12*(2), 403–417. https://doi.org/10.1007/S13399-020-00688-Z/TABLES/7

FDA. (2001). Spirulina extract. In *21 Code of federal legislations* (p. 73.1530). https://www.ecfr.gov/current/title-21/chapter-I/subchapter-A/part-73/subpart-B/section-73.1530

Feng, L., Wu, J., Song, J., Li, D., Zhang, Z., Xu, Y., Yang, R., Liu, C., & Zhang, M. (2021). Effect of particle size distribution on the carotenoids release, physicochemical properties and 3D printing characteristics of carrot pulp. *LWT*, *139*, 110576. https://doi.org/10.1016/J.LWT.2020.110576

Fernandes, A. S., Nass, P. P., Oliveira, Á., & Zepka, L. Q. (2020). Chlorophylls as food additives BT. In E. Jacob-Lopes, M. I. Queiroz, & L. Q. Zepka (Eds.), *Pigments from microalgae handbook* (pp. 391–420). Springer International Publishing. https://doi.org/10.1007/978-3-030-50971-2_16

Franco-Morgado, M., Amador-Espejo, G. G., Pérez-Cortés, M., & Gutiérrez-Uribe, J. A. (2023). Microalgae and cyanobacteria polysaccharides: Important link for nutrient recycling and revalorization of agro-industrial wastewater. *Applied Food Research*, *3*(1), 100296. https://doi.org/10.1016/j.afres.2023.100296

Fratelli, C., Burck, M., Amarante, M. C. A., & Braga, A. R. C. (2021). Antioxidant potential of nature's "something blue": Something new in the marriage of biological activity and extraction methods applied to C-phycocyanin. *Trends in Food Science & Technology*, *107*(October 2020), 309–323. https://doi.org/10.1016/j.tifs.2020.10.043

Fratelli, C., Bürck, M., Silva-Neto, A. F., Oyama, L. M., De Rosso, V. V., & Braga, A. R. C. (2022). Green extraction process of food grade C-phycocyanin: Biological effects and metabolic study in mice. *Processes*, *10*(9), 1793. https://doi.org/10.3390/pr10091793

Goh, B. H. H., Ong, H. C., Cheah, M. Y., Chen, W.-H., Yu, K. L., & Mahlia, T. M. I. (2019). Sustainability of direct biodiesel synthesis from microalgae biomass: A critical review. *Renewable and Sustainable Energy Reviews*, *107*, 59–74. https://doi.org/10.1016/j.rser.2019.02.012

Goshtasbi, H., Okolodkov, Y. B., Movafeghi, A., Awale, S., Safary, A., Barar, J., & Omidi, Y. (2023). Harnessing microalgae as sustainable cellular factories for biopharmaceutical production. *Algal Research*, *74*, 103237. https://doi.org/10.1016/j.algal.2023.103237

Goyal, S., Dhanker, R., Hussain, T., Ferreira, A., Gouveia, L., Kumar, K., & Mohamed, H. I. (2023). Modern advancement in biotechnological applications for wastewater treatment through microalgae: A review. *Water, Air, & Soil Pollution*, *234*(7), 417. https://doi.org/10.1007/s11270-023-06409-2

Hachicha, R., Elleuch, F., Ben Hlima, H., Dubessay, P., de Baynast, H., Delattre, C., Pierre, G., Hachicha, R., Abdelkafi, S., Michaud, P., & Fendri, I. (2022). Biomolecules from microalgae and cyanobacteria: Applications and market survey. *Applied Sciences*, *12*(4), 1924. https://doi.org/10.3390/app12041924

Hoang, A. T., Sirohi, R., Pandey, A., Nižetić, S., Lam, S. S., Chen, W.-H., Luque, R., Thomas, S., Arıcı, M., & Pham, V. V. (2023). Biofuel production from microalgae: Challenges and chances. *Phytochemistry Reviews*, *22*(4), 1089–1126. https://doi.org/10.1007/s11101-022-09819-y

Honda, M. (2020). Nutraceutical and pharmaceutical applications of carotenoids BT. In E. Jacob-Lopes, M. I. Queiroz, & L. Q. Zepka (Eds.), *Pigments from microalgae handbook* (pp. 449–469). Springer International Publishing. https://doi.org/10.1007/978-3-030-50971-2_18

Hoseini, S. M., Khosravi-Darani, K., & Mozafari, M. R. (2013). Nutritional and medical applications of spirulina microalgae. *Mini Reviews in Medicinal Chemistry*, *13*(8), 1231–1237. https://doi.org/10.2174/1389557511313080009

Hussein, H. A., Syamsumir, D. F., Radzi, S. A. M., Siong, J. Y. F., Zin, N. A. M., & Abdullah, M. A. (2020). Phytochemical screening, metabolite profiling and enhanced antimicrobial activities of microalgal crude extracts in co-application with silver nanoparticle. *Bioresources and Bioprocessing*, *7*(1), 39. https://doi.org/10.1186/s40643-020-00322-w

Huynh, A., Maktabi, B., Reddy, C. M., O'Neil, G. W., Chandler, M., & Baki, G. (2020). Evaluation of alkenones, a renewably sourced, plant-derived wax as a structuring agent for lipsticks. *International Journal of Cosmetic Science*, *42*(2), 146–155. https://doi.org/10.1111/ics.12597

Jeevanandam, J., Choudhary, V., Selvam, J. D., & Danquah, M. K. (2020). The bioeconomy of production of microalgal pigments BT. In E. Jacob-Lopes, M. I. Queiroz, & L. Q. Zepka (Eds.), *Pigments from microalgae handbook* (pp. 325–362). Springer International Publishing. https://doi.org/10.1007/978-3-030-50971-2_14

Jesionowska, M., Ovadia, J., Hockemeyer, K., Clews, A. C., & Xu, Y. (2023). EPA and DHA in microalgae: Health benefits, biosynthesis, and metabolic engineering advances. *Journal of the American Oil Chemists' Society*. https://doi.org/10.1002/aocs.12718

Kamm, B., & Kamm, M. (2004). Biorefinery – systems. *Chemical and Biochemical Engineering Quarterly*, *18*(1), 1–6. https://doi.org/10.1016/b978-0-444-59561-4.00014-0

Karaduman, F. R., Türk Çulha, S., & Horzum, N. (2022). Algal nanofibers: Current status and recent developments. *Materials Today Communications*, *33*, 104248. https://doi.org/10.1016/j.mtcomm.2022.104248

Kholssi, R., Ramos, P. V., Marks, E. A. N., Montero, O., & Rad, C. (2021). 2Biotechnological uses of microalgae: A review on the state of the art and challenges for the circular economy. *Biocatalysis and Agricultural Biotechnology*, *36*, 102114. https://doi.org/10.1016/J.BCAB.2021.102114

Kose, A., Ozen, M. O., Elibol, M., & Oncel, S. S. (2017). Investigation of in vitro digestibility of dietary microalga *Chlorella vulgaris* and cyanobacterium Spirulina platensis as a nutritional supplement. *3 Biotech*, *7*(3), 170. https://doi.org/10.1007/s13205-017-0832-4

Koyande, A. K., Chew, K. W., Rambabu, K., Tao, Y., Chu, D.-T., & Show, P.-L. (2019). Microalgae: A potential alternative to health supplementation for humans. *Food Science and Human Wellness*, *8*(1), 16–24. https://doi.org/10.1016/j.fshw.2019.03.001

Kuddus, M., Singh, P., Thomas, G., & Al-Hazimi, A. (2013). Recent developments in production and biotechnological applications of c-phycocyanin. *BioMed Research International*, *2013*. https://doi.org/10.1155/2013/742859

Kumar, V., Arora, N., Pandey, S., Jaiswal, K. K., Nanda, M., Vlaskin, M. S., & Chauhan, P. K. (2022). Microwave-assisted pretreatment of harmful algal blooms for microbial oil-centered biorefinery approach. *Biomass Conversion and Biorefinery*, *12*(8), 3097–3105. https://doi.org/10.1007/S13399-020-00941-5/TABLES/3

Kumari, A., Pabbi, S., & Tyagi, A. (2023). Recent advances in enhancing the production of long chain omega-3 fatty acids in microalgae. *Critical Reviews in Food Science and Nutrition*, 1–19. https://doi.org/10.1080/10408398.2023.2226720

Kuntzler, S. G., Costa, J. A. V., Brizio, A. P. D. R., & de Morais, M. G. (2020). Development of a colorimetric pH indicator using nanofibers containing Spirulina sp. LEB 18. *Food Chemistry*, *328*, 126768.

Kuntzler, S. G., de Almeida, A. C. A., Costa, J. A. V., & de Morais, M. G. (2018). Polyhydroxybutyrate and phenolic compounds microalgae electrospun nanofibers: A novel nanomaterial with antibacterial activity. *International Journal of Biological Macromolecules*, *113*, 1008–1014.

Lauritano, C., Andersen, J. H., Hansen, E., Albrigtsen, M., Escalera, L., Esposito, F., Helland, K., Hanssen, K., Romano, G., & Ianora, A. (2016). Bioactivity screening of microalgae for antioxidant, anti-inflammatory, anticancer, anti-diabetes, and antibacterial activities. *Frontiers in Marine Science*, *3*(May), 1–2. https://doi.org/10.3389/FMARS.2016.00068/BIBTEX

Leng, L., Wei, L., Xiong, Q., Xu, S., Li, W., Lv, S., Lu, Q., Wan, L., Wen, Z., & Zhou, W. (2020). Use of microalgae based technology for the removal of antibiotics from wastewater: A review. *Chemosphere*, *238*, 124680. https://doi.org/10.1016/j.chemosphere.2019.124680

Li, X., Liu, J., Chen, G., Zhang, J., Wang, C., & Liu, B. (2019). Extraction and purification of eicosapentaenoic acid and docosahexaenoic acid from microalgae: A critical review. *Algal Research*, *43*, 101619. https://doi.org/10.1016/j.algal.2019.101619

Li, Y., Horsman, M., Wang, B., Wu, N., & Lan, C. Q. (2008). Effects of nitrogen sources on cell growth and lipid accumulation of green alga Neochloris oleoabundans. *Applied Microbiology and Biotechnology*, *81*, 629–636.

Lin, Y.-H., Chen, G.-W., Yeh, C., Song, H., & Tsai, J.-S. (2018). Purification and identification of angiotensin I-converting enzyme inhibitory peptides and the antihypertensive effect of *Chlorella sorokiniana* protein hydrolysates. *Nutrients*, *10*(10), 1397. https://doi.org/10.3390/nu10101397

Lopes, L. C., da Costa, M. R. F., de Morais, M. G., Costa, J. A. V., & Moreira, J. B. (2022). Nanostructures in microalgae biorefinery: A potential approach in the blue economy design. *Biomass Conversion and Biorefinery*, *1*, 1–12. https://doi.org/10.1007/S13399-022-03400-5/TABLES/2

López-Hernández, J.-F., Kean-Meng, T., Asencio-Alcudia, G.-G., Asyraf-Kassim, M., Alvarez-González, C.-A., & Márquez-Rocha, F.-J. (2022). Sustainable microalgae and cyanobacteria biotechnology. *Applied Sciences*, *12*(14). https://doi.org/10.3390/app12146887

Ma, R., Wang, B., Chua, E. T., Zhao, X., Lu, K., Ho, S. H., Shi, X., Liu, L., Xie, Y., Lu, Y., & Chen, J. (2020). Comprehensive utilization of marine microalgae for enhanced co-production of multiple compounds. *Marine Drugs*, *18*(9). https://doi.org/10.3390/MD18090467

Maadane, A., Merghoub, N., El Mernissi, N., Ainane, T., Amzazi, S., Wahby, I., & Bakri, Y. (2017). Antimicrobial activity of marine microalgae isolated from Moroccan coastlines. *Journal of Microbiology, Biotechnology and Food Sciences*, *6*(6), 1257–1260. https://doi.org/10.15414/jmbfs.2017.6.6.1257-1260

Maity, S., & Mallick, N. (2022). Trends and advances in sustainable bioethanol production by marine microalgae: A critical review. *Journal of Cleaner Production*, *345*, 131153. https://doi.org/10.1016/j.jclepro.2022.131153

Mapelli-Brahm, P., Gómez-Villegas, P., Gonda, M. L., León-Vaz, A., León, R., Mildenberger, J., Rebours, C., Saravia, V., Vero, S., Vila, E., & Meléndez-Martínez, A. J. (2023). Microalgae, seaweeds and aquatic bacteria, archaea, and yeasts: Sources of carotenoids with potential antioxidant and anti-inflammatory health-promoting actions in the sustainability era. *Marine Drugs*, *21*(6), 340. https://doi.org/10.3390/md21060340

Marey, R. S., Abo-Shady, A. M., Khairy, H. M., Abd El-Moneim, A. M., & Abomohra, A. (2022). Enhanced lipid production and essential ω-fatty acids synthesis by the hypersaline biodiesel-promising microalga *Tetraselmis elliptica* through growth medium optimization. *Biomass Conversion and Biorefinery*, *1*, 1–14. https://doi.org/10.1007/S13399-022-03290-7/TABLES/7

Martínez-Ruiz, M., Martínez-González, C. A., Kim, D. H., Santiesteban-Romero, B., Reyes-Pardo, H., Villaseñor-Zepeda, K. R., Meléndez-Sánchez, E. R., Ramírez-Gamboa, D., Díaz-Zamorano, A. L., Sosa-Hernández, J. E., Coronado-Apodaca, K. G., Gámez-Méndez, A. M., Iqbal, H. M. N., & Parra-Saldivar, R. (2022). Microalgae bioactive compounds to topical applications products—a review. *Molecules (Basel, Switzerland)*, *27*(11). https://doi.org/10.3390/MOLECULES27113512

McIntosh, K., Smith, A., Young, L., Leitch, M., Tiwari, A., Reddy, C., O'Neil, G., Liberatore, M., Chandler, M., & Baki, G. (2018). Alkenones as a promising green alternative for waxes in cosmetics and personal care products. *Cosmetics*, *5*(2), 34. https://doi.org/10.3390/cosmetics5020034

Montalvo, G. E. B., Thomaz-Soccol, V., Vandenberghe, L. P. S., Carvalho, J. C., Faulds, C. B., Bertrand, E., Prado, M. R. M., Bonatto, S. J. R., & Soccol, C. R. (2019). *Arthrospira maxima* OF15 biomass cultivation at laboratory and pilot scale from sugarcane vinasse for potential biological new peptides production. *Bioresource Technology*, *273*, 103–113. https://doi.org/10.1016/j.biortech.2018.10.081

Moreira, J. B., Terra, A. L. M., Costa, J. A. V., & Morais, M. G. de. (2018). Development of pH indicator from PLA/PEO ultrafine fibers containing pigment of microalgae origin. *International Journal of Biological Macromolecules*, *118*, 1855–1862. https://doi.org/10.1016/J.IJBIOMAC.2018.07.028

Morocho-Jácome, A. L., dos Santos, B. B., de Carvalho, J. C. M., de Almeida, T. S., Rijo, P., Velasco, M. V. R., Rosado, C., & Baby, A. R. (2022). Microalgae as a sustainable, natural-oriented and vegan dermocosmetic bioactive ingredient: The case of *Neochloris oleoabundans*. *Cosmetics*, *9*(1), 9. https://doi.org/10.3390/cosmetics9010009

Mückschel, F., Ollo, E., Glaeser, S. P., Düring, R., Yan, F., Velten, H., Theilen, U., & Frei, M. (2023). Nitrogen use efficiency of microalgae application in wheat compared to mineral fertilizer. *Journal of Plant Nutrition and Soil Science*, *186*(5), 522–531. https://doi.org/10.1002/jpln.202300125

Muñoz-Miranda, L. A., & Iñiguez-Moreno, M. (2023). An extensive review of marine pigments: Sources, biotechnological applications, and sustainability. *Aquatic Sciences*, *85*(3), 68. https://doi.org/10.1007/s00027-023-00966-8

Muthusamy, G., Thangasamy, S., Raja, M., Chinnappan, S., & Kandasamy, S. (2017). Biosynthesis of silver nanoparticles from Spirulina microalgae and its antibacterial activity. *Environmental Science and Pollution Research*, *24*(23), 19459–19464. https://doi.org/10.1007/s11356-017-9772-0

Nakamoto, M. M., Assis, M., de Oliveira Filho, J. G., & Braga, A. R. C. (2023). Spirulina application in food packaging: Gaps of knowledge and future trends. *Trends in Food Science & Technology*, *133*, 138–147. https://doi.org/10.1016/J.TIFS.2023.02.001

Navarro, F., Forján, E., Vázquez, M., Toimil, A., Montero, Z., Ruiz-Domínguez, M. D. C., . . . & Vega, J. M. (2017). Antimicrobial activity of the acidophilic eukaryotic microalga Coccomyxa onubensis. *Phycological Research*, *65*(1), 38–43.

Novoveská, L., Ross, M. E., Stanley, M. S., Pradelles, R., Wasiolek, V., & Sassi, J.-F. (2019). Microalgal carotenoids: A review of production, current markets, regulations, and future direction. *Marine Drugs*, *17*(11), 640. https://doi.org/10.3390/md17110640

Parameswari, R. P., & Lakshmi, T. (2022). Microalgae as a potential therapeutic drug candidate for neurodegenerative diseases. *Journal of Biotechnology*, *358*, 128–139. https://doi.org/10.1016/J.JBIOTEC.2022.09.009

Patel, V., Berthold, D., Puranik, P., & Gantar, M. (2015). Screening of cyanobacteria and microalgae for their ability to synthesize silver nanoparticles with antibacterial activity. *Biotechnology Reports*, *5*, 112–119. https://doi.org/10.1016/j.btre.2014.12.001

Peltomaa, E., Hällfors, H., & Taipale, S. J. (2019). Comparison of diatoms and dinoflagellates from different habitats as sources of PUFAs. *Marine Drugs*, *17*(4), 233. https://doi.org/10.3390/md17040233

Pentón-Rol, G., Marín-Prida, J., & McCarty, M. F. (2021). C-Phycocyanin-derived phycocyanobilin as a potential nutraceutical approach for major neurodegenerative disorders and COVID-19-induced damage to the nervous system. *Current Neuropharmacology*, *19*(12), 2250–2275. https://doi.org/10.2174/1570159X19666210408123807

Pina-Pérez, M. C., Rivas, A., Martínez, A., & Rodrigo, D. (2017). Antimicrobial potential of macro and microalgae against pathogenic and spoilage microorganisms in food. *Food Chemistry*, *235*, 34–44. https://doi.org/10.1016/j.foodchem.2017.05.033

Pratita, A. T. K., & Fathurohman, M. (2020, June). Potential of autotrophic microalgae (spirulina plantensis) in decreased Coliform bacteria. In *2nd Bakti Tunas Husada-health science international conference (BTH-HSIC 2019)* (pp. 208–210). Atlantis Press.

Rai, P., Mehrotra, S., Priya, S., Gnansounou, E., & Sharma, S. K. (2021). Recent advances in the sustainable design and applications of biodegradable polymers. *Bioresource Technology*, *325*, 124739. https://doi.org/10.1016/J.BIORTECH.2021.124739

Rajesh, K., Rohit, M. V., & Venkata Mohan, S. (2017). Microalgae-based carotenoids production. In *Algal green chemistry: Recent progress in biotechnology* (pp. 139–147). https://doi.org/10.1016/B978-0-444-63784-0.00007-2

Ramesh Kumar, B., Deviram, G., Mathimani, T., Duc, P. A., & Pugazhendhi, A. (2019). Microalgae as rich source of polyunsaturated fatty acids. *Biocatalysis and Agricultural Biotechnology*, *17*, 583–588. https://doi.org/10.1016/j.bcab.2019.01.017

Ramos-Romero, S., Torrella, J. R., Pagès, T., Viscor, G., & Torres, J. L. (2021). Edible microalgae and their bioactive compounds in the prevention and treatment of metabolic alterations. *Nutrients*, *13*(2), 1–16. https://doi.org/10.3390/NU13020563

Ribeiro, M. C. M., Salles, T. S., Moreira, M. F., Barbarino, E., do Valle, A. F., & Couto, M. A. P. G. (2022). Antiviral activity of microalgae extracts against Mayaro virus. *Algal Research*, *61*, 102577. https://doi.org/10.1016/j.algal.2021.102577

Riccio, G., De Luca, D., & Lauritano, C. (2020). Monogalactosyldiacylglycerol and sulfolipid synthesis in microalgae. *Marine Drugs*, *18*(5). https://doi.org/10.3390/MD18050237

Saeed, M. U., Hussain, N., Shahbaz, A., Hameed, T., Iqbal, H. M. N., & Bilal, M. (2022). Bioprospecting microalgae and cyanobacteria for biopharmaceutical applications. *Journal of Basic Microbiology*, *62*(9), 1110–1124. https://doi.org/10.1002/JOBM.202100445

Saide, A., Martínez, K. A., Ianora, A., & Lauritano, C. (2021). Unlocking the health potential of microalgae as sustainable sources of bioactive compounds. *International Journal of Molecular Sciences*, *22*(9). https://doi.org/10.3390/IJMS22094383

Sajjadi, B., Chen, W.-Y., Raman, A. A. A., & Ibrahim, S. (2018). Microalgae lipid and biomass for biofuel production: A comprehensive review on lipid enhancement strategies and their effects on fatty acid composition. *Renewable and Sustainable Energy Reviews*, *97*, 200–232. https://doi.org/10.1016/j.rser.2018.07.050

Salgado, M. T. S. F., Fernandes e Silva, E., Matsumoto, A. M., Mattozo, F. H., Amarante, M. C. A. de, Kalil, S. J., & Votto, A. P. de S. (2022). C-phycocyanin decreases proliferation and migration of melanoma cells: In silico and in vitro evidences. *Bioorganic Chemistry*, *122*, 105757. https://doi.org/10.1016/j.bioorg.2022.105757

Sathasivam, R., Radhakrishnan, R., Hashem, A., & Abd Allah, E. F. (2019). Microalgae metabolites: A rich source for food and medicine. *Saudi Journal of Biological Sciences*, *26*(4), 709–722. https://doi.org/10.1016/j.sjbs.2017.11.003

Senhorinho, G., Ross, G., & Scott, J. (2015). Cyanobacteria and eukaryotic microalgae as potential sources of antibiotics. *Phycologua*, *54*, 271–282.

Shankar, P. D., Shobana, S., Karuppusamy, I., Pugazhendhi, A., Ramkumar, V. S., Arvindnarayan, S., & Kumar, G. (2016). A review on the biosynthesis of metallic nanoparticles (gold and silver) using bio-components of microalgae: Formation mechanism and applications. *Enzyme and Microbial Technology*, *95*, 28–44. https://doi.org/10.1016/j.enzmictec.2016.10.015

Sherratt, S. C. R., Lero, M., & Mason, R. P. (2020). Are dietary fish oil supplements appropriate for dyslipidemia management? A review of the evidence. *Current Opinion in Lipidology*, *31*(2), 94–100. https://doi.org/10.1097/MOL.0000000000000665

Shi, Q., Chen, C., Zhang, W., Wu, P., Sun, M., Wu, H., Wu, H., Fu, P., & Fan, J. (2021). Transgenic eukaryotic microalgae as green factories: providing new ideas for the production of biologically active substances. *Journal of Applied Phycology*, *33*(2), 705–728. https://doi.org/10.1007/s10811-020-02350-7

Sidari, R., & Tofalo, R. (2019). A comprehensive overview on microalgal-fortified/based food and beverages. *Food Reviews International*, *35*(8), 778–805. https://doi.org/10.1080/87559129.2019.1608557

Silva, S. C., Ferreira, I. C. F. R., Dias, M. M., & Filomena Barreiro, M. (2020). Microalgae-derived pigments: A 10-year bibliometric review and industry and market trend analysis. *Molecules*, *25*(15). MDPI AG. https://doi.org/10.3390/molecules25153406

Singh, K., Ansari, F. A., Ingle, K. N., Gupta, S. K., Ahirwal, J., Dhyani, S., Singh, S., Abhilash, P. C., Rawat, I., Byun, C., & Bux, F. (2023). Microalgae from wastewaters to wastelands: Leveraging microalgal research conducive to achieve the UN Sustainable Development Goals. *Renewable and Sustainable Energy Reviews*, *188*, 113773. https://doi.org/10.1016/j.rser.2023.113773

Sirohi, R., Ummalyma, S. B., Sagar, N. A., Sharma, P., Awasthi, M. K., Badgujar, P. C., Madhavan, A., Rajasekharan, R., Sindhu, R., Sim, S. J., & Pandey, A. (2021). Strategies and advances in the pre-treatment of microalgal biomass. *Journal of Biotechnology*, *341*, 63–75. https://doi.org/10.1016/j.jbiotec.2021.09.010

Srimongkol, P., Songserm, P., Kuptawach, K., Puthong, S., Sangtanoo, P., Thitiprasert, S., Thongchul, N., Phunpruch, S., & Karnchanatat, A. (2023). Sulfated polysaccharides derived from marine microalgae, *Synechococcus* sp. VDW, inhibit the human colon cancer cell line Caco-2 by promoting cell apoptosis via the JNK and p38 MAPK signaling pathway. *Algal Research*, *69*, 102919. https://doi.org/10.1016/j.algal.2022.102919

Srivastava, A., Kalwani, M., Chakdar, H., Pabbi, S., & Shukla, P. (2022). Biosynthesis and biotechnological interventions for commercial production of microalgal pigments: A review. *Bioresource Technology*, *352*(January). https://doi.org/10.1016/j.biortech.2022.127071

Srivastava, A. K., Dev, A., & Karmakar, S. (2018). Nanosensors and nanobiosensors in food and agriculture. *Environmental Chemistry Letters*, *16*(1), 161–182. https://doi.org/10.1007/s10311-017-0674-7

Stark, M., & O'Gara, I. (2012). An introduction to photosynthetic microalgae. *1*(2), 65–67. https://doi.org/10.1089/DST.2012.0017. https://Home.Liebertpub.Com/Dst.

Stirk, W. A., & van Staden, J. (2022). Bioprospecting for bioactive compounds in microalgae: Antimicrobial compounds. *Biotechnology Advances*, *59*. https://doi.org/10.1016/J.BIOTECHADV.2022.107977

Sukhikh, S., Prosekov, A., Ivanova, S., Maslennikov, P., Andreeva, A., Budenkova, E., Kashirskikh, E., Tcibulnikova, A., Zemliakova, E., Samusev, I., & Babich, O. (2022). Identification of metabolites with antibacterial activities by analyzing the FTIR spectra of microalgae. *Life (Basel, Switzerland)*, *12*(9). https://doi.org/10.3390/LIFE12091395

Suttisuwan, R., Phunpruch, S., Saisavoey, T., Sangtanoo, P., Thongchul, N., & Karnchanatat, A. (2019). Isolation and characterization of anti-inflammatory peptides derived from trypsin hydrolysis of microalgae protein (*Synechococcus* sp. VDW). *Food Biotechnology*, *33*(4), 303–324. https://doi.org/10.1080/08905436.2019.1673171

Terra, A. L. M., Moreira, J. B., Costa, J. A. V., & Morais, M. G. de. (2021). Development of pH indicators from nanofibers containing microalgal pigment for monitoring of food quality. *Food Bioscience*, *44*. https://doi.org/10.1016/j.fbio.2021.101387

Toshkova-Yotova, T., Georgieva, A., Iliev, I., Alexandrov, S., Ivanova, A., Pilarski, P., & Toshkova, R. (2022). Antitumor and antimicrobial activity of fatty acids from green microalga Coelastrella sp. BGV. *South African Journal of Botany*, *151*, 394–402.

Tounsi, L., Ben Hlima, H., Hentati, F., Hentati, O., Derbel, H., Michaud, P., & Abdelkafi, S. (2023). Microalgae: A promising source of bioactive phycobiliproteins. *Marine Drugs*, *21*(8), 1–27. https://doi.org/10.3390/md21080440

Tounsi, L., Hentati, F., Ben Hlima, H., Barkallah, M., Smaoui, S., Fendri, I., Michaud, P., & Abdelkafi, S. (2022). Microalgae as feedstock for bioactive polysaccharides. *International Journal of Biological Macromolecules*, *221*, 1238–1250. https://doi.org/10.1016/j.ijbiomac.2022.08.206

Verni, M., Demarinis, C., Rizzello, C. G., & Pontonio, E. (2023). Bioprocessing to preserve and improve microalgae nutritional and functional potential: Novel insight and perspectives. *Foods*, *12*(5), 983. https://doi.org/10.3390/foods12050983

Wang, M., Yin, Z., & Zeng, M. (2022). Microalgae as a promising structure ingredient in food: Obtained by simple thermal and high-speed shearing homogenization. *Food Hydrocolloids*, *131*, 107743. https://doi.org/10.1016/j.foodhyd.2022.107743

Wang, Y., Yang, S., Liu, J., Wang, J., Xiao, M., Liang, Q., Ren, X., Wang, Y., Mou, H., & Sun, H. (2023). Realization process of microalgal biorefinery: The optional approach toward carbon net-zero emission. *Science of the Total Environment*, *901*, 165546. https://doi.org/10.1016/J.SCITOTENV.2023.165546

Williamson, E., Ross, I. L., Wall, B. T., & Hankamer, B. (2023). Microalgae: Potential novel protein for sustainable human nutrition. *Trends in Plant Science*. https://doi.org/10.1016/j.tplants.2023.08.006

Wu, J., Gu, X., Yang, D., Xu, S., Wang, S., Chen, X., & Wang, Z. (2021). Bioactive substances and potentiality of marine microalgae. *Food Science & Nutrition*, *9*(9), 5279–5292. https://doi.org/https://doi.org/10.1002/fsn3.2471

Xie, Z., He, J., Peng, S., Zhang, X., & Kong, W. (2023). Biosynthesis of protein-based drugs using eukaryotic microalgae. *Algal Research*, *74*, 103219. https://doi.org/10.1016/j.algal.2023.103219

Zhang, S., Zhang, L., Xu, G., Li, F., & Li, X. (2022). A review on biodiesel production from microalgae: Influencing parameters and recent advanced technologies. *Frontiers in Microbiology*, *13*. https://doi.org/10.3389/fmicb.2022.970028

Zhang, T.-T., Xu, J., Wang, Y.-M., & Xue, C.-H. (2019). Health benefits of dietary marine DHA/EPA-enriched glycerophospholipids. *Progress in Lipid Research*, *75*, 100997. https://doi.org/10.1016/j.plipres.2019.100997

Zhou, L., Duan, X., Li, K., Hill, D. R. A., Martin, G. J. O., & Suleria, H. A. R. (2023). Extraction and characterization of bioactive compounds from diverse marine microalgae and their potential antioxidant activities. *Chemistry & Biodiversity*. https://doi.org/10.1002/CBDV.202300602

Zimmerman, J. B., Anastas, P. T., Erythropel, H. C., & Leitner, W. (2020). Designing for a green chemistry future. *Science*, *367*(6476), 397–400. https://doi.org/10.1126/SCIENCE.AAY3060/ASSET/8997170C-4499-4F22-A12F-B46943929ED6/ASSETS/GRAPHIC/367_397_F3.JPEG

Section 2

Supercritical Fluid Extraction

4 Supercritical Fluid Extraction as a Promising Technique for Agri-Food Products and By-Products

Maicon Sérgio N. dos Santos, Lissara Polano Ody, Carolina E. Demaman Oro, Bruno Douglas Kerber, Norton M. Silveira, Rogério M. Dallago, and Giovani L. Zabot

4.1 INTRODUCTION

In recent years, interest in research related to the recovery of bioactive compounds from agri-food waste sources for food and biofuel production has considerably grown. Supercritical fluid extraction (SFE) is characterized as an innovative and promising technological method for application in the food, pharmaceutical, cosmetic, and industrial sectors, as it provides high selectivity, reaction time reduction, and the minimization of toxic organic solvents (Serra et al., 2010). The use of supercritical fluids is an alternative to the use of conventional organic solvents in various applications, such as separation, extraction, sorption, chromatography, and drying processes. In the SFE process, the fluid can be used to extract specific compounds or to separate mixtures by varying the pressure and temperature of the system without the dynamics of the phase transition (Bartlett et al., 2014).

To be considered supercritical, a fluid must be above its critical pressure (*Pc*) and critical temperature (*Tc*). From the phase diagram of a pure compound (Figure 4.1), the distinct three physical phases of a fluid are verified: solid, liquid, and gas. These phases are separated by the melting curve (solid–liquid), sublimation (solid–vapor), and vaporization (liquid–vapor). In the diagram, the *triple point* is the condition where the distinct phases coexist, and the *critical point* is the condition where the gas and liquid phases maintain a homogeneous fluid.

These fluids are generally compared based on their reduced parameters, being reduced pressure ($P_r = P / P_c$), reduced temperature ($T_r = T / T_c$), reduced volume ($V_r = V / V_c$), and reduced density ($\rho_r = \rho / \rho_c$). V_c and ρ_c are critical *volume* and *density*, respectively. Van der Waals related the data to the fundamental equations of the corresponding states and the equation of state.

Corresponding equation of state:

$$(P_r + 3 / V_r^2)(V_r - 1/3) = 8T_r / 3 \tag{4.1}$$

Equation of state:

$$(\mathrm{P{+}a} / V^2)(V - b) = RT \tag{4.2}$$

Several compounds have already been studied to be used as solvents in supercritical processes. Nonetheless, most of these compounds cannot be applied as solvents due to characteristics such as

DOI: 10.1201/9781003462804-6

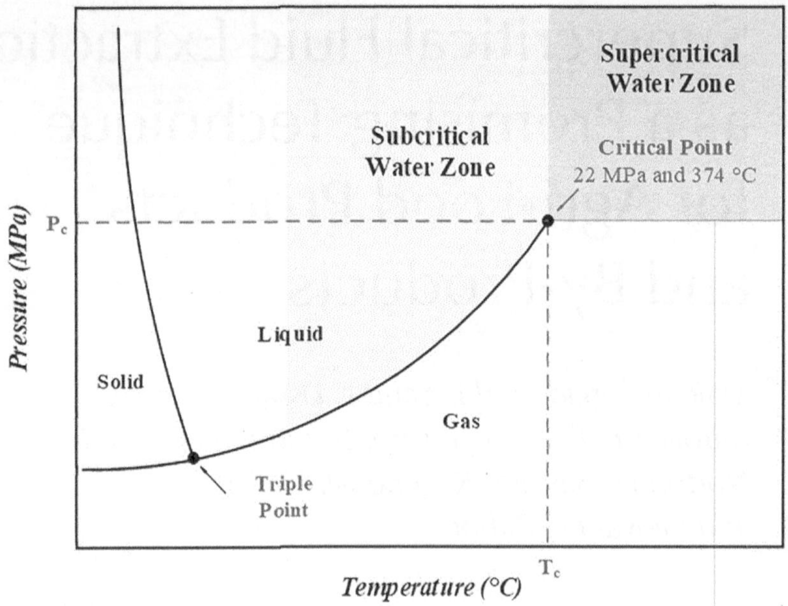

FIGURE 4.1 Schematic phase diagram approach with the phases of a pure substance involving the temperature and pressure.

TABLE 4.1
Critical Properties Characterization of Solvents Largely Applied in SFE Processes

Solvent	Critical Temperature (°C)	Critical Pressure (bar)
Carbon dioxide	31.1	73.7
Ethane	32.3	48.8
Ethylene	9.3	50.3
Propane	96.7	42.4
Methane	190.4	4.6
Propylene	91.9	46.2
Cyclohexane	280.3	40.7
Benzene	289	48.9
Toluene	318.6	41.1
Chlorotrifluoromethane	28.9	39.2
Acetone	235	47
Ammonia	132.5	112.7
Water	374.2	220.4

toxicity, reactivity, flammability, environmental damage, high cost, extreme supercritical conditions, and low purity (Temelli, 2009). In Table 4.1, observed is the critical characterization of the main solvents applied in a diversity of SFE processes.

Furthermore, the most used compound for supercritical processes is carbon dioxide, as it is a non-flammable, non-polluting, non-corrosive, non-toxic solvent, and it does not leave residues in the extracts. Additionally, carbon dioxide is a colorless, inert, relatively low-cost substance abundant when compared to classic organic solvents and has moderate critical temperature (approximately 31°C) and critical pressure (approximately 74 bar) (Jokić et al., 2011). Furthermore, CO_2 allows the extraction of temperature-sensitive compounds without compound degradation (Bozan and Temelli, 2002).

CO_2 is a non-polar substance, as it has a null dipole moment. In addition, it has similarity in polarity to other non-polar solvents that are commonly used in solid–liquid extraction processes. The solubility of the solute in supercritical CO_2 is related to polarity, molecular weight, and the presence of functional groups. Non-polar or low-molecular-weight organic compounds have high solubility in carbon dioxide. Polar and higher-molecular-weight compounds are moderately soluble, while highly polar and higher-molecular-weight compounds are hardly soluble in supercritical pure CO_2. Therefore, the addition of liquid co-solvents in low concentrations can increase solubility and modify the properties of the mixture, as well as the interactions between the solute and the solid matrix, improving the transport of solutes (Cardenas-Toro et al., 2014). In addition to reducing the limitations of CO_2 solvation power, co-solvents make it possible to control the selectivity of the extraction process, providing the use of milder pressures. The most used co-solvents are methanol, ethanol, and water, as they form hydrogen bonds and dipole–dipole interactions with solute molecules that have polar groups. The co-solvents make it possible to ensure the selectivity of the extraction process, providing the use of milder pressures (Brunner, 2005).

4.2 SUPERCRITICAL FLUIDS PROPERTIES

The physicochemical characterization, considering parameters such as temperature, pressure, density, diffusivity coefficient, and viscosity of liquids and gases, is combined in the supercritical phase, making them desirable solvents for numerous processes (Table 4.2) (Bartlett et al., 2014). At any critical pressure, it is not possible to distinguish liquid and gaseous fluid in equilibrium, with a gradual transition from high-density fluid to low-density fluid with an increase in temperature (Forooghi and Hooman, 2014).

4.2.1 Density

Some characteristics of a supercritical fluid, such as solvation strength and viscosity, are density-dependent. Moreover, by modifying the fluid density, it is possible to vary a fluid with gas characteristics to the characteristics of a liquid without crossing the gas–liquid equilibrium line (Bozan and Temelli, 2002). Accordingly, the density is close to liquids, which gives them a high solubilization capacity (Sahena et al., 2009).

In the supercritical zone, a high increase in fluid density is observed when the parameter pressure increases at constant temperature. Likewise, a reduction of the density is verified with the expansion of the temperature at constant pressure. In regions close to the critical point, with an increase in the temperature and pressure, the solvent density decreases. Above the critical point, a minimum increase in pressure results in a considerable increase in fluid density (Maxim et al., 2021). With increasing density, molecular interactions are stronger due to a decrease in intermolecular distances. This means that a supercritical fluid solvates in much the same way as a liquid (Yue et al., 2020).

TABLE 4.2
Density, Viscosity, and Diffusion Coefficient of Supercritical Fluids, Gases, and Liquids

Property	Gas	SFE Requirements	Liquid
Temperature (°C)	25	—	25
Pressure (bar)	1	—	1
Density (g/cm³)	0.6–2 × 10^{-3}	0.2–0.5	0.6–2.0
Diffusivity (cm²/s)	0.1–0.4	0.5–4 × 10^{-3}	0.2–2 × 10^{-5}
Viscosity (MPa s)	0.01–0.03	0.01–0.03	0.2–20.0

Source: Adapted from Bartlett et al. (2014).

4.2.2 Diffusivity

When considering a supercritical fluid, its diffusivity can be 100 times higher than in a conventional liquid and up to 10,000 times lower than in a conventional gas. A solute presents a higher diffusivity in a supercritical fluid than in a conventional liquid (due to the higher diffusivity of the supercritical fluid). Diffuse is proportional to temperature and inversely proportional to pressure. With the expansion of the pressure, the supercritical fluid molecules are attractive, decreasing diffusivity. The high diffusivity of supercritical fluid benefits them in being fast conveyors, and this characteristic is important for extractions using supercritical fluid (Cravotto et al., 2023).

4.2.3 Viscosity

Supercritical fluids have a viscosity similar to a gas and 10 to 100 times less than a liquid. The very low viscosity of those fluids allows very favorable flow characteristics for extraction. Therefore, the mass transfer of solutes in extractions is much faster with supercritical fluids than with liquids.

4.2.4 Solubility

The solubility performance of solutes in SFE conditions is extremely pertinent to observe the extraction profile. In an SFE process, the solute must be compatible with the solvent to be used, while in some cases, it is necessary to introduce a co-solvent to modify the polarity of the fluid before extraction. Conventional liquids present a solubility lower than that of supercritical fluids. Nonetheless, this parameter is significantly higher in gases. Furthermore, supercritical fluids can be used as very efficient solvents in the extraction process (Hodes et al., 2004). The solubility of supercritical fluid is directly affected by the pressure and temperature conditions (Wen et al., 2009).

The solubility of naphthalene in CO_2 at low pressure is practically null. When the system pressure is close to the critical pressure of CO_2 (73 bar), there is an increase in solubility. On the other hand, at pressures above the critical point, the solubility of naphthalene increases significantly, being higher than 8% at 300 bar. The main advantages and disadvantages of the SFE application in a large and pilot scale are presented in the following.

Benefits

- Generally, the solvents used are gaseous at atmospheric pressure and room temperature (near 25°C), which facilitates their removal from the extracted products.
- Solvent properties vary considerably, either by adjusting temperature and pressure or by using co-solvents.
- The diffusivity of solute is considerably higher than that of liquid solvents.
- As supercritical fluids have lower viscosity and higher diffusibility, mass transfer is much faster compared to that for liquid solvents, resulting in reduced extraction time.
- The use of fluids that have a low critical temperature produces thermally unstable solutes.
- It is environmentally beneficial.
- It has good scalability.
- It presents as simple and easy for continuous biofuel production.

Disadvantages

- High cost of equipment.
- Very polar compounds present a significant difficulty to extracting without a co-solvent introduction.

- Low-value-added and low-yield products cannot be economically extracted by this process.
- Hostile operating environment.
- Corrosion and salt deposition (for industrial applications).

4.3 POTENTIAL MOLECULES EXTRACTED BY SFE

Supercritical extraction has a wide range of applications in products from different raw material sources (Figure 4.2) due to its selectivity and efficiency in extracting high-quality compounds. SFE is used to obtain essential oils from grains, seeds, plants, flowers, and herbs, among others. Simultaneously, this technology can also be used to evaluate the antimicrobial activity of natural compounds against microorganisms considered as pathogens. Furthermore, SFE can help in the extraction of antioxidant compounds from fruits, vegetables, and plants, among others, capable of protecting cells against oxidative stress. Finally, it is worth highlighting that the extraction of SFE is increasingly being explored in the production of biofuels and bioenergy from vegetable oils and

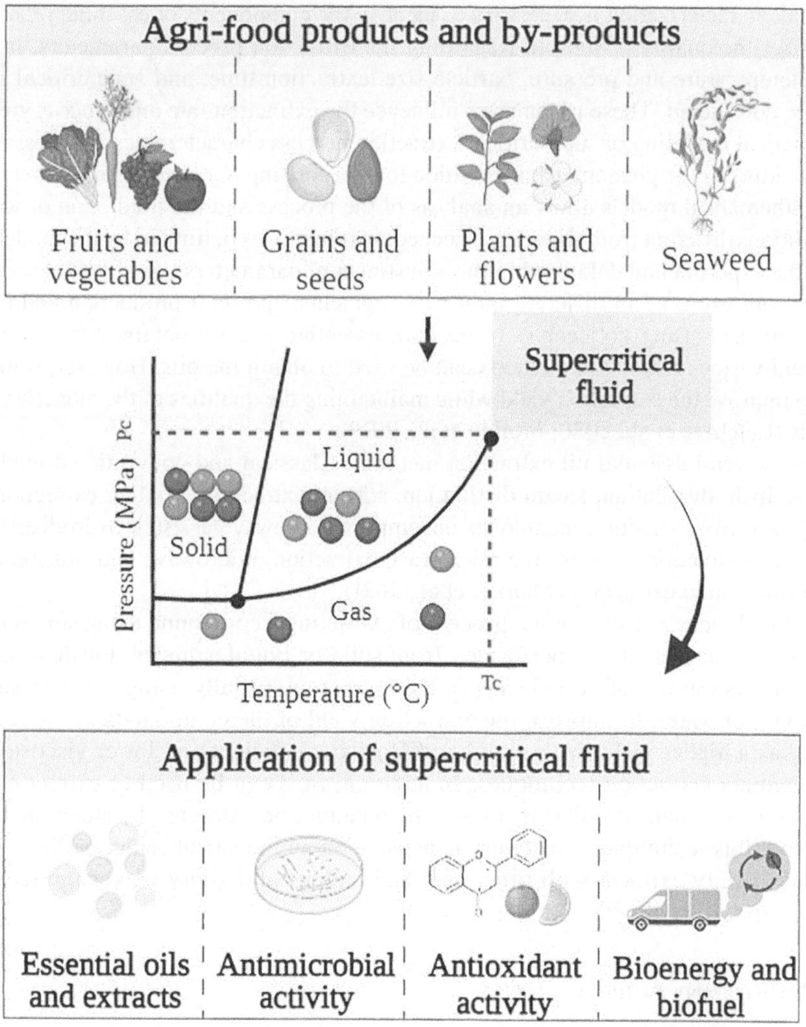

FIGURE 4.2 Application of supercritical fluid extraction in agri-food products and by-products.

lignocellulosic by-products. In the following sections, the applications of SFE technology in the agri-food sector are explained.

4.3.1 Essential Oils and Extracts

Essential oils present in plants are complex mixtures of organic compounds, most of which are insoluble in water. The composition and concentration of essential oils may differ from species to species and may even be different within the same species, or in the same population, at different times of the year (Jugreet et al., 2020).

Essential oils are volatile elements contained in various structures of plants and are named because of their lipid composition, unlike the glyceride composition of oils and fats. They are associated with several functions necessary for the survival of the plant in its ecosystem, playing a fundamental role in defense against microorganisms and predators, and also in attracting insects and other fertilizing agents (Campolo et al., 2020). Essential oil extraction studies can be divided into three parts: chemical characterization of extracts, evaluation of the influence of process parameters, and mathematical modeling.

Chemical characterization is necessary to identify the components of essential oils. These components affect the quality of the oil. Regarding the extraction process parameters, in addition to extraction temperature and pressure, particle size, extraction time, and supercritical solvent flow rate must be considered. These parameters influence the extraction rate and process yield.

Mathematical modeling of supercritical extraction process characteristics is necessary for a better understanding of the phenomena, in addition to interpolating or extrapolating experimental conditions. Mathematical models allow an analysis of the process and the prediction of what happens under conditions different from those experienced, within ranges delimited by the model, and by the quality of the experimental data used in the adjustment of parameters.

Essential oils are used in different industrial segments, and their production and consumption worldwide are increasing noticeably. In general, essential oils are obtained from vegetable raw material, and various extraction methods can be used to obtain the oils. However, some care must be taken to improve the extraction yield while maintaining the qualities of the bioactive and interest compounds (Lefebvre et al., 2020; Molino et al., 2019).

There are several essential oil extraction methods. Classical and conventional methods can be classified as hydrodistillation, steam distillation, solvent extraction, Soxhlet extraction, and cold-pressing. Innovative extraction techniques encompass microwave-assisted hydrodistillation, ultrasound-assisted extraction, solvent-free microwave extraction, microwave hydrofusion and gravity, and supercritical fluid extraction (Mankar et al., 2021).

The SFE technique consists of the process of extracting a compound using supercritical fluids as a solvent. Extraction can be performed from solid or liquid samples. Furthermore, the SFE technique allows the use of co-solvents in the extraction, usually using solvents such as ethanol, methanol, or water, to improve the extraction yield of the compounds of interest. The SFE technique has a higher yield, with a higher diffusion coefficient and lower viscosity compared to other common extraction techniques. In addition, SFE can be used to extract essential oils from heat-sensitive matrices, that is, those which cannot be extracted by steam distillation. On the one hand, this technique is still very expensive on an industrial scale; on the other, obtaining superior-quality extracts with functional and biological activity makes the technique very attractive (Capuzzo et al., 2013).

4.3.2 Antioxidant Activity

Supercritical CO_2 (sC-CO_2) has a non-polar solvent characteristic and is widely used in SFE extractions as it maintains the natural characteristics of the matrix used for extraction among all the other advantages already mentioned. As a result, extracts and oils obtained by the SFE–CO_2

technique are easily used in different industrial segments due to their high purity. Furthermore, for a higher yield of antioxidant compounds, the system depressurization rate must be fast, and each extract or oil must have its extraction parameters optimized (Ferro et al., 2019; Uwineza and Waśkiewicz, 2020; Rivas et al., 2021).

Essential oils can be obtained by the SFE technique with a high content of antioxidants (Table 4.3). For example, essential oils from *Thymus munbyanus* subsp. *coloratus* and subsp. *munbyanus* obtained by SFE–CO_2 have lipophilic α-tocopherol antioxidant content of 1,580 and 780 µg/g dry weight extract, respectively, and are rich in squalene and n-triacontane (Bendif et al., 2018).

In addition to essential oils, extracts can be obtained with or without the use of a co-solvent in the SFE technique. Chestnut (*Castanea sativa*) burs extracts obtained with SFE–CO_2 (45°C and 200 bar, 10% ethanol) showed approximately 40 mg Trolox/g extract for the analysis of Trolox equivalent antioxidant capacity (TEAC) and 15% inhibition against 2,2-diphenyl-1-picrylhydrazyl (DPPH) (Díaz-Reinoso et al., 2020). In another study, epigallocatechin, ellagic acid, epicatechin/catechin, and caffeic acid derivatives were identified in an extract of *C. sativa* shells obtained by SFE– (60°C, 350 bar, and 15% of ethanol as co-solvent), and the extract showed a 53.04% DPPH inhibition (Pinto et al., 2020). Extracts from tangerine (*C. reticulata* var. *Arrayana*) peels with antioxidant activity were obtained by SFE technique (220 bar, 80°C, and 5.0% w/w ethanol) and showed a positive effect on the protection of mayonnaise against lipid oxidation. The effect observed was similar to that found for synthetic antioxidants, showing that the extraction technique was efficient in obtaining an extract rich in bioactives and antioxidants (Franco-Arnedo et al., 2020).

SFE–CO_2 with an increase in ethanol concentration may show an increase in antioxidant activity, for example, the increase from 5% to 25% of ethanol in the extraction of mango leaves extracts

TABLE 4.3
Current Scientific Studies on the Application of the SFE Process in the Extraction of Antioxidants from Agri-Food Products

Source Material	Supercritical Fluid	Operation Conditions	Antioxidant Agents	Reference
Papaya (*Carica papaya*)	CO_2	35°C, 250 bar	Flavonoids	(Khor et al., 2021)
Pomegranate seed (*Punica granatum* L.)	CO_2	40-60°C, 200–300 bar	Phenolics, tocopherols, phytosterols	(Silva et al., 2019)
Basil (*Ocimum basilicum*)	CO_2	50°C, 300 bar	Phenolic, flavonoids	Romano et al., 2022a)
Citrus peels	CO_2	60°C, 300 bar	Flavonoids	Romano et al., 2022b)
Grape seeds	CO_2	60°C, 350 bar	Tocopherols	(Dimic et al., 2020)
Apple pomace	CO_2	40–60°C, 100–300 bar	Phenolics	(Ferrentino et al., 2020)
Potato peels (*Solanum tuberosum*)	CO_2	80°C, 350 bar	Phenolics	(Lima et al., 2021)
Mango peel	CO_2	60°C, 250 bar	Phenolic	(Sánchez-Camargo et al., 2019)
Pink pepper	CO_2	30°C, 300 bar	Phenolic	(Rebelatto et al., 2020)
Chestnut	CO_2	35–70°C, 100–300 bar	Phenolic	(Díaz-Reinoso et al., 2020)
Paeonia delavayi flowers	CO_2	40°C, 250 bar	Polyphenolics Flavonoids	(Yu et al., 2022)
Lamium album flowers	CO_2	50°C, 250 bar	Phenolic	(Uwineza et al., 2021)
Date (*Phoenix dactylifera*)	CO_2	35–70°C, 150–400 bar	Phenolic Flavonoids Carotenoids	(Ghafoor et al., 2022)

increased from 1.11 to 1.89 μg DPPH index/μg extract, respectively (Guamán-Balcázar et al., 2018). The authors also highlighted a higher amount of extraction of the compounds iriflophenone-3-C-β-D-glucoside, quercetin 3-D-galacto-side, gallic acid, and iriflophenone-3-C-(2-O-galloryl)-β-D-glucoside. This increase in yield and antioxidant activity by increasing the amount of ethanol in the extraction may be due to the polar characteristics of polysaccharides (Gong et al., 2021).

Another co-solvent that can be used to increase the extraction of compounds with high antioxidant activity is water through a binary system ($H_2O + CO_2 - SFE$) (Barbosa et al., 2020). The authors reported 80.83% antioxidant activity by DPPH in *Pleurotus ostreatus* extracts obtained by the binary system (250 bar and 20% H_2O). These results show that the SFE technique is very versatile, as it can be combined with different co-solvents, depending on the compound of interest to be extracted from the sample matrix.

The applications of compounds extracted by SFE in health care include mainly the food, cosmetics, and pharmaceutical industries. The compounds can be incorporated into foods for fortification, supplementation, or improvement of functional properties (Lončarević et al., 2018; Lorenzo et al., 2018), and in beauty creams and lotions to enhance the product's antioxidant action (Raza et al., 2020). In all cases, these compounds are usually encapsulated, as the encapsulation protects the bioactive compounds and preserves their functions. Encapsulation can be performed by different techniques, the most common being encapsulation by liposomes (Ji et al., 2019; Kalam et al., 2020), hydrogel (Munim and Raza, 2019), solvent evaporation (Ozkan et al., 2019), molecular inclusion (Fernandes et al., 2018), emulsions (Nooshkam and Varidi, 2020), freeze-drying (Čakarević et al., 2020), spray drying (Kaderides and Goula, 2019; Kim et al., 2019), electrospinning (Shao et al., 2018; Tavassoli-Kafrani et al., 2018), and supercritical antisolvent (SAS) (Machado et al., 2018), among other techniques.

4.3.3 Antimicrobial Activity

The direct application of supercritical fluids as antimicrobial action strategies has been reported as an extremely attractive alternative in recent years. The physicochemical characterization of a supercritical fluid, such as high density, reduced surface tension, and significant diffusivity, arouses interest in the development of materials predominantly with antimicrobial applications (Zizovic, 2020). The scientific literature reports several sources as potential alternatives for extracting compounds of antifungal and antimicrobial interest, such as vegetable sources (Mazzutti et al., 2012), marine and agricultural food wastes (Zhou et al., 2021), algae and cyanobacteria (Alsenani et al., 2020), etc. Table 4.4 delineates important scientific studies on the application of the SFE process in many sources, aiming to obtain compounds with significant antimicrobial action.

4.4 VALUE-ADDED LIGNOCELLULOSIC BY-PRODUCTS

4.4.1 Bioenergy and Biofuel Production

Conventionally, biofuel synthesis is based on two routes: thermochemical and biochemical conversions. The thermochemical route is widely applied to agri-food residual materials, such as wood, straw, and shells, through the gasification, liquefaction, and pyrolysis procedures. The biochemical route predominantly refers to the production of bioethanol and biodiesel through acidic and enzymatic hydrolysis and/or fermentation of different sets of raw materials. Considering an economic perspective, thermochemical conversion is economically viable when compared to biochemical routes. Accordingly, there is a global trend directly related to the application of alternative technologies with the direct use of a range of vegetable residual biomasses. In this context, the SFE technology is largely attractive and has been considered an innovative and visionary technique in recent years (Silva and Oliveira, 2014).

TABLE 4.4
Current Scientific Studies on the Application of the SFE Process Considering a Diversity of Compound Sources and Targets

Source Material	Supercritical Fluid	Operation Conditions	Target	Promising Conclusions	Reference
Lippia thymoides Mart. and Schauer	CO_2	40–50°C, 100–300 bar	*C. albicans*, *C. tropicalis*, *E. coli*, and *S. aureus*	High thymol yields and significant fungi and bacteria inhibition	(Silva et al., 2021)
Feijoa leaves	CO_2	55°C, 300 bar	*S. aureus*, *E. coli*, and *S. typhimurium*	High inhibitory concentration and effectiveness	(Dos Santos et al., 2009)
Artemisia arborescens, *Artemisia abyssinica*, *Pulicaria jaubertii*, and *Pulicaria petiolaris*	CO_2	40°C, 300 bar, 120 min	Gram-positive and Gram-negative bacteria	High antimicrobial compound concentrations and low bactericidal concentrations	(Al-Maqtari et al., 2021)
Red, brown, and green propolis	CO_2	80°C	*C. pseudotuberculosis*	High bacterial biofilm formation	(Dos Santos et al., 2009)
Fucus vesiculosus L.	CO_2	50°C, 300 bar	*F. culmorum* and *F. oxysporum*	Fungi inhibition and structural degradation up to 100%	(Tyskiewicz et al., 2019)
Baeckea frutescens L.	CO_2	40°C, 180 bar	*C. gloeosporioides* and *P. camelliae-sinensis*	High antimicrobial compound concentration and antifungal activity	(Jiang et al., 2020)
Butia catarinensis	CO_2	40–60°C, 100–300 bar	*E. coli* and *B. cereus*	Cinnamic acid and caprylic acids with high bacterial inhibition	(Cruz et al., 2017)
Ilex paraguariensis	Propane, CO_2	35–55°C, 40–250 bar	*E. coli* and *S. aureus*	High compound extraction and inhibitory concentration	(Fernandes et al., 2017)
Arctium lappa	CO_2, ethanol	40–80°C, 150–250 bar	*S. aureus*	Significant antimicrobial inhibitory concentration	(Souza et al., 2018)
Ag nanoparticles	CO_2	100°C, 200 bar	*E. coli* and *A. niger*	High antibacterial and antifungal properties inhibition	(Matsuyama et al., 2019)
Artemisia absinthium L.	CO_2	40°C, 300 bar, 60 min	*M. incognita*, *F. oxysporum*, *F. solani*, *P. infestans*, and *P. capsica*	Noticeable inhibitory concentration for nematodes and fungi	(Liu et al., 2019)
Ilex guayusa L.	CO_2, Ethanol	45–75°C, 150–250 bar	Gram-positive and Gram-negative bacteria, fungi, and yeast	High antimicrobial compound concentration and antimicrobial potentiality	(Cadena-Carrera et al., 2019)
Pelargonium graveolens	CO_2	40–50°C, 90–300 bar	*S. litoralis*	High antifungal activity and fungi growth inhibition	(Machalova et al., 2015)

Biofuels or bio-based fuels are liquid or gaseous materials synthesized primarily from vegetable biomasses for application in the transportation industry. Biofuels are renewable and environmentally friendly fuels and have been employed as attractive alternatives in comparison to conventional gasoline and diesel engines. Supercritical fluids have been considered a promising alternative for large-scale proportions and industrial approaches, mainly for biodiesel production from animal residual fats (Wen et al., 2009).

In comparison to traditional biofuel production strategies, the SFE technology is extremely advantageous; once the reaction time is accelerated, the kinetic profile is largely optimized, with significant fuel synthesis rate, performance simplicity, and minimization of the application of catalysts. Regarding the disadvantages of supercritical fluid technology, there is a harsh operating environment and high cost of the equipment (Wen et al., 2009).

The fuel quality of vegetable oils can be evaluated according to analytical determination and calorific value, viscosity, cloud point, distillation curve, and cetane index. The calorific value of the oil directly depends on the maximum power to be reached by the engine in operation. However, the cetane index defines the self-ignition power of oil combustion, since cetane increases the efficiency of fuel burning. Compared to diesel oil, vegetable oils have lower heat of combustion and a similar cetane index (Islam et al., 2013).

Nonetheless, the use of natural vegetable oils as strategic alternatives for diesel oil is possible due to several negative factors, and its use becomes unfeasible. Among these factors, one can highlight engine performance problems, filter clogging, injectors' carbonization, piston ring breakage, dryness, rupture of seals, and degradation of the engine lubricating oil.

This occurs due to heating at temperatures close to 250°C, causing complementary reactions of thermal decomposition, the results of which can lead to the formation of polymeric compounds through condensation reactions. The existence of polymeric compounds increases the distillation temperature and the level of engine smoke, reduces the viscosity of the lubricating oil, and causes a reduction in power due to the incomplete burning of secondary products. This behavior is not observed with methanolized or ethanolized derivatives (biodiesel); its mixture is fully distilled at temperatures below 350°C (Basha et al., 2009).

Biodiesel production can be understood as the chemical transformation of vegetable oil in nature into compounds that have similar characteristics to petroleum diesel. Generally, these compounds are fatty acid esters. Vegetable oils have high viscosity and lower volatility. These materials have high concentrations of polyunsaturated compounds, where physical or chemical alterations can contribute to reducing problems caused by viscosity, which include microemulsion, dilution and transesterification, esterification, and pyrolysis (Pinto et al., 2005). Table 4.5 presents some recent works regarding the use of supercritical transesterification technology for biodiesel production.

Transesterification or alcoholysis is the most commercially used technique for biodiesel synthesis. The reaction occurs when a source of triglycerides and alcohols, with or without a catalyst, produces esters and glycerol (Kusdiana and Saka, 2004). Biodiesel synthesis under supercritical conditions has other advantages over the common transesterification process. Fatty acids present in vegetable oils are esterified in supercritical alcohols. Significant concentrations of water in the reagent medium favor the formation of alkyl esters and facilitate the separation of reaction products, as glycerin is more soluble in water than in alcohol. These characteristics make the supercritical technology of transesterification favorable for the quality of vegetable oils applied for biofuel synthesis (Kusdiana and Saka, 2004).

4.5 CHALLENGES AND FUTURE OUTLOOKS

One of the biggest challenges in implementing SFE technology on an industrial scale is its cost of implementation and operation. Specific equipment for supercritical gases must be purchased, and special care must be taken. Additionally, the major limitation of the technique is the low flow of compost to be extracted due to the expansion of the gas, that is, the extraction process cannot be

TABLE 4.5
Recent Scientific Studies on the Direct Use of Supercritical Transesterification Technology for Biodiesel Production

Feedstock	Reactor	Temperature (°C)	Time (min)	Income (%)	Pressure (bar)	Reference
Soybean oil	Batch	255-300	80	98	90-170	(Cheng et al., 2015)
Soybean oil	Expanded milk	70	60	90	200	(Colombo et al., 2015)
Chlorella protothecoides	Batch	300–400	0.5–10	95.5	150–300	(Liu et al., 2015)
Soybean flakes	Continuous	350	180	86	200	(Xu et al., 2016)
Castor oil	Batch	300	90	96.5	210	(Román-Figueroa et al., 2016)
Jojoba (*Simmondsia chinensis*)	Continuous	350	90	96.8	420	(Andreo-Martínez et al., 2017)
Spirulina oil	Continuous	300	30	72	200	(Tobar and Núñez, 2018)
Soybean oil	Batch	280	10	95.19	130	(Costa et al., 2019)

accelerated by opening the extraction valve further, as the gas will expand and freeze the valve and connections. Furthermore, for larger volumes of extracted compost, it is recommended to increase the production line, adding more extracting equipment. The time taken to assemble, disassemble, and clean the equipment must always be taken into account. This time is often called "dead time" because the equipment is not performing the extraction. Nevertheless, all these parameters must be considered before implementing the technique in an industrial proportion. Also, the compounds extracted by the SFE technique have high purity and do not have solvent residues, since the gas is easily separated from the oil or extract.

REFERENCES

Al-Maqtari, Q. A., Al-Ansi, W., Mahdi, A. A., Al-Gheethi, A. A. S., Mushtaq, B. S., Al-Adeeb, A., Wei, M., Yao, W. (2021). Supercritical fluid extraction of four aromatic herbs and assessment of the volatile compositions, bioactive compounds, antibacterial, and anti-biofilm activity. *Environmental Science and Pollution Research*. 28(20):25479–25492. https://doi.org/10.1007/s11356-021-12346-6

Alsenani, F., Tupally, K. R., Chua, E. T., Eltanahy, E., Alsufyani, H., Parekh, H. S., Schenk, P. M. (2020). Evaluation of microalgae and cyanobacteria as potential sources of antimicrobial compounds. *Saudi Pharmaceutical Journal*. 28(12):1834–1841. https://doi.org/10.1016/j.jsps.2020.11.010

Andreo-Martínez, P., García-Martínez, N., Quesada-Medina, J., Román-Figueroa, C., Paneque, M. (2017). Synthesis of value added fatty alcohols and FAMEs by non-catalytic supercritical methanol transesterification of jojoba (*Simmondsia chinensis*) wax. *Energy Conversion and Management*. 154:430–439. https://doi.org/10.1016/j.enconman.2017.11.028

Barbosa, J. R., Maurício, M. M., Oliveira, L. C., Luiza, L. H., Almada-Vilhena, A. O., Oliveira, R. M., Pieczarka, J. C., Davi, S. D., Carvalho Junior, R. N. (2020). Obtaining extracts rich in antioxidant polysaccharides from the edible mushroom *Pleurotus ostreatus* using binary system with hot water and supercritical CO2. *Food Chemistry*. 330. https://doi.org/10.1016/j.foodchem.2020.127173

Bartlett, P. N., Cook, D. A., George, M. W., Hector, A. L., Ke, J., Levason, W., Reid, G., Smith, D. C., Zhang, W. (2014). Electrodeposition from supercritical fluids. *Physical Chemistry Chemical Physics*. 16(20):9202–9219. https://doi.org/10.1039/c3cp54955k

Basha, S. A., Gopal, K. R., Jebaraj, S. (2009). A review on biodiesel production, combustion, emissions and performance. *Renewable and Sustainable Energy Reviews*. 13(6–7):1628–1634. https://doi.org/10.1016/j.rser.2008.09.031

Bendif, H., Adouni, K., Miara, M. D., Baranauskienė, R., Kraujalis, P., Venskutonis, P. R., Nabavi, S. M., Maggi, F. (2018). Essential oils (EOs), pressurized liquid extracts (PLE) and carbon dioxide supercritical fluid extracts (SFE-CO2) from Algerian *Thymus munbyanus* as valuable sources of antioxidants to be used on an industrial level. *Food Chemistry*. 260:289–298. https://doi.org/10.1016/j.foodchem.2018.03.108

Bozan, B., Temelli, F. (2002). Supercritical CO2 extraction of flaxseed. *Journal of the American Oil Chemists' Society*. 79(3):231–235. https://doi.org/10.1007/s11746-002-0466-x

Brunner, G. (2005). Supercritical fluids: technology and application to food processing. *Journal of Food Engineering*. 67(1–2):21–33. https://doi.org/10.1016/j.jfoodeng.2004.05.060

Cadena-Carrera, S., Tramontin, D. P., Bella Cruz, A., Bella Cruz, R. C., Müller, J. M., Hense, H. (2019). Biological activity of extracts from guayusa leaves (*Ilex guayusa* Loes.) obtained by supercritical CO2 and ethanol as cosolvent. *Journal of Supercritical Fluids*. 152. https://doi.org/10.1016/j.supflu.2019.104543

Čakarević, J., Šeregelj, V., Tumbas Šaponjac, V., Ćetković, G., Čanadanović Brunet, J., Popović, S., Kostić, M. H., Popović, L. (2020). Encapsulation of beetroot juice: a study on the application of pumpkin oil cake protein as new carrier agent. *Journal of Microencapsulation*. https://doi.org/10.1080/02652048.2019.1705408

Campolo, O., Puglisi, I., Barbagallo, R. N., Cherif, A., Ricupero, M., Biondi, A., Palmeri, V., Baglieri, M., Zappala, L. (2020). Side effects of two citrus essential oil formulations on a generalist insect predator, plant and soil enzymatic activies. *Chemosphere*. 257:127252. https://doi.org/10.1016/j.chemosphere.2020.127252

Capuzzo, A., Maffei, M. E., Occhipinti, A. (2013). Supercritical fluid extraction of plant flavors and fragrances. *Molecules*. 18(6):7194–7238. https://doi.org/10.3390/molecules18067194

Cardenas-Toro, F. P., Forster-Carneiro, T., Rostagno, M. A., Petenate, A. J., Maugeri Filho, F., Meireles, M. A. A. (2014). Integrated supercritical fluid extraction and subcritical water hydrolysis for the recovery of bioactive compounds from pressed palm fiber. *Journal of Supercrit Fluids*. 93:42–48. https://doi.org/10.1016/j.supflu.2014.02.009

Cheng, J., Li, T., Peng, N., Huang, R., Zhou, J. H., Cen, K. F. (2015). Combustion dynamics of biodiesel produced by supercritical methanol transesterification. *Fuel Processing Technology*. 131:409–413. https://doi.org/10.1016/j.fuproc.2014.11.040

Colombo, T. S., Mazutti, M. A., Di Luccio, M., De Oliveira, D., Oliveira, J. V. (2015). Enzymatic synthesis of soybean biodiesel using supercritical carbon dioxide as solvent in a continuous expanded-bed reactor. *Journal of Supercritical Fluids*. 97:16–21. https://doi.org/10.1016/j.supflu.2014.11.002

Costa, W. A., Bezerra, F. W. F., de Oliveira, M. S., da Silva, M. P., Cunha, V. M. B., Andrade, E. H. de A., de Carvalho Júnior, R. N. (2019). Appliance of a high pressure semi-batch reactor: supercritical transesterification of soybean oil using methanol. *Food Science and Technology*. 39(3):754–773. https://doi.org/10.1590/fst.05118

Cravotto, C., Claux, O., Bartier, M., Fabioano-Tixier, A., Tabasso, S. (2023). Leading edge technologies and perspectives in industrial oilseed extraction. *Molecules*. (28):1–48. https://doi.org/10.3390/molecules28165973

Cruz, P. N., Pereira, T. C. S., Guindani, C., Oliveira, D. A., Rossi, M. J., Ferreira, S. R. S. (2017). Antioxidant and antibacterial potential of butia (*Butia catarinensis*) seed extracts obtained by supercritical fluid extraction. *Journal of Supercritical Fluids*. 119:229–237. https://doi.org/10.1016/j.supflu.2016.09.022

Díaz-Reinoso, B., Moure, A., Domínguez, H. (2020). Ethanol-modified supercritical CO2 extraction of chestnut burs antioxidants. *Chemical Engineering and Processing*. 156:108092. https://doi.org/10.1016/j.cep.2020.108092

Dimic, I., Teslic, N., Putnik, P., Kovacevic, D. B., Zekovic, Z., Sojic, B., Mrkonkic, Z., Colovic, D., Montesano, D., Pavlic, B. (2020). Innovative and conventional valorizations of grape seeds fromwinery by-products as sustainable source of lipophilic antioxidants. *Antioxidants*. (9):2–19. https://doi.org/10.3390/antiox9070568

Dos Santos, K. L., Peroni, N., Guries, R. P.; Nodari, R. O. (2009). Traditional knowledge and management of Feijoa (Acca sellowiana) in southern Brazil. *Economic Botany*. 63:204–214.

Fernandes, A., Rocha, M. A. A., Santos, L. M. N. B. F., Brás, J., Oliveira, J., Mateus, N., de Freitas, V. (2018). Blackberry anthocyanins: β-cyclodextrin fortification for thermal and gastrointestinal stabilization. *Food Chemistry*. 245:426–431. https://doi.org/10.1016/j.foodchem.2017.10.109

Fernandes, C. E. F., Scapinello, J., Bohn, A., Boligon, A. A., Athayde, M. L., Magro, J. D., Palliga, M., Oliveira, J. V., Tres, M. V. (2017). Phytochemical profile, antioxidant and antimicrobial activity of extracts obtained from erva-mate (*Ilex paraguariensis*) fruit using compressed propane and supercritical CO2. *Journal of Food Science and Technology*. 54(1):98–104. https://doi.org/10.1007/s13197-016-2440-4

Ferrentino, G., Giampiccolo, S., Morozova, K., Haman, N., Spilimbergo, S., Scampichio, M. (2020). Supercritical fluid extraction of oils from apple seeds: process optimization, chemical characterization and comparison with a conventional solvent extraction. *Innovative Food Science and Emerging Technologies*. 1–38. https://doi.org/10.1016/j.ifset.2020.102428

Ferro, D. M., Mazzutti, S., Vitali, L., Oliveira Müller, C. M., Ferreira, S. R. S. (2019). Integrated extraction approach to increase the recovery of antioxidant compounds from *Sida rhombifolia* leaves. *Journal of Supercritical Fluids*. 149:10–19. https://doi.org/10.1016/j.supflu.2019.03.013

Forooghi, P., Hooman, K. (2014). Experimental analysis of heat transfer of supercritical fluids in plate heat exchangers. *International Journal of Heat and Mass Transfer*. 74:448–459. https://doi.org/10.1016/j.ijheatmasstransfer.2014.03.052

Franco-Arnedo, G., Buelvas-Puello, L. M., Miranda-Lasprilla, D., Martínez-Correa, H. A., Parada-Alfonso, F. (2020). Obtaining antioxidant extracts from tangerine (*C. reticulata* var. Arrayana) peels by modified supercritical CO2 and their use as protective agent against the lipid oxidation of a mayonnaise. *Journal of Supercritical Fluids*. 165. https://doi.org/10.1016/j.supflu.2020.104957

Ghafoor, K., Sarker, M. Z. I., Al-Juhaimi, F. Y., Babiker, E. E., Alkaltham, M. S., Almubarak, A. K. (2022). Extraction and evaluation of bioactive compounds from date (*Phoenix dactylifera*) seed using supercritical and subcritical CO2 techniques. *Foods*. 11:1806. https://doi.org/10.3390/foods11121806

Gong, T., Liu, S., Wang, H., Zhang, M. (2021). Supercritical CO2 fluid extraction, physicochemical properties, antioxidant activities and hypoglycemic activity of polysaccharides derived from fallen Ginkgo leaves. *Food Bioscience*. 42(82):101153. https://doi.org/10.1016/j.fbio.2021.101153

Guamán-Balcázar, M. C., Montes, A., Fernández-Ponce, M. T., Casas, L., Mantell, C., Pereyra, C., Martínez de la Ossa, E. (2018). Generation of potent antioxidant nanoparticles from mango leaves by supercritical antisolvent extraction. *Journal of Supercritical Fluids*. 138:92–101. https://doi.org/10.1016/j.supflu.2018.04.005

Hodes, M., Marrone, P. A., Hong, G. T., Smith, K. A., Tester, J. W. (2004). Salt precipitation and scale control in supercritical water oxidation—part A: fundamentals and research. *Journal of Supercritical Fluids*. 29(3):265–288. https://doi.org/10.1016/S0896-8446(03)00093-7

Islam, M. A., Ayoko, G. A., Brown, R., Stuart, D., Heimann, K. (2013). Influence of fatty acid structure on fuel properties of algae derived biodiesel. *Procedia Engineering*. 56:591–596. https://doi.org/10.1016/j.proeng.2013.03.164

Ji, W., Zhang, T., Lu, Z., Shen, J., Xiao, Z., Zhang, X. (2019). Synthesis and characterization of novel biocompatible nanocapsules encapsulated lily fragrance. *Chinese Chemical Letters*. 30(3):739–742. https://doi.org/10.1016/j.cclet.2018.09.008

Jiang, H., Zhang, M., Qin, L., Wang, D., Yu, F., Liang, W., Song, C., Granato, D. (2020). Chemical composition of a supercritical fluid (SFE-CO2) extract from *Baeckea frutescens* L. leaves and its bioactivity against two pathogenic fungi isolated from the tea plant (*Camellia sinensis* (L.) O. Kuntze). *Plants*. 9(9):1–15. https://doi.org/10.3390/plants9091119

Jokić, S., Svilović, S., Zeković, Z., Vidović, S., Velić, D. (2011). Solubility and kinetics of soybean oil and fatty acids in supercritical CO2. *European Journal of Lipid Science and Technology*. 113(5):644–651. https://doi.org/10.1002/ejlt.201000403

Jugreet, B. S., Suroowan, S., Renhasamy, R. P., Mahomoodaly, M. F. (2020). Chemistry, bioactivities, mode of action and industrial applications of essential oils. *Trends in Food Science & Technology*. 101:89–105. https://doi.org/10.1016/j.tifs.2020.04.025

Kaderides, K., Goula, A. M. (2019). Encapsulation of pomegranate peel extract with a new carrier material from orange juice by-products. *Journal of Food Engineering*. 253:1–13. https://doi.org/10.1016/j.jfoodeng.2019.02.019

Kalam, M. A., Alkholief, M., Badran, M., Alshememry, A., Alshamsan, A. (2020). Co-encapsulation of metformin hydrochloride and reserpine into flexible liposomes: characterization and comparison of in vitro release profile. *Journal of Drug Delivery Science and Technology*. 57:101670. https://doi.org/10.1016/j.jddst.2020.101670

Khor, B. K., Chear, N. J., Azizi, J., Khaw, K. (2021). Chemical composition, antioxidant and cytoprotective potentials of *Carica papaya* leaf extracts: a comparison of supercritical fluid and conventional extraction methods. *Molecules*. 26:1–18. https://doi.org/10.3390/molecules26051489

Kim, E. H. J., Paredes, D., Motoi, L., Eckert, M., Wadamori, Y., Tartaglia, J., Green, C., Hedderley, D. I., Morgenstern, M. P. (2019). Dynamic flavor perception of encapsulated flavors in a soft chewable matrix. *Food Research International*. 241–250. https://doi.org/10.1016/j.foodres.2019.04.038

Kusdiana, D., Saka, S. (2004). Effects of water on biodiesel fuel production by supercritical methanol treatment. *Bioresource Technology*. 91(3):289–295. https://doi.org/10.1016/S0960-8524(03)00201-3

Lefebvre, T., Destandau, E., West, C., Lesellier, E. (2020). Supercritical fluid chromatography development of a predictive analytical tool to selectively extract bioactive compounds by supercritical fluid extraction and pressurized liquid extraction. *Journal of Chromatography*. 1632:461582. https://doi.org/10.1016/j.chroma.2020.461582

Lima, M., Andreou, R., Charalampopoulos, D., Chatzifragkou, A. (2021). Supercritical carbon dioxide extraction of phenolic compounds from potato (*Solanum tuberosum*) peels. *Applied Scienses*. 1–18. https://doi.org/10.3390/app11083410

Liu, J., Lin, R., Nan, Y., Tavlarides, L. L. (2015). Production of biodiesel from microalgae oil (*Chlorella protothecoides*) by non-catalytic transesterification: evaluation of reaction kinetic models and phase behavior. *Journal of Supercritical Fluids*. 99:38–50. https://doi.org/10.1016/j.supflu.2015.01.010

Liu, T., Wu, H., Wu, H., Zhang, J. (2019). Wormwood (*Artemisia absinthium* L.) as a promising nematicidal and antifungal agent: chemical composition, comparison of extraction techniques and bioassay-guided isolation. *Industrial Crops and Products*. 133:295–303. https://doi.org/10.1016/j.indcrop.2019.03.039

Lončarević, I., Pajin, B., Fišteš, A., Tumbas Šaponjac, V., Petrović, J., Jovanović, P., Vulić, J., Zarić, D. (2018). Enrichment of white chocolate with blackberry juice encapsulate: impact on physical properties, sensory characteristics and polyphenol content. *Lebensmittel-Wissenschaft & Technologie*. 92:458–464. https://doi.org/10.1016/j.lwt.2018.03.002

Lorenzo, J. M., Pateiro, M., Domínguez, R., Barba, F. J., Putnik, P., Kovačević, D. B., Shpigelman, A., Granato, D., Franco, D. (2018). Berries extracts as natural antioxidants in meat products: a review. *Food Research International*. 106:1095–1104. https://doi.org/10.1016/j.foodres.2017.12.005

Machado, A. P. da F., Rezende, C. A., Rodrigues, R. A., Barbero, G. F., Rosa, P. de T. V., Martínez, J. (2018). Encapsulation of anthocyanin-rich extract from blackberry residues by spray-drying, freeze-drying and supercritical antisolvent. *Powder Technology*. 340:553–562. https://doi.org/10.1016/j.powtec.2018.09.063

Machalova, Z., Sajfrtova, M., Pavela, R., Topiar, M. (2015). Extraction of botanical pesticides from *Pelargonium graveolens* using supercritical carbon dioxide. *Industrial Crops and Products*. 67:310–317. https://doi.org/10.1016/j.indcrop.2015.01.070

Mankar, A. R., Pandey, A., Modak, A., Pant, K. K. (2021). Pretreatment of lignocellulosic biomass: a review on recent advances. *Bioresource Technology*. 334:125235. https://doi.org/10.1016/j.biortech.2021.125235

Matsuyama, K., Morotomi, K., Inoue, S., Nakashima, M., Nakashima, H., Okuyama, T., Kato, T., Muto, H., Sugiyama, H. (2019). Antibacterial and antifungal properties of Ag nanoparticle-loaded cellulose nanofiber aerogels prepared by supercritical CO2 drying. *Journal of Supercritical Fluids*. 143:1–7. https://doi.org/10.1016/j.supflu.2018.08.008

Maxim F., Karalis K., Boillat P., Banuti D. T., Damian J. I., Niceno B., Ludwig C. (2021). Thermodynamics and dynamics of supercritical waterpseudo-boiling. *Advance Science*. 8(3):2002312. https://doi.org/10.1002/advs.202002312

Mazzutti, S., Ferreira, S. R. S., Riehl, C. A. S., Smania, A., Smania, F. A., Martínez, J. (2012). Supercritical fluid extraction of *Agaricus brasiliensis*: antioxidant and antimicrobial activities. *Journal of Supercritical Fluids*. 70:48–56. https://doi.org/10.1016/j.supflu.2012.06.010

Molino, A., Larocca, V., Di Sanzo, G., Martino, M., Casella, P., Marino, T., Laratza, D., Musmarra, D. (2019). Extraction of bioactive compounds using supercritical carbon dioxide. 24(4):782. *Molecules*. https://doi.org/10.3390/molecules24040782

Munim, S. A., Raza, Z. A. (2019). Poly(lactic acid) based hydrogels: formation, characteristics and biomedical applications. *Journal of Porous Materials*. 26(3):881–901. https://doi.org/10.1007/s10934-018-0687-z

Nooshkam, M., Varidi, M. (2020). Food hydrocolloids Maillard conjugate-based delivery systems for the encapsulation, protection, and controlled release of nutraceuticals and food bioactive ingredients: a review. *Food Hydrocolloids*. 100:105389. https://doi.org/10.1016/j.foodhyd.2019.105389

Ozkan, G., Franco, P., De Marco, I., Xiao, J., Capanoglu, E. (2019). A review of microencapsulation methods for food antioxidants: principles, advantages, drawbacks and applications. *Food Chemistry*. 272:494–506. https://doi.org/10.1016/j.foodchem.2018.07.205

Pinto, A. C., Guarieiro, L. L. N., Rezende, M. J. C., Ribeiro, N. M., Torres, E. A., Lopes, W. A., De Pereira, P. A. P., De Andrade, J. B. (2005). Biodiesel: an overview. *Journal of the Brazilian Chemical Society*. 16:1313–1330. https://doi.org/10.1590/S0103-50532005000800003

Pinto, D., De La Luz Cádiz-Gurrea, M., Sut, S., Ferreira, A. S., Leyva-Jimenez, F. J., Dall'acqua, S., Segura-Carretero, A., Delerue-Matos, C., Rodrigues, F. (2020). Valorisation of underexploited *Castanea sativa* shells bioactive compounds recovered by supercritical fluid extraction with CO2: a response surface methodology approach. *Journal of CO2 Utilization*. 40. https://doi.org/10.1016/j.jcou.2020.101194

Raza, Z. A., Khalil, S., Ayub, A., Banat, I. M. (2020). Recent developments in chitosan encapsulation of various active ingredients for multifunctional applications. *Carbohydrate Research*. 108004. https://doi.org/10.1016/j.carres.2020.108004

Rebelatto, E., Rodrigues, L. G., Rudke, A., Andrade, K., Ferreira, S. (2020). Sequential green-based extraction processes applied to recover antioxidant extracts from pink pepper fruits. *Journal of Supercritical Fluids*. 1–9. https://doi.org/10.1016/j.supflu.2020.105034

Rivas, M. Á., Casquete, R., de Guía Córdoba, M., Benito, M. J., Hernández, A., Ruiz-Moyano, S., Martín, A. (2021). Functional properties of extracts and residual dietary fibre from pomegranate (*Punica granatum* L.) peel obtained with different supercritical fluid conditions. *Lebensmittel-Wissenschaft & Technologie*. 145. https://doi.org/10.1016/j.lwt.2021.111305

Román-Figueroa, C., Olivares-Carrillo, P., Paneque, M., Palacios-Nereo, F. J., Quesada-Medina, J. (2016). High-yield production of biodiesel by non-catalytic supercritical methanol transesterification of crude castor oil (*Ricinus communis*). *Energy*. 107:165–171. https://doi.org/10.1016/j.energy.2016.03.136

Romano, R., De Luca, L., Aiello, A., Pagano, R., Di Pierro, P., Pizzolongo, F., Masi, P. (2022a). Basil (*Ocimum basilicum* L.) leaves as a source of bioactive compounds. *Foods*. 11(20):1–13. https://doi.org/10.3390/foods11203212

Romano, R., Luca, L., Aiello, A., Rossi, D., Pizzolongo, F., Masi, P. (2022b). Bioactive compounds extracted by liquid and supercritical carbon dioxide from citrus peels. *Food Science and Technology*. 57:3826–3837. https://doi.org/10.1111/ijfs.15712

Sahena, F., Zaidul, I. S. M., Jinap, S., Karim, A. A., Abbas, K. A., Norulaini, N. A. N., Omar, A. K. M. (2009). Application of supercritical CO2 in lipid extraction—a review. *Journal of Food Engineering*. 95(2):240–253. https://doi.org/10.1016/j.jfoodeng.2009.06.026

Sánchez-Camargo, A., Gutiérrez, L., Vargas, S. M., Martinez-Correa, H., Parada-Afonso, F., Narváez-Cuenca, C. (2019). Valorisation of mango peel: proximate composition, supercritical fluidextraction of carotenoids, and application as an antioxidant additivefor an edible oil. *Journal of Supercritical Fluids*. 1–9. https://doi.org/10.1016/j.supflu.2019.104574

Serra, A. T., Seabra, I. J., Braga, M. E. M., Bronze, M. R., De Sousa, H. C., Duarte, C. M. M. (2010). Processing cherries (*Prunus avium*) using supercritical fluid technology. Part 1: recovery of extract fractions rich in bioactive compounds. *Journal of Supercritical Fluids*. 55(1):184–191. https://doi.org/10.1016/j.supflu.2010.06.005

Shao, P., Niu, B., Chen, H., Sun, P. (2018). Fabrication and characterization of tea polyphenols loaded pullulan-CMC electrospun nanofiber for fruit preservation. *International Journal of Biological Macromolecules*. 107:1908–1914. https://doi.org/10.1016/j.ijbiomac.2017.10.054

Silva, C., Oliveira, J. V. (2014). Biodiesel production through non-catalytic supercritical transesterification: current state and perspectives. *Brazilian Journal of Chemical Engineering*. 31(2):271–285. https://doi.org/10.1590/0104-6632.20140312s00002616

Silva, L. O., Ranquine, L. G., Monteiro, M., Torres, A. G. (2019). Pomegranate (*Punica granatum* L.) seed oil enriched with conjugated linolenic acid (cLnA), phenolic compounds and tocopherols: improved extraction of a specialty oil by supercritical CO2. *Journal of Supercriticial Fluids*. 126–137. https://doi.org/10.1016/j.supflu.2019.02.019

Silva, S. G., de Oliveira, M. S., Cruz, J. N., da Costa, W. A., da Silva, S. H. M., Barreto Maia, A. A., de Sousa, R. L., Carvalho Junior, R. N., de Aguiar Andrade, E. H. (2021). Supercritical CO2 extraction to obtain *Lippia thymoides* Mart. & Schauer (Verbenaceae) essential oil rich in thymol and evaluation of its antimicrobial activity. *Journal of Supercritical Fluids*. 168. https://doi.org/10.1016/j.supflu.2020.105064

Souza, A. R. C., Guedes, A. R., Folador Rodriguez, J. M., Bombardelli, M. C. M., Corazza, M. L. (2018). Extraction of *Arctium lappa* leaves using supercritical CO2 + ethanol: kinetics, chemical composition, and bioactivity assessments. *Journal of Supercritical Fluids*. 140:137–146. https://doi.org/10.1016/j.supflu.2018.06.011

Tavassoli-Kafrani, E., Goli, S. A. H., Fathi, M. (2018). Encapsulation of orange essential oil using cross-linked electrospun gelatin nanofibers. *Food and Bioprocess Technology*. 11(2):427–434. https://doi.org/10.1007/s11947-017-2026-9

Temelli, F. (2009). Perspectives on supercritical fluid processing of fats and oils. *Journal of Supercritical Fluids*. 47(3):583–590. https://doi.org/10.1016/j.supflu.2008.10.014

Tobar, M., Núñez, G. A. (2018). Supercritical transesterification of microalgae triglycerides for biodiesel production: effect of alcohol type and co-solvent. *Journal of Supercritical Fluids*. 137:50–56. https://doi.org/10.1016/j.supflu.2018.03.008

Tyskiewicz, K., Tyskiewicz, R., Konkol, M., Rój, E., Jaroszuk-Sciseł, J., Skalicka-Wozniak, K. (2019). Antifungal properties of *Fucus vesiculosus* L. supercritical fluid extract against *Fusarium culmorum* and *Fusarium oxysporum*. *Molecules*. 24(19). https://doi.org/10.3390/molecules24193518

Uwineza, P. A., Gramza-Michałowska, A., Bryła, M., Waśkiewicz, A. (2021). Antioxidant activity and bioactive compounds of *Lamium album* flower extracts obtained by supercritical fluid extraction. *Applied Sciences*. 11:1–12. https://doi.org/10.3390/app11167419

Uwineza, P. A., Waśkiewicz, A. (2020). Recent advances in supercritical fluid extraction of natural bioactive compounds from natural plant materials. *Molecules*. 25(17). https://doi.org/10.3390/molecules25173847

Wen, D., Jiang, H., Zhang, K. (2009). Supercritical fluids technology for clean biofuel production. *Progress in Natural Science*. 19(3):273–284. https://doi.org/10.1016/j.pnsc.2008.09.001

Xu, Q. Q., Li, Q., Yin, J. Z., Guo, D., Qiao, B. Q. (2016). Continuous production of biodiesel from soybean flakes by extraction coupling with transesterification under supercritical conditions. *Fuel Processing Technology*. 144:37–41. https://doi.org/10.1016/j.fuproc.2015.12.018

Yu, X., Zhang, H., Wang, J., Wang, J., Wang, Z., Li, J. (2022). Phytochemical compositions and antioxidant activities of essential oils extracted from the flowers of *Paeonia delavayi* using supercritical carbon dioxide fluid. *Molecules*. 2022; (27) 1–16. https://doi.org/10.3390/molecules27093000

Yue, L., Zhang, L., Lv, L., Yang, M. (2020). Phase separation in the supercritical mixtures of N2, H2O and CO2 through a molecular dynamics study. *Journal of Molecular Liquids*. https://doi.org/10.1016/j.molliq.2020.113811

Zhou, J., Gullón, B., Wang, M., Gullón, P., Lorenzo, J. M., Barba, F. J. (2021). The application of supercritical fluids technology to recover healthy valuable compounds from marine and agricultural food processing by-products: a review. *Processes*. 9(2):1–23. https://doi.org/10.3390/pr9020357

Zizovic, I. (2020). Supercritical fluid applications in the design of novel antimicrobial materials. *Molecules*. 25(11). https://doi.org/10.3390/molecules25112491.

5 Last Innovations in Supercritical Fluid Extraction Processes for Metabolite Extraction from Agriculture and Agroindustrial Residues in the New Food Waste Biorefineries Context

Daissy Lorena Restrepo Serna, Mariana Ortiz Sánchez, Sara Piedrahita-Rodríguez, and Carlos Ariel Cardona Alzate

5.1 OVERVIEW

Nature presents many bioactive compounds in different parts of a wide variety of plants (Bernhoft, 2010). Parts such as leaves, seeds, flowers, and stems have been shown to contain bioactive compounds of interest to humans (Elgndi et al., 2017; Favareto et al., 2017; Sánchez-Vicente et al., 2009). This type of compound has been found in some residues of different agricultural and agroindustrial processes (Bitencourt et al., 2020; Dávila et al., 2017; Mamma & Christakopoulos, 2008). Among the bioactive compounds are terpenes (i.e., carotenoids and essential oils), phenolic compounds (i.e., flavonoids, lignin, and tannins), glycosides, alkaloids, carbohydrates, proteins, lipids, and amino acids (Croteau et al., 2000), which have applications in different industries, such as food, pharmaceuticals, and cosmetics. Several technologies have been used to obtain these compounds. The selection of one technique or another considers factors such as the source of the bioactive compound, the yield of the extraction process (global or total), productivity, and selectivity. The first three factors are linked with the economic aspects of the process, explicitly concerning economic viability. In comparison, selectivity relates to the final product and to product quality (Pereira & Meireles, 2010).

Extraction technologies can be classified into conventional and non-conventional (Kamran et al., 2020; Putnik et al., 2019). Conventional technologies include liquid–liquid extraction, evaporation, maceration, hydrodistillation, and Soxhlet extraction (Hasmida et al., 2014; Mazzutti et al., 2017), which are used in many studies as reference methods for comparison with non-conventional technologies (Azmir et al., 2013). Techniques such as accelerated solvents (Christen & Kaufmann, 2002), enzymatic digestion (Gai et al., 2013; Zhang et al., 2013), microwave heating (Routray & Orsat, 2012), ohmic heating (Lakkakula et al., 2004), pulsed electric field (Toepfl et al., 2006), supercritical fluids (Goyeneche et al., 2020), and ultrasonic (Deng et al., 2015; Taofiq et al., 2019) are considered among the non-conventional technologies implemented to obtain bioactive compounds. These last technologies have been focused on more efficient processes, so it employs less chemicals and lowers energy consumption (Bursać Kovačević et al., 2018). Each technology developed intends

DOI: 10.1201/9781003462804-7

to be a "green" technology, for which Chemat et al. (2012) have developed the six principles of a process that satisfies the denomination of being a "green" technology: (1) innovation in renewable plant waste (innovative raw material); (2) alternative solvents (agrosolvents); (3) energy recovery system; (4) obtaining co-products by implementing bio- and agro-refinery concepts (waste minimization); (5) short operation times (safe processes); (6) biodegradable extracts, without contaminants and not denaturalized (obtaining natural and safe extracts) (Chemat et al., 2012; De Melo et al., 2014; Giacometti et al., 2018). In this context, compared to conventional extraction methods and other non-conventional technologies, supercritical fluid extraction (SFE) is presented as a viable alternative. SFE is a technology that uses gases as solvents, which are found to be above their critical temperature and pressure (Knez et al., 2019). By modifying the operating conditions (maintaining the supercritical conditions of the solvent), it is possible to obtain different varieties of the same extract (variation in the final product), with different characteristics (Klein et al., 2018; Manjare & Dhingra, 2019). This characteristic has converted SFE in a technology very interesting for the scientific community.

5.2 INTRODUCTION TO SUPERCRITICAL FLUIDS

5.2.1 Subcritical, Near-Critical, and Supercritical Fluid

Traditionally, three phases can be identified in a phase diagram: solid, liquid, and gas. However, additional phases of matter can also be identified, which are of interest in industrial processes. Regions such as subcritical, near-critical, and supercritical are used both at the laboratory and industrial levels for these regions' characteristics. Figure 5.1 shows the location of these last three regions in a phase diagram.

5.2.1.1 Subcritical Region

Located in the liquid region, the *subcritical region* presents temperature and pressure conditions below the critical values and above the boiling temperature and vapor pressure obtained at the critical temperature (Haghighi & Khajenoori, 2013). According to Figure 5.1, in a diagram, the subcritical region is located above the vapor–liquid equilibrium line up to the critical pressure line, and from the boiling temperature line to the critical temperature line (Okajima & Sako, 2014). In an enthalpy–pressure diagram (see Figure 5.2), this region is located at the bottom of the dome, below the saturated liquid line, and to the left of the critical line (Pieters, 2017). Under critical conditions, water becomes a suitable solvent for organic compound hydrolysis, even without additives. Under subcritical conditions, water has properties as a potential solvent and catalyst for organic reactions (Quitain et al., 2015).

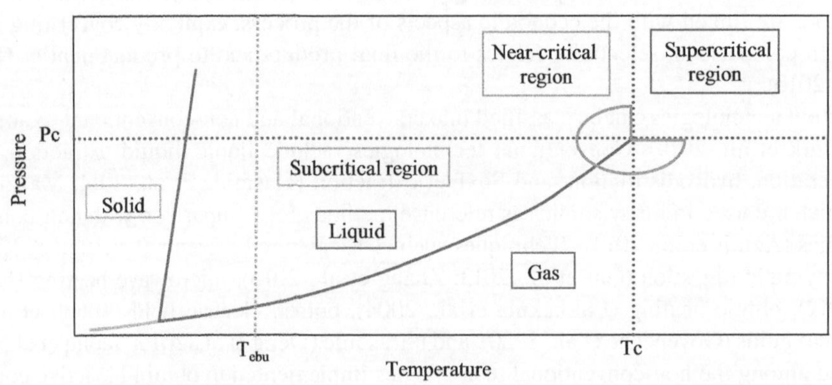

FIGURE 5.1 Position of subcritical, near-critical, and supercritical regions in a phase diagram.

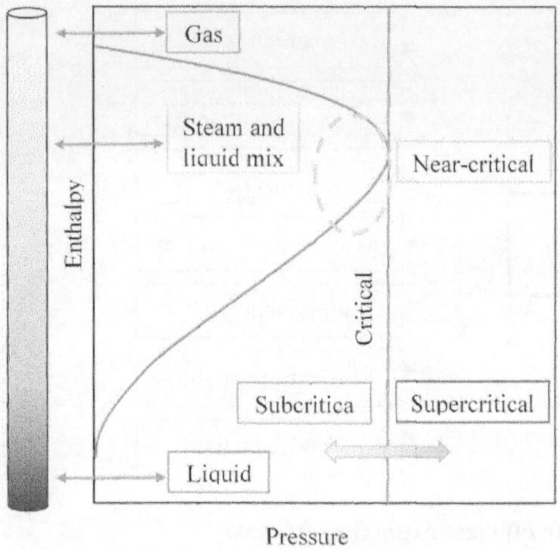

FIGURE 5.2 Diagram: pressure–enthalpy, localization of phases.

5.2.1.2 Near-Critical Region

The *near-critical region* occurs when a fluid is below its critical temperature and pressure but exhibits one of two behaviors: (1) the density of a supercritical fluid and (2) the solvating properties of a supercritical fluid (Castor, 2005; Okajima & Sako, 2014). In Figure 5.1, this region comprises three-quarters of a circumference whose center is located at the critical point and does not consider the region to the right of the critical temperature and above the critical pressure line. In a pressure–enthalpy diagram, the limitation between the critical and near-critical region can be seen. A vertical line passing through the critical point (top of the dome) divides both regions (see Figure 5.2).

5.2.1.3 Supercritical Region

A fluid is considered in this region when it presents temperature and pressure conditions higher than its critical values. As shown in Figure 5.1, there is not a gas–liquid interface in this region. The characteristics of this region have made it the subject of different studies focused on its use for extraction and reaction processes (Aristizábal-Marulanda & Cardona Alzate, 2019; Azmir et al., 2013; Bitencourt et al., 2020; Daza Serna et al., 2016; G. Huang et al., 2009; Mitra et al., 2009; Squillace et al., 2020); characteristics such as the following are of interest in these processes: (1) The viscosity is much lower than that of liquids, which confers very favorable hydrodynamic properties. (2) Low surface tension allows high penetrability through porous solids and packed beds. (3) Higher diffusion coefficients (diffusivity) than in liquids so that the transfer of mass is favorable. (4) Density above the critical point depends basically on pressure and temperature, but in any case, it is closer to that of liquids than that of gases. Density increases with increasing pressure at constant temperature and decreasing temperature at constant pressure.

5.2.2 Supercritical Fluid Extraction (SFE or SCFE)

The SFE is a method used to obtain different compounds of interest for the human through a solvent. At supercritical conditions, the solvent presents characteristics such as density similar to the liquid, high solubility, low viscosity (approximately equal to that of gas), and high diffusivity (favoring the transport properties) (Brunner, 2010; Díaz-Reinoso et al., 2006; Herrero et al., 2006). Compared with other technologies, SFE employs shorter residence times and requires a smaller amount of solvent. In addition, the alteration of its operating parameters allows obtaining extracts with different properties and combinations that improve the selectivity of different interest compounds (Pronyk &

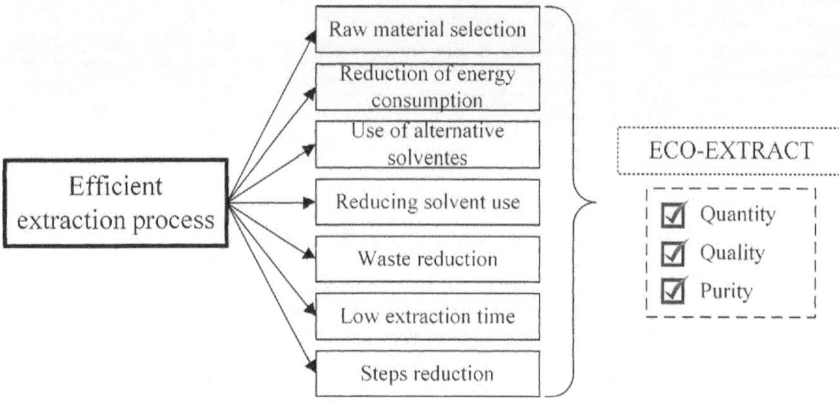

FIGURE 5.3 Principles of an efficient extraction process.

Source: Adapted from Chemat et al. (2019).

Mazza, 2009). Thus, the versatility of SFE provides a promising technology for the efficient production of extracts. In addition, it is possible to modify the extraction conditions and the process to meet the objectives for the efficient extraction of extracts (see Figure 5.3). Among the changes that can be made to the SFE process are the addition of a co-solvent (Ludwig et al., 2021), use of enzymes (Mushtaq et al., 2015), ultrasound (Mazzutti et al., 2017), solvent recirculation (Caballero Galván, 2017), and sequential extraction (Rubashvili et al., 2018), among others. The changes in the solvent type and concentration are the most manipulated variables in SFE.

Different solvents have been used as supercritical fluids; acetone, ammonia, benzene, carbon dioxide (CO_2), fluorocarbon compounds, and ethanol have been found among these solvents, ethylene, methane, methanol, among others. Table 5.1 presents some of the solvents used as supercritical fluids and their molecular weight and critical point properties.

The energy required to reach the supercritical state is considered as one of the parameters to be considered for the selection of a solvent in SFE. Figure 5.4 shows the energy required by each solvent presented in Table 5.1 to achieve the supercritical state. Dichlorofluoromethane presents the lowest energy consumption, with 0.25 MJ per kg of the solvents analyzed. On the other hand, ammonia has the highest consumption, with 3.14 MJ per kg. However, in selecting the solvent, other aspects, such as the nature of the solvent, must be considered.

Due to some of these solvents' nature, these are prohibited or used as supercritical fluids that must meet high operational standards for health and safety during operation. These limitations are due to some of these solvents' harmful or hazardous effects on humans (Khaw et al., 2017). These solvents include chlorofluorocarbons (e.g., chlorotrifluoromethane and dichlorofluoromethane), which have high densities at a critical point harmful to the environment. Compounds such as acetone, benzene, dimethyl ether, ethane, ethanol, methane, and methanol have low densities at the critical point and are hazardous to the environment. Fluids such as air, nitrogen, ammonia, and water are considered green technology fluids and have relatively high critical point density values. However, a solvent has a high density at the critical point and is considered a green technology (see **Figure 5.5**). This solvent is CO_2. Additionally, it is a re-usable gas; has relatively low critical point conditions ($T_c = 304.1K(31.1°C)$ (and $P_c = 7.38MPa$) (Shilpi et al., 2013); is non-toxic, non-flammable, non-corrosive, and easy to operate; has a low cost; and is available in large quantities with high levels of purity (Girotra et al., 2013; Z. Huang et al., 2012; Smith, 1999). Although CO_2 is a greenhouse gas, it is obtained from nature, so when released (returned to the atmosphere), it does not change its levels (the net change is zero). In this context, CO_2 is the most common supercritical fluid used in different processes (Knez et al., 2019). The CO_2 can be easily removed from the extract, allowing a

TABLE 5.1
Critical Conditions for Some Solvents Used as Supercritical Fluids

Supercritical Fluid	Molecular Weight (g/mol)	Critical Temperature (K)	Critical Pressure (MPa)	Density at a Critical Point (kg/m³)
Acetone	58.08	508.1	4.7	0.278
Ammonia	17.03	405.35	11.33	225
Benzene	78.11	562	4.9	30.9
Carbon dioxide	44.01	304.1	7.38	0.469
Chlorotrifluoromethane	104.5	301.8	3.88	582.9
Dichlorofluoromethane	102.9	451	5.18	526.1
Dimethyl ether	46.1	400.1	5.34	277
Ethane	30.07	305.3	4.87	0.203
Ethanol	46.07	513.9	6.14	0.276
Ethylene	28.05	282.4	5.04	0.215
Methane	16.04	190.4	4.6	0.162
Methanol	32.04	512.6	8.09	0.272
Nitrogen	28.01	126	3.4	313.3
Propane	44.09	369.8	4.25	0.217
Propylene	42.08	364.9	4.6	0.232
Water	18.01	647.09	22.06	0.322

Source: Khaw et al. (2017); Sapkale et al. (2010).

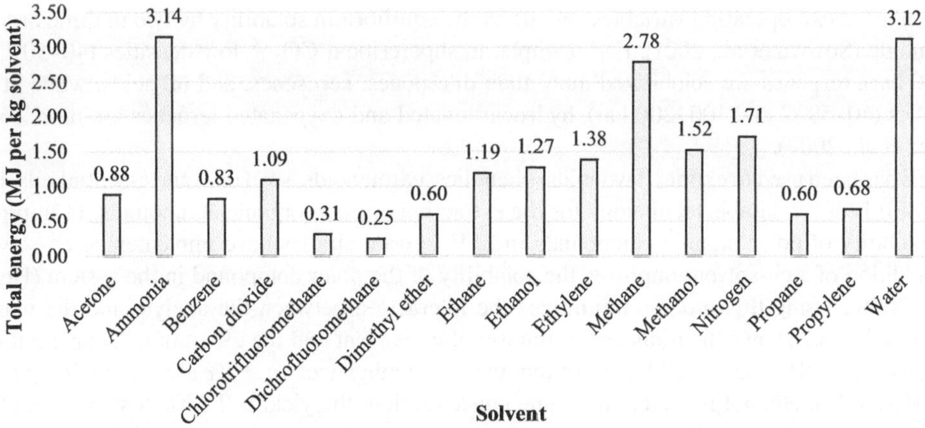

FIGURE 5.4 Different solvents consume energy to reach the supercritical state. *Estimated values using Aspen plus v.9.*

solvent-free extract to be obtained. These properties make supercritical CO_2 a favorable alternative in many applications—for example, extraction of seeds, peels, and leaves of plants; pretreatment of lignocellulosic materials (Daza Serna et al., 2016); drug transport in the pharmaceutical field (Davies et al., 2008; Temtem et al., 2009); virus inactivation (Fages et al., 1998); sterilization of biomaterials and medical devices (Chang et al., 2011); and synthesis of polymers (Kazarian, 2000; Nalawade et al., 2006).

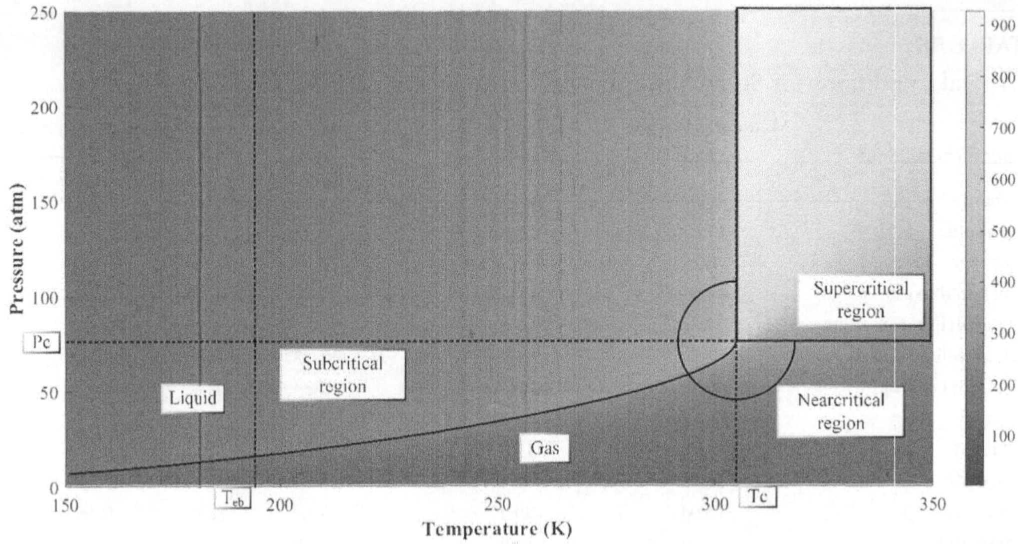

FIGURE 5.5 The density of CO_2 as a function of temperature and pressure, and location of the aggregation states of mass.

Source: Data obtained resolving the Peng–Robinson equation in MATLAB (Peng & Robinson, 1976).

Temperature and pressure are the most important process variables that can be manipulated in supercritical fluid processes. The solvating power (dissolving capacity) of supercritical fluids and CO_2 is sensitive to these two variables' changes. Variations in the temperature and pressure of a supercritical fluid allow it to be adjusted according to particular needs (Shi et al., 2007). Modifications in these two operating variables can affect the equilibrium solubility by two to three orders of magnitude (Sovova et al., 2001). For example, in supercritical CO_2 at low densities (40–50°C and 80–90 bar), terpenes are solubilized more than diterpenes, kerosenes, and oil acids, while at high densities (40–50°C and 100–200 bar), hydrocarbonated and oxygenated terpenes are more soluble (Zorca et al., 2007).

SFE has been used to extract flavonoids, phenolics, carotenoids, seed oils, and essential oils. However, SFE is not a suitable technology for the extraction of polar organic compounds. In improving the solubility of polar organic compounds in SFE, several studies have implemented co-solvents. The addition of a co-solvent improves the solubility of the polar compound in the system (Herrero et al., 2006). Using this process minimizes the interaction between the analyte and the material matrix. Table 5.2 shows the influence of the use of co-solvent and the effect of its concentration in SFE processes. Klein et al. (2021) report that the use of ethyl acetate at 5% allows obtaining a yield of 2.04%. When ethanol is used at the same concentration, the yield is 2.27%. It shows the effects of the solvent used. The affinity of the metabolite for the co-solvent is a determining factor of the yield to be obtained in each case. The work of Sun et al. (2021) shows that adding a co-solvent such as ethanol decreases the extraction yield of tocopherol. However, the extraction yield of canola oil shows a change from 3.55% to 5.43%. The works of Ludwig et al. (2021) and M. de A. Lima et al. (2021) show that increasing the co-solvent concentration improves extraction yields. However, the solubility limit of the extracted metabolite in the co-solvent used must be taken into account.

As shown in Table 5.2, co-solvents such as ethanol, ethyl acetate, and methanol have been used. Chai et al. (2021) report that ethanol is the most used co-solvent in SFE, with 53%. Figure 5.6 presents the most used co-solvents in SFE. Ethanol is followed by methanol with 21%. These two co-solvents are the most used in SFE. Among the least-used co-solvents are vegetable oils, water, dichloromethane, n-hexane, and combinations.

TABLE 5.2
Effect of the Use of Co-Solvents on SFE

Raw Material	Component	Co-Solvent	Yield	Reference
Uvaia leaves (Eugenia pyriformis)	α-amyrin and β-amyrin	0	2.06%	(Klein et al., 2021)
		5% ethyl acetate	2.04%	
		5% ethanol	2.27%	
Canola seeds	Tocopherol	0	6.61 mg/g oil	(Sun et al., 2021)
		10% (v/w) ethanol	1.05 mg/g oil	
	Canola oil	0	3.55%	
		10% (v/w) ethanol	5.43%	
Dunaliella salina	β-carotene	0	4.3%	(Ludwig et al., 2021)
		4 wt% ethanol	12.2%	
		10 wt% ethanol	25.0%	
Potato (*Solanum tuberosum*) peels	Caffeic acid	7% methanol	6.0%	(M. de A. Lima et al., 2021)
	Chlorogenic acid	15% methanol	11.1%	

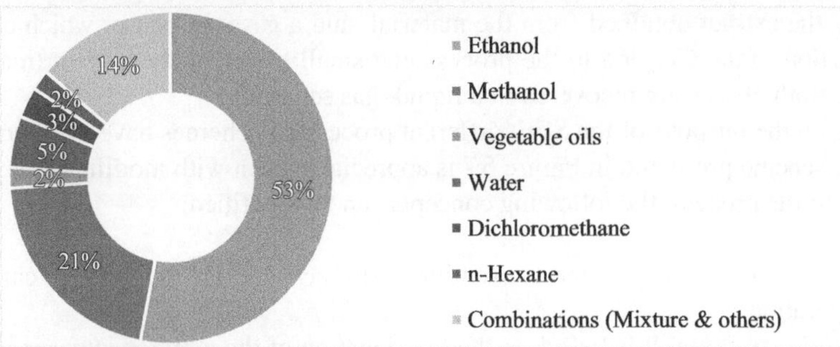

FIGURE 5.6 Types of co-solvents used in SFE.
Source: Adapted from Chai et al. (2021).

5.3 SUPERCRITICAL FLUID TECHNOLOGY

The supercritical state of the solvent is reached when the critical point of the solvent is exceeded. For this purpose, a semi-continuous reactor is used where the raw material is placed and the solvent (CO_2) is continuously fed. In SFE processes, five main pieces of equipment are involved: condenser (heat exchanger), pump for CO_2, reactor–extractor (high-pressure extractor unit), valve (back pressure regulator), and liquid–gas separator (collector), as shown in Figure 5.7. At the beginning of the process, the CO_2 is contained in a tank from which it is driven by the same tank pressure to a condenser, reaching a liquid state at the output. Then, the liquid CO_2 is subjected to a pressure increase above its critical point, which is the extraction pressure. For this, it is recommended to use at least a factor of 1.3 times the critical pressure of the CO_2. Finally, the CO_2 at high pressure is injected into a closed tank, where the extraction occurs (reactor–extractor). The material to be extracted has been previously introduced in the extractor, which is contained in a cloth bag to facilitate the recovery of the material. Additionally, it allows the separation of the solid and liquid phases at the end of the extraction process. This extractor consists of a temperature control system utilizing a water bath heated by electric resistance. The temperature of the bath corresponds to the extraction temperature. Therefore, mainly processes involving supercritical fluids operate in batches. Once the extractor has been loaded with the extraction material, it is closed and begins to pump pressurized CO_2,

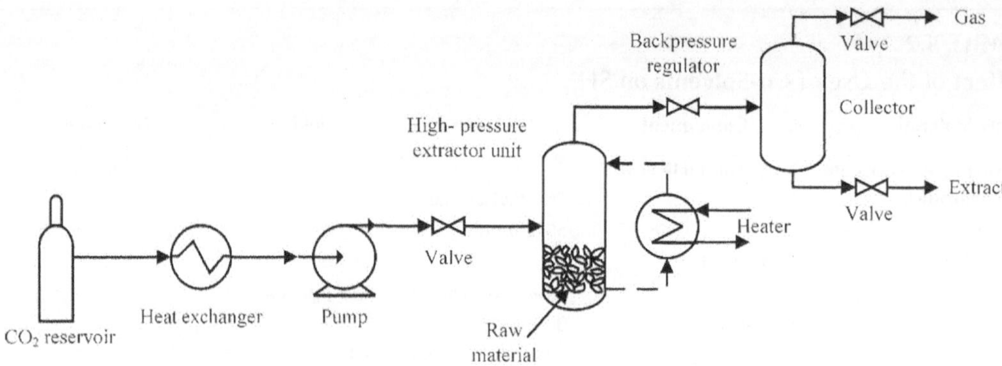

FIGURE 5.7 Main equipment involved in the conventional SFE.

controlling the extraction pressure that is desired to work. The extraction time starts counting from the moment the extractor reaches the temperature and pressure conditions selected for the process. When the extraction time has elapsed, the valve located at the extractor's outlet is opened to start the depressurization process. With the decrease of pressure, two phases are generated: a liquid phase, constituted by the extract obtained from the material, and a gaseous phase, which corresponds in greater proportion to the CO_2 fed to the process and small traces of the extract (mainly the most volatile ones). Both phases are recovered in a liquid–gas separator.

Depending on the purpose of the SFE, different processing schemes have been proposed. However, the basic scheme presented in Figure 5.7 is appreciable even with modifications. Based on the modifications to the process, the following concepts can be identified:

- *Simple extraction.* Refers to the conventional SFE scheme. Extractions are carried out in batch operation.
- *Sequential extraction.* It is based on the fractionation of the components using sequential extractions.
- *Co-integrated extraction.* Considers the mixture of ultrasound-assisted extraction (UAE) with SFE.
- *Enzyme-assisted extraction.* Incorporates the use of enzymes in SFE.
- *Extraction with energy recovery.* Takes advantage of the conditions of the streams to implement an energy recovery system.

5.3.1 Single and Sequential Extractions Concept

A *simple extraction* uses a batch reactor for extraction, which occurs in a single stage. The mode of operation requires simple extraction to be carried out at low scales. Among the products obtained from solid biomass are essential oils, phenolic compounds, lipids, and sterols. Since CO_2 has a solvation power similar to hexane, it is possible to obtain better yields than those presented in a Soxhlet extraction. Therefore, it is an alternative to non-polar organic solvents, which have negative effects on the environment.

The batch operation presents relatively easy operation, high yields, and a lower investment cost than a continuous operation. However, there may be variations in product compositions. In a batch operation, specific temperature, pressure, and solvent concentration are used so that certain compounds present higher solubility and are extracted in higher proportions. If changes in the operating conditions are made in the process, it is possible to obtain changes in the density of CO_2 and therefore solubilize other compounds of interest. A sequential extraction can be implemented in SFE to take advantage of this principle. This process makes it possible to increase the number of

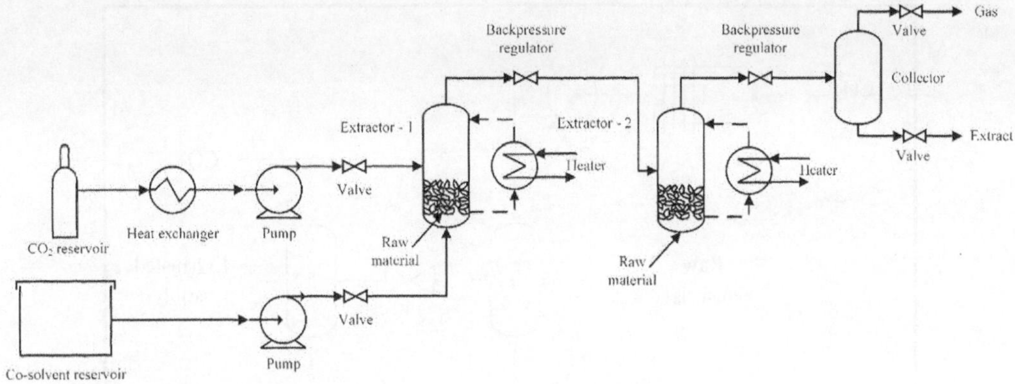

FIGURE 5.8 Process flow diagram of the sequential extraction concept.

compounds extracted from the biomass by varying the pressure, temperature, or solvent used in the subsequent stages. Figure 5.8 highlights the process flow diagram of a sequential extraction concept.

Depending on the nature of the bioactive compound to be extracted, the process described earlier has been modified to improve both the yield and the selectivity of the process, among the main modifications that can be found in using a co-solvent. For using a co-solvent in supercritical fluids, there are two alternatives to feed it into the process. First, the selected co-solvent is added to the extractor along with the solid material. Second, it can be pumped to the extractor at the selected extraction pressure. However, this alternative has disadvantages, such as high co-solvent consumption and diluting extracted bioactive compounds in the co-solvent. This is because the co-solvent in the depressurization process remains in the liquid phase—the scheme in Figure 5.9A. Among the modifications that have been made to this scheme to reduce costs in solvent and co-solvent are: First, the liquid phase recovered in the liquid–gas separator is taken to an evaporator, which allows the recovery of the co-solvent used and obtaining a more concentrated extract. Second, a recycling system of the co-solvent that has been recovered in the evaporator; this system considers a purge system to feed a clean co-solvent and not to lose its properties by excessive contamination with any of the compounds with which it comes into contact during the process. Third, this third modification now focuses on the recovery of the solvent used, considering the same premises as in the case of the co-solvent. These three modifications are presented in Figure 5.9B.

5.3.2 Enzyme-Assisted Extraction

Some enzymes can operate in the presence of subcritical or supercritical fluids. Such characteristic, combined with the possibility of being reused without any change in selectivity, makes them attractive from an economic point of view. The appropriate reaction parameters must be estimated for each substrate/enzyme/dense fluid system for their use. The process parameters influence the stability of the enzymes, and there is a very narrow margin for their operation. High-pressure and high-temperature conditions and low pH lead to enzyme deactivation (Clifford, 1994). Additionally, it must be considered that the active centers of the enzyme can be blocked by both the substrate and the reaction product. As a result, the catalytic activity of the enzyme is decreased or even completely deactivated. In reactions with subcritical or supercritical solvents, the enzymes can be used in their native form or immobilized. Immobilized enzymes are expected to be more stable in subcritical and supercritical media and to have no changes in their enzymatic activity (Brunner, 2005; Stephenson et al., 2006; Wahyudiono et al., 2007).

The formation of carbamates or carbonaceous acid during enzyme-assisted extraction can lead to changes in the catalytic activity of enzymes (Knez et al., 2015). Different reaction media than

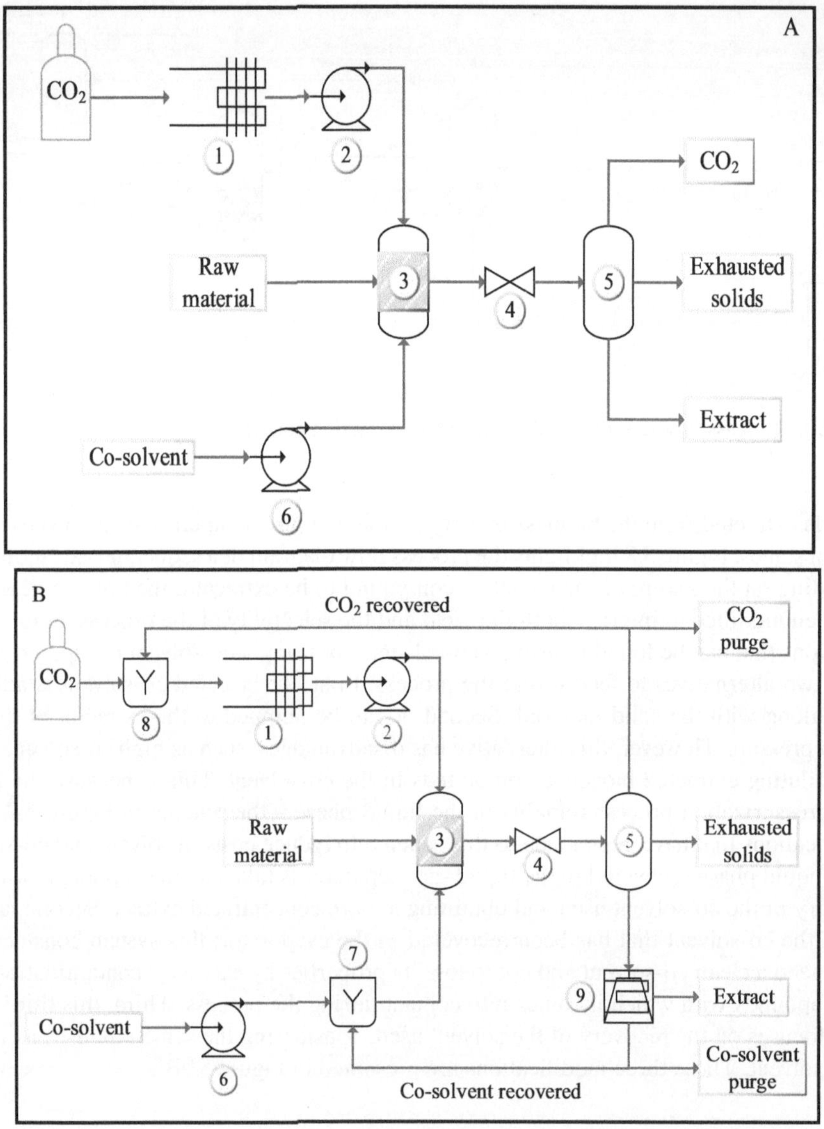

FIGURE 5.9 SFE process flowchart: (A) SFE with co-solvent. (B) Recycled co-solvent and CO_2. (1) Condenser, (2) pump for CO_2, (3) reactor–extractor, (4) valve, (5) liquid–gas separator, (6) pump for co-solvent, (7) co-solvent mixer, (8) CO_2 mixer, (9) evaporator.

CO_2 can be used as a component to avoid deactivation. Alternatives include fluoroform, ethylene, propane, and sulfur hexafluoride (Cantero et al., 2015).

5.3.3 SFE Combined with UAE

The UAE provides advantages over conventional extraction techniques, such as liquid–liquid extraction and Soxhlet extraction (Caballero-Galván et al., 2018). UAE is based on the application of a large negative pressure to a liquid, producing a chemical phenomenon associated with acoustic cavitation and the formation of microbubbles (Capelo-Martínez, 2009). When these bubbles reach an unstable size, they collapse violently. This fact is associated with the release of intense local energy with important chemical (i.e., free radicals) and mechanical (i.e., microjets) effects. The collapse of the bubbles causes the temperature to increase by thousands of degrees and to pressure changes of more than 1,000 atmospheres. In addition to the operating variables of a solvent extraction, the UAE additionally considers the effects of power, frequency, sonication time, and type of material treated (Pokhrel et al., 2016).

Liquid solvents such as water, ethanol, methanol, or solutions thereof are used in the UAE (Orio et al., 2011; Parniakov et al., 2015). However, the use of supercritical fluids as solvents in UAE has long been studied (Riera et al., 2004; Wei & Yang, 2015). It has allowed the obtaining of a pure extract and the reduction of one step in the extraction process: the removal of the extraction solvent.

5.3.4 Energy Recovery

In SFE, processes have three pieces of equipment that operate at high pressure and therefore have a high energy consumption. When analyzing the costs of the process, it has been found that the costs associated with this energy represent a high percentage. Therefore, different alternatives have been proposed to reduce manufacturing costs that allow greater use of the energy present in the process. A first alternative proposed is to use the energy contained in the exhaust stream, which is at high pressure and, when in a gaseous state, can be passed through a turbine instead of a valve to take advantage of its energy content. The change of the valve by the turbine makes it seem that, by the principle of operation of this one, instead of consuming energy in the form of work, this one is generating. Depending on the operating conditions and the efficiency of the turbine operation, two situations may arise. In the first case, the work generated will partly supply the demand presented in terms of power by the process, and a smaller amount will have to be fed to complete the demand. In the second case, the work generated will supply all the demand of the process, and it can be that over a certain amount which can be sold or, in its defect, sent to other parts of the same plant. An outline of this alternative is presented in Figure 5.10a.

A second alternative is to supply this work by incorporating other stages to the process, which have as an objective the generation of energy in the form of work. A technology that fulfills this is gasification, a thermochemical process in which solid waste is used and transformed into energy and chemical products. Now the question is: What solid waste should be used for gasification? Should the amount of raw material to be processed be reduced to use this technology? During SFE, solid waste, an exhausted solid, is generated, which presents a minimum amount of bioactive compounds but a high amount of lignocellulosic compounds, depending on the waste. With this waste in gasification, the amount of raw material to be processed is not compromised. On the contrary, one of the main wastes generated during an SFE is being used. Additionally, it must be considered that during the gasification process, chemical products are generated which have a value in the market or can be subjected to other transformation processes to increase the product portfolio that the company would be offering.

5.4 SUPERCRITICAL FLUIDS IN EXTRACTION PROCESS

One of the most abundant sources of bioactive compounds are plants, in which a great variety of classes of compounds, such as carotenoids, oils, and polyphenols, among others, can be found. When talking about plants, we can consider three main parts: the fruits, the leaves, and the root. Each of these parts is characterized by the accumulation of certain compounds of interest that can be extracted using a variety of technologies. However, among the most commonly used technologies are SFE, microwave-assisted extraction (MAE), pressurized liquid extraction (PLE), and ultrasound-assisted extraction (UAE). These techniques are known for their short extraction time, reduced volume of organic and hazardous solvents, and simplicity to perform, achieving a higher extraction performance with lower energy consumption, so they are classified within the techniques of "green extraction" (Khoddami et al., 2013). In Table 5.3, some of the chemical compounds that

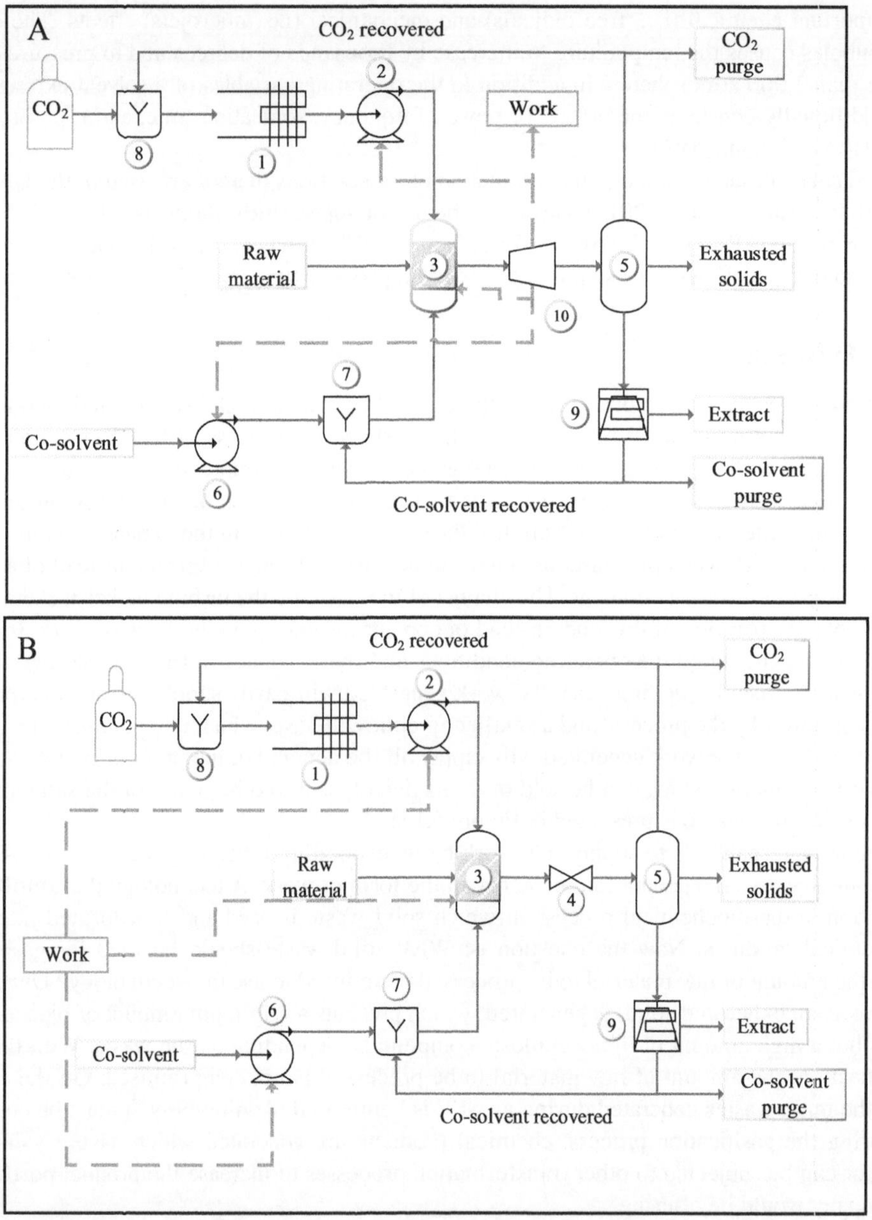

FIGURE 5.10 SFE process flowchart: (A) SFE with energy recovery employed in a turbine. (B) SFE energy recovery. (1) Condenser, (2) pump for CO_2, (3) reactor–extractor, (4) valve, (5) liquid–gas separator, (6) pump for co-solvent, (7) co-solvent mixer, (8) CO_2 mixer, (9) evaporator, (10) turbine.

can be obtained from plants are presented, the part of the plant where they have the highest concentration, and the most recommended technologies for their extraction. With the exception of polysaccharides, the other chemical compounds considered in Table 5.3 can be obtained using SFE, showing the great applicability that this technology presents for the extraction of a great variety of bioactive compounds.

During the last decade, publications have been found dedicated to the study of obtaining bioactive compounds from different plants and their residues using SFE. Mainly studies dedicated to the

TABLE 5.3
Parts of Plants Used for Obtaining Bioactive Compounds and The Preferential Technologies for These Processes

	Plant Part			Extraction Method			
Chemical Class	**Fruits**	**Leaves**	**Roots**	**SFE**	**MAE**	**UAE**	**PLE**
Alkaloids		x	x	x	x		
Anthocyanins	x	x		x	x		
Carotenoids	x	x		x			x
Flavonoids	x	x		x	x		
Oils	x	x		x	x		
Polyphenolics	x	x		x	x		
Polysaccharides	x		x		x	x	
Saponins		x	x	x	x		

Source: Adapted from Belwal et al. (2018).

obtaining of carotenoids, oils, and extracts in general can be found. CO_2 is the solvent used in all cases, with variations in the use or not of a co-solvent. Among the most commonly used co-solvents are ethanol, methanol, and isopropanol, considering that, in the case of oil extraction, some authors report the use of hexane as a co-solvent. The different authors had a common objective, to identify the extraction conditions that favor the obtaining of a determined compound or, in its defect, the conditions that maximize the concentration of the total extract. In the latter case, the focus was mainly on the identification of the compounds present in the extracts obtained and on showing an additional use for the raw materials used. Among the investigations focused on the use of SFE to obtain different bioactive compounds in recent years are the studies presented in Table 5.4.

5.4.1 Carotenoids

Carotenoids are hydrocarbons with 40 carbon atoms and two terminal rings, to which the red, orange, and yellow colorations are attributed. These compounds can be of animal or plant origin, mainly synthesized from photosynthetic organisms (plants, algae, and cyanobacteria), some bacteria, and non-photosynthetic fungi. Carotenoids are classified into two groups: The first group consists of carotenes, such as β-carotene; the second group is composed of oxygenated carotene derivatives, such as lutein and xanthophylls (e.g. violaxanthin, neoxanthin, and zeaxanthin).

Table 5.5 summarizes the SFE methods for carotenoids.

5.5 ECONOMIC SUSTAINABILITY OF SUPERCRITICAL EXTRACTION PROCESSES

The economic factor is highly influential in industrial processes. When talking about SFE using different biomasses, the productivity and selectivity of the process should be considered (Posada et al., 2012; Sánchez-Camargo et al., 2019). In addition, the type of compound to be extracted needs to be considered for the value it presents in the market. Manjare and Dhingra (2019) report that SFE is a technology with high utility and flexibility. However, it highlighted the energy requirements and high capital investment cost. Since it is a process that employs high pressures, additional operating costs are present compared to solvent extraction; despite this disadvantage, Manjare and Dhingra (2019) highlight different economic benefits provided by the use of SFE. The main benefits

TABLE 5.4
Compounds Obtained through SFE in the Last Decade

Raw Material	Temperature (°C)	Pressure (MPa)	CO_2 Flow (mL/min)	Compound	Concentration	Co-Solvent	Time (min)	Reference
Iberis amara seeds	55	25	2.5	Cucurbitacin E	7.08 mg/g	12% ethanol	100	(Liu et al., 2020)
Turmeric (*Curcuma longa* L.)	50	25	3	Curcumin	1.69 % w/w	10% ethanol	90	(Chhouk et al., 2017)
Pumpkin (*Cucurbita máxima*)	68.1	32.14	250	Seed oil	98%	Hexane	94.6	(Mitra et al., 2009)
Chia (*Salvia hispánica* L.)	80	45	1.82 kg/h	Seed oil	88.1%	ND	240	(Ixtaina et al., 2010)
Spent coffee grounds	60	40	0.9 kg/h	Oil	25%	10% ethanol	—	(Bitencourt et al., 2020)
Dry leaves of *Piper hispidum* (ASE 29962)	353	22	—	Cinnamoyl pyrrolidine amides	3.6%	5% methanol	120	(R. N. Lima et al., 2020)
	313	20			3.3%	5% ethanol		
	333	18			2.6	5% isopropanol		
	353	20			1.9%	—		
Leaves of *Morus alba* L. (*Moraceae*)	333	20	2.0 × 10^{-3} kg/min	β-sitosterol	155.74 mg/100 g leaves	—	120	(Santos et al., 2020)
Tetraclinis articulata	313	9	20 g/min	Volatile oil	1.63 mg/kg	—	30	(Herzi et al., 2013)
Candeia (*Eremanthus erythropappus*)	343	24	1.96 × 10^{-3} kg/min	α-bisadolol	9.84g/kg wood	—	60	(Santos et al., 2016)
Yarrow (*Achillea millefolium* L.)	313	14	70 g/min	Extracts	0.79 g extract/g plant material	—	180	(García-Risco et al., 2017)
Marigold (*Calendula officinalis*)					2.78 g extract/g plant material	—		
Marjoram (*Origanum majorana*)					1.51 g extract/g plant material	—		
Balm (*Melissa officianlis*)					0.65 g extract/g plant material	—		
Marigold (*Calendula officinalis*)	313	34	5.0 × 10^{-4} kg/s	Extract	6.28%	—	270	(López-Padilla et al., 2017)
		14	7.5 × 10^{-4} kg/s		6.11%	—		
Marigold (*Calendula officinalis*)	313	14	70 g/min	Extract	2.1%	—	180	(Martin et al., 2016)
					7.5%	10% ethanol		
Greek sage leaves (*Salvia fruticosa*)	333	28	3 kg/h	Extract	10.3%	—	150	(Kavoura et al., 2019)

TABLE 5.5
Some Carotenoids Obtained under SFE 2010–2020

Raw Material	Compound	Conditions					Carotenoid		Ref.
		T (°C)	P (bar)	CO_2 Flow	Time (min)	Co-Solvent	Recovery	Extract	
Tomato	Lycopene	80	380	15 kg/h	—	—	—	1,719 µg/g	(Squillace et al., 2020)
		90	400	4 mL/min	180	—	44%	—	(MacHmudah et al., 2012)
		50	300	3 mL/min	60	Hazelnut oil	21.6%	—	(Watanabe et al., 2018)
		100	400	1.5 mL/min	90	—	—	31.25 µg/g	(Yi et al., 2009)
		52	550		180		62%	—	(Urbonaviciene & Viskelis, 2017)
	Carotenoids	59	350	15 g/min	30	15.5% ethanol	91%	—	(de Andrade Lima et al., 2019)
Gac	Lycopene	50	300	3 mL/min	480	—	—	76.6 mg/g	(Honda et al., 2018)
		55	250	—	120	—	97.5%	11.01 mg/g	(Akkarachaneeyakorn et al., 2017)
		86	503	4 mL/min	60	—	63%	—	(Bruno et al., 2018)
		50	200	2 L/min	120	—	95%	508 mg/100 mL oil	(Kha et al., 2014)
Mango peel	Carotenoids	60	250	—	—	15% ethanol	6.12 g/g	1.9 mg/g RM	(Sánchez-Camargo et al., 2019)
		40	300	1.1 L/min	450	—	3.15g/100g	5.60 mg/g	(Garcia-Mendoza et al., 2015)
Avocado	Lycopene	50	400	1.5 L/min	210	—	32.3%	131µg/g	(Barros et al., 2017)
	Capsanthin	50	400	2.475 g/min	—	—	57%	—	(Barros et al., 2016)
Red pepper	Capsanthin	50	400	2.475 g/min	—	—	15%	1,1000 µg/g	(Barros et al., 2016)
		80	150	—	17	20% methanol	49%	—	(Zoccali et al., 2017)
	Carotenoids	59	350	15 g/min	30	15.5% ethanol	98.1	—	(de Andrade Lima et al., 2019)
Tangerine peel	β-carotene	40	150	2 mL/min	90	7% acetone	—	26.65–32.18 µg/g	(Rubashvili et al., 2018)
	Lycopene						—	11.12–17.91 µg/g	
Orange peel	β-carotene	40	150	2 mL/min	90	7% acetone	—	41.66–59.16 µg/g	
	Lycopene	40	150	2 mL/min	90	7% acetone	—	9.65–10.65 µg/g	
Murici pulp (*Byrsonima crassifolia*)	Lutein	50	350	8.85×10^{-5} kg/s	300	Ethanol	—	209.73 µg/g	(Pires et al., 2019)
Sweet corn	Lutein	60	350	15 g/min	60	15% ethanol	—	3.81 mg/kg	(Lau et al., 2019)
	Zeaxanthin						—	8.47 mg/kg	
	β-carotene						—	177.29 mg/kg	
Scenedesmus almeriensis	Lutein	50	550	14.48 g/min	120	—	—	2.81 mg/g	(Mehariya et al., 2019)
Haematococcus pluvialis	Lutein	65	550	3.62 g/min	40	12.5% ethanol	92.9%	7.15 mg/g	(Molino et al., 2018)
	Astaxanthin				20		92.4%	18.5 mg/g	
	Lutein	50	550	3.62 g/min	120	—	52.32%	4.03 mg7g	(Sanzo et al., 2018)
	Astaxanthin					—	98.62%	19.72 mg/g	

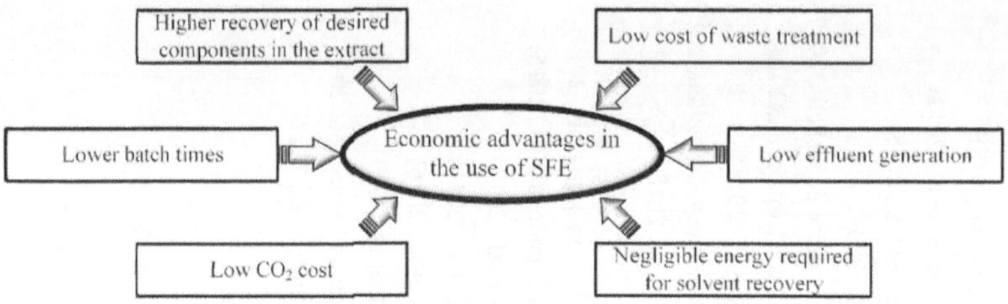

FIGURE 5.11 Economic benefits with the use of SFE.

Source: Manjare and Dhingra (2019).

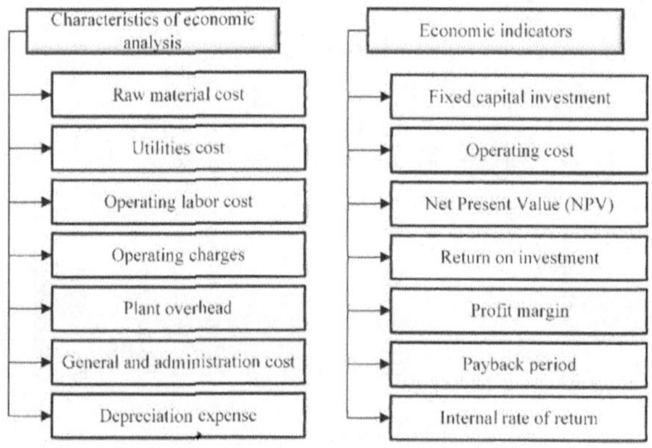

FIGURE 5.12 Features and indicators of economic analysis.

are presented in Figure 5.11. These economic benefits can be seen in production costs. With the decrease in production costs, investment costs can be mitigated.

In estimating the cost of production, Daza Serna et al. (2016) mention seven features that should be considered (see Figure 5.12). Raw material costs are one of the highest costs, representing up to 70% of production costs. Consequently, the use of low-cost raw materials and inputs contributes to reducing production costs. However, these costs depend on the requirements of the technology to be used. Therefore, technologies such as SFE presenting a low-cost input, such as CO_2, are projected as an economically promising technology (Manjare & Dhingra, 2019). The purpose of evaluating the impact of each of the characteristics of production costs (Aristizábal-Marulanda & Cardona Alzate, 2019) mentions seven indicators (see Figure 5.12). The estimation of these indicators allows identifying the economic viability of a process. Additionally, when evaluating SFE, its indicators allow checking the economic benefits presented in Figure 5.11.

Different authors have estimated production costs for their proposed SFE processes. Some of these studies for five different raw materials are shown in Table 5.6. Náthia-Neves et al. (2020) studied the extraction of fatty acids and genipin from *Genipa americana* L. The main fatty acid in the extract is linoleic acid, with a composition of 76%. In the market, linoleic acid presents an average price of 22 USD/kg (*Alibaba.Com*, 2021). Genipin is an iridoid glucoside; in the market, it can present a cost of over 100 USD/kg (*Alibaba.Com*, 2021). Obtaining the extract of *Genipa americana* L. presents a cost of 86.33 USD/kg, where 65% corresponds to raw material. de Aguiar et al. (2018) developed a cost estimation study for obtaining capsaicinoid-rich oleoresin using SFE. The economic analysis obtained a production cost of 125.41 USD/kg, a profit margin of 27.9%, and an internal rate of return of 23.52%. Around 63% of the production cost is related to raw material costs.

These values were obtained considering a reference price of 104 USD/kg of oleoresin, as reported by Ademe et al. (2014). Johner et al. (2018) studied the SFE of pequi oil and conducted an economic study of the process, as main results concluded that the raw material cost represents 85% of the total production costs. With a total production cost of 37.03 USD/kg, it obtained a profit margin of 43% and an internal rate of return of 149%. For these indicators, the authors considered an oil selling price of 75 USD/kg. Carvalho et al. (2015) highlight that oil extraction from turmeric presents a cost for the plant matrix of 91% of the total production cost. Osorio-Tobón et al. (2016) considered the data obtained by Carvalho et al. (2015) to estimate the production costs of turmeric oil, as results (Osorio-Tobón et al., 2016) obtained a production cost of 112.7 USD/kg, a profit margin of 27.37%, and an internal rate of return of 14.92%. These values consider a product cost of 7.27 USD/kg. Additionally, they analyze the influence of scale on the cost of production. The cost of raw materials has the greatest influence. When considering a volume of 5 L, the raw material cost corresponds to 28%. If this volume is increased up to 500 L, the raw material cost becomes 78%. Therefore, this shows the influence of scale on production costs and, specifically, the contribution to the total cost of raw material costs. Rodrigues et al. (2019) analyzed technically and economically the extraction of artemisinin. The technical results are presented in **Table 5.6**. As the main economic results, the negative profitability of the process under the analyzed schemes stands out. The profit margin obtained was −442%, and an internal rate of return of −38.02%. When the influence of the operating volume was analyzed, it was found that the raw material costs increased from 19.3% to 63.84% when the volume was increased from 5 L to 500 L.

When comparing the results obtained in the studies presented in Table 5.6, the following factors influence production costs: (1) the extraction volume and (2) the extract cost in the market. The combination of these two factors can lead to both positive and negative profitability. For this reason, it is

TABLE 5.6
Production Cost Obtained for the Extraction of Some Compounds

Raw Material	Compound	Conditions	Concentration	Production Cost	Reference
Genipa americana L.	Genipin and fatty acid	30 MPa 333 K 5.6 g/min CO_2 400 min	2.2 mg/g 16.6 mg/g	86.83 USD/kg	(Náthia-Neves et al., 2020)
Malagueta peppers (*Capsicum frutescens*)	Capsaicinoid-rich oleoresin	15 MPa 40°C 240 min	4.18 mg/g	125.41 USD/kg	(de Aguiar et al., 2018)
Pequi (*Caryocar brasiliense*)	Oil	40 MPa 60°C 2.17×10^{-4} kg/s of CO_2	49 mg/g	37.03 USD/kg	(Johner et al., 2018)
Turmeric (*Curcuma longa* L.)	Oil	25 MPa 333 K 8.4×10^{-3} kg/min of CO_2	6.4%	178.8 USD/kg	(Carvalho et al., 2015)
Artemisia annua L.	*Artemisinina*	250 bar 60°C 5.0 g of CO_2/kg raw material 6 hours	11.54 mg/g	93.49 USD/kg	(Rodrigues et al., 2019)

necessary to consider which compound is to be extracted and from which biomass. If the extraction yield is low, the sales costs will be lower than the production costs.

5.6 BIOREFINERIES COUPLED WITH SFE

Moncada & Aristizábal (2016) discussed the concept of biorefinery. The following definition is proposed:

> A biorefinery is a complex system, where biomass is integrally processed or fractionated to obtain more than one product including bioenergy, biofuels, chemicals and high value-added compounds that only can be extracted from bio-based sources. The latter after a comprehensive study of the raw materials to be used and a sustainable design based on the latest state of the art technologies and approaches, including aspects of the three pillars of sustainability.

The IEA Bioenergy Task 42 classified biorefineries according to four main characteristics: platforms, products, feedstocks, and conversion processes (Sonnenberg et al., 2009). *Platforms* are intermediate or final products of feedstock transformation, which could be used in other transformation processes to obtain final products. The platforms allow connecting the biorefinery systems and their processes (Gnansounou & Pandey, 2017). The platforms are the basis for the formulation of different biorefinery schemes to make obtaining a certain product viable. The most prominent platforms include C5 and C6 sugars, organic solution, syngas, lignin, oil, hydrogen, pyrolytic liquid, and biogas. In terms of the products to be obtained, biorefineries can be classified into two groups: energy-driven biorefineries and product-driven biorefineries (Ng et al., 2017). Figure 5.13 presents some products that can be obtained under each biorefinery group. However, products obtained under a biorefinery scheme are classified into two groups of biorefineries. Bioethanol and hydrogen can be considered chemical products and biofuels (Aristizábal-Marulanda & Cardona Alzate, 2019).

The variety of processes and raw materials that can be used in biorefineries makes them highly versatile. As a result of this characteristic, biorefineries can be designed to meet different needs or objectives. Generally, biorefineries are designed to meet three objectives: (1) *Energy security*. Energy production from biomass is carried out in developed or underdeveloped countries using thermochemical processes (Martinez-hernandez et al., 2013). Additionally, biomass can be transformed into biofuels (Moncada & Aristizábal, 2016). (2) *Reduction in the environmental impact*

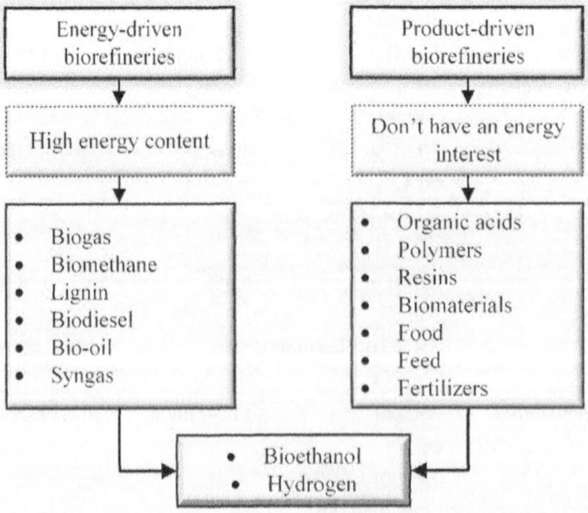

FIGURE 5.13 Biorefineries classification based on the products to be obtained.

of chemical products. Biochemical production generates less waste than conventional processes. Additionally, the amount of greenhouse gases is lower (Cherubini & Ulgiati, 2009). (3) *Rural development.* Biorefineries are designed to improve all the people involved in the supply chain (Cherubini, 2010).

The use of biomasses in SFE processes provides metabolites with high added value. If the processes are coupled at biorefineries, it is possible to valorize the post-extraction residue. The residue is solid, rich in the polysaccharide constituents of the extracted material. When disposed of, these polysaccharides would have an environmental impact. However, when it is transformed under biorefinery schemes, the second objective of biorefineries is achieved. Additionally, these polysaccharides can be used to obtain different biofuels (Aristizabal M et al., 2016; Botero et al., 2017; Moncada et al., 2014). Under this consideration, it is also possible to fulfill the first objective by implementing biorefineries.

Gao et al. (2021) propose the use of industrial ginger waste under a biorefinery scheme (see Figure 5.14). The first step is the SFE to obtain ginger oil. The extraction residue (ginger residue in Figure 5.14) can be transformed under two routes: hydrolysis or pyrolysis. Hydrolysis of the ginger residue yields starch, sugar and acids, and microfibrillated cellulose. In comparison, pyrolysis allows obtaining ginger char and bio-oil.

Scaglia et al. (2020) propose a biorefinery for the utilization of pomace tomato. The first stage consists of an SFE to obtain an extract rich in lycopene (619.3 mg/kg). In a second stage, they used an anerobic digestion for the production of biogas and digestate. With the scheme presented in Figure 5.15, a profit of €787.9 per ton of tomato pomace can be obtained.

Espada et al. (2020) propose a biorefinery for the utilization of *Dunaliella salina* biomass (Figure 5.16). In the biorefinery, the main purpose is the extraction of β-carotene using supercritical fluids. First, the microalgal biomass is cultured and recovered using centrifugation. The extraction is then carried out for 15 min with CO_2 at 60°C and 300 bar. The depressurization process is based on the use of an instant distillation unit. The process allows the elimination of CO_2 and the recovery of an extraction (liquid fraction) rich in β-carotene (90% by weight). An anaerobic digestion process carries out the recovery of the post-extraction solid residue. The biogas obtained is fed to

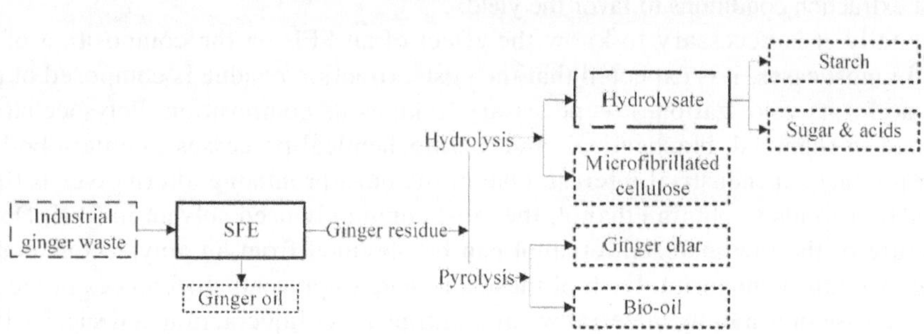

FIGURE 5.14 Process diagram for the SFE coupled to a biorefinery of industrial ginger waste.

Source: Adapted from Gao et al. (2021).

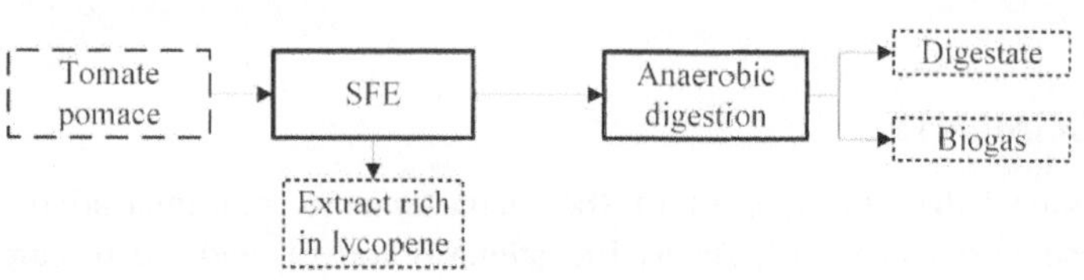

FIGURE 5.15 Biorefinery scheme for SFE and subsequent anaerobic digestion of tomato pomace.

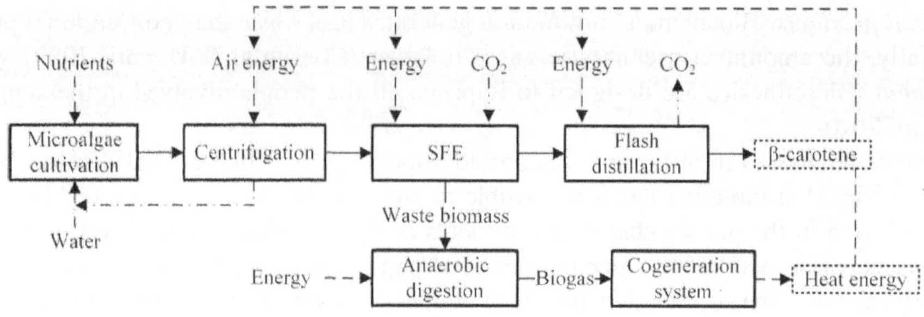

FIGURE 5.16 Biorefinery for the utilization of *Dunaliella salina* biomass.

Source: Adapted from Espada et al. (2020).

a cogeneration process to produce electricity and heat. The energy obtained is used to supply the energy requirements of the process.

5.7 FUTURE PERSPECTIVES OF SFE

As highlighted throughout this work, fruit residues are an interesting source of bioactive compounds. The compounds obtained have different applications in industries such as food and pharmaceuticals. SFE presents itself as a promising technology for these extractions because of its low-residue emissions. However, it is still necessary to expand the information available on the valorization of the post-extraction residue. The information available on this subject is limited and focuses on the use of few transformation processes. If SFE is combined with the biorefinery concept, it is possible to design processes that allow an integral use of the raw material. In order to achieve this, it is necessary to have a complete characterization of the feedstock. This is a limiting factor for many studies. Mainly regarding the identification of the different bioactive compounds present in the materials. The correct identification and quantification of the bioactive compounds would allow the selection of the best extraction conditions to favor the yield.

Additionally, it is necessary to know the effect of an SFE on the composition of the raw material. In most cases, it is expected that the post-extraction residue is composed of polysaccharides, so for its valorization, it is necessary to know its composition. Polysaccharides can be subjected to chemical, biochemical, and thermochemical processes to obtain both energy and other products of industrial interest. One of the most promising alternatives is the use of post-extraction solids to obtain ethanol, the most commonly used solvent in SFE. Depending on the nature of the raw material, ethanol can be obtained from an amylaceous material or from a lignocellulosic material. Each of these alternatives presents differences in the process. While in the case of a starchy material we are talking about liquefaction and saccharification, in the case of lignocellulosic materials, a pretreatment must be used for subsequent enzymatic hydrolysis.

The preceding text shows the great number of study possibilities that are presented around the use of SFE. It is still necessary to carry out studies to improve the potential of SFE for each material of interest.

ACKNOWLEDGMENTS

The authors acknowledge the support of the Departamento Administrativo de Ciencia, Tecnología e Innovación (Colciencias), Becas Bicentenario corte I, and the research program titled

"Reconstrucción del tejido social en zonas posconflicto en Colombia," SIGP code: 57579, with the project titled "Competencias empresariales y de innovación para el desarrollo económico y la inclusión productiva de las regiones afectadas por el conflicto colombiano," SIGP code 58907. Contract number: FP44842-213-2018.

REFERENCES

Ademe, Y., Axtell, B., Fellows, P., Gedi, L., Harcourt, D., La Grenade, C., Lubowa, L., & Hounhouigan, J. (2014). *A handbook for setting up and running a small-scale business producing high-value foods* (P. Fellows & B. Axtell, Eds.). CTA. www.fab.be

Akkarachaneeyakorn, S., Boonrattanakom, A., Pukpin, P., Rattanawaraha, S., & Mattaweewong, N. (2017). Extraction of aril oil from gac (*Momordica cochinchinensis* spreng) using supercritical carbon dioxide. *Journal of Food Processing and Preservation*, *41*(5), e13122. https://doi.org/10.1111/jfpp.13122

Alibaba.com. (2021). https://spanish.alibaba.com/

Aristizabal, M. V., Garcıa, V. C. A., & Cardona, A. C. A. (2016). Integrated production of different types of bioenergy from oil palm through biorefinery concept. *Waste and Biomass Valorization*, *7*(4), 737–745. https://doi.org/10.1007/s12649-016-9564-7

Aristizábal-Marulanda, V., & Cardona Alzate, C. A. (2019). Methods for designing and assessing biorefineries: Review. *Biofuels, Bioproducts and Biorefining*, *13*(3), 789–808. https://doi.org/10.1002/bbb.1961

Azmir, J., Zaidul, I. S. M. S. M., Rahman, M. M. M., Sharif, K. M. M., Mohamed, A., Sahena, F., Jahurul, M. H. A. H. A., Ghafoor, K., Norulaini, N. A. N. A. N., & Omar, A. K. M. K. M. (2013). Techniques for extraction of bioactive compounds from plant materials: A review. *Journal of Food Engineering*, *117*(4), 426–436. https://doi.org/10.1016/j.jfoodeng.2013.01.014

Barros, H. D. F. Q., Coutinho, J. P., Grimaldi, R., Godoy, H. T., & Cabral, F. A. (2016). Simultaneous extraction of edible oil from avocado and capsanthin from red bell pepper using supercritical carbon dioxide as solvent. *Journal of Supercritical Fluids*, *107*, 315–320. https://doi.org/10.1016/j.supflu.2015.09.025

Barros, H. D. F. Q., Grimaldi, R., & Cabral, F. A. (2017). Lycopene-rich avocado oil obtained by simultaneous supercritical extraction from avocado pulp and tomato pomace. *Journal of Supercritical Fluids*, *120*, 1–6. https://doi.org/10.1016/j.supflu.2016.09.021

Belwal, T., Ezzat, S. M., Rastrelli, L., Bhatt, I. D., Daglia, M., Baldi, A., Devkota, H. P., Orhan, I. E., Patra, J. K., Das, G., Anandharamakrishnan, C., Gomez-Gomez, L., Nabavi, S. F., Nabavi, S. M., & Atanasov, A. G. (2018). A critical analysis of extraction techniques used for botanicals: Trends, priorities, industrial uses and optimization strategies. *TrAC—Trends in Analytical Chemistry*, *100*, 82–102. Elsevier B.V. https://doi.org/10.1016/j.trac.2017.12.018

Bernhoft, A. (2010). A brief review on bioactive compounds in plants. In A. Bernhoft (Ed.), *Bioactive compounds in plants—benefits and risks for man and animals* (pp. 11–17). Det Norske Videnskaps-Akademi.

Bitencourt, R. G., Mello, F. M. P. A., Cabral, F. A., & Meirelles, A. J. A. (2020). High-pressure fractionation of spent coffee grounds oil using green solvents. *The Journal of Supercritical Fluids*, *157*, 104689. https://doi.org/10.1016/j.supflu.2019.104689

Botero, C. D., Restrepo, D. L., & Cardona, C. A. (2017). A comprehensive review on the implementation of the biorefinery concept in biodiesel production plants. *Biofuel Research Journal*, *4*(3), 691–703. https://doi.org/10.18331/BRJ2017.4.3.6

Brunner, G. (2005). Supercritical fluids: Technology and application to food processing. *Journal of Food Engineering*, *67*, 21–33. https://doi.org/10.1016/j.jfoodeng.2004.05.060

Brunner, G. (2010). Applications of supercritical fluids. *Annual Review of Chemical and Biomolecular Engineering*, *1*(1), 321–342. https://doi.org/10.1146/annurev-chembioeng-073009-101311

Bruno, A., Durante, M., Marrese, P. P., Migoni, D., Laus, M. N., Pace, E., Pastore, D., Mita, G., Piro, G., & Lenucci, M. S. (2018). Shades of red: Comparative study on supercritical CO2 extraction of lycopene-rich oleoresins from Gac, tomato and watermelon fruits and effect of the α-cyclodextrin clathrated extracts on cultured lung adenocarcinoma cells' viability. *Journal of Food Composition and Analysis*, *65*, 23–32. https://doi.org/10.1016/j.jfca.2017.08.007

Bursać Kovačević, D., Maras, M., Barba, F. J., Granato, D., Roohinejad, S., Mallikarjunan, K., Montesano, D., Lorenzo, J. M., & Putnik, P. (2018). Innovative technologies for the recovery of phytochemicals from *Stevia rebaudiana* Bertoni leaves: A review. *Food Chemistry*, *268*, 513–521. Elsevier Ltd. https://doi.org/10.1016/j.foodchem.2018.06.091

Caballero Galván, A. S. (2017). *Evaluation of the influence of ultrasound and supercritical fluids in processes of polyphenolic compounds extraction from agroindustrial wastes*. Universidad Nacional de Colombia.

Caballero-Galván, A. S., Restrepo-Serna, D. L., Ortiz-Sánchez, M., & Carlos-Alzate, C. A. (2018). Analysis of extraction kinetics of bioactive compounds from spent coffee grounds (*Coffea arábica*). *Waste and Biomass Valorization*, *9*(12), 2381–2389. https://doi.org/10.1007/s12649-018-0332-8

Cantero, D. A., Bermejo, M. D., & Cocero, M. J. (2015). Governing chemistry of cellulose hydrolysis in supercritical water. *ChemSusChem*, *8*(6), 1026–1033. https://doi.org/10.1002/cssc.201403385

Capelo-Martínez, J.-L. (2009). *Ultrasound in chemistry: Analytical applications*. Wiley. https://books.google.com.co/books?hl=es&lr=&id=G-N5goiAFMcC&oi=fnd&pg=PR11&ots=5ZbGVFWufZ&sig=91nl6MGk-lZN_NnTUed1FrROgSc&redir_esc=y#v=onepage&q&f=false

Carvalho, P. I. N., Osorio-Tobón, J. F., Rostagno, M. A., Petenate, A. J., & Meireles, M. A. A. (2015). Techno-economic evaluation of the extraction of turmeric (*Curcuma longa* L.) oil and ar-turmerone using supercritical carbon dioxide. *Journal of Supercritical Fluids*, *105*, 44–54. https://doi.org/10.1016/j.supflu.2015.03.020

Castor, T. (2005). *Compositions, methods and apparatus for supercritical fluid virus inactivation* (Patent No. US20060269928A1).

Chai, Y. H., Yusup, S., Kadir, W. N. A., Wong, C. Y., Rosli, S. S., Ruslan, M. S. H., Chin, B. L. F., & Yiin, C. L. (2021). Valorization of tropical biomass waste by supercritical fluid extraction technology. *Sustainability (Switzerland)*, *13*(1), 1–24. https://doi.org/10.3390/su13010233

Chang, L., Chen, Y. J., Chen, Y. P., Chen, C. T., & Yu, W. H. (2011). Biocompatibility of human bone allograft powder processed by supercritical CO2. *Formosan Journal of Musculoskeletal Disorders*, *2*(2), 55–61. https://doi.org/10.1016/j.fjmd.2011.03.004

Chemat, F., Abert-Vian, M., Fabiano-Tixier, A. S., Strube, J., Uhlenbrock, L., Gunjevic, V., & Cravotto, G. (2019). Green extraction of natural products. Origins, current status, and future challenges. *TrAC Trends in Analytical Chemistry*, *118*, 248–263. https://doi.org/10.1016/J.TRAC.2019.05.037

Chemat, F., Vian, M. A., & Cravotto, G. (2012). Green extraction of natural products: Concept and principles. *International Journal of Molecular Sciences*, *13*(7), 8615–8627. https://doi.org/10.3390/ijms13078615

Cherubini, F. (2010). The biorefinery concept: Using biomass instead of oil for producing energy and chemicals. *Energy Conversion and Management*, *51*(7), 1412–1421. https://doi.org/10.1016/j.enconman.2010.01.015

Cherubini, F., & Ulgiati, S. (2009). Crop residues as raw materials for biorefinery systems—a LCA case study. *Applied Energy*, *87*, 47–57. https://doi.org/10.1016/j.apenergy.2009.08.024

Chhouk, K., Kanda, H., & Goto, M. (2017). Comparison of conventional and ultrasound assisted supercritical carbon dioxide extraction of curcumin from turmeric (*Curcuma longa* L.). *Engineering Journal*, *21*(5), 53–65. https://doi.org/10.4186/ej.2017.21.5.53

Christen, P., & Kaufmann, B. (2002). Recent extraction techniques for natural products: Microwave-assisted extraction and pressurised solvent extraction. *Phytochemical Analysis*, *113*(13), 105–113. University of Geneva, School of Pharmacy, Laboratory of Pharmaceutical Analytical Chemistry, 20 bd d'Yvoy, CH-1211 Geneva 4, Switzerland.

Clifford, A. A. (1994). Reactions in supercritical fluids. In *Supercritical fluids* (pp. 449–479). Springer Netherlands. https://doi.org/10.1007/978-94-015-8295-7_18

Croteau, R., Kutchan, T. M., & Lewis, N. G. (2000). Natural products (secondary metabolites). In *Biochemistry & molecular biology of plants* (pp. 1250–1318). American Society of Plant Physiologists.

Davies, O. R., Lewis, A. L., Whitaker, M. J., Tai, H., Shakesheff, K. M., & Howdle, S. M. (2008). Applications of supercritical CO2 in the fabrication of polymer systems for drug delivery and tissue engineering. *Advanced Drug Delivery Reviews*, *60*(3), 373–387. Elsevier. https://doi.org/10.1016/j.addr.2006.12.001

Dávila, J. A., Rosenberg, M., Castro, E., & Cardona, C. A. (2017). A model biorefinery for avocado (*Persea americana* mill.) processing. *Bioresource Technology*, *243*, 17–29. https://doi.org/10.1016/j.biortech.2017.06.063

Daza Serna, L. V., Orrego Alzate, C. E., & Cardona Alzate, C. A. (2016). Supercritical fluids as a green technology for the pretreatment of lignocellulosic biomass. *Bioresource Technology*, *199*, 113–120. https://doi.org/10.1016/j.biortech.2015.09.078

de Aguiar, A. C., Osorio-Tobón, J. F., Silva, L. P. S., Barbero, G. F., & Martínez, J. (2018). Economic analysis of oleoresin production from malagueta peppers (*Capsicum frutescens*) by supercritical fluid extraction. *Journal of Supercritical Fluids*, *133*, 86–93. https://doi.org/10.1016/j.supflu.2017.09.031

de Andrade Lima, M., Kestekoglou, I., Charalampopoulos, D., & Chatzifragkou, A. (2019). Supercritical fluid extraction of carotenoids from vegetable waste matrices. *Molecules*, *24*(3). https://doi.org/10.3390/molecules24030466

De Melo, M. M. R., Silvestre, A. J. D., & Silva, C. M. (2014). Supercritical fluid extraction of vegetable matrices: Applications, trends and future perspectives of a convincing green technology. *Journal of Supercritical Fluids*, *92*, 115–176. Elsevier. https://doi.org/10.1016/j.supflu.2014.04.007

Deng, Q., Zinoviadou, K. G., Galanakis, C. M., Orlien, V., Grimi, N., Vorobiev, E., Lebovka, N., & Barba, F. J. (2015). The effects of conventional and non-conventional processing on glucosinolates and its derived forms, isothiocyanates: Extraction, degradation, and applications. *Food Engineering Reviews*, *7*(3), 357–381. https://doi.org/10.1007/s12393-014-9104-9

Díaz-Reinoso, B., Moure, A., Domínguez, H., & Parajó, J. C. (2006). Supercritical CO2 extraction and purification of compounds with antioxidant activity. *Journal of Agricultural and Food Chemistry*, *54*(7), 2441–2469. https://doi.org/10.1021/jf052858j

Elgndi, M. A., Filip, S., Pavlić, B., Vladić, J., Stanojković, T., Žižak, Ž., & Zeković, Z. (2017). Antioxidative and cytotoxic activity of essential oils and extracts of *Satureja montana* L., *Coriandrum sativum* L. and *Ocimum basilicum* L. obtained by supercritical fluid extraction. *Journal of Supercritical Fluids*, *128*, 128–137. https://doi.org/10.1016/j.supflu.2017.05.025

Espada, J. J., Pérez-Antolín, D., Vicente, G., Bautista, L. F., Morales, V., & Rodríguez, R. (2020). Environmental and techno-economic evaluation of β-carotene production from *Dunaliella salina*. A biorefinery approach. *Biofuels, Bioproducts and Biorefining*, *14*(1), 43–54. https://doi.org/10.1002/BBB.2012

Fages, J., Jean, E., Frayssinet, P., Mathon, D., Poirier, B., Autefage, A., & Larzul, D. (1998). Bone allografts and supercritical processing: Effects on osteointegration and viral safety. *Journal of Supercritical Fluids*, *13*(1–3), 351–356. https://doi.org/10.1016/S0896-8446(98)00071-0

Favareto, R., Teixeira, M. B., Soares, F. A. L., Belisário, C. M., Corazza, M. L., & Cardozo-Filho, L. (2017). Study of the supercritical extraction of Pterodon fruits (Fabaceae). *Journal of Supercritical Fluids*, *128*, 159–165. https://doi.org/10.1016/j.supflu.2017.05.028

Gai, Q. Y., Jiao, J., Mu, P. S., Wang, W., Luo, M., Li, C. Y., Zu, Y. G., Wei, F. Y., & Fu, Y. J. (2013). Microwave-assisted aqueous enzymatic extraction of oil from *Isatis indigotica* seeds and its evaluation of physicochemical properties, fatty acid compositions and antioxidant activities. *Industrial Crops and Products*, *45*, 303–311. https://doi.org/10.1016/j.indcrop.2012.12.050

Gao, Y., Ozel, M. Z., Dugmore, T., Sulaeman, A., & Matharu, A. S. (2021). A biorefinery strategy for spent industrial ginger waste. *Journal of Hazardous Materials*, *401*, 123400. https://doi.org/10.1016/J.JHAZMAT.2020.123400

Garcia-Mendoza, M. P., Paula, J. T., Paviani, L. C., Cabral, F. A., & Martinez-Correa, H. A. (2015). Extracts from mango peel by-product obtained by supercritical CO2 and pressurized solvent processes. *LWT—Food Science and Technology*, *62*(1), 131–137. https://doi.org/10.1016/j.lwt.2015.01.026

García-Risco, M. R., Mouhid, L., Salas-Pérez, L., López-Padilla, A., Santoyo, S., Jaime, L., Ramírez de Molina, A., Reglero, G., & Fornari, T. (2017). Biological activities of asteraceae (*Achillea millefolium* and *Calendula officinalis*) and lamiaceae (*Melissa officinalis* and *Origanum majorana*) plant extracts. *Plant Foods for Human Nutrition*, *72*(1), 96–102. https://doi.org/10.1007/s11130-016-0596-8

Giacometti, J., Bursać Kovačević, D., Putnik, P., Gabrić, D., Bilušić, T., Krešić, G., Stulić, V., Barba, F. J., Chemat, F., Barbosa-Cánovas, G., & Režek Jambrak, A. (2018). Extraction of bioactive compounds and essential oils from mediterranean herbs by conventional and green innovative techniques: A review. *Food Research International*, *113*, 245–262. https://doi.org/10.1016/j.foodres.2018.06.036

Girotra, P., Singh, S. K., & Nagpal, K. (2013). Supercritical fluid technology: A promising approach in pharmaceutical research. *Pharmaceutical Development and Technology*, *18*(1), 22–38. https://doi.org/10.3109/10837450.2012.726998

Gnansounou, E., & Pandey, A. (2017). Classification of biorefineries taking into account sustainability potentials and flexibility. In *Life-cycle assessment of biorefineries* (pp. 1–39). Elsevier Inc. https://doi.org/10.1016/B978-0-444-63585-3.00001-2

Goyeneche, R., Di Scala, K., Ramirez, C. L., & Fanovich, M. A. (2020). Recovery of bioactive compounds from beetroot leaves by supercritical CO2 extraction as a promising bioresource. *Journal of Supercritical Fluids*, *155*, 104658. https://doi.org/10.1016/j.supflu.2019.104658

Haghighi, A., & Khajenoori, M. (2013). Subcritical water extraction. In H. Nakajima (Ed.), *Mass transfer—advances in sustainable energy and environment oriented numerical modeling*. InTechOpen. https://doi.org/10.5772/54993

Hasmida, M. N., Nur Syukriah, A. R., Liza, M. S., & Mohod Azizi, C. Y. (2014). Effect of different extraction techniques on total phenolic content and antioxidant activity of *Quercus infectoria* galls. *International Food Research Journal*, *21*(3), 1039–1043. https://web.a.ebscohost.com/abstract?direct=true&profile=ehost&scope=site&authtype=crawler&jrnl=19854668&AN=96165028&h=ZDjp1oh4SuBpntS%2BQ0e9%2FrAq4RN1fwGBhUfkLhsen836rD7ThKqrOpql%2B5EoewnxwmI1%2BgzeGG8W9kWiijrpSA%3

D%3D&crl=c&resultNs=AdminWebAuth&resultLocal=ErrCrlNotAuth&crlhashurl=login.aspx%3Fdirect%3Dtrue%26profile%3Dehost%26scope%3Dsite%26authtype%3Dcrawler%26jrnl%3D19854668%26AN%3D96165028

Herrero, M., Cifuentes, A., & Ibañez, E. (2006). Sub- and supercritical fluid extraction of functional ingredients from different natural sources: Plants, food-by-products, algae and microalgae—a review. *Food Chemistry*, *98*(1), 136–148. https://doi.org/10.1016/j.foodchem.2005.05.058

Herzi, N., Camy, S., Bouajila, J., Destrac, P., Romdhane, M., & Condoret, J. S. (2013). Supercritical CO2 extraction of *Tetraclinis articulata*: Chemical composition, antioxidant activity and mathematical modeling. *Journal of Supercritical Fluids*, *82*, 72–82. https://doi.org/10.1016/j.supflu.2013.06.007

Honda, M., Watanabe, Y., Murakami, K., Hoang, N. N., diono, W., Kanda, H., & Goto, M. (2018). Enhanced lycopene extraction from Gac (*Momordica cochinchinensis* Spreng.) by the Z-isomerization induced with microwave irradiation pre-treatment. *European Journal of Lipid Science and Technology*, *120*(2), 1700293. https://doi.org/10.1002/ejlt.201700293

Huang, G., Chen, F., Wei, D., Zhang, X., & Chen, G. (2009). Biodiesel production by microalgal biotechnology. *Applied Energy*, *87*, 38–46. https://doi.org/10.1016/j.apenergy.2009.06.016

Huang, Z., Shi, X. H., & Jiang, W. J. (2012). Theoretical models for supercritical fluid extraction. *Journal of Chromatography A*, *1250*, 2–26. Elsevier. https://doi.org/10.1016/j.chroma.2012.04.032

Ixtaina, V. Y., Vega, A., Nolasco, S. M., Tomás, M. C., Gimeno, M., Bárzana, E., & Tecante, A. (2010). Supercritical carbon dioxide extraction of oil from Mexican chia seed (*Salvia hispanica* L.): Characterization and process optimization. *Journal of Supercritical Fluids*, *55*(1), 192–199. https://doi.org/10.1016/j.supflu.2010.06.003

Johner, J. C. F., Hatami, T., Zabot, G. L., & Meireles, M. A. A. (2018). Kinetic behavior and economic evaluation of supercritical fluid extraction of oil from pequi (*Caryocar brasiliense*) for various grinding times and solvent flow rates. *Journal of Supercritical Fluids*, *140*, 188–195. https://doi.org/10.1016/j.supflu.2018.06.016

Kamran, M., Chemat, F., Sattar, S., & Imran, M. (2020). Extraction of bioactive molecules: Conventional versus novel methods. In H. A. Rasul Suleria & C. Barrow (Eds.), *Bioactive compounds in plant origin. Extraction, applications, and potential health benefits* (pp. 3–44). Taylor & Francis Group. https://doi.org/10.1201/b10300-9

Kavoura, D., Kyriakopoulou, K., Papaefstathiou, G., Spanidi, E., Gardikis, K., Louli, V., Aligiannis, N., Krokida, M., & Magoulas, K. (2019). Supercritical CO2 extraction of *Salvia fruticosa*. *Journal of Supercritical Fluids*, *146*, 159–164. https://doi.org/10.1016/j.supflu.2019.01.010

Kazarian, S. G. (2000). Polymer processing with supercritical fluids. *Polymer Science—Series C*, *42*(1), 78–101.

Kha, T. C., Phan-Tai, H., & Nguyen, M. H. (2014). Effects of pre-treatments on the yield and carotenoid content of Gac oil using supercritical carbon dioxide extraction. *Journal of Food Engineering*, *120*(1), 44–49. https://doi.org/10.1016/j.jfoodeng.2013.07.018

Khaw, K. Y., Parat, M. O., Shaw, P. N., & Falconer, J. R. (2017). Solvent supercritical fluid technologies to extract bioactive compounds from natural sources: A review. *Molecules*, *22*(7). https://doi.org/10.3390/molecules22071186

Khoddami, A., Wilkes, M. A., & Roberts, T. H. (2013). Techniques for analysis of plant phenolic compounds. *Molecules*, *18*(2), 2328–2375. https://doi.org/10.3390/molecules18022328

Klein, E. J., Carvalho, P. I. N., Náthia-Neves, G., Vardanega, R., Meireles, M. A. A., da Silva, E. A., & Vieira, M. G. A. (2021). Techno-economical optimization of uvaia (*Eugenia pyriformis*) extraction using supercritical fluid technology. *The Journal of Supercritical Fluids*, *174*, 105239. https://doi.org/10.1016/J.SUPFLU.2021.105239

Klein, E. J., Santos, K. A., Palú, F., Vieira, M. G. A., & da Silva, E. A. (2018). Use of supercritical CO2 and ultrasound-assisted extractions to obtain A/B-amyrin-rich extracts from uvaia leaves (*Eugenia pyriformis* Cambess.). *Journal of Supercritical Fluids*, *137*, 1–8. https://doi.org/10.1016/j.supflu.2018.02.019

Knez, Ž., Leitgeb, M., & Primožič, M. (2015). Enzymatic reactions in supercritical fluids. In *Food engineering series* (pp. 185–215). Springer. https://doi.org/10.1007/978-3-319-10611-3_6

Knez, Ž., Pantić, M., Cör, D., Novak, Z., & Knez Hrnčič, M. (2019). Are supercritical fluids solvents for the future? *Chemical Engineering and Processing—Process Intensification*, *141*, 107532. Elsevier B.V. https://doi.org/10.1016/j.cep.2019.107532

Lakkakula, N. R., Lima, M., & Walker, T. (2004). Rice bran stabilization and rice bran oil extraction using ohmic heating. *Bioresource Technology*, *92*(2), 157–161. https://doi.org/10.1016/j.biortech.2003.08.010

Lau, T., Harbourne, N., & Oruña-Concha, M. J. (2019). Valorisation of sweet corn (*Zea mays*) cob by extraction of valuable compounds. *International Journal of Food Science and Technology*, *54*(4), 1240–1246. https://doi.org/10.1111/ijfs.14092

Lima, M. de A., Andreou, R., Charalampopoulos, D., & Chatzifragkou, A. (2021). Supercritical carbon dioxide extraction of phenolic compounds from potato (*Solanum tuberosum*) peels. *Applied Sciences*, *11*(8), 3410. https://doi.org/10.3390/APP11083410

Lima, R. N., Santos, A. D. C., Ribeiro, A. S., Cardozo-Filho, L., Freitas, L. S., Barison, A., Costa, E. V., & Alves, P. B. (2020). Selective amides extraction and biological activity from *Piper hispidum* leaves using the supercritical extraction. *Journal of Supercritical Fluids*, *157*, 104712. https://doi.org/10.1016/j.supflu.2019.104712

Liu, X., Ou, H., & Gregersen, H. (2020). Ultrasound-assisted supercritical CO2 extraction of cucurbitacin E from *Iberis amara* seeds. *Industrial Crops and Products*, *145*. https://doi.org/10.1016/j.indcrop.2020.112093

López-Padilla, A., Ruiz-Rodriguez, A., Reglero, G., & Fornari, T. (2017). Supercritical carbon dioxide extraction of *Calendula officinalis*: Kinetic modeling and scaling up study. *Journal of Supercritical Fluids*, *130*, 292–300. https://doi.org/10.1016/j.supflu.2017.03.033

Ludwig, K., Rihko-Struckmann, L., Brinitzer, G., Unkelbach, G., & Sundmacher, K. (2021). β-Carotene extraction from *Dunaliella salina* by supercritical CO2. *Journal of Applied Phycology*, *33*(3), 1435–1445. https://doi.org/10.1007/S10811-021-02399-Y

MacHmudah, S., Zakaria, Winardi, S., Sasaki, M., Goto, M., Kusumoto, N., & Hayakawa, K. (2012). Lycopene extraction from tomato peel by-product containing tomato seed using supercritical carbon dioxide. *Journal of Food Engineering*, *108*(2), 290–296. https://doi.org/10.1016/j.jfoodeng.2011.08.012

Mamma, D., & Christakopoulos, P. (2008). Citrus peels: An excellent raw material for the bioconversion into value-added products. *Tree and Forestry Science and Biotechnology*, *2*(1), 83–97. http://www.globalsciencebooks.info/Online/GSBOnline/images/0812/TFSB_2(SI1)/TFSB_2(SI1)83-97o.pdf

Manjare, S. D., & Dhingra, K. (2019). Supercritical fluids in separation and purification: A review. *Materials Science for Energy Technologies*, *2*(3), 463–484.

Martin, D., Navarro Del Hierro, J., Villanueva Bermejo, D., Fernández-Ruiz, R., Fornari, T., & Reglero, G. (2016). Bioaccessibility and antioxidant activity of *Calendula officinalis* supercritical extract as affected by in vitro codigestion with olive oil. *Journal of Agricultural and Food Chemistry*, *64*(46), 8828–8837. https://doi.org/10.1021/acs.jafc.6b04313

Martinez-hernandez, E., Sadhukhan, J., & Campbell, G. M. (2013). Integration of bioethanol as an in-process material in biorefineries using mass pinch analysis. *Applied Energy*, *104*, 517–526. https://doi.org/10.1016/j.apenergy.2012.11.054

Mazzutti, S., Riehl, C. A. S., Ibañez, E., & Ferreira, S. R. S. (2017). Green-based methods to obtain bioactive extracts from *Plantago major* and *Plantago lanceolata*. *Journal of Supercritical Fluids*, *119*, 211–220. https://doi.org/10.1016/j.supflu.2016.09.018

Mehariya, S., Iovine, A., Di Sanzo, G., Larocca, V., Martino, M., Leone, G., Casella, P., Karatza, D., Marino, T., Musmarra, D., & Molino, A. (2019). Supercritical fluid extraction of lutein from *Scenedesmus almeriensis*. *Molecules*, *24*(7), 1324. https://doi.org/10.3390/molecules24071324

Mitra, P., Ramaswamy, H. S., & Chang, K. S. (2009). Pumpkin (*Cucurbita maxima*) seed oil extraction using supercritical carbon dioxide and physicochemical properties of the oil. *Journal of Food Engineering*. https://doi.org/10.1016/j.jfoodeng.2009.04.033

Molino, A., Mehariya, S., Iovine, A., Larocca, V., Di Sanzo, G., Martino, M., Casella, P., Chianese, S., & Musmarra, D. (2018). Extraction of astaxanthin and lutein from microalga *Haematococcus pluvialis* in the red phase using CO2 supercritical fluid extraction technology with ethanol as co-solvent. *Marine Drugs*, *16*(11), 432. https://doi.org/10.3390/md16110432

Moncada, J., & Aristizábal, V. (2016). Design strategies for sustainable biorefineries. *Biochemical Engineering Journal*, *116*, 122–134. https://doi.org/10.1016/j.bej.2016.06.009

Moncada, J., Tamayo, J., & Cardona, C. A. (2014). Evolution from biofuels to integrated biorefineries: Techno-economic and environmental assessment of oil palm in Colombia. *Journal of Cleaner Production*, *81*, 51–59. https://doi.org/10.1016/j.jclepro.2014.06.021

Mushtaq, M., Sultana, B., Anwar, F., Adnan, A., & Rizvi, S. S. H. (2015). Enzyme-assisted supercritical fluid extraction of phenolic antioxidants from pomegranate peel. *Journal of Supercritical Fluids*. https://doi.org/10.1016/j.supflu.2015.05.020

Nalawade, S. P., Picchioni, F., & Janssen, L. P. B. M. (2006). Supercritical carbon dioxide as a green solvent for processing polymer melts: Processing aspects and applications. *Progress in Polymer Science*, *31*(1), 19–43. https://doi.org/10.1016/j.progpolymsci.2005.08.002

Náthia-Neves, G., Vardanega, R., Hatami, T., & Meireles, M. A. A. (2020). Process integration for recovering high added-value products from *Genipa americana* L.: Process optimization and economic evaluation. *Journal of Supercritical Fluids*, *164*, 104897. https://doi.org/10.1016/j.supflu.2020.104897

Ng, D. K. S., Ng, K. S., & Ng, R. T. L. (2017). Integrated biorefineries. In *Encyclopedia of sustainable technologies* (Vol. 4). Elsevier. https://doi.org/10.1016/B978-0-12-409548-9.10138-1

Okajima, I., & Sako, T. (2014). Energy conversion of biomass with supercritical and subcritical water using large-scale plants. *Journal of Bioscience and Bioengineering*, *117*(1), 1–9. Elsevier. https://doi.org/10.1016/j.jbiosc.2013.06.010

Orio, L., Alexandru, L., Cravotto, G., Mantegna, S., & Barge, A. (2011). UAE, MAE, SFE-CO2 and classical methods for the extraction of *Mitragyna speciosa* leaves. https://doi.org/10.1016/j.ultsonch.2011.10.001

Osorio-Tobón, J. F., Carvalho, P., Rostagno, M., & Meireles, A. (2016). Process integration for turmeric products extraction using supercritical fluids and pressurized liquids: Economic evaluation. *Food and Bioproducts Processing*. https://doi.org/10.1016/j.fbp.2016.02.001

Parniakov, O., Apicella, E., Koubaa, M., Barba, F. J., Grimi, N., Lebovka, N., Pataro, G., Ferrari, G., & Vorobiev, E. (2015). Ultrasound-assisted green solvent extraction of high-added value compounds from microalgae *Nannochloropsis* spp. *Bioresource Technology*, *198*, 262–267. https://doi.org/10.1016/j.biortech.2015.09.020

Peng, D.-Y., & Robinson, D. B. (1976). A new two-constant equation of state. *Industrial & Engineering Chemistry Fundamentals*, *15*(1), 59–64. https://doi.org/10.1021/i160057a011

Pereira, C. G., & Meireles, M. A. A. (2010). Supercritical fluid extraction of bioactive compounds: Fundamentals, applications and economic perspectives. *Food and Bioprocess Technology*, *3*(3), 340–372. https://doi.org/10.1007/s11947-009-0263-2

Pieters, J. (2017). *Evaluation of sub- and supercritical Rankine cycle optimisation criteria Potchefstroom Campus of the North-West University Supervisor* (Issue March). North-West University. https://doi.org/10.13140/RG.2.2.30220.41608

Pires, F. C. S., Silva, A. P. de S. e., Salazar, M. de los A. R., Costa, W. A. da, Costa, H. S. C. da, Lopes, A. S., Rogez, H., & Carvalho Junior, R. N. de. (2019). Determination of process parameters and bioactive properties of the murici pulp (*Byrsonima crassifolia*) extracts obtained by supercritical extraction. *Journal of Supercritical Fluids*, *146*, 128–135. https://doi.org/10.1016/j.supflu.2019.01.014

Pokhrel, N., Vabbina, P. K., & Pala, N. (2016). Sonochemistry: Science and engineering. *Ultrasonics Sonochemistry*, *29*, 104–128. Elsevier B.V. https://doi.org/10.1016/j.ultsonch.2015.07.023

Posada, J. A., Rincón, L. E., & Cardona, C. A. (2012). Design and analysis of biorefineries based on raw glycerol: Addressing the glycerol problem. *Bioresource Technology*, *111*, 282–293. https://doi.org/10.1016/j.biortech.2012.01.151

Pronyk, C., & Mazza, G. (2009). Design and scale-up of pressurized fluid extractors for food and bioproducts. *Journal of Food Engineering*, *95*(2), 215–226. Elsevier. https://doi.org/10.1016/j.jfoodeng.2009.06.002

Putnik, P., Barba, F. J., Lucini, L., Rocchetti, G., & Montesano, D. (2019). Conventional, non-conventional extraction techniques and new strategies for the recovery of bioactive compounds from plant material for human nutrition. *Food Research International*, *123*, 516–517. Elsevier Ltd. https://doi.org/10.1016/j.foodres.2019.05.010

Quitain, A. T., Herng, C. Y., Yusup, S., Sasaki, M., & Uemura, Y. (2015). Conversion of biomass to bio-oil in sub- and supercritical water. In K. Biernat (Ed.), *Biofuels—status and perspective*. InTechOpen. https://doi.org/10.5772/59947

Riera, E., Golás, Y., Blanco, A., Gallego, J. A., Blasco, M., & Mulet, A. (2004). Mass transfer enhancement in supercritical fluids extraction by means of power ultrasound. *Ultrasonics Sonochemistry*, *11*(3–4), 241–244. https://doi.org/10.1016/j.ultsonch.2004.01.019

Rodrigues, M. F. F., Sousa, I. M. O., Vardanega, R., Nogueira, G. C., Meireles, M. A. A., Foglio, M. A., & Marchese, J. A. (2019). Techno-economic evaluation of artemisinin extraction from *Artemisia annua* L. using supercritical carbon dioxide. *Industrial Crops and Products*, *132*, 336–343. https://doi.org/10.1016/j.indcrop.2019.02.049

Routray, W., & Orsat, V. (2012). Microwave-assisted extraction of flavonoids: A review. *Food and Bioprocess Technology*, *5*(2), 409–424. Springer. https://doi.org/10.1007/s11947-011-0573-z

Rubashvili, I., Tsitsagi, M., Ebralidze, K., Tsitsishvili, V., Eprikashvili, L., Chkhaidze, M., & Zautashvili, M. (2018). Extraction and analysis of the major carotenoids of agro-industrial waste materials using sequential extraction techniques and high performance liquid chromatography. *Eurasian Journal of Analytical Chemistry*, *13*(2). https://doi.org/10.29333/ejac/82931

Sánchez-Camargo, A. del P., Gutiérrez, L. F., Vargas, S. M., Martinez-Correa, H. A., Parada-Alfonso, F., & Narváez-Cuenca, C. E. (2019). Valorisation of mango peel: Proximate composition, supercritical fluid extraction of carotenoids, and application as an antioxidant additive for an edible oil. *Journal of Supercritical Fluids*, *152*, 104574. https://doi.org/10.1016/j.supflu.2019.104574

Sánchez-Vicente, Y., Cabañas, A., Renuncio, J. A. R., & Pando, C. (2009). Supercritical fluid extraction of peach (*Prunus persica*) seed oil using carbon dioxide and ethanol. *The Journal of Supercritical Fluids*, *49*(2), 167–173. https://doi.org/10.1016/j.supflu.2009.01.001

Santos, K. A., Klein, E. J., Fiorese, M. L., Palú, F., da Silva, C., & da Silva, E. A. (2020). Extraction of *Morus alba* leaves using supercritical CO2 and ultrasound-assisted solvent: Evaluation of β-sitosterol content. *Journal of Supercritical Fluids*, *159*, 104752. https://doi.org/10.1016/j.supflu.2020.104752

Santos, K. A., Klein, E. J., Gazim, Z. C., Gonçalves, J. E., Cardozo-Filho, L., Corazza, M. L., & Da Silva, E. A. (2016). Wood and industrial residue of candeia (*Eremanthus erythropappus*): Supercritical CO2 oil extraction, composition, antioxidant activity and mathematical modeling. *Journal of Supercritical Fluids*, *114*, 1–8. https://doi.org/10.1016/j.supflu.2016.02.015

Sanzo, G., Mehariya, S., Martino, M., Larocca, V., Casella, P., Chianese, S., Musmarra, D., Balducchi, R., & Molino, A. (2018). Supercritical carbon dioxide extraction of astaxanthin, lutein, and fatty acids from *Haematococcus pluvialis* microalgae. *Marine Drugs*, *16*(9), 334. https://doi.org/10.3390/md16090334

Sapkale, G. N., Patil, S. M., Surwase, U. S., & Bhatbhage, P. K. (2010). Supercritical fluid extraction—a review. *International Journal of Chemical Science*, *8*(2), 729–743.

Scaglia, B., D'Incecco, P., Squillace, P., Dell'Orto, M., De Nisi, P., Pellegrino, L., Botto, A., Cavicchi, C., & Adani, F. (2020). Development of a tomato pomace biorefinery based on a CO2-supercritical extraction process for the production of a high value lycopene product, bioenergy and digestate. *Journal of Cleaner Production*, *243*, 118650. https://doi.org/10.1016/J.JCLEPRO.2019.118650

Shi, J., Mittal, G., Kim, E., Xue, S. J., Shi, J., Mittal, G., Kim, E., & Xue, S. J. U. N. (2007). Solubility of carotenoids in supercritical solubility of carotenoids in supercritical CO2. *Food Reviews International*, *23*, 341–371. https://doi.org/10.1080/87559120701593806

Shilpi, A., Shivhare, U. S., & Basu, S. (2013). Supercritical CO2 extraction of compounds with antioxidant activity from fruits and vegetables waste—a review. *Focusing on Modern Food Industry*, *2*(1), 43–62.

Smith, R. M. (1999). Supercritical fluids in separation science—the dreams, the reality and the future. *Journal of Chromatography A*, *856*(1–2), 83–115. Elsevier. https://doi.org/10.1016/S0021-9673(99)00617-2

Sonnenberg, A., Baars, J., & Hendrickx, P. (2009). IEA bioenergy task 42 biorefinery. *International Energy Agency*. www.cropenergies.com.

Sovova, H., Stateva, R. P., & Galushko, A. A. (2001). Solubility of β-carotene in supercritical CO2 and the effect of entrainers. *21*(3), 195–203.

Squillace, P., Adani, F., & Scaglia, B. (2020). Supercritical CO2 extraction of tomato pomace: Evaluation of the solubility of lycopene in tomato oil as limiting factor of the process performance. *Food Chemistry*, *315*, 126224. https://doi.org/10.1016/j.foodchem.2020.126224

Stephenson, P., Kondor, B., Licence, P., Scovell, K., Ross, S. K., & Poliakoff, M. (2006). Continuous asymmetric hydrogenation in supercritical carbon dioxide using an immobilised homogeneous catalyst. *Advanced Synthesis & Catalysis*, *348*(12–13), 1605–1610. https://doi.org/10.1002/adsc.200606172

Sun, Q., Shi, J., Scanlon, M., Xue, S. J., & Lu, J. (2021). Optimization of supercritical-CO2 process for extraction of tocopherol-rich oil from canola seeds. *LWT*, *145*, 111435. https://doi.org/10.1016/J.LWT.2021.111435

Taofiq, O., Corrêa, R. C. G., Barros, L., Prieto, M. A., Bracht, A., Peralta, R. M., González-Paramás, A. M., Barreiro, M. F., & Ferreira, I. C. F. R. (2019). A comparative study between conventional and non-conventional extraction techniques for the recovery of ergosterol from *Agaricus blazei* Murrill. *Food Research International*, *125*, 108541. https://doi.org/10.1016/j.foodres.2019.108541

Temtem, M., Silva, L. M. C., Andrade, P. Z., dos Santos, F., da Silva, C. L., Cabral, J. M. S., Abecasis, M. M., & Aguiar-Ricardo, A. (2009). Supercritical CO2 generating chitosan devices with controlled morphology. Potential application for drug delivery and mesenchymal stem cell culture. *Journal of Supercritical Fluids*, *48*(3), 269–277. https://doi.org/10.1016/j.supflu.2008.10.020

Toepfl, S., Mathys, A., Heinz, V., & Knorr, D. (2006). Review: Potential of high hydrostatic pressure and pulsed electric fields for energy efficient and environmentally friendly food processing. *Food Reviews International*, *22*(4), 405–423. https://doi.org/10.1080/87559120600865164

Urbonaviciene, D., & Viskelis, P. (2017). The cis-lycopene isomers composition in supercritical CO2 extracted tomato by-products. *LWT—Food Science and Technology*, *85*, 517–523. https://doi.org/10.1016/j.lwt.2017.03.034

Wahyudiono, Kanetake, T., Sasaki, M., & Goto, M. (2007). Decomposition of a lignin model compound under hydrothermal conditions. *Chemical Engineering and Technology*, *30*(8), 1113–1122. https://doi.org/10.1002/ceat.200700066

Watanabe, Y., Honda, M., Higashiura, T., Fukaya, T., Machmudah, S., Wahyudiono, Kanda, H., & Goto, M. (2018). Rapid and selective concentration of lycopene Z-isomers from tomato pulp by supercritical CO2 with co-solvents. *Solvent Extraction Research and Development, Japan*, *25*(1), 47–57. https://doi.org/10.15261/serdj.25.47

Wei, M.-C., & Yang, Y.-C. (2015). Kinetic studies for ultrasound-assisted supercritical carbon dioxide extraction of triterpenic acids from healthy tea ingredient *Hedyotis diffusa* and *Hedyotis corymbosa*. https://doi.org/10.1016/j.seppur.2015.01.008

Yi, C., Shi, J., Xue, S. J., Jiang, Y., & Li, D. (2009). Effects of supercritical fluid extraction parameters on lycopene yield and antioxidant activity. *Food Chemistry*, *113*(4), 1088–1094. https://doi.org/10.1016/j.foodchem.2008.08.083

Zhang, G., Hu, M., He, L., Fu, P., Wang, L., & Zhou, J. (2013). Optimization of microwave-assisted enzymatic extraction of polyphenols from waste peanut shells and evaluation of its antioxidant and antibacterial activities in vitro. *Food and Bioproducts Processing*, *91*(2), 158–168. https://doi.org/10.1016/j.fbp.2012.09.003

Zoccali, M., Giuffrida, D., Dugo, P., & Mondello, L. (2017). Direct online extraction and determination by supercritical fluid extraction with chromatography and mass spectrometry of targeted carotenoids from red Habanero peppers (*Capsicum chinense* Jacq.). *Journal of Separation Science*, *40*(19), 3905–3913. https://doi.org/10.1002/jssc.201700669

Zorca, M., Găinar, I., & Bala, D. (2007). The CO2 density variation in the supercritical extraction of anet essential oils. *Analele Universitatii Bucuresti Chimie*, *I*, 43–47. http://193.226.51.177/anunivch/analele_univ_chimie.html

6 Prefeasibility Analysis of Supercritical Fluid Extraction of Polyphenols from Orange as Part of Food Waste Biorefineries

Mariana Ortiz-Sánchez, Sara Piedrahita-Rodríguez, Daissy Lorena Restrepo Serna, and Carlos Ariel Cardona Alzate

6.1 INTRODUCTION

Globally, one-third of the food produced for human consumption is lost. Food waste can vary from region to region, depending on the present production. Fruits and vegetables, roots, and tubers have the highest waste disposal rates among all food products. As reported by the Food and Agriculture Organization of the United Nations (FAO), these foods present waste of 40–50% [1]. It generates an environmental and economic problem for farmers. The adequate disposal of these wastes implies a high cost that cannot be supplied in many regions. As a result, these wastes do not receive adequate treatment and are dumped in sanitary landfills or neighboring areas.

Food waste is considered a complex raw material with several properties [1]. However, the proper use of the fractions that make up food waste is of great interest, especially in approaches such as biorefinery design [2]. Biorefineries based on food waste are currently being studied to obtain value-added products comprehensively. The extraction stages then become a suitable process to get involved in designing food waste biorefineries [3]. The prefeasibility of applying one of these designs must be evaluated from technical and economic perspectives [4]. Finally, the results determine whether extracting polyphenolic compounds is feasible on an industrial scale. For this purpose, two alternatives are presented: either a stand-alone process or a comprehensive approach to biorefineries [5]. Among the industries that can take advantage of these molecules are the food, cosmetics, pharmaceutical, and chemical industries. The compounds of interest present in these wastes include polyphenolic compounds. Molecules such as ferulic acid, gallic acid, caffeic acid, chlorogenic acid, quercetin, and hesperidin can be obtained from different food wastes. One of the main properties of polyphenols is their antioxidant capacity. Therefore, polyphenols are among the most desired and required properties in food, pharmaceuticals, and chemicals. To extract polyphenols from food waste, conventional and non-conventional technologies can be used. Conventional technologies include solvent extraction and Soxhlet extraction. Supercritical fluid extraction (SFE) and ultrasound-assisted extraction (UAE) are non-conventional technologies. The selection of the extraction technology depends on a combination of technical, energetic, economic, and environmental aspects.

Polyphenolic compounds benefit human health, making them tempting for several applications, and are present in the food, pharmaceutical, medical, and chemical industries [6]. These compounds are important due to the growing interest in having a healthy and natural lifestyle. The extraction of

DOI: 10.1201/9781003462804-8

polyphenolic compounds is, therefore, a topic worth investigating. Some authors assure that with the COVID-19 pandemic, more people are becoming aware of how they eat, take care of themselves, and know the components they consume [7]. Plants, fruits, and vegetables contain multiple bioactive components [8]. It is possible to find fatty acids, pigments [9], flavonoids, essential oils [10], and alkaloids within these compounds [11]. They have been reported as beneficial agents for health, either due to their antioxidant, anti-inflammatory [12], or even antimicrobial capacity [13].

Due to this, several technologies for the extraction of bioactive components of biomass have been studied [14, 15]. The most common ones (i.e., conventional extraction) [16] use solvents and substances highly harmful to human health since they alter the properties and chemical nature of the component [7]. In this way, the application is restricted, as it would not be viable for human consumption [17]. On the other hand, the generation of contamination due to solvents is significant [17]. Therefore, the most efficient, economical, environmentally friendly, and safe technologies have been studied. Supercritical fluid extraction (SFE) is recognized, is safe, and yields better than other techniques. The use of supercritical fluids allows the efficient extraction of polyphenolic compounds. It breaks the complex network of polysaccharides (mainly composed of cellulose, hemicellulose, lignin, pectin, and starch, among others) present in the biomass [6].

The use of biomass to obtain bioactive compounds has interesting results. However, the high energy and economic requirements have been reported as challenging to carry out these processes on an industrial scale [18]. This chapter evaluates orange residues as a potential biomass for extracting polyphenolic compounds through supercritical fluid extraction. The evaluation considers orange residues as food waste for valorization using biorefineries. Initially, a theoretical context on polyphenolic compounds, economic perspectives, and raw materials will be provided. Subsequently, the analysis of the solubility of hesperidin in CO_2, acetone, ethanol, hexane, methanol, and water is presented. The technology used for the extraction is supercritical fluid extraction (SFE). In this sense, CO_2 is considered the supercritical fluid used for the analysis.

Additionally, the effect of using a co-solvent in the process is considered. Finally, the conceptual design of a biorefinery with SFE will be proposed based on orange peel. Computational tools such as Aspen Plus V9.0 and Aspen Process Economic Analyzer V9.0 will be used. Additionally, the data will be analyzed as part of sustainable process design.

6.2 POLYPHENOLS AND EXTRACTION METHODS

Polyphenolic compounds are antioxidant compounds that plants, fruits, and vegetables biosynthesize. They are considered highly beneficial compounds for diseases such as cellular and subcellular lesions, help activate liver detoxification systems, and inhibit tumor growth by blocking metabolic pathways that cause cancer, among many other benefits [19]. They generally act as scavengers and stabilizers of free radicals, producing chelation of metals with carboxylic groups in their structure. Polyphenolic compounds can generally be classified as phenolic acids, flavonoids, and tannins [20]. Depending on the raw materials' origin and composition, these substances can be found in different proportions. Additionally, the complexity of the structures in terms of the primary bioactive components (carbohydrates, for example) limits the efficiency of extracting polyphenolic compounds [21]. In fact, in some cases, pretreatment steps are required to condition the raw materials to facilitate access to the components of interest [22]. Polyphenolic compounds of the flavonoid type are polyphenolic compounds with antioxidant capacity and immunomodulatory activity. It means that these components are applied mainly in the pharmaceutical industry. Currently, there are about 9,000 flavonoids available [23].

Table 6.1 shows the polyphenolic compound extraction techniques most applied by several authors. It also includes the characterization techniques applied according to the compound of interest.

SFE has been one of the most studied of the technologies mentioned earlier. Additionally, it is considered a green process capable of improving the yields of conventional extractions [29]. The SFE is a method that involves the transfer of mass around the critical point of the solvent used in the process [25]. Properties such as temperature and pressure increase the solvent power of the fluid and

TABLE 6.1
Technology Applied to Polyphenolic Compounds Extraction

Technology	Principle	References
Microwave-assisted extraction (MAE)	It is based on heating the solid material with a solvent by microwave irradiation. Its foundation is dialectical heating.	[15, 24]
Ultrasound-assisted extraction (UAE)	It is based on the cavitation process induced by the compression and expansion of cycles associated with the passage of ultrasound through the sample. The implosion of the cavitation bubbles originates the rupture of the particles, making the extraction of the component by the solvent more accessible.	[15, 21]
Supercritical fluid extraction (SFE)	It is based on using supercritical fluids (especially supercritical CO_2) to extract the compounds of interest.	[15, 25]
Deep eutectic solvent (DES) extraction	It is based on the dissolution of polysaccharides present in the material (cellulose, hemicellulose, starch, and lignin), facilitating the extraction of polyphenolic compounds.	[15, 26]
Enzyme-assisted extraction (EAE)	It is based on hydrolytic enzymes that facilitate the access of solvents to the complex structure of the raw material and obtain the bioactive component.	[26, 27]
Pressurized liquid extraction (PLE)	It is based on applying pressure to heat the solvent in contact with the raw material to its boiling point.	[15, 28]

provide better penetration to the material. The greatest advantage of the SFE is its lower negative impact on the environment, mainly because the generated compounds are not harmful [26]. The most widely used fluid in this technology is supercritical CO_2. This solvent is more advantageous than conventional solvents in terms of controlling variables in the process, contamination, and stability of the bioactive compounds, among other aspects.

Furthermore, supercritical CO_2 can be combined using co-solvents (such as ethanol) to increase extraction yields [30]. Besides, this allows it to be used in several industries without generating toxic traces in the extracts. Regarding the operating conditions, this solvent allows the process temperatures to be adjusted to avoid the degradation of thermolabile components. Also, in energy terms, it has been shown that the extraction process of polyphenolic compounds requires less energy than conventional extractions [31]. Moreover, the solvent can be easily recovered and reused, decreasing the overall economic impact [29].

The extraction of polyphenolic compounds by SFE is based on solubility. This aspect affects the extraction rate and depends on conditions such as temperature and pressure. For example, increasing the pressure causes an improvement in the density of the supercritical fluid, which, in turn, improves the solvation capacity [15]. Additionally, the high pressure applied to the biomass can break down plant cells, facilitating the solubility of biocomponents. On the other hand, temperature behaves in a more complex way. If the pressure is constant, for example, an increase in temperature improves the vapor pressure of the solute. In this sense, the solubility in the extraction fluid is improved [32]. Another important factor to consider in the SFE process is the addition of co-solvent.

In some cases, the use of a co-solvent is studied when molecules of interest have higher polarity. Co-solvents increase the density around the solute due to intermolecular forces (physical and chemical) affectation. In this way, the extraction benefits, achieving high yields [33].

However, a good SFE process depends not only on the nature of the supercritical fluid, the temperature, the pressure, or the use or not of co-solvent [34]. The matrix of the raw material conditions the extraction from the need to apply a previous pretreatment stage. Therefore, the SFE process can be divided into the sections shown in Figure 6.1. Each section must be carefully designed and optimized to obtain the best possible extraction yields and to be applied to any biomass matrix [33]. This strategy would be interesting to apply, especially for raw materials as complex as food waste. Additionally, the characterization of the residual raw material can be valued by integrating

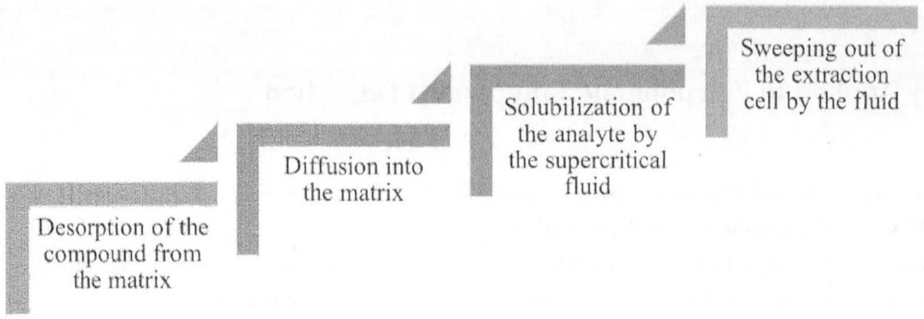

FIGURE 6.1 SFE, a four-stage process.

Source: Adapted from [33].

other processes through the design of biorefineries [35]. In this way, SFE can be accompanied by other technologies, fully exploiting the potential of the raw material and making the process globally viable.

6.3 POLYPHENOLS FROM FOOD WASTES

Mercado-Mercado et al. [36] found that plants' polyphenolic compounds are considered biosynthetic antioxidants. These are classified by Pathak et al. [37] as phenolic acids, flavonoids, and tannins. Therefore, these three can be considered the first classification of polyphenolic compounds. From each of these classifications, divisions are presented in which the different polyphenolic compounds are classified. Figure 6.2 shows the hierarchy of classification of polyphenolic compounds. Additionally, examples for each of the classifications are shown.

Several large cross-sectional studies suggest that polyphenolic compounds benefit the body against diseases, such as cellular and sub-cellular lesions, inhibition of tumor growth, activation of liver detoxification systems, and blockage of metabolic pathways that can cause carcinogenesis [36]. Some studies have begun to examine the content of polyphenolic compounds in different fruits. However, most researchers investigating phenolic compounds have utilized peels as raw material for the high content of these compounds by comparison with the content present in the pulp [38, 39].

Studies performed by Özbek et al. and Torti and Dearing [41, 42] have shown that polyphenolic compounds can be extracted by different technologies, including liquid–liquid extraction (also known as solvent extraction), which is remarkable. Although solvent extraction is one of the most used for its effectiveness and simplicity, other extraction methods are available. SFE is characterized by selectivity, low consumption of organic solvents, higher rates, better reproducibility, environmental safety, and obtained extracts enriched in phenolic compounds [43]. Another method is ultrasound-assisted extraction (UAE). The UAE has a higher yield of antioxidant activity in the obtained products than those obtained through solvent extraction [44, 45]. When the Soxhlet extraction method is used, it is common to recognize contact with the fresh solvent throughout the extraction step.

Different raw materials can be found as sources of polyphenolic compounds. Therefore, it is possible to find different studies focused on quantifying bioactive compounds present in fruits or their residues in the literature. [2] shows raw materials, the polyphenolic compounds present, and the technology used for extraction. There are three main stages for extracting bioactive compounds: selection of the raw material, selection of the technology, and implementation of the process (final stage) (see Figure 6.3). In recent years, research has focused on using SFE as the main technology. However, regardless of the technology used, there are three streams at the end of the process: product (extract), liquids and/or gases, and solid residues (exhausted solids).

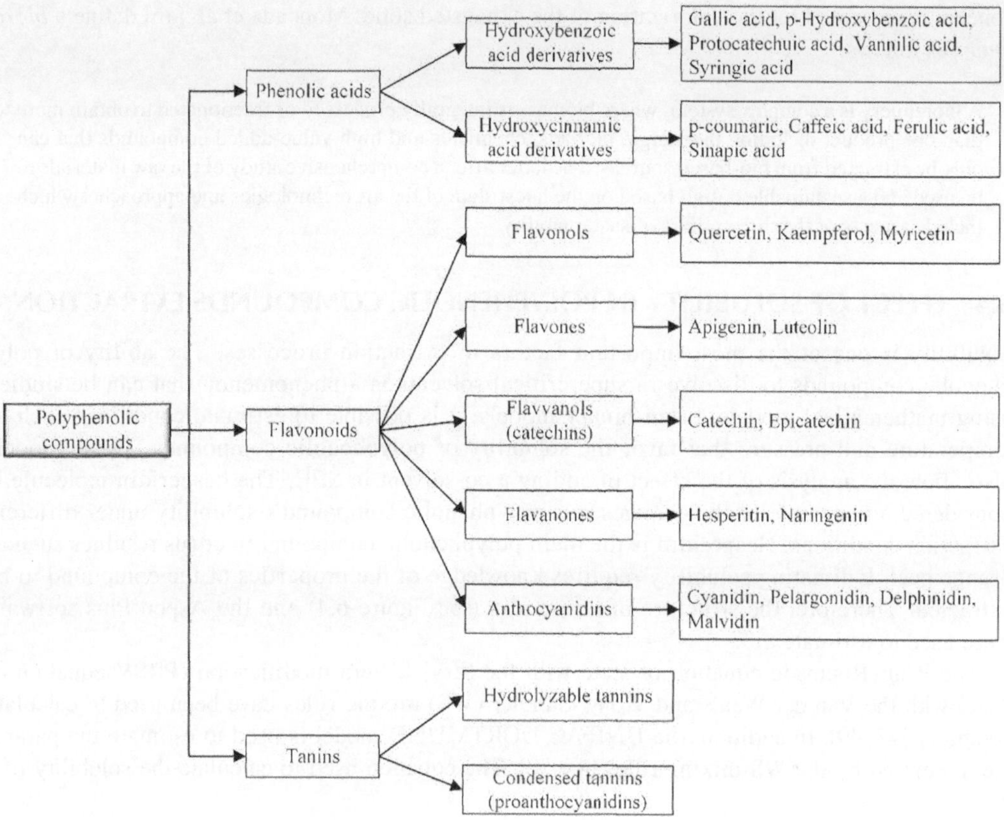

FIGURE 6.2 Polyphenolic compounds classification.

Source: Adapted from [40].

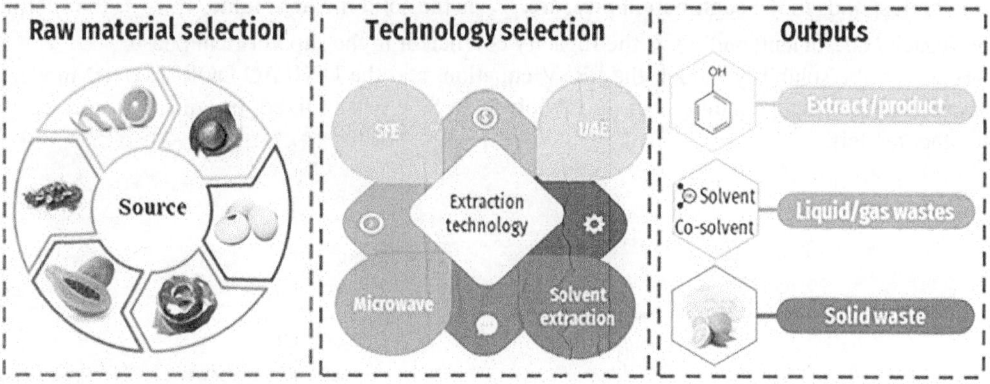

FIGURE 6.3 Steps in an extraction process.

Two factors must be considered in the extraction processes to achieve an integral utilization of the raw material and obtain good yields: the extraction conditions and the possible valorization of the exhausted solid. The extraction conditions can be estimated from a theoretical point of view using mathematical models based on solubility calculations. The possibility of using the biorefinery concept is presented for the valorization of the exhausted solid. Moncada et al. [46] define a *biorefinery* as follows:

> A biorefinery is a complex system, where biomass is integrally processed or fractionated to obtain more than one product including bioenergy, biofuels, chemicals and high value-added compounds that can only be extracted from bio-based sources. The latter after a comprehensive study of the raw materials to be used and a sustainable design based on the latest state of the art technologies and approaches which include aspects of the three pillars of sustainability.

6.4 EFFECT OF SOLUBILITY IN POLYPHENOLIC COMPOUNDS EXTRACTION

Solubility is one of the most important factors in extraction processes. The ability of polyphenolic compounds to dissolve in supercritical solvents is a phenomenon that can be studied using mathematical models. From binary models, it is possible to estimate conditions such as temperature and pressure that favor the solubility of polyphenolic compounds. These models also allow the analysis of the effect of adding a co-solvent in SFE. The hesperidin molecule is considered an example of the changes in a polyphenolic compound's solubility under different extraction conditions. Hesperidin is the main polyphenolic compound in citrus residues such as orange peel. Estimating solubility requires knowledge of the properties of the compound to be extracted. Therefore, the structure of hesperidin (see Figure 6.4) and the Aspen Plus software were used to estimate.

The Peng–Robinson equation of state with the Stryjek Vera modification (PRSV equation of state) with the Van der Waals and Wong–Sandler (WS) mixing rules have been used to calculate solubility [47–49]. In addition, the UNIFAC DORTMUND model is used to estimate the parameters required by the WS mixing rules [47, 50]. The equation used to calculate the solubility of a solid in gas is:

$$y_2 = \frac{P_2^{subli}\varphi_2^{subli}}{P\varphi_2}\exp\left[\frac{V_2^s}{RT}\left(P - P_2^{subli}\right)\right]$$

P_2^{subli} corresponds to the sublimation pressure, V_2^s is the solid molar volume, φ_2^{subli} is the sublimation fugacity coefficient, and φ_2 is the fugacity coefficient in the supercritical phase.

However, the solubility model, the PRSV equation, and the UNIFAC DORTMUND model do not provide the critical conditions of the co-solvent in SFE when mixed. For this, it is necessary to use other models.

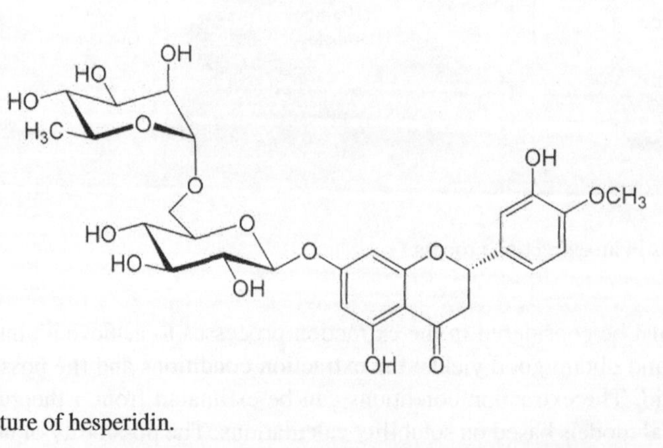

FIGURE 6.4 Structure of hesperidin.

6.4.1 Estimation of Critical Properties of the Co-Solvent Mixture

Nichita [51] proposes a method for estimating the critical conditions based on the method proposed by Heidemann and Khalil [52]. The method is based on solving two equations simultaneously. The first one corresponds to the spinodal equation, which is considered the critical point of the proposed model. The second criterion allows locating the critical points in the spinodal. The equations must consider the number of components (*nc*). The main equations that compose the model are presented in the following (Table 6.2).

The critical conditions of four solvents in a solution at different concentrations have been estimated in this work. Solving the equations presented in [4] and considering the critical properties of the pure solvents (see Table 6.3), the results presented in Figure 6.5 are obtained. The highest critical point presented corresponds to water (647.3 K and 224.13 bar). The lowest critical conditions among the solvents analyzed are presented in the hexane mixtures. A 70% hexane composition presents a critical temperature of 503.20 K, being the lowest value in this work. The lowest critical pressure corresponds to pure hexane (30.49 bar). On the other hand, the lowest value is presented for pure hexane (30.49 bar) regarding the critical pressure. Methanol has the highest critical pressure as a pure component.

However, the difference in critical temperature and critical pressure values for the pure components must be considered. These solvents' average critical temperature and critical pressure correspond to 510.53 K and 55.6 bar, respectively, with a difference between the highest and the lowest value of 6.40 K for the critical temperature. For the critical pressure, the range is 31.7 bar. The most appreciable differences for these two properties are evident at concentrations of 30% of each solvent. The difference corresponds to 61.29 K. The critical pressure corresponds to a value of 8.29 bar for the critical temperature. This concentration of solvents would be expected to have more significant differences in energy and economics. When considering the critical temperature change, the energetic part requires a higher energy consumption to reach this point. With a higher energy consumption, the utility costs in the processes increase. Therefore, selecting a solvent or co-solvent must consider different aspects and the energetic and economic aspects. The affinity of the compound of interest to be extracted by the solvent has a key role in the technical aspects. Therefore, the solubility of the compound of interest in different solvents and solvent concentrations must be evaluated.

TABLE 6.2
Main Equations of the Nichita Model for Critical Properties Estimation

Equation	Criterium
$\det(Q_s) = 0$ $Q_{Sij} = \left(\frac{\delta^2 A}{\delta n_i \delta n_j}\right)_{T,V,n_k \neq i,j} = \left(\frac{\delta \ln f_i}{\delta n_j}\right)_{T,V,n_k \neq i,j}$	1
$c = \sum_{k=1}^{nc}\sum_{j=1}^{nc}\sum_{i=1}^{nc}\left(\frac{\delta^3 A}{\delta n_i \delta n_j \delta n_k}\right)_{T,V,n_k \neq i,j} \quad \Delta n_i \Delta n_j \Delta n_k = 0$	2

Source: Based on [51].

TABLE 6.3
Critical Properties Used in This Work

Component	Tc (K)	Pc (bar)	w
Acetone	508.1	47.62	0.304
Ethanol	513.9	62.21	0.644
Hexane	507.5	30.49	0.299
Methanol	512.6	81.97	0.556
Water	647.3	224.13	0.344
CO2	304.12	73.74	0.225
Hesperidin	1,066.45	14.65	0.790

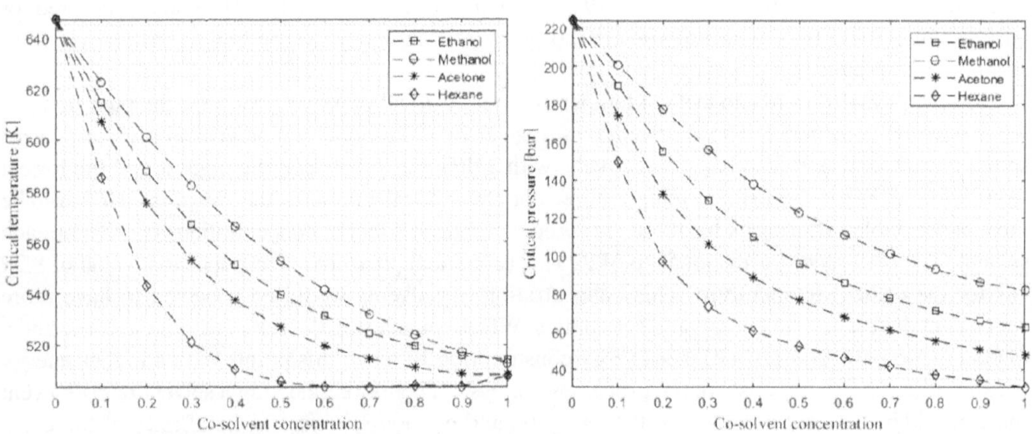

a b

FIGURE 6.5 Critical properties estimated using the proposed methodology for co-solvent mass concentration into 0 at 1. a. Critical temperature b. Critical pressure.

6.5 SOLUBILITY OF HESPERIDIN UNDER SUPERCRITICAL CONDITIONS

It uses the methodology described for the estimation of the solubility of hesperidin. As a result, the results presented in Figure 6.6 are obtained. *Hesperidin* is a flavonoid with a polar behavior. It is evident from its low solubility in CO_2, a non-polar compound. In alcohol and other polar organic solvents, hesperidin presents higher solubility values. However, hesperidin is not very soluble in water (see Figure 6.6.f). The solubility of hesperidin has been estimated for temperatures between 300 K and 340 K. Because the range considered is of low temperatures, the hesperidin exhibits low solubility in water. However, as reported by López Luengo [53], hesperidin is soluble in hot water.

When acetone, ethanol, methanol, or hexane is used as co-solvent, it is observed that solubility reaches its highest values at a pressure of 80 bar. Pressures higher than 80 bar lead to a decrease in the solubility of hesperidin in the different co-solvents tested, showing that to extract hesperidin using SFE, the extraction pressure should be 80 bar. With the purpose observed in greater detail, the changes in solubility concerning temperature (Figure 6.7) are presented. In this figure, it is possible to observe how the solubility of hesperidin decreases with increasing temperature, regardless of the

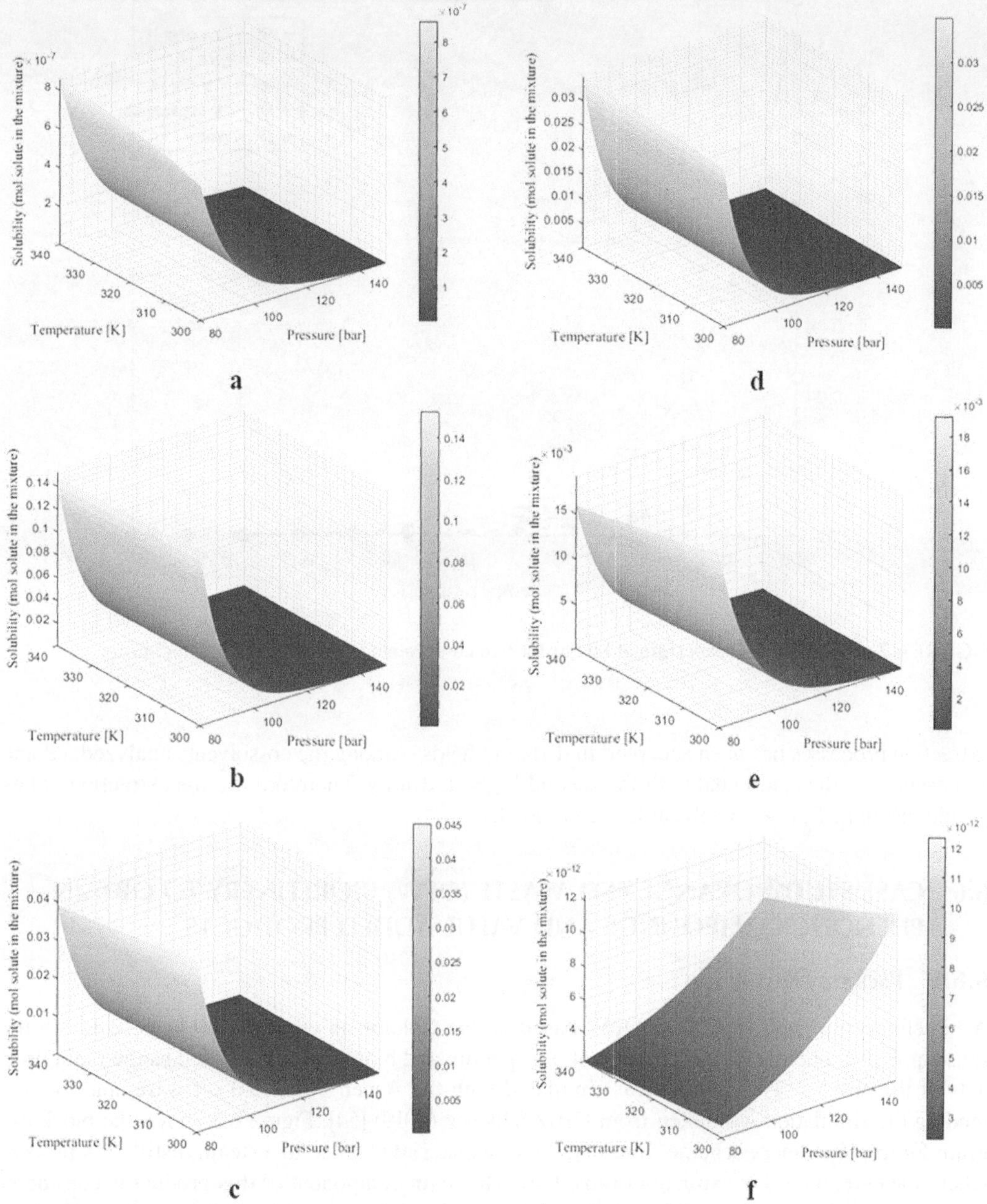

FIGURE 6.6 Solubility of hesperidin in CO2 and some co-solvents: (a) without co-solvent, (b) acetone, (c) ethanol, (d) hexane, (e) methanol, (f) water.

co-solvent used. The co-solvent with the highest affinity for hesperidin is acetone. However, acetone is a liquid that evaporates easily and is flammable. These characteristics mean that its handling may present greater restrictions than other co-solvents.

Moreover, it is not an accepted compound in pharmaceutical and/or food products. On the other hand, hesperidin is a compound used in the pharmaceutical industry for its antioxidant properties. Therefore, it cannot present traces of compounds undesired by this industry. The use of ethanol in

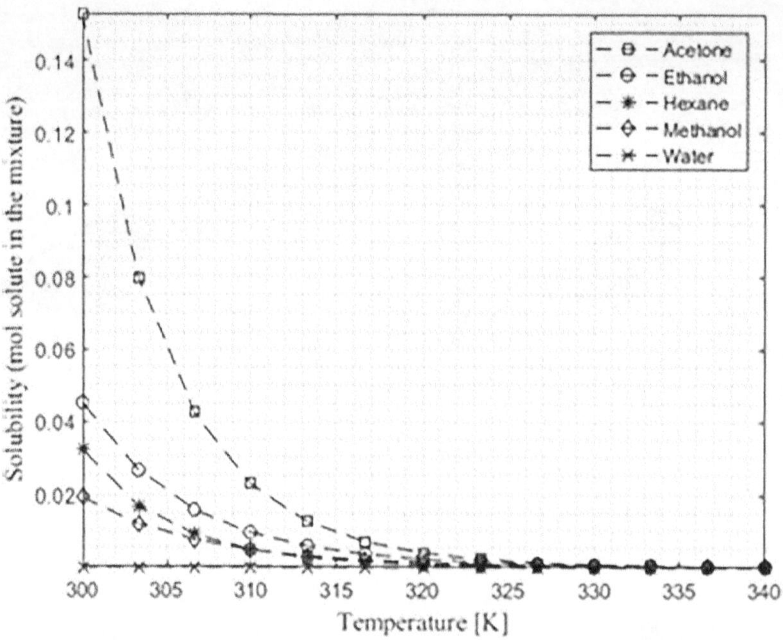

FIGURE 6.7 Solubility of hesperidin at 80 bar for five co-solvents as a function of temperature.

extraction processes has been accepted in different fields. Among the co-solvents analyzed, ethanol is presented as the compound with the second-highest affinity. Therefore, for the extraction of hesperidin using SFE, this co-solvent is recommended.

6.6 CASE STUDY: ORANGE PEEL WASTE (OPW) BIOREFINERY TO OBTAIN PHENOLIC COMPOUNDS AND VALUE-ADDED PRODUCTS

6.6.1 Process Description

A biorefinery was proposed from OPW based on the Colombian context (140 kg/h scale). In this sense, polyphenolic compounds, essential oils, pectin, and biogas were proposed as the main products of the process. The process was simulated with the Aspen Plus V9.0 software, and the data feeding the simulation was taken from Ortiz-Sánchez (2019) [54]. Figure 6.8 shows the block diagram for the biorefinery scheme. Initially, the raw material is taken to a steam distillation process, where the essential oil is extracted (Table 6.4). The main component of this product is considered as limonene. The remaining OPW is then taken to the SFE stage, where the polyphenolic compounds are obtained. In this case, ethanol is used as a co-solvent with supercritical CO_2 to have a better yield and a wide portfolio of applications. Next, a pectin extraction process is carried out with ethanol and citric acid, allowing the generation of a stream of OPW that is more available and appropriate for the anaerobic digestion stage. Besides, the pectin market is growing in different countries, such as Colombia. Finally, the anaerobic digestion stage will allow biogas produced by anaerobic bacteria present in the inoculum.

6.6.2 Techno-Economic and Energy Assessment Methodology

The evaluation of the proposed biorefinery considers the analysis of three aspects: technical, energetic, and economical. The first is related to the mass balance, in which indicators such as product

TABLE 6.4
Raw Material and Technology Used to Extract Polyphenolic Compounds

Raw Material (Waste)	Polyphenolic Compounds	Extraction Techniques	Reference
Onion waste	Phenolics	Solid–liquid extraction	[55]
	Flavonoids		
Apple pomace	Phenolics	Solvent extraction	[56]
Macadamia skin	Phenolics	UAE	[57]
	Flavonoids		
	Proanthocyanidins		
Potato peels	Phenolics	Hydroalcoholic solution extraction	[58]
	Flavonoids	Solvent extraction	[59]
	Ferulic acid		
	Chlorogenic acid		
Pomegranate peels	Phenolics	Solvent extraction	[60]
	Flavonoids		
	Carotenoids	UAE	[61]
Mango seed	Mangiferin	SFE	[62]
	Gallic acid		
	Ellagic acid		
	Quercetin		
Guava seeds and pomace	Phenolics	Solvent extraction	[63]
	Phenolics	SFE	[64]
		Soxhlet extraction	
Spent ground coffee	Phenolics	SFE	[64]
		Soxhlet extraction	
	Phenolics	Solvent extraction	[65]
	Chlorogenic acid	Soxhlet extraction	
	Ferulic acid	UAE	
	Vanillin		
	Quercetin		
	Vanillic acid		
	Caffeic acid		
Olive leaves and pomace	Phenolics	Solvent extraction	[59]
Tangerine peels	Phenolics	SFE	[66]
	Naringin		
	Hesperidin		
	Naringenin		
	Tangeretin		
Papaya	Phenolics	SFE	[67]
	Chlorogenic acid		
	Caffeic acid		
	Ferulic acid		
	Quercetin		
Orange peel	Limonene	SFE	[68]
	Phenolics	UAE	
	Hesperidin	SFE	[69]
	Naringin		
	Gallic acid		

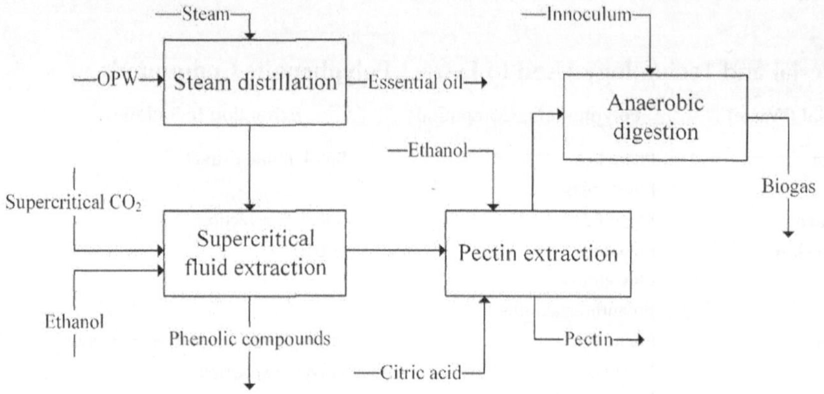

FIGURE 6.8 Block diagram for case study.

Product yield (Y_P)

$$Y_P = \frac{\sum \dot{m}_{Product,i}}{\dot{m}_{Raw\ material}}$$

Process mass intensity index (PMI)

$$PMI = \frac{\sum_{i=1}^{N} \dot{m}_i^{in}}{\sum \dot{m}_{Product,i}}\ kg\ of\ raw\ material/\ kg\ product$$

Mass loss index (MLI)

$$MLI = \frac{\sum_{i=1}^{N} \dot{m}_i^{in} - \dot{m}_{Product}}{\sum \dot{m}_{Product,i}}\ kg\ waste\ streams/\ kg\ products$$

Overall energy efficiency (η)

$$\eta = \frac{\sum \dot{m}_{Product} * LHV_{products}}{(\dot{m}_{Raw\ material} * LHV_{Raw\ material}) + \dot{Q}_{Total} + \dot{W}_{Total}}$$

Specific energy consumption (S_{EC})

$$S_{EC} = \frac{\dot{Q}_{Total} + \dot{W}_{Total}}{\dot{m}_{Raw\ material}}$$

Resource energy efficiency (η_E)

$$\eta_E = \frac{\sum \dot{m}_{Product} * LHV_{products}}{(\dot{m}_{Raw\ material} * LHV_{Raw\ material})}$$

Self-generation index (SGI)

$$SGI = \frac{(\sum \dot{m}_{Product} * LHV_{products}) * \eta_{conversion}}{\dot{Q}_{Total} + \dot{W}_{Total}}$$

FIGURE 6.9 Mass and energy indicators for technical evaluation.

yield (*Yp*), process mass intensity index (*PMI*), and mass loss index (*MLI*) are used. Secondly, the energy aspects of the process are analyzed using both mass and energy balances, considering indicators such as overall energy efficiency, specific energy consumption, resource energy efficiency, and self-generation index (SGI). The equations used to estimate the technical and energy indicators are presented in Figure 6.9. These indicators serve as a measure of the sustainability of the biorefinery proposed in this work. In addition, it provides results compared to other biorefineries configurations, even though they are based on different contexts. The mass and energy balances were obtained using Aspen Plus V.9 software. The simulation of the biorefinery in context was performed in this software.

The Aspen Process Economic Analyzer was used for economic analysis. This software is a tool that allows evaluating the capital cost of industrial plants. This purpose has a basis on design, costs, methods, and models to generate preliminary equipment designs and simulate the supplier costs. Also, it has a cost for pipelines, valves, and control instrumentation in the process [70]. The economic analysis was estimated in dollars for a 20-year period, in which 8,000 h/year was assumed [71], considering Colombian economic conditions. Table 6.5 shows the data used in the analysis. The economic indicators maintenance costs, labor costs, fixed charges, general and administrative costs, plant overhead, and capital depreciation were calculated based on Peters and Timmerhaus (1991) [72]. Then, with the annual cash flows in the project lifetime, the net present value (NPV) was obtained.

6.6.3 Biorefinery Assessment

Fort the proposed scheme, the product yields reached were 0.6, 6.8, 3.8, and 4.9% (kg product/kg OPW) of essential oil, polyphenolic compounds, pectin, and biogas. In this sense, the most productive product is biogas, and the product with the lowest yield is essential oil. On the other hand, the PMI obtained was 407.7 kg raw materials/kg products (globally). Finally, the global mass indicator MLI was 151.2 kg waste streams/kg products. Therefore, the proposed biorefinery scheme has an overall energy efficiency of 0.78% regarding the energy indicators. This result is explained due to the high utilities in the process. Therefore, biogas could be considered to be taken to a cogeneration system that allows partially supplying the energy requirements (thermal and electrical) of the biorefinery [80]. As a result, the specific energy consumption is 5.16%, and the self-generation index is 0.47%. However, these results can be improved if the mass integration concept is applied in the biorefinery, if the use of waste streams is proposed or any of the products is considered a platform to obtain value-added products or process requirements (biogas case).

Regarding economic assessment, the CAPEX of the proposed biorefinery is 15.9 MUSD, and the biorefinery process sections that contribute the most representative equipment cost value are biogas and oil extraction. Mainly, this behavior is because both plants require large equipment to maintain continuous operating conditions (i.e., tank jackets). Besides, Figure 6.10 shows the cost distribution for the economic analysis of the proposed biorefinery. It can be evidenced that the raw material costs are the most representative, so mass integration within the process could be considered an alternative to recover and reuse the streams that can still be processed and thus

TABLE 6.5
Values Considered for the Economic Evaluation of the Biorefinery Scheme

Item	Value	Unit	Reference
Raw materials, reagents, and supplies			
OPW	0.02	USD/kg	[54]
Water	0.33	USD/cum	[73]
Carbon dioxide	0.03	USD/kg	[74]
Ethanol	0.98	USD/kg	[75]
Citric acid	0.19	USD/kg	[76]
Product prices			
Essential oil	10	USD/kg	[54]
Phenolic compounds	45	USD/kg	[54]
Pectin	11	USD/kg	[77]
Biogas	0.26	USD/kg	[78]
Utilities			
Low-pressure steam	7.89	USD/ton	[79]
Cooling water	0.04	USD/cum	[73]
Electricity	0.02	USD/MJ	[54]
Labor cost			
Operator labor cost	1.2	USD/h	[75]

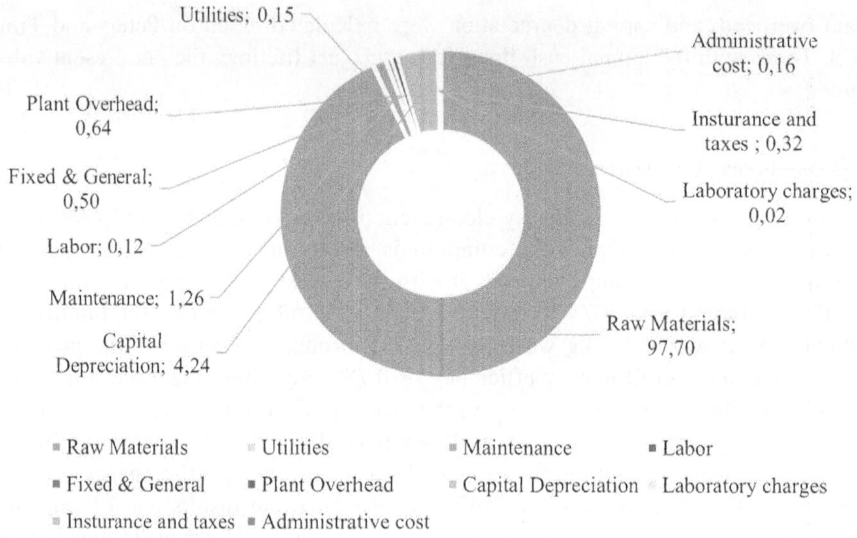

FIGURE 6.10 Distribution cost of biorefinery scheme.

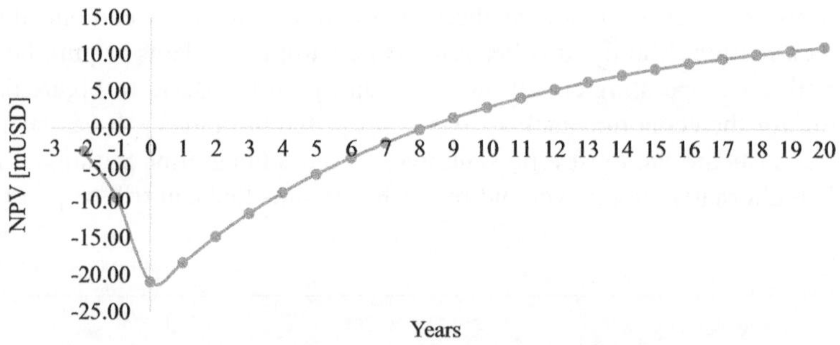

FIGURE 6.11 Net present value over project lifetime.

require a fewer amount of reagents. In addition, this improvement is reflected economically, but it is also possible that it could be reflected in process yields, final waste disposal, and environmental impact.

Finally, Figure 6.11 shows the NPV behavior over 20 years of project life. It indicates that the process proposed under the conditions and contexts is economically viable, and the investment is recovered in approximately the eighth year. Despite being a small-scale process, it is evident that the integral use of biomasses (especially those belonging to the food waste group) is feasible in regions such as Colombia.

6.7 CONCLUSIONS

Food wastes are renewable sources of molecules of industrial interest that can be extracted and valorized in different fields. Among the industries that can take advantage of these molecules are the food, cosmetics, pharmaceutical, and chemical industries. The compounds of interest present in these wastes include polyphenolic compounds. To extract polyphenols from food waste, conventional and non-conventional technologies can be used. Conventional technologies include solvent extraction and Soxhlet extraction. Supercritical fluid extraction (SFE) and ultrasound-assisted extraction (UAE) are non-conventional technologies. The selection of the extraction technology depends on a combination of technical, energetic, economic, and environmental aspects.

From this work, it is possible to conclude that the extraction processes of polyphenolic compounds in food waste utilization schemes are important, viable, and worth studying. Furthermore, high-value-added products from comprehensive bioprocesses such as biorefineries complement and valorize the biomass available. As future work, the analysis of the other pillars of sustainability can be proposed to demonstrate the sustainable viability of the project (environmental and social dimension).

ACKNOWLEDGMENTS

The authors acknowledge the support of the Departamento Administrativo de Ciencia, Tecnología e Innovación (Colciencias), Becas Bicentenario corte I, and the research program titled "Reconstrucción del tejido social en zonas posconflicto en Colombia," SIGP code: 57579, with the project titled "Competencias empresariales y de innovación para el desarrollo económico y la inclusión productiva de las regiones afectadas por el conflicto colombiano," SIGP code 58907. Contract number: FP44842-213-2018.

REFERENCES

1. K. Bandyopadhyay, S. Ganguly, C. Chakraborty, and A. Modak, "Studies on extraction of polyphenols from food wastes and its utilization for fortification of polyphenols," *International Journal of Agricultural Engineering*, vol. 11, no. 1, pp. 216–219, Apr. 2018, doi: 10.15740/HAS/IJAE/11.1/216-219.
2. I. Mourtzinos and A. Goula, "Polyphenols in agricultural byproducts and food waste," in *Polyphenols in Plants*, 2nd ed., Elsevier, 2019, pp. 23–44, doi: 10.1016/B978-0-12-813768-0.00002-5.
3. B. Tsegaye, S. Jaiswal, and A. K. Jaiswal, "Food waste biorefinery: Pathway towards circular bioeconomy," *Foods*, vol. 10, no. 6, p. 1174, May 2021, doi: 10.3390/foods10061174.
4. C. Caldeira, A. Vlysidis, G. Fiore, V. De Laurentiis, G. Vignali, and S. Sala, "Sustainability of food waste biorefinery: A review on valorisation pathways, techno-economic constraints, and environmental assessment," *Bioresource Technology*, vol. 312, p. 123575, Sep. 2020, doi: 10.1016/j.biortech.2020.123575.
5. S. A. Socaci *et al.*, "Antioxidant compounds recovered from food wastes," in *Functional Food—Improve Health through Adequate Food*, InTech, 2017, pp. 3–22, doi: 10.5772/intechopen.69124.
6. K. Niaz and F. Khan, "Analysis of polyphenolics," in *Recent Advances in Natural Products Analysis*, Elsevier, 2020, pp. 39–197, doi: 10.1016/B978-0-12-816455-6.00003-2.
7. T. Arumugham, R. K, S. W. Hasan, P. L. Show, J. Rinklebe, and F. Banat, "Supercritical carbon dioxide extraction of plant phytochemicals for biological and environmental applications—a review," *Chemosphere*, vol. 271, p. 129525, May 2021, doi: 10.1016/j.chemosphere.2020.129525.
8. J. Cao *et al.*, "Pinene-rich essential oils from *Haplophyllum dauricum* (L.) G. Don display anti-insect activity on two stored-product insects," *International Biodeterioration & Biodegradation*, vol. 140, no. August 2018, pp. 1–8, May 2019, doi: 10.1016/j.ibiod.2019.03.007.
9. S. A. Petropoulos, Â. Fernandes, R. C. Calhelha, N. Danalatos, L. Barros, and I. C. F. R. Ferreira, "How extraction method affects yield, fatty acids composition and bioactive properties of cardoon seed oil?" *Industrial Crops and Products*, vol. 124, no. August, pp. 459–465, Nov. 2018, doi: 10.1016/j.indcrop.2018.08.027.
10. G. Maniet, N. Jacquet, and A. Richel, "Recovery of sterols from vegetable oil distillate by enzymatic and non-enzymatic processes," *Comptes Rendus Chimie*, vol. 22, no. 4, pp. 347–353, Apr. 2019, doi: 10.1016/j.crci.2019.01.004.
11. J. Tian *et al.*, "Flavanoids derivatives from the root bark of *Broussonetia papyrifera* as a tyrosinase inhibitor," *Industrial Crops and Products*, vol. 138, no. May, p. 111445, Oct. 2019, doi: 10.1016/j.indcrop.2019.06.008.
12. S. A. Villarreal-Soto *et al.*, "Impact of fermentation conditions on the production of bioactive compounds with anticancer, anti-inflammatory and antioxidant properties in kombucha tea extracts," *Process Biochemistry*, vol. 83, no. April, pp. 44–54, Aug. 2019, doi: 10.1016/j.procbio.2019.05.004.
13. L. P. Cappato *et al.*, "Whey acerola-flavoured drink submitted Ohmic Heating: Bioactive compounds, antioxidant capacity, thermal behavior, water mobility, fatty acid profile and volatile compounds," *Food Chemistry*, vol. 263, no. April, pp. 81–88, Oct. 2018, doi: 10.1016/j.foodchem.2018.04.115.

14. E. M. C. Alexandre *et al.*, "Antimicrobial activity of pomegranate peel extracts performed by high pressure and enzymatic assisted extraction," *Food Research International*, vol. 115, no. August 2018, pp. 167–176, Jan. 2019, doi: 10.1016/j.foodres.2018.08.044.
15. L. Panzella, F. Moccia, R. Nasti, S. Marzorati, L. Verotta, and A. Napolitano, "Bioactive phenolic compounds from agri-food wastes: An update on green and sustainable extraction methodologies," *Frontiers in Nutrition*, vol. 7, no. 60, pp. 1–27, May 2020, doi: 10.3389/fnut.2020.00060.
16. M. Roy, H. Dutta, R. Jaganmohan, M. Choudhury, N. Kumar, and A. Kumar, "Effect of steam parboiling and hot soaking treatments on milling yield, physical, physicochemical, bioactive and digestibility properties of buckwheat (*Fagopyrum esculentum* L.)," *Journal of Food Science and Technology*, vol. 56, no. 7, pp. 3524–3533, Jul. 2019, doi: 10.1007/s13197-019-03849-9.
17. S. Mukhopadhyay, R. Dutta, and P. Das, "A critical review on plant biomonitors for determination of polycyclic aromatic hydrocarbons (PAHs) in air through solvent extraction techniques," *Chemosphere*, vol. 251, p. 126441, Jul. 2020, doi: 10.1016/j.chemosphere.2020.126441.
18. P. A. Marín-Valencia, E. Carmona-Garcia, J. A. Poveda Giraldo, N. P. Arias Duque, L. F. Ballesteros, and C. A. Cardona Alzate, "The integral use of aromatic plants: prefeasibility comparison of stand-alone and biorefinery processes using thyme (*Thymus vulgaris*) as base case," *Biomass Conversion and Biorefinery*, vol. 11, no. 2, pp. 681–691, 2021, doi: 10.1007/s13399-020-00734-w.
19. B. Ozturk, C. Parkinson, and M. Gonzalez-Miquel, "Extraction of polyphenolic antioxidants from orange peel waste using deep eutectic solvents," *Separation and Purification Technology*, vol. 206, pp. 1–13, Nov. 2018, doi: 10.1016/j.seppur.2018.05.052.
20. A. Kosińska, M. Karamać, I. Estrella, T. Hernández, B. Bartolomé, and G. A. Dykes, "Phenolic compound profiles and antioxidant capacity of *Persea americana* Mill. Peels and seeds of two varieties," *Journal of Agricultural and Food Chemistry*, vol. 60, no. 18, pp. 4613–4619, May 2012, doi: 10.1021/jf300090p.
21. P. Garcia-Salas, A. Morales-Soto, A. Segura-Carretero, and A. Fernández-Gutiérrez, "Phenolic-compound-extraction systems for fruit and vegetable samples," *Molecules*, vol. 15, no. 12, pp. 8813–8826, Dec. 2010, doi: 10.3390/molecules15128813.
22. J. A. Poveda-Giraldo and C. A. Cardona Alzate, "A biorefinery for the valorization of marigold (*Calendula officinalis*) residues to produce biogas and phenolic compounds," *Food and Bioproducts Processing*, vol. 125, pp. 91–104, Jan. 2021, doi: 10.1016/j.fbp.2020.10.015.
23. F. S. Al Amri and M. A. Hossain, "Comparison of total phenols, flavonoids and antioxidant potential of local and imported ripe bananas," *Egyptian Journal of Basic and Applied Sciences*, vol. 5, no. 4, pp. 245–251, Dec. 2018, doi: 10.1016/j.ejbas.2018.09.002.
24. X. Ran, M. Zhang, Y. Wang, and B. Adhikari, "Novel technologies applied for recovery and value addition of high value compounds from plant byproducts: A review," *Critical Reviews in Food Science and Nutrition*, vol. 59, no. 3, pp. 450–461, Feb. 2019, doi: 10.1080/10408398.2017.1377149.
25. R. P. F. F. da Silva, T. A. P. Rocha-Santos, and A. C. Duarte, "Supercritical fluid extraction of bioactive compounds," *TrAC Trends in Analytical Chemistry*, vol. 76, pp. 40–51, Feb. 2016, doi: 10.1016/j.trac.2015.11.013.
26. A. Mena-García, A. I. Ruiz-Matute, A. C. Soria, and M. L. Sanz, "Green techniques for extraction of bioactive carbohydrates," *TrAC Trends in Analytical Chemistry*, vol. 119, p. 115612, Oct. 2019, doi: 10.1016/j.trac.2019.07.023.
27. P. D. Patil, S. P. Patil, R. K. Kelkar, N. P. Patil, P. V. Pise, and S. S. Nadar, "Enzyme-assisted supercritical fluid extraction: An integral approach to extract bioactive compounds," *Trends in Food Science & Technology*, vol. 116, no. July, pp. 357–369, Oct. 2021, doi: 10.1016/j.tifs.2021.07.032.
28. L. Rodrigues, S. Alcázar-Alay, M. Gomes, and A. Meireles, "Pressurized Liquid Extraction (PLE) of Bixin from defatted Annatto seeds," *III Iberoamerican Conference on Supercritical Fluids Cartagena de Indias (Colombia), 2013*, pp. 1–6, 2013.
29. H. Ahangari, J. W. King, A. Ehsani, and M. Yousefi, "Supercritical fluid extraction of seed oils—a short review of current trends," *Trends in Food Science & Technology*, vol. 111, no. February, pp. 249–260, May 2021, doi: 10.1016/j.tifs.2021.02.066.
30. A. B. Zanqui *et al.*, "Extraction and assessment of oil and bioactive compounds from cashew nut (*Anacardium occidentale*) using pressurized n-propane and ethanol as cosolvent," *The Journal of Supercritical Fluids*, vol. 157, no. November, p. 104686, Mar. 2020, doi: 10.1016/j.supflu.2019.104686.
31. M. González Prieto, F. A. Sánchez, and S. Pereda, "Thermodynamic model for biomass processing in pressure intensified technologies," *Journal of Supercritical Fluids*, vol. 96, pp. 53–67, 2015, doi: 10.1016/j.supflu.2014.08.024.

32. J. Martínez and A. Aguiar, "Extraction of triacylglycerols and fatty acids using supercritical fluids—review," *Current Analytical Chemistry*, vol. 10, no. 1, pp. 67–77, Oct. 2013, doi: 10.2174/1573411011410010006.
33. S. M. Pourmortazavi and S. S. Hajimirsadeghi, "Supercritical fluid extraction in plant essential and volatile oil analysis," *Journal of Chromatography A*, vol. 1163, no. 1–2, pp. 2–24, Sep. 2007, doi: 10.1016/j.chroma.2007.06.021.
34. Y. Li *et al.*, "Towards a zero-waste biorefinery using edible oils as solvents for the green extraction of volatile and non-volatile bioactive compounds from rosemary," *Antioxidants*, vol. 8, no. 5, p. 140, May 2019, doi: 10.3390/antiox8050140.
35. C. A. Cardona-Alzate, S. Serna-Loaiza, and M. Ortiz-Sanchez, "Sustainable biorefineries: What was learned from the design, analysis and implementation," *Journal of Sustainable Development of Energy, Water and Environment Systems*, vol. 8, no. 1, pp. 88–117, 2020, doi: 10.13044/j.sdewes.d7.0268.
36. G. Mercado-Mercado, L. de la R. Carrillo, A. Wall-Medrano, J. A. L. Díaz, and E. Álvarez-Parrilla, "Compuestos polifenólicos y capacidad antioxidante de especias típicas consumidas en México," *Nutricion Hospitalaria*, vol. 28, no. 1, pp. 36–46, 2013, doi: 10.3305/nh.2013.28.1.6298.
37. P. D. Pathak, S. A. Mandavgane, and B. D. Kulkarni, "Valorization of pomegranate peels: A biorefinery approach," *Waste and Biomass Valorization*, vol. 8, no. 4, pp. 1127–1137, 2017, doi: 10.1007/s12649-016-9668-0.
38. J. Contreras-calderón, L. Calderón-jaimes, E. Guerra-hernández, and B. García-villanova, "Antioxidant capacity, phenolic content and vitamin C in pulp, peel and seed from 24 exotic fruits from Colombia," *FRIN*, vol. 44, no. 7, pp. 2047–2053, 2011, doi: 10.1016/j.foodres.2010.11.003.
39. A.-L. Gancel, P. Alter, C. Dhuique-Mayer, J. Ruales, and F. Vaillant, "Identifying carotenoids and phenolic compounds in naranjilla (*Solanum quitoense* Lam. var. Puyo hybrid), an Andean fruit," *Journal of Agricultural and Food Chemistry*, vol. 56, pp. 11890–11899, 2008.
40. R. H. Liu, "Health-promoting components of fruits and vegetables in the diet," *Advances in Nutrition*, vol. 4, no. 3, pp. 384S–392S, May 2013, doi: 10.3945/AN.112.003517.
41. H. N. Özbek, F. Halahlih, F. Göğüş, D. Koçak Yanık, and H. Azaizeh, "Pistachio (*Pistacia vera* L.) hull as a potential source of phenolic compounds: Evaluation of ethanol–water binary solvent extraction on antioxidant activity and phenolic content of pistachio hull extracts," *Waste and Biomass Valorization*, 2018, doi: 10.1007/s12649-018-0512-6.
42. S. D. Torti and M. D. Dearing, "Extraction of phenolic compounds from fesh leaves: A comparison of methods," *Journal of Chemical Ecology*, vol. 21, no. 2, pp. 117–125, 1995.
43. L. Valadez-Carmona, A. Ortiz-Moreno, G. Ceballos-Reyes, J. A. Mendiola, and E. Ibáñez, "Valorization of cacao pod husk through supercritical fluid extraction of phenolic compounds," *Journal of Supercritical Fluids*, vol. 131, no. June 2017, pp. 99–105, 2018, doi: 10.1016/j.supflu.2017.09.011.
44. J. Živković, K. Šavikin, T. Janković, N. Ćujić, and N. Menković, "Optimization of ultrasound-assisted extraction of polyphenolic compounds from pomegranate peel using response surface methodology," *Separation and Purification Technology*, vol. 194, no. November 2017, pp. 40–47, 2017, doi: 10.1016/j.seppur.2017.11.032.
45. M. I. Hakimi, Z. Ilham, and R. A. Abdul Kohar, "Enhancement of agro-industrial copra residue oil yield using microwave-assisted extraction," *Waste and Biomass Valorization*, pp. 1–8, 2018, doi: 10.1007/s12649-018-0274-1.
46. J. Moncada B, V. Aristizábal M, and C. A. Cardona A, "Design strategies for sustainable biorefineries," *Biochemical Engineering Journal*, vol. 116, pp. 122–134, 2016, doi: 10.1016/j.bej.2016.06.009.
47. J. Gmehling, J. Li, and N. Schiller, "A modified UNIFAC model. 2. Present parameter matrix and results for different thermodynamic properties," *Industrial & Engineering Chemistry Research*, vol. 32, pp. 178–193, 1993.
48. W. W. Focke, "Excess Gibbs energy of binary liquid mixtures and the Wong–Sandler postulates," *AIChE Journal*, vol. 50, no. 12, pp. 3284–3289, Dec. 2004, doi: 10.1002/AIC.10222.
49. R. Stryjek and J. H. Vera, "PRSV: An improved Peng-Robinson equation of state for pure compounds and mixtures," *The Canadian Journal of Chemical Engineering*, vol. 64, no. 2, pp. 323–333, 1986.
50. D. Shan and H. Wong, "A theoretically correct equations mixing rule for cubic of state," *AIChE*, vol. 38, no. 5, pp. 671–680, 1992.
51. D. V. Nichita, "Calculation of critical points using a reduction method," *Fluid Phase Equilibria*, pp. 223–231, 2005, doi: 10.1016/j.fluid.2004.09.036.
52. T. A. Heidemann and A. M. Khalil, "The calculation of critical points," *AIChE Journal*, vol. 26, no. 5, pp. 769–779, Sep. 1980.
53. M. T. López Luengo, "Flavonoides," *Offarm*, vol. 21, no. 4, pp. 108–113, Apr. 2002.

54. M. Ortiz-Sanchez, "Evaluation of different biorefinery configurations for orange peel waste processing—Master's Thesis," Universidad Nacional de Colombia sede Manizales, 2019.
55. V. Benítez *et al.*, "Characterization of industrial onion wastes (*Allium cepa* L.): Dietary fibre and bioactive compounds," *Plant Foods for Human Nutrition (Dordrecht, Netherlands)*, vol. 66, no. 1, pp. 48–57, Mar. 2011, doi: 10.1007/S11130-011-0212-X.
56. S. Rana, S. Gupta, A. Rana, and S. Bhushan, "Functional properties, phenolic constituents and antioxidant potential of industrial apple pomace for utilization as active food ingredient," *Food Science and Human Wellness*, vol. 4, no. 4, pp. 180–187, Dec. 2015, doi: 10.1016/J.FSHW.2015.10.001.
57. A. Dailey and Q. Vuong, "Effect of extraction solvents on recovery of bioactive compounds and antioxidant properties from macadamia (*Macadamia tetraphylla*) skin waste," *Cogent Food & Agriculture*, vol. 1, no. 1, 2015, doi: 10.1080/23311932.2015.1115646.
58. I. R. Amado, D. Franco, M. Sánchez, C. Zapata, and J. A. Vázquez, "Optimisation of antioxidant extraction from *Solanum tuberosum* potato peel waste by surface response methodology," *Food Chemistry*, vol. 165, pp. 290–299, Dec. 2014, doi: 10.1016/J.FOODCHEM.2014.05.103.
59. I. Khalifa, H. Barakat, H. A. El-Mansy, and S. A. Soliman, "Optimizing bioactive substances extraction procedures from guava, olive and potato processing wastes and evaluating their antioxidant capacity," *Journal of Food Chemistry and Nanotechnology*, vol. 2, no. 4, pp. 170–177, 2016, doi: 10.17756/jfcn.2016-027.
60. A. Sood and M. Gupta, "Extraction process optimization for bioactive compounds in pomegranate peel," *Food Bioscience*, vol. 12, pp. 100–106, Dec. 2015, doi: 10.1016/J.FBIO.2015.09.004.
61. A. M. Goula, M. Ververi, A. Adamopoulou, and K. Kaderides, "Green ultrasound-assisted extraction of carotenoids from pomegranate wastes using vegetable oils," *Ultrasonics Sonochemistry*, vol. 34, pp. 821–830, Jan. 2017, doi: 10.1016/J.ULTSONCH.2016.07.022.
62. N. Sánchez-Mesa, J. U. Sepúlveda-Valencia, H. J. Ciro-Velásquez, and M. A. Meireles, "Bioactive compounds from mango peel (*Mangifera indica* L. var. Tommy Atkins) obtained by supercritical fluids and pressurized liquids extraction," *Revista Mexicana de Ingeniería Química*, vol. 19, no. 2, pp. 755–766, 2020.
63. T. I. Lafka, V. Sinanoglou, and E. S. Lazos, "On the extraction and antioxidant activity of phenolic compounds from winery wastes," *Food Chemistry*, vol. 104, no. 3, pp. 1206–1214, Jan. 2007, doi: 10.1016/J.FOODCHEM.2007.01.068.
64. L. Manna, C. A. Bugnone, and M. Banchero, "Valorization of hazelnut, coffee and grape wastes through supercritical fluid extraction of triglycerides and polyphenols," *The Journal of Supercritical Fluids*, vol. 104, pp. 204–211, Sep. 2015, doi: 10.1016/J.SUPFLU.2015.06.012.
65. A. S. Caballero-Galván, D. L. Restrepo-Serna, M. Ortiz-Sánchez, and C. Cardona-Alzate, "Analysis of extraction kinetics of bioactive compounds from spent coffee grounds (*Coffea arábica*)," *Waste and Biomass Valorization*, vol. 9, pp. 2381–2389, 2018, doi: 10.1007/s12649-018-0332-8.
66. G. Franco-Arnedo, L. M. Buelvas-Puello, D. Miranda-Lasprilla, H. A. Martínez-Correa, and F. Parada-Alfonso, "Obtaining antioxidant extracts from tangerine (*C. reticulata* var. Arrayana) peels by modified supercritical CO2 and their use as protective agent against the lipid oxidation of a mayonnaise," *Journal of Supercritical Fluids*, vol. 165, p. 104957, Nov. 2020, doi: 10.1016/j.supflu.2020.104957.
67. H. I. Castro-Vargas, W. Baumann, S. R. S. Ferreira, and F. Parada-Alfonso, "Valorization of papaya (*Carica papaya* L.) agroindustrial waste through the recovery of phenolic antioxidants by supercritical fluid extraction," *Journal of Food Science and Technology*, vol. 56, no. 6, pp. 3055–3066, 2019, doi: 10.1007/s13197-019-03795-6.
68. S. Jokić, M. Molnar, A. M. Cikoš, M. Jakovljević, S. Šafranko, and I. Jerković, "Separation of selected bioactive compounds from orange peel using the sequence of supercritical CO2 extraction and ultrasound solvent extraction: optimization of limonene and hesperidin content," *Separation Science and Technology (Philadelphia)*, vol. 55, no. 15, pp. 2799–2811, Oct. 2020, doi: 10.1080/01496395.2019.1647245.
69. F. M. Barrales *et al.*, "Recovery of phenolic compounds from citrus by-products using pressurized liquids—an application to orange peel," *Food and Bioproducts Processing*, vol. 112, pp. 9–21, Nov. 2018, doi: 10.1016/J.FBP.2018.08.006.
70. Aspen Technology Inc, "Aspen capital cost estimator user's guide | Tapas Bhattacharyya—Academia.edu," Burlington, 2012.
71. C. Pereira and A. Meireles, "Economic analysis of rosemary, fennel and anise essential oils obtained by supercritical fluid extraction," *Flavour and Fragance Journal*, vol. 22, pp. 407–413, 2007, doi: 10.1002/ffj.1813.

72. J. Moncada, J. Tamayo, and C. A. Cardona, "Evolution from biofuels to integrated biorefineries: Techno-economic and environmental assessment of oil palm in Colombia," *Journal of Cleaner Production*, vol. 81, pp. 51–59, 2014, doi: 10.1016/j.jclepro.2014.06.021.
73. M. S. Peters and K. D. Timmerhaus, *Plant design and economics for chemical engineers*, Singapore: McGraw-Hill, 1991.
74. J. Moncada, J. A. Tamayo, and C. A. Cardona, "Integrating first, second, and third generation biorefineries: Incorporating microalgae into the sugarcane biorefinery," *Chemical Engineering Science*, vol. 118, pp. 126–140, 2014, doi: 10.1016/j.ces.2014.07.035.
75. J. Moncada, "Design and evaluation of sustainable biorefineries from feedstocks in tropical regions—Master's Thesis," Universidad Nacional de Colombia sede Manizales, 2012.
76. C. A. García-Velásquez and C. A. Cardona, "Comparison of the biochemical and thermochemical routes for bioenergy production: A techno-economic (TEA), energetic and environmental assessment," *Energy*, vol. 172, pp. 232–242, Apr. 2019, doi: 10.1016/j.energy.2019.01.073.
77. Intratec Solutions, "Chemical pricing data subscriptions. Historical prices & forecast of chemical commodities," 2019. https://www.intratec.us/ (accessed Nov. 12, 2020).
78. R. Ciriminna, A. Fidalgo, R. Delisi, L. M. Ilharco, and M. Pagliaro, "Pectin production and global market," *Agro Food Industry Hi-Tech*, vol. 27, no. 5, pp. 17–20, 2016.
79. International Renewable Energy Agency, "Biogas cost reductions to boost sustainable transport," *IRENA*, 2017.
80. A. Dávila, M. Rosenberg, and C. A. Cardona, "A biorefinery for efficient processing and utilization of spent pulp of Colombian Andes Berry (*Rubus glaucus* Benth.): Experimental, techno-economic and environmental assessment," *Bioresource Technology*, vol. 223, pp. 227–236, Jan. 2017, doi: 10.1016/j.biortech.2016.10.050.

Section 3

Essential Oils

7 Application of Essential Oils in the Food Industry

Challenges and Innovation

Kanza Saeed, Imran Pasha,
Muhammad Farhan Jahangir Chughtai,
Zaryab Ali, Hina Bukhari, and Muhammad Zuhair

7.1 INTRODUCTION

Natural plant materials like herbs and spices are gaining consumer attraction due to their safety status in comparison to synthetic chemical additives (1). With increasing consumer awareness in the past few years, people are more inclined toward buying high-quality processed food with minimal chemical additives, thus raising the need to explore natural alternatives for commercial purposes (Sachithananthan et al., 2017).

For centuries, herbs and spices have been utilized in folk medications because of their curative properties. For culinary purposes, herbs were used to create distinctive flavors of food and to enhance their shelf life (2). Around the world, spices still hold a preeminent place in tribal culinary traditions. Different plant parts can be used for the extraction of essential oils, that is, leaves of peppermint, cinnamon bark, flower buds of clove, ginger rhizomes/roots, pepper fruit, and cardamom seeds (3).

Essential oils are aromatic plant secondary metabolites of volatile nature, usually synthesized by single-cell metabolic activities or involved in groups of special cells working in coordinated manner to produce these volatile liquids with special characteristics (4). Essential oils are complex formulations of multiple compounds, but the principal fraction is of monoterpenes, which contain the highest proportion of hydrocarbons (5). Essential oils are synthesized and accumulate in specialized structures near cavities, glandular trichomes, or epidermal cells, canals, secretory cells, or other propagating parts. Essential oils are usually associated with significant storage structures, like the anise fruit, cinnamon bark, wild bergamot flower, nutmeg seeds, mint leaves, vetiver roots, ginger rhizome, and sandalwood (6).

Environmental factors like photoperiod, temperature, cultivation practices, relative humidity, and irradiance have a dominant effect on essential oil composition. In addition to all these factors, essential oil extraction techniques impact not only its composition but also the propensity of essential oil constituents (7). Compositionally, essential oils are a highly volatile and complex mixtures of compounds like terpenoids and aromatic compounds. These metabolic by-products are used in beverages as primary flavoring agents (8).

Researchers and food processors are focusing on the application of essential oils in processed food products due to their ability to control food spoilages and pathogenic microorganisms. Vast research in the field has established the fact that these marvelous properties of essential oils are due to complex bioactive compounds like glycosides, phenols, alkaloids, steroids, tannins, and coumarins (8). Different herbs and spices have their own uniqueness with respect to their functional characteristics due to their diverse composition of bioactive compounds, like carvacrol and thymol in oregano, vanillin in vanilla, eugenol in cloves, allyl isothiocyanate in mustard, menthol in peppermint, allicin in garlic, etc. (9).

DOI: 10.1201/9781003462804-10

This data provides researchers with conclusive scientific evidences that these bioactive compounds have the ability to shield against multitudinous food spoilage and poisoning issues. Secondary metabolites are also regarded as competitive weapons of plants that enable them to combat against invaders, like bacteria, fungi, insect, as well as herbivorous animals. Secondary metabolites like saponins, anthraquinones, glucosinolates, glycosides, tannins, polyacetylenes, anthraquinones, cyanogen, and terpenes, when released in the environment, act as allelochemicals, having a detrimental physiological impact on competitor species (10). These essential oils also attract insects for pollination, thus positively affecting growth and the development of proximate plants (11). The current review is an attempt to explore the utilization of essential oils in commercial food processing as a safer alternative to hazardous chemical additives.

7.2 MECHANISM OF ACTION OF ESSENTIAL OILS

Past research data regarding antimicrobial potential of essential oils and their bioactive compounds provide various different concepts about their working mechanism against microbial cells. The most convincing concept regarding the antimicrobial potential of essential oil is diversity in its composition (12). Every functional group has a peerless way of interaction with the components of target bacterial cell. So a single mechanism cannot define the exact mode of action of essential oils against bacterial cell (13).

All target operations within the cell are interlinked; among them, some cellular mechanisms are directly disrupted by the bioactive compounds of essential oils. Disturbance in one system consequently interferes in other cellular activities, that is, indirect disruption (14). Among these, the most delicate biochemical activities are cytoplasmic membrane remodeling or modification, bacterial cell wall disintegration, cell content extravasation, cytoplasmic coagulation, membrane protein degeneration, and diminution of proton motive force (15).

Bioactive compounds of essential oil are hydrophobic in nature. This hydrophobicity is the key reason for bacterial cell membrane lipid segmentation, resulting in configurational disfigurement, which leads to exacerbated permeability. This structural tempering causes cellular ionic loss and content leakage (16). Minor content and ion loss do not significantly affect the bacterial cell viability, whereas extensive content loss, particularly indispensable ionic and molecular exhaustion, leads to cellular death. For example, numerous studies have provided clear evidence that tea tree essential oil causes lysis or cell death of *Escherichia coli* strains (17).

Hostility of essential oils against microorganisms depends on the concentration of its phenolic content, excessively due to a high percentage of phenolic content, like thymol, menthol, caravel, and eugenol(2-methoxy-4-(2-propenyl)), which results in a stronger antimicrobial potential of essential oils (18). The mode of action of all these phenolic compounds is more or less similar to each other, including destruction of active electron transport flow, cytoplasmic membrane disruption, disturbance in proton motive forces, and cell content coagulation (15).

The meticulous microbial control mechanism of essential oils depends on the versatility of the chemical framework of individual essential oils. Scientific studies provide evidence that the presence of certain functional groups in phenolic compounds strongly impacts its antimicrobial potential, like the hydroxyl group of menthol being the key reason of its antimicrobial strength (19). However, the location of the -OH group on the phenolic ring does not influence its activity. The non-phenolic components of essential oil also play an active role in restraining microbial activity, for example, alkyl groups. 1-methyl-4-(1-methylethenyl)-cyclohexene or limonene has far more vigorous antimicrobial potential than *p*-cymene (20).

The bioactive compounds of essential oil ambush cellular proteins embedded in the cytoplasmic membrane. Bacterial cell membrane embeds some indispensable enzymes, for example, ATPase enzyme cushioned and hedged by lipid molecules. These protected enzymes can be susceptible to essential oils in two different ways. First, they are prone to cyclic hydrocarbon attack. These cyclic hydrocarbons are of lipophilic nature, making it easier for them to accumulate in the

TABLE 7.1
Mode of Action of Essential Oil against Different Bacterial Cell Component

Cell Component	Action	Reference
Cell membrane	Inhibition of integrity results in leakage of ions	(23)
DNA	DNA damage resulting in disruption of cell cycle regulation	(24)
Ribosomes	Ribosome dysfunction due to oxidation of protein	(25)
Mitochondria	Mitochondrial perturbation resulting in: • Membrane damage • Apoptosis • Cytotoxicity • Uncoupling of proton translocation • Interrupt ADP phosphorylation	(26)
QS factor	Interferes with QS factors and reduces: • Biofilm-forming potential • Proteolytic activity • Swarming	(27)
Membrane-associated enzymes	Protein denaturation and inhibition of ergosterol biosynthesis	(28)

cytoplasmic membrane lipid bilayer, prompting lipid–protein interaction disfigurement (21). The second attacking mechanism is based on head-on interaction between the hydrophobic enzyme portion with the lipophilic hydrocarbons. Some essential oils trigger pseudomycelia formation, that is, cellular growth in series with end-to-end adhesion. The antimicrobial agents of essential oil cause partial separation between neonate yeast cells by interaction with the enzymes responsible for energy regulation and synthesizing its cellular components. This is the prime mode of interaction between vital cellular enzymes and the bioactive compounds of essential oils (22). An overview about the mode of action of essential oil against different bacterial cell components is presented in Table 7.1.

The antimicrobial potential of essential oil can be determined through various techniques, like disc diffusion assay, broth dilution method, bioautography, etc. (29). Scientific literature also substantiates that these principles and techniques are used to evaluate antimicrobial potential, but to date, no authenticated procedure has been designed to anticipate the results of the effectiveness of essential oils against foodborne pathogens and spoilage agents (30). The National Committee for Clinical Laboratory Standards provided the scientific community with antibacterial susceptibility tests for antibiotics and essential oils. Researchers are focusing on selecting better experimental techniques to broaden their scope of application in various fields of life sciences (31). But factors like essential oil extraction method, inoculum size, volume of growth phase, nutrient medium, pH, and incubation time–temperature combination influence the selection of procedure. Extensive research data in the field makes it even harder to compile the outcomes and compare all the results together, as a large number of scientists are still investigating the antimicrobial capability of essential oils (32).

In the 1990s, research-based data regarding essential oil characterization and its potential application was insufficient. In that era, due to less consumer awareness, commercial processors were using cheaper synthetic preservatives. Improved research in the field increased consumer awareness and concerns regarding the health risks of synthetic chemical food additives, raising the urge to find natural and safer alternatives (33). At that time, natural preservatives gained scientific attention. After extensive research, the application of extracts and essential oils from herbs and spices in food preparation gained importance (34).

Initially, *in vitro* testing revealed that essential oils are required in higher concentrations in food preparations to achieve comparable results to that of synthetic preservatives (35). An old study conducted in 2001 provided the data regarding food and essential oil ratio to attain the desired results for commercial application, that is, 2-fold in skimmed milk, 100-fold in soft cheese, 10-fold in sausage, whereas 50-fold for soup preparations. *Aeromonas hydrophila* was an exception, as very low essential oil concentrations effectively controlled microbial proliferation in lettuce and pork samples (36).

Various scientists investigated and unveiled the numerous characteristics of essential oils as natural preservatives. In addition, researchers focused on the factors supporting microbial resistance in food systems. The interaction between food constituents and microbes enables them to withstand the detrimental effects of essential oils (37). Microbes grown on nutrient-rich medium under favorable laboratory conditions have a stronger ability of instantaneous repair. In food systems, both intrinsic (preservatives, salt, fats, water, antioxidants, H+, and OH-concentration) and extrinsic (packaging condition, that is, vacuum, air, gaseous combinations, temperature) factors greatly impact bacterial sensitivity (38).

It was observed that food that is packed in acidic and oxygen-deficit environment and kept in cold storage supports the antimicrobial potential of essential oils (39). This is because a low pH boosts the hydrophobicity of essential oil, which in turn improves essential oil solubility in the cytoplasmic membrane lipid matrix of the target microbial cell. However, research studies have provided evidence that eminent fat and protein content retard the activity of essential oil against microbes (40). Further research supports the concept that essential oil solubilization in food lipid matrix depresses its potential to assail microorganism residing in the aqueous phase of the food matrix. Food medium with lower water content does not support essential oil progression to target microbial cell site (41).

With further research in the field, scientists designed improved technologies to reinforce essential oil delivery mechanism to the target cell. Immobilized cilantro essential oil gel exhibited far more intense attacking ability against *Listeria monocytogenes* (42). Encapsulated rosemary essential oil has better antimicrobial potential against *L. monocytogenes* than non-encapsulated essential oil (43). The reason behind this improvement is that encapsulation forbids loss of volatile compounds, thus maintaining high essential oil concentration (44).

7.3 ESSENTIAL OILS IN DIFFERENT FOOD PRODUCTS

7.3.1 Meat and Meat Products

Autochthonous spoilage of meat microflora can be controlled using essential oil from thyme, oregano, eugenol, clove, and coriander; essential oils cause notable initial viable cell number reduction (45). According to research, meat fat affects essential oil bioactivity (46). The implementation of hurdle technology for meat preservation involving the utilization of essential oils in combination with novel packaging techniques improves the preservation potential of essential oils (47).

A study proved the application of oregano essential oil in modified atmosphere packaging for shelf life extension of fresh chicken breast meat stored at refrigeration temperature. Two different essential oil concentrations, that is, 0.1% and 1% (weight by weight) of essential oil, were tested in two different modified atmosphere packaging designs, that is, (30:70, CO_2:N_2) and (70:30, CO_2:N_2). Results depicted that 1% oregano essential oil gave better results in both modified atmospheric conditions in a 25-day trial study. During the trial period, 1% essential oil kept the cell count below critical levels (7 log cfu/g) (48). *Satureja horvattii* essential oil (thymol and p-cymene) in a 96 h storage study showed effective results of pork preservation against *L. monocytogenes* (49). Another study proved that carnosic acid and thymol, when applied to meat and meat products, penetrate and accumulate in muscle fiber, thus providing long-term microbial protection to meat (50).

7.3.2 Seafoods

Fats reduce the bioactive potential of essential oils in seafood in a similar manner to meat and meat products. According to a report, 0.05% volume by weight oregano essential oil showed better results against *Photobacterium phosphoreum* in cod fillets than salmon due to the higher fatty profile of salmon (45). Numerous studies proved that the synergetic effect of essential oil combinations improves the microbial stability of fat-rich seafood products (51).

A storage study of fish burger proved that the mixture of thymol, grapefruit seed, and lemon essential oil in a concentration 0.11%, 0.10%, and 0.12%, respectively, enhanced its shelf life up to 40%. Further combined effect of these essential oils in modified atmospheric packaging was evaluated. For this purpose, three different gas mixtures (5:95 $O_2 : CO_2$, 50:50 $CO_2 : O_2$, and 30:40:30 $N_2 : CO_2 : O_2$) were used and found to be more effective against mesophilic bacteria than the product packaged in air bag (39).

7.3.3 Fruits and Vegetables

In case of pasteurized fruit juices, bacterial enzymes are the prime reason of spoilage. The addition of essential oils of aromatic plants like lemongrass, ginger, etc. in melon, pear, and apple juice has been found to be effective against *E. coli*, *Salmonella* spp., and *Listeria* spp. (52). Upon further research, Raybaudi-Massilia et al. (53) found that keeping the quality of fresh "Piel de Sapo" melon can be enhanced using malic acid in combination with gingerol, citral, and eugenol.

Edible food coatings containing essential oils incorporated in an alginate base have been found to be effective in a 21-day study against *Salmonella enteritidis* (54). In another study, apple puree–alginate edible coating impregnated with oregano, lemongrass, and vanillin essential oils enhanced Fuji apples' shelf life: 0.3% w/w vanillin conserve their organoleptic characteristics for 21 days at refrigeration temperature (55). Cinnamaldehyde and thymol protect alfalfa seeds against *Salmonella* contamination (56). According to another study, oregano essential oil–treated (7– 21 μL/g) eggplant salad impeded *E. coli* O157:H7 contamination, while 0.75 μL/g carvacrol protect rice seed against *Bacillus cereus* (57).

7.3.4 Dairy Products

In dairy products, *Pseudomonas* spp., *E. coli*, *Salmonella senftenberg*, and *Staphylococcus aureus* can be controlled using D-limonene, terpeneless lime, geraniol, and terpinol. In an *in vitro* study, terpineol was used in combination with orange essential oil to extend milk keeping quality. However, the fat content of milk remarkably reduces the antimicrobial potential of terpineol. The results proved that terpineol was most effective in skimmed milk (7 log cfu/mL), then in low-fat milk (4 log cfu/mL), while it was least effective in full cream milk (3 log cfu/mL). Direct application of thyme essential oil and its nanoemulsions has been found to be effective in controlling Gram-positive bacteria (*Enterococcus hirae*, *S. aureus*, and *Bacillus licheniformis*) in contaminated milk samples. In addition, the physicochemical characteristics of milk were also meliorated (46). A similar study proved that thyme, clove, and cinnamon essential oil at a concentration of 0.5% tremendously controlled *S. enteritidis* growth statistics in low-fat cheese; however, the addition of these aromatic essential oils altered the cheese sensory profile (58).

Yangilar and Yildiz (59) conducted chemical, microbiological, and sensorial assessment of probiotic yogurt-infused ginger essential oil at a concentration of 0.2% and chamomile essential oil at a concentration of 0.4%. The results exhibited that yogurt with 0.4% ginger essential oil produced 8.01 cfu/mL, while control samples without essential oil had 8.32 cfu/mL. The results depicted that chamomile and ginger essential oil incorporation in yogurt enhanced preservation impact due to their strong antimicrobial and antioxidant properties.

In another study, Jemaa et al. (60) implemented nanotechnology for milk preservation by preparing *Thymus capitatus* essential oil nanoemulsions and incorporating them in semi-skimmed ultra-high-temperature (UHT) treated milk; the resultant product had improved physiological and microbiological properties, controlled the growth and proliferation of spoilage organisms, and thus improved the oxidative and fermentative stability of UHT-treated semi-skimmed milk. This study proves the fact that natural preservatives like essential oils give better results in comparison to artificial preservatives.

7.3.5 Cereal-Based Products

In bread, the major shelf life–limiting factor is stalling due to fungal spoilage, particularly caused by molds. In order to improve its quality, different natural compounds are added to it. According to Rehman et al. (61), the application of citrus peel essential oils in bread impacts its sensory characteristics and impedes microbial growth. The preservative effect of orange peel essential oil can be enhanced by spraying it on bread. A study on fresh amaranth-based pasta proved that natural bioactive compounds limonene, chitosan, and thymol ameliorate microbial stability at concentrations of 2,000 ppm and 4,000 ppm under refrigerated conditions (39).

Authors in (62) worked on oregano, thyme, and basil essential oils to retard the activity of *B. cereus* in rice-based food products. A study conducted by (63) on active packing designed for modified atmosphere packaging with cinnamon essential oil to extend the shelf life of gluten-free sliced bread resulted in improved sensory quality of bread. Active packaging also inhibited microbial growth in bread slices. The antimicrobial and antioxidant potentials of herbs and spices like sage, rosemary, anise, and black cumin essential oil and phenolic compounds have been found to be effective against Gram-positive and Gram-negative bacteria in increasing the shelf life of some bakery products (64). Table 7.2 provides a summary of the application of essential oils in different food products.

7.4 ORGANOLEPTIC ASPECTS OF ESSENTIAL OIL

While using essential oil in edible commodities as a preservative, its strong organoleptic characteristics should be considered first (65). The sensory characteristics of the foods strongly seasoned with herbs and spices remain unaffected (66). According to a study, oregano essential oil in small quantity was added in processed fish product without disturbing its sensory characteristics. However, 0.8% oregano essential oil in beef fillet resulted in undesirable flavor (67), whereas 0.05% concentration oregano essential oil gave cod fish fillet a notable pleasant flavor (68).

Thyme essential oil coating at a concentration of 0.9% (of total mass of shrimp) on cooked shrimp does not affect its organoleptic characteristics. However, when this essential oil concentration is elevated to 1.8% to improve antimicrobial properties, the sensory acceptability of shrimps was curtailed (43). Essential oils are commercially used as a flavoring agent in processed food commodities due to their characteristic aromatic properties (32). However, concentration of the oil should be adjusted at an optimum level to attain maximum preservation effect without disturbing the sensory food characteristics (69). Each essential oil has distinct aromatic properties, depending on its predominant aromatic compound. For example, geranial has rose-like fragrance, citral imparts a lemon-like aroma, and caravol gives a pungent aroma (70).

7.5 LEGAL ASPECTS OF ESSENTIAL OIL

The European Commission has registered essential oil bioactive components as a commercial flavoring agent (71). Some of the most commonly used natural flavoring agents are cinnamaldehyde, citral, thymol, menthol, caravol, limonene, eugenol, and carron, with no antagonistic human health effects (72). The European Union (EU) validates the safety profile of natural agents through toxicological and microbial tests before registrations (Kiewnick, 2007). These registered flavors are then added to the Everything Added to Food in the United States (EAFUS) list. This is a Food and Drug

TABLE 7.2
Summary of the Application of Essential Oils in Different Food Products

Product	Essential Oil	Target Microorganism	Modern Technology	Examples of Food Products	Reference
Meat and meat products	Thyme, oregano, eugenol, clove, and coriander essential oils; winter savory essential oil	*Listeria monocytogenes*	Modified atmosphere packaging	Fresh chicken, breast meat, pork preservation	(45)(46) (47) (48)(49)(50)
Seafoods	Oregano, thymol, grapefruit seed, and lemon essential oil	*Photobacterium phosphoreum*; mesophilic bacteria	Modified atmosphere packaging	Cod fillets, salmon	(45)(51)(39)
Fruits and vegetables	Lemongrass, ginger, oregano, lemongrass, and vanillin essential oils	*Salmonella* spp., *Listeria* spp., *Salmonella enteritidis*, *Escherichia coli* O157:H7, *Bacillus cereus*	Edible food coatings	Pear and apple juice, Piel de Sapo melon, Fuji apples, eggplant salad	(52)(53)(54) (55)(56)(57)
Dairy products	Orange, thyme, clove, and cinnamon essential oil	*Pseudomonas* spp., *Escherichia coli*, *Salmonella senftenberg*, *Staphylococcus aureus*, *Enterococcus hirae*, *Staphylococcus aureus*, *Bacillus licheniformis*, *Salmonella enteritidis*	Nanoencapsulation	Skimmed milk, low-fat milk, full cream milk, low-fat cheese	(46)(58)
Cereal-based products	Citrus peel, oregano, thyme, basil, cinnamon, sage, rosemary, anise, and black cumin essential oil	*Bacillus cereus*, fungi	Active packaging	Bread, fresh amaranth-based pasta, rice-based food products, gluten-free sliced bread, bakery products	(61),(39)(64), (62)

Administration–approved list that also includes food additives with "generally recognized as safe" (GRAS) status. There are a few exceptions, like estragole, which is in the EAFUS list but prohibited on the EU list due to its genotoxic and carcinogenic nature (73).

7.6 ESSENTIAL OIL NANOENCAPSULATION

Water-insoluble, hydrophobic essential oils do not perform efficiently in real food systems in comparison to *in vitro* designed models (12). This performance diminution is more pronounced in high-fat foods like butter, margarine, mayonnaise, milk, etc. (74). A study proved that to attain comparable efficacy of essential oil in cheese to that of *in vitro* models, the concentration of essential oil needs to be increased 100-fold (75). The key reason behind this inefficacy is the high dissolution of essential oils in the lipid phase, which decreases their concentration in the aqueous phase (76), while bacteria prefer an aqueous environment for proliferation (77). In order to attain similar antimicrobial potential in *in vivo* and *in vitro* systems, essential oils should be delivered to maintain a direct contact with microbes (78).

To achieve this purpose, essential oil nanoencapsulation is designed with a hydrophilic coating to encapsulate the hydrophobic essential oil, which is harbored in the hydrophilic capsule core (79). This nanoencapsulation precludes interactions between essential oils and food constituents (80). As a major proportion of essential oil binds with hydrophobic food components, it makes it unavailable for combat against target microbes (81), whereas nanodispersed essential oil is uniformly spread in food matrix and released locally to maintain its concentration to a level sufficient to obliterate spoilage bacteria (82). Entrapped essential oil molecules due to their limited interaction with food components do not impact organoleptic food characteristics negatively. So nanoencapsulation also resolves this issue, which limits the commercial use of essential oil as a food preservative (44).

The antibacterial potential of essential oils is their pivotal attribute. In order to intensify their potency, different methods were tested; among these, nanoencapsulated essential oils showed prominent enhancement in antibacterial efficiency (83). In order to achieve better results, we need to adopt nanoencapsulation techniques which boost the pathogenic bacteria–fighting potential of essential oil (84). Nanoencapsulation acts as an essential oil carrier across microbial cell membrane; these carriers overcome the hydrophobic limitations of essential oils (79). Nanoencapsulated essential oil easily crosses the lipid bilayer and wields their antimicrobial activities from within the cell. As discussed earlier, essential oils disrupt the bacterial membrane, resulting in the loss of vital ions, enzymes, and nutrient and compromising bacterial cell viability, leading to cellular death (85). The antibacterial potential of essential oils is restricted due to its uneven distribution in the cellular cytoplasm because of poor water solubility (83).

As a matter of fact, essential oil nanoencapsulations ameliorate their antimicrobial potential to a level that surpasses its antibiotic efficacy (86). Moreover, nanoencapsulated essential oils can be a promising solution to the global issue of controlling drug-resistant bacterial strains (87). The mode of action of antibiotics includes ribosomal blocking that halts protein synthesis necessary for cellular metabolic and survival activities, resulting in cell death (88). However, resistant strains mutate their drug reception sites in such a way that even in the presence of drugs, cellular metabolic activities and ribosomal formation continue (89). However, the mode of action of nanoencapsulated essential oils is based on bacterial cytoplasmic membrane disruption. So bactericidal active sites remain uninterrupted even in case of bacterial genetic mutations (12). Accordingly, the nanoencapsulation-based delivery system of essential oils improves its efficacy against food spoiling and pathogenic bacteria and does not support resistant strain development, making it an ideal remedy to the global risks posed by drug-resistant bacteria formed due to antibiotic abuse (90).

According to a study, *Chamaemelum* and *Lavandala* essential oils altered yogurt sensory characters to an unacceptable level for the sensory panelists (91). The essential oil binds with the fat content of yogurt in such a way that it negatively alters the yogurt's organoleptic profile (60). In comparison, nanoencapsulated *T. capitatus* essential oil does not interfere with the sensory characteristics of milk, as fat droplets entrap essential oil, restricting its interaction with milk ingredients. Thus, essential oil incorporation does not alter milk fat (92). Figure 7.1 enlists the benefits of nanoencapsulated essential oils.

7.7 ANTIMICROBIAL PACKAGING

Essential oils have proved their significance in microbial population control procedures. In the food industry, essential oils usually have a specified target microorganism in order to achieve enhanced efficiency, resulting in safe and high-quality products with maximum durability (93). Numerous studies have proved the efficiency of essential oils as part of active packaging design to extend the shelf life of processed food (94). The demand for active packaging material with an antimicrobial agent increased with time, as consumer preferences are more inclined toward the usage of minimally processed food with natural additives. Hence, food processors started using essential oils as active antimicrobial agents in food packaging material (95). As a result, bioactive molecules of

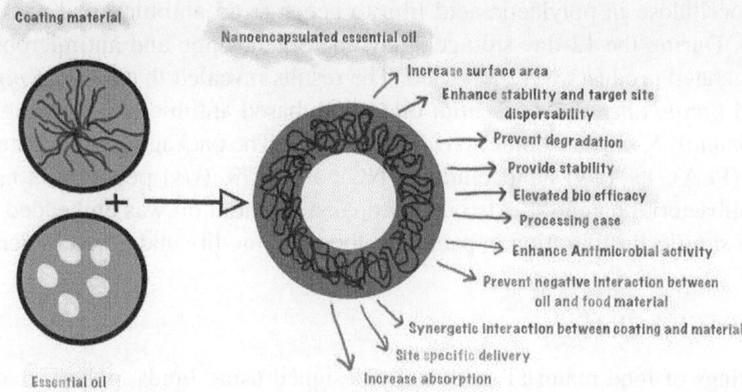

FIGURE 7.1 Benefits of nanoencapsulated essential oils.

essential oils are being comprehensively studied to establish their antibacterial, antioxidant, antiviral, antifungal, and insecticidal attributes (96).

In order to incorporate essential oil in packaging material, usually three different methods are adapted: (1) direct loading of essential oil in antibacterial pouch (97), (2) mixing of essential oil in packaging material (98), and (3) essential oil coating in food packages (99). Among these three techniques, essential oil integration in packaging material is the best technique for commercial applications (Vasile et al.) (100). Biodegradable packaging polymers (chitosan, polylactic acid, skate skin gelatin) with antimicrobial agents have been found to be an effective solution to increasing issues of foodborne diseases and global environmental concerns (101).

Authors in (102) used various concentrations of *Origanum vulgare* essential oil in polylactic acid film in order to assess the extent of microbial protection potential of ready-to-eat salad. This active material has been found to be effective in controlling yeast and mold when 5% and 10% essential oils are incorporated in the packaging film. Polylactic acid (PLA) films with 5% and 10% essential oil has acceptable physical and mechanical stability.

Authors in (103) worked on the antimicrobial, antioxidant, physical, and thermal characteristics of quince seed mucilage film with 1% *O. vulgare* essential oil. This film has been found to be ineffective against *Pseudomonas aeruginosa* and *Salmonella typhimurium* but exhibited exponential results against *S. aureus* and *E. coli*. Montmorillonite nanoclay–based nanocomposite films with thyme essential oil have been found to be effective against *E. coli* and *Salmonella typhi*. Skate skin gelatin films with 1% thyme essential oil protect chicken tenderloin against *E. coli* 0157: 47 and *L. monocytogenes*.

Similarly, authors in (104) subsumed rosemary essential oil in chitosan films. Chitosan films have aspirating and antimicrobial properties, and the rosemary essential oil incorporation further enhanced the antimicrobial characteristics of chitosan active packaging. Essential oils have also been found compatible with non-biodegradable polymers used for the packaging of commercial food products. But the conventional techniques for essential oil incorporation in polymeric packaging material had their own limitations, whereas techniques like electrospinning have been found to be effective in essential oil incorporation in polymeric packaging films. Electrospinning techniques generate nanofibers by generating an electric field to melt polymers to form solutions for nanofiber preparation. Nanofibers produced using this technique are highly porous, with diameter in the submicron range and a high surface area.

Authors in (105) implemented the electrospinning technique to create antimicrobial food packaging material by integrating tea tree essential oil and beta-cyclodextrin in polyethylene oxide film. Authors in (106) used the same technique to pin klein grass and peppermint essential oil along with nanocellulose in polylactic acid film to generate an antimicrobial packaging material for ground beef. During the 12-day-storage study, the organoleptic and antimicrobial characteristics of the refrigerated products were assessed. The results revealed that *Mentha piperita* essential oil (MPO)– and *Bunium persicum* essential oil (BPO)–based antimicrobial packaging is effective against *Pseudomonas*, *S. aureus*, and *Enterobacteriaceae*. The packaging film containing 1% (w/v) polylactic acid (PLA), 1% (v/v) nanocellulose (NC), and 0.5% (v/v) peppermint essential oil has the strongest antibacterial potential. Also, 20% copaiba essential oil was embedded in PLA film to control *Bacillus subtilis* proliferation in packaged food commodity and gave efficient results (107).

7.8 EDIBLE PACKAGING

The edible coatings of food material are usually designed using lipids, polysaccharides, and proteins (108). Polysaccharide- and protein-based films have good mechanical strength, but because of its hydrophilic nature, these films are water-permeable (109). To enhance their hydrophilic potential, these films are impregnated with calcium caseinate (110). A water-soluble polymer was commercially obtained through acid precipitation of casein. Sodium caseinate (SC)–based packaging materials also add to the nutritional and sensory profile of packaged food. Casein-based edible films also provide an adequate climatic shield for packaged food products. However, this polymer does not provide a strong and reliable water barrier (111). In order to enhance vapor permeability of sodium caseinate films, oleic acid, along with beeswax, is mixed into it (112). Essential oils like ginger and cinnamon essential oil improve the antimicrobial characteristics of SC-based edible packaging (113). Cinnamon and ginger essential oils remain homogeneously distributed into the SC matrix, providing a textural evenness to the film. The inclusion of essential oil affects the optical characteristics of SC films. In drying procedures due to lipid aggregation, ginger essential oil escalates surface cragginess, which leads to the loss of surface sheen. Due to low oxygen and humidity permeability, these films were found appropriate to protect sunflower oil against oxidative environmental stress (114, 115).

7.9 CONCLUSIONS

Essential oils are plant secondary metabolites that enable them to combat against invaders like bacteria, fungi, insect, as well as herbivorous animals. Essential oils have been found to be effective against both pathogenic and food spoilage agents. Bioactive compounds of essential oils disrupt the cellular mechanisms directly or indirectly, and disturbance in one system consequently interrupts other cellular activities, making them suitable for application in edible commodities as preservative. But the strong organoleptic characteristics of aromatic compounds in essential oils limit their incorporation in processed food as a preservative. This problem can be overcome by the use of nanoencapsulated essential oils that provide far better results in commercial food preservation. Essential oil–based antimicrobial and edible packaging not only exalt product shelf life but also strengthen consumer confidence in processed food products. This scientific approach reinforces the Green Earth concept.

DISCLOSURE STATEMENT

No potential conflict of interest was reported by the author(s).

REFERENCES

1. S.C. Lourenço, M. Moldão-Martins and V.D. Alves. Antioxidants of natural plant origins: from sources to food industry applications. *Molecules*, 24(22), 4132 (2019). Crossref. PubMed. ISI.

2. D. Diksha, S. Upadhyay, S. Singh and P. Gururani. Enhancing the flavour of oil by infusion of different herbs-review. *The Pharma Innovation Journal*, 8(5), 626–632 (2019).
3. R.K. Singh, K.B. Pandey and S.I. Rizvi. Medicinal properties of some Indian spices. *Annual Phytomedicine*, 1(1), 29–33 (2012).
4. D.M. Pott, S. Osorio and J.G. Vallarino. From central to specialized metabolism: an overview of some secondary compounds derived from the primary metabolism for their role in conferring nutritional and organoleptic characteristics to fruit. *Frontiers in Plant Science*, 10 (2019). Crossref.
5. G. Ben Salha, R. Herrera Díaz, O. Lengliz, M. Abderrabba and J. Labidi. Effect of the chemical composition of free-terpene hydrocarbons essential oils on antifungal activity. *Molecules*, 24(19), 3532 (2019). Crossref.
6. M.A. Hanif, S. Nisar, G.S. Khan, Z. Mushtaq and M. Zubair. Essential oils. *Essential Oil Research*. Edits., Malik, Sonia. pp. 3–17, Cham: Springer (2019). Crossref.
7. R. Mickiene, B. Bakutis and V. Baliukoniene. Antimicrobial activity of two essential oils. *Annuals of Agricultural and Environmental Medicine*, 18(1), 139–144 (2011). PubMed.
8. W.Y. Huang, Y.Z. Cai and Y. Zhang. Natural phenolic compounds from medicinal herbs and dietary plants: potential use for cancer prevention. *Nutrition and Cancer*, 62(1), 1–20 (2009). Crossref. ISI.
9. H.A. Shaaban, A.H. El-Ghorab and T. Shibamoto. Bioactivity of essential oils and their volatile aroma components. *Journal of Essential Oil Research*, 24(2), 203–212 (2012). Crossref. ISI.
10. P. Mambanzulua Ngoma, S. Hiligsmann, E. Sumbu Zola, M. Ongena and P. Thonart. Impact of different plant secondary metabolites addition: saponin, tannic acid, salicin and aloin on glucose anaerobic co-digestion. *Fermentation Technology*, 4(1), 1–11 (2015).
11. P. Satyal, H.V. Hieu, N.T.H. Chuong, N.H. Hung, T.A. Tai, V.T. Hien and W.N. Setzer. Chemical composition, Aedes mosquito larvicidal activity, and repellent activity against *Triatoma rubrofasciata* of *Severinia monophylla* leaf essential oil. *Parasitology Research*, 118(3), 733–742 (2019). Crossref. PubMed.
12. S. Chouhan, K. Sharma and S. Guleria. Antimicrobial activity of some essential oils—present status and future perspectives. *Medicines*, 4(3), 58 (2017). Crossref.
13. H. Sakkas, P. Gousia, V. Economou, V. Sakkas, S. Petsios and C. Papadopoulou. In vitro antimicrobial activity of five essential oils on multidrug resistant Gram-negative clinical isolates. *Journal of Intercultural Ethnopharmacology*, 5(3), 212 (2016). Crossref. PubMed.
14. A.A.A. Almarie. Roles of terpenoids in essential oils and its potential as natural weed killers: recent developments. *Essential Oils-Bioactive Compounds, New Perspectives and Applications*. Edits., de Oliveira, Mozaniel Santana. pp. 189–210, IntechOpen (2020). Crossref.
15. F. Nazzaro, F. Fratianni, L. De Martino, R. Coppola and V. De Feo. Effect of essential oils on pathogenic bacteria. *Pharmaceuticals*, 6(12), 1451–1474 (2013). Crossref.
16. Y. Wu, J. Bai, K. Zhong, Y. Huang, H. Qi, Y. Jiang and H. Gao. Antibacterial activity and membrane-disruptive mechanism of 3-p-trans-coumaroyl-2-hydroxyquinic acid, a novel phenolic compound from pine needles of Cedrus deodara, against *Staphylococcus aureus*. *Molecules*, 21(8), 1084 (2016). Crossref. ISI.
17. L. Wang, C. Hu and L. Shao. The antimicrobial activity of nanoparticles: present situation and prospects for the future. *International Journal of Nanomedicine*, 12, 1227 (2017). Crossref. PubMed. ISI.
18. T.H. Wang, S.M. Hsia, C.H. Wu, S.Y. Ko, M.Y. Chen, Y.H. Shih . . . C.Y. Wu. Evaluation of the antibacterial potential of liquid and vapor phase phenolic essential oil compounds against oral microorganisms. *PLoS One*, 11(9), e0163147 (2016). Crossref. PubMed. ISI.
19. A.C. Guimarães, L.M. Meireles, M.F. Lemos, M.C.C. Guimarães, D.C. Endringer, M. Fronza and R. Scherer. Antibacterial activity of terpenes and terpenoids present in essential oils. *Molecules*, 24(13), 2471 (2019). Crossref. PubMed. ISI.
20. H.D. Dorman and S.G. Deans. Antimicrobial agents from plants: antibacterial activity of plant volatile oils. *Journal of Applied Microbiology*, 88(2), 308–316 (2000). Crossref. PubMed. ISI.
21. G. Fabre. *Molecular interaction of natural compounds with lipid bilayer membranes: Towards a better understanding of their biological and pharmaceutical actions* (Doctoral dissertation) (2015).
22. K.T. Gabriel, L. Kartforosh, S.A. Crow and C.T. Cornelison. Antimicrobial activity of essential oils against the fungal pathogens ascosphaera apis and pseudogymnoascus destructans. *Mycopathologia*, 183(6), 921–934 (2018). Crossref. PubMed. ISI.
23. A. Bouyahya, J. Abrini, N. Dakka and Y. Bakri. Essential oils of *Origanum compactum* increase membrane permeability, disturb cell membrane integrity, and suppress quorum-sensing phenotype in bacteria. *Journal of Pharmaceutical Analysis*, 9(5), 301–311 (2019). Crossref. PubMed. ISI.
24. L.H. Li, P. Wu, J.Y. Lee, P.R. Li, W.Y. Hsieh, C.C. Ho . . . S.M. Yang. Hinokitiol induces DNA damage and autophagy followed by cell cycle arrest and senescence in gefitinib-resistant lung adenocarcinoma cells. *PloS One*, 9(8), e104203 (2014). Crossref. PubMed.

25. S. Pattnaik and N. Behera. Interaction between aromatic oil components and bacterial targets. *Plant Biotic Interactions*. Edits., Varma, Ajit, Tripathi, Swati, and Prasad, Ram. pp. 211–226, Cham: Springer (2019). ISBN: 978-3-030-26656-1. Crossref.
26. L. Wang, N. Jiang, D. Wang and M. Wang. Effects of essential oil citral on the growth, mycotoxin biosynthesis and transcriptomic profile of alternaria alternata. *Toxins*, 11(10), 553 (2019). Crossref. ISI.
27. A. Bouyahya, N. Dakka, A. Et-Touys, J. Abrini and Y. Bakri. Medicinal plant products targeting quorum sensing for combating bacterial infections. *Asian Pacific Journal of Tropical Medicine*, 10(8), 729–743 (2017). Crossref. PubMed. ISI.
28. J. He, D. Wu, Q. Zhang, H. Chen, H. Li, Q. Han . . . W. Qin. Efficacy and mechanism of cinnamon essential oil on inhibition of *Colletotrichum acutatum* isolated from 'Hongyang' kiwifruit. *Frontiers in Microbiology*, 9, 1288 (2018). Crossref. PubMed. ISI.
29. G. Horváth, N. Jámbor, A. Végh, A. Böszörményi, É. Lemberkovics, É. Héthelyi . . . B. Kocsis. Antimicrobial activity of essential oils: the possibilities of TLC–bioautography. *Flavour and Fragrance Journal*, 25(3), 178–182 (2010). Crossref. ISI.
30. G. Luo, J. Tang, W. He, J. Wu, B. Ma, X. Wang and M. Fitzgerald. Methods for dilution antimicrobial susceptibility tests for bacteria that grow aerobically. Approved standard-methods for dilution antimicrobial susceptibility tests for bacteria that grow aerobically. Approved standard-2000. *Wound Repair and Regeneration*, 16(6), 800–804 (2008). Crossref. PubMed.
31. M. Balouiri, M. Sadiki and S.K. Ibnsouda. Methods for in vitro evaluating antimicrobial activity: a review. *Journal of Pharmaceutical Analysis*, 6(2), 71–79 (2016). Crossref. PubMed. ISI.
32. H.A. Shaaban. Essential oil as antimicrobial agents: efficacy, stability, and safety issues for food application. *Essential Oils-Bioactive Compounds, New Perspectives and Applications*. Edits., de Oliveira, Mozaniel Santana. pp. 153–186, IntechOpen (2020). Crossref.
33. M. Mitterer-Daltoé, J. Bordim, C. Lise, L. Breda, M. Casagrande and V. Lima. Consumer awareness of food antioxidants. *Synthetic vs. Natural. Food Science and Technology (AHEAD)*, 41(Suppl. 1), 208–212 (2020).
34. A. Bearth, M.E. Cousin and M. Siegrist. The consumer's perception of artificial food additives: influences on acceptance, risk and benefit perceptions. *Food Quality and Preference*, 38, 14–23 (2014). Crossref. ISI.
35. F. Saeed, M. Afzaal, T. Tufail and A. Ahmad. Use of natural antimicrobial agents: a safe preservation approach. *Active Antimicrobial Food Packaging*, 18(0), (2019).
36. A.K. Karatzas, E.P.W. Kets, E.J. Smid and M.H.J. Bennik. The combined action of carvacrol and high hydrostatic pressure on *Listeria monocytogenes* Scott A. *Journal of Applied Microbiology*, 90(3), 463–469 (2001). Crossref. PubMed.
37. J. Krisch, R. Tserennadmid and C. Vágvölgyi. Interaction of essential oils and food components: consequences in food preservation. *Essential Oils as Natural Food Additives: Composition, Applications, Antioxidant and Antimicrobial Properties*. Edits., Valgimigli, Luca. pp. 361–376, UK: Nova Science Publishers, Inc (2012).
38. C. Rolfe and H. Daryaei. Intrinsic and extrinsic factors affecting microbial growth in food systems. *Food Safety Engineering*. pp. 3–24, Cham: Springer (2020). Crossref.
39. A. Lucera, M.A. Del Nobile, C. Costa and A. Conte. Food applications of natural antimicrobial compounds. *Frontiers in Microbiology*, 3, 287 (2012). Crossref. PubMed.
40. L. Wang, J. Heising, V. Fogliano and M. Dekker. Fat content and storage conditions are key factors on the partitioning and activity of carvacrol in antimicrobial packaging. *Food Packaging and Shelf Life*, 24, 100500 (2020). Crossref. ISI.
41. M. Hyldgaard, T. Mygind and R.L. Meyer. Essential oils in food preservation: mode of action, synergies, and interactions with food matrix components. *Frontiers in Microbiology*, 3, 12 (2012). Crossref. PubMed. ISI.
42. M.M.M.D. De-Oliveira, D.F. Brugnera and R.H. Piccoli. Essential oils of thyme and rosemary in the control of *Listeria monocytogenes* in raw beef. *Brazilian Journal of Microbiology*, 44(4), 1181–1188 (2013). Crossref. PubMed.
43. B. Ouattara, S.F. Sabato and M. Lacroix. Combined effect of antimicrobial coating and gamma irradiation on shelf life extension of pre-cooked shrimp (*Penaeus* spp.). *International Journal of Food Microbiology*, 68(1–2), 1–9 (2001). Crossref. PubMed. ISI.
44. R. Becerril, C. Nerín and F. Silva. Encapsulation systems for antimicrobial food packaging components: an update. *Molecules*, 25(5), 1134 (2020). Crossref.
45. B. Speranza, M. R. Corbo and M. Sinigaglia. Essential oils for preserving perishable foods: possibilities and limitations. *Application of Alternative Food-Preservation Technologies to Enhance Food Safety and Stability*, 23, 35–37 (2010).

46. M. Perricone, E. Arace, M.R. Corbo, M. Sinigaglia and A. Bevilacqua. Bioactivity of essential oils: a review on their interaction with food components. *Frontiers in Microbiology*, 6, 76 (2015). Crossref. PubMed. ISI.
47. H.L. Degala, A.K. Mahapatra, A. Demirci and G. Kannan. Evaluation of non-thermal hurdle technology for ultraviolet-light to inactivate *Escherichia coli* K12 on goat meat surfaces. *Food Control*, 90, 113–120 (2018). Crossref.
48. E. Chouliara, A. Karatapanis, I.N. Savvaidis and M.G. Kontominas. Combined effect of oregano essential oil and modified atmosphere packaging on shelf-life extension of fresh chicken breast meat, stored at 4 C. *Food Microbiology*, 24(6), 607–617 (2007). Crossref. PubMed. ISI.
49. D. Bukvički, D. Stojković, M. Soković, L. Vannini, C. Montanari, B. Pejin . . . P.D. Marin. *Satureja horvatii* essential oil: in vitro antimicrobial and antiradical properties and in situ control of *Listeria monocytogenes* in pork meat. *Meat Science*, 96(3), 1355–1360 (2014). Crossref. PubMed. ISI.
50. M. Al-Hijazeen and M. Al-Rawashdeh. Preservative effects of rosemary extract (*Rosmarinus officinalis* L.) on quality and storage stability of chicken meat patties. *Food Science and Technology*, 39(1), 27–34 (2017). Crossref.
51. A. Hassoun and Ö.E. Çoban. Essential oils for antimicrobial and antioxidant applications in fish and other seafood products. *Trends in Food Science & Technology*, 68, 26–36 (2017). Crossref. ISI.
52. R.M. Raybaudi-Massilia, J. Mosqueda-Melgar and O. Martin-Belloso. Antimicrobial activity of essential oils on *Salmonella enteritidis*, *Escherichia coli*, and *Listeria innocua* in fruit juices. *Journal of Food Protection*, 69(7), 1579–1586 (2006). Crossref. PubMed. ISI.
53. R.M. Raybaudi-Massilia, J. Mosqueda-Melgar, R. Soliva-Fortuny and O. Martín-Belloso. Control of pathogenic and spoilage microorganisms in fresh-cut fruits and fruit juices by traditional and alternative natural antimicrobials. *Comprehensive Reviews in Food Science and Food Safety*, 8(3), 157–180 (2009). Crossref. PubMed.
54. J.F. Ayala-Zavala, G.A. González-Aguilar and L. Del-Toro-Sánchez. Enhancing safety and aroma appealing of fresh-cut fruits and vegetables using the antimicrobial and aromatic power of essential oils. *Journal of Food Science*, 74(7), R84–R91 (2009). Crossref. PubMed. ISI.
55. M.A. Rojas-Graü, R.M. Raybaudi-Massilia, R.C. Soliva-Fortuny, R.J. Avena-Bustillos, T.H. McHugh and O. Martín-Belloso. Apple puree-alginate edible coating as carrier of antimicrobial agents to prolong shelf-life of fresh-cut apples. *Postharvest Biology and Technology*, 45(2), 254–264 (2007). Crossref. ISI.
56. M. Friedman. Chemistry, antimicrobial mechanisms, and antibiotic activities of cinnamaldehyde against pathogenic bacteria in animal feeds and human foods. *Journal of Agricultural and Food Chemistry*, 65(48), 10406–10423 (2017). Crossref. PubMed. ISI.
57. S.A. Burt. *Antibacterial Activity of Essential Oils: Potential Applications in Food* (2007). https://dspace.library.uu.nl/bitstream/handle/1874/24273/full.pdf?sequence=6
58. G. Kavas, N. Kavas and D. Saygili. The effects of thyme and clove essential oil fortified edible films on the physical, chemical and microbiological characteristics of Kashar cheese. *Journal of Food Quality*, 38(6), 405–412 (2015). Crossref. ISI.
59. F. Yangilar and P.O. Yildiz. Effects of using combined essential oils on quality parameters of bio-yogurt. *Journal of Food Processing and Preservation*, 42(1), e13332 (2018). Crossref.
60. M.B. Jemaa, H. Falleh, M. Saada, M. Oueslati, M. Snoussi and R. Ksouri. *Thymus capitatus* essential oil ameliorates pasteurization efficiency. *Journal of Food Science and Technology*, 55(9), 3446–3452 (2018). Crossref. PubMed.
61. S.H. Salim-ur-Rehman, H. Nawaz, M.M. Ahmad, M.A. Murtaza and A.J. Rizvi. Inhibitory effect of citrus peel essential oils on the microbial growth of bread. *Pakistan Journal of Nutrition*, 6(6), 558–561 (2007). Crossref.
62. D. Budka and N.A. Khan. The effect of ocimum basilicum, thymus vulgaris, origanum vulgare essential oils on *Bacillus cereus* in rice-based foods. *Ejbs*, 2(1), 17–20 (2010).
63. L. Gutiérrez, R. Batlle, S. Andújar, C. Sánchez and C. Nerín. Evaluation of antimicrobial active packaging to increase shelf life of gluten-free sliced bread. *Packaging Technology and Science*, 24(8), 485–494 (2011). Crossref. ISI.
64. M. Basuny Amany, L. Nasef Shereen, A.M. Mahmoud Eman and M.A. Shaker. Use of medicinal and aromatic plants for increasing quality of some bakery products. *International Science and Investigation Journal*, 54, 820–828 (2012).
65. H. Sharma, S.K. Mendiratta, R.K. Agrawal, K. Gurunathan, S. Kumar and T.P. Singh. Use of various essential oils as bio preservatives and their effect on the quality of vacuum packaged fresh chicken sausages under frozen conditions. *LWT-Food Science and Technology*, 81, 118–127 (2017). Crossref.

66. A.G. Osman, V. Raman, S. Haider, Z. Ali, A.G. Chittiboyina and I.A. Khan. Overview of analytical tools for the identification of adulterants in commonly traded herbs and spices. *Journal of AOAC International*, 102(2), 376–385 (2019). Crossref. PubMed. ISI.
67. I. Cabarkapa, M. Skrinjar, N. Blagojev, J. Gubic, D. Plavsic, B. Kokic and T. Radusin. Effect of *Origanum heracleoticum* L. essential oil on marinated chicken meat shelf-life. *Journal of Pure and Applied Microbiology*, 7(1), 221–228 (2013).
68. P. Ramroop and H. Neetoo. Antilisterial activity of *Cymbopogon citratus* on crabsticks. *AIMS Microbiology*, 4(1), 67 (2018). Crossref. PubMed.
69. M. Pateiro, F.J. Barba, R. Domínguez, A.S. Sant'Ana, A.M. Khaneghah, M. Gavahian . . . J.M. Lorenzo. Essential oils as natural additives to prevent oxidation reactions in meat and meat products: a review. *Food Research International*, 113, 156–166 (2018). Crossref. PubMed. ISI.
70. R. Chizzola. Natural products. *Regular Monoterpenes and Sesquiterpenes (Essential Oils)*. Edits., Ramawat, K., and Mérillon, J.M. p. 10, Berlin, Heidelberg: Springer (2013). Crossref.
71. S. Gupta, S. Khan, M. Muzafar, M. Kushwaha, A.K. Yadav and A.P. Gupta, Encapsulation: entrapping essential oil/flavors/aromas in food. *Encapsulations*. Edits., Grumezescu, Mihai, and Grumezescu, Alexandru Mihai. pp. 229–268, Elsevier Inc. (2016). Crossref.
72. N. Hayatgheib, C. Fournel, S. Calvez, H. Pouliquen and E. Moreau. In vitro antimicrobial effect of various commercial essential oils and their chemical constituents on *Aeromonas salmonicida* subsp. salmonicida. *Journal of Applied Microbiology*, 129, 137–145 (2020). Crossref. PubMed.
73. M. Bošković, M.Ž. Baltić, J. Ivanović, J. Đurić, J. Lončina, M. Dokmanović and R. Marković. Use of essential oils in order to prevent foodborne illnesses caused by pathogens in meat. *Tehnologija Mesa*, 54(1), 14–20 (2013). Crossref.
74. A.E. Edris. Pharmaceutical and therapeutic potentials of essential oils and their individual volatile constituents: a review. *Phytotherapy Research: An International Journal Devoted to Pharmacological and Toxicological Evaluation of Natural Product Derivatives*, 21(4), 308–323 (2007). Crossref.
75. F. Salaün, Ed. *Microencapsulation: Processes, Technologies and Industrial Applications*. Intechopen (2019). Crossref.
76. S. Asadinezhad, F. Khodaiyan, M. Salami, H. Hosseini and B. Ghanbarzadeh. Effect of different parameters on orange oil nanoemulsion particle size: combination of low energy and high energy methods. *Journal of Food Measurement and Characterization*, 13(4), 2501–2509 (2019). Crossref.
77. X.F. Xin, K. Nomura, K. Aung, A.C. Velásquez, J. Yao, F. Boutrot . . . S.Y. He. Bacteria establish an aqueous living space in plants crucial for virulence. *Nature*, 539(7630), 524–529 (2016). Crossref. PubMed. ISI.
78. F. Donsì, A. Cuomo, E. Marchese and G. Ferrari. Infusion of essential oils for food stabilization: unraveling the role of nanoemulsion-based delivery systems on mass transfer and antimicrobial activity. *Innovative Food Science & Emerging Technologies*, 22, 212–220 (2014). Crossref. ISI.
79. N. Lammari, O. Louaer, A.H. Meniai and A. Elaissari. Encapsulation of essential oils via nanoprecipitation process: overview, progress, challenges and prospects. *Pharmaceutics*, 12(5), 431 (2020). Crossref. PubMed. ISI.
80. L. Jin, J. Teng, L. Hu, X. Lan, Y. Xu, J. Sheng . . . M. Wang. Pepper fragrant essential oil (PFEO) and functionalized MCM-41 nanoparticles: formation, characterization, and bactericidal activity. *Journal of the Science of Food and Agriculture*, 99(11), 5168–5175 (2019). Crossref. PubMed.
81. J. Xue. *Essential oil nanoemulsions prepared with natural emulsifiers for improved food safety* (Doctoral dissertations) (2015).
82. C.D. Ferreira and I.L. Nunes. Oil nanoencapsulation: development, application, and incorporation into the food market. *Nanoscale Research Letters*, 14(1), 9 (2019). Crossref. PubMed.
83. N. Basavegowda, J.K. Patra and K.H. Baek. Essential oils and mono/bi/tri-metallic nanocomposites as alternative sources of antimicrobial agents to combat multidrug-resistant pathogenic microorganisms: an overview. *Molecules*, 25(5), 1058 (2020). Crossref.
84. L. Fu, Z. Wang, O.P. Dhankher and B. Xing. Nanotechnology as a new sustainable approach for controlling crop diseases and increasing agricultural production. *Journal of Experimental Botany*, 71(2), 507–519 (2020). Crossref. PubMed.
85. F.A. Omonijo, L. Ni, J. Gong, Q. Wang, L. Lahaye and C. Yang. Essential oils as alternatives to antibiotics in swine production. *Animal Nutrition*, 4(2), 126–136 (2018). Crossref. ISI.
86. M.B. Jemaa, H. Falleh and R. Ksouri. Encapsulation of natural bioactive compounds: nanoemulsion formulation to enhance essential oils activities. *Microencapsulation-Processes, Technologies and Industrial Applications*. Edits., Salaün, Fabien, IntechOpen (2019). Crossref.

87. K. Ssekatawa, D.K. Byarugaba, C.D. Kato, F. Ejobi, R. Tweyongyere, M. Lubwama and E.M. Wampande. Nanotechnological solutions for controlling transmission and emergence of antimicrobial-resistant bacteria, future prospects, and challenges: a systematic review. *Journal of Nanoparticle Research*, 22(5), (2020). Crossref.
88. J. Lin, D. Zhou, T.A. Steitz, Y.S. Polikanov and M.G. Gagnon. Ribosome-targeting antibiotics: modes of action, mechanisms of resistance, and implications for drug design. *Annual Review of Biochemistry*, 87, 451–478 (2018). Crossref. PubMed. ISI.
89. D.T. Hoagland, J. Liu, R.B. Lee and R.E. Lee. New agents for the treatment of drug-resistant *Mycobacterium tuberculosis*. *Advanced Drug Delivery Reviews*, 102, 55–72 (2016). Crossref. PubMed. ISI.
90. D.G. Barrera-Ruiz, G.C. Cuestas-Rosas, R.I. Sánchez-Mariñez, M.L. Álvarez-Ainza, G.M. Moreno-Ibarra, A.K. López-Meneses and M.O. Cortez-Rocha. Antibacterial activity of essential oils encapsulated in chitosan nanoparticles. *Food Science and Technology*, 40, 568–573 (2020).
91. B.R. Ghalem and B. Zouaoui. Evaluation of the quality of steamed yogurt treated by Lavandula and Chamaemelum species essential oils. *Journal of Medicinal Plants Research*, 7(42), 3121–3126 (2013).
92. H. Majeed, Y.Y. Bian, B. Ali, A. Jamil, U. Majeed, Q.F. Khan and Z. Fang. Essential oil encapsulations: uses, procedures, and trends. *RSC Advances*, 5(72), 58449–58463 (2015). Crossref. ISI.
93. R.C. McGregor, K.A. Parker, J.M. Hornby and L.C. Latta. Microbial population dynamics under microdoses of the essential oil arborvitae. *BMC Complementary and Alternative Medicine*, 19(1), 1–9 (2019). Crossref. PubMed.
94. R. Ribeiro-Santos, M. Andrade, N.R. de Melo and A. Sanches-Silva. Use of essential oils in active food packaging: recent advances and future trends. *Trends in Food Science & Technology*, 61, 132–140 (2017). Crossref. ISI.
95. M.Z. Akram, S.Y. Fırıncıoğlu, H. Jalal and S.C. Doğan. The use of essential oils in active food packaging: a review of recent studies. *Turkish Journal of Agriculture-Food Science and Technology*, 7(11), 1799–1804 (2019). Crossref.
96. S. Tariq, S. Wani, W. Rasool, K. Shafi, M.A. Bhat, A. Prabhakar . . . M.A. Rather. A comprehensive review of the antibacterial, antifungal and antiviral potential of essential oils and their chemical constituents against drug-resistant microbial pathogens. *Microbial Pathogenesis*, 134, 103580 (2019). Crossref. PubMed. ISI.
97. S.J. Kwon, Y. Chang and J. Han. Oregano essential oil-based natural antimicrobial packaging film to inactivate *Salmonella enterica* and yeasts/molds in the atmosphere surrounding cherry tomatoes. *Food Microbiology*, 65, 114–121 (2017). Crossref. PubMed. ISI.
98. S. Ataei, P. Azari, A. Hassan, B. Pingguan-Murphy, R. Yahya and F. Muhamad. Essential oils-loaded electrospun biopolymers: a future perspective for active food packaging. *Advances in Polymer Technology*, 2020(1), 9040535 (2020). Crossref.
99. J. Ju, Y. Xie, Y. Guo, Y. Cheng, H. Qian and W. Yao. Application of edible coating with essential oil in food preservation. *Critical Reviews in Food Science and Nutrition*, 59(15), 2467–2480 (2019). Crossref. PubMed. ISI.
100. C. Vasile, M. Sivertsvik, A.C. Miteluţ, M.A. Brebu, E. Stoleru, J.T. Rosnes . . . A. Irimia. Comparative analysis of the composition and active property evaluation of certain essential oils to assess their potential applications in active food packaging. *Materials*, 10(1), 45 (2017). Crossref. PubMed. ISI.
101. T. Huang, Y. Qian, J. Wei and C. Zhou. Polymeric antimicrobial food packaging and its applications. *Polymers*, 11(3), 560 (2019). Crossref. PubMed. ISI.
102. M. Llana-Ruiz-Cabello, S. Pichardo, S. Maisanaba, M. Puerto, A.I. Prieto, D. Gutierrez-Praena . . . A.M. Camean. In vitro toxicological evaluation of essential oils and their main compounds used in active food packaging: a review. *Food and Chemical Toxicology*, 81, 9–27 (2015). Crossref. PubMed. ISI.
103. M. Jouki, S.A. Mortazavi, F.T. Yazdi and A. Koocheki. Characterization of antioxidant–antibacterial quince seed mucilage films containing thyme essential oil. *Carbohydrate Polymers*, 99, 537–546 (2014). Crossref. PubMed. ISI.
104. M. Abdollahi, M. Rezaei and G. Farzi. Improvement of active chitosan film properties with rosemary essential oil for food packaging. *International Journal of Food Science & Technology*, 47(4), 847–853 (2012). Crossref.
105. H. Cui, M. Bai and L. Lin. Plasma-treated poly (ethylene oxide) nanofibers containing tea tree oil/beta-cyclodextrin inclusion complex for antibacterial packaging. *Carbohydrate Polymers*, 179, 360–369 (2018). Crossref. PubMed.
106. F. Talebi, A. Misaghi, A. Khanjari, A. Kamkar, H. Gandomi and M. Rezaeigolestani. Incorporation of spice essential oils into poly-lactic acid film matrix with the aim of extending microbiological and sensorial shelf life of ground beef. *LWT*, 96, 482–490 (2018). Crossref. ISI.

107. C.L. Morelli, M. Mahrous, M.N. Belgacem, M.C. Branciforti, R.E.S. Bretas and J. Bras. Natural copaiba oil as antibacterial agent for bio-based active packaging. *Industrial Crops and Products*, 70, 134–141 (2015). Crossref.
108. P.K. Raghav, N. Agarwal and M. Saini. Edible coating of fruits and vegetables: a review. *International Journal of Scientific Research and Modern Education*, 1(1), 188–204 (2016).
109. P. Cazón, G. Velazquez, J.A. Ramírez and M. Vázquez. Polysaccharide-based films and coatings for food packaging: a review. *Food Hydrocolloids*, 68, 136–148 (2017). Crossref. ISI.
110. C.V. Dhumal, K. Pal and P. Sarkar. Synthesis, characterization, and antimicrobial efficacy of composite films from guar gum/sago starch/whey protein isolate loaded with carvacrol, citral and carvacrol-citral mixture. *Journal of Materials Science: Materials in Medicine*, 30(10), 117 (2019). Crossref. PubMed.
111. H. Chen, J. Wang, Y. Cheng, C. Wang, H. Liu, H. Bian and W. Han. Application of protein-based films and coatings for food packaging: a review. *Polymers*, 11(12), 2039 (2019). Crossref. PubMed. ISI.
112. T.A. Ochoa, B.E.G. Almendárez, A.A. Reyes, D.M.R. Pastrana, G.F.G. López, O.M. Belloso and C. Regalado-González. Design and characterization of corn starch edible films including beeswax and natural antimicrobials. *Food and Bioprocess Technology*, 10(1), 103–114 (2017). Crossref. ISI.
113. M. Saikia and L.S. Badwaik. Characterization and antimicrobial property of casein, gelatin and pectin based active composite films. *Journal of Packaging Technology and Research*, 2(3), 233–242 (2018). Crossref.
114. B.S. Álvarez, N. Damiani, M. Czerner, J.F. Martucci and L.B. Gende. Application of active and edible films of sodium caseinate to extend the shelf-life of argentine pategrás cheese. *Journal of Applied Packaging Research*, 11(3), 3 (2019).
115. V. Sachithananthan. A study on the consumer awareness of food additives in packaged food and their effects on health in Abha region, Saudi Arabia. *Journal Food Technolology Press*, 1(3), 14–27 (2017).

8 Application of Essential Oil Nanoemulsions in Food Preservation

Mohammad Aslam, Georgeos Deeb, Mohd. Amir Mirza, Javed Ahmad, and Leo M.L. Nollet

8.1 INTRODUCTION

Nowadays, the whole world shrinks into a small platform due to the political and economic ties between countries. Due to globalization and advanced technology, everything is available from one nook and corner of the world to another. Food is an essential part of our daily lives; hence, advanced and organized packaging of food materials is required to reduce food waste. This has given rise to a novel challenge of food packaging to protect the nutritional and organoleptic value of packaged foods.

Food degradation occurs mainly due to microbial growth and chemical degradation. Microbial growth in food not only reduces the quality of food but also causes foodborne diseases (Centers for Disease Control and Prevention 2015), while, chemically, food degrades through the process of oxidation and peroxidation. Lipid is an important part of food, and food with a high content of lipid is more prone to chemical degradation.

In the oxidation process, the unsaturated fatty acids present in the food react with the atmospheric oxygen present in the surroundings of the food, leading to the production of peroxide radicals. These peroxide radicals start the peroxidation process and hasten the oxidation of lipids present in the food [1–3]. This chemical process accelerates in the presence of environmental conditions such as humidity, high temperature, and radiations [2]. These chemical reactions decrease the food nutritional values and alter the organoleptic properties of foods by changing the flavor, odor, taste, and texture. The final products of these chemical degradations are one of the major reasons leading to cellular mutation, cancer, atherosclerosis, and cardiac problems [2, 4, 5].

With the evolution of human civilization, humans are using some conventional ways to preserve foods. The concepts of conventional food preservation methods are still in use in different parts of the world, especially in the Indian (India, Bangladesh, and Pakistan) and Levantine (Syria, Lebanon, and Iraq) civilizations in the form of salt brine and various condiments as food preservatives. Salt brine mixed with different condiments is renowned for its antimicrobial and antioxidant assets. The World Health Organization suggests to reduce the daily intake of salt due to the increasing risk of cardiovascular diseases.

Antioxidants are molecules which bind to the free oxygen radicals and can be used to prevent food degradation by oxidation and peroxidation chemical degradation. Generally used antioxidants in food containing lipids are BHT (butylated hydroxytoluene) and BHA (butylated hydroxyanisole) [6–8]. While many synthetic antibacterials are also used to control food spoilage by pathogenic microorganisms, these artificial preservatives, such as BHT, BHA, sulfite benzoic acid, potassium sorbate, sodium nitrite, and sodium acetate, have shown to cause various behavioral diseases, such as neuronal toxicity and tumorigenicity, in numerous research studies [9]. BHT and BHA have demonstrated to cause attention deficit hyperactivity disorder [10]. N-nitroso compounds are found

DOI: 10.1201/9781003462804-11

to be related with the different types of cancers, including ovarian [11], renal [12], colorectal, and pancreatic [13]. As far as the adverse effects of the aforementioned synthetic antioxidants, antibacterials, and chemical preservatives are concerned, nowadays consumers are switching to natural alternatives for food additives [14].

8.2 OVERVIEW OF ESSENTIAL OILS AND THEIR USE AS ALTERNATIVES TO SYNTHETIC PRESERVATIVES

The main phytochemical bioactive compounds are the secondary metabolic products or secondary metabolites. The secondary metabolites are produced from specific precursors after following their biosynthetic pathways [15]. They are generally classified as nitrogen-containing compounds, phenolics, polyacetates, and terpenoids. Terpenes, also called isoprenoids, are one of the largest classes of natural chemicals formed by head-to-tail arrangement of two or more isoprene molecules. Indeed, the structural diversity associated with at least 40,000 compounds makes the class of terpenoids exemplary amongst all plant chemicals. The evolutionary success of this class of compounds is based on the simplicity of constructing different-size molecules. They are important constituents of essential oils [16–18].

Essential oils are the terpenoid aromatic substances produced by plants that can be very well used by several food industries for different purposes, including food preservation and imparting flavor as well as aroma to food products [19]. Generally, essential oils are composed of 20–60 bioactive compounds at variable concentrations (commonly ranging from 20 to 70%, and some available in trace amounts). The entities which are present in higher proportions (terpenes, terpenoids, and aromatic compounds) are responsible for biological/pharmacological activities [20]. Minor entities/components can also play a significant role in the exhibition of various biologically functional activities through agonistic and antagonistic effects [21]. So far, the activity of essential oils with respect to its chemical nature is attributed to aldehydes or phenols as major entities (e.g., cinnamaldehyde, citral, carvacrol, eugenol, or thymol), followed by terpene alcohols, ketones, or esters (β-myrcene, α-thujone, or geranyl acetate) [22, 23].

The use of salt as alternative for synthetic preservatives is not advisable nowadays, while the use of different condiments is still in the choices as alternatives for synthetic preservatives. The active ingredient in these condiments is an essential oil, which is responsible for its food preservation action. The pathogens which enter the packaged food with environmental oxygen are the real cause of scourging of foodborne illness, and this needs to be addressed. The essential oils having antibacterial and antioxidant properties are of great interest to researchers. The successful incorporation of essential oils into raw uncut pieces, minced meat, patties, sausages, fish fillets, and broth has been demonstrated to be advantageous in various research studies. While the importance of modified atmosphere packaging (MAP) in enhancing food safety is being widely recognized, its combination with essential oil yields far more superior results. Essential oils have been widely used in food packaging and are found to exert enhancement in the shelf life of packaged food products.

8.3 APPLICATION OF ESSENTIAL OILS IN DIFFERENT FOOD INDUSTRIES

8.3.1 Meat Industry

Meat and meat products packed using chemical preservatives lead to sensory deterioration mainly due to the chemical changes that occur in lipids and proteins. These changes are called post-mortem changes, leading to high risks of foodborne illness [24]. Therefore, the uses of natural preservatives, such as essential oils, are encouraged in preserving meat products.

Hydrophobic extracts are extracted from various parts of the plant, such as roots, leaves, stems, barks, fruits, flowers, and seeds. These extracts possess desirable properties, like antioxidant, food pathogen–inhibition, shelf life–enhancing, and organoleptic characteristics, enhancing attributes

TABLE 8.1
Application of Essential Oils as Preservatives in Meat

S. No.	Essential Oils	Application	Outcome	Reference
1.	Coriander, clove, oregano, and thyme oils	Control of pathogens and autochthonous spoilage flora in meat	Cause a marked initial reduction in the viable cell numbers	[26]
2.	Thyme and cinnamon oils	Ham	Significantly decrease *L. monocytogenes*	[27]
3.	Rosemary/thyme oils	Mortadella	Extend the shelf life of mortadella	[28]
4.	Oregano oil	Bologna sausage	Extend the shelf life of bologna sausages	[29]
5.	Garlic and oregano oils	Against *Salmonella* spp., *L. monocytogenes*, and *S. aureus*	Improve safety of dry-cured sausages	[30]
6.	Clove oil (concentrations of 0.5% and 1%)	Restriction of the growth of *L. monocytogenes* in meat	Restrict the growth of *L. monocytogenes* in meat at both 30°C and 7°C	[31]

for preserving meat and meat products [25]. Essential oils having such desirable properties in food preservation and recognized as safe (GRAS) by regulating authorities can be used in the preservation of meat products. Examples of some of these natural preservatives for meat and meat-like products include oregano, mint, rosemary, orange, sage, cinnamon, lavender, bay, clove, cumin, fennel, coriander, peppermint, thyme, balm, tea tree, coriander, laurel, summer savory, eucalyptus, verbena, and hyssop (as elaborated in Table 8.1).

8.3.2 Seafood Products

Seafood, usually including mollusks (oysters, clams, and mussels), finfish, marine mammals, fish eggs (roe), and crustaceans (shrimp, crab, and lobster), is a highly perishable commodity due to various factors. Worldwide, seafood consumption has been growing exponentially. To fulfill the requirements of a growing market, the seafood industry has progressively been modernizing through different ways to keep their products fresh and safe. The growth of various bacteria depends on environmental conditions and the microbiological quality of the water. The temperature of the water and its salt content, methods of catching, and storage conditions are also important factors [32]. Fresh fish are more liable to spoilage because of its high moisture and nutrient content. Seafood spoilage agents are mainly bacteria of the *Vibrio* spp., *Listeria* spp., *Aeromonas* spp., and many more, which can exploit specific metabolites of fish tissues during storage. These bacteria represent an important food safety concern both for fresh and slightly preserved farmed seafood.

Seafood spoilage is any change that makes the product unacceptable for human consumption, and it commences just after fatality due to autolysis, microbial invasion, and autoxidation [33].

The preservation of overly sensitive seafood products by natural means is an important issue that needs consideration.

Therefore, seafood requires efficient preservation methods. Besides traditional preservation methods, such as chilling, salting, and drying, the potentially safe preservatives of natural origin are required to extend the shelf life of seafood, such as essential oils. Various essential oils have been used for the preservation of seafood in different research studies. Kostaki et al. have evaluated the combined effect of modified atmosphere packaging (MAP) condition and 0.2% of thyme oil as a natural preservative to enhance the shelf life of fresh filleted sea bass (*Dicentrarchus labrax*) [34]. It was found that the presence of thyme oil improved the sensory quality of sea bass fillets and improved shelf life with 17 days as compared to 6 days of the control samples. In another study, Corbo et al. have investigated the synergistic effect of natural compounds (mixture of three essential oils containing 0.11% of thymol, 0.10% of grapefruit seed extract, and 0.012% of lemon extracts) on the microbial/sensorial quality decay of packed fish hamburger against the main fish spoilage microorganisms (*Pseudomonas fluorescens*, *Photobacterium phosphoreum*, and *Shewanella putrefaciens*) [35]. A significant enhancement in the shelf life of the packed fish hamburger was reported, and also a delay of the sensorial quality decay without compromising the flavor of the fish hamburgers.

8.3.3 Fruits and Vegetables

Fruit preservation and packaging have been practiced since the ages to maintain the constant supply of seasonal fruits over lengthened periods around the year. The growing demand for minimally or unprocessed packaged fruits has further aggravated safety concerns, which fueled an extensive research with the objective to develop novel techniques of food processing, preservation, and packaging. There is a need for rapid, accurate, and early detection of contaminant products/microbes.

Fruits and vegetables are an important part of our recommended diet; the dietary guidelines for Americans recommend that daily half-plate of fruit and vegetables is essential for every individual. The vitamins and minerals present in the food are called phytochemicals and are a source of immunity and provide a good source of fiber to the diet and a diverse array of nonessential nutrients.

These fruits, before reaching consumers' home, need up to five days in transit post-harvest to the distribution center and one to three days in display in the retail store. Fruits such as tomatoes, apples, melons, and peaches (known as climacterics) are harvested before they ripen, and they continue to ripen until they reach their peak color after being detached from the mother plant. There are considerable losses of vitamin C in such process compared to that found if the product had been freshly picked at its peak of maturity [36–38].

Microorganisms may be controlled through the use of heat, cold, dehydration, acid, sugar, salt, smoke, atmospheric composition, and radiation. Mild heat treatments in the range of 82–93°C are commonly used to kill bacteria in foods with pH $\geq$ 4.6, but to ensure spore destruction, temperatures of 121°C wet heat for 15 min or longer are required. Highly acidic foods with pH $<$ 4.6 require less heat, and often, a treatment of 93°C for 15 min will ensure commercial sterility. Water activity (*a*w) of a food of 0.85 or below requires no thermal process, regardless of the pH of food. Most fruits are highly acidic, except for some fruits which are less acidic, such as bananas, figs, mangoes, and some mature stone fruit. Vegetables, on the other hand, are primarily low-acidic or alkaline in pH, apart from some "fruit vegetables," such as tomatoes, whose pH values $<$ 4.6.

Another main consideration in choosing the most appropriate method of food preservation is the intended shelf life required for the product. This will dictate to a large extent the method of preservation selected. If the product is meant for consumption within a week or two, fresh-cut or minimal processing may be sufficient, but refrigeration and other means of preventing microbial growth will be required. The main reason for the perishing of these cultivated fruits or vegetable foodstuffs is decay due to fungal infections. Use of synthetic chemicals post-harvesting is able to destroy fungal and bacterial infections and controls its deterioration, but due to the restriction of the chemicals in the food industry, alternative ways are always in demand. The use of controlled atmosphere (CA) or modified atmosphere (MAP) with different essential oils helps in increasing the shelf life of fruits and vegetables to be preserved. Various essential oils have been used in the preservation of various fruits and vegetables, as shown in Table 8.2.

8.3.4 Cereal-Based Products

Agriculture-based crop products basically consist of barley, wheat, corn, and rice, and they are highly susceptible to fungal infections. The mycotoxins infect various crops to a great extent, which is a matter of serious concern. The microorganisms are mainly present in the pericarp of the grains, which covers the endosperm. Agricultural crops or cereals are also contaminated by bacteria.

TABLE 8.2
Application of Essential Oils as Preservatives in Fruits and Vegetables

S. No.	Essential Oils (EOs)	Application	Outcome	Reference
1.	Lemongrass and geraniol	Have been found effective against *E. coli*, *Salmonella* sp., and *Listeria* spp. In apple, pear, and melon juices	Shelf life of unpasteurized fruit juices limited by microbial enzymatic spoilage	[39]
2.	Malic acid and EOs extracted from cinnamon, palmarosa, and lemongrass (0.3 and 0.7%) or their major compounds (eugenol, geraniol, and citral)	*Salmonella enteritidis* (108 cfu/ml) used as the target microorganism	To prolong the shelf life of fresh-cut "Piel de Sapo" melon (*Cucumis melo* L.)	[40]
3.	Lemongrass, oregano oil, and vanillin	In apples	In an apple puree–alginate edible coating to prolong the shelf life of fresh-cut Fuji apples	[41]
4.	Vanillin (0.3%w/w)	In general fruits	Preserve sensory quality for at least 21 days at 4°C	[42]
5.	Thyme and sage EOs	In strawberries	Decreasing the amount of fungi	[43]
6.	*Echinophora platyloba* EO	All fruits	Antifungal power	[44]
7.	Palmarosa (*Cymbopogon martini*), red thyme (*Thymus zygis*), cinnamon leaf (*Cinnamomum zeylanicum*), and clove buds (*Eugenia caryophyllata*)	Testing numerous extracts from plants and EOs	Evaluated for their antifungal activity against *Botrytis cinerea*	[45]
8.	Thyme, oregano, and lemon	Strawberry, tomato, and cucumber	Reduction of propagation of different fungi	[46]
9.	Caraway, cumin, savory thyme, and peppermint	Strawberries	Check fungus *Botrytis cinerea*	[47]
10.	Fennel, black caraway, peppermint, and thyme	Fruits	Fungicidal effect *on Botrytis cinerea* and *Rhizopus stolonifer* rot fungus	[48]
11.	Eugenol, thymol, and carvacrol	Grapes berry decay		[49]
12.	Eugenol, thymol, or menthol	"Crimson seedless" table grapes		[49]
13.	Eugenol and menthol EOs	"Crimson seedless" table grapes	Better results during storage time compared to grapes without such treatments	[50]
14.	Lemongrass, cinnamon, and oregano oils	Pomegranate fruit	Antifungal activity against *Botrytis* sp., *Penicillium* sp., and *Pilidiella granati* pathogens	[51]

The main categories of bacteria which infect cereals come under the species of *Bacillaceae*, *Lactobacillaceae*, *Pseudomonadaceae*, and *Micrococcaceae* [52]. Only a few bacteria can survive during storage, but their spores can survive for a longer period of time. For example, the spore of the bacterial species of lactic acid bacteria can survive and infect the dough made from those cereals [53]. Several research studies have been conducted to get rid of these contaminations, and the methods mainly applied include gamma radiation, steam treatment, heat, and various fungicides and bactericides. However, at present, the use of natural-origin preservatives is given preference over all other methods. That is why the antimicrobial effect of essential oil is profoundly explored. The antimicrobial and antifungal activities of these essential oils help in the preservation of cereal-based products. In a recent research study, thymol mixed with lemon extract and chitosan was tested against mesophilic and psychrotrophic bacteria, total coliforms, *Staphylococcus* spp., yeasts, and molds. The study results have shown an improved microbiological stability of refrigerated amaranth-based fresh pasta [54].

8.3.5 Chicken Industry

Due to the wide variety of commercial availability, low-cost productions, and higher nutritional values, the production of fresh chicken and their products is prevalent all over the world. The limited shelf life of fresh chicken and their products is the main hurdle in its commercialization. Chicken products are a rich source of protein and nutrients, and the presence of moisture makes them vulnerable to microbial spoilage [55]. The growth of microorganisms can cause a change in the organoleptic characteristic and make the quality of chicken low, with shorter shelf life and more susceptibility toward putrefaction [56].

New preservation techniques are being developed to maintain the freshness of chicken and likely products. Essential oils are used for the purpose of chicken preservation to increase flavor. Essential oils have antimicrobial activity against various pathogenic microorganisms, and therefore, they are used for the preservation of chicken and related food products [57].

Synthetic antioxidants are common preservatives, but their carcinogenicity is debatable [58]; therefore, essential oils are preferred by food technologists as additives for their antioxidant property. The antimicrobial activities of essential oils help improve the shelf life and taste of fresh chicken and its products.

In chicken meat, putrefaction occurs mainly due to biogenic amines, such as putrescine, cadaverine, and histamine. Bacteria such as *Pseudomonas* species and endobacteria are also the reason for chicken meat deterioration. Therefore, active packagings with essential oils help in chicken preservation [59].

In a recent study, oregano was used to maintain the freshness of chicken, and it was observed that the shelf life of fresh chicken was improved in modified atmospheric packaging [60]. Further, rosemary was used as an active packaging system for chicken meat, and it was shown to inhibit the increase of biogenic amines, namely, putrescine, cadaverine, and histamine, as well as enterobacteria, *Pseudomonas* spp., and *Brochothrix thermosphacta* [59]. These research study findings boost the application of essential oils as preservatives for chicken and similar products.

8.3.6 Dairy Products and the Cheese Industry

Contamination of milk and its products can occur either from endogenous (itself from animal infection or disease of mastitis) or exogenous (during the process of milking via the udders and teats, milking machine, bucket, or milk storage tank, and finally, during storage and transport) sources.

The species of pathogenic bacteria that are present in raw milk and milk products such as cheese are mainly *Listeria monocytogenes*, verocytotoxin-producing *Escherichia coli*, *Staphylococcus aureus*, *Salmonella* sp., and *Campylobacter* sp. [61, 62]. During the ripening of cheese, physicochemical changes such as an increase in pH make the conditions more favorable for the growth of

TABLE 8.3
Application of Essential Oils as Preservatives in Dairy Products

S. No.	Essential Oils	Application	Outcome	Reference
1.	Orange, lemon, grapefruit, mandarin, terpeneless lime, orange, D-limonene, terpineol, and geraniol	Tested against *Salmonella senftenberg*, *E. coli*, *S. aureus*, and *Pseudomonas* spp.	In different types of milk	[64]
2.	The combination of thymol dipping and MAP (modified atmosphere packaging)		Prolonged shelf life by 8 days without negative effects on the sensory quality and on the growth kinetics of LAB	[65]
3.	Clove, cinnamon, bay, and thyme EOs (concentrations of 0.1%, 0.5%, and 1%)	Low-fat and full-fat soft cheese	Antimicrobial effect on *Listeria monocytogenes*, *Salmonella enteritidis*	[66]
4.	Oregano and thyme EOs	Feta cheese	*L. monocytogenes* and *Escherichia coli* O157: H7	[67]
5.	Oregano and rosemary EOs	Cream cheese	Affected the number of mesophilic microorganisms	[68]
6.	Marjoram and rosemary EOs	In full cream cheese	Mesophilic bacteria	[69]
7.	Clove EO (concentrations of 0.5% and 1%)	In cheese	Restricted the growth of *L. monocytogenes* in meat at both 30°C and 7°C	[31]
8.	*Zataria multiflora* (thyme-like plant) EO	Gouda cheese	Reduction in the production of histamine and tyramine	[70]

pathogenic bacteria. The addition of plant or herbal products influences the growth of microorganisms in cheese. Many countries use these plant products during the production of cheese for their organoleptic properties or for decorative or preservation purposes. The commonly used products for these purposes include red, black, and green peppers and thyme, cloves, cumin, parsley, paprika, onion/garlic, and many more.

Essential oils have been added to milk and cheese products to promote shelf life, confer functional properties, and maintain sensory features, with very promising conclusions [63].

Essential oils can increase the shelf life of dairy products, not only by eliminating unwanted microorganisms, but also by decreasing the degree of chemical deterioration during storage and marketing periods.

The antimicrobial effect of essential oils is associated with the composition and characteristics of each food product and specific microorganisms to be eliminated. Additionally, several other factors may affect the antimicrobial effect of essential oils in food products, namely, heat treatments, smoking, chemical preservatives, and packaging. The use of essential oils as natural preservatives in food industries meets the current consumer trends of "green," "biological," "natural," and "no chemicals added" labels. The cheese is more susceptible to different types of bacterial contamination. Nowadays, the antimicrobial effect of essential oils is explored with proper packaging for preserving milk and its products, as exemplified in Table 8.3.

8.4 NANOEMULSION-BASED APPROACH TO IMPROVE THE PROPERTIES OF ESSENTIAL OILS FOR APPLICATION IN FOOD SYSTEMS

The use of essential oils has changed the prospects of the preservation of food, but the direct addition of essential oils into the food is not feasible, attributed to low water solubility, high volatility, and objectionable odor. These factors lead to a poor bioavailability, and the odor of oils alter the aesthetic value of the food itself. When essential oils are mixed with food and the food goes through various food elaboration processes, such as heat treatment or air and light exposition, the volatile nature of essential oils causes the degradation with higher rates. That is why several methods for the fortification of essential oils are used, so that they can exert their antimicrobial effect efficiently for a longer time. Encapsulation techniques of essential oils within polymeric particles, liposomes, solid lipid nanoparticles, and nanolipid carriers have been widely studied to increase stability and efficacy.

8.4.1 Application of Essential Oil Nanoemulsions in Food Systems

Nanoemulsions are a colloidal delivery system in the size range of 10–200 nm and are used as carriers of bioactive molecules. Nanoemulsions help in the solubilization of lipophilic biomolecules and can be changed into various formulations, such as foams, creams, liquids, and sprays. They are helpful in taste masking and have greater surface area, providing greater absorption, which causes improved bioavailability of bioactive molecules. In formulations aspect, they can be used as substitute to liposomes and vesicles [71], and they require less amount of energy in its preparation. Essential oils have also been used as additives in biodegradable films and coatings for active food packaging [72, 73]. Essential oils can provide films and coatings with antioxidant and/or antimicrobial properties.

The antioxidant activity of oil also depends upon the film's oxygen permeability. Incorporation into edible films can promote the antimicrobial capacity of essential oils. The efficacy of the edible film against microbial growth will depend on the oil's nature and the type of microorganism. Essential oils' controlled release from edible films is another aspect that positively affects their effectiveness. Although nanoencapsulation is a promising tool for the effective delivery of essential oils into food, still the toxicological aspects of most of the nanocarriers at their molecular target site must be further investigated. The recent advances in nanotechnology have made possible the development of novel carrier agents for the delivery and control release of essential oils in food systems with enhanced chemical, oxidative, and thermal stability [74, 75]. The edible coating of food products is another area where essential oils have proved their potential [76]. For example, nanoemulsions containing lemongrass essential oil as an antimicrobial agent have been used as edible coatings to control the safety and quality parameters of fresh-cut Fuji apples during storage. The performance of essential oil–based products is comparable to that of synthetic molecules and is concentration-dependent. In a study, it was demonstrated that the encapsulation of carvacrol, limonene, and cinnamaldehyde in sunflower oil droplets of nanoemulsions has shown concentration-reliant antimicrobial activity that was dependent on the concentration of essential oil in the aqueous phase [77]. Many food scientists have encapsulated essential oils in a nanoemulsions to protect and improve their organoleptic properties and biological activities. This encapsulation into a nanoemulsion system increases their solubility and bioavailability and limits their oxidation process [78]. Li and Chiang prepared a d-limonene nanoemulsion by ultrasonic emulsification technique having a particle size of <100 nm and improved stability against Ostwald ripening at a storage duration of eight weeks [79]. β-carotene is an important antioxidant, and consumption of a sufficient amount of it can reduce the risk of various chronic diseases and cataracts [80]. However, its lipophilic nature and low stability against high temperature, oxygen, and light limit its use as fortified food. O/W nanoemulsion of β-carotene was prepared by high-pressure homogenization technique. It was observed that bioaccessibility was increased from 3.1 to 35% [81]. Shah et al. reported that eugenol-containing essential oil nanoemulsion was more active against *Escherichia coli* O157:H7 and *L. monocytogenes* in bovine milk compared to free essential oil due to their uniform distribution in aqueous-based food products [82]. The limited solubility of free essential oils in aqueous systems makes it challenging to disperse it uniformly in food matrices, ultimately influencing food quality and antimicrobial efficacy against spoilage. In another study, Donsi et al. investigated the potential of encapsulation of essential oils (terpenes mixture and D-limonene) into nanoemulsion-based delivery systems for integration into fruit juices (orange juice and pear juice) in order to increase their antimicrobial activity and the shelf life of the product [83]. The investigator has reported higher antimicrobial activity of the D-limonene containing the essential oil nanoemulsion system, with minimal alteration of the organoleptic characteristics of the fruit juice.

8.5 CONCLUSION AND FUTURE DIRECTIONS

The formed nanoemulsions of essential oils are kinetically stable and could have longer shelf life, but the higher volatility of the oils and instability in an acidic environment may cause stability disruption in the gastrointestinal tract milieu [84]. The formed small droplets of a nanoemulsion may go through Ostwald ripening and may lead to problems of instability, such as flocculation, coalescence, and sedimentation. Further, essential oil processing as nanoemulsion in the food industry is associated with serious concerns of stability, cost, and safety of ingredients. These are certain issues with which researchers are dealing at present. Future perspectives of nanoemulsions are very promising for different applications in food products, especially their incorporation in the formulation of functional foods. An example of future application of nanoemulsions as delivery platform is the encapsulation of probiotics by means of nanoemulsions. Indeed, such bioactives are usually encapsulated in liposomes. Moreover, it will be interesting to investigate *in vivo* performance after the consumption of functional foods fortified with encapsulated bioactives via nanoemulsions.

CONFLICT OF INTEREST

The authors declare no conflict of interest.

REFERENCES

1. Ferrari, C.K.B., 1998. Oxidação lipídica em alimentos e sistemas biológicos: mecanismos gerais e implicações nutricionais e patológicas. *Revista de Nutrição*, *11*(1), pp. 3–14.
2. McClements, D.J. and Decker, E.A., 2000. Lipid oxidation in oil-in-water emulsions: Impact of molecular environment on chemical reactions in heterogeneous food systems. *Journal of Food Science*, *65*(8), pp. 1270–1282.
3. Frankel, E.N., 2005. *Lipid oxidation*, 2 (pp. 1–14). Oily Press Lipid Library.
4. Chanwitheesuk, A., Teerawutgulrag, A. and Rakariyatham, N., 2005. Screening of antioxidant activity and antioxidant compounds of some edible plants of Thailand. *Food Chemistry*, *92*(3), pp. 491–497.
5. Márquez-Ruiz, G., Garcia-Martinez, M.C. and Holgado, F., 2008. Changes and effects of dietary oxidized lipids in the gastrointestinal tract. *Lipid Insights*, *2*, pp. 11–19.
6. Ultee, A., Kets, E.P., Alberda, M., Hoekstra, F.A. and Smid, E.J., 2000. Adaptation of the food-borne pathogen *Bacillus cereus* to carvacrol. *Archives of Microbiology*, *174*(4), pp. 233–238.
7. Moleyar, V. and Narasimham, P., 1992. Antibacterial activity of essential oil components. *International Journal of Food Microbiology*, *16*(4), pp. 337–342.
8. André, C., Castanheira, I., Cruz, J.M., Paseiro, P. and Sanches-Silva, A., 2010. Analytical strategies to evaluate antioxidants in food: A review. *Trends in Food Science & Technology*, *21*(5), pp. 229–246.
9. Parke, D.V. and Lewis, D.F.V., 1992. Safety aspects of food preservatives. *Food Additives & Contaminants*, *9*(5), pp. 561–577.
10. Feingold, B.F., 1982. The role of diet in behaviour. *Ecology of Disease*, *1*(2–3), pp. 153–165.
11. Aschebrook-Kilfoy, B., Heltshe, S.L., Nuckols, J.R., Sabra, M.M., Shuldiner, A.R., Mitchell, B.D., Airola, M., Holford, T.R., Zhang, Y. and Ward, M.H., 2012. Modeled nitrate levels in well water supplies and prevalence of abnormal thyroid conditions among the Old Order Amish in Pennsylvania. *Environmental Health*, *11*(1), p. 6.
12. Dellavalle, C.T., Daniel, C.R., Aschebrook-Kilfoy, B., Hollenbeck, A.R., Cross, A.J., Sinha, R. and Ward, M.H., 2013. Dietary intake of nitrate and nitrite and risk of renal cell carcinoma in the NIH-AARP Diet and Health Study. *British Journal of Cancer*, *108*(1), pp. 205–212.

13. DellaValle, C.T., Xiao, Q., Yang, G., Shu, X.O., Aschebrook-Kilfoy, B., Zheng, W., Lan Li, H., Ji, B.T., Rothman, N., Chow, W.H. and Gao, Y.T., 2014. Dietary nitrate and nitrite intake and risk of colorectal cancer in the Shanghai Women's Health Study. *International Journal of Cancer*, *134*(12), pp. 2917–2926.
14. Yap, P.S.X., Yiap, B.C., Ping, H.C. and Lim, S.H.E., 2014. Essential oils, a new horizon in combating bacterial antibiotic resistance. *The Open Microbiology Journal*, *8*, p. 6.
15. Buchanan, B.B., Gruissem, W. and Jones, R.L. eds., 2015. *Biochemistry and molecular biology of plants*. John Wiley & Sons.
16. Rohmer, M., 1999. The discovery of a mevalonate-independent pathway for isoprenoid biosynthesis in bacteria, algae and higher plants. *Natural Product Reports*, *16*(5), pp. 565–574.
17. Boucher, Y. and Doolittle, W.F., 2000. The role of lateral gene transfer in the evolution of isoprenoid biosynthesis pathways. *Molecular Microbiology*, *37*(4), pp. 703–716.
18. Lange, B.M., Rujan, T., Martin, W. and Croteau, R., 2000. Isoprenoid biosynthesis: The evolution of two ancient and distinct pathways across genomes. *Proceedings of the National Academy of Sciences*, *97*(24), pp. 13172–13177.
19. Pavela, R., 2015. Essential oils for the development of eco-friendly mosquito larvicides: A review. *Industrial Crops and Products*, *76*, pp. 174–187.
20. Bakkali, F., Averbeck, S., Averbeck, D. and Idaomar, M. 2008. Biological effects of essential oils—a review. *Food and Chemical Toxicology*, *46*, pp. 446–475.
21. Bassolé, I.H.N. and Juliani, H.R., 2012. Essential oils in combination and their antimicrobial properties. *Molecules*, *17*(4), pp. 3989–4006.
22. Dorman, H.D. and Deans, S.G., 2000. Antimicrobial agents from plants: Antibacterial activity of plant volatile oils. *Journal of Applied Microbiology*, *88*(2), pp. 308–316.
23. de Barros, J.C., da Conceição, M.L., Neto, N.J.G., da Costa, A.C.V., Júnior, J.P.S., Junior, I.D.B. and de Souza, E.L., 2009. Interference of *Origanum vulgare* L. essential oil on the growth and some physiological characteristics of *Staphylococcus aureus* strains isolated from foods. *LWT—Food Science and Technology*, *42*(6), pp. 1139–1143.
24. Casaburi, A., Nasi, A., Ferrocino, I., Di Monaco, R., Mauriello, G., Villani, F. and Ercolini, D., 2011. Spoilage-related activity of *Carnobacterium maltaromaticum* strains in air-stored and vacuum-packed meat. *Applied and Environmental Microbiology*, *77*(20), pp. 7382–7393.
25. Del Nobile, M.A., Lucera, A., Costa, C. and Conte, A., 2012. Food applications of natural antimicrobial compounds. *Frontiers in Microbiology*, *3*, p. 287.
26. Speranza, B. and Corbo, M.R., 2010. Essential oils for preserving perishable foods: Possibilities and limitations. *Application of Alternative Food-Preservation Technologies to Enhance Food Safety and Stability*, *23*, pp. 35–37.
27. Dussault, D., Vu, K.D. and Lacroix, M., 2014. In vitro evaluation of antimicrobial activities of various commercial essential oils, oleoresin and pure compounds against food pathogens and application in ham. *Meat Science*, *96*(1), pp. 514–520.
28. Viuda-Martos, M., Ruiz-Navajas, Y., Fernández-López, J. and Pérez-Álvarez, J.A., 2010. Effect of added citrus fibre and spice essential oils on quality characteristics and shelf-life of mortadella. *Meat Science*, *85*(3), pp. 568–576.
29. Viuda-Martos, M., Ruiz-Navajas, Y., Fernández-López, J. and Pérez-Álvarez, J.A., 2010. Effect of orange dietary fibre, oregano essential oil and packaging conditions on shelf-life of bologna sausages. *Food Control*, *21*(4), pp. 436–443.
30. García-Díez, J., Alheiro, J., Pinto, A.L., Soares, L., Falco, V., Fraqueza, M.J. and Patarata, L., 2016. Behaviour of food-borne pathogens on dry cured sausage manufactured with herbs and spices essential oils and their sensorial acceptability. *Food Control*, *59*, pp. 262–270.
31. Menon, K.V. and Garg, S.R., 2001. Inhibitory effect of clove oil on *Listeria monocytogenes* in meat and cheese. *Food Microbiology*, *18*(6), pp. 647–650.
32. Giuffrida, A., 2003. Application of risk management to the production chain of intensively reared fish. *Veterinary Research Communications*, *27*(1), pp. 491–496.
33. Gokoglu, N. and Yerlikaya, P., 2015. *Seafood chilling, refrigeration and freezing: Science and technology*. John Wiley & Sons.
34. Kostaki, M., Giatrakou, V., Savvaidis, I.N. and Kontominas, M.G., 2009. Combined effect of MAP and thyme essential oil on the microbiological, chemical and sensory attributes of organically aquacultured sea bass (*Dicentrarchus labrax*) fillets. *Food Microbiology*, *26*(5), pp. 475–482.
35. Corbo, M.R., Speranza, B., Filippone, A., Granatiero, S., Conte, A., Sinigaglia, M. and Del Nobile, M.A., 2008. Study on the synergic effect of natural compounds on the microbial quality decay of packed fish hamburger. *International Journal of Food Microbiology*, *127*(3), pp. 261–267.

36. Ahmad, A., Mukherjee, P., Senapati, S., Mandal, D., Khan, M.I., Kumar, R. and Sastry, M., 2003. Extracellular biosynthesis of silver nanoparticles using the fungus *Fusarium oxysporum*. *Colloids and Surfaces B: Biointerfaces*, *28*(4), pp. 313–318.
37. Shrivastava, S., Bera, T., Roy, A., Singh, G., Ramachandrarao, P. and Dash, D., 2007. Characterization of enhanced antibacterial effects of novel silver nanoparticles. *Nanotechnology*, *18*(22), p. 225103.
38. Bradford, M.M., 1976. A rapid and sensitive method for the quantitation of microgram quantities of protein utilizing the principle of protein-dye binding. *Analytical Biochemistry*, *72*(1–2), pp. 248–254.
39. Raybaudi-Massilia, R.M., Mosqueda-Melgar, J. and Martin-Belloso, O., 2006. Antimicrobial activity of essential oils on *Salmonella enteritidis*, *Escherichia coli*, and *Listeria innocua* in fruit juices. *Journal of Food Protection*, *69*(7), pp. 1579–1586.
40. Raybaudi-Massilia, R.M., Mosqueda-Melgar, J., Sobrino-López, A., Soliva-Fortuny, R. and Martín-Belloso, O., 2009. Use of malic acid and other quality stabilizing compounds to assure the safety of fresh-cut "Fuji" apples by inactivation of *Listeria monocytogenes*, *Salmonella enteritidis* and *Escherichia coli* O157: H7. *Journal of Food Safety*, *29*(2), pp. 236–252.
41. Rojas-Graü, M.A., Raybaudi-Massilia, R.M., Soliva-Fortuny, R.C., Avena-Bustillos, R.J., McHugh, T.H. and Martín-Belloso, O., 2007. Apple puree-alginate edible coating as carrier of antimicrobial agents to prolong shelf-life of fresh-cut apples. *Postharvest Biology and Technology*, *45*(2), pp. 254–264.
42. Ayala-Zavala, J.F., González-Aguilar, G.A. and Del-Toro-Sánchez, L., 2009. Enhancing safety and aroma appealing of fresh-cut fruits and vegetables using the antimicrobial and aromatic power of essential oils. *Journal of Food Science*, *74*(7), pp. R84–R91.
43. Campos, T., Barreto, S., Queirós, R., Ricardo-Rodrigues, S., Félix, M.R., Laranjo, M. and Agulheiro-Santos, A.C., 2016. Conservação de morangos com utilização de óleos essenciais. *AGROTEC*, *18*, pp. 90–96.
44. Moghaddam, M., Taheri, P., Pirbalouti, A.G. and Mehdizadeh, L., 2015. Chemical composition and antifungal activity of essential oil from the seed of *Echinophora platyloba* DC. against phytopathogens fungi by two different screening methods. *LWT—Food Science and Technology*, *61*(2), pp. 536–542.
45. Wilson, C.L., Solar, J.M., El Ghaouth, A. and Wisniewski, M.E., 1997. Rapid evaluation of plant extracts and essential oils for antifungal activity against *Botrytis cinerea*. *Plant Disease*, *81*(2), pp. 204–210.
46. Vitoratos, A., Bilalis, D., Karkanis, A. and Efthimiadou, A., 2013. Antifungal activity of plant essential oils against *Botrytis cinerea*, *Penicillium italicum* and *Penicillium digitatum*. *Notulae botanicae horti agrobotanici Cluj-Napoca*, *41*(1), pp. 86–92.
47. Beikzadeh, N. and Afzali, H. 2020. Impact of six essential oils on strawberry gray mold. *University of Yasouj Journals System Plant Pathology Science*, *9*(1), pp. 129–140.
48. Mohammadi, S., Aroiee, H., Aminifard, M.H., Tehranifar, A. and Jahanbakhsh, V., 2014. Effect of fungicidal essential oils against *Botrytis cinerea* and *Rhizopus stolonifer* rot fungus in vitro conditions. *Archives of Phytopathology and Plant Protection*, *47*(13), pp. 1603–1610.
49. Valero, D., Valverde, J.M., Martínez-Romero, D., Guillén, F., Castillo, S. and Serrano, M., 2006. The combination of modified atmosphere packaging with eugenol or thymol to maintain quality, safety and functional properties of table grapes. *Postharvest Biology and Technology*, *41*(3), pp. 317–327.
50. Ricardo-Rodrigues, S., Tim-Tim, S., Véstia, J., Agostinho, J., Derreado, A., Rato, A.E., and Santos, A.C.A. 2018. Avaliação das potencialidades da aplicação de Atmosfera Modificada em combinação com eugenol na conservação de uva de mesa 'Crimson Seedless'. In *IX Congresso Ibérico de Agroengenharia: Livro de Atas* (pp. 716–724). Instituto Politécnico de Bragança.
51. Munhuweyi, K., Caleb, O.J., Lennox, C.L., van Reenen, A.J. and Opara, U.L., 2017. In vitro and in vivo antifungal activity of chitosan-essential oils against pomegranate fruit pathogens. *Postharvest Biology and Technology*, *129*, pp. 9–22.
52. Laca, A., Mousia, Z., DÍaz, M., Webb, C. and Pandiella, S.S., 2006. Distribution of microbial contamination within cereal grains. *Journal of Food Engineering*, *72*(4), pp. 332–338.
53. Sabillón, L. and Bianchini, A., 2016. From field to table: A review on the microbiological quality and safety of wheat-based products. *Cereal Chemistry*, *93*(2), pp. 105–115.
54. Del Nobile, M.A., Di Benedetto, N., Suriano, N., Conte, A., Lamacchia, C., Corbo, M.R. and Sinigaglia, M., 2009. Use of natural compounds to improve the microbial stability of Amaranth-based homemade fresh pasta. *Food Microbiology*, *26*(2), pp. 151–156.
55. Aymerich, T., Picouet, P.A. and Monfort, J.M., 2008. Decontamination technologies for meat products. *Meat Science*, *78*(1–2), pp. 114–129.
56. Petrou, S., Tsiraki, M., Giatrakou, V. and Savvaidis, I.N., 2012. Chitosan dipping or oregano oil treatments, singly or combined on modified atmosphere packaged chicken breast meat. *International Journal of Food Microbiology*, *156*(3), pp. 264–271.

57. De Martino, L., De Feo, V. and Nazzaro, F., 2009. Chemical composition and in vitro antimicrobial and mutagenic activities of seven Lamiaceae essential oils. *Molecules*, *14*(10), pp. 4213–4230.
58. Chen, C., Pearson, A.M. and Gray, J.I., 1992. Effects of synthetic antioxidants (BHA, BHT and PG) on the mutagenicity of IQ-like compounds. *Food Chemistry*, *43*(3), pp. 177–183.
59. Sirocchi, V., Caprioli, G., Cecchini, C., Coman, M.M., Cresci, A., Maggi, F., Papa, F., Ricciutelli, M., Vittori, S. and Sagratini, G., 2013. Biogenic amines as freshness index of meat wrapped in a new active packaging system formulated with essential oils of *Rosmarinus officinalis*. *International Journal of Food Sciences and Nutrition*, *64*(8), pp. 921–928.
60. Chouliara, E., Karatapanis, A., Savvaidis, I.N. and Kontominas, M.G., 2007. Combined effect of oregano essential oil and modified atmosphere packaging on shelf-life extension of fresh chicken breast meat, stored at 4°C. *Food Microbiology*, *24*(6), pp. 607–617.
61. Hammad, A.M., Hassan, H.A. and Shimamoto, T., 2015. Prevalence, antibiotic resistance and virulence of *Enterococcus* spp. in Egyptian fresh raw milk cheese. *Food Control*, *50*, pp. 815–820.
62. Verraes, C., Vlaemynck, G., Van Weyenberg, S., De Zutter, L., Daube, G., Sindic, M., Uyttendaele, M. and Herman, L., 2015. A review of the microbiological hazards of dairy products made from raw milk. *International Dairy Journal*, *50*, pp. 32–44.
63. Carocho, M., Barreira, J.C., Antonio, A.L., Bento, A., Morales, P. and Ferreira, I.C., 2015. The incorporation of plant materials in "Serra da Estrela" cheese improves antioxidant activity without changing the fatty acid profile and visual appearance. *European Journal of Lipid Science and Technology*, *117*(10), pp. 1607–1614.
64. Fisher, K. and Phillips, C., 2008. Potential antimicrobial uses of essential oils in food: Is citrus the answer? *Trends in Food Science & Technology*, *19*(3), pp. 156–164.
65. Bevilacqua, A., Corbo, M.R. and Sinigaglia, M., 2007. Combined effects of modified atmosphere packaging and thymol for prolonging the shelf life of caprese salad. *Journal of Food Protection*, *70*(3), pp. 722–728.
66. Smith-Palmer, A., Stewart, J. and Fyfe, L., 2001. The potential application of plant essential oils as natural food preservatives in soft cheese. *Food Microbiology*, *18*(4), pp. 463–470.
67. Govaris, A., Botsoglou, E., Sergelidis, D. and Chatzopoulou, P.S., 2011. Antibacterial activity of oregano and thyme essential oils against *Listeria monocytogenes* and *Escherichia coli* O157: H7 in feta cheese packaged under modified atmosphere. *LWT—Food Science and Technology*, *44*(4), pp. 1240–1244.
68. Olmedo, R.H., Nepote, V. and Grosso, N.R., 2013. Preservation of sensory and chemical properties in flavoured cheese prepared with cream cheese base using oregano and rosemary essential oils. *LWT—Food Science and Technology*, *53*(2), pp. 409–417.
69. Asensio, C.M., Grosso, N.R. and Juliani, H.R., 2015. Quality preservation of organic cottage cheese using oregano essential oils. *LWT—Food Science and Technology*, *60*(2), pp. 664–671.
70. Laranjo, M., Fernandez-Leon, A.M., Potes, M.E., Agulheiro-Santos, A.C. and Elias, M., 2017. Use of essential oils in food preservation. In *Antimicrobial research: Novel bioknowledge and educational programs* (pp. 177–188). Formatex Research Centre.
71. Bouchemal, K., Briançon, S., Perrier, E. and Fessi, H., 2004. Nano-emulsion formulation using spontaneous emulsification: Solvent, oil and surfactant optimisation. *International Journal of Pharmaceutics*, *280*(1–2), pp. 241–251.
72. Atarés, L. and Chiralt, A., 2016. Essential oils as additives in biodegradable films and coatings for active food packaging. *Trends in Food Science & Technology*, *48*, pp. 51–62.
73. Ribeiro-Santos, R., Andrade, M., de Melo, N.R. and Sanches-Silva, A., 2017. Use of essential oils in active food packaging: Recent advances and future trends. *Trends in Food Science & Technology*, *61*, pp. 132–140.
74. Prakash, B., Kujur, A., Yadav, A., Kumar, A., Singh, P.P. and Dubey, N.K., 2018. Nanoencapsulation: An efficient technology to boost the antimicrobial potential of plant essential oils in food system. *Food Control*, *89*, pp. 1–11.
75. Aureli, P., Costantini, A. and Zolea, S., 1992. Antimicrobial activity of some plant essential oils against *Listeria monocytogenes*. *Journal of Food Protection*, *55*(5), pp. 344–348.
76. Salvia-Trujillo, L., Rojas-Graü, M.A., Soliva-Fortuny, R. and Martín-Belloso, O., 2015. Use of antimicrobial nanoemulsions as edible coatings: Impact on safety and quality attributes of fresh-cut Fuji apples. *Postharvest Biology and Technology*, *105*, pp. 8–16.
77. Donsì, F., Annunziata, M., Vincensi, M. and Ferrari, G., 2012. Design of nanoemulsion-based delivery systems of natural antimicrobials: Effect of the emulsifier. *Journal of Biotechnology*, *159*(4), pp. 342–350.

78. Mahfoudhi, N., Ksouri, R. and Hamdi, S., 2016. Nanoemulsions as potential delivery systems for bioactive compounds in food systems: Preparation, characterization, and applications in food industry. In *Emulsions, nanotechnology in the agri-food industry*, 3 (pp. 365–403). Academic Press.
79. Li, P.H. and Chiang, B.H., 2012. Process optimization and stability of D-limonene- in-water nanoemulsions prepared by ultrasonic emulsification using response surface methodology. *Ultrasonics Sonochemistry*, *19*(1), pp. 192–197.
80. Failla, M.L., Huo, T. and Thakkar, S.K., 2008. In vitro screening of relative bioaccessibility of carotenoids from foods. *Asia Pacific Journal of Clinical Nutrition*, *17* (S1), pp. 200–203.
81. Liang, R., Shoemaker, C.F., Yang, X., Zhong, F. and Huang, Q., 2013. Stability and bioaccessibility of β-carotene in nanoemulsions stabilized by modified starches. *Journal of Agricultural and Food Chemistry*, *61*(6), pp. 1249–1257.
82. Shah, B., Davidson, P.M. and Zhong, Q., 2013. Nanodispersed eugenol has improved antimicrobial activity against *Escherichia coli* O157: H7 and *Listeria monocytogenes* in bovine milk. *International Journal of Food Microbiology*, *161*(1), pp. 53–59.
83. Donsì, F., Annunziata, M., Sessa, M. and Ferrari, G., 2011. Nanoencapsulation of essential oils to enhance their antimicrobial activity in foods. *LWT—Food Science and Technology*, *44*(9), pp. 1908–1914.
84. Klinkesorn, U. and McClements, D.J., 2009. Influence of chitosan on stability and lipase digestibility of lecithin-stabilized tuna oil-in-water emulsions. *Food Chemistry*, *114*(4), pp. 1308–1315.

9 Traditional Knowledge and the Development of Botanical Bioactive Compounds in Brazil for the Control of *Aedes aegypti*

Paula Maria Correa de Oliveira, João Paulo Barreto Sousa, Lorena C. Albernaz, Ellem Suane Ferreira Alves, Laila Salmen Espindola, and Márlia Coelho-Ferreira

9.1 INTRODUCTION

Records from 2000 BC on Hindu populations already point to the use of plants to control insects considered pests, either because they act to destroy crops or to transmit diseases. The Egyptians during the time of the pharaohs, and the Chinese around 1200 BC, also used plants as sources of insecticidal compounds [1]. Travel between the Old and New Worlds allowed for the exchange of plants, animals, and consequently, insects that caused disease. However, these trips also led to the discovery of new ways to control these organisms, as many native cultures already used plant extracts to control pests [2]. The seeds of the *Sabadilla officinarum* Brandt lily, for example, were used by Venezuelan Indians for this purpose in the mid-twentieth century, while in North America, nicotine extracted from *Nicotiana tabacum* L. was the main insecticidal compound used. Another example is the use of cinchona (*Quassia amara* L.) by Central American Indigenous people against pathogenic insects. In China and India, *Acorus calamus* L. was the main species used in this sense. From the exchange of knowledge during these trips, the chemical control of arthropods, based on plants, spread throughout the world [1].

Despite the advent of organo-synthetic insecticides after World War II, plants remain promising candidates for the formulation of insecticides and biocides. It is due to the increase in problems related to insect resistance to synthetic/organo-synthetic insecticides, the resurgence and eruption of pests, and the problems arising from the indiscriminate use of these substances in the environment [3].

There are different ways of selecting plants for the discovery and development of new drugs, which can directly influence the time, effectiveness, costs, and other aspects involve in the preparation of a ready-to-use drug. The pathways used can be random selection followed by chemical screening, random selection followed by one or more biological assays, monitoring biological activity reports, use of databases and ethnobiological studies of plants [4].

This last approach, in general, is based on studies in the area of ethnobotany and ethnopharmacology, which have contributed substantially to the discovery of drugs since the nineteenth century [5]. Several studies in these areas have investigated the conception of local populations about insecticidal plants, as well as diseases transmitted by mosquitoes, and have shown as promising species for the control of vectors such as *Aedes aegypti* [6–9].

DOI: 10.1201/9781003462804-12

Aedes aegypti is one of the most important mosquito species vectors of the viruses that cause dengue, Zika, chikungunya, and urban yellow fever arboviruses, which together cause thousands of deaths annually in most of the world [10]. The control of this mosquito has been a public health challenge, since its distribution currently occurs in all continents, except Antarctica. Studies indicate that global climate change can boost the adaptation of this species to habitats that were previously inhospitable and may become favorable [11]. In view of this, this study presents a bibliographic review of plants traditionally documented in Brazil as repellents and/or insecticides, in order to highlight potential species to advance research on the control of *Aedes aegypti.*

9.2 MATERIALS AND METHODS

The compiled studies were found in Scopus, Google Academic, Scielo, Pubmed, Science Direct, FioCruz, and CAPES Portal de Periódicos databases using the keywords *ethnobotanical studies, ethnopharmacology, traditional medicine, medicinal plants, repellent plants,* and *plant insecticides.*

The selection research followed the criteria: (1) ethnobotanical studies developed in Brazil from 2018 to 2022, published in national and international journals, in Portuguese, English, and Spanish, (2) that presented one or more reports that the use of plants by local and/or traditional populations as repellents, as insecticides, or for "insect bites" (3) works with the presentation of sufficiently clear methodological guidelines regarding the precision in the identification of species.

9.3 RESULTS AND DISCUSSION

Among the 424 works consulted, 45 recorded plant species were traditionally used against mosquitoes (Table 9.1). This research, developed in the field of ethnobotany and ethnopharmacology, was realized in municipalities from 16 states located in the five regions of Brazil. These compiled represent a heterogeneous sample of the different sociocultural contexts, such as indigenous lands, extractive reserves, urban areas, riverside communities, quilombolas, and small farmers.

The higher number of species was recorded in the North region (39%), followed by the Northeast (21%), Southeast (14%), South (13%), and Midwest (12%). In recent years, the North region has appeared as a stage for the endemism of diseases, such as dengue, malaria, leishmaniasis, and others classified among the neglected diseases [12]. The epidemiological profile of a given region reflects the use of resources as a control alternative by local populations. In addition, this is an issue that is neglected by health authorities, especially in areas where live the most vulnerable populations [13].

9.3.1 Insecticide and Repellent Plants Traditionally Used in Brazil

We documented the use of the 103 species belonging to 79 genera and 43 botanical families (Table 9.1). Asteraceae, Fabaceae, and Poaceae are the most representative families (Figure 7.2). These families also stand out in reviews of species traditionally used as repellents in Brazil and other parts of the world [14, 15]. Asteraceae and Fabaceae are among the families whose species provide essential oils that have been extensively tested against *A. aegypti.* Asteraceae was the most important for adulticidal, repellent, and oviposition effects [16]. Among the substances of insecticidal importance in this family are the pyrethrins isolated from the dried flowers of *Chrysanthemum cinerariaefolium* Vis., different monoterpenes, sesquiterpenes, and phototoxic thiophenes, such as tertienyl (2,2':5',2"-tertiophene) present in the genera *Tagetes* and *Porophyllum* [17].

Several species of Fabaceae have rotenone, a potent insecticide, which gives them high commercial value [18]. Essential oils and extracts of the plants Poaceae, especially those of the genus *Cymbopogon*, are used as ingredients in herbal mosquito repellents [19]. Citral and geranyl acetate constitute the majority components of the essential oils of the Poaceae plants, to which different insecticidal activities are attributed [20].

Recorded species are predominantly trees (47%) native to Brazil (65%) (Table 9.1, Figure 7.3a). Woody plants stand out mainly in the North, Northeast, and Midwest regions. In the South and Southeast regions, exotic herbs are more frequent. There are several herbs whose biological potential is known throughout the country. Regardless, the expressiveness of tree and native species may be related to the availability of these resources throughout the year, their versatility, and their cultural factors. Tree resources have stood out among ethnobotanical and ethnopharmacological research in the Northeast, both because herbs and leaves of deciduous woody plants are only available for a short period (generally two to four months) in habitats such as the Brazilian Caatinga and because supply is often unpredictable due to annual variation in climatic conditions. This result points to the importance of the sustainable exploitation of tree resources, since native plants are sources of substances of economic, technological, and social importance, especially for the control of disease-carrying mosquitoes.

9.3.2 Plant Parts Used and Methods of Preparation and Application of Botanical Repellents and Insecticides

Some of the evaluated works did not present information about the plant parts used or details about the methods of preparation and use of the mentioned species as repellents or insecticides. Different plant organs were mentioned in traditional preparations, with leaves (43%) most frequently used (Figure 7.3b). Fumigation and infusion are the most commonly used preparation methods (Figure 7.3c). Concerning application, topical use (50%) and smoking (32%) stood out (Figure 7.3d) (Table 9.1). Topical use often occurs in the application of vegetable oils and juices throughout the body, as well as through macerated or infusions baths. Sample plants with topical application are effective in repelling mosquitoes and, in addition, are a little or not toxic to humans [21]. Fumigation occurs from the burning of plant parts so that the resulting smoke is spread around the residences in order to ward off mosquitoes. This process mostly involves woody parts of trees or shrubs, or even residues resulting from processing or plant use. Examples are the bark of coconut fruits (*Cocus nucifera* L.), residues of andiroba seeds, bark of species such as *Bertholletia excelsa* Bonpl., *Hymenaea courbaril* L. Expulsion of smoke and/or heat stood out among the methods used in the preparation of herbal repellents on the African continent [15]. Dube et al. (2011) evaluated the repellent potential of species traditionally used in Ethiopia to ward off mosquitoes and observed that volatile extracts from the smoke of burning dried leaves were found to be more repellent than those from fresh leaves.

9.3.3 Species with the Highest Citation Frequency

Carapa guianensis stood out among the species recorded in this review, followed by *Ruta graveolens* L., *Cymbopogon nardus* (L.) Rendle, *Petiveria alliacea* L., *Azadirachta indica* A. Juss. and *Cymbopogon winterianus* Jowitt ex Bor (Table 9.1). These species, except *C. guianensis*, are aromatic, release substances capable of repelling insects, including mosquitoes, due to their neurotoxic activities, through GABA–octopamine synapses and acetylcholinesterase inhibition [18].

Andiroba is historically used in the North region as a repellent and an insecticide. The oil extracted from its seeds is commonly applied as a topical repellent or through fumigation with smoke burn of the seed or stem bark. This oil, however, did not show significant repellency results against *A. aegypti* [22]. On the other hand, the hexane/ethyl acetate/dichloromethane seed shell extract showed larvicidal activity as well as a residual effect of approximately ten days against L3 larvae of these mosquitoes [9].

R. graveolens use occurs in all regions of Brazil (Table 9.1). This plant, known for its odorous characteristic, is used as a spatial repellent, topical, and hair insecticide. Its essential oil shows significant larvicidal and repellent activity against *A. aegypti* [23].

The *Cymbopogon* species mentioned earlier are known worldwide as citronella. These herbs are employed as a repellent and/or insecticide extending throughout the country. Beyond essential

TABLE 9.1
Plant Species Traditionally Indicated as Repellents or Insecticides in Ethnobotanical Studies Conducted in Brazil

Family/Species	Traditional Name	Use	H	Or	UP	Mode of Preparation and Application	Brazilian States	Reference
Acanthaceae								
Avicennia germinans (L.) L.	Siriúba	Repellent	Tree	Nat	—	Fumigation/smoke	PA	[30]
Amaranthaceae								
Amaranthus viridis L.	Bredo-de-porco	Repellent	Herb	Ex	—	—	PE	[31]
Amaranthus spinosus L.	Bredo-de-espinho	Repellent	Herb	Ex	—	—	PE	[31]
Dysphania ambrosioides (L.) Mosyakin and Clemants	Santa maria, herb-de-santa maria	Repellent of fleas	Herb	Ex	—	—	SC	[32]
	Mastruz, mastruço	Insecticide Repellent	Shrub	Ex	Leaves	Fumigation/ smoke	AP	[33]
Anacardiaceae								
Anacardium occidentale L.	Caju	Repellent	Tree	Nat	Fruits	Infusion bath	PA	[34]
Annona exsucca DC.	Envira-preta	Repellent	Tree	Nat	Bark	Maceration bath	PA	[9]
Annona squamosa L.	Ata	Insecticide	Tree	Ex	Leaves	Maceration	PE	[35]
Astronium urundeuva (M.Allemão) Engl.	Aroeira	Picada de inseto	Tree	Nat	Stem bark	—	PE	[36]
Annonaceae			Tree					
Annona coriacea Mart.	Araticum	Insecticide	Tree	Nat	Seed	*In natura* topical	PI	[37]
Annona crassifla Mart.	Araticum	Insecticide	Tree	Nat	—	—	GO	[38]
Apocynaceae			Tree					
Aspidosperma nitidum Benth. ex Müll.Arg.	Carapanã	Repellent	Tree	Nat	Stem bark	Maceration bath	PA	[9]
Tabernaemontana catharinensis A.DC.	Cobrina	Repel insects	Tree	Nat	Leaves	Alcoholic infusion, oral	RS	[39]
Geissospermum argenteum Woodson	Taquarirana	Repellent	Tree	Nat	Stem bark	Maceration bath	PA	[9]
Araceae								
Montrichardia linifera (Arruda) Schott	Aninga	Repellent	Shrub	Nat	Leaves	Fumigation/smoke	PA	[9]

(Continued)

TABLE 9.1 (*Continued*)
Plant Species Traditionally Indicated as Repellents or Insecticides in Ethnobotanical Studies Conducted in Brazil

Family/Species	Traditional Name	Use	H	Or	UP	Mode of Preparation and Application	Brazilian States	Reference
Araucariaceae								
Columbea angustifolia Bertol.	Pinheiro-do-paraná	Repellent	Ar	Nat	Leaves	Infusion, topical	PR	[40]
Arecaceae								
Astrocaryum vulgare Mart.	Tucum	Repellent	Ar	Nat	Leaves, fruits	Fumigation/smoke	PI	[37]
Cocos nucifera L.	Coco	Repellent	Herb	Ex	Fruits	Fumigation/smoke	PA	[30]
	Coco	Repellent	Herb	Ex	Fruits	Fumigation/smoke	BA	[41]
	Coco	Repellent	Herb	Ex	Fruits	Fumigation/smoke	BA	[41]
Copernicia prunifera (Mill.) H.E.Moore	Carnaúba	Repellent	Tree	Nat	Leaves	Fumigation/smoke	PI	[37]
Astrocaryum aculeatum G.Mey.	Tucumã	Repellent	Tree	Nat	—	*In natura*	RO	[42]
		Repellent	Tree	Nat	—	Fumigation/smoke	PA	[30]
Aristolochiaceae								
Aristolochia triangularis Cham. and Schltdl.	Cipó-mil-homens	Repellent	Liana	Nat	Leaves, stem bark	Infusion, topical	PR	[40]
Asparagaceae								
Aloe arborescens Mill.	—	Picada de mosquito	Herb	Ex	—	—	SC	[43]
Aloe vera (L.) Burm.f.	Babosa	Insecticide, repellent	Herb	Ex	Leaves	—	MG	[44]
Asteraceae								
Artemisia alba Turra	Canforinha	Repellent	Tree	Ex	—	—	RJ	[45]
Bidens pilosa L.	Picão, pião-branco	Repel insects	Herb	Ex	—	—	SP	[46]
Cabobanthus polysphaerus (Baker) H.Rob.	Assa-peixe	Repellent	Herb	Ex	Leaves, roots	Infusion, topical	PR	[40]
Calea unifla Less.	Arnica-da-praia, arnica-do campo	Repel insects	Herb	Nat	Flowers	Juice, topical	SC	[47]
Lychnophora ericoides Mart.	—	Repellent	Tree	Nat	All plant	Alcoholic infusion	GO	[48]
Mikania glomerata Spreng.	Guaco	Repel insects	Lia	Nat	—	—	SP	[49]
Rolandra fruticosa (L.) Kuntze	Carica-á	Repellent	Tree	Nat	Roots	Infusion bath	PA	[9]
Solidago chilensis Meyen		Repellent	Shrub	Nat	All plant	Alcoholic infusion	GO	[48]
	Arnica	Picada de inseto	Shrub	Nat	—	—	SP	[46]

Sphagneticola trilobata (L.) Pruski	—	Picada de mosquito	Herb	Nat	—	—	SC	[43]
Symphyopappus casarettoi B.L.Rob.	Vassoura-de-bicho, vassourinha, vassourão-braba, vassoura-mansa	Evitar picada de inseto	Shrub	Nat	—	Tea, oral	SC	[32]
Tagetes erecta L.	Cravo-de-defunto	Insecticide ou repellent	Herb	Ex	Flowers	—	MG	[44]
Crescentia cujete L.	Cuieira	Repellent	Shrub	Ex	Fruits	*In natura*, topical	PA	[50]
		Repellent	Shrub		—	Fumigation/smoke	PA	[30]
Mansoa alliacea (Lam.) A.H.Gentry	Cipó-d'alho	Repellent	Liana	Nat	Leaves	Juice, topical	PA	[50]
Bixaceae								
Bixa orellana L.	Coloral	Repellent	Tree	Nat	Fruits, seeds	Infusion	MT	[51]
Burseraceae								
Protium heptaphyllum (Aubl.) Marchand	Breu branco	Repellent	Tree	Nat	Stem bark	Fumigation/smoke	PA	[30]
	Breu branco	Repellent	Tree	Nat	—	—	PA	[30]
Caryophyllaceae								
Drymaria cordata (L.) Willd. ex Roem. and Schult.	Herb-sapinho	Insecticide, antipulga	Herb	Ex	—	—	RJ	[45]
Calophyllaceae								
Calophyllum brasiliense Cambess.	Jacareúba	Repellent	Tree	Nat	Resin	—	AM	[52]
Convolvulaceae								
Ipomoea asarifolia (Desr.) Roem. and Schult.	Salsa	Repellent	Liana	Nat	Leaves	Alcoholic infusion bath	PA	[50]
Ipomoea carnea Jacq.	Algodão-bravo	Repellent	Liana	Nat	Leaves	Infusion bath	PA	[50]
Cannabaceae								
Trema micrantha (L.) Blume	Grandiuva, periquiteiro	Repel insect	Tree	Nat	—	—	MT	[53]
Combretaceae								
Laguncularia racemosa (L.) C.F.Gaertn.	Tinteiro	Repellent	Tree	Nat	—	Fumigation/smoke	PA	[30]
Commelinaceae								
Murdannia nudifla (L.) Brenan	Trapoeraba	Picada de inseto	Herb	Ex	Leaves	—	RJ	[54]
Chrysobalanaceae								

(Continued)

TABLE 9.1 (*Continued*)
Plant Species Traditionally Indicated as Repellents or Insecticides in Ethnobotanical Studies Conducted in Brazil

Family/Species	Traditional Name	Use	H	Or	UP	Mode of Preparation and Application	Brazilian States	Reference
Hymenopus heteromorphus (Benth.) Sotherbs and Prance	Caripé	Repellent	Tree	Nat	—	—	PA	[55]
Cucurbitaceae								
Cayaponia botryocarpa C.Jeffrey	Cipó-alho-bravo	Repellent	Liana	Nat	Leaves	Topical	AM	[56]
Momordica charantia L.	Melão-de-são-caetano	Repellent	Liana	Ex	Leaves	Juice, tea	PA	[50]
	Melão-de-são-caetano	Repellent	Liana	Ex	Leaves	Decoction, fumigation	GO	[48]
	Melão-são-caetano	Insecticide	Liana	Ex	Leaves	*In natura*, topical, spatial	PI	[37]
Ebenaceae								
Diospyros guianensis L. (Aubl.) Gürke	Maria preta	Repel insects from fungal wounds	Tree	Nat	Fruits	*In natura* topical: minced fruits applied as poultice	PA	[9]
Euphorbiaceae								
Croton campestris A.St.-Hil.	Velame	Repellent, insecticide	Tree	Nat	Leaves	*In natura* topical	PI	[37]
Croton tricolor Klotzsch ex Baill.	Sacatinga, marmeleiro-branco	Repellent	Tree	Nat	—	—	PE	[31]
Jatropha gossypiifolia L.	Peão-roxo	Repellent	Tree	Nat	Leaves	Juice	MA	[57]
Manihot esculenta Crantz	Mandioca	Insecticide	Tree	Ex	Roots	Topical	PI	[37]
Ricinus communis L.	Mamona	Repellent	Tree	Ex	Seed	Fumigation/smoke	PA	[50]
Sapium marmieri Huber.	Burra-leiteira	Repellent	Tree	Nat	Látex	Fumigation/smoke	PA	[58]
Fabaceae								
Amburana acreana (Ducke) A.C.Sm.	Cumaru-de-cheiro	Repellent	Tree	Nat	Stem bark	Decoction, fumigation	AM	6[1]
	Imburana, amburana-decheiro, cerejeira-do-nordeste	Repellent	Tree	Nat	Stem bark	—	MT	[59]
Amburana cearensis (Allemão) A.C.Sm.	Imburana	Repellent	Tree	Nat	Stem bark	Fumigation/smoke	PI	[37]
Copaifera langsdorffii Desf.	Copaíba, paud'óleo, podoi	Repellent	Tree	Nat	Stem bark, oilseed	Decocção, maceração, *in natura*	MT	[60]

Derris sp.	Timbó	Repellent	Tree	—	Roots	—	RO	[42]
Hymenaea courbaril L.	Jatobá	Repellent	Tree	Nat	Stem bark	Decoction, fumigation	PI	[37]
	Jatobá/breu jutaicíca	Repellent	Tree	Nat	Stem bark	Maceration/fumigation	PA	[9]
Libidibia ferrea (Mart. ex Tul.) L. P. Queiroz	Jucá	Picada de inseto	Tree	Nat	Fruits	—	PA	[61]
Pterodon emarginatus Vogel	Sucupira	Repellent	Tree	Nat	Leaves	Spatial repellent	PI	[37]
Lamiaceae								
Hyptidendron canum (Pohl ex Benth.) Harley	Hortelã-do-campo	Repellent	Herb	Nat	—	—	MT	[62]
Plectranthus barbatus Andrews	Boldo	Insecticide	Herb	Ex	Leaves	Topical	PI	[37]
Lecythidaceae								
Bertholletia excelsa Bonpl.	Castanheira	Repellent	Tree	Nat	Stem bark, fruits	Fumigation/smoke	AM	[63]
Couratari sp.	Tauari	Repellent			Stem bark	—	PA	[58]
Malvaceae								
Sida spinosa L.	Malva-lanceta, malva-relógio	Repel insect	Shrub	Nat	Leaves	Topical	PB	[64]
Meliaceae								
Carapa guianensis Aubl.	Andiroba	Repellent	Tree	Nat	Roots	Fumigation/smoke	AM	[63]
		Repellent	Tree		Seeds	Fumigation/smoke	PA	[30]
		Repellent	Tree		Seeds	*In natura*	AM	[56]
		Repellent	Tree		Seeds	*In natura* topical	PA	[65]
		Repellent	Tree		Oilseed	*In natura* topical	AM	[66]
		Repellent	Tree		Oilseed	*In natura* topical	AM	[67]
		Repellent, insecticide	Tree		Oilseed	*In natura* topical	AM	[68]
		Repellent	Tree		Oilseed	*In natura* topical	PA	[50]
		Repellent	Tree		Oilseed	*In natura* topical	AP	[32]
		Repellent	Tree		Oilseed	Tea, juice, topical	PA	[69]
		Picada de inseto	Tree		Oilseed, stem bark	*In natura* topical	PA	[61]
		Repellent	Tree		—	—	PA	[55]
		Repellent	Tree		Oilseed	*In natura*/fumigation	PA	[9]
Melia azedarach L.	Cinamão	Repellent	Tree	Ex	Leaves	Infusion, topical	Pr	[40]

(Continued)

TABLE 9.1 (*Continued*)
Plant Species Traditionally Indicated as Repellents or Insecticides in Ethnobotanical Studies Conducted in Brazil

Family/Species	Traditional Name	Use	H	Or	UP	Mode of Preparation and Application	Brazilian States	Reference
Azadirachta indica A.Juss.	Nin	Repellent	Tree	Ex	Leaves	Juice, topical	PA	[50]
		Insecticide e repellent	Tree	Ex		—	PA	[70]
		Repellent	Tree	Ex	Leaves	Infusion, topical	SP	[71]
		Repellent, insecticide			Leaves, all plant	Spatial	PI	[37]
Moringaceae								
Moringa oleifera Lam.	Noni	—	Tree	Ex	—	—	MT	[60]
Myrtaceae								
Corymbia citriodora (Hook.) K. D. Hill and L. A. S. Johnson	—	Repellent	Tree	Ex	Leaves	Infusion	GO	[48]
Eucalyptus cinerea F.Muell. ex Benth.	Eucalipto	Repel insects	Tree	Ex	Leaves	Alcoholic infusion	RJ	[72]
Eucalyptus sp.	Eucalipto	Repellent	Tree	Ex	Leaves	Fumigation/smoke	PI	[37]
Syzygium sp.		Repel insects	Tree	Ex	—	—	SP	[46]
Piperaceae								
Piper callosum Ruiz and Pav.	Óleo elétrico	Repel insects	Shrub	Nat	Leaves	Tea	AM	[73]
	Elixir paregórico	Repellent	Shrub	Nat	Roots	Infusion, topical	PA	[34]
Piper klotzschianum (Kunth) C. DC.	Jambú do mato	Repellent	Shrub	Nat	Leaves	Topical: crushed and applied	PA	[9]
Phytolaccaceae					Leaves			
Petiveria alliacea L.	Mucura-caá	Repellent	Herb	Ex	Leaves	Juice	PA	[50]
		Repellent	Herb	Ex	Roots Leaves	Juice, topical	PA	[69]
		Repellent	Herb	Ex	Leaves	Fumigation/smoke infusion/ topical	AP	[32]
	Guiné	Repellent	Herb	Ex	Leaves	Infusion	Pr	[40]
Plantaginaceae								
Stemodia maritima L.	Meladinha	Repellent	Herb	Nat	All plant	Juice	PB	[64]
Poaceae								
Cymbopogon citratus (DC.) Stapf	Capim santo, capim cheiroso, capim marinho, capim limão	Repellent	Herb	Ex	All plant	Fumigation/smoke	AP	[32]

Cymbopogon nardus (L.) Rendle	—	Repellent	Herb	Ex	Leaves	Alcoholic infusion	GO	[48]
	Citronela	Repellent	Herb	Ex	Leaves	Alcoholic infusion	SP	[74]
	Citronela	Repellent	Herb	Ex	Leaves	Maceration, topical	SP	[74]
	Citronela	Insecticide ou repellent	Herb	Ex	Leaves	—	MG	[44]
	Capim-citronela	Repellent	Herb	Ex	Leaves	—	BA	[41]
Cymbopogon winterianus Jowitt ex Bor	—	Repellent	Herb	Ex	Leaves	—	BA	[41]
	Citronela	Repellent	Herb	Ex	Leaves	Alcoholic infusion	MT	[59]
	Citronela	Repellent	Herb	Ex	—	—	RJ	[45]
Cymbopogon sp.	Citronela	Repellent	Herb	Ex	All plant	Spatial	PI	[37]
Guadua superba Huber	Taboca	Repellent	Tree	Nat	—	Fumigation/smoke	PA	[30]
Guadua weberbaueri Pilg.	Tabuqui	Repel insects	Herb	Nat	Leaves	*In natura*	AM	[73]
Melinis minutifla P.Beauv.	Capim-gordura	Repel insects	Herb	Ex	All plant	Infusion	RJ	[72]
Polygonaceae								
Muehlenbeckia sagittifolia (Ortega) Meisn.	Salsaparrilha	Repel insects	Shrub	Nat	—	—	RS	[34]
Rubiaceae								
Hamelia patens Jacq.	Falsa-erva-de-rato; amélia	Repel insects	Tree	Nat	—	—	—	[75]
Ruta graveolens L	Arruda	Insecticide	Herb	Ex	—	—	PA	[76]
		Insecticide and repellent	Herb	Ex	Leaves, flower	Tea, maceration	AP	[32]
	Arruda	Insecticide and repellent	Herb	Ex	Leaves, flower	—	MG	[44]
Ruta graveolens L	Arruda	Repellent	Herb	Ex	Leaves	Alcoholic infusion	MT	[59]
	Arruda	Insecticide, antipulga	Herb	Ex	—	—	RJ	[45]
	Arruda	Repellent	Herb	Ex	Leaves, roots	Infusion	Pr	[40]
	Arruda	Insecticide	Herb	Ex	Leaves	—	PI	[37]
Rutaceae								
Citrus limon (L.) Osbeck	Limão-azedo	Insecticide	Tree	Ex	Fruit	—	PI	[37]
Rhamnaceae								
Ziziphus joazeiro Mart	Juá	Repellent	Tree	Nat	Leaves	Juice, topical	PA	[50]

(*Continued*)

TABLE 9.1 ***(Continued)***
Plant Species Traditionally Indicated as Repellents or Insecticides in Ethnobotanical Studies Conducted in Brazil

Family/Species	Traditional Name	Use	H	Or	UP	Mode of Preparation and Application	Brazilian States	Reference
Rhizophoraceae								
Rhizophora racemosa G.Mey.	Mangue	Repellent	Tree	Nat	—	Fumigation/smoke	PA	[30]
Sapindaceae								
Magonia pubescens A.St.-Hil.	Tingui	Insecticide, mata piolho	Tree	Nat	—	—	GO	[38]
Sapindus saponária L.	Saboneteira	Insecticide	Tree	Nat	Seed	*In natura* topical	—	[77]
Simaroubaceae								
Homalolepis maiana (Casar.) Devecchi and Piran	Prá-tudo	Insecticide	Tree	Nat	Stem bark	—	PI	[37]
Homalolepis cedron (Planch.) Devecchi and Piran	Pau-para-tudo	Repellent	Tree	Nat	Stem bark	Maceration bath	PA	[78]
Siparunaceae								
Siparuna guianensis Aubl.	Negramina	Repellent	Shrub	Nat	Leaves	Maceration	MT	[60]
Solanaceae								
Capsicum frutescens L.	Pimenta malagueta	Repellent	Shrub	Ex	—	—	PA	[79]
	Pimenta malagueta	Repellent	Shrub	Ex	Seeds	Fumigation/smoke	PI	[37]
Nicotiana tabacum L.	Tabaco	Repellent	Shrub	Ex	—	—	RO	[42]
	Fumo	Insecticide	Shrub	Ex	—	—	MT	[80]
	Fumo	Repellent	Shrub	Ex	Fruit	Infusion topical	PR	[40]
Solanum lycopersicum L.	Tomate	Picada de inseto	Shrub	Ex	Fruit	—	RJ	[54]
Verbenaceae								
Lippia alba (Mill.) N.E.Br. ex Britton and P.Wilson	Salva	Picada de inseto	Shrub	Nat	—	—	RS	[34]
Lippia grata Schauer	Alecrim	Repellent	Shrub	Nat	Leaves	Fumigation/smoke	PI	[37]

Note: H: Habit; *Or*: origin; *Nat*: native to Brazil; *Ex*: exotic or introduced; *UP*: used part; *PA*: Pará; *PE*: Pernambuco; *SC*: Santa Catarina; *AP*: Amapá; *PI*: Piauí; *GO*: Goiás; *RS*: Rio Grande do Sul; *PR*: Paraná; *BA*: Bahia; *RO*: Roraima; *MG*: Minas Gerais; *RJ*: Rio de Janeiro; *SP*: São Paulo; *MT*: Mato Grosso; *AM*: Amazonas; *MA*: Maranhão; *PB*: Paraíba.

oil being traded internationally, Brazilian populations cultivate it in backyards to promote spatial repellency. They also use the maceration or alcoholic infusion of its leaves as a topical repellent. The insecticidal and repellent activities of *C. nardus* and *C. winterianus* have been recorded in different studies against insect pests and mosquito vectors of infectious diseases [24].

Petiveria alliacea, known as mucura-caá, is a naturalized species, widely distributed in Central and South America. Studies report the use of its crushed or burned leaves as repellents. The insecticidal and acaricidal activities of this species were related to the compound dibenzyl-trisulfide (DBTS) isolated from its roots [25]. Later studies showed the significant toxic effect of *P. alliacea* on the whitefly *Trialeurodes vaporariorum* [26], as well as the cytotoxic activity of the hydroalcoholic leaf extract against *A. aegypti* larvae [27].

Azadirachta indica, a tree historically known as neem, originates from India and is widely cultivated in Brazil, especially for its insecticidal and repellent properties. This plant is cultivated around homes to promote spatial repellency, and the juice and infusions of its leaves are applied as topical repellents. Extracts obtained with different solvents, as well as its essential oil, showed insecticidal and repellent activity against insects, including various disease vector mosquitoes, and this activity is attributed to the abundant presence of limonoids in this plant, especially azadiactin [28]. Neem derivatives have taken the industrial market by storm, especially for agricultural pest control, with an estimated market value of US$16.2 billion, and are projected to grow to more than US$27 billion by 2025 [29].

9.3.4 Species Evaluated against the Vector *Aedes aegypti*

Crude extracts and essential and fixed oils of 67 of the 103 species recorded have already been submitted for biological assays against immature and adult forms of *A. aegypti* (Table 9.2). Larvicidal tests are the most frequent (65 spp.), followed by repellent (12 spp.) (Figure 7.02).

The literature specializing in tests against *A. aegypti* considers that a $LC_{50} < 50$ µg/mL is very active, a $LC_{50} = 50–100$ µg/mL is active, and a $LC_{50} > 100$ µg/mL is weak/inactive [16]. The repellency results ranged from 47 to 100%, and the protection period was not disclosed for all species.

Crude extracts are obtained by different extraction methods with organic solvents or water. Only two extract plants show repellency against *A. aegypti*. *C. winterianus* extract at a concentration of 75% showed 92.26% protection against this mosquito during 5 h of exposure [81].

The study that evaluated the topical repellency of extracts obtained from the seeds of *Bixa orellana* L. using three different solvents showed that the dry hexane extract resuspended in acetone was the most active, followed by the hydroalcoholic extract, with 90% and 73% (113.8 µg/mL) of protection, respectively [82].

The extracts recorded in Table 9.2 were mostly submitted to larvicidal activity assays. Of the 39 extracts, 10 showed high larvicidal activity ($LC_{50} < 50$ µg/mL), 4 were active (LC_{50} 50–100 µg/mL), and 18 showed weak activity ($LC_{50} > 100$ µg/mL).

The ethanolic extract of the stem bark of *Annona crassiflora* Mart. showed greater larvicidal activity (LC_{50} 0.71 µg/ml) in a study that evaluated the larvicidal activity of 84 extracts obtained from 51 Brazilian plant species. Ethanol extracts from the fruit and bark of *Anacadium occidentale* also stand out, with LC_{50} values 7.84 µg/mL and 4.01 µg/mL, respectively. In the cashew fruit extract, several phenolic compounds were identified, such as anacardic acids, cardol, and cardanol, which in turn also showed significant larvicidal activity against the same vector (LC_{50} 5.93, 6.14, and 20.32 µg/mL) [83].

The species from which the essential oils were evaluated are predominantly herbaceous and/or aromatic and are the main samples used in repellency studies. Essential oils have been widely investigated for insect control, as they concentrate compounds capable of scaring away different species, which, throughout history, classify them as insecticides, anti-ovipositional, and repellents of great importance, both for the control of insect pests of crops like mosquitoes. In addition, these oils are composed of volatile substances with fast degradation, thus causing little or no damage to other organisms in the environment.

TABLE 9.2
Plant Species Used as Repellents and/or Insecticides, with Their Respective Biological Activity Information against the *Aedes aegypti* Vector

Plant Species	Part Used	Extraction Solvent	Activity	Result	Reference
Aloe vera (L.) Burm.f.	Leaves	Petroleum ether	Larvicidal	LC_{50} 253 μg/mL	[89]
Amburana cearensis (Allemão) A.C.Sm.	Seed	Water	Larvicidal	LC_{50} 8,100 μg/mL	[90]
Anacardium occidentale L.	Fruit	Ethanol	Larvicidal	LC_{50} 7.84 μg/mL	[89]
	Fruit	Ethanol	Pupicidal	LC_{50} 338.28 μg/mL	[83]
	Fruit	Ethanol	Larvicidal	LC_{50} 6.55 μg/mL	[91]
	Oil fruit	—	Larvicidal	LC_{50} 14.5 μg/mL	[92]
	Bark	Hexane	Larvicidal	LC_{50} 4.01 μg/mL	[93]
Annona coriacea Mart.	Seed	Ethanol	Larvicidal	LC_{50} 50 μg/mL	[94]
Annona crassiflora Mart.	Bark root, bark stem	Ethanol	Larvicidal	LC_{50} 0.71 μg/mL	[95]
Annona exsucca DC.	Wood	Hexane:ethyl acetate:dichloromethane (45:45:10)	Larvicidal	Not active	[9]
Annona squamosa L.	Root, leaves, seed	Ethanol	Larvicidal	LC_{50} 31.9 μg/mL	[95]
Aspidosperma nitidum Benth. ex Müll.Arg.	Wood	Hexane:ethyl acetate:dichloromethane	Larvicidal	LC_{50} 137 μg/mL	[9]
Aristolochia triangularis Cham. and Schltdl.	Not informed	Methanol; dichloromethane	Larvicidal	LC_{50} 285.5; 188.1 μg/mL	[96]
Astrocaryum aculeatum G.Mey.	Fruit	Methanol	Larvicidal	LC_{50} 341.5 μg/mL	[97]
Astronium urundeuva (M.Allemão) Engl.	Seed	Ethanol	Larvicidal	LC_{50} 15.2 μg/mL	[98]
Astronium urundeuva (M.Allemão) Engl.	Seed	Water	Larvicidal	LC_{50} 370 μg/mL	[99]
	Seed	Sodium phosphate buffer	Oviposition	98.4% [20%]	[99]
	Wood	NaCl	Larvicidal	LC_{50} 40 μg/mL	[100]
Avicennia germinans (L.) L.	Leaves	Water	Larvicidal	LC_{50} 10.465 μg/mL	[101]
Azadirachta indica A.Juss.	Oil fruit	—	Repellent	84.21% [5%]	[102]
	Leaves	Acetone	Larvicidal	LC_{50} 50 μg/mL	[103]
Bidens pilosa L.	Leaves	Water	Oviposition	37. 5% [not informed]	[104]
Bixa orellana L.	Seed	Hexane	Repellent	90% [113.2 μg/mL]	[82]
	Seed	Water	Larvicidal	Not active	[105]
Calophyllum brasiliense Cambess.	Leaves	Dichloromethane	Larvicidal	Not active	[106]
Capsicum frutescens L.	Fruit	Ethanol	Larvicidal	LC_{50} 141.92 μg/mL	[106]
Carapa guianensis Aubl.	Oilseed	—	Larvicidal	LC_{50} 136 μg/mL	[107]
	Oilseed	—	Repellent	Not active	[102]
	Stem wood	Hexane:ethyl acetate:dichloromethane (45:45:10)	Not active	Not active	[9]

TABLE 9.2 (*Continued*)
Plant Species Used as Repellents and/or Insecticides, with Their Respective Biological Activity Information against the *Aedes aegypti* Vector

Plant Species	Part Used	Extraction Solvent	Activity	Result	Reference
	Seed shells	Hexane:ethyl acetate:dichloromethane (45:45:10)	Larvicidal	LC_{50} 70 μg/mL	[9]
***Citrus limon* (L.) Osbeck**	Essential oil	—	Larvicidal	LC_{50} 1.48 μg/mL	[108]
***Copaifera langsdorffii* Desf.**	Oil bark	—	Larvicidal	LC_{50} 41 μL/mL	[92]
***Corymbia citriodora* (Hook.) K.D. Hill and L.A.S. Johnson**	Essential oil leaves	—	Larvicidal	LC_{50} 38.7 μg/mL	[109]
	Essential oil	—	Repellent	99.34% [1%]	[102]
	Essential oil	—	Repellent	59.4% [0.1 mL/30 cm²] 2 h e 30 min	[85]
***Crescentia cujete* L.**	Leaves	Ethanol	Larvicidal	not active	[110]
***Cymbopogon citratus* (DC.) Stapf**	Essential oil bark	—	Adulticidal	87.3% [5%]	[111]
***Cymbopogon nardus* (L.) Rendle**	Essential oil leaves	—	Repellent	47.03% 6 h	[87]
	Essential oil leaves	—	Larvicidal	LC_{50} 35.133 μg/mL	[112]
	Commercial essential oil	—	Repellent	60% [50%] 2 h	[113]
***Cymbopogon winterianus* Jowitt ex Bor**	Essential oil	—	Larvicidal	LC_{50} 38.8 μg/mL	[114]
	—	Crude extract	Repellent	92.26% [5%]	[81]
	Essential oil	—	Repellent	75.7% [0.1 mL/30 cm²] 2 h	[85]
***Drymaria cordata* (L.) Willd. ex Roem. and Schult.**	Leaves	Petroleum ether, methanol, acetone, water	Larvicidal	Not active	[115]
***Dysphania ambrosioides* (L.) Mosyakin and L Cemants**	Aerial parts	N-hexane	Larvicidal	LC_{50} 50 μg/mL	[116]
***Diospyros guianensis* L. (Aubl.) Gürke**	Fruits	Hexane:ethyl acetate:dichloromethane (45:45:10)	Larvicidal	LC_{50} 62 μg/mL	[9]
***Eucalyptus cinerea* F.Muell. ex Benth.**	Essential oil	—	Larvicidal	LC_{50} 0.38 μg/mL	[117]
***Geissospermum argenteum* Woodson**	Stem bark, wood, leaves	Hexane:ethyl acetate:dichloromethane (45:45:10)	Larvicidal	Not active	[9]
***Hymenaea courbaril* L.**	Essential oil fruit	—	Larvicidal	LC_{50} 14.8 μg/mL	[118]
***Homalolepis cedron* (Planch.) Devecchi and Piran**	Stem, leaves, wood	Hexane:ethyl acetate:dichloromethane (45:45:10)	Larvicidal	Not active	[9]

(*Continued*)

TABLE 9.2 (*Continued*)
Plant Species Used as Repellents and/or Insecticides, with Their Respective Biological Activity Information against the *Aedes aegypti* Vector

Plant Species	Part Used	Extraction Solvent	Activity	Result	Reference
Ipomoea carnea Jacq.	Stem	Ethyl acetate	Larvicidal	LC_{50} 119.61 µg/mL	[119]
***Libidibia ferrea* (Mart. ex Tul.) L.P.Queiroz**	Fruit	Ethanol 70%	Larvicidal	LC_{50} 14,900 µg/mL	[120]
	Seed	Sodium phosphate (50 mm	Larvicidal	LC_{50} 12,020 µg/mL	[99]
***Lippia alba (Mill.) N.E.Br. ex Britton and* P.Wilson**	Essential oil aerial parts	—	Larvicidal	LC_{50} 113.99 µg/mL	[121]
	Oil essential leaves	—	Oviposition	50.4% at 1,000 µg/mL	[122]
	Oil essential oil	—	Adulticidal	Not active	[122]
***Lippia alba (Mill.) N.E.Br. ex Britton and* P.Wilson**	Oil essential	—	Repellent	50% [1 mg/cm²]	[122]
Lippia grata Schauer	Essential oil	—	Larvicidal	LC_{50} 22.79 µg/mL	[123]
Magonia pubescens A.St.-Hil.	Bark stem	Ethanol	Larvicidal	LC_{50} 60,000 µg/mL	[124]
Manihot esculenta Crantz	Leaves	Water	Larvicidal	LC_{50} 66.14 µg/mL	[125]
Melia azedarach L.	Leaves	Ethanol	Larvicidal	LC_{50} 760 µg/mL	[126]
	Leaves	Ethanol	Oviposition	1g/L	[126]
	Seed	Ethanol	Larvicidal	LC_{50} 34 µg/mL	[127]
Momordica charantia L.	Fruit	N-hexane	Larvicidal	LC_{50} 122.45 µg/mL	[128]
	Flower, fruit	Hexane ethyl acetate methanol (400 ml; 1:1).	Larvicidal	LC_{50} 37.2 µg/mL	[129]
	Leaves	Water	Oviposition	87.7%	[104]
Moringa oleifera Lam.	Seed	Water	Larvicidal	LC_{50} 1,260 µg/mL	[130]
Nicotiana tabacum L.	Leaves	Ethanol	Adulticidal	90% [5%] 4 h	[131]
	Leaves	Ethanol	Larvicidal	LC_{50} 1,022.97 µg/mL	[131]
Petiveria alliacea L.	Leaves	Water	Larvicidal	LC_{50} 25% *v/v*	[27]
Plectranthus barbatus Andrews	Leaves	Acetone	Larvicidal	LC_{50} 610 µg/mL	[132]
***Protium heptaphyllum (Aubl.)* Marchand**	Leaves, bark	Ethanol	Larvicidal	Not active	[133]
***Protium heptaphyllum (Aubl.)* Marchand**	Essential oil nanoemulsion	—	Adulticidal	200 µg/mL	[134]
	Essential oil nanoemulsion	—	Repellent	77.67%. [200 µg/mL] 3 h	[134]
Pterodon emarginatus Vogel	Essential oil fruit	—	Larvicidal	LC_{50} 134.90 g/mL	[135]
Pterodon emarginatus Vogel	Fruit	N-Hexane	Larvicidal	LC_{50} 23.99 g/mL	[135]
Ricinus communis L.	Oilseed	—	Oviposition	LC_{50} 0.26 µL/mL	[136]

TABLE 9.2 (*Continued*)

Plant Species Used as Repellents and/or Insecticides, with Their Respective Biological Activity Information against the *Aedes aegypti* Vector

Plant Species	Part Used	Extraction Solvent	Activity	Result	Reference
	Oilseed	—	Larvicidal	LC_{50} 0.029 μL/mL	[136]
	Oilseed	—	Repellent	86% [100%] 3 h	[137]
	Leaves	Methanol	Larvicidal	LC_{50} 191.54 μL/mL	[138]
	Seed	Methanol	Larvicidal	LC_{50} 15.52 μL/mL	[138]
Rolandra fruticosa (L.) Kuntze	Aerial parts, roots	Hexane:ethyl acetate:dichloromethane (45:45:10)	Larvicidal	Not active	[9]
Ruta graveolens L	Essential oil leaves	—	Larvicidal	LC_{50} 14.37 μL/mL	[23]
	Essential oil leaves	—	Repellent	100% [0.187 mg/cm²]	[23]
	Essential oil leaves	—	Larvicidal	LC_{50} 61.641 μL/mL	[139]
Sapindus saponária L.	Fruit	Methanol	Larvicidal	LC_{50} 601.81 μL/mL	[140]
Siparuna guianensis Aubl.	Essential oil leaves	—	Oviposition	80% [0.5 μg/mL]	[88]
	Essential oil leaves	—	Oviposition	1 μg/mL	[88]
	Stem, leaf, and fruit essential oils		Larvicidal	LC_{50} 1.76, 0.98, 2.46 μg/mL	[88]
	Essential oil leaves	—	Pupicidal	LC_{50} 1.69 μg/mL	[88]
	Essential oil leaves	—	Repellent	100% [0.450 μg/cm²] 2 h	8[88]
Solanum lycopersicum L.	Leaves	Methanol	Larvicidal	LC_{50} 20,323 μL/mL	[141]
Solidago chilensis Meyen	Aerial parts	Water	Larvicidal	Not active	[142]
Stemodia maritima L.	Essential oil leaves	—	Larvicidal	LC_{50} 55.4 μL/mL	[143]
	Essential oil bark		Larvicidal	LC_{50} 22.9 μL/mL	[143]
Tagetes erecta L.	Essential oil steam		Larvicidal	LC_{50} 79.78 μg/mL	[144]
Ziziphus joazeiro Mart				Not active	[145]

Source: According to the studies in Table 9.1.

Cymbopogon species are the most studied regarding repellency. These plants have repellent activity not only against *A. aegypti* but also against other vector mosquitoes, such as *Anopheles stephensi* and *Anopheles dirus* [84].

Cymbopogon winterianus Jowitt ex Bor essential oil applied at a rate of 0.1 mL/30 cm² provided 75.7% protection for 2 h against *A. aegypti.* This result was more effective than that observed with the application of standard DEET 20%, a commercial repellent [85]. The main compounds of the essential oil of *C. winterianus*, citronellal, citrolenol, and geraniol, were tested separately and showed repellency >70% for 1 h against *A. aegypti* [86]. The essential oil of *C. nardus* showed 47.03% protection for 6 h against this vector [87] (Figures 9.1 and 9.2).

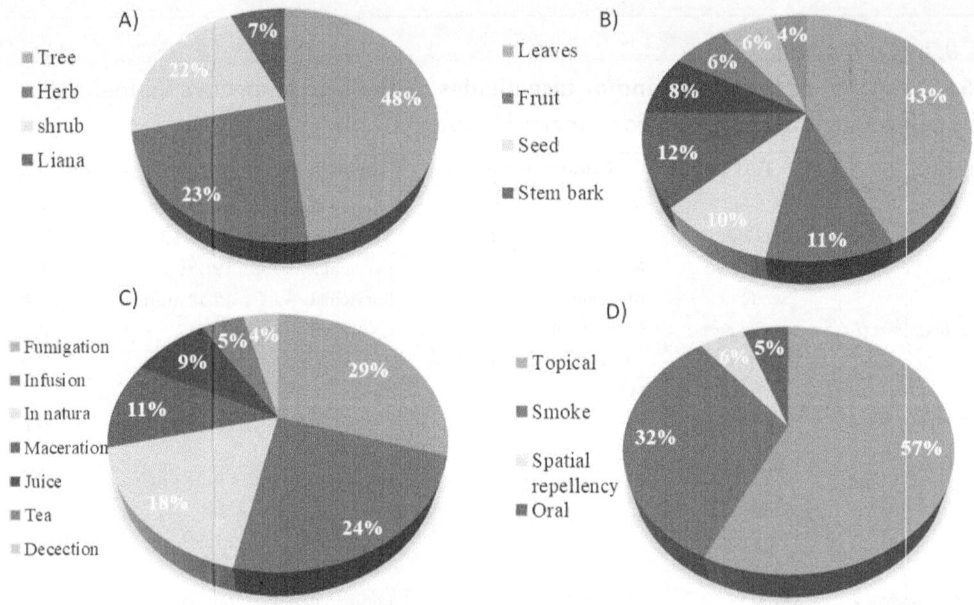

FIGURE 9.1 Ethnobotanical characteristics of the species traditionally used in Brazil as insecticides and repellents: (a) habits; (b) used plant parts; (c) forms of preparation; (d) modes of application.

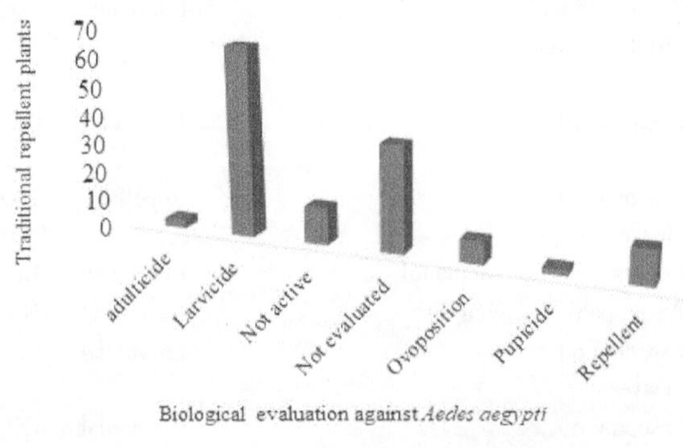

FIGURE 9.2 Status of biological evaluation studies of plants traditionally used as repellents against *Aedes aegypti.*

The essential oils obtained from the leaves, stems, and fruits of the native species *Siparuna guianensis* Aubl. showed significant ovicidal (80% [0.5 μg/mL), larvicidal (LC_{50} 1.76 μg/mL), pupicidal (LC_{50} 1.69 μg /mL), anti-oviposition (1 μg/mL), and repellent (100% [0.450 μg/cm² 2 h]). The main components of the leaf oils were β-myrcene (79.71%) and 2-undecanone (14.58%); from the stem, β-myrcene (26.91%), δ-elemene (20.92%), germacrene D (9.42%), α-limonene (7.91%), and bicyclo-germacrene (7.79%); and from the fruit, 2-tridecanone (38.75%), 2-undecanone (26.5%), and β-myrcene (16.42%) [88].

9.4 CONCLUSION

This result points to two important issues. The first refers to the progress in research carried out with plants widely used by traditional populations in Brazil, bringing to light species that can act as sources of effective products in the control of the *A. aegypti*. On the other hand, almost half of the species recorded for their traditional use have not yet been the subject of studies related to the control of this vector, which may direct further research in the area of biopesticides.

Therefore, ethnobotany and ethnopharmacology studies encourage the development of research regarding mosquito vectors using plant species. This parallel between the traditional use of repellent plants and their potential against this mosquito reinforces the importance of developing multidisciplinary studies that seek to value traditional local knowledge as well as strategies for the development of different plant-based insecticides.

ACKNOWLEDGMENTS

We would like to thank the CNPq–Brasil for financial support.

CONFLICT OF INTEREST

The authors declare no conflict of interest.

REFERENCES

1. Thacker JRM. *An introduction to arthropod pest control*. Cambridge University Press. 2002. Tng DY, Apgaua DMG, Lisboa MM, El-Hani, CN. Gender differences in plant use knowledge within a traditional fishing community in northeastern Brazil. *Ethnobotany Research and Applications*, 2016, 21, 1–36.
2. Casida JE, Quistad GB. Golden age of insecticide research: past, present, or future? *Annual Review of Entomology*, 1998, 43, 1–16.
3. Bellinato DF, Viana-Medeiros PF, Araújo SC, Martins AJ, Lima JBP, Valle D. Resistance status to the insecticides temephos, deltamethrin, and diflubenzuron in Brazilian *Aedes aegypti* populations. *BioMed Research International*, 2016, 1–12.
4. Rout SP, Choudary KA, Kar DM, Das L, Jain A. Plants in traditional medicinal system—future source of new drugs. *International Journal of Pharmacy and Pharmaceutical Sciences*, 2009, 1, 1.
5. Heinrich M, Gibbons S. Ethnopharmacology in drug discovery: an analysis of its role and potential contribution. *Journal of Pharmacy and Pharmacology*, 2001, 53, 425–432.
6. Orozco OL, Lentz DL. Poisonous plants and their uses as insecticides in Cajamarca, Peru. *Economic Botany*, 2005, 59, 2, 166–173.
7. Luna JDS, dos Santos AF, De Lima MRF, De Omena MC, De Mendonça FAC, Bieber LW. A study of the lavicidal and molluscicidal activities of some medicinal plants from northeast Brazil. *Journal of Ethnopharmacology*, 2005, 97, 199–206.
8. dos Santos EA, De Carvalho CM, Costa AL, Conceição AS, Moura FB, Santana AE. Bioactivity evaluation of plant extracts used in indigenous medicine against the snail, *Biomphalaria glabrata*, and the larvae of *Aedes aegypti*. *Evidence-Based Complementary and Alternative Medicine*, 2012, 1–9.
9. de Oliveira-Melo PMC, Lima PGC, Costa JC, Coelho-Ferreira MR. Ethnobotanical study in a rural settlement in Amazon: contribution of local knowledge to public health policies. *Research, Society and Development*, 2022, 11, 1, e56911125258–e56911125258.
10. HO. *Neglected tropical diseases*. Accessed August 20, 2020, at https://www.who.int/neglected-tropical-diseases-progress-dashboard-2011-2020.
11. Laporta GZ, Potter AM, Oliveira JF, Bourke BP, Pecor DB, Linton YM. Global distribution of *Aedes aegypti* and *Aedes albopictus* in a climate change scenario of regional rivalry. *Insects*, 2023, 14, 1, 49.
12. Sousa FDCA, Soares HVA, Lemos LEAS, Reis DM, Da Silva WC, De Sousa Rodrigues LA. Perfil epidemiológico de doenças negligenciadas de notificação compulsória no Brasil com análise dos investimentos governamentais nessa área. *Research, Society and Development*, 2020, 9, 1, e62911610.
13. Fonseca BDP, Albuquerque PC, Zicker F. Neglected tropical diseases in Brazil: lack of correlation between disease burden, research funding and output. *Tropical Medicine & International Health*, 2020, 25, 11, 1373–1384.
14. Rehman JU, Ali A, Khan IA. Plant based products: Use and development as repellents against mosquitoes: A review. *Fitoterapia*, 2014, 95, 65–74.
15. Pavela R, Benelli G. Ethnobotanical knowledge on botanical repellents employed in the African region against mosquito vectors—a review. *Experimental Parasitology*, 2016, 167, 103–108.

16. Silvério MRS, Espindola LS, Lopes NP, Vieira PC. Plant natural products for the control of *Aedes aegypti*: The main vector of important arboviruses. *Molécules*, 2020, 25.
17. Sukh D, Opender K. *Insecticides of natural origin* (1st ed.). Routledge, 1997. https://doi.org/10.1201/9780203750759.
18. Regnault-Roger C, Philogène BJR. Past and current prospects for the use of botanicals and plant allelochemicals in integrated pest management. *Pharmaceutical Biology*, 2008, 46, 41–52. https://doi.org/10.1080/13880200701729794.
19. Dias CN, Moraes DFC. Essential oils and their compounds as *Aedes aegypti* L. (Diptera: Culicidae) larvicides. *Parasitology Research*, 2014, 113, 2, 565–592.
20. Plata-Rueda A, Martínez LC, Da Silva Rolim G, Coelho RP, Santos MH, De Souza Tavares W. Insecticidal and repellent activities of *Cymbopogon citratus* (Poaceae) essential oil and its terpenoids (citral and geranyl acetate) against *Ulomoides dermestoides*. *Crop Protection*, 2020, 137, 105299.
21. Wangai LN, Kamau KK, Munyekenye G, Nderu D, Maina E, Gitau W, Otieno F. Efficacy of plant-based repellents against Anopheles mosquitoes: a systematic review. *Biomedical Sciences*, 2020, 6, 44.
22. Miot HA, Batistella RF, Batista KDA, Volpato DEC, Augusto LST, Madeira NG, . . . Miot LDB. Comparative study of the topical effectiveness of the andiroba oil (*Carapa guianensis*) and DEET 50% as repellent for *Aedes* sp. *Revista do Instituto de Medicina Tropical de São Paulo*, 2004, 46, 253–256.
23. Tabanca N, Demirci B, Kiyan HT, Ali A, Bernier UR, Wedge DE, Başer KHC. Repellent and larvicidal activity of *Ruta graveolens* essential oil and its major individual constituents against *Aedes aegypti*. *Planta Medica*, 2012, 78, 05, 90.
24. Devi MA, Sahoo D, Singh TB, Rajashekar Y. Toxicity, repellency and chemical composition of essential oils from *Cymbopogon* species against red flour beetle *Tribolium castaneum* Herbst (Coleoptera: Tenebrionidae). *Journal of Consumer Protection and Food Safety*, 2020, 15, 2, 181–191.
25. Williams LAD, Rosner H, Levy HG, Barton EN. A critical review of the therapeutic potential of dibenzyl trisulphide isolated from *Petiveria alliacea* L (guinea hen weed, anamu). *West Indian Medical Journal*, 2007, 56, 1, 17.
26. García-Mateos MR, Elizalde Sánchez E, Espinosa-Robles P, Álvarez-Sánchez ME. Toxicidad de *Petiveria alliacea* L.: Sobre la mosca blanca (*Trialeurodes vaporariorum* west.). *Interciencia*, 2007, 32, 2, 121–124.
27. Hartmann I, Silva AD, Walter EM, Jeremias WDJ. Investigação do efeito larvicida de *Petiveria alliacea* (Guiné) sobre as larvas de mosquitos da espécie *Aedes aegypti*. *Revista virtual de Química*, 2018, 10, 3, 529–541.
28. Ferreira LS, Alves SN. Neem (*Azadirachta indica*): larvicidal properties—a review. *Revista Conexão Ciência*, 2021, 16, 1.
29. Globe Newswire. *Pest control market projected to reach $27.5 billion by 2025—report by marketsand marketsTM. Marketsand Markets*. 2019. Accessed February 10, 2022, at https://www.globenewswire.com/news-release/2019/10/07/1925966/0/en/Pest-Control-Market-Projected-to-Reach-27-5-Billion-by-2025-Report-by-MarketsandMarkets.html.
30. Rocha TT, Tavares-Martins ACC, Lucas FCA. Traditional populations in environmentally protected areas: an ethnobotanical study in the Soure Marine Extractive Reserve of Brazil. *Boletín Latinoamericano y del Caribe de Plantas Medicinales y Aromáticas*, 2017, 16, 4, 410–427.
31. Albuquerque UP, Andrade LHC. Conhecimento botânico tradicional e conservação em uma área de caatinga no estado de Pernambuco, Nordeste do Brasil. *Acta Botanica Brasilica*, 2002, 16, 3, 273–285.
32. Melo S, Lacerda VD, Hanazaki N. Espécies de restinga conhecidas pela comunidade do Pântano do Sul, Florianópolis, Santa Catarina, Brasil. *Rodriguésia*, 2008, 59, 4, 799–812.
33. Lima RB, dos Santos JUM, Da Luz Freitas J, Souto RNP. Caracterização agroecológica e socioeconômica dos moradores da comunidade quilombola do Curiaú, Macapá-AP, Brasil. *Biota Amazônia (Biote Amazonie, Biota Amazonia, Amazonian Biota)*, 2013, 3, 3, 113–138.
34. Ritter RA, Monteiro MVB, Monteiro FOB, Rodrigues ST, Soares ML, Silva JCR, . . . Tourinho MM. Ethnoveterinary knowledge and practices at Colares island, Pará state, eastern Amazon, Brazil. *Journal of Ethnopharmacology*, 2012, 144, 2, 346–352.
35. Magalhães SIR, Oliveira SA, Araújo LM, Costa PRC, Araújo Roland I, Borrás MRL. Determination of Cu, Fe, Mn, and Zn in the leaves and tea of *Arrabidaea chica* (Humb. & Bompl.) Verl. *Biological Trace Element Research*, 2019, 132, 1, 239–246.
36. Albuquerque UP, Monteiro JM, Ramos MA, Amorim ELC. Medicinal and magic plants from a public market in northeastern Brazil. *Journal of Ethnopharmacology*, 2007, 110, 1, 76–91.
37. Almeida Neto JR, dos Santos KPP, Chaves ME, de Morais RF, Neto EMC, Silva PRR, Barros RFM. Conhecimento sobre uso de plantas repelentes e inseticidas em duas comunidades rurais do Complexo vegetacional de Campo Maior, Nordeste do Brasil. *Gaia Scientia*, 2017, 11, 1.

38. Souza CDD, Felfili JM. Uso de plantas medicinais na região de Alto Paraíso de Goiás, GO, Brasil. *Acta Botanica Brasilica*, 2006, 20, 135–142.
39. Garlet TMB, Irgang BE. Plantas medicinais utilizadas na medicina popular por mulheres trabalhadoras rurais de Cruz Alta, Rio Grande do Sul, Brasil. *Revista Brasileira de plantas medicinais*, 2001, 4, 1, 9–18.
40. Burei ST, Santana DAD, Lopez BB, Sotomaior CS, Acra LA, Weber SH, Ollhoff RD. Ethnoveterinary knowledge and practice applied to domestic animals raised in the Ukraine Colonization Community of Palmital, Paraná State, Brazil. *Economic Botany*, 2022, 1–12.
41. Tng DY, Apgaua DMG, Lisboa MM, El-Hani CN. Gender differences in plant use knowledge within a traditional fishing community in northeastern Brazil. *Ethnobotany Research and Applications*, 2016, 21, 1–36.
42. Pinho RCD, Pedreira JL, Rocha JC, Perez IU, Hada AR, Miller RP, Alfaia SS. Agroextrativismo indígena no Lavrado: o caso da Terra Indígena Araçá, Roraima, Amazônia. *Terra e Água-degradação e desenvolvimento sustentável*, 2012, 143–164.
43. Merétika AHC, Peroni N, Hanazaki N. Local knowledge of medicinal plants in three artisanal fishing communities (Itapoá, Southern Brazil), according to gender, age, and urbanization. *Acta Botanica Brasilica*, 2010, 24, 2, 386–394.
44. Prado AC, Rangel EB, Sousa HCD, Messias MCT. Etnobotânica como subsídio à gestão socioambiental de uma unidade de conservação de uso sustentável. *Rodriguésia*, 2019, 70.
45. Horta BO, Neto GM. Levantamento etnobotânico de plantas medicinais em duas comunidades da região serrana do Rio de Janeiro, Brasil. *Revista Fitos*, 2019, 13, 3, 212–231.
46. Pilla MAC, Amorozo, MCDM, Furlan A. Obtenção e uso das plantas medicinais no distrito de Martim Francisco, Município de Mogi-Mirim, SP, Brasil. *Acta Botanica Brasilica*, 2006, 20, 789–802.
47. Ramos LS, Cardoso PS, Freitas MD, Paghan R, Borges MS, Citadini-Zanette V, . . . Dalbo S. Popular medicinal uses of *Calea uniflora* Less. (Asteraceae) and its contribution to the study of Brazilian medicinal plants. *Anais da Academia Brasileira de Ciências*, 2016, 88, 2319–2330.
48. Guimarães OB, Morais LI, Oliveira PA. Medicinal plants and their popular use in Boa Esperança Settlement, Piracanjuba, Goiás, Brazil. *Boletín Latinoamericano y del Caribe de Plantas Medicinales y Aromáticas*, 2022, 21, 4.
49. Macedo AF, Oshiiwa M, Guarido CF. Ocorrência do uso de plantas medicinais por moradores de um bairro do município de Marília-SP. *Revista de Ciências Farmacêuticas Básica e Aplicada*, 2007, 28, 1.
50. Monteiro MVB, Bevilaqua CML, Palha MDDC, Braga RR, Schwanke K, Rodrigues ST, Lameira OA. Ethnoveterinary knowledge of the inhabitants of Marajó Island, eastern Amazonia, Brazil. *Acta Amazonica*, 2011, 41, 2, 233–242.
51. Pinto JDS, Oliveira AKMD, Fernandes V, Matias R. Ethnobotany and popular culture in the use of plants in settlements on the southern edge of southern Pantanal Mato Grosso. *Journal of Biosciences* (Online), 2017, 193–203.
52. Santos J, Pagani E, Ramos J, Rodrigues E. Observations on the therapeutic practices of riverine communities of the Unini River, AM, Brazil. *Journal of Ethnopharmacology*, 2012, 142, 2, 503–515. https://doi.org/10.1016/j.jep.2012.05.027.
53. Bieski IGC, Leonti M, Arnason JT, Ferrier J, Rapinski M, Violante IMP, . . . de Oliveira Martins DT. Ethnobotanical study of medicinal plants by population of Valley of Juruena Region, Legal Amazon, Mato Grosso, Brazil. *Journal of Ethnopharmacology*, 2015, 173, 383–423.
54. Boscolo OH, Senna Valle L. Plantas de uso medicinal em Quissamã, Rio de Janeiro, Brasil. *Iheringia, Série Botânica*, 2008, 63, 2, 263–278.
55. Shanley P, Rosa NA. Eroding knowledge: an ethnobotanical inventory in eastern Amazonia's logging frontier. *Economic Botany*, 2004, 58, 2, 135–160.
56. Pedrollo CT, Kinupp VF, Shepard Jr G, Heinrich M. Medicinal plants at Rio Jauaperi, Brazilian Amazon: ethnobotanical survey and environmental conservation. *Journal of Ethnopharmacology*, 2016, 186, 111–124.
57. Scudeller VV, Veiga JD, Araújo-Jorge, LD. Etnoconhecimento de plantas de uso medicinal nas comunidades São João do Tupé e Central (Reserva de Desenvolvimento Sustentável do Tupé). *Biotupé: meio físico, diversidade biológica e sociocultural do Baixo Rio Negro, Amazônia Central*, 2009, 2, 185–189.
58. Santos JX, Reis ARS, Parry SM, Leão FM, Carvalho JC. Caracterização etnobotânica de essências florestais com fins medicinais utilizados pela Etnia Xipaya, no município de Altamira-PA. *Biota Amazônia (Biote Amazonie, Biota Amazonia, Amazonian Biota)*, 2016, 6, 2, 1–8.
59 Cavalheiro L, Guarim-Neto G. Ethnobotany and regional knowledge: combining popular knowledge with the biotechnological potential of plants in the Aldeia Velha community, Chapada dos Guimarães,

Mato Grosso, Brazil. *Boletín Latinoamericano y del Caribe de Plantas Medicinales y Aromáticas*, 2018, 17, 2.

60. Ribeiro RV, Bieski IGC, Balogun SO, Oliveira DTM. Ethnobotanical study of medicinal plants used by Ribeirinhos in the North Araguaia microregion, Mato Grosso, Brazil. *Journal of Ethnopharmacology*, 2017, 205, 69–102.
61. Magno-Silva RE, Rocha TT, Tavares-Martins CAC. Ethnobotany and ethnopharmacology of medicinal plants used in communities of the Soure Marine Extractive Reserve, Pará State, Brazil. *Boletín Latinoamericano y del Caribe de Plantas Medicinales y Aromáticas*, 2020, 19, 1.
62. Fiebig DA, Pasa GMC. A Etnobotânica na Comunidade Passagem da Conceição em Várzea Grande, Mato Grosso, Brasil. *Biodiversidade*, 2016, 15, 2.
63. Costa PSP. *Estudo Etnobotânico e Farmacognóstico de plantas antimaláricas de uso popular na comunidade Céu do Mapiá, Pauini-AM*. Dissertação (Mestrado em ciências farmacêuticas), UFAM, Manaus, 2013.
64. Agra MF, Baracho GS, Nurit K, Basílio IJLD, Coelho VPM. Medicinal and poisonous diversity of the flora of "Cariri Paraibano", Brazil. *Journal of Ethnopharmacology*, 2007, 111, 2, 383–395.
65. Shanley P, Luz L. The impacts of forest degradation on medicinal plant use and implications for health care in eastern Amazonia. *BioScience*, 2003, 53, 6, 573–584.
66. Silva AL, Tamashiro J, Begossi A. Ethnobotany of riverine populations from the Rio Negro, Amazonia (Brazil). *Journal of Ethnobiology*, 2007, 27, 1, 46–72.
67. Frausin G, Hidalgo A, Lima RB, Kinupp VF, Ming LC, Pohlit AM, Milliken W. An ethnobotanical study of anti-malarial plants among indigenous people on the upper Negro River in the Brazilian Amazon. *Journal of Ethnopharmacology*, 2015, 174, 238–252. https://doi.org/10.1016/j.jep.2015.07.033.
68. Mendonça AP, Ferraz IDK. Óleo de andiroba: processo tradicional da extração, uso e aspectos sociais no estado do Amazonas, Brasil. *Acta amazônica*, 2007, 37, 3, 353–364.
69. Carvalho TLGS. *Etnofarmacologia e fisiologia de plantas medicinais do quilombo Tiningú, Santarém, Pará, Brasil*. Dissertação (Mestrado em Ciências Ambientais), UFOPA, Santarém, 2015.
70. Flor ASSO, Barbosa WLR. *Sabedoria popular no uso de plantas medicinais pelos moradores do bairro do sossego no distrito de Marudá-PA*. Revista Brasileira de Plantas Medicinais, 2015.
71. Antonio RL, Souza RM, Furlan MR, Pedro CR, Cassas F, Honda S, Rodrigues E. Investigation of urban ethnoveterinary in three veterinary clinics at east zone of São Paulo city, Brazil. *Journal of Ethnopharmacology*, 2015, 173, 183–190.
72. Parente CET, Da Rosa MMT. Plantas comercializadas como medicinais no Município de Barra do Piraí, RJ. *Rodriguésia*, 2001, 52, 47–59.
73. Filho PGX, Ribeiro AF, Moraes AF, Penha WF, Borges, WL, Santos RHS. *Ethnobotanical knowledge on non-conventional food and medicinal plants in Rio Cajari Extractivist Reserve, Amazon, Brazil*. Research Square, 2020.
74. Sauini T, Fonseca-Kruel VS, Baptistela Yazbek P, Matta P, Cassas F, Da Cruz C, . . . Rodrigues E. Participatory methods on the recording of traditional knowledge about medicinal plants in Atlantic forest, Ubatuba, São Paulo, Brazil. *Plos One*, 2020, 15, 5, e0232288.
75. Dias SRK, Alcantara MMAC, Pessoa SMA. Aspectos etnobotánicos, fitoquímicos y farmacológicos de especies de Rubiaceae en Brasil. *Revista Cubana de Plantas Medicinales*, 2013, 18, 1, 140–156.
76. Martins AG, Rosário DLD, Barros MND, Jardim MAG. Levantamento etnobotânico de plantas medicinais, alimentares e tóxicas da Ilha do Combu, Município de Belém, Estado do Pará, Brasil. *Revista Brasileira de Farmacologia*, 2005, 86, 1, 21–30.
77. Guarin Neto G, Santana SR, Silva, J. Notas etnobotânicas de espécies de Sapindaceae jussieu. *Acta Botânica Brasílica*, 2000, 14, 327.
78. Farias DF, Cavalheiro MG, Viana MP, Queiroz VA, Rocha-Bezerra LC, Vasconcelos IM. Water extracts of Brazilian leguminous seeds as rich sources of larvicidal compounds against *Aedes aegypti* L. *Anais da Academia Brasileira de Ciências*, 2010, 82, 3, 585–594.
79. Roman ALC, Ming LC, Carvalho ID, Sablayrolles MDGP. Uso medicinal da pimenta malagueta (*Capsicum frutescens* L.) em uma comunidade de várzea à margem do rio Amazonas, Santarém, Pará, Brasil. *Boletim do Museu Paraense Emílio Goeldi. Ciências Humanas*, 2011, 543–557.
80. Carniello MA, Silva RDS, Cruz MA, Guarim Neto G. Quintais urbanos de Mirassol D'Oeste-MT, Brasil: uma abordagem etnobotânica. *Acta amazonica*, 2010, 40, 3, 451–470.
81. Yanti CA, Sari M, Triana ADPSW. (*Cymbopogon winterianus* Jowitt) sebagai Repelen dari Nyamuk Aedes aegypti. *Jurnal Vektor Penyakit*, 2021, 15, 2, 99–106.
82. Giorgi A, De Marinis P, Granelli G, Chiesa LM, Panseri S. Secondary metabolite profile, antioxidant capacity, and mosquito repellent activity of *Bixa orellana* from Brazilian Amazon region. *Journal of Chemistry*, 2013, 2013.

83. Carvalho GHF, Andrade MA, Araújo CN, Santos ML, Castro NA, Charneau S, . . . Bastos IMD. Larvicidal and pupicidal activities of eco-friendly phenolic lipid products from Anacardium occidentale nutshell against arbovirus vectors. *Environmental Science and Pollution Research*, 2019, 26, 6, 5514–5523.
84. Asadollahi A, Khoobdel M, Zahraei-Ramazani A, Azarmi S, Mosawi SH. Effectiveness of plant-based repellents against different Anopheles species: a systematic review. *Malaria Journal*, 2019, 18, 1, 1–20.
85. Amer A, Mehlhorn, H. Repellency effect of forty-one essential oils against *Aedes*, *Anopheles*, and *Culex* mosquitoes. *Parasitology Research*, 2006, 99, 4, 478–490.
86. Eden WT, Alighiri D, Supardi KI, Cahyono, E. The mosquito repellent activity of the active component of air freshener gel from Java Citronella oil (*Cymbopogon winterianus*). *Journal of Parasitology Research*, 2020, 2020.
87. Arpiwi NL, Muksin IK, Kartini NL. Essential oil from *Cymbopogon nardus* and repellant activity against *Aedes aegypti*. *Biodiversitas Journal of Biological Diversity*, 2020, 21, 8.
88. Aguiar RWS, Santos SF, Morgado SF, Ascencio SD, Mendonça LM, Viana KF, . . . Ribeiro BM. Insecticidal and repellent activity of *Siparuna Guianensis* Aubl. (Negramina) against *Aedes Aegypti* and *Culex Quinquefasciatus*. *Plos One*, 2015, 10, 2, E0116765.
89. Subramaniam J, Kovendan K, Kumar PM, Murugan K, Walton W. Mosquito larvicidal activity of *Aloe vera* (Family: Liliaceae) leaf extract and *Bacillus sphaericus*, against Chikungunya vector, *Aedes aegypti*. *Saudi Journal of Biological Sciences*, 2012, 19, 4, 503–509.
90. Guissoni ACP, Silva IG, Geris R, Cunha LCD, Silva, HHGD. Atividade larvicida de *Anacardium occidentale* como alternativa ao controle de Aedes aegypti e sua toxicidade em *Rattus norvegicus*. *Revista Brasileira de Plantas Medicinais*, 2013, 15, 363–367.
91. Mendonça FA, Da Silva KFS, dos Santos KK, Júnior KR, Sant'ana AEG. Activities of some Brazilian plants against larvae of the mosquito *Aedes aegypti*. *Fitoterapia*, 2005, 76, 629–636.
92. Torres RC, Garbo AG, Walde RZML. Characterization and bioassay for larvicidal activity of *Anacardium occidentale* (cashew) shell waste fractions against dengue vector *Aedes aegypti*. *Parasitology Research*, 2015, 114, 10, 3699–3702.
93. Dill EM. Pereira MJB, Costa MS. Efeito residual do extrato de *Annona coriacea* sobre *Aedes aegypti*. *Arquivos do Instituto Biológico*, 2012, 79, 595–602.
94. Omena MC, Navarro DMAF, De Paula JE, Luna JS, De Lima M.F, Sant'ana AEG. Larvicidal activities against *Aedes aegypti* of some Brazilian medicinal plants. *Bioresource Technology*, 2007, 98, 2549–2556.
95. Ciccia G, Coussio J, Mongelli E. Insecticidal activity against *Aedes aegypti* larvae of some medicinal South American plants. *Journal of Ethnopharmacology*, 2000, 72, 185–189.
96. Hidalgo PDSP, Nunomura RDCS, Nunomura, SM, Pinto SAC, Tadei WP. Larvicidal Activities against *Aedes aegypti* and *Culex quinquefasciatus* of some extracts from Amazon edible fruits. *Journal of Mosquito Research*, 2017, 6.
97. Souza TM, Farias DV, Soares BM, Viana MP, Lima GPG, Machado LKA, . . . Carvalho AFU. Toxicity of Brazilian plant seed extracts to two strains of *Aedes aegypti* (Diptera: Culicidae) and non-target animals. *Journal of Medical Entomology*, 2011, 48, 846–851.
98. Barbosa PBBM, De Oliveira JM, Chagas JM, Rabelo LMA, Medeiros GF, Giodani RB, . . . Ximenes MFFM. Evaluation of seed extracts from plants found in the Caatinga biome for the control of *Aedes aegypti*. *Parasitology Research*, 2014, 113, 10, 3565–3580.
99. Sá RA, De Lima Santos ND, Da Silva CSB, Napoleão TH, Gomes FS, Cavada BS, . . . Paiva PMG. Larvicidal activity of lectins from *Myracrodruon urundeuva* on *Aedes aegypti*. *Comparative Biochemistry and Physiology Part C: Toxicology & Pharmacology*, 2009, 149, 3, 300–306.
100. Mohammed A, Chadee DD. An evaluation of some Trinidadian plant extracts against larvae of *Aedes aegypti* mosquitoes. *Journal of the American Mosquito Control Association*, 2007, 23, 2, 172–176.
101. Kiplang'at KP, Mwangi RW. Repellent activities of *Ocimum basilicum, Azadirachta indica* and *Eucalyptus citriodora* extracts on rabbit skin against *Aedes aegypti*. *Journal of Entomology and Zoology Studies*, 2013, 84.
102. Nour AH, Jessinta D, Nour AH. Larvicidal activity of extracts from different parts of neem (*Azadirachta indica*) against *Aedes Aegypti* mosquitoes larvae. *Scientific Research and Essays*, 2012, 7, 31, 2810–2815.
103. Ajileye AY, Muse WA. Biological effect of some plant extractions on the reproductive development of the yellow fever mosquito *Aedes aegypti* Linn. *Ife Journal of Science*, 2006, 8, 1, 23–27.
104. Sanabria L, Segovia EA, González N, Alcaraz P, Bilbao VN. Larvicidal activity of aqueous plants extracts on *Aedes aegypti* larva (first trials). *Memorias del Instituto de Investigaciones en Ciencias de la Salud*, 2009, 26–31.
105. Alvarez MRS, Heralde Iii FM. Quiming NS. Potent larvicidal activities of *Capsicum frutescens* (L.) fruit ethanolic and partially purified extracts against *Aedes aegypti* (L.) and *Aedes albopictus* (S.). *Der Pharmacia Lettre*, 2015, 7, 11, 94–99.

106. Prophiro JS, Da Silva MAN, Kanis LA, Da Rocha LCB, Duque-Luna JE, Da Silva OS. First report on susceptibility of wild *Aedes aegypti* (Diptera: Culicidae) using *Carapa guianensis* (Meliaceae) and *Copaifera* sp. (Leguminosae). *Parasitology Research*, 2012, 110, 699–705.
107. Miot HA, Batistella RF, Batista KDA, Volpato DEC, Augusto LST, Madeira NG, . . . Miot LDB. Comparative study of the topical effectiveness of the andiroba oil (*Carapa guianensis*) and DEET 50% as repellent for *Aedes* sp. *Revista do Instituto de Medicina Tropical de São Paulo*, 2004, 46, 253–256.
108. Gomes PRB, Oliveira MB, De Sousa DA, Silva JC, Fernandes R P, Louzeiro HC, . . . Mouchrek Filho VE. Larvicidal activity, molluscicide and toxicity of the essential oil of *Citrus limon* peels against, respectively, *Aedes aegypti, Biomphalaria glabrata* and *Artemia salina. Eclética Química*, 2019, 44, 4, 85–95.
109. Massebo F, Tadesse M, Bekele T, Balkew M, Gebre-Michael T. Evaluation on larvicidal effects of essential oils of some local plants against *Anopheles arabiensis* Patton and *Aedes aegypti* Linnaeus (Diptera, Culicidae) in Ethiopia. *African Journal of Biotechnology*, 2009, 8, 17, 4183–4188.
110. Porto KR, Motti PR, Yano M, Roel AR, Cardoso CA, Matias R. Screening of plant extracts and fractions on *Aedes aegypti* larvae found in the state of Mato Grosso do Sul (linnaeus, 1762) (culicidae). *Anais da Academia Brasileira de Ciências*, 2017, 89, 895–906.
111. Soonwera M, Sittichok S. Adulticidal activities of *Cymbopogon citratus* (Stapf.) and *Eucalyptus globulus* (Labill.) essential oils and of their synergistic combinations against *Aedes aegypti* (L.), *Aedes albopictus* (Skuse), and *Musca domestica* (L.). *Environmental Science and Pollution Research*, 2020, 27, 16, 20201–20214.
112. Kurniasih N, Nuryadin W, Harahap MN, Supriadin A, Kinasih I. Toxicity of essential oils from orange (*Citrus sinesis* L. Obbeck) and lemongrass (*Cymbopogon nardus* L. Rendle) on *Aedes aegypti* a vector of Dengue Hemorrhagic Fever (DHF). *Journal of Physics: Conference Series. IOP Publishing*, 2021, 012015.
113. Trongtokit Y, Rongsriyam Y, Komalamisra N, Apiwathnasorn C. Comparative repellency of 38 essential oils against mosquito bites. *Phytotherapy Research: An International Journal Devoted to Pharmacological and Toxicological Evaluation of Natural Product Derivatives*, 2005, 19, 4, 303–309.
114. Manh HD, Hue DT, Hieu NTT, Tuyen DTT, Tuyet OT. The mosquito larvicidal activity of essential oils from *Cymbopogon* and *Eucalyptus* species in Vietnam. *Insects*, 2020, 11, 2, 128.
115. Anoopkumar AN, Puthur S, Rebello S, Aneesh EM. Screening of a few traditionally used medicinal plants for their larvicidal efficacy against *Aedes aegypti* Linn (Diptera: Culicidae), a dengue fever vector. *Symbiosis Microbiology & Infectious Diseases*, 2017, 5, 1–5.
116. Pinto LPDAP, da Cruz ILS, Dias TD, Honório, NA, da Cunha Gonçalves SJ, Maleck M. Extratos de Erva-de-Santa-Maria na saúde pública: controle do vetor de arboviroses. *Revista Pró-univerSUS*, 2019, 10, 1, 102–105.
117. Cavalca PAM, Lolis MIGDA, Reis B, Bonato CM. Homeopathic and larvicide effect of *Eucalyptus cinerea* essential oil against *Aedes aegypti. Brazilian Archives of Biology and Technology*, 2010, 53, 835–843.
118. Aguiar JCD, Santiago GM, Lavor PL, Veras HN, Ferreira YS, Lima MA, . . . Braz-Filho R. Chemical constituents and larvicidal activity of *Hymenaea courbaril* fruit peel. *Natural Product Communications*, 2010, 5, 12.
119. Khatiwora E, Adsul V.B, Pawar P, Joseph M, Deshpande NR, Kashalkar RV. Larvicidal activity of *Ipomoea carnea* stem extracts and its active ingredient dibutyl phthalate against *Aedes aegypti* and *Culex quinquefasciatus. Der Pharma Chemica*, 2014, 6, 1, 155–161.
120. Andrade TAS, Santana IM, Jimenez GC, Farias ETN, Macedo LO, Alves LC, . . . de Carvalho GA. Larvicidal activity of *Caesalpinia ferrea* Mart. and *Lippia origanoides* Cham. against *Aedes aegypti* (LINNAEUS, 1762) (Diptera: Culicidae). *Revista de Patologia Tropical/Journal of Tropical Pathology*, 2020, 49, 1, 45–54,
121. Sobrinho ACN, Morais, SM, Marinho MM, Souza NV, Lima DM. Antiviral activity on the Zika virus and larvicidal activity on the *Aedes* spp. of *Lippia alba* essential oil and β-caryophyllene. *Industrial Crops and Products*, 2021, 162, 113281.
122. Mahanta S, Sarma R, Khanikor, B. The essential oil of *Lippia alba* Mill (Lamiales: Verbenaceae) as mosquitocidal and repellent agent against *Culex quinquefasciatus* Say (Diptera: Culicidae) and *Aedes aegypti* Linn (Diptera: Culicidae). *The Journal of Basic and Applied Zoology*, 2019, 80, 1, 1–7.
123. Felix SF, Rodrigues AM, Rodrigues ALM, Freitas JCC, Alves DR, Silva AA, . . . de Morais SM. Chemical composition, larvicidal activity, and enzyme inhibition of the essential oil of *Lippia grata* Schauer from the Caatinga Biome against dengue vectors. *Pharmaceuticals*, 2021, 14, 3, 250.
124. Silva IG, Guimarães VP, Lima CG, Silva HHG, Elias CN, Mady CM, . . . Isac E. Efeito larvicida e toxicológico do extrato bruto etanólico da casca do caule de *Magonia pubescens* sobre *Aedes aegypti* (Diptera, Culicidae) em criadouros artificiais. *Revista de Patologia Tropical*, 2003, 32, 73–86.

125. Velayutham K, Ramanibai R, Umadevi M. Green synthesis of silver nanoparticles using Manihot esculenta leaves against *Aedes aegypti* and *Culex quinquefasciatus*. *The Journal of Basic & Applied Zoology*, 2016, 74, 37–40.
126. Coria C, Almiron W, Valladares G, Carpinella C, Ludueña F, Defago M, Palacios S. Larvicide and oviposition deterrent effects of fruit and leaf extracts from *Melia azedarach* L. on *Aedes aegypti* (L.) (Diptera: Culicidae). *Bioresource Technology*, 2008, 99, 8, 3066–3070.
127. Wandscheer CB, Duque JE, Da Silva MA, Fukuyama Y, Wohlke JL, Adelmann J, Fontana JD. Larvicidal action of ethanolic extracts from fruit endocarps of *Melia azedarach* and *Azadirachta indica* against the dengue mosquito *Aedes aegypti*. *Toxicon*, 2004, 44, 8, 829–835.
128. Singh RK, Dhiman RC, Mittal PK. Mosquito larvicidal properties of *Momordica charantia* Linn (family: Cucurbitaceae). *Journal of Vector Borne Diseases*, 2006, 43, 2, 88.
129. Mituiassu LMP, Serdeiro MT, Vieira RRBT, Oliveira LS, Maleck M. *Momordica charantia* L. extracts against *Aedes aegypti* larvae. *Brazilian Journal of Biology*, 2021, 82.
130. Ferreira PMP, Carvalho AF, Farias DF, Cariolano NG, Melo VM, Queiroz MG. Larvicidal activity of the water extract of *Moringa oleifera* seeds against *Aedes aegypti* and its toxicity upon laboratory animals. *Anais da Academia Brasileira de Ciências*, 2009, 81, 2, 207–216.
131. Sandralintang TC, Fauzantoro A, Hermansyah H, Jufri M, Gozan M. Use of *Nicotiana tabacum* L extract for anti-*Aedes Aegypti* mosquito paint. In: *AIP Conference Proceedings*. AIP Publishing LLC, 2018, 030010.
132. Musau JK, Mbaria JM, Nguta JM, Mathiu M, Kiama SG. Phytochemical composition and larvicidal properties of plants used for mosquito control in Kwale County, Kenya. *International Journal of Mosquito Research*, 2016, 3, 3, 12–17.
133. Oliveira PV, Ferreira JC, Moura FS, Lima GS, Oliveira FM, Oliveira PES, . . . Lemos, RPL. Larvicidal activity of 94 extracts from ten plant species of northeastern of Brazil against *Aedes aegypti* L. (Diptera: Culicidae). *Parasitology Research*, 2010, 107, 2, 403–407.
134. Faustino CG, Medeiros FA, Galardo AKR, Rodrigues ABL, Da Costa ALP, Martins RL, . . . da Silva de Almeida SSM. Biocidal activity of a nanoemulsion containing essential oil from *Protium heptaphyllum* Resin against *Aedes aegypti* (Diptera: Culicidae). *Molecules*, 2021, 26, 21, 6439.
135. Pimenta AT, Santiago GM, Arriaga Â, Menezes GH, Bezerra SB. Estudo fitoquímico e avaliação da atividade larvicida de *Pterodon polygalaeflorus* Benth (Leguminosae) sobre *Aedes aegypti*. *Revista brasileira de farmacognosia*, 2006, 16, 501–505.
136. Candido LP, Beserra EB. Repellent activity of *Cnidoscolus phyllacanthus* Mart. and *Ricinus communis* L. extracts against *Aedes aegypti* L. oviposition behavior. *Biotemas*, 2015, 28, 4, 105–112.
137. Cerna GO, Rodríguez JR. Actividad repelente y tiempo de protección experimental del aceite del endospermo de *Ricinus communis* (Euphorbiaceae) en *Aedes aegypti*. *Revista Rebiolest*, 2014, 2, 2, 48–60.
138. Sogan N, Kapoor N, Singh H, Kala S, Nayak A, Nagpal BN. Larvicidal activity of *Ricinus communis* extract against mosquitoes. *Journal of Vector Borne Diseases*, 2018, 55, 4, 282.
139. Orlanda JFF, Mouchrek AN. Efeito larvicida do óleo essencial das folhas de *Ruta graveolens* L. no controle de *Aedes aegypti* (L, 1762) (Diptera: Culicidae). *Research, Society and Development*, 2021, 10, 12, e115101220028.
140. Amariles-Barrera S, García-Pajón C, Parra-Henao G. Actividad insecticida de extractos vegetales sobre larvas de *Aedes aegypti*, Diptera: Culicidae. *CES Medicina*, 2013, 27, 2, 193–203.
141. Nityasree BR, Chalannavar RK, Ghosh SK, Divakar MS, Sowmyashree K. Effect of *Solanum lycopersicum* leaf extracts against larvicidal activity of *Aedes aegypti* L. *Biomedicine*, 2020, 40, 4, 467–473.
142. Guarda C, Lutinski JA, Roman-Junior WA, Busato MA. Atividade larvicida de produtos naturais e avaliação da susceptibilidade ao inseticida temefós no controle do *Aedes aegypti* (Diptera: culicidae). *Interciência*, 2016, 41, 4, 243–247.
143. Arriaga AM, Rodrigues FE, Lemos TL, Oliveira MDC, Lima JQ, Santiago GM. Composition and larvicidal activity of essential oil from *Stemodia maritima* L. *Natural Product Communications*, 2007, 2, 12, 934578X0700201209.
144. Marques MM, Moraes SM, Vieira ÍG, Vieira MG, Silva ARA, Almeida RR, Guedes MIF. Larvicidal activity of *Tagetes erecta* against *Aedes aegypti*. *Journal of the American Mosquito Control Association*, 2011, 27, 2, 156–158.
145. *Medeiros VF. Potencial larvicida de extratos de plantas regionais no controle de larvas de Aedes aegypti (Diptera: Clucidae)*. Dissertação de Mestrado. Programa de pós-graduação em Ciências Biológicas (Universidade Federal do Rio Grande do Norte), 2007.

10 Essential Oil from Medicinal and Food Plants

A Pharmacological Approach

Devyani Rajput, Rakhi Khabiya, Akanksha Dwivedi, Vishal Soni, and Priyanka Soni

10.1 INTRODUCTION

Essential oils are highly concentrated, volatile compounds extracted from various parts of plants, such as leaves, flowers, stems, bark, or roots. These oils contain the characteristic aroma and flavor of the plant they are derived from, and they are used for a wide range of purposes, including aromatherapy, cosmetics, perfumery, and traditional medicine. Essential oils are obtained through processes such as steam distillation, cold-pressing, or solvent extraction, depending on the plant material (Bakkali et al., 2008).

Each essential oil is composed of a complex mixture of chemical constituents, primarily including terpenes and other aromatic compounds. These compounds are responsible for the distinctive therapeutic and aromatic properties of essential oils. Some of the common terpenes found in essential oils include limonene, pinene, myrcene, and linalool, among others (Nazzaro et al., 2013).

Essential oils have been used for centuries in various cultures around the world for their potential health benefits. While the traditional uses of essential oils have been primarily rooted in folk medicine, in recent years, there has been a growing body of scientific research exploring their pharmacological properties and potential therapeutic applications (Guenther et al., 1959). Some of the well-studied effects and uses of essential oils include:

- *Aromatherapy.* Essential oils are often used in aromatherapy, a holistic healing practice that uses the inhalation of aromatic compounds to promote physical and psychological well-being. Different essential oils are believed to have various effects on mood, stress, relaxation, and cognitive function (Cooke & Ernst, 2000).
- *Antimicrobial properties.* Many essential oils exhibit antimicrobial activity against bacteria, viruses, and fungi. They have been studied for their potential role in supporting the immune system and combating infections (Murbach Teles Andrade et al., 2014).
- *Anti-inflammatory effects.* Certain essential oils have shown anti-inflammatory properties and are being investigated for their potential use in managing inflammatory conditions and promoting tissue healing (Chao et al., 2005).
- *Analgesic and pain relief.* Some essential oils, such as lavender and peppermint oil, have been studied for their potential to alleviate pain and discomfort, often through topical application or massage (De Sousa, 2011).
- *Skin health.* Essential oils are commonly used in skin care products for their potential benefits in promoting healthy skin, reducing the appearance of blemishes, and soothing skin irritations (Wei & Shibamoto, 2007).
- *Digestive health.* Certain essential oils are believed to support digestion and relieve digestive discomfort when used appropriately (Jang et al., 2007).

DOI: 10.1201/9781003462804-13

- *Relaxation and stress reduction.* Aromatherapy with specific essential oils is commonly used to promote relaxation, reduce stress, and improve sleep quality (Souza et al., 2019).

The use of essential oils derived from medicinal and food plants has been a cornerstone of traditional medicine and culinary practices for centuries. These aromatic and volatile compounds, extracted from various plant parts, such as leaves, flowers, stems, and roots, have gained significant attention due to their potential pharmacological properties and therapeutic applications. The rich tapestry of plant species, each harboring its unique combination of bioactive constituents, has intrigued researchers, pharmacologists, and health enthusiasts alike (Guenther, 1950).

In recent years, there has been a resurgence of interest in harnessing the power of essential oils for their diverse medicinal benefits, ranging from antimicrobial and anti-inflammatory effects to psychological and emotional well-being. This resurgence is rooted in both an increased understanding of the complex interactions between essential oil compounds and the human body and as well as an expanding body of scientific evidence supporting their potential efficacy (Hanif et al., 2019).

This chapter aims to provide a comprehensive pharmacological exploration of essential oils derived from medicinal and food plants. By delving into the biochemistry and mechanisms of action of these natural compounds, we aim to elucidate their therapeutic potential and shed light on their role in modern health care. Additionally, we will explore the methods of extraction and quality control to ensure the potency and safety of essential oils for various applications (Djilani & Dicko, 2012).

10.2 ETHNOBOTANICAL AND TRADITIONAL USES

10.2.1 Historical Usage of Essential Oils in Different Cultures

The historical usage of essential oils spans across various cultures and civilizations, each with its unique traditions, beliefs, and practices. Essential oils have been valued for their aromatic, medicinal, and spiritual properties, and their usage can be traced back thousands of years (Tomić et al., 2023). Here are some notable examples of how essential oils have been used in different cultures:

- *Ancient Egypt.* Egyptians are renowned for their extensive use of essential oils in various aspects of life. They used aromatic oils in embalming practices, cosmetics, perfumery, and religious rituals. Frankincense and myrrh were highly valued and were often used in religious ceremonies and burial rituals. These resins were considered precious gifts fit for kings (Sadgrove & Jones, 2015).
- *Ancient Greece and Rome.* The Greeks and Romans used essential oils for their therapeutic properties. Hippocrates, often referred to as the father of medicine, documented the use of aromatic plants for their healing effects. Lavender, rosemary, and thyme were commonly used for their antiseptic and aromatic qualities. They were also used in bathing, massage, and skin care (Burdock & Carabin, 2009).
- *Traditional Chinese Medicine.* Aromatic herbs and oils have been a part of traditional Chinese medicine (TCM) for centuries. Aromatherapy was used to balance the body's energies and promote healing. Camphor, ginger, and cinnamon were among the oils used in TCM to address various health concerns and support wellness (Carson & Hammer, 2011).
- *Ayurveda.* In Ayurvedic medicine, which originated in ancient India, aromatic oils were used for both therapeutic and spiritual purposes. Oils were applied in massage, in inhalation, and as ingredients in herbal remedies. Sandalwood, patchouli, and basil were frequently used in Ayurveda for their balancing and rejuvenating effects on the body and the mind (Sadiki et al., 2017).

- *Native American cultures.* Many Indigenous tribes in North America used aromatic plants for their medicinal and spiritual properties. Smoke from burning herbs was believed to carry prayers and intentions to the spirit world. Sage, cedarwood, and sweetgrass were commonly used for smudging rituals to cleanse and purify spaces and individuals.
- *Middle Eastern traditions.* The Middle East has a rich history of using essential oils, particularly resins, like frankincense and myrrh, in religious ceremonies, perfumery, and incense burning. These aromatic substances were prized for their connection to spirituality and were offered as gifts to deities and honored guests.
- *European medieval period.* During the Middle Ages, essential oils were used for both practical and spiritual purposes. They were often incorporated into apothecary practices to create remedies for various ailments. Clove, nutmeg, and rose were among the essential oils used during this time for their medicinal benefits (Li et al., 2014).

10.2.2 Case Studies of Traditional Applications

Certainly, here are a few case studies showcasing the traditional applications of essential oils in various cultures:

- *Lavender oil in ancient Greece and Rome.* Lavender oil (*Lavandula angustifolia*) was used by the ancient Greeks and Romans for its aromatic and medicinal properties. It was often added to baths for its calming effects and pleasant scent. The Romans recognized its antiseptic qualities and used lavender oil to clean wounds and prevent infections. Today, modern aromatherapy still utilizes lavender oil for relaxation, for stress reduction, and as a natural sleep aid (Cavanagh & Wilkinson, 2005; Pazyar et al., 2013).
- *Tea tree oil in Aboriginal Australian medicine.* Tea tree oil (*Melaleuca alternifolia*) has a long history of use among the Aboriginal communities in Australia. The leaves were crushed to extract the oil, which was used to treat cuts, wounds, and skin infections. Tea tree oil's antiseptic and anti-inflammatory properties made it an effective remedy for various skin ailments, including insect bites and fungal infections. Today, tea tree oil is a well-known natural remedy for skin issues and is used in skin care products and natural antiseptics (Pazyar et al., 2013).
- *Frankincense and myrrh in ancient Egypt.* Frankincense (*Boswellia*) and myrrh (*Commiphora myrrha*) were highly valued in ancient Egypt and were used for religious ceremonies, embalming, and medicinal purposes. Frankincense was burned as incense during rituals and was believed to have spiritual and healing properties. Myrrh was used in cosmetics, in perfumes, and as an analgesic. These resins continue to be used in spiritual practices and aromatherapy for relaxation and emotional well-being (El Ashry et al., 2003).
- *Patchouli oil in traditional Chinese medicine (TCM).* Patchouli oil (*Pogostemoncablin*) has been used in traditional Chinese medicine (TCM) to address various health issues, including skin conditions, digestive discomfort, and emotional imbalances. In TCM, patchouli oil was believed to have a grounding and harmonizing effect on the body and mind, making it a popular choice for balancing energy (van Beek & Joulain, 2018).
- *Eucalyptus oil in Australian Aboriginal medicine.* Eucalyptus oil (*Eucalyptus* spp.) has been used for centuries by Australian Aboriginal peoples for its respiratory benefits. The leaves were steam-distilled to extract the oil, which was used in inhalations to clear congestion and ease breathing. The antiseptic properties of eucalyptus oil were also utilized in treating wounds and infections (Sadlon & Lamson, 2010).
- *Sandalwood oil in Ayurvedic medicine.* Sandalwood oil (*Santalum album*) is an integral part of Ayurvedic medicine and is valued for its cooling and soothing properties. It has

been used to balance doshas (energetic forces) and promote relaxation. Sandalwood oil has been used in Ayurvedic skin care to alleviate skin irritations and promote a clear complexion (Burdock & Carabin, 2008).

These case studies provide a glimpse into how essential oils have been an integral part of traditional healing practices in various cultures. The historical use of essential oils for their aromatic, medicinal, and spiritual qualities has laid the foundation for their continued popularity and exploration in modern holistic and natural health approaches.

10.3 CHEMICAL COMPOSITION OF ESSENTIAL OILS

Essential oils (EOs) are highly concentrated, volatile compounds derived from various parts of plants, such as leaves, flowers, stems, bark, and roots. These oils capture the distinct aroma and flavor of the plant they are extracted from and contain a complex mixture of bioactive constituents, including terpenes, phenols, alcohols, esters, and more (Figure 10.1). There are several methods used to extract essential oils from plant material, each suited for different types of plants and compounds, as mentioned in Table 10.1 and Figure 10.2.

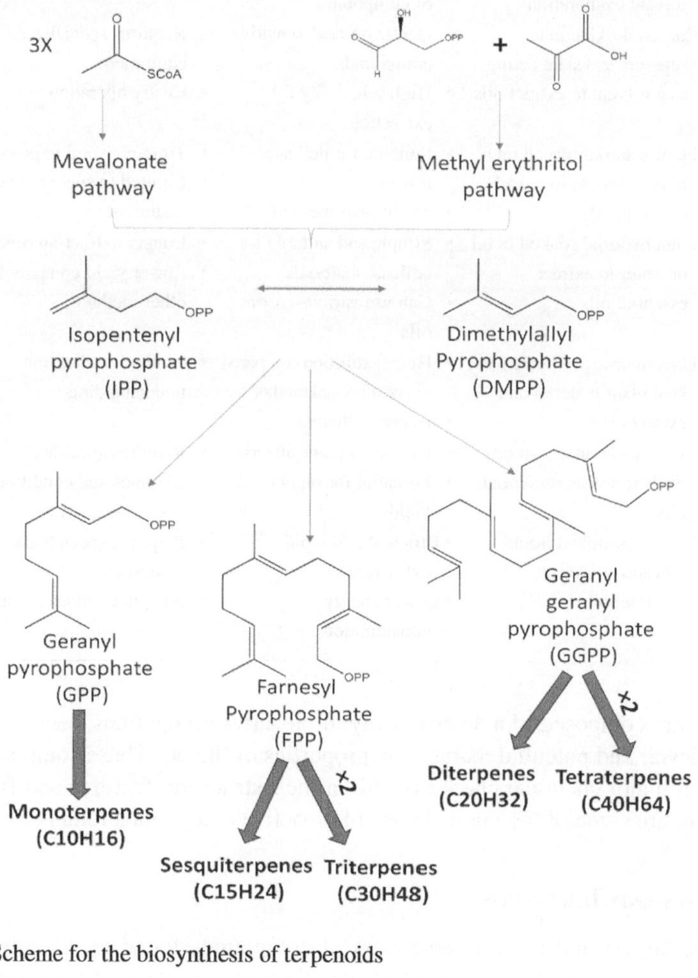

FIGURE 10.1 Scheme for the biosynthesis of terpenoids

TABLE 10.1
Method of Extraction of EOs

Extraction Method	Principle and Description	Pros	Cons	Reference
Steam distillation (see Figure 10.2)	Uses steam to volatilize essential oils, which are then condensed and collected	• Suitable for heat-sensitive compounds • Simple and widely used • Yields high-quality oils	• Requires significant quantities of plant material • Long extraction times • High energy consumption	(Masango, 2005)
Hydrodistillation	Variation of steam distillation where water is directly added to plant material	• Suitable for water-insoluble compounds • Similar advantages to steam distillation	• Similar disadvantages to steam distillation	(Sovová & Aleksovski, 2006)
Expression (cold-press)	Mechanical pressing to extract oils from citrus peels	• Simple and suitable for citrus oils • No heat involved	• Limited to citrus fruits	(Russo et al., 2021)
Solvent extraction	Uses solvents like hexane to dissolve oils from plant material, followed by solvent evaporation	• Efficient for oil extraction • Can extract a wide range of compounds	• May leave traces of solvents in the final product	(Zhang et al., 2012)
Supercritical CO_2 extraction	Carbon dioxide in a supercritical state acting as a solvent to extract oils	• Gentle on heat-sensitive compounds • High selectivity for extraction	• Requires specialized equipment • Costly operation	(Mira et al., 1996)
Enfleurage	Plant material placed on a fatty substrate to absorb essential oils	• Suitable for delicate flowers • Traditional method	• Time-consuming process • Limited to specific plant materials	(Oktavianawati et al., 2019)
Maceration	Plant material soaked in oil or water to extract essential oils	• Simple and suitable for delicate materials • Can use various carrier oils	• Longer extraction time • Lower yield compared to other methods	(Dobreva et al., 2011)
Microwave-assisted extraction	Uses microwave energy to heat plant material and extract oils	• Faster extraction compared to conventional methods • Energy-efficient	• Limited studies and understanding	(Tran et al., 2018)
Enzymatic extraction	Enzymes break down cell walls to release essential oils	• Gentle on plant material • Potential for higher yields	• Requires specific enzymes and conditions	(Hosni et al., 2013)
Pressurized liquid extraction	Uses pressurized liquid solvents to extract essential oils	• Efficient and rapid extraction • Lower energy consumption	• Requires specialized equipment • Solvent residues possible	(Mustafa & Turner, 2011)

Essential oils are composed of a diverse array of bioactive compounds, each contributing to the unique aroma, flavor, and potential therapeutic properties of the oil. These compounds can be classified into several major chemical classes based on their structural features and functional groups (Table 10.2). Here are some of the main classes of bioactive compounds found in essential oils.

10.3.1 Terpenes and Terpenoids

Terpenes are the largest and most diverse class of compounds found in essential oils. They are responsible for the characteristic aromas of plants. Terpenes are derived from isoprene units and

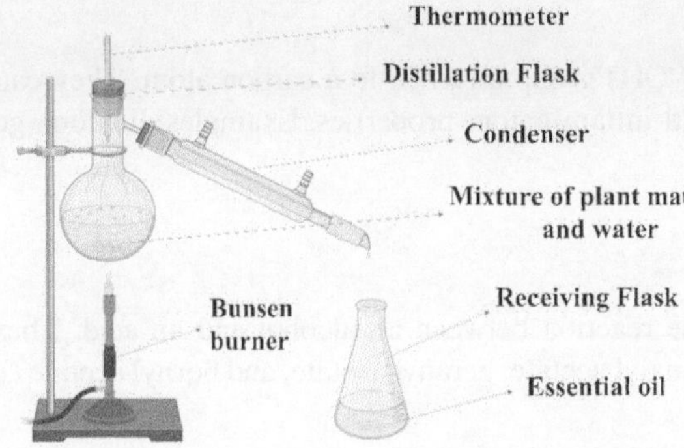

FIGURE 10.2 Extraction of EOs.

TABLE 10.2
Chemical Constituents of EOs

Essential Oil	Major Chemical Constituents
Lavender	Linalool, linalyl acetate, 1,8-cineole
Peppermint	Menthol, menthone, menthyl acetate
Tea tree	Terpinen-4-ol, cineole, α-terpineol
Eucalyptus	1,8-cineole, α-pinene, limonene
Lemon	Limonene, β-pinene, γ-terpinene
Frankincense	α-pinene, limonene, α-thujene
Rosemary	1,8-cineole, α-pinene, camphor
Chamomile	α-bisabolol, chamazulene, farnesene
Bergamot	Limonene, linalyl acetate, linalool
Orange	Limonene, β-myrcene, α-pinene

can be classified into monoterpenes (containing two isoprene units) and sesquiterpenes (containing three isoprene units), among others. Examples: limonene, pinene, myrcene (monoterpenes), β-caryophyllene, and farnesene (sesquiterpenes) (Guimarães et al., 2019).

10.3.2 Phenols

Phenols are aromatic compounds with a hydroxyl (-OH) group attached to an aromatic ring. They often have strong antimicrobial and antioxidant properties. Examples: thymol, carvacrol, and eugenol (Duarte et al., 2010).

10.3.3 Aldehydes

Aldehydes are compounds containing a carbonyl group (C=O) at the end of a carbon chain. They can have a fresh and citrusy aroma. Examples: citral, citronellal, or benzaldehyde (Djilani & Dicko, 2012).

10.3.4 Ketones

Ketones contain a carbonyl group (C=O) within the carbon chain. They are often present in smaller quantities due to their potentially toxic nature. Examples: camphor, menthone, and pulegone (Herrera et al., 2014).

10.3.5 Alcohols

Alcohols have a hydroxyl (-OH) group attached to a carbon atom. They can have various effects, including antiseptic and anti-inflammatory properties. Examples: linalool, geraniol, and citronellol (Kurita et al., 1981).

10.3.6 Esters

Esters are formed from the reaction between an alcohol and an acid. They often have pleasant, fruity aromas. Examples: linalyl acetate, geranyl acetate, and bornyl acetate (Radulović et al., 2015).

10.3.7 Oxides

Oxides contain an oxygen atom bound to at least one other atom. They can contribute to respiratory benefits and have a fresh aroma. Examples: 1,8-cineole (eucalyptol) or ascaridole (Božović et al., 2015).

10.3.8 Sulfur-Containing Compounds

Some essential oils contain sulfur compounds that contribute to their distinct aromas. Examples: allyl sulfides (found in garlic and onion oils) or sulfides in *Allium* species (Fogang et al., 2014).

10.3.9 Lactones

Lactones contain a cyclic ester group. They are often found in oils derived from fruits and nuts. Examples: γ-nonalactone or δ-decalactone (Neerman, 2003).

It is important to note that essential oils are complex mixtures of these and other compounds, and the presence and proportions of these compounds can vary based on factors such as plant species, geographical origin, growing conditions, and extraction methods. The interactions between these bioactive compounds contribute to the overall therapeutic effects and sensory experiences associated with essential oils.

10.4 PHARMACOLOGICAL ACTIVITIES OF ESSENTIAL OILS

Essential oils exhibit a wide range of pharmacological activities (Figure 10.3) which have been the subject of scientific investigation and traditional use for various health and wellness purposes, summarized in Table 10.3. Here are some of the notable pharmacological activities of essential oils:

10.4.1 Antimicrobial Activity

There are numerous biological effects of essential oils. They are utilized for their antiseptic characteristics in herbal therapy to treat infectious disorders with bacterial and fungal origins. They do, however, also have cytotoxic qualities that are comparable to those of antiseptics and disinfectants, which are employed as antimicrobial agents with a wide range of activity. Essential oils or their bioactive components may potentially be employed as protective agents against phytopathogenic fungi and pathogenic bacteria in food in the two fields of health and food safety (Hammer et al., 1999). Actually, there is a wealth of information describing the antibacterial characteristics of various types and origins of essential oils. However, many articles simply state whether or not the investigated agent (an essential oil or one of its purified components) is bioactive against Gram-positive or Gram-negative bacteria or fungus in order to provide broad results. The majority of the time, the authors provide no information regarding the scope and degree of engagement (Janssen et al., 1987). Similar to this, some publications merely compare the activity of various essential oils tested

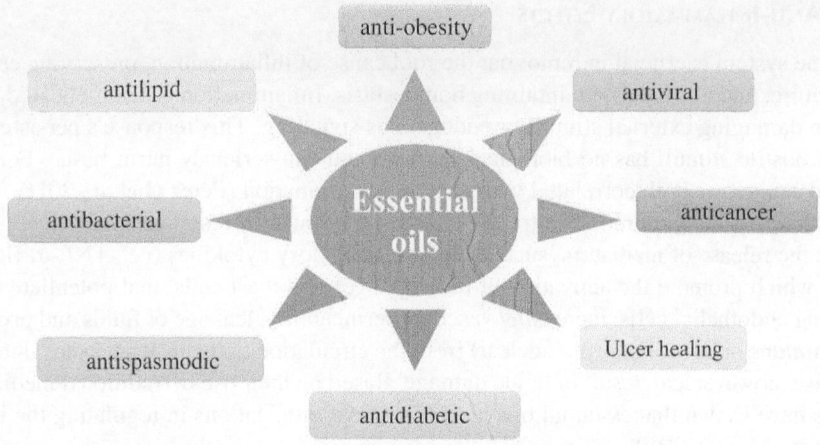

FIGURE 10.3 Pharmacological properties exhibited by essential oils.

TABLE 10.3
Essential Oil, Pharmacological Activity, and Common Use

Essential Oil	Pharmacological Activity	Common Uses
Lavender	Analgesic, anxiolytic, antimicrobial	Pain relief, anxiety, skin infections
Peppermint	Analgesic, anti-inflammatory	Headaches, muscle pain, digestive issues
Tea tree	Antimicrobial, antifungal	Acne treatment, fungal infections
Eucalyptus	Decongestant, anti-inflammatory	Respiratory congestion, sore muscles
Chamomile	Anti-inflammatory, sedative	Skin irritations, anxiety, sleep aid
Lemon	Antioxidant, antibacterial	Immune support, cleaning, mood elevation
Frankincense	Anti-inflammatory, analgesic	Inflammation, joint pain, spiritual use
Rosemary	Cognitive enhancer, antioxidant	Memory, mental clarity, hair care
Geranium	Anti-inflammatory, astringent	Skin care, hormonal balance
Ginger	Digestive, antinausea	Nausea, indigestion, muscle pain

against the same microorganisms to assess the relative activity of an essential oil. It should be noted that there may be a large difference in the outcomes when comparing the bioactivity of essential oils with the same common name and the same primary bioactive component. This is a result of the overall compositional differences between them, which have a significant impact on the bioactivity of the essential oil. Climate and environmental circumstances are the primary causes of this variation in the composition of essential oils. Additionally, the approach taken to assess the antibacterial activity and the selection of tested microorganisms differs from laboratory to laboratory (Murbach Teles Andrade et al., 2014).

Finally, since different parameters may change between research, findings are frequently unpredictable. These include variations in microbial growth, inoculum concentration, oil or component solubility, or the application of an emulsifier and its concentration. To investigate the antimicrobial activities of essential oils, a trustworthy standard approach is therefore urgently needed. Similar to this, in vivo research is required to validate in vitro findings (Elgayyar et al., 2001; Knobloch et al., 1986).

10.4.2 Anti-Inflammatory Effects

The immune system is crucial in removing the root cause of inflammation, preserving cellular and tissue structure, and ultimately, maintaining homeostasis. Inflammation is a complicated protective response to damaging external stimuli or endogenous signaling. This response's persistence in the absence of hostile stimuli has no biological purpose and can seriously harm tissue. For instance, obesity and cancer are both correlated with chronic inflammation (Pérez G et al., 2011). Therefore, clinical intervention is required. The triggers of the inflammatory response involve cellular stimulation with the release of mediators, such as pro-inflammatory cytokines (i.e., TNF-α, IL-1β, IL-6, and IL-8), which promote the activation of macrophages and mast cells, that potentiate this effect by activating endothelial cells, increasing vascular permeability, leakage of fluids and proteins, and influx of immune cells (polymorphonuclear) from the circulation to the inflamed site. Uncontrolled, this response, however, can result in tissue damage. Based on their use in traditional medicine, various studies have shown that essential oils have biological implications in regulating the inflammatory response (Miguel, 2010).

Geranium essential oil possesses anti-inflammatory, antiseptic, antifungal, and antibacterial properties and has been used in traditional medicine to treat inflammatory skin disorders. Kobayashi and colleagues (2016) demonstrated that geranium essential oil significantly and dose-dependently reduced β-hexosaminidase secretion but did not suppress the enzyme's activity in cultured mast cells (CMC) activated with IgE. When mast cells, which are innate cells involved in the tissue allergy and inflammatory response, are stimulated and then degranulate, this enzyme is produced (Higaki et al., 2016). This finding so implies that geranium essential oil inhibits mast cell degranulation to reduce IgE-induced allergic reaction. The chemical components of this essential oil were also described by the authors, who also showed that citronellol (54.6% inhibition), citronellylformate (33.8% inhibition), and geraniol (32.2% inhibition) have an inhibitory impact on mast cell degranulation (Pérez G et al., 2011). A crucial pro-inflammatory cytokine of inflammation, TNF-α is produced in CMCs when IgE is present. TNF mediates edema, neutrophil recruitment, and activation of T cells to the inflamed region. These results support the application of geranium essential oil and especially its major constituent, citronellol, in the treatment of allergic conditions. α-phellandrene is a monoterpene found in essential oils of a variety of plants, such as *Schinusmolle* L., *Monodoramyristica* (Gaertn.) Dunal, and *Anethum graveolens* L. Studies have demonstrated the protective effect of α-phellandrene against ifosfamide-induced hemorrhagic cystitis, as well as wound healing, anti-hyperalgesic, and antidepressive properties (Bayala et al., 2014). Regarding the modulatory role on the inflammatory response, a study showed that α-phellandrene inhibited leukocyte influx and a number of rolling neutrophils or adhered to vascular endothelium, associated with carrageenan-induced acute inflammation. These inflammatory processes start with the production of signaling molecules from resistant cells, such as cytokines and chemokines, which play a crucial part in activating the vascular endothelium and raising the expression of adhesion molecules. In this sense, the anti-inflammatory effect of the α-phellandrene can be explained, at least in part, by the reduction of the pro-inflammatory cytokines TNF-α and IL-6. Consistent with these results, recently, Gonçalves and collaborators (2020) reported the modulatory effect of α-phellandrene on TNF-α and IL-1β secretion using an experimental model of hemorrhagic cystitis in mice (Alves-Silva et al., 2020). Additionally, when mesenteric tissues were treated with compound 48/80 (a stimulator of mast cell degranulation), the treatment with phellandrene decreased the amount of degranulated mast cells. Accordingly, our findings imply that phellandrene encourages mast cell stability by limiting the release of pro-inflammatory mediators and moderating the inflammatory response. *Citrus bergamia* Risso and Poit., the source of bergamot essential oil, and the plant *Cnidiummonnieri* (L.) Cuss, a member of the Umbelliferae family, combine to form bergapten, a furocoumarin. Bergapten has been used to treat skin conditions such psoriasis, dermatitis, mycosis, and vitiligo because of its pharmacological activity. Recent experimental research have documented its anti-inflammatory and anti-allergic properties, as well as its ability to prevent osteoporosis (Zuzarte et al., 2013).

10.4.3 Analgesic and Pain Relief

For their ability to relieve pain and act as analgesics, essential oils have been researched. These oils include chemical elements that have therapeutic potential for treating a range of disorders and producing effects akin to analgesics. It has been shown that combining solid coconut oil with essential oils like lavender, lemongrass, and rosemary can temporarily relieve pain brought on by both physical and psychological illnesses like arthritis, migraines, and anxiety. A combination of essential oils with *Chamaemelum nobile* present in a specific quantity has also been shown to have anti-inflammatory and analgesic activity (Sarmento-Neto et al., 2015). These oils are eucalyptus *Citriodora, Citronellafera, Rosmarinus officinalis, Chamaemelum nobile, Helichrysum italicum,* and *Lavandula angustifolia*. Studies have also shown that essential oils from different plant species have analgesic properties, emphasizing their antinociceptive action and chemical makeup characteristics. In addition, a blend of ginger oil, coriander oil, copaiba balsam, and juniper berry oil has been created as an essential oil for pain management that focuses on musculoskeletal illnesses (De Sousa, 2011).

The growing body of research on the antinociceptive properties of components found in essential oils demonstrates this chemical class's medicinal promise. These substances are an intriguing molecules for research in clinical methods since they are effective in numerous animal models of pain and function through various modes of action. Despite the small amount of various antinociceptive constituents in essential oils, it is possible, using low-cost reactions, to easily synthesize some of these compounds, such as monoterpenes α, β-epoxy-carvone, and hydroxydihydrocarvone, which were obtained via organic synthesis (Ou et al., 2012). Another intriguing step toward the creation of novel analgesic medicines is the utilization of these bioactive ingredients as prototypes for the synthesis of comparable molecules. Additionally, it is important to look at the toxicological side of essential oils. A small number of publications have examined the potential toxicological effects of essential oils on people. For instance, only a few toxicological studies in humans have been conducted on compounds like linalool, which has well-documented antinociceptive action in animals (Hajhashemi et al., 2009). To determine better dosing and administration routes, experimental protocols must be standardized. In this manner, a more accurate comparative examination of the oils is feasible. To support the pharmacological and toxicological approach, it is also critical to investigate the chemical makeup of essential oils. The structural diversity of the bioactive components prevents the identification of a specific chemical property underlying antinociceptive activity. The use of computational medicinal chemistry in conjunction with advanced studies on these constituents' mechanisms of action may be a more effective technique to comprehend the molecular prerequisites for this pharmacological activity (Lenardão et al., 2016).

10.4.4 Antioxidant Activity

Numerous research assessed the antioxidant properties of essential oils. *Rosmarinus officinalis* essential oil had a 28.76% ± 2.68% antioxidant activity. In tests for total phenolic content, β-carotene linoleic acid bleaching, and DPPH radical scavenging, the essential oil from *Reinwardtiodendron cinereum* only showed marginal antioxidant activity. Coumarins and furanocoumarins are the main non-volatile ingredients that contribute to the antioxidant properties of pomelo essential oil, which displayed in vitro antioxidant activity (Amorati et al., 2013). The essential oil of *Mentha pulegium* has shown strong antioxidant properties as well as high DPPH radical scavenging and reduction capability. In DPPH radical and β-carotene/linoleic acid bleaching experiments, *Nepeta faassenii* essential oil demonstrated significant antioxidant activity, with effective radical scavenging and relative antioxidative activity (Teixeira et al., 2013).

In tests for phenolic content, β-carotene linoleic acid bleaching, and DPPH radical scavenging, the essential oil from *Reinwardtiodendron cinereum* only displayed marginal antioxidant activity. The study advances knowledge of the non-volatile oxygenated heterocyclic compounds

(OHCs) found in pomelo essential oil, particularly coumarins and furanocoumarins, which make up the majority of the oil's non-volatile components. Thirteen OHCs in pomelo cold-pressed essential oil (CPEO) have been discovered and quantified, including some previously unreported substances, such as auraptenol, 6',7-dihydroxybergamottin (6',7'-DHB), imperatorin, isoimperatorin, and 8-geranyloxypsoralen (Wei & Shibamoto, 2007). Additionally, they emphasize the antioxidant properties of pomelo essential oil, with CPEO demonstrating greater antioxidant properties than distilled essential oil (DEO). According to the study, pomelo CPEO may be used as a natural antioxidant and has the potential to be a robust, long-lasting antioxidant. *Mentha pulegium* essential oil demonstrated strong antioxidant activity, including the ability to reduce and scavenge free radicals. In numerous experiments, the *Nepeta faassenii* essential oil shows significant antioxidant activity. Thymol and *Trachyspermum ammi* essential oil were discovered to have significant antioxidant activity. In a DPPH experiment, the essential oils from four species of orchids have shown dose-dependent antioxidant activity (Chua et al., 2019). The high polyphenolic concentration of *Cymbopogon flexuosus* essential oil is what gives it its antioxidant effects. Growing interest is being shown in the use of essential oils as natural antioxidants, particularly in complementary medicine and food science. However, in order to provide relevant data that can be compared amongst research teams and easily applied to the practical use of EOs for the preservation of food as well as the production of health-conscious products, a more logical and uniform approach to experimental design is required (Misharina et al., 2009).

10.4.5 Cognitive and Mood Effects

It has been discovered that essential oils have an impact on mood and cognition. Inhaling lemongrass essential oil helped healthy female participants' mood and cognitive function. In healthy people, inhaling *Litsea cubeba* essential oil also lowered salivary cortisol levels and improved mood states. While rosemary essential oil improved overall memory quality but slowed reaction times and working memory, lavender essential oil damaged both. These results imply that essential oils can influence mood and cognition objectively and subjectively (Lizarraga-Valderrama, 2021). To comprehend the underlying mechanisms and confirm the impact of essential oils on the human central nervous system, more research is required. In healthy women, lemongrass essential oil increased mood and cognitive function (Benny & Thomas, 2019). Inhaling *Litsea cubeba* essential oil led to better mood states and less disorientation in healthy individuals, according to the study. In preclinical and clinical research, essential oils have been found to have effects on the nervous system that include anxiolytic, depressive, sedative, and anticonvulsant. The study discovered that rosemary essential oil improved memory but decreased speed, whereas lavender essential oil reduced working memory and reaction times (Moss et al., 2003). Mood was subjectively impacted by both oils. In comparison to the control group, the experimental group experienced better focus and general mood states after inhaling wood essential oil (*Chamaecyparisobtusa*). Prefrontal brain activity was decreased by olfactory stimulation with perilla essential oil, although sensations of comfort or relaxation were not significantly altered. Although it has been demonstrated that essential oils can reduce anxiety when inhaled, the underlying neurochemical pathways are not fully understood (Halder et al., 2012).

10.4.6 Respiratory Support

Some essential oils have qualities that help improve respiratory health by easing respiratory infection symptoms, removing congestion, and calming airways. Research has been done on the respiratory effects of essential oils. When combined with medications, ajowan essential oil (AjEO) and thymol exhibited synergistic benefits against respiratory bacteria such methicillin-resistant *Staphylococcus aureus* and penicillin-resistant *Streptococcus pneumoniae* (Fung et al., 2021). Another essential oil that contains d-limonene, eudesmol, sabinene, among other ingredients, has been discovered to have impacts on haze clearing, allergy resistance, and lung facilitation. *Haemophilus influenzae* was susceptible to the antibacterial actions of *Artemisia asiatica* essential oil, which resulted in harm to the bacterial cell wall and shape. Inducing morphological alterations and ion leakage in bacterial cultures, the essential oil of *Artemisia capillaris* and its primary components demonstrated broad-spectrum antibacterial action against respiratory infections (Leigh-de Rapper et al., 2021). Different essential oils' volatile components, including those from rosemary, tea tree, and cassia, have shown strong antibacterial action against both Gram-positive and Gram-negative bacteria. These results imply that essential oils may have respiratory activity and may be effective against respiratory illnesses. Antibacterial action against respiratory bacteria was demonstrated by ajowan essential oil. Combinations of antibiotics and thymol-containing essential oils have been found to have synergistic effects (Vladimir-Knežević et al., 2012). *Haemophilus influenzae* and other common bacteria that cause respiratory infections were all resistant to the essential oil of *Artemisia asiatica*. *Abies holophylla* essential oil showed significant antibacterial action against respiratory tract pathogens. Caryophyllene oxide was shown to be a key component with potent antibacterial properties. According to the study, *Eucalyptus globulus* essential oil showed notable antibacterial action against infections of the respiratory tract, especially when used in vapor form. Against respiratory pathogens, *Thymus longicaulis* essential oil possesses antibacterial properties. The oil might work against germs that are resistant to medication. *Monarda punctata* essential oil and the majority of its constituents have antibacterial activity against microorganisms that cause respiratory problems. In bacterial strains, the essential oil causes membrane damage (Amat et al., 2017).

10.4.7 Skin Health

Due to their potential to treat a variety of skin issues, such as acne, irritability, and dryness, many essential oils are used in skin care products. They might be anti-inflammatory and antimicrobial. Numerous essential oils have positive effects on skin health. They can support skin cell renewal, increase skin suppleness, combat germs, and preserve the proper balance of oil and water. Some essential oils can relieve allergies and have brightening benefits (Michalak, 2022). Additionally, essential oils can boost blood circulation, delay the signs of ageing, repair and moisturize the skin, and increase skin moisture. They effectively moisturize dry skin and soothe itching, reducing skin issues brought on by dryness. Certain essential oils, particularly those that include active ingredients from traditional Chinese medicines, can nourish the skin, regulate qi, and invigorate blood, all of which have excellent skin care effects (Guzmán & Lucia, 2021). Essential oil–based skin care products provide nourishing, hydrating, and acne-clearing properties in addition to the ability to lighten and balance skin color. The skin-whitening, spot-banishing *Angelica sinensis* essential oil and sweet almond essential oil promote cell regeneration, preserve moisture, nourish and enhance skin elasticity and smoothness, and whiten and banish spots. Lemon essential oil has properties that benefit the human skin. Continuous application of lemon essential oil can prevent skin diseases. Essential oil–containing skin care products have a natural and charming smell, good safety, and no irritation. It has nourishing, moisturizing, and conditioning effects on the skin (Zheng et al., 2020).

10.4.8 Immune System Support

Some essential oils have immune-enhancing properties and can help strengthen the body's defense mechanisms against infections. Essential oils have been found to have immunomodulatory effects, enhancing both cellular and humoral immune responses. Various essential oils, including those from *Foeniculum vulgare*, *Satureacuneifolia*, and *Origanum munitiflorum*, have been shown to inhibit reactive oxygen species (ROS) production from phagocytes and polymorphonuclear neutrophils (PMNs). Geranial, geranial acetate, gingerol, and eugenol essential oils have also been found to enhance humoral and cell-mediated immune responses in rats, increasing hemagglutinating antibody titer (HAT) and delayed-type hypersensitivity (DTH) response (Serafino et al.,

2008). A composition of five essential oils has been developed as an immune enhancement product, showing excellent immune-enhancing action in both cellular and humoral immunity. Additionally, an essential oil blend has been shown to have biological activities that model the molecular biology of autoimmune diseases and chronic inflammation (Li et al., 2012). Essential oils from needle *Juniperrus* seed and trunk and *Olibanum* resin have been observed to enhance the growth of human immune cells and inhibit human cancer cell lines. The essential oils (geranial, geranial acetate, gingerol, and eugenol) showed significant immunostimulant activity on both specific and non-specific immune mechanisms. The essential oil from *Olibanum* resin enhanced the growth of human immune T cells and showed inhibitory effects on human cancer cell lines (Baba et al., 2016; Bhalla et al., 2013).

10.4.9 Hormonal and Reproductive Health

Certain essential oils may have hormonal effects and can be used to support hormonal balance, menstrual health, and reproductive wellness. Essential oils extracted from various plants have been studied for their hormonal and reproductive activity. The essential oil of *Rosemarinus officinalis* showed inhibitory effects on the asexual reproductive stages of *Aspergillus niger*, *A. flavus*, and *A. awamor I* (Sakhteman et al., 2020). The essential oil from *Zhumeriamajdae* exhibited potential antifungal activity against pathogenic fungal species, including *Candida albicans* and *Aspergillus flavus*. Soybean, peachy, and buckthorn oil injections were found to inhibit ovary function activity in rats, leading to changes in the estrous cycle and fertility. These findings suggest that essential oils can have an impact on hormonal and reproductive processes, potentially affecting fertility and the estrous cycle. However, further research is needed to fully understand the mechanisms and potential applications of essential oils in this context. The study found that plant oils, such as soybean, peachy, and buckthorn oil, inhibited ovary function and had negative effects on fertility and reproductive activity in rats. Plant oils inhibit ovary function and prolong the estrous cycle. Plant oils have anti-implantative and abortive effects (Ahmadniaye Motlagh et al., 2023).

10.4.10 Antitumor Activity

Numerous studies have documented the EOs' in vivo anticancer and in vitro cytotoxic properties. Numerous hypothesized methods of action in particular. These include activation of the MAPK pathway, inhibition of the AKT/mTOR and NF-κB pathways, and induction of apoptotic cell death that was linked to an increase in the quantity of reactive oxygen species (ROS) (Sobral et al., 2014). In contrast to *Zataria multiflora* Boiss, EO from the bulb of *Allium sativum* L. produced cytotoxicity in leukemia cells by increasing ROS and inducing apoptosis and differentiation. In colon cancer cells, EO caused apoptosis and elevated ROS. By inhibiting the AKT/mTOR pathway, the EO derived from *Litsea cubeba* (Lour.) Pers. seeds caused apoptotic cell death. On the other hand, NF-κB pathway suppression was linked to the cytotoxicity of *Euphorbia intisy* Drake stem EO and *Cedrus deodara* (Roxb. ex D. Don) G. Don bark EO (Machado et al., 2022). Additionally, by activating MAPK, *Artemisia capillaris* Thunb. EO caused apoptotic cell death. These cytotoxic action mechanisms are connected to the chemical makeup of each EO. For instance, in vivo research on perillyl alcohol revealed that this monoterpene can prevent type I and type II geranylgeranyl–protein transferases from prenylating a particular protein (Dahham et al., 2016). In NIH3T3 cells, it has been demonstrated that it has no effect on the farnesyl–protein transferase enzyme. Additionally, the cytotoxic effects of perillyl alcohol on pancreatic cancer cells were studied. It is advised to prevent growth-regulatory proteins like K-Ras and H-Ras from being prenylated. Additionally, research on thymoquinone, an element of the *Nigella sativa* L. plant's essential oil, revealed that it triggers p53-independent apoptosis, as was demonstrated in experiments with human osteosarcoma cells (Abu Zarga et al., 2013). The essential oil from the leaves of *Psidium guajava* L. has shown notable cytotoxic action against human oral epidermal cancer. This oil's two primary chemical components are limonene (38.1%) and β-caryophyllene (27.98%). Numerous mechanisms of action contribute to limonene's anticancer effects. For instance, it encourages autophagy in lung cancer cells, which causes them to undergo apoptosis. The antitumor action of β-caryophyllene has also been reported. In addition to other methods of action in various tumor cell lines, this substance increases the anticancer activity of cisplatin in lung cancer cell lines via controlling the cell cycle and apoptosis-signaling molecules. As a result, limonene and β-caryophyllene may help explain why *Psidium guajava* L. leaf essential oil and other anticancer oils with high concentrations of these components are effective (Lin et al., 2012; Xiang et al., 2017)

10.5 MECHANISMS OF ACTION

In numerous experimental research, essential oils (EOs) have demonstrated outstanding pharmacological potentials in addition to being investigated for their medicinal effects (Table 10.4). They have a reputation for being antibacterial, anti-inflammatory, and anticancer, as well as having the power to alleviate problems related to cardiovascular and brain disorders (Saad et al., 2013). EOs work in a variety of ways, changing bacterial cells' permeability and integrity as well as breaking the cell membrane and releasing vital cell components. Studies on a variety of pathogens, including *Escherichia coli* and enterotoxigenic *E. coli*, have noted these effects (ETEC). The main elements included in EOs, like limonene, have been recognized and are involved in their antibacterial activity (Verdeguer et al., 2020). Brazilian orange terpenes (BOT) and *Cymbopogon khasianus* essential oil (CKEO) are two EOs that have demonstrated potential as secure substitutes for the treatment of a variety of pathogenic bacterial strains. Through interactions with diverse cellular pathways and molecular targets, essential oils have pharmacological effects on the body. The various medicinal qualities of essential oils can be attributed to these interactions, which can modify physiological processes. Here are some typical molecular targets and cellular pathways that essential oils may affect, though the precise targets and pathways will depend on the oil and its composition (Mani-López et al., 2021).

The two asarone isomers that were discovered in the essential oil of *Acorus calamus* L. rhizomes (Araceae) in 2018 by Koul et al. have various mechanisms of action. Cis-asarone is poisonous in addition to having considerable antifeedant activity, but the trans-isomer works exclusively

TABLE 10.4
Essential Oils and Their Pharmacotherapeutics Applications

Pharmacological Activity	Examples of Essential Oils	Mechanisms of Action
Antimicrobial	Tea tree, oregano, thyme	Disrupts cell membranes, enzymes, biofilms
Anti-inflammatory	Lavender, frankincense, chamomile	Modulates inflammatory pathways
Analgesic	Peppermint, eucalyptus, clove	Blocks pain receptors, reduces inflammation
Antioxidant	Rosemary, lemon, geranium	Neutralizes free radicals, reduces oxidative stress
Anxiolytic	Lavender, bergamot, chamomile	Interacts with neurotransmitter receptors
Respiratory support	Eucalyptus, tea tree, peppermint	Clears congestion, soothes airways
Digestive support	Ginger, peppermint, fennel	Supports healthy digestion, reduces discomfort
Skin health	Tea tree, lavender, chamomile	Antimicrobial, anti-inflammatory, soothing
Immune system support	Lemon, oregano, cinnamon	Enhances immune responses
Neuroprotection	Rosemary, lemon balm, frankincense	Supports cognitive health
Hormonal balance	Clary sage, ylang-ylang, fennel	Modulates hormonal pathways
Mood enhancement	Bergamot, citrus oils, patchouli	Influences mood and emotions
Relaxation	Lavender, chamomile, cedarwood	Promotes relaxation and stress reduction

as an antifeedant with no noticeable toxicity. Every time an insecticide (or molecules acting as an insecticide) penetrates the insect cuticle, it acts at the appropriate binding site(s), particularly on the target insect's receptor proteins (neurologically active substances), such as the antagonist of NMDA-sensitive glutamate receptors (Stappen et al., 2018).

Two primary responses that differ in their mode of action as molluscicides have been used to demonstrate the toxicity of essential oils on *B. truncates* snails. Following the discharge of the hemolymph, the mollusk either retracts inside the shell or expands and protrudes from the shell. A molluscicide's activity might theoretically involve multiple systems and multiple components. In the literature, a number of similar reactions suggestive of this have been described, including decreasing heart rate (a cardiotoxic effect). In this instance, the contraction and concomitant suppression of the heartbeat demonstrated the toxicity of several natural compounds on mollusks (Oussalah et al., 2006). The stiffness in the cardiac muscle fibers that is seen during the contraction is evidence of a potential disruption in the interaction between the molluscicide and calcium in the contracting myocardium tissue swelling and changes to the water balance brought on by the breakdown of osmotic equilibrium, which is controlled by neurohormones. Chloride levels in neurons are changed. These cells' bicarbonate transport and pH are just two of the characteristics that this event alone changes. These changes cause a little increase in excitability and a group discharge response (Bouyahya et al., 2017).

There are not many scientific studies on the processes through which essential oils affect bacteria. Others have discovered that essential oils cause a deterioration of the cytoplasmic membrane, and Belkamel et al. have demonstrated an effect of essential oil of thyme on the cell life cycle (Belkamel et al., 2018). Some authors have attributed this action to the interaction of their functional groups (phenol primarily) with the microbial cell envelop (Lahlou, 2004). The effect of this oil was amply demonstrated by the development of severe lesions in a variety of microorganisms, including *Escherichia coli*, *Bacillus subtilis*, and *Saccharomyces cerevisiae.* Linalool decreased spore germination and fungal growth, while linalyl acetate suppressed spore production. The suppression of aerial mycelia's respiratory activity appeared to be the cause of the sporulation inhibition (Li & Yu, 2015).

Due to the obvious parallels between the lipophilic, water-insoluble, and extremely volatile essential oils and inhalation drugs, Hof-Mussier theorizes an unspecific action principle which more or less operates upon the peripheral nerve system and relaxes the smooth muscles. This notion appears to be supported by experiments showing a relationship between particular essential oil compounds' increasing efficacy and their decreasing water solubility (Al-Harrasi et al., 2022). Furthermore, according to Buchbauer, smooth muscle contractions can be induced by either musculotropic or neurotropic pathways (Buchbauer et al., 1992). However, some monoterpene derivatives, such as fenchone or carvacrol, inhibit enzymes like acetylcholinesterase which inactivate the spasmogenic acetylcholine, even if a neurotropic action of scent compounds on the excitation transmission by neurotransmitters is theoretically feasible. For instance, the essential oil of *Croton* sp. caused depressions in the smooth muscle fibers' ability to contract (Xu et al., 2016). Additionally, these investigations demonstrated that the potassium channel activation and adjustment of the transmembrane resting potential result in a primary neuronal impact that causes the spasmolytic effect of essential oils, which is highly unlikely to be primarily indirect. More evidence, however, supports the existence of a musculotropic inhibition mechanism. This is because the smooth muscle that lines the walls of cavity organs like the stomach, bronchi, and intestines, and which is primarily in charge of maintaining a continuous potential, exhibits a spasmolytic effect (Diao et al., 2014). By releasing Ca2+ from the sarcoplasmic reticulum, several essential oils and their components, such as those from *Croton zehntneri* and *Mentha* × *villosa*, were able to cause muscular contraction (or prevent excitation–contraction coupling) (active principles estragole, anethole, and eugenol). Additionally, it was believed that cAMP was a mediator of the spasmolytic effects of lavender (*Lavendula angustifolia*). Additionally, it was discovered that linalool, a key component of real lavender, has a comparable mode of action to the whole oil (Al-Harrasi et al., 2022).

A thorough study on how essential oils affect the nervous system has been released. Both linalyl acetate and linalyl linalool have the potential to depress the central nervous system. Additionally, *L. angustifolia* oil, linalyl acetate, and linalool have been shown by Ghelardini et al. to exhibit local anesthetic effect both in vitro and in vivo (Ghelardini et al., 1999). According to these writers, the mechanism of action involves ion (Na+ or Ca2+) channel blocking or antimuscarinic activity. Seeman had investigated inducing narcosis or local anesthesia by a series of reactions (Rao et al., 2019). It was also described how these membrane-active substances interact with protein kinase C, an enzyme that is thought to affect the spontaneous rhythm of cardiac and neuronal cells. Additionally, Buchbauer discovered that the aroma compounds, or essential oil ingredients, build up in the nerve cell membrane and sterically block the ion channels, which are embedded function proteins. The membrane's physical characteristics are altered as a result of this blockage, particularly its ion permeability. The inability of calcium ions to enter causes spasmolysis. Local anesthesia is produced by inhibiting the formation of action potentials, and an anti-inflammatory effect is produced by inhibiting mast cells' irritability, which are the cause of inflammation. Two mechanisms—pharmacological, or involving a direct contact between smell molecules and receptors or nerve endings, and psychological, or involving a person's subjective perception of odor—are addressed for how scents affect human autonomic functioning and behavior (Shirole et al., 2014). The findings of this study show that both mechanisms are involved, albeit to variable degrees. However, both pharmacological and psychological mechanisms might be used to understand the stimulating effects of a substance under test. The same outcome has been reported for chiral perfumes' effects on self-evaluation and the parameters of the human autonomic nervous system (enantiomers of limonene and carvone). The wide range of biological activities of scent compounds is mostly caused by their lipophilic nature. Their lipophilic nature also contributes to their simple transit through the blood–brain barrier. They make it easier for scent molecules from the circulation to enter lipid-rich tissues like those of the central nervous system (CNS) due to their affinity for these tissues. The central and peripheral parts of the *Psidium guyanensis* essential oil's action, however, were attributed to processes reliant on endogenous adenosine and acetylcholine, respectively (blockade of nicotinic cholinergic transmission at neuromuscular junction) (Angane et al., 2022).

Santos et al. (2021) claim that whereas α-pinene is exclusively active at the spinal level in the modulation of nociceptive transmission, the entire essential oil may be useful at both peripheral and spinal sites. It is widely accepted that the hot-plate assay is only sensitive to medications operating supraspinally. As they failed to significantly alter the pain threshold in the hot-plate test, the essential oil and its constituents do not appear to contain such a site of action. Since naloxone, an opioid antagonist, was unable to counteract the essential oil's effects in the formalin test, it suggests that the antinociceptive impact of the oil is caused by processes other than the activation of opioid receptors. Adenosine has been linked to antinociception in several studies, and adenosine antagonists may prevent this effect. In this instance, coffee greatly counteracted the essential oil's antinociceptive effects, indicating that adenosine receptors likely played a role in the mechanism (Batista et al., 2008; dos Santos Cavalcante et al., 2021).

Essential oils and aroma compounds have been, and are still, utilized as medicinal agents in both phytotherapy and folk medicine. These natural substances have pleasing aromas in addition to having important biological and pharmacological impacts on many experimental animals, mostly because of their lipophilic makeup. Therefore, a better and more carefully targeted application is made possible by intimate understanding of essential oil composition. It is also obvious that the successful application of such chemicals across a range of application domains depends on a thorough understanding of the mechanisms behind their activity (Yang et al., 2020).

10.6 CONCLUSION

In conclusion, the exploration of essential oils derived from medicinal and food plants through a pharmacological approach reveals their significant potential for various health and wellness applications. Throughout history, these oils have been valued for their aromatic, therapeutic, and cultural significance. As science advances, our understanding of the intricate biochemical composition of essential oils and their interactions with the body deepens, shedding light on their diverse pharmacological activities.

Essential oils possess a wide range of pharmacological properties, including antimicrobial, anti-inflammatory, analgesic, and antioxidative effects. These oils have shown promise in supporting respiratory health, skin care, emotional well-being, and more. Their versatility is attributed to the intricate blend of bioactive compounds within each oil, working synergistically to produce multifaceted benefits.

Furthermore, the mechanisms of action through which essential oils exert their effects are complex and multifaceted. These mechanisms involve interactions with receptors, enzymes, gene expression, and cellular pathways. Essential oils exhibit specificity in their actions, allowing them to target various systems within the body, including the nervous, immune, and digestive systems.

However, it is essential to approach the use of essential oils with caution and responsibility. Proper dilution, understanding of contraindications, and consideration of individual sensitivities are crucial aspects of their safe and effective application. Collaboration between traditional knowledge and scientific research provides a balanced perspective on the potential benefits and limitations of essential oils.

As essential oils continue to gain popularity in the realm of holistic health and alternative medicine, further research is needed to unravel their mechanisms in greater detail, determine optimal dosages, and validate their efficacy through rigorous clinical studies. This will enable us to harness the full potential of these natural compounds for promoting well-being, while respecting the rich cultural and historical contexts that have shaped their use throughout time. In embracing the intersection of tradition and modern science, we open the door to a world of possibilities for enhancing our health and quality of life through the remarkable properties of essential oils.

REFERENCES

Abu Zarga, M. H., Al-Jaber, H. I., Baba Amer, Z. Y., Sakhrib, L., Al-Qudah, M. A., Al-humaidi, J. Y., Abaza, I. F., & Afifi, F. U. (2013). Chemical composition, antimicrobial and antitumor activities of essential oil of *Ammodaucus leucotrichus* growing in Algeria. *Journal of Biologically Active Products from Nature*, *3*(3), 224–231.

Ahmadniaye Motlagh, H., Horie, Y., Rashid, H., Banaee, M., Multisanti, C. R., & Faggio, C. (2023). Unveiling the effects of fennel (*Foeniculum vulgare*) seed essential oil as a diet supplement on the biochemical parameters and reproductive function in female common carps (*Cyprinus carpio*). *Water*, *15*(16), 2978.

Al-Harrasi, A., Bhatia, S., Behl, T., Kaushik, D., Ahmed, M. M., & Anwer, K. (2022). Antibacterial mechanism of action of essential oils. In *Role of essential oils in the management of COVID-19* (pp. 227–237). CRC Press, Boca Raton, FL.

Alves-Silva, J. M., Guerra, I., Gonçalves, M. J., Cavaleiro, C., Cruz, M. T., Figueirinha, A., & Salgueiro, L. (2020). Chemical composition of *Crithmum maritimum* L. essential oil and hydrodistillation residual water by GC-MS and HPLC-DAD-MS/MS, and their biological activities. *Industrial Crops and Products*, *149*, 112329.

Amat, S., Baines, D., & Alexander, T. (2017). A vapour phase assay for evaluating the antimicrobial activities of essential oils against bovine respiratory bacterial pathogens. *Letters in Applied Microbiology*, *65*(6), 489–495.

Amorati, R., Foti, M. C., & Valgimigli, L. (2013). Antioxidant activity of essential oils. *Journal of Agricultural and Food Chemistry*, *61*(46), 10835–10847.

Angane, M., Swift, S., Huang, K., Butts, C. A., & Quek, S. Y. (2022). Essential oils and their major components: An updated review on antimicrobial activities, mechanism of action and their potential application in the food industry. *Foods*, *11*(3), 464.

Baba, E., Acar, Ü., Öntaş, C., Kesbiç, O. S., & Yılmaz, S. (2016). Evaluation of *Citrus limon* peels essential oil on growth performance, immune response of *Mozambique tilapia Oreochromis mossambicus* challenged with *Edwardsiella tarda*. *Aquaculture*, *465*, 13–18.

Bakkali, F., Averbeck, S., Averbeck, D., & Idaomar, M. (2008). Biological effects of essential oils—a review. *Food and Chemical Toxicology*, *46*(2), 446–475.

Batista, P. A., de Paula Werner, M. F., Oliveira, E. C., Burgos, L., Pereira, P., da Silva Brum, L. F., & dos Santos, A. R. S. (2008). Evidence for the involvement of ionotropic glutamatergic receptors on the antinociceptive effect of (–)-linalool in mice. *Neuroscience Letters*, *440*(3), 299–303.

Bayala, B., Bassole, I. H. N., Gnoula, C., Nebie, R., Yonli, A., Morel, L., Figueredo, G., Nikiema, J.-B., Lobaccaro, J.-M. A., & Simpore, J. (2014). Chemical composition, antioxidant, anti-inflammatory and anti-proliferative activities of essential oils of plants from Burkina Faso. *PloS One*, *9*(3), e92122.

Belkamel, A., Bammi, J., Janneot, V., Belkamel, A., Dehbi, Y., & Douira, A. (2018). Contribution to the study of the chemical composition of Lemon Verbena: *Aloysia triphylla* (Hert). Britt. cultivated in Morocco. *International Journal of Environment, Agriculture and Biotechnology*, *3*(2), 239070.

Benny, A., & Thomas, J. (2019). Essential oils as treatment strategy for Alzheimer's disease: Current and future perspectives. *Planta Medica*, *85*(03), 239–248.

Bhalla, Y., Gupta, V. K., & Jaitak, V. (2013). Anticancer activity of essential oils: A review. *Journal of the Science of Food and Agriculture*, *93*(15), 3643–3653.

Bouyahya, A., Bakri, Y., Et-Touys, A., Talbaoui, A., Khouchlaa, A., Charfi, S., Abrini, J., & Dakka, N. (2017). Resistance to antibiotics and mechanisms of action of essential oils against bacteria. *Phytothérapie*, 1–11.

Božović, M., Pirolli, A., & Ragno, R. (2015). *Mentha suaveolens* Ehrh. (Lamiaceae) essential oil and its main constituent piperitenone oxide: Biological activities and chemistry. *Molecules*, *20*(5), 8605–8633.

Buchbauer, G., Jirovetz, L., Remberg, B., Remberg, G., & Nikiforov, A. (1992). Headspace analysis of the dried herb of passion flower (Herba Passiflorae) and dried flowers of lime tree (Flores Tiliae). *Flavour and Fragrance Journal*, *7*(6), 329–332.

Burdock, G. A., & Carabin, I. G. (2008). Safety assessment of sandalwood oil (*Santalum album* L.). *Food and Chemical Toxicology*, *46*(2), 421–432.

Burdock, G. A., & Carabin, I. G. (2009). Safety assessment of coriander (*Coriandrum sativum* L.) essential oil as a food ingredient. *Food and Chemical Toxicology*, *47*(1), 22–34.

Carson, C. F., & Hammer, K. A. (2011). Chemistry and bioactivity of essential oils. In *Lipids and essential oils as antimicrobial agents* (pp. 203–238). Wiley.

Cavanagh, H. M., & Wilkinson, J. M. (2005). Lavender essential oil: a review. *Australian Infection Control*, *10*(1), 35–37.

Chao, L. K., Hua, K.-F., Hsu, H.-Y., Cheng, S.-S., Liu, J.-Y., & Chang, S.-T. (2005). Study on the antiinflammatory activity of essential oil from leaves of *Cinnamomum osmophloeum*. *Journal of Agricultural and Food Chemistry*, *53*(18), 7274–7278.

Chua, L. Y., Chong, C. H., Chua, B. L., & Figiel, A. (2019). Influence of drying methods on the antibacterial, antioxidant and essential oil volatile composition of herbs: A review. *Food and Bioprocess Technology*, *12*, 450–476.

Cooke, B., & Ernst, E. (2000). Aromatherapy: A systematic review. *British Journal of General Practice*, *50*(455), 493–496.

Dahham, S. S., Hassan, L. E. A., Ahamed, M. B. K., Majid, A. S. A., Majid, A. M. S. A., & Zulkepli, N. N. (2016). In vivo toxicity and antitumor activity of essential oils extract from agarwood (*Aquilaria crassna*). *BMC Complementary and Alternative Medicine*, *16*(1), 1–11.

De Sousa, D. P. (2011). Analgesic-like activity of essential oils constituents. *Molecules*, *16*(3), 2233–2252.

Diao, W.-R., Hu, Q.-P., Zhang, H., & Xu, J.-G. (2014). Chemical composition, antibacterial activity and mechanism of action of essential oil from seeds of fennel (*Foeniculum vulgare* Mill.). *Food Control*, *35*(1), 109–116.

Djilani, A., & Dicko, A. (2012). The therapeutic benefits of essential oils. *Nutrition, Well-Being and Health*, *7*, 155–179.

Dobreva, A., Kovatcheva, N., Astatkie, T., & Zheljazkov, V. D. (2011). Improvement of essential oil yield of oil-bearing (*Rosa damascena* Mill.) due to surfactant and maceration. *Industrial Crops and Products*, *34*(3), 1649–1651.

dos Santos Cavalcante, J., Júnior, F. A. N., Jorge, R. J. B., & Almeida, C. (2021). Pain modulated by Bothrops snake venoms: Mechanisms of nociceptive signaling and therapeutic perspectives. *Toxicon*, *201*, 105–114.

Duarte, A. R., Santos, S. C., Seraphin, J. C., & Ferri, P. H. (2010). Environmental influence on phenols and essential oils of *Myrciaria cauliflora* leaves. *Journal of the Brazilian Chemical Society*, *21*, 1672–1680.

El Ashry, E., Rashed, N., Salama, O., & Saleh, A. (2003). Components, therapeutic value and uses of myrrh. *Die Pharmazie—An International Journal of Pharmaceutical Sciences*, *58*(3), 163–168.

Elgayyar, M., Draughon, F., Golden, D., & Mount, J. (2001). Antimicrobial activity of essential oils from plants against selected pathogenic and saprophytic microorganisms. *Journal of Food Protection*, *64*(7), 1019–1024.

Fogang, H. P., Maggi, F., Tapondjou, L. A., Womeni, H. M., Papa, F., Quassinti, L., Bramucci, M., Vitali, L. A., Petrelli, D., & Lupidi, G. (2014). In vitro biological activities of seed essential oils from the cameroonian spices *Afrostyrax lepidophyllus* mildbr. and *Scorodophloeus zenkeri* harms rich in sulfur-containing compounds. *Chemistry & Biodiversity*, *11*(1), 161–169.

Fung, T. K., Lau, B. W., Ngai, S. P., & Tsang, H. W. (2021). Therapeutic effect and mechanisms of essential oils in mood disorders: Interaction between the nervous and respiratory systems. *International Journal of Molecular Sciences*, *22*(9), 4844.

Ghelardini, C., Galeotti, N., Salvatore, G., & Mazzanti, G. (1999). Local anaesthetic activity of the essential oil of *Lavandula angustifolia*. *Planta Medica*, *65*(08), 700–703.

Gonçalves, R. L. G., Cunha, F. V. M., Sousa-Neto, B. P. S., Oliveira, L. S. A., Lopes, M. E., Rezende, D. C., . . . & Oliveira, F. D. A. (2020). α-Phellandrene attenuates tissular damage, oxidative stress, and TNF-α levels on acute model ifosfamide-induced hemorrhagic cystitis in mice. *Naunyn-Schmiedeberg's Archives of Pharmacology*, *393*, 1835–1848.

Guenther, E. (1950). *The essential oils* (Vol. IV). D. Van Nostrade Company, Inc., Toronto, New York and London.

Guenther, E., Kulka, K., & Rogers, J. A. (1959). Essential oils. *Analytical Chemistry*, *31*(4), 679–687.

Guimarães, A. C., Meireles, L. M., Lemos, M. F., Guimarães, M. C. C., Endringer, D. C., Fronza, M., & Scherer, R. (2019). Antibacterial activity of terpenes and terpenoids present in essential oils. *Molecules*, *24*(13), 2471.

Guzmán, E., & Lucia, A. (2021). Essential oils and their individual components in cosmetic products. *Cosmetics*, *8*(4), 114.

Hajhashemi, V., Sajjadi, S. E., & Heshmati, M. (2009). Anti-inflammatory and analgesic properties of *Heracleum persicum* essential oil and hydroalcoholic extract in animal models. *Journal of Ethnopharmacology*, *124*(3), 475–480.

Halder, S., Mehta, A. K., Mediratta, P. K., & Sharma, K. K. (2012). Acute effect of essential oil of *Eugenia caryophyllata* on cognition and pain in mice. *Naunyn-Schmiedeberg's Archives of Pharmacology*, *385*, 587–593.

Hammer, K. A., Carson, C. F., & Riley, T. V. (1999). Antimicrobial activity of essential oils and other plant extracts. *Journal of Applied Microbiology*, *86*(6), 985–990.

Hanif, M. A., Nisar, S., Khan, G. S., Mushtaq, Z., & Zubair, M. (2019). Essential oils. In Malik, S. (Ed.), *Essential oil research: Trends in biosynthesis, analytics, industrial applications and biotechnological production* (pp. 3–17). Springer.

Herrera, J. M., Zunino, M. P., Massuh, Y., Pizzollito, R., Dambolena, J. S., Gañan, N. A., & Zygadlo, J. A. (2014). Fumigant toxicity of five essential oils rich in ketones against *Sitophilus zeamais* (Motschulsky). *Agriscientia*, *31*(1), 35–41.

Higaki, Y., Kobayashi, M., Murakami, D., & Takahara, A. (2016). Anti-fouling behavior of polymer brush immobilized surfaces. *Polymer Journal*, *48*(4), 325–331.

Hosni, K., Hassen, I., Chaâbane, H., Jemli, M., Dallali, S., Sebei, H., & Casabianca, H. (2013). Enzyme-assisted extraction of essential oils from thyme (*Thymus capitatus* L.) and rosemary (*Rosmarinus officinalis* L.): Impact on yield, chemical composition and antimicrobial activity. *Industrial Crops and Products*, *47*, 291–299.

Jang, I., Ko, Y., Kang, S., & Lee, C. (2007). Effect of a commercial essential oil on growth performance, digestive enzyme activity and intestinal microflora population in broiler chickens. *Animal Feed Science and Technology*, *134*(3–4), 304–315.

Janssen, A., Scheffer, J., & Svendsen, A. B. (1987). Antimicrobial activity of essential oils: a 1976–1986 literature review. Aspects of the test methods. *Planta Medica*, *53*(05), 395–398.

Knobloch, K., Weis, N., & Weigand, H. (1986). Mechanism of antimicrobial activity of essential oils. *Planta Medica*, *52*(06), 556–556.

Kobayashi, Y., Sato, H., Yorita, M., Nakayama, H., Miyazato, H., Sugimoto, K., & Jippo, T. (2016). Inhibitory effects of geranium essential oil and its major component, citronellol, on degranulation and cytokine production by mast cells. *Bioscience, Biotechnology, and Biochemistry*, *80*(6), 1172–1178.

Kurita, N., Miyaji, M., Kurane, R., & Takahara, Y. (1981). Antifungal activity of components of essential oils. *Agricultural and Biological Chemistry*, *45*(4), 945–952.

Lahlou, M. (2004). Essential oils and fragrance compounds: Bioactivity and mechanisms of action. *Flavour and Fragrance Journal*, *19*(2), 159–165.

Leigh-de Rapper, S., Viljoen, A., & van Vuuren, S. (2021). Essential oil blends: The potential of combined use for respiratory tract infections. *Antibiotics*, *10*(12), 1517.

Lenardão, E. J., Savegnago, L., Jacob, R. G., Victoria, F. N., & Martinez, D. M. (2016). Antinociceptive effect of essential oils and their constituents: An update review. *Journal of the Brazilian Chemical Society*, *27*, 435–474.

Li, C. M., & Yu, J. P. (2015). Chemical composition, antimicrobial activity and mechanism of action of essential oil from the leaves of *M. acleaya* Cordata (Willd.) R. Br. *Journal of Food Safety*, *35*(2), 227–236.

Li, P., Piao, X., Ru, Y., Han, X., Xue, L., & Zhang, H. (2012). Effects of adding essential oil to the diet of weaned pigs on performance, nutrient utilization, immune response and intestinal health. *Asian-Australasian Journal of Animal Sciences*, *25*(11), 1617.

Li, Y., Fabiano-Tixier, A.-S., & Chemat, F. (2014). *Essential oils as reagents in green chemistry* (Vol. 1). Springer.

Lin, J., Dou, J., Xu, J., & Aisa, H. A. (2012). Chemical composition, antimicrobial and antitumor activities of the essential oils and crude extracts of Euphorbia macrorrhiza. *Molecules*, *17*(5), 5030–5039.

Lizarraga-Valderrama, L. R. (2021). Effects of essential oils on central nervous system: Focus on mental health. *Phytotherapy Research*, *35*(2), 657–679.

Machado, T. Q., da Fonseca, A. C., Duarte, A., Robbs, B. K., & de Sousa, D. P. (2022). A narrative review of the antitumor activity of monoterpenes from essential oils: An update. *BioMed Research International*, *2022*.

Mani-López, E., Cortés-Zavaleta, O., & López-Malo, A. (2021). A review of the methods used to determine the target site or the mechanism of action of essential oils and their components against fungi. *SN Applied Sciences*, *3*, 1–25.

Masango, P. (2005). Cleaner production of essential oils by steam distillation. *Journal of Cleaner Production*, *13*(8), 833–839.

Michalak, M. (2022). Plant-derived antioxidants: Significance in skin health and the ageing process. *International Journal of Molecular Sciences*, *23*(2), 585.

Miguel, M. G. (2010). Antioxidant and anti-inflammatory activities of essential oils: A short review. *Molecules*, *15*(12), 9252–9287.

Mira, B., Blasco, M., Subirats, S., & Berna, A. (1996). Supercritical CO2 extraction of essential oils from orange peel. *The Journal of Supercritical Fluids*, *9*(4), 238–243.

Misharina, T., Terenina, M., & Krikunova, N. (2009). Antioxidant properties of essential oils. *Applied Biochemistry and Microbiology*, *45*, 642–647.

Moss, M., Cook, J., Wesnes, K., & Duckett, P. (2003). Aromas of rosemary and lavender essential oils differentially affect cognition and mood in healthy adults. *International Journal of Neuroscience*, *113*(1), 15–38.

Murbach Teles Andrade, B. F., Nunes Barbosa, L., da Silva Probst, I., & Fernandes Júnior, A. (2014). Antimicrobial activity of essential oils. *Journal of Essential Oil Research*, *26*(1), 34–40.

Mustafa, A., & Turner, C. (2011). Pressurized liquid extraction as a green approach in food and herbal plants extraction: A review. *Analytica Chimica Acta*, *703*(1), 8–18.

Nazzaro, F., Fratianni, F., De Martino, L., Coppola, R., & De Feo, V. (2013). Effect of essential oils on pathogenic bacteria. *Pharmaceuticals*, *6*(12), 1451–1474.

Neerman, M. F. (2003). Sesquiterpene lactones: A diverse class of compounds found in essential oils possessing antibacterial and antifungal properties. *International Journal of Aromatherapy*, *13*(2–3), 114–120.

Oktavianawati, I., Letisya, N., Citra, P., Utari, D. P., Winata, I. N. A., Handayani, W., & Nugraha, A. S. (2019). Essential oil composition of rose flowers from Karangpring Village Jember district extracted by distillation and enfleurage. *Jurnal Ilmu Dasar*, *20*(2), 67–74.

Ou, M. C., Hsu, T. F., Lai, A. C., Lin, Y. T., & Lin, C. C. (2012). Pain relief assessment by aromatic essential oil massage on outpatients with primary dysmenorrhea: A randomized, double-blind clinical trial. *Journal of Obstetrics and Gynaecology Research*, *38*(5), 817–822.

Oussalah, M., Caillet, S., & Lacroix, M. (2006). Mechanism of action of Spanish oregano, Chinese cinnamon, and savory essential oils against cell membranes and walls of *Escherichia coli* O157: H7 and *Listeria monocytogenes*. *Journal of Food Protection*, *69*(5), 1046–1055.

Pazyar, N., Yaghoobi, R., Bagherani, N., & Kazerouni, A. (2013). A review of applications of tea tree oil in dermatology. *International Journal of Dermatology*, *52*(7), 784–790.

Pérez, G. S., Zavala, S. M., Arias, G. L., & Ramos L. M. (2011). Anti-inflammatory activity of some essential oils. *Journal of Essential Oil Research*, *23*(5), 38–44.

Radulović, N. S., Mladenović, M. Z., Randjelovic, P. J., Stojanović, N. M., Dekić, M. S., & Blagojević, P. D. (2015). Toxic essential oils. Part IV: The essential oil of *Achillea falcata* L. as a source of biologically/pharmacologically active trans-sabinyl esters. *Food and Chemical Toxicology*, *80*, 114–129.

Rao, J., Chen, B., & McClements, D. J. (2019). Improving the efficacy of essential oils as antimicrobials in foods: Mechanisms of action. *Annual Review of Food Science and Technology*, *10*, 365–387.

Russo, M., Rigano, F., Arigò, A., Dugo, P., & Mondello, L. (2021). Coumarins, psoralens and polymethoxyflavones in cold-pressed Citrus essential oils: A review. *Journal of Essential Oil Research*, *33*(3), 221–239.

Saad, N. Y., Muller, C. D., & Lobstein, A. (2013). Major bioactivities and mechanism of action of essential oils and their components. *Flavour and Fragrance Journal*, *28*(5), 269–279.

Sadgrove, N., & Jones, G. (2015). A contemporary introduction to essential oils: Chemistry, bioactivity and prospects for Australian agriculture. *Agriculture*, *5*(1), 48–102.

Sadiki, M., El Abed, S., Balouiri, M., Barkai, H., El Bergadi, F., El Farricha, O., & Ibnsouda Koraichi, S. (2017). Combined effect of essential oils against bacteria associated with deterioration of historical wood. *Journal of Materials and Environmental Science*, *8*, 594–602.

Sadlon, A. E., & Lamson, D. W. (2010). Immune-modifying and antimicrobial effects of Eucalyptus oil and simple inhalation devices. *Alternative Medicine Review*, *15*(1), 33–43.

Sakhteman, A., Pasdaran, A., Afifi, M., & Hamedi, A. (2020). An assay on the possible effect of essential oil constituents on receptors involved in women's hormonal health and reproductive system diseases. *Journal of Evidence-Based Integrative Medicine*, *25*, 2515690X20932527.

Sarmento-Neto, J. F., Do Nascimento, L. G., Felipe, C. F. B., & De Sousa, D. P. (2015). Analgesic potential of essential oils. *Molecules*, *21*(1), 20.

Serafino, A., Vallebona, P. S., Andreola, F., Zonfrillo, M., Mercuri, L., Federici, M., Rasi, G., Garaci, E., & Pierimarchi, P. (2008). Stimulatory effect of Eucalyptus essential oil on innate cell-mediated immune response. *BMC Immunology*, *9*, 1–16.

Shirole, R., Shirole, N., Kshatriya, A., Kulkarni, R., & Saraf, M. (2014). Investigation into the mechanism of action of essential oil of *Pistacia integerrima* for its antiasthmatic activity. *Journal of Ethnopharmacology*, *153*(3), 541–551.

Sobral, M. V., Xavier, A. L., Lima, T. C., & de Sousa, D. P. (2014). Antitumor activity of monoterpenes found in essential oils. *The Scientific World Journal*, *2014*.

Souza, C. d. F., Baldissera, M. D., Baldisserotto, B., Heinzmann, B. M., Martos-Sitcha, J. A., & Mancera, J. M. (2019). Essential oils as stress-reducing agents for fish aquaculture: A review. *Frontiers in Physiology*, *10*, 785.

Sovová, H., & Aleksovski, S. A. (2006). Mathematical model for hydrodistillation of essential oils. *Flavour and Fragrance Journal*, *21*(6), 881–889.

Stappen, I., Tabanca, N., Ali, A., Wanner, J., Lal, B., Jaitak, V., Wedge, D. E., Kaul, V. K., Schmidt, E., & Jirovetz, L. (2018). Antifungal and repellent activities of the essential oils from three aromatic herbs from western Himalaya. *Open Chemistry*, *16*(1), 306–316.

Teixeira, B., Marques, A., Ramos, C., Neng, N. R., Nogueira, J. M., Saraiva, J. A., & Nunes, M. L. (2013). Chemical composition and antibacterial and antioxidant properties of commercial essential oils. *Industrial Crops and Products*, *43*, 587–595.

Tomić, A., Šovljanski, O., Nikolić, V., Pezo, L., Aćimović, M., Cvetković, M., Stanojev, J., Kuzmanović, N., & Markov, S. (2023). Screening of antifungal activity of essential oils in controlling biocontamination of historical papers in archives. *Antibiotics*, *12*(1), 103.

Tran, T. H., Nguyen, H. H. H., Nguyen, D. C., Nguyen, T. Q., Tan, H., Nhan, L. T. H., Nguyen, D. H., Tran, L. D., Do, S. T., & Nguyen, T. D. (2018). Optimization of microwave-assisted extraction of essential oil from vietnamese basil (*Ocimum basilicum* L.) using response surface methodology. *Processes*, *6*(11), 206.

van Beek, T. A., & Joulain, D. (2018). The essential oil of patchouli, *Pogostemon cablin*: A review. *Flavour and Fragrance Journal*, *33*(1), 6–51.

Verdeguer, M., Sánchez-Moreiras, A. M., & Araniti, F. (2020). Phytotoxic effects and mechanism of action of essential oils and terpenoids. *Plants*, *9*(11), 1571.

Vladimir-Knežević, S., Kosalec, I., Babac, M., Petrović, M., Ralić, J., Matica, B., & Blažeković, B. (2012). Antimicrobial activity of *Thymus longicaulis* C. Presl essential oil against respiratory pathogens. *Central European Journal of Biology*, *7*, 1109–1115.

Wei, A., & Shibamoto, T. (2007). Antioxidant activities and volatile constituents of various essential oils. *Journal of Agricultural and Food Chemistry*, *55*(5), 1737–1742.

Xiang, H., Zhang, L., Yang, Z., Chen, F., Zheng, X., & Liu, X. (2017). Chemical compositions, antioxidative, antimicrobial, anti-inflammatory and antitumor activities of *Curcuma aromatica* Salisb. essential oils. *Industrial Crops and Products*, *108*, 6–16.

Xu, J.-G., Liu, T., Hu, Q.-P., & Cao, X.-M. (2016). Chemical composition, antibacterial properties and mechanism of action of essential oil from clove buds against *Staphylococcus aureus*. *Molecules*, *21*(9), 1194.

Yang, Y. J., Lin, M. Y., Feng, S. Y., Gu, Q., Chen, Y. C., Wang, Y. D., Song, D. f., & Gao, M. (2020). Chemical composition, antibacterial activity, and mechanism of action of essential oil from *Litsea cubeba* against foodborne bacteria. *Journal of Food Processing and Preservation*, *44*(9), e14724.

Zhang, X., Gao, H., Zhang, L., Liu, D., & Ye, X. (2012). Extraction of essential oil from discarded tobacco leaves by solvent extraction and steam distillation, and identification of its chemical composition. *Industrial Crops and Products*, *39*, 162–169.

Zheng, Y., Pan, C., Zhang, Z., Luo, W., Liang, X., Shi, Y., Liang, L., Zheng, X., Zhang, L., & Du, Z. (2020). Antiaging effect of *Curcuma longa* L. essential oil on ultraviolet-irradiated skin. *Microchemical Journal*, *154*, 104608.

Zuzarte, M., Gonçalves, M., Cavaleiro, C., Cruz, M., Benzarti, A., Marongiu, B., Maxia, A., Piras, A., & Salgueiro, L. (2013). Antifungal and anti-inflammatory potential of *Lavandula stoechas* and *Thymus herba-barona* essential oils. *Industrial Crops and Products*, *44*, 97–103.

Section 4

Fixed Oils

11 Carotenoid-Rich Vegetable Oils of Palm Trees from the Amazonia and Their Potential Food Applications

Fernanda Wariss Figueiredo Bezerra,
Marielba de Los Angeles Rodriguez Salazar,
Marcilene Paiva da Silva, Saulo Edgar Gomes de Souza,
Elivaldo Nunes Modesto Junior,
Luiza Helena Da Silva Martins, and Renan Campos Chisté

11.1 INTRODUCTION

Carotenoids are bioactive compounds belonging to the isoprenoid class, mostly C_{40} tetraterpenes, which are widely widespread in nature as they are produced by all photosynthetic organisms, as well as some non-photosynthetic organisms (archaea, bacteria, fungi, and animals) (Sun et al., 2022; Torres-Montilla & Rodriguez-Concepcion, 2021). In plants, carotenoids are specialized secondary metabolites acting as light-collecting pigments and structural components of photosystems, playing an indispensable role in the survival of these organisms (Torres-Montilla & Rodriguez-Concepcion, 2021). The chemical structures of carotenoids consist of a long polyene chain with several conjugated double bonds responsible for their high reactivity, light-absorption properties (chromophore), and typical noticeable colors, yet some important colorless carotenoids can be found in nature (Maoka, 2020). These compounds are lipophilic and can be classified as carotenes and xanthophylls. *Carotenes* are hydrocarbons, with β-carotene being the most-known carotenoid, in addition to α-carotene, lycopene, and γ-carotene. Xanthophylls are oxygenated structures derived from carotenes and include relevant carotenoids, such as β-cryptoxanthin, lutein, zeaxanthin, and astaxanthin.

In fruits and vegetables, carotenoids are highlighted as natural pigments that provide yellow, orange, and red colors, showing high potential as replacements for artificial dyes in food, cosmetic, and pharmaceutical products. Due to their antioxidant properties, carotenoids effectively scavenge/quench reactive pro-oxidant species, and their frequent consumption promotes overall health as they are usually associated with the prevention of chronic degenerative diseases, such as age-related macular degeneration (ARMD), cardiovascular and neurodegenerative diseases, and cancer, among others (Edge & Truscott, 2018; Igielska-Kalwat et al., 2015).

The Amazonia biome is a source of great biodiversity, where various species of palm trees with oleaginous fruits rich in carotenoids can be found. These fruits are consumed by the local population, but the majority of them are still poorly known to the scientific community (Matos et al., 2019). Oils from fruits of palm trees belonging to the Arecaceae family, such as *buriti* (*Mauritia flexuosa* L.), *tucumã* (*Astrocaryum vulgare* Mart., *A. aculeatum* G.Mey., and *A. giganteum* Barb. Rodr.), peach palm (*Bactris gasipaes* Kunth), *murumuru* (*Astrocaryum murumuru* Mart.), *inajá* (*Maximiliana maripa* (Aubl.) Mart.), and oil palm (*Elaeis oleifera* (Kunth) Cortés), which are shown in Figure 11.1, are sources of carotenoids that may be associated with biological activities, such

DOI: 10.1201/9781003462804-15

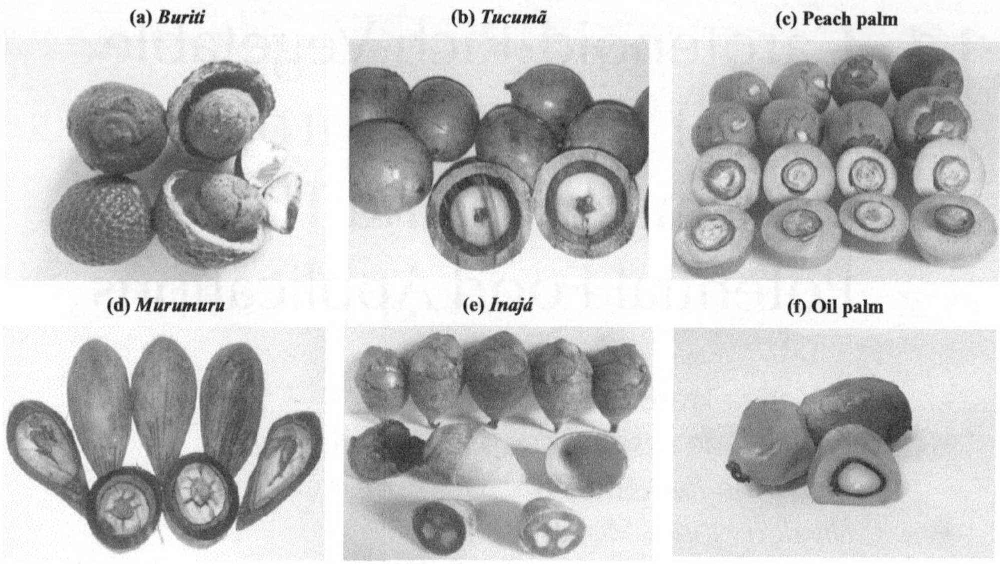

FIGURE 11.1 Pictures of (a) *buriti*, (b) *tucumã*, (c) peach palm, (d) *murumuru*, (e) *inajá*, and (f) oil palm.

Source: Authors.

as controlling phytopathogens (Abreu et al., 2014), preventing cardiovascular diseases, combating various types of cancer, and preventing vision-related diseases, such as ARMD, night blindness (nyctalopia), and dry eyes (xerophthalmia) (Gul et al., 2015). Therefore, in this chapter, scientific information focused on these selected plant species is gathered by approaching the relevant agronomic aspects, the carotenoid composition reported for these fruits and derived oils, along with the reported biological properties whenever available in the literature.

11.2 *BURITI (MAURITIA FLEXUOSA* L.)

11.2.1 Main Agronomic and Physiological Aspects

The *buriti* (*Mauritia flexuosa* L.) is a palm tree belonging to the Arecaceae family and the *Mauritia* genus. The tree is tall, reaching up to 25 m in height, with a diameter ranging from 30 to 50 cm, and at its apex, it bears 20 to 25 leaves. *Buriti* can be found distributed along Northern South America, and in Brazil, the geographical distribution encompasses mainly the states in the North (Amazonia biome), but they can also be found in other Brazilian states (Maranhão, Piauí, Bahia, Ceará, Distrito Federal, Minas Gerais e Mato Grosso) (Filho & Aleluia, 2015; Pereira et al., 2022). Depending on the cultivation region, the plant species may produce fruits with different chemical compositions as influenced by local climatic conditions, as reviewed elsewhere (Barboza et al., 2022).

The edible fruits of *buriti* palm are generally large, with hard, thick, and scaly brown-red skins, ranging in diameter from 3 to 5 cm and weighing around 25 to 40 g (Figure 11.1a). These skins encase a soft and oily pulp. The fruit pulp (mesocarp) has a vibrant yellow-orange color, accounting for about 38% of the fruit weight (Barboza et al., 2022), and it is the most exploited part of the fruit due to its pleasant taste, nutrient, and phytochemical composition. Additionally, the fruits consist of an endocarp that encloses the seed inside (Freitas et al., 2018).

The oil extracted from the fruit pulp is composed mainly by monounsaturated (oleic acid, ≈62–78%) and saturated fatty acids (palmitic acid, ≈16–25%), but some polyunsaturated fatty acids are also present in less extent (linoleic, linolenic, gadolinic, arachidonic, and eicosapentaenoic acids)

(Barboza et al., 2022). Other lipophilic compounds are expected to be present in the oil, as the pulp is a promising source of bioactive compounds that contain very high levels of carotenoids (23–53 mg/100 g of fresh pulp), tocopherols (117 mg/100 g dry weight), and phenolic compounds (23,800 mg/100 g pulp dry weight) (Chisté & Fernandes, 2016). These characteristics certainly make *buriti* an interesting option for the development of sustainable agribusiness stimulated by the food and pharmaceutical industries, owing to the bioactive compounds that may act as potential therapeutic agents for various diseases (Pereira, 2014; Bovi et al., 2017).

The production of *buriti* fruits still comes from extractive activities by small rural communities, characterized mainly by sustainable management, being *buriti* is sometimes seen as the sole source of income for many local farming families for the production of pulps and few derived products (beverages, flours, and sweets). In another scenario, the cosmetics industry seems to dominate the annual purchase of *buriti* fruits extracted by small cooperatives in the Amazonia to produce various products, mainly lotions, creams, soaps, shampoos, and more sophisticated products with *buriti* seed oil, which is rich in lauric acid, to be exported worldwide (Koolen et al., 2018). Moreover, *buriti* was investigated as an alternative source of energy for biodiesel production due to its physicochemical properties that meet desirable specifications for renewable fuel production (Pereira Lima et al., 2017).

11.2.2 *Buriti* Oil: Main Carotenoids and Extraction Techniques

Buriti oil is reported to exhibit antioxidant properties due to its rich composition in carotenoids, mainly β-carotene, α-carotene, and γ-carotene, which are provitamin A compounds, and tocopherols (Table 11.1). Importantly, the contents of carotenoids in *buriti* oil can vary depending on the region of origin and the oil itself.

Several extraction techniques have been reported for the extraction of carotenoids from the lipid fraction of *buriti* fruits. Leite et al. (2021) carried out the extraction of carotenoids from the pulp and peels of *buriti* fruits by solid–liquid extraction or by ultrasound-assisted extraction using ethanol or a supramolecular solvent system (SUPRA) formed by octanoic acid aggregates. According to the authors, both the solvents were able to extract β-carotene, whereas SUPRA showed the highest efficiency (26.7 mg/g of pulp and 5.82 mg/g of peel), which can be justified, as SUPRA composition can favor an equilibrium of hydrophobic and electrostatic interactions between β-carotene and the formed micellar system.

TABLE 11.1
Composition of Carotenoids and Tocopherols of *Buriti* Oils

Compound	Oil (%)	Reference
β-carotene	76.8	Ribeiro et al. (2012)
α-carotene	8.8	
γ-carotene	4.5	
β-carotene	55	Santos et al. (2015)
Z-β-carotene	30.63	
Lutein	5.94	
α-carotene	45.53	Mesquita et al. (2020)
β-carotene	54.56	
β-tocopherol	48.13	Freitas et al. (2018)
α-tocopherol	43.11	
β-tocopherol	44.22	Serra et al. (2018)
α-tocopherol	26.20	

Silva et al. (2022) proposed an optimal extraction method of carotenoids from *buriti* pulp by high-intensity ultrasound (frequency at 40 kHz and power at 80 W), combined with multivariate sequential optimization approach. These authors reported that acetone/ethanol (75:25 v/v) was selected as the optimal extracting solution, achieving 1,026 µg carotenoids/g fresh pulp after 30 min of extraction, with β-carotene as the major compound, which resulted in contents about twofold higher than the methods known in the literature for *buriti*.

Rudke et al. (2019) optimized the pressurized liquid extraction (PLE) method to extract carotenoids from *buriti* shells using ethanol/water mixtures and found total carotenoids ranging from 23 to 1,056 µg/g of dried pulps, being the global optimum response obtained at 71.21°C and 91.58% of ethanol content.

Another example of an efficient technique to extract *buriti* oil with high contents of carotenoids is extraction with supercritical CO_2. França et al. (1999) reported that the oil extracted with supercritical CO_2 at 20 MPa and 313 K showed 1,043 µg/g oil, while hexane by Soxhlet presented 77.7 mg β-carotene/100 g dried material (shell and pulp).

11.2.3 Antioxidant Potential and Other Biological Properties of *Buriti* Oil

Buriti oil was considered an important source of vitamins A and E, as investigated in young rats fed with diet including refined *buriti* oil, which exhibited reduced total cholesterol (up to 60%), LDL (~64%), triglycerides (~55%), and aspartate transaminase activity (~21%) compared to those fed with diet added with crude oil (Aquino et al., 2015). Additionally, the frequent intake of dietary β -carotene revealed that the consumption of rich sources of this compound has been correlated with the prevention of the development of chronicle degenerative diseases, such as various types of cancer, cardiovascular diseases, and the prevention of vision-related conditions, such as night blindness (nyctalopia) and dry eyes (xerophthalmia) (Gul et al., 2015).

In addition to the potential biological functions for human health, *buriti* oil is mainly investigated for its antioxidant properties, due to the presence of valuable bioactive compounds (Table 11.1), which may contribute to protecting cells against oxidative stress by reactive oxidant species, such as biological free radicals that are associated with various diseases and premature cell aging (Speranza et al., 2016; Shahidi & Zhong, 2015).

Speranza et al. (2016) assessed the antioxidant capacity of *buriti* oil against peroxyl radicals ($ROO^{\bullet}$) by the oxygen radical absorbance capacity (ORAC) method, and the results demonstrated its high potential for cosmetic, food, and pharmaceutical purposes due to the high tocopherol contents found in the oil. In another study, Morais et al. (2022) evaluated the effect of gelatin nanoparticles containing *buriti* oil on antioxidant capacity as determined by the ABTS assay, *in vivo* cytotoxicity, and antimicrobial activity. These authors showed that *buriti* oil is a viable alternative product to enhance functionality and add commercial value to *buriti*.

Lage et al. (2018) assessed the effect of flour of *buriti* pulp, which contains expressive levels of antioxidant compounds, on oxidative damage of biomarkers in the liver, heart, and pancreas (*in vivo* assays). In this study, *buriti* flour exhibited high capacity to scavenge DPPH radical and reduced oxidative damage in the studied organs, suggesting high *in vivo* antioxidant effects for the bioactive compounds present in the *buriti* flour. Cândido et al. (2015) reported the antioxidant capacity of *buriti* pulp by different methods (ABTS, DPPH, FRAP, and ORAC), and high antioxidant capacities were observed in all the tests.

The information discussed in this section highlights the promising prospects for future exploitation of *buriti* fruits as potential sources of bioactive compounds with antioxidant properties, which may contribute as a strategy to the prevention of various diseases associated with oxidative stress. Therefore, scientific studies related to the biological properties of *buriti* fruits, such as antioxidant capacity, should be better stimulated and financed to provide valuable information related to the potential of *buriti* oil and its by-products from various regions, including the Amazonia biome.

11.2.4 Food Applications and Future Prospects for *Buriti* Oil

The products derived from *buriti* fruits have wide technological potential to be used by food industries, being that its oil is used to obtain high contents of β-carotene to be used as natural food colorant and antioxidant, incorporating interesting physical and biological properties to food products (Singh & Sambyal, 2022).

According to Ibiapina et al. (2022), the essential and fixed oils from *buriti* fruits have relevant physicochemical, nutritional, and pharmacological properties, along with potential economic value for various biotechnological and industrial applications. However, further systematic research concerning the technological, biotechnological, and biological properties and applications of *buriti* is still needed to add commercial value to produce high-quality food products.

Encapsulation technology has gained prominence in recent years, entering various industrial sectors to protect bioactive compounds from degradation and loss of functional properties. In the food sector, encapsulated *buriti* oil can be found to be incorporated into the food matrix or as part of industrial processes. Nevertheless, further studies concerning encapsulation of *buriti* oil are required, leading to solutions applicable on industrial scales (Barroso et al., 2021; Pereira de Oliveira et al., 2022).

The study by de Souza et al. (2022) developed and characterized edible films produced with *buriti* oil, showing suitable mechanical, thermal, and optical properties for various applications, such as edible packaging. Therefore, *buriti* fruits have promising technological applications to be used by the food industry. However, the fruit is still underexploited and needs to expand its production and processing chain by sustainable means to stimulate consumption and application.

11.3 *Tucumã (Astrocaryum vulgare* Mart., *A. aculeatum* G.Mey., and *A. giganteum* Barb.Rodr.)

11.3.1 Main Agronomic and Physiological Aspects

The genus *Astrocaryum* (Arecaceae family) is commonly found in South America, in Pacific forests, and in the Atlantic forests of Central America. The species *A. vulgare* and *A. aculeatum* are the most known, and both have high economic potential (Kahn, 2008).

A. vulgare, popularly known in Portuguese as "*tucumã-do-Pará*," is highly valued due to the chemical composition of its fruits, considered a good source of nutritional and bioactive compounds (Menezes et al., 2022). In Brazil, the species is mainly found in the states of Pará, Amapá, Goiás, Maranhão, and Tocantins. It has a clumped stem reaching 20 m in height × 20 cm in diameter, pinnate leaves with 8 to 16 contemporaneous leaflets, and these parts of the palm are covered with spiky black thorns. The fruits are ellipsoidal/globose (Figure 11.1b), measuring 3.5–5 cm in length × 2.5–4 cm in diameter, with an orange- or red-colored peel and orange pulp (Forzaa, 2020).

A. aculeatum is also known in Portuguese as "*tucumã-do-Amazonas*" or "*tucumã-açu*." The species is native to the Amazon, mainly found in the Brazilian state named Amazonas. It is a large palm with unbranched stems ranging from 10 to 30 m, featuring feather-shaped leaves measuring 4–5 m in length, covered with black or brown spiky thorns. The fruit has a diameter of 4.5–6.0 cm, with a greenish-yellow peel, and the fruit pulp varies in color from yellow to orange (Santos et al., 2018).

The oil extracted from the fruit pulp has high contents of unsaturated fatty acids and carotenoids, mainly β-carotene. The almond oil is also rich in saturated fatty acids, mainly lauric acid (Ferreira et al., 2022). Concerning the evaluation of bioactive compounds in the peel and pulp of *A. aculeatum* fruits, Sagrillo et al. (2015) showed higher contents of bioactive compounds in the peel than in the pulp: For the peel, the carotenoid content was 6.29%, rutin 3.05%, and quercetin 1.27%, while in the pulp, the carotenoid content was 2.75%, rutin 1.91%, and quercetin 0.65%. Similar findings were reported by Matos et al. (2019) for the species *A. vulgare*, where the total carotenoid contents in the fruit peel (18 mg/100 g) were higher than in the pulp (8.39 mg/100 g). Additionally, A. *vulgare* was reported to contain ascorbic acid (19 mg/100 g), yellow flavonoids (31 mg/100 g), total phenolic compounds (159 mg gallic acid equivalent/100 g), and high antioxidant capacity as compared to other palm trees (dos Santos et al., 2015). In the oils extracted from the fruit pulp, the major fatty acids were palmitic (C16:0), oleic (C18:1), and linoleic (C18:2). For *A. aculeatum* fruit oils, oleic acid was predominant (73.81%), and in the oil of *A. vulgare* fruit, oleic acid prevailed (65.38%) in a lesser proportion (Costa et al., 2016).

Another promising *tucumã* species is *Astrocaryum giganteum* Barb. Rodr. ("*tucumã-i-da-várzea*" in Portuguese), which is found in the eastern region of the Amazon basin, in Pará State, Brazil. It is a large-leaved palm with leaves up to 7 m long, a short stem of up to 4 m in height and 22 cm in diameter, and obovoid-shaped fruits (Kahn, 2008). The species is known only regionally in the Amazonia biome and requires further studies for its valorization. The pulp oil of *A. giganteum* was reported to be rich in carotenoids (409 μg β-carotene/g), in addition to the presence of unsaturated fatty acids, mainly oleic (62%) and linoleic (11.8%), and total phenolic compounds (158 mg gallic acid equivalents/100 g), considered a nutritionally valuable oil to be used as a substitute for traditional oils (Santos et al., 2022).

Carotenoids and phenolic compounds are the main bioactive compounds reported for the species *A. vulgare*, *A. aculeatum*, and *A. giganteum*. Based on the fact that these bioactive compounds exhibit high antioxidant activity, the oils from different parts of *tucumã* fruits are expected to be promising sources of antioxidant compounds that can provide various health benefits.

11.3.2 *Tucumã* Oil: Main Carotenoids and Extraction Techniques

The carotenoid composition in oils of fruits from different palm trees is highly variable. Studies comparing the levels of total carotenoids among different species highlighted that *tucumã* fruits (*A. vulgare*) were the richest source of carotenoids (1,222 mg/kg), followed by peach palm (*Bactris gasipaes*) (357 mg/kg), and *buriti* (*Mauritia flexuosa*) (540 mg/kg) (Santos et al., 2015). Table 11.2 shows the carotenoid composition reported for different *tucumã* species (*A. vulgare*, *A. aculeatum*, and *A. giganteum*).

The carotenoids in the pulp of *A. vulgare* include β-carotene, neoxanthin, γ-carotene, lutein, α-carotene, β-cryptoxanthin, luteoxanthin, and violaxanthin (Santos et al., 2015). In a study by Matos et al. (2019), the carotenoid composition of *A. vulgare* fruit peels was determined for the first time, with (all-*E*)-β-carotene the major compound (7.8 mg/100 g fresh weight), followed by (all-*E*)-γ-carotene (2.6 mg/100 g), *Z*-γ-carotene (0.8 mg/100 g), and (all-*E*)-δ-carotene (0.3 mg/10 g), yielding a vitamin A value of 7.9 μg retinol activity equivalent/g fresh peel. Additionally, in another study, the oily extract from *A. vulgare* almonds coproducts, obtained by green extraction using palm oil by the ultrasound method, showed high total carotenoid contents (3.3 mg/100 g), which enabled the oily extract to be considered as a carotenoid-rich source for encapsulation purposes for application in functional foods (Ferreira et al., 2021).

The carotenoid composition of *A. aculeatum* fruits is similar to that found in *A. vulgare*. One of the early studies on the carotenoid composition of *A. aculeatum*, carried out by de Rosso and Mercadante (2007), identified the same previous carotenoids of *A. vulgare*, in addition to α-cryptoxanthin and zeaxanthin, and reported vitamin A value of 850 μg retinol equivalent/100 g. Moreover, β-carotene was also identified at high proportions in the fruit peel (Sagrillo et al., 2015; de Souza Filho et al., 2013). Thus, β-carotene is the predominant carotenoid in the pulp and peels of both fruit species (*A. vulgare* and *A. aculeatum*), and as far as our knowledge is concerned, the carotenoid composition of *A. giganteum* fruits is not available in the literature.

The oil extracted from *A. vulgare* fruits (pulps + peel) obtained through conventional solid–liquid extraction reached total carotenoids of 7.2 mg/100 g (dos Santos et al., 2015), while Matos

TABLE 11.2
Carotenoid Composition and Extraction Techniques for *Astrocaryum vulgare, Astrocaryum aculeatum*, and *Astrocaryum giganteum*

Part of the Fruit	Extraction Process	Analytical Methods	Carotenoids	Reference
		***Astrocaryum vulgare* Mart.**		
Freeze-dried pulp	Soxhlet extraction Time: 6 h Solvent: diethyl ether	HPLC-DAD C18 Mobile phase: water, methanol, and acetone λ: 430–450 nm	β-carotene (567 mg/kg oil), Z-β-carotene (231); neoxanthin (76), Z-γ-carotene (75.9), γ-carotene (68), lutein (44); α-carotene (29), 5,8 epoxy β-carotene (27), β-cryptoxanthin (14), luteoxanthin (16), violaxanthin (13), Z-violaxanthin (12.8), Z-lutein (12)	Santos et al. (2015)
Freeze-dried pulp	Supercritical fluid extraction, 40, 50, and 60°C/300 bar Flow rate: 15 g/min, oil collection every 10 min Solvent: CO_2	Total carotenoid contents by spectrophotometry at 450 nm Solvent: hexane/acetone (3:7, v/v)	40°C: 2,101 mg/kg 50°C: 2,065 mg/kg 60°C: 2,077 mg/kg	Costa et al. (2016)
Freeze-dried pulp	Supercritical fluid extraction, 40°C/400 bar Flow rate: 5.31 g/min Dynamic time: 30 min Solvent: CO_2	Total carotenoid contents by spectrophotometry at 450 nm Solvent: hexane/acetone (3:7, v/v)	1,351.8 μg/g	Menezes et al. (2022)
Pulp dried in an oven	Soxhlet extraction, 60°C Time: 6 h Solvent: hexane Supercritical fluid extraction (SFE), 60°C/300 bar Flow rate: 8.8×10^{-5} kg/s Dynamic time: 3 h Solvent: CO_2	Total carotenoid contents by spectrophotometry at 450 nm Solvent: petroleum ether	Soxhlet: 877 μg β-carotene/100 g SFE: 1,082 μg β-carotene/100 g	dos Santos et al. (2018)
Peel *in natura*	Solid–liquid extraction at 1:7 (w/v) ratio Time: 80 min of agitation Solvent: ethyl acetate and acetone	HPLC-DAD-MS/MS C30 column Mobile phases: methanol and MTBE Flow rate: 0.9 mL/min λ: 450 nm	(all-*E*)-β-carotene (7.8 mg/100 g peel); (all-*E*)-γ-carotene (2.6); Z-γ-carotene (0.8); and (all-*E*)-δ-carotene (0.3).	Matos et al. (2019)
Dry almond residue in the oven	Ultrasound-assisted extraction, 35°C Time: 4 h Solvent: palm oil Frequency: 40 KHz Power: 154 W	Total carotenoid contents by spectrophotometry at 470 nm Solvent: hexane	3.3 mg/g	Ferreira et al. (2022)

(*Continued*)

TABLE 11.2 (*Continued*)
Carotenoid Composition and Extraction Techniques for *Astrocaryum vulgare, Astrocaryum aculeatum,* and *Astrocaryum giganteum*

Part of the Fruit	Extraction Process	Analytical Methods	Carotenoids	Reference
		***Astrocaryum aculeatum* Meyer**		
Pulp	Conventional exhaustive extraction Solvent: acetone, petroleum ether, and diethyl ether, saponified with 10% KOH in methanol; oil recovered with acetone	HPLC-DAD-MS/MS C30 column Mobile phase: methanol with 0.1% triethylamine and MTBE Flow rate: 0.9 mL/min λ: 450 nm	(all-*E*)-β-carotene (47.36 μg/g), (all-*E*)-α-carotene (1.68), (all-*E*)-β-cryptoxanthin (1.64), 13*Z*-β-carotene (1.60), (all-*E*)-α-cryptoxanthin (1.30), zeinoxanthin (1.02), (all-*E*)-lutein (0.79), Z-γ-carotene 3 (0.89), 15*Z*-β-carotene (0.80), 5,8-epoxy-β-carotene (0.76), Z-β-zeacarotene 2 (0.65), Z-α-zeacarotene 1 (0.60), (all-*E*)-δ-carotene (0.52), (all-*E*)-β-zeacarotene (0.44), (all-*E*)-γ-carotene (0.35), (all-*E*)-neoxanthin (0.26), Z-violaxanthin (0.24), Z-neoxanthin (0.18), (all-*E*)-zeaxanthin (0.16), (all-*E*)-ξ-carotene (0.14), Z-lutein (0.04).	de Rosso and Mercadante (2007)
Peel and pulp *in natura*	Solid–liquid extraction, ratio of 1:5 (w/v) Time: 96 h (4 days) Solvent: ethanol	HPLC-DAD C18 column Mobile phase: water with 2% acetic acid and methanol Flow rate: 0.8 mL/min λ: 450 nm	β-carotene: Peel > 60 mg/g Pulp > 25 mg/g	de Souza Filho et al. (2013)
Peel and pulp *in natura*	Solid–liquid extraction, 1:5 (w/v) Time: 96 h (4 days) Solvent: ethanol	HPLC-DAD Column: C18 Mobile phase: water with 2% acetic acid and methanol Flow rate: 0.8 mL/min λ: 450 nm	β-carotene peel: 6.29% pulp: 2.75%	Sagrillo et al. (2015)
Freeze-dried pulp	Supercritical fluid extraction: 40, 50, and 60°C/300 bar Flow rate: 15 g/min, oil collection every 10 min Solvent: CO_2	Total carotenoid contents by spectrophotometry at 450 nm Solvent: hexane/acetone (3:7 v/v)	40°C: 1,065 mg/kg 50°C: 1,006 mg/kg 60°C: 1,021 mg/kg	Costa et al. (2016)
Peel and pulp *in natura*	Solid–liquid extraction ratio of 1:5 (w/v) Time: 96 h (4 days) Solvent: ethanol	HPLC-DAD Column: C18 Mobile phase: 1% phosphoric acid in water and methanol Flow rate: 0.6 mL/min λ: 450 nm	β-carotene 26.15 mg/g	Cabral et al. (2020)

TABLE 11.2 (*Continued*)
Carotenoid Composition and Extraction Techniques for *Astrocaryum vulgare, Astrocaryum aculeatum*, and *Astrocaryum giganteum*

Part of the Fruit	Extraction Process	Analytical Methods	Carotenoids	Reference
		***Astrocaryum giganteum* Barb. Rodr.**		
Autoclaved pulp	Enzymatic aqueous extraction Enzymes: Celluclast® and Pectinex® (1.7% concentration) at 50°C	Total carotenoid contents by spectrophotometry at 450 nm Solvent: petroleum ether	409.85 μg β-carotene/g	dos Santos et al. (2018)

Note: *MS* = mass spectra; *SFE* = supercritical fluid extraction; *HPLC* = high-performance liquid chromatography; *DAD* = diode array detector.

et al. (2019), also using conventional solid–liquid extraction, reported total carotenoid contents about 8 mg/100 g for the fresh pulp and 18 mg/100 g for the fresh peel, being that β-carotene is the major compound (~68%) in both parts.

The extraction of carotenoid-rich oils from Amazonian palm fruits through green extraction processes has stimulated increased interest by the scientific community, as these processes are environmentally friendly and may be efficiently designed to prevent compound degradation (Santos et al., 2022). A successful comparative study regarding the quality of *A. vulgare* and *A. aculeatum* oils using supercritical fluid extraction (SFE) with CO_2 showed that *A. vulgare* exhibited higher carotenoid contents than *A. aculeatum* (da Fonseca Machado et al., 2022).

Recently, the effects of two extraction techniques, Soxhlet (hexane) and SFE (CO_2), were evaluated concerning the bioactive compound contents of oils extracted from *A. vulgare* fruits (dos Santos et al., 2018). Although conventional Soxhlet extraction presented a slightly higher oil yield (39.63%) than SFE (36.75%), the SFE–CO_2 oil showed higher carotenoid and fatty acid contents. The authors stated that the superior contents of bioactive compounds in the oil by SFE could be attributed to the lower extraction temperature and shorter extraction time (dos Santos et al., 2018).

During oil extraction, the carotenoid composition of the fruit may undergo chemical modifications, such as transformation reactions, degradation, concentration, and/or selective transfer of carotenoids (Santos et al., 2018). Thus, suitable extraction and determination methods are needed for the characterization of carotenoids from any source. Moreover, further studies should characterize different parts of *tucumã* fruits focusing on carotenoids extraction, especially by-products, such as fruit peels. Furthermore, identifying the best extraction approaches for obtaining bioactive compounds, such as conventional extraction with green solvents, SFE, and ultrasound-assisted extraction, among others, is a promising strategy for extracting carotenoids by sustainable means.

11.3.3 Antioxidant Potential and Other Biological Properties of *Tucumã* Oil

The reported biological properties of oils from *A. vulgare* and *A. aculeatum* fruits include anti-inflammatory, antimicrobial, antioxidant, antidiabetic, and anticancer properties, while the oil from *A. giganteum* was only investigated regarding its antioxidant potential (Table 11.3). These effects are frequently associated with the presence of carotenoids in the fruits, particularly the high concentration of β-carotene.

TABLE 11.3
Biological Properties Reported for *Astrocaryum vulgare, Astrocaryum aculeatum*, and *Astrocaryum giganteum*

Fruit	Biological Activity/Main Results	Reference
	Anti-Inflammatory	
Astrocaryum vulgare	Oil from pulp extracted by Soxhlet with hexane, at a dose of 5 mL/kg, decreased pro-inflammatory cytokines and increased anti-inflammatory cytokines. In a rat model of lung inflammation, it efficiently neutralized both acute and chronic inflammatory responses *in vivo*.	Bony et al. (2012)
Astrocaryum aculeatum	Bark and pulp extracts were obtained by solid–liquid extraction with ethanol. At 20–30 μg/mL, the extract increased antioxidant defenses, effectively reduced the inflammatory process, and increased anti-inflammatory cytokines, suggesting its potential in preventing inflammatory diseases.	Cabral et al. (2020)
	Oil from almond extracted by Soxhlet using hexane exhibited *in vitro* anti-inflammatory effect by inhibiting the production of nitric oxide, at 50 μg/mL.	Aguiar et al. (2022)
	Anticancer	
Astrocaryum vulgare	The oil extracted from the pulp by cold-pressing and nanoencapsulation showed promising results on the antiproliferative effect in breast adenocarcinoma cells (MCF-7).	Nascimento et al. (2021)
Astrocaryum aculeatum	Nanoemulsion loaded with extract exhibited better antileukemic results than free extract.	Copetti et al. (2022)
	Antimicrobial	
Astrocaryum vulgare	The oils at 0.14–4.5 mg/mL were active against *Clostridium perfringens*, *Enterococcus cecorum*, *Listeria monocytogenes*, and *Staphylococcus aureus*.	Hovorková et al. (2018)
Astrocaryum aculeatum	Pulp and peel extracts obtained by conventional ethanol extraction presented antibacterial effects on three Gram-positive bacteria (*Enterococcus faecalis*, *Bacillus cereus*, and *Listeria monocytogenes*), and antifungal effects on *Candida albicans*.	Jobim et al. (2014)
	Antidiabetic	
Astrocaryum vulgare	The oil from the pulp extracted by cold-pressing showed protective action against diabetes induced in mice and was able to modulate changes caused by hyperglycemia, maintaining normal levels of ATP, ADP, AMP, and adenosine.	Baldissera et al. (2017)
Astrocaryum aculeatum	The fruit was reported to exhibit potential antidiabetic properties, with inhibitory activity against α-glucosidase and α-amylase.	Gonçalves et al. (2011)
	Antioxidant	
Astrocaryum vulgare	Oils extracted from pulp and peel by conventional extraction scavenged DPPH radical (3,343 g/g DPPH, EC50), peroxyl radicals (ORAC assay: 64 μM·Trolox/g), and inhibited β-carotene bleaching (92%, β-carotene/linoleic acid system).	dos Santos et al. (2015)
	The oil extracted from pulp by cold-pressing showed efficiently scavenged DPPH (IC_{50} = 8.59 μg/mL) and ABTS (2.99 μmol Trolox/mg) radicals.	Nascimento et al. (2021)
	The almond oil extracted by ultrasound-assisted extraction efficiently scavenged ABTS radical (584.75 μM Trolox equivalent/g) and inhibited β-carotene bleaching (49.9%, β-carotene/linoleic acid system).	Ferreira et al. (2022)

TABLE 11.3 (*Continued*)
Biological Properties Reported for *Astrocaryum vulgare, Astrocaryum aculeatum,* and *Astrocaryum giganteum*

Fruit	Biological Activity/Main Results	Reference
	Anti-Inflammatory	
Astrocaryum aculeatum	The almond oil extracted by Soxhlet, using hexane, showed low antioxidant capacity at 1 mg/mL by scavenging DPPH (4.62%) and ABTS (1.51%) radicals.	Aguiar et al. (2021)
Astrocaryum giganteum	The oil extracted from pulp by enzymatic aqueous extraction was able to scavenge ABTS radicals (146.98 μM Trolox/g).	Santos et al. (2022)

In a study conducted by Jantsch et al. (2021), the extract of the species was assessed for its impact on memory and the redox balance of the cerebral cortex in hyperlipidemic rats. The extract prevented memory loss and oxidative damage in the cerebral cortex, suggesting neuroprotective and potential nutraceutical activity of the species. *A. vulgare* seed oil, when compared to palm kernel oil (*Elaeis guineensis*), coconut oil (*Cocos nucifera*), babassu oil (*Attalea speciosa*), and *murumuru* oil (*Astrocaryum murumuru*), stands out for having the highest antimicrobial activity (Hovorková et al., 2018).

Due to its broad biological application and traditional use by the population for treating and/or preventing various diseases, a study by Guex et al. (2022) was conducted to evaluate acute and repeated dose toxicity over 28 days of the crude extract from the pulp of *A. aculeatum* fruit in Wistar rats. The data obtained suggest that *A. aculeatum* fruits did not induce toxicity after exposure to single doses at 2,000 mg/kg. In people, it may be considered safe when administered repeatedly at low doses.

11.3.4 Food Applications and Future Prospects for *Tucumã* Oil

All the parts of *A. aculeatum* and *A. vulgare* fruits are used by technological approaches. For instance, the fruit peels are used to produce flour and for the extraction of pigments and oil, while the fruit pulps are used for human consumption, either fresh or in the preparation of ice cream, creams, sweets, jams, nectar, alcoholic beverages, animal feed, and extraction of pigments and oil, and the almonds are used for extracting oil to produce biodiesel or cosmetic products (Martins et al., 2022).

Mota et al. (2022) produced olein from the pulp of *A. vulgare* fruits with potential use as salad oil, rich in carotenoids and mono- and polyunsaturated fatty acids. Additionally, these authors also prepared a medium fraction with high contents of monounsaturated fatty acids, stable at room temperature, with potential application in the food industry to replace palm oils and *trans*-fats. The oil from the pulp of *A. vulgare* fruits was used for the production of microparticles, by spray chilling method, which showed high contents of β-carotene, with great potential as natural additive for the development of food products (Santos et al., 2023). In another study, the oil extracted from the almonds of *A. vulgare* fruits was also used as a rich source of carotenoids, and the oil microencapsulated by atomization (spray-drying) presented high carotenoid contents (2.5 mg/100 g), spherical and heterogeneous structures, and high encapsulation efficiency using malt dextrin and gum arabic as wall materials (Ferreira et al., 2021).

From the pulp of *A. aculeatum* fruits, microparticles were produced by freeze-drying using different biopolymers (gum arabic, maltodextrin, dextrin, and modified starch) at 85:15 ratio (pulp/wall material solution, w/w), and the highest retention of bioactive compounds (phenolic compounds and carotenoids) was observed for microparticles coated with modified starches (Silva et al., 2018).

Regarding the applications of *tucumã* fruits in other industrial sectors, in the cosmetic industry, formulations of bar soaps were developed from the peel and pulp of *A. aculeatum* (Oliveira et al., 2023). According to the authors, the produced soaps exhibited homogeneity, hard consistency, low shine, the characteristic odor of *tucumã*, and a pleasant appearance. In another example, due to the chemical composition of palm fruit, *tucumã* fruits were considered as promising oil sources with great potential for biodiesel production. Lima et al. (2007) reported the use of oil from *A. vulgare* fruits as a raw material for biodiesel production through transesterification in anhydrous ethanol or methanol (NaOH as a catalyst), and they observed similar properties for the two types of biodiesels when compared to diesel oil. Furthermore, biodiesel production from the oil of *tucumã* almonds (*Astrocaryum aculeatum*) was studied by applying the electrolyte paste from used batteries as a catalyst. These authors observed that the extracted oil presented a high percentage of saturated fatty acids (75%) and acidity (9.25 KOH/g oil) and made available an alternative potentially efficient and environmentally friendly catalyst for biodiesel production (de Freitas et al., 2022).

11.4 PEACH PALM (*BACTRIS GASIPAES* KUNTH)

11.4.1 Main Agronomic and Physiological Aspects

Bactris gasipaes Kunth (Arecaceae family), mainly known as peach palm and popularly known in the Brazilian Amazon as "*pupunha*," is a palm tree with starchy and oily fruits with an ovoid shape (Figure 11.1c) that exhibits a high concentration of carotenoids, which may vary according to different varieties (Soares et al., 2022). Native to the neotropical regions of Central and South America, especially in countries covering the legal Amazonia, the palm trees are characterized as producing drupe-shaped fruits and thorny stems, and among the fruits from genus *Bactris*, the species *B. gasipaes* Kunth is one of the most used for human consumption (González-Jaramillo et al., 2022).

Due to the significant genetic variability resulting from successive natural hybridization processes, peach palm exhibits various genotypes, leading to different morphological and nutritional characteristics (Clement et al., 2017). This is evident when comparing *Bactris gasipaes* Kunth var. *chichagui* (wild genotype) and *B. gasipaes* Kunth var. *gasipaes* (domesticated variety), where the former is characterized by smaller (~2 cm diameter) and more oily fruits, while the latter features larger fruits (~4 cm diameter), less oily pulp, and is starchier, resulting in pulps with "dry" taste (González-Jaramillo et al., 2022; Clement et al., 2017).

In general, peach palm fruits have fibrous exocarp, which can vary in color shades of green, yellow, orange, red, but white-colored varieties (albino fruits) were recently reported (dos Santos et al., 2018; Soares et al., 2022). According to Chisté et al. (2021), the peel color is directly related to the type and concentration of carotenoids accumulated by the fruit. Apart from the peel, the fruits had starchy mesocarps, which can be more or less soft, depending on the lipid content (Neri-Numa et al., 2018). Peach palm fruits have excellent nutritional profile, containing carbohydrates (mainly starch), lipids, proteins, fibers, minerals, and essential amino acids (de Cássia Spacki et al., 2022). The endocarp consists of a hard seed (pit), with an internal white mass containing high contents of polyunsaturated fatty acids (González-Jaramillo et al., 2022). Interestingly, some genotypes do not have seeds (parthenocarpic), and this variety can occur in cases of self-fertilization (Nascim et al., 2010).

11.4.2 Main Carotenoids of Peach Palm

Due to being an oilseed fruit with high levels of carotenoids, studies have been conducted with peach palm in an effort to determine the carotenoid composition in each fruit part (peel and pulp), considering the various existing varieties of peach palm (Chisté et al., 2021; Menezes Silva et al., 2023).

In peach palm, as in other fruits, the types of carotenoids and their contents are related to each fruit part and may also vary depending on various existing genotypes or breeds. The reported carotenoid composition of peach palm fruits is shown in Table 11.4.

TABLE 11.4
Main Carotenoid Reported in Different Parts of *Bactris gasipaes* Kunth Fruits

Fruit Part	Variety/Color	Carotenoid	Reference
Peel	Orange	(all-*E*)-β-carotene and *Z*-isomers	
		(all-*E*)-γ-carotene and *Z*-isomers	
		(all-*E*)-δ-carotene and *Z*-isomers	
Pulp	Orange/yellow	(all-*E*)-phytoene (all-*E*)-phytofluene	Chisté et al., (2021); de Rosso and Mercadante (2007)
	Red/yellow/orange	(all-*E*)-lutein	Chisté et al., (2021); de Rosso and Mercadante (2007); Basto et al. (2016); Santos et al. (2015)
		(all-*E*)-zeaxantina	Chisté et al., (2021); de Rosso and Mercadante (2007); Basto et al. (2016)
	Red/yellow	(all-*E*)-α-carotene	Furtado et al. (2004); de Rosso et al. (2007); Jatunov et al. (2010); Quesada et al. (2011)
	Red/orange/yellow	(all-*E*)-α-cryptoxanthin	de Rosso and Mercadante (2007); Rojas-Garbanzo et al. (2011)
		(all-*E*)-β-carotene and *Z*-isomers	Furtado et al. (2004); de Rosso et al. (2007); Jatunov et al. (2010); Quesada et al. (2011)
		(all-*E*)-β-cryptoxanthin and *Z*-isomers	Chisté et al. (2021); de Rosso et al. (2007); Basto et al. (2016)
		(all-*E*)-γ-carotene and *Z*-isomers	de Rosso et al. (2007); Santos et al., 2015); Quesada et al. (2011)
	Yellow/orange	(all-*E*)-δ-carotene and *Z*-isomers	de Rosso et al. (2007); Santos et al., 2015); Chisté et al. (2021); Rojas-Garbanzo et al. (2011)
	Red/orange/yellow	(all-*E*)-lycopene and *Z*-isomers	Furtado et al. (2004); Jatunov et al. (2010); Quesada et al. (2011)

11.4.3 Antioxidant Potential and Other Biological Properties of Peach Palm Oil

Considering the wide variety of bioactive compounds found in peach palm fruits, such as carotenoids and phenolic compounds, its frequent intake can be considered an important food source that may contribute to health maintenance and the prevention of chronic and degenerative diseases for people from the Amazon region (Chisté et al., 2021; Meléndez-Martínez et al., 2019). However, few studies are available in the literature regarding the biological properties of peach palm, including its derived products. Investigations conducted with dietary supplements enriched with peach palm flour, implemented in the diet of broiler chickens, showed a more favorable hemostatic pattern, along with notable improvement in the LDL/HDL ratio, suggesting a decreased risks of vascular-related diseases in individuals with diets supplemented with peach palm flour (Baldizán et al., 2010).

These health benefits are mainly related to the capacity of carotenoids to scavenge reactive oxygen species, and other pro-oxidant species, by preventing oxidative reactions in low-density lipoproteins, and as consequence, the development of atherosclerosis, enabling a more regular blood flow, with improvements in lipid regulation (dos Santos et al., 2020; Pui Sing Ng et al., 2018). Additionally, peach palm fruits show excellent profile of unsaturated fatty acids, including oleic (ω-9), linoleic (ω-6), and linolenic (ω-3) acids (dos Santos et al., 2020), contributing to the reduction of blood lipid levels and, consequently, lowering the risks of cardiovascular diseases (Pui Sing Ng et al., 2018).

In addition to the expected antioxidant effect provided by the carotenoids found at high concentrations in peach palm fruits, other relevant biological functions were reported (Amengual, 2019). For instance, provitamin A carotenoids, such as β-carotene and β-cryptoxanthin, which are found in the peel and pulp of peach palm fruits (Noronha Matos et al., 2019), are able to act as important regulators in various body functions (Bakac et al., 2023) and contribute to maintaining ocular (Li et al., 2023), skin, and various organ tissues' health (Meléndez-Martínez, 2019).

Another important biological benefit related to the consumption of peach palm fruits is the noticeable nutritional composition (Leterme et al., 2005), presenting high energy value due to its carbohydrate content, with the advantage of having a low glycemic index resulting from the slow metabolism of glucose in the body (Jiménez et al., 2012). Furthermore, peach palm fruits are considered as fiber-rich foods (Soares et al., 2022), inducing the regulation of digestive functions, which result in health benefits (Bernaud et al., 2013). Additionally, peach palm fruits have the excellent profile of unsaturated fatty acids (Graefe et al., 2012) and a protein composition with the presence of all essential amino acids, particularly leucine, valine, threonine, phenylalanine, isoleucine, and lysine, found at high percentages (Yuyama et al., 2003).

11.4.4 Food Applications and Future Prospects for Peach Palm Oil

Based on the reported nutritional aspects, bioactive compounds composition, and expected biological properties, numerous possibilities for the application of peach palm fruits and its derived products, such as the oil fraction, are available for application by interested food industries.

For example, carotenoids from peach palm fruits were extracted using sunflower oil and incorporated in the formulation of an oil-in-water (O/W) food emulsion similar to commercial mayonnaise, aiming to investigate carotenoid's bioaccessibility, which was about 11-fold more bioaccessible in the mayonnaise than in lyophilized fruit (de Souza Mesquita et al., 2020).

In another study, Pires et al. (2019) investigated the potential application of peach palm flour as ingredient for food industries, highlighting its high nutritional value and promising functional properties. As peach palm fruits are starchy, gluten-free, characterized as good sources of fibers (≥6%) with high energy value (≥300 kcal/100 g) (Pires et al., 2019), they can be seen as an excellent alternative for the development of food products for specific consumers, such as people with celiac disease. Other related studies indicated that peach palm flour has a low glycemic index due to the presence of slowly digestible starch, allowing a delay in the energy release, which is ideal for energy supplementation (Pereira, 2007; Ferrari Felisberto et al., 2020). Additionally, the starchy components of peach palm fruits present a low retrogradation index that is adequate for formulating food products without the use of emulsifiers, such as bread and pasta (Ferrari Felisberto et al., 2020).

Other interesting technological applications attributed to peach palm fruits, such as part of food formulations (de Souza Mesquita et al., 2020), are due to the high contents of β-carotene in its peel and pulp, which can be used directly (Martínez-Girón et al., 2017), or even as by-products, in biotechnological processes for obtaining high-value-added products (Carvalho et al., 2018). Despite significant advances in scientific research, additional and systematic investigations are needed to stimulate the sustainable production chain of peach palm fruits focused on optimizing its nutritional and functional properties for application by the food industry and related areas.

11.5 *MURUMURU (ASTROCARYUM MURUMURU* MART.)

11.5.1 Main Agronomic and Physiological Aspects

Astrocaryum murumuru Mart. (synonym: *Astrocaryum yauaperyense* Barb.Rodr), belonging to the Arecaceae family, is popularly known in Brazil as "*murumuru*," "*mumbaca*," "*murmur*," "*murumuru-da-terra-firme*," "*murumuruí*," and "*murumuruzeiro*." *Murumuru* is a typical palm tree native to the Amazonia biome (Brazil North, French Guiana, Guyana, Suriname, Venezuela), and in Northern Brazil, its occurrence is confirmed in swampy forests in the Amazonas, Amapá, Rondônia, and Pará States (MMA, 2016; Coradin et al., 2022).

Murumuru has a cespitous stem, meaning, it grows perpendicular to the ground and can reach up to 15 m in length, with pinnate or composed leaves where leaflets emanate from the rachis (MMA, 2020). The fruits are conical and laterally flattened, measuring 6–8.5 cm in length (Figure 11.1d), with an epicarp covered by brownish bristles, a yellow or reddish mesocarp due to the presence of carotenoids, a hard black endocarp containing a rigid white almond rich in saturated fatty acids (MMA, 2020). The stems, leaves, and fruits are covered with black spines, which can make handling the fruits challenging.

Despite the existing challenges, the extractive exploitation of *murumuru* is becoming increasingly promising due to the widespread use of the fats and oils obtained from the fruits by the food and cosmetics industries, being that *murumuru* butter is the main commercial derived product (Cruz et al., 2017). According to Vidal et al. (2021), the value addition, the use of appropriate technologies, the sustainable forest management, and the development of appropriate research converge in a strategic plan to restructure the marketing of Amazonian oleaginous matrices such as *murumuru*.

Therefore, there is a clear need to boost scientific research focused on characterizing bioactive compounds from *murumuru* fruits, considering all the fruit parts, not just the almond. This includes systematic studies concerning the chemical composition and nutritional and biological properties, with the perspective of developing new food products, co-products, and sustainable processes for the utilization and valorization of the species.

11.5.2 Main Chemical Compounds of *Murumuru* Oils

The chemical composition of *murumuru* fruits can vary due to cultivation conditions, ripening stages, analysis methods, and the specific plant part being investigated. As far as our knowledge is concerned, no prior literature described the carotenoid composition of *murumuru* fruits, although their yellowish/reddish fruit pulps are expected to contain a promising composition. One report stated that the fruit pulp of *murumuru* fruits may contain higher contents of β-carotene than carrots (Duke et al., 1994), which were reported to be in the range of 68–76 mg/100 g (dry basis) (Hiranvarachat et al., 2008).

However, the fatty acid composition of *murumuru* oil has been extensively studied, which is predominantly characterized by saturated fatty acids with concentrations about 47–49% lauric acid (C12:0), ~28% myristic acid (C14:0), and 6–7% palmitic acid (C16:0) (Serra et al., 2018; Pereira et al., 2018). Likewise, most fruits from palm trees, *murumuru* provides oil from both the pulp (2.6%) and the almond (27-44%), where the total oil yields in the entire fruit can be considered lower than those found in *tucumã* fruits (*A. aculeatum*), which present 21.25–61.60% in the pulps and 28.40–45.50% in the almonds (Teixeira et al., 2022). These values may explain the usual preference for the utilization of *murumuru* almonds to extract oil over the fruit pulps.

In general, the oil extraction process for *murumuru* fruits involves drying, heating, grinding, and pressing, and the separation of the husk and almond facilitates the extraction process of the butter and oil (MMA, 2016). Vidal et al. (2021) studied the oil extraction from almonds of *murumuru* fruits from the Amazonas state (Brazil), and these authors found 23% yield of oil with significant waste production, suggesting the need for adopting sustainable techniques to add value to the *murumuru* processing chain. Considering the promising composition of carotenoids in *murumuru* fruits, the adoption of sustainable extraction methods for carotenoids, focusing on green chemistry approaches, is expected to reach high popularity and add a number of benefits to industries interested to include *murumuru* carotenoids in food formulations (Yara-Varón et al., 2017; Chisté et al., 2014).

According to Mambrim and Barrera-Arellano (1997), the *murumuru* oil extracted from the fruit almonds showed carotenoid contents at 5 μg/g, which was similar to the contents found in *tucumã* almonds (3 μg/g), while in the same study, the oil from *tucumã* pulp showed very high contents of carotenoids (2,420 μg/g). Contrastingly, in another study, the *murumuru* fat, produced by neutralization and bleaching, presented higher total carotenoid contents of 21.8 μg/g than in the previous study (Falcão et al., 2017), but Serra et al. (2019) and Lozano-Garzón et al. (2023) did not detect the total carotenoids in the butter from *murumuru* almonds and reported that saturated fatty acids and β-sitosterol were the main detected compounds. Therefore, further studies are needed to determine the carotenoid composition of *murumuru* oil, as extracted from both the pulp and almonds, to stimulate the sustainable exploitation of this oleaginous plant species from the Amazonia.

11.5.3 Antioxidant Potential and Other Biological Functions of *Murumuru* Oil

Murumuru oil was reported to exhibit biological properties, such as antioxidant, antibacterial, and antifungal properties. These effects were mostly attributed to the presence of bioactive compounds, mainly carotenoids (β-carotene), phenolic compounds, and fatty acids.

In a study conducted by Falcão et al. (2017), the antioxidant capacity of semi-refined *murumuru* fat was determined by the reduction of 2,2-diphenyl-1-picrylhydrazyl radicals (DPPH•) and peroxyl radicals (ROO•, ORAC assay). According to these authors, *murumuru* fat was able to scavenge ROO• (ORAC value of 61.2), and the DPPH assay showed an increase in radical scavenging capacity at the highest fat concentrations (50 and 100 mg/mL).

Concerning the antimicrobial activity of *murumuru* oils, Speranza et al. (2015) demonstrated that the emulsions formulated with interesterified oils have bactericidal effect by inhibiting the growth of *Bacillus cereus* and *Escherichia coli*. In another study, Abreu et al. (2014) demonstrated the antimicrobial effect of *murumuru* essential oil against *Colletotrichum* sp., as the studied fungi was inhibited at 1 mL of essential oil, with more expressive results at higher doses (15 and 30 mL), which highlighted the promising applications of *murumuru* essential oil in controlling phytopathogens.

In addition to the mentioned *in vitro* biological properties, *murumuru* butter can potentially be used for the controlled release of bioactive compounds. Gomes et al. (2019) produced nanostructured lipid carriers (NLC) (35 nm) loaded with β-carotene by low emulsification using *murumuru* butter, and α-tocopherol was required to protect the encapsulated carotenoid. These authors observed 42% of bioaccessibility for β-carotene in the dynamic *in vitro* digestion, and the tested NLC decreased cell viability of Caco-2 (human colon adenocarcinoma) and HEPG (human hepatoma) cells probably due to the surfactants used and to the extremely reduced particle size, which may have led to an intense and fast permeation of the NLC through the cells (Gomes et al., 2019).

11.5.4 Food Applications and Future Prospects for *Murumuru* Oil

Murumuru fruits contain edible kernels and pulps that allow high technological potential for food applications. Cruz et al. (2017) conducted a study valuing *murumuru* plant in the high and low floodplain forests of Pará State (Brazil), highlighting its high potential as a non-timber forest product, and the main reported uses of *murumuru* fruits include human and animal feed, as ingredient in cosmetic formulations, biodiesel production, and the use of leaf fibers for crafting and paper production.

As an example of animal feed, Silva et al. (2021) studied the effects of concentrated supplementation with *murumuru* by-products on milk production, milk composition, and milk fatty acid profile of dairy buffaloes. The authors showed that concentrate supplementation with *murumuru* cake slightly increased the levels of saturated fatty acids (lauric and myristic acids) in the milk and slightly reduced the linoleic acid contents as compared with control diet. Although concentrate supplementation with *murumuru* could be considered adequate for inclusion in dairy water buffalo feed, this supplementation requires caution and should be further investigated as the secretion of lauric and myristic fatty acids were induced (Silva et al., 2021).

Pereira et al. (2019) investigated the composition and physicochemical properties of *murumuru* fat, and considering its similarity to *tucumã*, they suggested its alternative use as replacement for cocoa butter in the food industry. *Murumuru* fats have also been explored in Brazilian exports, with studies on its chemical composition, management, distribution areas, and use in beauty and hygiene products (Frare et al., 2022).

Regarding other interested applications, according to literature, *murumuru* butter can be used as an alternative source to produce biodiesel, as it exhibits physical and chemical properties, such as good thermal stability and volatility, which is in accordance with the Brazilian, American, and European standards (Lima et al., 2017; Neves et al., 2018). Furthermore, *murumuru* leaf fibers revealed lignified, hyaline, and rarely undulating vascular fibers, and due to the micrometric characteristics of fibers located in the petiole and rachis of the leaves, they can be seen as a promising material to be used by the paper industry (Rocha & Potiguara, 2007).

11.6 *INAJÁ (MAXIMILIANA MARIPA* (AUBL.) MART.)

11.6.1 Main Agronomic and Physiological Aspects

Inajá (*Maximiliana maripa* (Aubl.) Mart.) is a palm tree that belongs to the family Arecaceae (Palmae) and is found throughout the Amazonia biome. The flowering occurs from October to November, with fruit production on the eastern side of the Amazonia concentrated between January and March, while in the western side, flowering occurs in mid-July, with fruits in early November (Bezerra, 2011).

Inajá palm tree produces clusters containing edible fruits that are widely commercialized in street markets in northern Brazil (Passos et al., 2014). The palm tree has a solitary and erect stem, reaching up to 25 m in height, and it is common in clearings with well-drained soil. *Inajá* can be found in various types of environments, from clay-sandy to dry and nutrient-poor soils, and the palm tree exhibits rapid growth in clearings and can spontaneously regenerate in secondary forests (Viana et al., 2014). Both the kernel and the fresh or cooked edible pulps of *inajá* fruits (Figure 11.1e) can be used by the food and cosmetic industries, as the whole fruit is considered a source of phosphorus, magnesium, and fatty acids (Bezerra, 2011; Shanley & Medina, 2005).

The technological potential of *inajá* relies on the edible oil extracted from the fruit kernel, as its oil processing by-products were considered promising sources of phenolic compounds (mainly catechins and procyanidins), to be used as nutraceutical and/or functional ingredients for the food, pharmaceutical, and cosmetic industries (de Souza Silva et al., 2021).

11.6.2 Main Carotenoids of *Inajá* Oil

The *inajá* oil can be extracted from both the pulp fruits and kernel through mechanical pressing and by solvent extraction (Antoniassi et al., 2015). Relevant unsaturated fatty acids of nutritional importance are present at high contents in *inajá* oil, namely, α-linolenic acid (ω-3), linoleic acid (ω-6), and oleic acid (ω-9), which demonstrate its high energy and nutritional potential for food utilization (Lobato et al., 2021). *Inajá* oil was also reported to contain high contents of ω-9 fatty acids (58–60%) and palmitic acid (20–21%) (Lobato et al., 2021).

The oil extracted from the pulp of *inajá* fruits (*Maximiliana maripa* (Aubl) Drude)) was reported to contain high contents of total carotenoids, varying from 76 (unripe fruits) to 98 mg/100 g oil (ripe fruits) (Barbi et al., 2020). Vegetable oils from palm trees are frequently reported as promising sources of carotenoids that provide an alternative for revaluing low-cost renewable resources, reducing industrial wastes, and delivering positive economic and environmental impacts (Contreras-Gámez et al., 2020). Given this framework, *inajá* fruits stand out as an underexploited native source of high contents of carotenoids.

The carotenoid composition was described for *inajá* oil in the literature, with β-carotene as the major compound (23 μg/g oil), followed by Z-lycopene (15.49 μg/g), lycopene (15.33 μg/g), γ-carotene (14.85 μg/g), Z-γ-carotene (9.81 μg/g), Z-β-carotene (6.01 μg/g), and α-carotene (0.51 μg/g) (Santos et al., 2015). In another study, *inajá* oil showed total carotenoid contents of 4 μg/g (dos

Santos et al., 2021), while only (all-*E*)-β-carotene was identified in pulps (13.7 μg/g, fresh weight) of *inajá* fruits from Amapá (Brazil) (Anunciação et al., 2019).

11.6.3 Antioxidant Potential and Other Biological Functions of *Inajá* Oil

The biological properties associated with *inajá* oil are a topic underexploited by the scientific community, and both the food and pharmaceutical areas of knowledge would benefit Amazonian people with the development of natural food ingredients or nutraceuticals, despite the limited research regarding the antioxidant capacity of the existing bioactive compounds in the fruits (Lobato et al., 2021).

Concerning antioxidant capacity studies, Barbi et al. (2018) reported the *in vitro* scavenging capacity of unripe and ripe pulp oils from *inajá* fruits (DPPH, ABTS, and FRAP assays) and the oil from the unripe pulp, which showed the highest total phenolic compound and the lowest total carotenoid contents and presented the highest antioxidant capacity for all the assays (Barbi et al., 2018). Thus, the highest contents of total phenolic compounds in the oil extracted from unripe *inajá* fruits might be responsible for the highest *in vitro* antioxidant efficiency of the oil over the oil extracted from ripe fruits. However, deeper *in vitro* and *in vivo* studies are needed to understand the antioxidant mechanisms of the bioactive compounds found in *inajá* fruits as potential antioxidants to scavenge physiologically relevant reactive species responsible for oxidative stress.

Inajá oil was able to inhibit the acetylcholinesterase enzyme (ACh) by 63.75%, demonstrating its high efficiency and promising potential as an alternative approach to synthetic molecules for the therapeutic treatment of various neurodegenerative pathologies related to the modulation of ACh (Fernandez et al., 2016). The inhibition of ACh is relevant in the treatment of neurodegenerative diseases, such as Alzheimer's and Parkinson's, as inhibitors prevent the enzyme from breaking down the neurotransmitter acetylcholine, as low levels of acetylcholine are associated with memory issues and muscle disorders (Jamwal & Kumar, 2018).

11.6.4 Food Applications and Future Prospects for *Inajá* Oil

The *inajá* fruits are edible but still poorly investigated for food application. Barbi et al. (2019) evaluated the potential of the oil extracted from pulps of ripe *inajá* fruits using an unconventional pressurized fluid extraction technique, and they stated that *inajá* oil can be considered a good source of bioactive phenolic compounds and essential fatty acids to be incorporated in food formulations. Moreover, ripe pulps of *inajá* fruits were studied by Santos et al. (2017), and the presence of relevant bioactive compounds with antioxidant properties was reported, such as carotenoids, flavonoids, and ascorbic acid.

Therefore, considering the lack of studies regarding the use of *inajá* fruits and its derived products in the food science and technology area of knowledge, there is a clear need to stimulate scientific research focused on the investigation of chemical, nutritional, and biological properties of the fruit parts, with the perspective of developing new food products or improving existing ones. Conducting such research would be essential for adding value to the derived food products and also to expand our understanding about the full potential of *inajá* fruits for the food industry.

11.7 OIL PALM (*ELAEIS OLEIFERA* (KUNTH) CORTÉS)

11.7.1 Main Agronomic and Physiological Aspects

Oil palm is one of the most studied palm trees in the world mainly due to its high industrial relevance. The genus *Elaeis* belongs to the class Liliopsida (Monocotyledoneae), order Arecales, family Arecaceae (formerly Palmaceae), and tribe Cocoseae. The main species are *E. guineensis*, native to Africa, and *E. oleifera* (also known as *E. melanococca*), native to America. Two other reported species are *E. odora* (previously *Barcella odora*) and *E. madagascariensis*, but only limited information about them is available in the literature. The oil palm or African oil palm (*E. guineensis*) presents high productivity potential, with average yield ten times higher than soybeans (Godswill et al., 2016; de Almeida et al., 2012). According to the United States Department of Agriculture (USDA) (USDA, 2023), the global palm oil production in 2022–2023 was ~79 million tons, with the following as the top ten producing countries: Indonesia (59%), Malaysia (24%), Thailand (4%), Colombia (2%), Nigeria (2%), Guatemala (1%), Papua New Guinea (1%), Cote d'Ivoire (1%), Honduras (1%), and Brazil (1%).

The species *E. oleifera* (Kunth) Cortés, as known as "*caiaué*," American oil palm, Amazon oil palm, Pará oil palm, nolí, ojon, or corozo, is an oleaginous species native to northern Latin America and Central America and primarily in the Amazon rainforest of Brazil (Lima et al., 2020; España et al., 2018). The plant is perennial, with a solitary stem, slow growth, and leaflets arranged in a single plane. The fruit bunches are cone-shaped, and the unripe fruits show a light-green or olive-green exocarp that turns to intense yellow or orange when fully ripe. The mesocarp presents intense yellow to orange colors (Figure 11.1f), and the endocarp is dark brown, woody, and fibrous (Hernández et al., 2012; Mendoza et al., 2022).

The American oil palm tree has anatomical, morphological, and physiological similarities to the African oil palm. The species possesses relevant agronomic characteristics, such as slow growth, high resistance to diseases and pests like fusarium wilt and lethal yellowing, and the fruits show oil with high quality. These features make it attractive for the genetic improvement of the African oil palm, one of the main sources of vegetable oil globally (Ithnin et al., 2021; Rey et al., 2004; Hormaza et al., 2012; Tezara et al., 2021).

The fruits of the *Elaeis* genus produce two types of oils: palm oil from the mesocarp (pulp) of the fruit and palm kernel oil from the endosperm (almond) of the fruit. Unlike the oil from *E. guineensis*, which is highly important in the food, cosmetic, and petrochemical industries, the *E. oleifera* species has not yet achieved the same global impact due to its lower yield. However, studies reported that *E. oleifera* oil has interesting characteristics, such as higher concentration of unsaturated fatty acids and low iodine value, often considered of superior quality compared to oil from *E. guineensis* fruits (Mendoza et al., 2022; Ithnin et al., 2021; Rey et al., 2004; Almeida et al., 2019; Dal Prá et al., 2016; Chaves et al., 2018).

11.7.2 Palm Oil: Main Carotenoids and Extraction Techniques

The chemical composition of palm oil from the American oil palm is predominantly composed by oleic and linoleic fatty acids, provitamin A carotenoids, tocopherols, ubiquinones, squalene, and sterols. The oil exhibits a lower rate of deterioration compared to the African oil palm due to the reduced hydrolytic action of fruit mesocarp lipases (triacylglycerol acylhydrolase, EC 3.1.1.3), which play a role in releasing free fatty acids (FFA) and are indicative of oil quality deterioration (España et al., 2018; Cadena et al., 2012).

Similar to the African oil palm, the American palm is also recognized as a natural source of carotenoids. This compound is responsible for the orange-reddish color and the stability and nutritional value of the edible oil, along with tocopherols. The main carotenoids found are α- and β-carotenes, which account for about 80% of the carotenoid composition, and the total contents may vary based on fruit species, extraction method, cultivation conditions, and location. The American species was reported to show total carotenoid contents up to 4,600 mg/kg, surpassing *E. guineensis*, which contains 500–1,500 mg/kg (Vidoca et al., 2020; Rasid et al., 2009; Sampaio et al., 2013; Mayamol et al., 2007).

Fruits of *E. oleifera*, *E. guineensis* varieties (*pisifera* and *dura*), and their hybrids (*E. o.* × *E. g.*), collected in Johor, Malaysia, were investigated for their total carotenoid contents and composition. *E. oleifera* fruits showed higher total carotenoid contents (4,600 mg/kg) as compared to

E. guineensis varieties pisifera (428 mg/kg), dura (997 mg/kg), and their hybrids *E. oleifera* × pisifera (1,430 mg/kg), *E. oleifera* × dura (2,324 mg/kg), and the backcross *E. oleifera*–dura × pisifera (896 mg/kg). The main carotenoids reported in the species and hybrids were α- and β-carotene, followed by phytoene, Z-β-carotene, phytofluene, Z-α-carotene, ζ-carotene, γ-carotene, δ-carotene, neurosporene, β-zeacarotene, α-zeacarotene, and lycopene (Yap et al., 1991).

Trujillo-Quijano et al. (1990) studied oils extracted from fresh and sterilized palm fruits of *E. guineensis* species, including dura dumpy, psifera, tenera, and *E. oleifera* collected in Una, Bahia, Brazil. The oils had total carotenoid contents of 1,120.7, 283.2, 660.5, and 1,576.8 mg/kg, respectively. The main identified carotenoids were Z-phytofluene, 13Z-α-carotene, α-carotene, 13Z-β-carotene, β-carotene, 9Z-β-carotene, ζ-carotene, zeaxanthin, β-cryptoxanthin, poly-Z-lycopene, mono-Z-lycopene, and lycopene, with α- and β-carotene accounting for 80% of the total carotenoid contents. This study also highlighted that *E. oleifera* fruit oil had the highest carotenoid content among the species.

In a comparative study of the genotypes of *E. oleifera* Coarí (Brazil), Peru (Peru), and Sinú (Colombia), the total carotenoid contents were 861, 785.2, and 1,524.7 mg/kg, respectively. The contents of α-carotene were 298.2, 190.6, and 557.5 mg/kg, while β-carotenes were 562.8, 594.6, and 967.2 mg/kg, respectively. The study attributed the differences in the total contents to the quality and production of each population's fruit bunches (Chaves et al., 2018).

Five different genotypes of *E. oleifera* from the American oil palm germplasm bank maintained by *Embrapa Amazônia Ocidental* in the Rio Urubu Experimental Station, Amazonas State, Brazil, were studied by España et al. (2018). The total carotenoid contents varied from 1,526.68 to 3,343.56 mg/kg, indicating high influence of genetic factors over environmental factors on the phenotypic correlations between the studied traits. Mohd et al. (2000) studied *oleiferas* collected from open-pollinated families in Panama, Costa Rica, Honduras, and Colombia, obtaining a maximum total carotenoid content of 3,455.5 mg/kg oil from palm fruits from Costa Rica.

The efficient extraction of carotenoids and other bioactive compounds from *Elaeis* fruits begins with initial treatment stages, such as sterilization using pressurized steam or boiling, pulping, followed by drying. These steps are essential to prevent degradation, isomerization, and reduced yield of the target constituents, allowing them to be later used as colorants, dietary additives, or health-promoting products (Mortensen, 2005; Abd Rashid et al., 2021).

The most commonly used methods for oil extraction include hot or cold processes, such as cold-pressing and organic solvent extraction. However, unconventional methods are being studied to increase oil yield and protect sensitive compounds, including low-heat enzyme-assisted aqueous extraction, pressurized liquid extraction, and supercritical carbon dioxide extraction. Table 11.5 presents the reported studies for *Elaeis* species regarding the collection origin, oil extraction method, analytical determination of carotenoids, and the contents and main identified carotenoids (Chaves et al., 2018; da Silva et al., 2018; Szydłowska-Czerniak et al., 2011; Joshua, 2022; Rodríguez et al., 2016; Lau et al., 2008).

11.7.3 Antioxidant Potential of Palm Oil

The oils obtained from the fruits of the genus *Elaeis* and the fruits of the interspecific hybrid palm (*E. oleifera* × *E. guineensis*) are reported to exhibit relevant biological properties, such as reducing oxidative stress, which can be a precursor to heart diseases, neurodegenerative diseases, tumors, diabetes, and chronic diseases. These activities are related to the antioxidant properties provided by bioactive compounds, such as carotenoids, phytosterols, tocopherols, tocotrienols, and phenolic compounds in the oil. Additionally, these oils contribute to the reduction of cardiovascular risk by favorably affecting plasma lipid profiles, comparable to the effects of olive oil on the body. Other biological activities are associated with the high concentration of oleic acid and low concentration of saturated fatty acids (Abd Rashid et al., 2021; Szydłowska-Czerniak et al., 2011; Rodríguez et al., 2016; Ojeda et al., 2017).

TABLE 11.5
Carotenoid Composition and Extraction Techniques for *Elaeis* Fruit Oils

Species	Origin	Extraction Process	Analytical Determination of Carotenoids	Carotenoid Composition	Reference
Genotypes of *E. oleifera* **and their interspecific hybrids O × G with** *E. guineensis*	Brazil, Peru, and Colombia	Organic solvent extraction	HPLC using a fluorescence detector and C18 column	Total carotenoids: 785–1,524 mg/kg α-carotene: 190–577 mg/kg β-carotene: 321–967 mg/kg	Chaves et al. (2018)
E. guineensis dura, pisifera and tenera	Nigeria	Hot and cold process	HPLC-UV with C18 column	Total carotenoids: 54–123 mg/100 g Main carotenoids: β-cryptoxanthin (1.97–4.20); lycopene (1.57–7.89); α-carotene (1.16–2.00); β-carotene (28.54–84.18); lutein (2.07–4.86); zeaxanthin (4.27–9.40); antheraxanthine (1.42–9.86); violaxanthin (1.69–4.07); neoxanthin (2.33–6.37)	Joshua (2022)
Five accessions of *E. oleifera*	Brazil	Dionex-accelerated solvent extractor using petroleum ether	UV-visible spectrophotometry using acetone	Total carotenoids: 1,526–3,343 mg/kg	España et al. (2018)
E. oleifera ***E. guineenses pisifera*** ***E. guineenses dura*** ***E. oleifera × pisifera*** ***E. oleifera × dura*** ***E. oleifera, dura × pisifera***	Malaysia	Soxhlet (hexane)	HPLC-UV-visible	Total carotenoids: 428–4,592 mg/kg Main carotenoids: phytoene (1.12–2.49); *Z*-β-carotene (0.10–0.68); phytofluene (Tr.–1.24); β-carotene (51.64–60.53); α-carotene (33.11–54.35); *Z*-α-carotene (0.86–2.49); ζ-carotene (0.36–2.31); γ-carotene (0.09–1.10); δ-carotene (0.09–2.00); neurosporene (0.04–0.77); β-zeacarotene (0.56–1.53); α-zeacarotene (0.21–0.52); lycopene (0.02–7.81).	Yap et al. (1991)

(Continued)

TABLE 11.5 (*Continued*)
Carotenoid Composition and Extraction Techniques for *Elaeis*Fruit Oils

Species	Origin	Extraction Process	Analytical Determination of Carotenoids	Carotenoid Composition	Reference
E. guineensis	Malaysia	Commercial process	HPLC-UV-visible	Total carotenoids: 673 mg/kg Main carotenoids: phytoene (1.3); phytofluene (0.1); phytofluene (0.7); β-carotene (56); α-carotene (35.1); *Z*-α-carotene (2.5); ζ-carotene (0.7); γ-carotene (0.3); δ-carotene (0.8); neurosporene (0.3); β-zeacarotene (0.7); α-zeacarotene (0.2); lycopene (1.3)	Choo et al. (1992)
E. guineensis	Malaysia	Supercritical CO_2 at 40°C and 10–30 MPa	UV-visible spectrophotometry using hexane	Total carotenoids: 237–5,498 mg/kg	Lau et al. (2008)
E. guineensis	Malaysia	Low-heat enzyme-assisted aqueous extraction method	UV-visible spectrophotometry using n-hexane	Total carotenoids: 39.14 mg/100 g oil	Abd Rashid et al. (2021)
E. guineensis	—	Local vegetable oil factory	UV-visible spectrophotometry using n-hexane	Total carotenoids: 45.8–32.7 mg/100 g oil	Szydłowska-Czerniak et al. (2011)
E. guineensis	Brazil	Crude palm oil	HPLC coupled to diode array and fluorescence detectors	Total carotenoids: 851 mg/kg	Sampaio et al.***E. guineensis*** (2013)
E. guineensis	Brazil	Crude palm oil	UV-visible spectrophotometry using n-hexane	0.54 g β-carotene/L oil	Chiu et al. (2009)

The functional effect of interspecific hybrid (*E. oleifera* × *E. guineensis*) fruit oil was studied concerning its action on human antioxidant plasma/serum. The evaluation included total phenolic compound contents and antioxidant capacity (ORAC and ABTS assays). The study revealed that administering the oil at a dosage of 25 mL/day for three months significantly increased the total phenolic contents (19.2%) and the antioxidant capacity of human antioxidant plasma/serum in adults aged 50–77 years (ORAC, 92.1%, and ABTS, 42.9%). The results suggest that the oil may be comparable to olive oil, a key ingredient in the renowned Mediterranean diet, showing potential as a promising supplement for reducing the risk of cardiovascular diseases and other health benefits (Ojeda et al., 2017).

The concentration of phytonutrients, such as carotenoids and phenolic compounds, and the antioxidant capacity of oils obtained by low-heat, enzyme-assisted aqueous extraction method from palm fruits with and without hydrolysis treatment, with the pectinase enzyme, was compared in the study by Abd Rashid et al. (2021). These authors showed that the enzyme influenced the increase in the oil yield (26.11 to 52.70%) with an increase, but without statistically significant difference ($p < 0.05$), in the total phenolic compound contents (4.31 to 4.48 mg gallic acid equivalent/100 g oil) and in the total carotenoid contents (35.51 to 39.13 mg/100 g oil). The antioxidant capacity of the oil increased by 21% with enzymatic treatment of the fibers (86.81 to 105.41 Trolox equivalent/100 g oil), and the results were attributed to the enzyme's breakdown of the cell wall of the mesocarp tissue, facilitating the release of antioxidant components into the oil. The authors highlighted that fruit treatment methods combined with unconventional extraction techniques can contribute to obtaining a product with good yield and high antioxidant properties.

Changes in antioxidant capacity, total phenolic compound contents, and polyphenol composition of interspecific hybrid palm oil extracted from *E. oleifera* × *E. guineensis* (*O* × *G*, Coari × La Mé cultivar) were evaluated during fruit ripening at 18, 20, 22, and 24 weeks after anthesis. The results showed that the oils from fruits at the early maturation stage (18 weeks) presented the highest phenolic compound contents and antioxidant capacities (ORAC, FRAP, ABTS, and DPPH assays). Thus, the ripening stage had relevant influence on obtaining oils with high potential as food ingredient with antioxidant properties (Rodríguez et al., 2016).

11.7.4 Food Applications and Future Prospects for Palm Oil

The palm oil culture holds great relevance and impact on the global stage due to its extensive use in the market and versatility in various industrial sectors, such as petrochemical, chemical, pharmaceutical, biochemical, nutraceutical, and food (Rodríguez et al., 2016; Ojeda et al., 2017; Bezerra et al., 2018; Montoya et al., 2013). The biomass resulting from the processing of *Elaeis* fruits also represents a potential for future applications, including use in the diets of ruminants and non-ruminants, as a source of residual oil, biofertilizers, biogas, and biocomposites (Bezerra et al., 2018; Onoja et al., 2019).

The primary application of *E. oleifera* in literature is found in the study of its genetic sequencing and its application for improvement and interspecific hybrid production with *E. guineensis*. Despite being underutilized due to its lower oil yield (about 25% lower than African oil palm), *E. oleifera* has advantages over *E. guineensis* in terms of agronomic factors and its chemical composition, which contains a higher concentration of unsaturated fatty acids (oleic and linoleic) and phytochemical compounds, such as carotenoids, some of which are precursors of vitamin A, in addition to phytosterols, tocopherols, tocotrienols, and phenolic compounds, all of them exhibiting antioxidant properties with potential for preventing diseases (Chaves et al., 2018; Yap et al., 1991; Rodríguez et al., 2016; Ojeda et al., 2017).

11.8 CONCLUSION

The frequent intake of foods rich in carotenoids can significantly impact health, and the oils from Arecaceae palm fruits such as *buriti*, *tucumã*, peach palm, *murumuru*, *inajá*, and American oil palm, as shown in this chapter, are promising sources of carotenoids with biological properties. In this sense, carotenoid-rich vegetable oils of palm trees from the Amazonia may help enhance the antioxidant defenses in the body and act as anti-inflammatory, antimicrobial, and anticarcinogenic agents, thereby exerting beneficial effects on health.

In this regard, the future perspective on the use of these oils from the Amazon should include sustainable harvesting practices and responsible management to ensure their long-term availability. Furthermore, encouraging more research regarding their chemical composition and nutritional and biological properties would open new opportunities for their use in the food, pharmaceutical, cosmetic, and other interested industries. Therefore, further studies on these Amazonian palm trees species are necessary, as they can be employed in different food formulations, improving the nutritional and sensory composition of food products, valuing Amazonia, and promoting the bioeconomy around the countries covered by this biome.

REFERENCES

Abd Rashid, S.N.A.; Ab. Malik, S.; Embi, K.; Mohd Ropi, N.A.; Yaakob, H.; Cheng, K.K.; Sarmidi, M.R.; Leong, H.Y. Carotenoids and Antioxidant Activity in Virgin Palm Oil (VPO) Produced from Palm Mesocarp with Low Heat Aqueous-Enzyme Extraction Techniques. In *Proceedings of the Materials Today: Proceedings*; Elsevier Ltd., 2021; Vol. 42, pp. 148–152.

Abreu, M.G.P.; Ferreira, J.B.; Neves, y y b; Araujo, M.L.; Souza, R.B. Efeito Fungitóxico de Óleos Essenciais de Palmeiras Amazônicas Sobre *Colletotrichum* Sp. *Enciclopédia Biosfera* **2014**, *10*, 897–905.

Aguiar, L. M.; Bicas, J. L.; Fuentes, E.; Alarcón, M.; Gonzalez, I. P.; Pastore, G. M.; Cazarin, C. B. B. (2021). Non-nutrients and Nutrients from Latin American Fruits for the Prevention of Cardiovascular Diseases. *Food Research International* **2021**, *139*, 109844.

Aguiar, P.; Silva, M.; Ximenes, J.; Paes, E.; Lima, E. Avaliação Preliminar Do Potencial Antioxidante e Anti-Inflamatório Da Amêndoa Do Tucumã_em_(*Astrocaryum aculeatum*). *Brazilian Journal of Health Review* **2022**, *5*.

Almeida, E.S.; Carvalho, A.C.B.; Soares, I.O. de S.; Valadares, L.F.; Mendonça, A.R.V.; Silva, I.J.; Monteiro, S. Elucidating How Two Different Types of Bleaching Earths Widely Used in Vegetable Oils Industry Remove Carotenes from Palm Oil: Equilibrium, Kinetics and Thermodynamic Parameters. *Food Research International* **2019**, *121*, 785–797, doi:10.1016/j.foodres.2018.12.061.

Amengual, J. Bioactive Properties of Carotenoids in Human Health. *Nutrients* **2019**, *11*, 2388, doi:10.3390/NU11102388.

Antoniassi, R.; Schwengber, D.R.; Souza, R.C.; Duarte, I.D.; Wilhelm, A.E.; Machado, A.D.F.; Duarte, O.R. Rendimento Em Óleo e Composição Em Ácidos Graxos de Frutos de *Attaela maripa*. In *Proceedings of the Congresso Brasileiro de Processamento de Frutas e Hortaliças*; Búzios, 2015; p. 6.

Anunciação, P.C.; Giuffrida, D.; Murador, D.C.; de Paula Filho, G.X.; Dugo, G.; Pinheiro-Sant'Ana, H.M. Identification and Quantification of the Native Carotenoid Composition in Fruits from the Brazilian Amazon by HPLC–DAD–APCI/MS. *Journal of Food Composition and Analysis* **2019**, *83*, doi:10.1016/j.jfca.2019.103296.

Aquino, J.; Soares, J.; Magnani, M.; Stamford, T.; Mascarenhas, R.; Tavares, R.; Stamford, T. Effects of Dietary Brazilian Palm Oil (*Mauritia flexuosa* L.) on Cholesterol Profile and Vitamin A and E Status of Rats. *Molecules* **2015**, *20*, 9054–9070, doi:10.3390/molecules20059054.

Bakac, E.R.; Percin, E.; Gunes-Bayir, A.; Dadak, A. A Narrative Review: The Effect and Importance of Carotenoids on Aging and Aging-Related Diseases. *International Journal of Molecular Sciences* **2023**, *24*, 15199, doi:10.3390/IJMS242015199.

Baldissera, M.D.; Souza, C.F.; Doleski, P.H.; Grando, T.H.; Sagrillo, M.R.; da Silva, A.S.; Leal, D.B.R.; Monteiro, S.G. Treatment with Tucumã Oil (*Astrocaryum vulgare*) for Diabetic Mice Prevents Changes in Seric Enzymes of the Purinergic System: Improvement of Immune System. *Biomedicine and Pharmacotherapy* **2017**, *94*, 374–379, doi:10.1016/j.biopha.2017.07.113.

Baldizán, G.; Oviedo, M.; Michelangeli, C.; Vargas, R.E. Effects of Peach Palm Oil on Performance, Serum Lipoproteins and Haemostasis in Broilers. *British Poultry Science* **2010**, *51*, 784–790, doi:10.1080/00071668.2010.526925.

Barbi, R.C.; Teixeira, G.L.; Hornung, P.S.; Ávila, S.; Hoffmann-Ribani, R. *Eriobotrya Japonica* Seed as a New Source of Starch: Assessment of Phenolic Compounds, Antioxidant Activity, Thermal, Rheological and Morphological Properties. *Food Hydrocolloids* **2018**, *77*, 646–658, doi:10.1016/j.foodhyd.2017.11.006.

Barbi, R.C.T.; de Souza, A.R.C.; Hamerski, F.; Teixeira, G.L.; Corazza, M.L.; Ribani, R.H. Subcritical Propane Extraction of High-Quality Inajá (*Maximiliana maripa*) Pulp Oil. *Journal of Supercritical Fluids* **2019**, *153*, 1–11, doi:10.1016/j.supflu.2019.104576.

Barbi, R.C.T.; Hornung, P.S.; Ávila, S.; Alves, F.E.S.B.; Beta, T.; Ribani, R.H. Ripe and Unripe Inajá (*Maximilia maripa*) Fruit: A New High Source of Added Value Bioactive Compounds. *Food Chemistry* **2020**, *331*, doi:10.1016/j.foodchem.2020.127333.

Barboza, N.L.; Cruz, J.M. dos A.; Corrêa, R.F.; Lamarão, C.V.; Lima, A.R.; Inada, N.M.; Sanches, E.A.; Bezerra, J. de A.; Campelo, P.H. Buriti (*Mauritia flexuosa* L. f.): An Amazonian Fruit with Potential Health Benefits. *Food Research International* **2022**, *159*, 111654, doi:10.1016/j.foodres.2022.111654.

Barroso, T.L.C.T.; Alexandre, J. de B.; Silva, L.C. da; Castelo, R.M.; Ribeiro, L.B.; Furtado, R.F.; Zambelli, R.A. Tecnologia de Encapsulamento Na Área de Alimentos: Uma Revisão. *Research, Society and Development* **2021**, *10*, e6210716240, doi:10.33448/rsd-v10i7.16240.

Basto, G.J.; Carvalho, C.W.P.; Soares, A.G.; Costa, H.T.G.B.; Chávez, D.W.H.; Godoy, R.L. de O.; Pacheco, S. Physicochemical Properties and Carotenoid Content of Extruded and Non-Extruded Corn and Peach Palm (*Bactris gasipaes*, Kunth). *LWT – Food Science and Technology* **2016**, *69*, 312–318, doi:10.1016/J.LWT.2015.12.065.

Bernaud, F.S.R.; Rodrigues, T.C. Fibra Alimentar: Ingestão Adequada e Efeitos Sobre a Saúde Do Metabolismo. *Arquivos Brasileiros de Endocrinologia & Metabologia* **2013**, *57*, 397–405, doi:10.1590/S0004-27302013000600001.

Bezerra, F.W.F.; Costa, W.A. da; Oliveira, M.S. de; Aguiar Andrade, E.H. de; Carvalho, R.N. de Transesterification of Palm Pressed-Fibers (*Elaeis guineensis* Jacq.) Oil by Supercritical Fluid Carbon Dioxide with Entrainer Ethanol. *Journal of Supercritical Fluids* **2018**, *136*, 136–143, doi:10.1016/j.supflu.2018.02.020.

Bezerra, V.S. *Comunicado Técnico – Embrapa*; Amapá, 2011; pp. 1–6.

Bony, E.; Boudard, F.; Brat, P.; Dussossoy, E.; Portet, K.; Poucheret, P.; Giaimis, J.; Michel, A. Awara (*Astrocaryum vulgare* M.) Pulp Oil: Chemical Characterization, and Anti-Inflammatory Properties in a Mice Model of Endotoxic Shock and a Rat Model of Pulmonary Inflammation. *Fitoterapia* **2012**, *83*, 33–43, doi:10.1016/j.fitote.2011.09.007.

Bovi, G.G.; Petrus, R.R.; Pinho, S.C. Feasibility of Incorporating Buriti (*Mauritia flexuosa* L.) Oil Nanoemulsions in Isotonic Sports Drink. *International Journal of Food Science & Technology* **2017**, *52*, 2201–2209, doi:10.1111/ijfs.13499.

Cabral, F.L.; Bernardes, V.M.; Passos, D.F.; de Oliveira, J.S.; Doleski, P.H.; Silveira, K.L.; Horvarth, M.C.; Bremm, J.M.; Barbisan, F.; Azzolin, V.F.; et al. *Astrocaryum aculeatum* Fruit Improves Inflammation and Redox Balance in Phytohemagglutinin-Stimulated Macrophages. *Journal of Ethnopharmacology* **2020**, *247*, 112274, doi:10.1016/j.jep.2019.112274.

Cadena, T.; Prada, F.; Perea, A.; Romero, H.M. Lipase Activity, Mesocarp Oil Content, and Iodine Value in Oil Palm Fruits of *Elaeis guineensis*, *Elaeis oleifera*, and the Interspecific Hybrid O×G (*E. oleifera* × *E. guineensis*). *Journal of the Science of Food and Agriculture* **2012**, *93*, 674–680, doi:10.1002/jsfa.5940.

Cândido, T.L.N.; Silva, M.R.; Agostini-Costa, T.S. Bioactive Compounds and Antioxidant Capacity of Buriti (*Mauritia flexuosa* L.f.) from the Cerrado and Amazon Biomes. *Food Chemistry* **2015**, *177*, 313–319, doi:10.1016/j.foodchem.2015.01.041.

Carvalho, E.A.; Nunes, L.V.; Goes, L.M. dos S.; Silva, E.G.P. da; Franco, M.; Gross, E.; Uetanabaro, A.P.T.; Costa, A.M. da Peach-Palm (*Bactris gasipaes* Kunth.) Waste as Substrate for Xylanase Production by *Trichoderma stromaticum* AM7. *Chemical Engineering Communications* **2018**, *205*, 975–985, doi:10.1080/00986445.2018.1425208.

Chaves, G.; Ligarreto-Moreno, G.A.; Cayon-Salinas, D.G. Physicochemical Characterization of Bunches from American Oil Palm (*Elaeis oleifera* H.B.K. Cortes) and Their Hybrids with African Oil Palm (*Elaeis guineensis* Jacq.). *Acta Agronomica* **2018**, *67*, doi:10.15446/acag.v67n1.62028.

Chisté, R.C.; Benassi, M.T.; Mercadante, A.Z. Efficiency of Different Solvents on the Extraction of Bioactive Compounds from the Amazonian Fruit *Caryocar villosum* and the Effect on Its Antioxidant and Colour Properties. *Phytochemical Analysis* **2014**, *25*, 364–372, doi:10.1002/pca.2489.

Chisté, R.C.; Costa, E.L.N.; Monteiro, S.F.; Mercadante, A.Z. Carotenoid and Phenolic Compound Profiles of Cooked Pulps of Orange and Yellow Peach Palm Fruits (*Bactris gasipaes*) from the Brazilian Amazonia. *Journal of Food Composition and Analysis* **2021**, *99*, 103873, doi:10.1016/j.jfca.2021.103873.

Chisté, R.C.; Fernandes, E. In Silva, L.R.; Silva, B., Eds.; *Bioactive Compounds from Amazonian Fruits and Their Antioxidant Properties*; Bentham Science Publishers, Sharjah, UAE, 2016; pp. 244–264, doi:10.2174/9781681082394116010011.

Chiu, M.C.; de Morais Coutinho, C.; Gonçalves, L.A.G. Carotenoids Concentration of Palm Oil Using Membrane Technology. *Desalination* **2009**, *245*, 783–786, doi:10.1016/j.desal.2009.03.002.

Choo, Y.M.; Yap, S.C.; Ooi, A.S.H.; Goh, S.H. Production of Palm Oil Carotenoid Concentrate and its Potential Application in Nutrition. In Ong, A.S.H.; Packer, L., Eds.; *Molekularbiologie*; Switzerland, Bern: Bohlstr, 1992; pp. 243–255.

Clement, C.R.; de Cristo-Araújo, M.; Coppens d'Eeckenbrugge, G.; dos Reis, V.M.; Lehnebach, R.; Picanço-Rodrigues, D. Origin and Dispersal of Domesticated Peach Palm. *Frontiers in Ecology and Evolution* **2017**, *5*, 279143, doi:10.3389/FEVO.2017.00148/BIBTEX.

Contreras-Gámez, M. del M.; Gómez-Caravaca, A.M. Underutilized Sources of Carotenoids. In *Carotenoids: Properties, Processing and Applications*; Elsevier, 2020; pp. 107–147; ISBN 9780128170670.

Copetti, P.M.; Gündel, S. da S.; de Oliveira, P.S.B.; Favarin, F.R.; Ramos, A.P.; Pintos, F.G.; Pappis, L.; Gündel, A.; Machado, A.K.; Ourique, A.F.; et al. Development, Characterisation, Stability Study and Antileukemic Evaluation of Nanoemulsions Containing *Astrocaryum aculeatum* Extract. *Natural Product Research* **2022**, *36*, 1321–1326, doi:10.1080/14786419.2020.1862830.

Coradin, L.; Camillo, J.; Vieira, I.C.G. *Espécies Nativas Da Flora Brasileira de Valor Econômico Atual Ou Potencial: Plantas Para o Futuro: Região Norte*; Brasília: MMA, 2022.

Costa, B.E.T.; Santos, O.V. Dos; Corrêa, N.C.F.; França, L.F. De Comparative Study on the Quality of Oil Extracted from Two Tucumá Varieties Using Supercritical Carbon Dioxide. *Food Science and Technology (Brazil)* **2016**, *36*, 322–328, doi:10.1590/1678-457X.0094.

Cruz, G.S.; Gama, J.R. V; Ribeiro, R.B.S.; Santos, L.E.; Melo, L.O.; Coelho, A.A. Estrutura e Valoração de *Astrocaryum murumuru* Mart. Na Região Do Estuário Amazônico. *Nativa* **2017**, *5*, 581–587.

da Fonseca Machado, A. P.; do Nascimento, R. D. P.; da Rocha Alves, M.; Reguengo, L. M.; Junior, M. R. M. Brazilian Tucumã-do-Amazonas (Astrocaryum aculeatum) and Tucumã-do-Pará (Astrocaryum Vulgare) Fruits: Bioactive Composition, Health Benefits, and Technological Potential. *Food Research International* **2022**, *151*, 110902.

Dal Prá, V.; Soares, J.F.; Monego, D.L.; Vendruscolo, R.G.; Freire, D.M.G.; Alexandri, M.; Koutinas, A.; Wagner, R.; Mazutti, M.A.; Da Rosa, M.B. Extraction of Bioactive Compounds from Palm (*Elaeis guineensis*) Pressed Fiber Using Different Compressed Fluids. *Journal of Supercritical Fluids* **2016**, *112*, 51–56, doi:10.1016/j.supflu.2016.02.011.

da Silva, C.M.; Zanqui, A.B.; da Silva, E.A.; Gomes, S.T.M.; Filho, L.C.; Matsushita, M. Extraction of Oil from *Elaeis* Spp. Using Subcritical Propane and Cosolvent: Experimental and Modeling. *Journal of Supercritical Fluids* **2018**, *133*, 401–410, doi:10.1016/j.supflu.2017.11.006.

de Almeida, S.; Raimundo, R.; Vieira Da Cunha, N.; Lopes, R.; Barcelos, E. Empresa Brasileira de Pesquisa Agropecuária Embrapa Amazônia Ocidental Ministério Da Agricultura, Pecuária e Abastecimento. **2012**.

de Cássia Spacki, K.; Corrêa, R.C.G.; Uber, T.M.; Barros, L.; Ferreira, I.C.F.R.; Peralta, R.A.; de Fátima Peralta Muniz Moreira, R.; Helm, C.V.; de Lima, E.A.; Bracht, A.; et al. Full Exploitation of Peach Palm (*Bactris gasipaes* Kunth): State of the Art and Perspectives. *Plants* **2022**, *11*, 3175, doi:10.3390/PLANTS11223175.

de Freitas, F.A.; Mendonça, I.R.S.; Barros, S. de S.; Pessoa, W.G.A.; Sá, I.S.C.; Gato, L.B.; Silva, E.P.; Farias, M.A.S.; Nobre, F.X.; Maia, P.J.S.; et al. Biodiesel Production from Tucumã (*Astrocaryum aculeatum* Meyer) Almond Oil Applying the Electrolytic Paste of Spent Batteries as a Catalyst. *Renew Energy* **2022**, *191*, 919–931, doi:10.1016/j.renene.2022.04.083.

de Rosso, V.V.; Mercadante, A.Z. Identification and Quantification of Carotenoids, by HPLC-PDA-MS/MS, from Amazonian Fruits. *Journal of Agricultural and Food Chemistry* **2007**, *55*, 5062–5072, doi:10.1021/jf0705421.

de Souza, M.N.; Piedade, A.B.S.; Santos, T.P.; de Britto, V.F.; Nascimento, E.; de Faria, J.L.B.; de Faria, R.A.P.G. Interaction Effect of Cassava Starch × Buriti Oil on the Physical Properties of Edible Films. *eFood* **2022**, *3*, doi:10.1002/efd2.53.

de Souza Filho, O.C.; Sagrillo, M.R.; Garcia, L.F.M.; Machado, A.K.; Cadoná, F.; Ribeiro, E.E.; Duarte, M.M.M.F.; Morel, A.F.; Da Cruz, I.B.M. The in Vitro Genotoxic Effect of Tucuma (*Astrocaryum aculeatum*), an Amazonian Fruit Rich in Carotenoids. *Journal of Medicinal Food* **2013**, *16*, 1013–1021, doi:10.1089/jmf.2012.0287.

de Souza Mesquita, L.M.; Neves, B.V.; Pisani, L.P.; de Rosso, V.V. Mayonnaise as a Model Food for Improving the Bioaccessibility of Carotenoids from *Bactris gasipaes* Fruits. *LWT* **2020**, *122*, doi:10.1016/j.lwt.2020.109022.

de Souza Silva, A.P.; Rosalen, P.L.; de Camargo, A.C.; Lazarini, J.G.; Rocha, G.; Shahidi, F.; Franchin, M.; de Alencar, S.M. Inajá Oil Processing By-Product: A Novel Source of Bioactive Catechins and

Procyanidins from a Brazilian Native Fruit. *Food Research International* **2021**, *144*, doi:10.1016/j.foodres.2021.110353.

dos Santos, M.F.G.; Mamede, R.V.S.; Rufino, M. do S.M.; De Brito, E.S.; Alves, R.E. Amazonian Native Palm Fruits as Sources of Antioxidant Bioactive Compounds. *Antioxidants* **2015**, *4*, 591–602, doi:10.3390/antiox4030591.

dos Santos, M.T.; Morgavi, P.; Le Roux, G.A. (2018). Exploring Amazonian Fats and Oils Blends by Computational Predictions of Solid Fat Content. *OCL* **2018**, *25*(1), D107.

dos Santos, O.V.; Soares, S.D.; Dias, P.C.S.; de Paula de Almeida Duarte, S.; dos Santos, M.P.L.; das Chagas Alves do Nascimento, F. Chromatographic Profile and Bioactive Compounds Found in the Composition of Pupunha Oil (*Bactris gasipaes* Kunth): Implications for Human Health. *Revista de Nutrição* **2020**, *33*, e190146, doi:10.1590/1678-9805202033E190146.

Duke, J.A.; Vásquez, R.; Vásquez, M. *Amazonian Ethnobotanical Dictionary*; CRC Press, 1994; ISBN 0849336643.

Edge, R.; Truscott, T.G. Singlet Oxygen and Free Radical Reactions of Retinoids and Carotenoids—a Review. *Antioxidants* **2018**, *7*, 5, doi:10.3390/ANTIOX7010005.

España, M.D.; Mendonça, S.; Carmona, P.A.O.; Guimarães, M.B.; da Cunha, R.N.V.; Souza, M.T. Chemical Characterization of the American Oil Palm from the Brazilian Amazon Forest. *Crop Science* **2018**, *58*, 1982–1990, doi:10.2135/cropsci2018.04.0231.

Falcão, A. de O.; Speranza, P.; Ueta, T.; Martins, I.M.; Macedo, G.A.; Macedo, J.A. Antioxidant Potential and Modulatory Effects of Restructured Lipids from the Amazonian Palms on Liver Cells. *Food Technology and Biotechnology* **2017**, *55*, 553–561, doi:10.17113/ftb.55.04.17.5157.

Fernandez, I.M.; Mozombite, D.M.S.; Santos, R.C.; De Melo Filho, A.A.; Ribeiro, P.R.E.; Chagas, E.A.; Takahashi, J.A.; Ferraz, V.P.; Melo, A.C.G.R.; Maldonado, S.A.S. Oil in Inajá Pulp (*Maximiliana maripa*): Fatty Acid Profile and Anti-Acetylcholinesterase Activity. *Orbital – The Electronic Journal of Chemistry* **2016**, *8*, doi:10.17807/orbital.v7i4.769.

Ferrari Felisberto, M.H.; Souza Costa, M.; Villas Boas, F.; Lopes Leivas, C.; Maria Landi Franco, C.; Michielon de Souza, S.; Pedrosa Silva Clerici, M.T.; Mach Côrtes Cordeiro, L. Characterization and Technological Properties of Peach Palm (*Bactris gasipaes* Var. Gasipaes) Fruit Starch. *Food Research International* **2020**, *136*, 109569, doi:10.1016/J.FOODRES.2020.109569.

Ferreira, L.M. de M.C.; Pereira, R.R.; de Carvalho, F.B.; Santos, A.S.; Ribeiro-Costa, R.M.; Carréra Silva Júnior, J.O. Green Extraction by Ultrasound, Microencapsulation by Spray Drying and Antioxidant Activity of the Tucuma Coproduct (*Astrocaryum vulgare* Mart.) Almonds. *Biomolecules* **2021**, *11*, doi:10.3390/biom11040545.

Ferreira, M.J.A.; Mota, M.F.S.; Mariano, R.G.B.; Freitas, S.P. Current Scenario and Recent Advancements from Tucumã Pulp Oil and Kernel Fat Processing. *European Journal of Lipid Science and Technology* **2022**, *124*.

Filho, R.P.L.; Aleluia, R.L. *Caracterização de Frutos de Buritizeiros (Mauritia flexuosa L.) Objetivando a Seleção de Progênies Para Fins de Aproveitamento Do Óleo Da Polpa e Das Sementes*; Macapá, 2015; ISBN 1517-4867, 88.

Forzaa, R.C. *Coleção Flora Do Brasil*; Rio de Janeiro: Ministro do Meio Ambiente (MMA), 2020.

França, L.F.; Reber, G.; Meireles, M.A.A.; Machado, N.T.; Brunner, G. Supercritical extraction of carotenoids and lipids from buriti (*Mauritia flexuosa*), a fruit from the Amazon region. *The Journal of Supercritical Fluids* **1999**, *14*(3), 247–256, doi:10.1016/S0896-8446(98)00122-3.

Frare, A.B.; Milena, C.; Fernandes, G.; Capuano Da Cruz, A.P.; Gomes Barbosa, M.A. Biopiracy and Legitimacy: Strategy of Brazilian Company Natura S.A. in the Murumuru Case. *REUNIR: Revista de Administração, Contabilidade e Sustentabilidade* **2022**, *12*, 1–13.

Freitas, M.L.F.; Chisté, R.C.; Polachini, T.C.; Sardella, L.A.C.Z.; Aranha, C.P.M.; Ribeiro, A.P.B.; Nicoletti, V.R. Quality Characteristics and Thermal Behavior of Buriti (*Mauritia Flexuosa* L.) Oil. *Grasas y Aceites* **2018**, *68*, 220, doi:10.3989/gya.0557171.

Furtado, J.; Siles, X.; Campos, H. Carotenoid Concentrations in Vegetables and Fruits Common to the Costa Rican Diet. *International Journal of Food Sciences and Nutrition* **2004**, *55*, 101–113, doi:10.1080/09637480410001666522.

Godswill, N.N.; Frank, N.E.G.; Walter, A.N.; Edson, M.Y.J.; Kingsley, T.M.; Arondel, V.; Martin, B.J.; Emmanuel, Y. Oil Palm. In *Breeding Oilseed Crops for Sustainable Production: Opportunities and Constraints*; Elsevier Inc., 2016; pp. 217–273; ISBN 9780128014691.

Gomes, G.V.L.; Sola, M.R.; Rochetti, A.L.; Fukumasu, H.; Vicente, A.A.; Pinho, S.C. β-Carotene and α-Tocopherol Coencapsulated in Nanostructured Lipid Carriers of Murumuru (*Astrocaryum murumuru*) Butter Produced by Phase Inversion Temperature Method: Characterisation, Dynamic in Vitro Digestion and Cell Viability Study. *Journal of Microencapsulation* **2019**, *36*, 43–52, doi:10.1080/02652048.2019.1585982.

Gonçalves, O.; Pereira, R.; Gonçalves, F.; Mendo, S.; Coimbra, M.A.; Rocha, S.M. Evaluation of the Mutagenicity of Sesquiterpenic Compounds and Their Influence on the Susceptibility Towards Antibiotics of Two Clinically Relevant Bacterial Strains. *Mutation Research—Genetic Toxicology and Environmental Mutagenesis* **2011**, *723*, 18–25, doi:10.1016/j.mrgentox.2011.03.010.

González-Jaramillo, N.; Bailon-Moscoso, N.; Duarte-Casar, R.; Romero-Benavides, J.C. Peach Palm (*Bactris gasipaes* Kunth.): Ancestral Tropical Staple with Future Potential. *Plants* **2022**, *11*, 3134, doi:10.3390/PLANTS11223134.

Graefe, S.; Dufour, D.; Van Zonneveld, M.; Rodriguez, F.; Gonzalez, A.; Graefe, S.; Dufour, Á.D.; Rodriguez, Á.F.; Gonzalez, Á.A.; Dufour, D.; et al. Peach Palm (*Bactris gasipaes*) in Tropical Latin America: Implications for Biodiversity Conservation, Natural Resource Management and Human Nutrition. *Biodiversity and Conservation* **2012**, *22*, 269–300, doi:10.1007/S10531-012-0402-3.

Guex, C.G.; Cassanego, G.B.; Dornelles, R.C.; Casoti, R.; Engelmann, A.M.; Somacal, S.; Maciel, R.M.; Duarte, T.; Borges, W. de S.; Andrade, C.M. de; et al. Tucumã (*Astrocaryum Aculeatum*) Extract: Phytochemical Characterization, Acute and Subacute Oral Toxicity Studies in Wistar Rats. *Drug and Chemical Toxicology* **2022**, *45*, 810–821, doi:10.1080/01480545.2020.1777151.

Gul, K.; Tak, A.; Singh, A.K.; Singh, P.; Yousuf, B.; Wani, A.A. Chemistry, Encapsulation, and Health Benefits of β-Carotene – a Review. *Cogent Food & Agriculture* **2015**, *1*, doi:10.1080/23311932.2015.1018696.

Hernández, D.C.F.; Martínez, P.A.H.; Caicedo, L.P.M.; Romero, R.R. *Generalidades Sobre La Morfología y Fenología de La Palma de Aceite*; Angulo, H.M.R., Ed.; 2012; Vol. 1.

Hiranvarachat, B.; Suvarnakuta, P.; Devahastin, S. Isomerisation Kinetics and Antioxidant Activities of β-Carotene in Carrots Undergoing Different Drying Techniques and Conditions. *Food Chemistry* **2008**, *107*, 1538–1546, doi:10.1016/j.foodchem.2007.10.026.

Hormaza, P.; Fuquen, E.M.; Romero, H.M. Phenology of the Oil Palm Interspecific Hybrid *Elaeis oleifera* × *Elaeis guineensis*. *Scientia Agricola* **2012**, *69*, 275–280.

Hovorková, P.; Laloučková, K.; Skřivanová, E. Determination of in Vitro Antibacterial Activity of Plant Oils Containing Medium-Chain Fatty Acids Against Gram-Positive Pathogenic and Gut Commensal Bacteria. *Czech Journal of Animal Science* **2018**, *63*, 119–125, doi:10.17221/70/2017-CJAS.

Ibiapina, A.; Gualberto, L. da S.; Dias, B.B.; Freitas, B.C.B.; Martins, G.A. de S.; Melo Filho, A.A. Essential and Fixed Oils from Amazonian Fruits: Proprieties and Applications. *Critical Reviews in Food Science and Nutrition* **2022**, *62*, 8842–8854, doi:10.1080/10408398.2021.1935702.

Igielska-Kalwat, J.; Gościańska, J.; Nowak, I. Karotenoidy Jako Naturalne Antyoksydanty Carotenoids as Natural Antioxidants. *Advances in Hygiene and Experimental Medicine* **2015**, *69*, 418–428.

Ithnin, M.; Vu, W.T.; Shin, M.G.; Suryawanshi, V.; Sherbina, K.; Zolkafli, S.H.; Serdari, N.M.; Amiruddin, M.D.; Abdullah, N.; Mustaffa, S.; et al. Genomic Diversity and Genome-Wide Association Analysis Related to Yield and Fatty Acid Composition of Wild American Oil Palm. *Plant Science* **2021**, *304*, doi:10.1016/j.plantsci.2020.110731.

Jamwal, S.; Kumar, P. Insight into the Emerging Role of Striatal Neurotransmitters in the Pathophysiology of Parkinson's Disease and Huntington's Disease: A Review. *Current Neuropharmacology* **2018**, *17*, 165–175, doi:10.2174/1570159x16666180302115032.

Jantsch, M.H.; Bernardes, V.M.; Oliveira, J.S.; Passos, D.F.; Dornelles, G.L.; Manzoni, A.G.; Cabral, F.L.; da Silva, J.L.G.; Schetinger, M.R.C.; Leal, D.B.R. Tucumã (*Astrocaryum aculeatum*) Prevents Memory Loss and Oxidative Imbalance in the Brain of Rats with Hyperlipidemia. *Journl of Food Biochemistry* **2021**, *45*, 1–9, doi:10.1111/jfbc.13636.

Jatunov, S.; Quesada, S.; Diaz, C.; Murillo, Enrique. Composição de Carotenoides e Atividade Antioxidante Do Mesocarpo de Frutos Crus e Cozidos de Seis Variedades de *Bactris gasipaes*. **2010**, 99–104.

Jiménez, G.; Gómez, G.; Pérez, A.M.; Blanco-Metzler, A. Estimation of Glycaemic Index of Peach Palm (*Bactris gasipaes*) Cooked Fruits and Chips, and Pitahaya (*Hylocereus* Spp.) Pulp. *Archivos latinoamericanos de nutricion* **2012**, *62*, 242–248.

Jobim, M.L.; Santos, R.C.V.; dos Santos Alves, C.F.; Oliveira, R.M.; Mostardeiro, C.P.; Sagrillo, M.R.; de Souza Filho, O.C.; Garcia, L.F.M.; Manica-Cattani, M.F.; Ribeiro, E.E.; et al. Antimicrobial Activity of Amazon *Astrocaryum aculeatum* Extracts and Its Association to Oxidative Metabolism. *Microbiological Research* **2014**, *169*, 314–323, doi:10.1016/j.micres.2013.06.006.

Joshua, O. Effect of Processing Method on Carotenoid Profiles of Oils from Three Varieties of Nigerian Palm Oil (*Elaise guinensis*). *International Journal of Natural Sciences* **2022**, *3*, 1–15.

Kahn, F. The Genus *Astrocaryum* (Arecaceae). *Rev Peru Biol* **2008**, *15*, 31–48.

Koolen, H.H.F.; da Silva, F.M.A.; da Silva, V.S.V.; Paz, W.H.P.; Bataglion, G.A. Buriti Fruit—*Mauritia flexuosa*. In *Exotic Fruits*; Elsevier, 2018; pp. 61–67.

Lage, N.N.; Márcia, J.; Lopes, M.; Pereira, R.R.; Ferreira Da Costa Guerra, J.; De Fátima, M.; Pereira, A.; Silva, M.; De Freitas Bonomo, L.; Geraldo De Lima, W.; et al. Antioxidant Potential of Buriti (Mauritia Flexuosa) Pulp Flour in Diabetic Rats. *Archivos Latinoamericanos de Nutrición (ALAN)*, *68*(1), 59–70.

Lau, H.L.N.; Choo, Y.M.; Ma, A.N.; Chuah, C.H. Selective Extraction of Palm Carotene and Vitamin E from Fresh Palm-Pressed Mesocarp Fiber (*Elaeis guineensis*) Using Supercritical CO2. *Journal of Food Engineering* **2008**, *84*, 289–296, doi:10.1016/j.jfoodeng.2007.05.018.

Leite, P.I.P.; Barreto, S.M.A.G.; Freitas, P.R.; de Araújo, A.C.J.; Paulo, C.L.R.; de Almeida, R.S.; de Assis, C.F.; Padilha, C.E.A.; Ferrari, M.; de Sousa Junior, F.C. Extraction of Bioactive Compounds from Buriti (*Mauritia flexuosa* L.) Fruit by Eco-Friendly Solvents: Chemical and Functional Characterization. *Sustainable Chemistry and Pharmacy* **2021**, *22*, doi:10.1016/j.scp.2021.100489.

Leterme, P.; García, M.F.; Londoño, A.M.; Rojas, M.G.; Buldgen, A.; Souffrant, W.B. Chemical Composition and Nutritive Value of Peach Palm (*Bactris gasipaes* Kunth) in Rats. *Journal of the Science of Food and Agriculture* **2005**, *85*, 1505–1512, doi:10.1002/JSFA.2146.

Li, Z.; Li, Y.; Hou, Y.; Fan, Y.; Jiang, H.; Li, B.; Zhu, H.; Liu, Y.; Zhang, L.; Zhang, J.; et al. Association of Plasma Vitamins and Carotenoids, DNA Methylation of LCAT, and Risk of Age-Related Macular Degeneration. *Nutrients* **2023**, *15*, 2985, doi:10.3390/NU15132985/S1.

Lima, J.R. de O.; da Silva, R.B.; Miranda de Moura, E.; Rodarte de Moura, C.V. Biodiesel of Tucum Oil, Synthesized by Methanolic and Ethanolic Routes. *Fuel* **2008**, *87*, 1718–1723, doi:10.1016/j.fuel.2007.09.007.

Lima, R.P.; Souza da Luz, P.T.; Braga, M.; dos Santos Batista, P.R.; Ferreira da Costa, C.E.; Zamian, J.R.; Santos do Nascimento, L.A.; da Rocha Filho, G.N. Murumuru (*Astrocaryum murumuru* Mart.) Butter and Oils of Buriti (*Mauritia flexuosa* Mart.) and Pracaxi (*Pentaclethra macroloba* (Willd.) Kuntze) Can Be Used for Biodiesel Production: Physico-Chemical Properties and Thermal and Kinetic Studies. *Industrial Crops and Products* **2017**, *97*, 536–544, doi:10.1016/j.indcrop.2016.12.052.

Lima, S.C.D.S.; Genovese-Marcomini, P.R.; Quisen, R.C.; Mendonça, M.S. Morphoanatomic and Histochemical Aspects of *Elaeis oleifera* (Kunth) Cortés Seed. *Journal of Seed Science* **2020**, *42*, 1–12, doi:10.1590/2317.

Lobato, J.M.; Stamford, T.C.M.; Lobato, L.M.; Lobato, G.M.; Sousa, A.R.S.; Stamford, T.L.M. Caracterização Físico-Química e Propriedades Do Óleo de Inajá (*Maximiliana maripa* (Aubl.) Drude). In *Avanços em Ciência e Tecnologia de Alimentos – Volume 3*; Editora Científica Digital, 2021; pp. 35–47.

Lozano-Garzón, K.; Orduz-Díaz, L.L.; Guerrero-Perilla, C.; Quintero-Mendoza, W.; Carrillo, M.P.; Cardona-Jaramillo, J.E.C. Comprehensive Characterization of Oils and Fats of Six Species from the Colombian Amazon Region with Industrial Potential. *Biomolecules* **2023**, *13*, doi:10.3390/biom13060985.

Mambrim, M.C.T.; Barrera-Arellano, D. Caracterización de Aceites de Frutos de Palmeras de La Región Amazónica Del Brasil. *Grasas y Aceites* **1997**, *48*, 154–158.

Maoka, T. Carotenoids as Natural Functional Pigments. *Journal of Natural Medicines* **2020**, *74*.

Martínez-Girón, J.; Figueroa-Molano, A.M.; Ordóñez-Santos, L.E. Effect of the Addition of Peach Palm (*Bactris gasipaes*) Peel Flour on the Color and Sensory Properties of Cakes. *Food Science and Technology* **2017**, *37*, 418–424, doi:10.1590/1678-457X.14916.

Martins, M.G.; Brito, B.N.C.; Lima, D.; Matos, K.A.N.; Chiste, R.C. *Astrocaryum aculeatum* e *A. vulgare*-Tucumã-Açu e Tucumã-Do-Pará. In *Espécies nativas da flora brasileira de valor econômico atual ou potencial: plantas para o futuro: região Norte*; Coradin, L.; Camillo, J.; Vieira, I.C.G., Eds.; Ministério do Meio Ambiente – MMA, 2022; ISBN 978-65-88265-16-1.

Matos, K.A.N.; Praia Lima, D.; Pereira Barbosa, A.P.; Zerlotti Mercadante, A.; Campos Chisté, R. Peels of Tucumã (*Astrocaryum vulgare*) and Peach Palm (*Bactris gasipaes*) Are by-Products Classified as Very High Carotenoid Sources. *Food Chemistry* **2019**, 272, 216–221, doi:10.1016/j.foodchem.2018.08.053.

Mayamol, P.N.; Balachandran, C.; Samuel, T.; Sundaresan, A.; Arumughan, C. Process Technology for the Production of Micronutrient Rich Red Palm Olein. *JAOCS, Journal of the American Oil Chemists' Society* **2007**, *84*, 587–596, doi:10.1007/s11746-007-1078-9.

Meléndez-Martínez, A.J. An Overview of Carotenoids, Apocarotenoids, and Vitamin A in Agro-Food, Nutrition, Health, and Disease. *Molecular Nutrition & Food Research* **2019**, *63*.

Meléndez-Martínez, A. J., Esquivel, P., & Rodriguez-Amaya, D. B. Comprehensive review on carotenoid composition: Transformations during processing and storage of foods. *Food Research International* **2023**, *169*, 112773.

Mendoza, L.; Barba, J.; Ligarreto, G. Growth Parameters of Backcross-Derived *Elaeis oleifera* (Kunth) Cortés from Taisha – Ecuador and Their Potential for Plant Breeding in Oil Palm. *Pesquisa Agropecuária Tropical* **2022**, *52*, 1–10, doi:10.1590/1983-40632022v5273633.

Menezes Silva, J.V.; Silva Santos, A.; Araujo Pereira, G.; Campos Chisté, R. Ultrasound-Assisted Extraction Using Ethanol Efficiently Extracted Carotenoids from Peels of Peach Palm Fruits (*Bactris gasipaes* Kunth) Without Altering Qualitative Carotenoid Profile. *Heliyon* **2023**, *9*, doi:10.1016/j.heliyon.2023.e14933.

Menezes, E.G.O.; Barbosa, J.R.; Pires, F.C.S.; Ferreira, M.C.R.; Silva, A.P. de S.; Siqueira, L.M.M.; de Carvalho Junior, R.N. Development of a New Scale-up Equation to Obtain Tucumã-of-Pará (*Astrocaryum vulgare* Mart.) Oil Rich in Carotenoids Using Supercritical CO2 as Solvent. *The Journal of Supercritical Fluids* **2022**, *181*, 105481, doi:10.1016/j.supflu.2021.105481.

Mesquita, J. de A.; Oliveira, T.T. da S.; Santos, J.G. da S. dos; Gaspar, M.R.G.R. do C.; Vieira, V. de A.; Rodrigues, E.C.; Nascimento, E.; Faria, P.B.; Faria, R.A.P.G. de Fatty Acid Profile and Physicochemical Characterization of Buriti Oil During Storage. *Ciência Rural* **2020**, *50*, doi:10.1590/0103-8478cr20190997.

MMA. *Coleção Flora Do Brasil*; Rio de Janeiro, 2020.

MMA. *Murumuru: Boas Práticas Para o Extrativismo Sustentável Orgânico*. Secretaria de Extrativismo e Desenvolvimento Rural Sustentável. Departamento de Extrativismo, Ed.; Brasília, DF, 2016.

Mohd, D.; Rajanaidu, N.; Jalani, B.S. Performance of *Elaeis oleifera* from Panama, Costa Rica, Colombia and Honduras in Malaysia. *Journal of Oil Palm Research* **2000**, *12*, 71–80.

Montoya, C.; Lopes, R.; Flori, A.; Cros, D.; Cuellar, T.; Summo, M.; Espeout, S.; Rivallan, R.; Risterucci, A.M.; Bittencourt, D.; et al. Quantitative Trait Loci (QTLs) Analysis of Palm Oil Fatty Acid Composition in an Interspecific Pseudo-Backcross from *Elaeis oleifera* (H.B.K.) Cortés and Oil Palm (*Elaeis guineensis* Jacq.). *Tree Genetics & Genomes* **2013**, *9*, 1207–1225, doi:10.1007/s11295-013-0629-5.

Morais, N.S.; Passos, T.S.; Ramos, G.R.; Ferreira, V.A.F.; Moreira, S.M.G.; Chaves Filho, G.P.; Barreto, A.P.G.; Leite, P.I.P.; Almeida, R.S.; Paulo, C.L.R.; et al. Nanoencapsulation of Buriti Oil (*Mauritia flexuosa* L.f.) in Porcine Gelatin Enhances the Antioxidant Potential and Improves the Effect on the Antibiotic Activity Modulation. *PLoS One* **2022**, *17*, e0265649, doi:10.1371/journal.pone.0265649.

Mortensen, A. Analysis of a Complex Mixture of Carotenes from Oil Palm (*Elaeis guineensis*) Fruit Extract. *Food Research International* **2005**, *38*, 847–853, doi:10.1016/j.foodres.2005.01.009.

Mota, M.F.S.; Ferreira, M.J.A.; Novaes, F.J.M.; Marriott, P.J.; Rezende, C.M.; Freitas, S.P. Natural Crystallisation of Tucuma (*Astrocaryum vulgare* Mart.) Pulp Olein. *Journal of Food Composition and Analysis* **2022**, *114*, doi:10.1016/j.jfca.2022.104795.

Nascim, A.; Filho, K.; Roland, C.; Marcos, C.; Vilela De Resende, D.; Tomé, J.; Neto, F.; Luiz, C.; Gilberto, B.; Yokomizo, K.-I.; et al. *Empresa Brasileira de Pesquisa Agropecuária Embrapa Florestas Ministério Da Agricultura, Pecuária e Abastecimento Programa de Melhoramento Genético de Pupunha Na Embrapa, IAC e Inpa*. **2010**.

Nascimento, K.; Copetti, P.M.; Fernandes, A.; Klein, B.; Fogaça, A.; Zepka, L.Q.; Wagner, R.; Ourique, A.F.; Sagrillo, M.R.; da Silva, J.E.P. Phytochemical Analysis and Evaluation of the Antioxidant and Antiproliferative Effects of Tucumã Oil Nanocapsules in Breast Adenocarcinoma Cells (MCF-7). *Natural Product Research* **2021**, *35*, 2060–2065, doi:10.1080/14786419.2019.1648460.

Neri-Numa, I.A.; Soriano Sancho, R.A.; Pereira, A.P.A.; Pastore, G.M. Small Brazilian Wild Fruits: Nutrients, Bioactive Compounds, Health-Promotion Properties and Commercial Interest. *Food Research International* **2018**, *103*, 345–360, doi:10.1016/J.FOODRES.2017.10.053.

Neves, M.C.T.; Lopes, A.; de Oliveira, M.C.J.; Iamaguti, P.S.; Lira, T.A.M.; Moreti, T.C.F.; de Lima, L.P.; Koike, G.H.A. Effects of Murumuru (*Astrocaryum murumuru* Mart.) and Soybean Biodiesel Blends on Tractor Performance and Smoke Density. Australian *Journal of Crop Science* **2018**, *12*, 878–885, doi:10.21475/ajcs.18.12.06.PNE634.

Noronha Matos, K.A.; Praia Lima, D.; Pereira Barbosa, A.P.; Zerlotti Mercadante, A.; Campos Chisté, R. Peels of Tucumã (*Astrocaryum vulgare*) and Peach Palm (*Bactris gasipaes*) Are by-Products Classified as Very High Carotenoid Sources. *Food Chemistry* **2019**, 272, 216–221, doi:10.1016/j.foodchem.2018.08.053.

Ojeda, M.; Borrero, M.; Sequeda, G.; Diez, O.; Castro, V.; García, Á.; Ruiz, Á.; Pacetti, D.; Frega, N.; Gagliardi, R.; et al. Hybrid Palm Oil (*Elaeis oleifera* × *Elaeis guineensis*) Supplementation Improves Plasma Antioxidant Capacity in Humans. *European Journal of Lipid Science and Technology* **2017**, *119*, doi:10.1002/ejlt.201600070.

Oliveira, E.V.; Laranjeira, R.K. da S.; Yamaguchi, K.K. de L. Uso Do Tucumã (*Astrocaryum aculeatum* Meyer) Em Cosmético: Caracterização Química e Elaboração de Sabonetes Artesanais. *Pesquisa e Ensino em Ciências Exatas e da Natureza* **2023**, *7*, 177–191, doi:10.56814/pecen.v7i1.1925.

Onoja, E.; Chandren, S.; Abdul Razak, F.I.; Mahat, N.A.; Wahab, R.A. Oil Palm (*Elaeis guineensis*) Biomass in Malaysia: The Present and Future Prospects. *Waste Biomass Valorization* **2019**, *10*, 2099–2117.

Passos, M.A.B.; Duarte, O.R.; Yuyama, K. Morphometric Characterization of *Attalea maripa* (Arecaceae) from Roraima State, Brazil. *Journal of Agricultural Science* **2014**, *6*, doi:10.5539/jas.v6n12p115.

Pereira, E.; Ferreira, M.C.; Sampaio, K.A.; Grimaldi, R.; Meirelles, A.J. de A.; Maximo, G.J. Physical Properties of Amazonian Fats and Oils and Their Blends. *Food Chemistry* **2019**, *278*, 208–215, doi:10.1016/j.foodchem.2018.11.016.

Pereira, J. Rally Universitário patrocinado pelo UniCEUB e organizado pelo Buriti Rally Clube: competidores chegam ao final da prova, 2014.

Pereira, K.D. Amido Resistente, a Última Geração No Controle de Energia e Digestão Saudável. *Food Science and Technology* **2007**, *27*, 88–92, doi:10.1590/S0101-20612007000500016.

Pereira, W.F.S.; de Figueiredo Furtado, G.; Feltre, G.; Hubinger, M.D. Oleosomes from Buriti *(Mauritia flexuosa* L. f.): Extraction, Characterization and Stability Study. *Innovative Food Science and Emerging Technologies* **2022**, *82*, doi:10.1016/j.ifset.2022.103183.

Pereira de Oliveira, J.; Almeida, O.P.; Campelo, P.H.; Carneiro, G.; de Oliveira Ferreira Rocha, L.; Santos, J.H.P.M.; Gomes da Costa, J.M. Tailoring the Physicochemical Properties of Freeze-Dried Buriti Oil Microparticles by Combining Inulin and Gum Arabic as Encapsulation Agents. *LWT* **2022**, *161*, 113372, doi:10.1016/j.lwt.2022.113372.

Pereira Lima, R.; Souza da Luz, P.T.; Braga, M.; dos Santos Batista, P.R.; Ferreira da Costa, C.E.; Zamian, J.R.; Santos do Nascimento, L.A.; da Rocha Filho, G.N. Murumuru (*Astrocaryum murumuru* Mart.) Butter and Oils of Buriti (*Mauritia Flexuosa* Mart.) and Pracaxi (*Pentaclethra macroloba* (Willd.) Kuntze) Can Be Used for Biodiesel Production: Physico-Chemical Properties and Thermal and Kinetic Studies. *Industrial Crops and Products* **2017**, *97*, 536–544, doi:10.1016/j.indcrop.2016.12.052.

Pires, M.B.; Amante, E.R.; Lopes, A.S.; Rodrigues, A.M. da C.; da Silva, L.H.M. Peach Palm Flour (*Bactris gasipae* KUNTH): Potential Application in the Food Industry. *Food Science and Technology* **2019**, *39*, 613–619, doi:10.1590/FST.34617.

Pui Sing Ng, C.; Sui Lam Kwok, J.; Zhou, J.; Yuen Belinda Cheung, P.; Chi Hung Tzang, L.; Sundaram Prenetics Limited, S.; Kong, H.; Farhan Muhamad Sarkan, D.; Chia Hsin Chew, D.; Wong, Z.; et al. IDDF2018-ABS-0204 Effect of Whole Grains and Unsaturated Fatty Acids Dietary Intervention on Health in Hong Kong Overweight Population. *Gut* **2018**, *67*, A71, doi:10.1136/GUTJNL-2018-IDDFAB STRACTS.153.

Quesada, S.; Azofeifa, G.; Jatunov, S.; Jiménez, G.; Navarro, L.; Gó mez, G. Carotenoids Composition, Antioxidant Activity and Glycemic Index of Two Varieties of *Bactris gasipaes. Emirates Journal of Food and Agriculture* **2011**, *23*, 482–489.

Rasid, O.A.; Parveez, A.; Kadir, G.; Ho, C.-L.; Napis, S. Plant Carotenoids: Molecular Genetics and Regulation Malaysian Palm Oil Board. *Journal of Oil Palm Research* **2009**, *21*, 588–601.

Rey, L.; Gómez, P.L.; Ayala, I.; Delgado, W.; Rocha, P. Colecciones Genéticas de Palma de Aceite *Elaeis guineensis* Jacq. Y *Elaeis oleífera* (HBK) de Cenipalma: Características de Importancia Para El Sector Palmicultor. *Revista Palmas* **2004**, *25*, 39–48.

Ribeiro, B.D.; Coelho, M.A.Z.; Barreto, D.W. Production of Concentrated Natural Beta-Carotene from Buriti (*Mauritia vinifera*) Oil by Enzymatic Hydrolysis. *Food and Bioproducts Processing* **2012**, *90*, 141–147, doi:10.1016/j.fbp.2011.02.003.

Rocha, C.B.R.; Potiguara, R.C. V Morfometria Das Fibras Das Folhas de *Astrocaryum murumuru* Var. Murumuru Mart. (Arecaceae). *Acta Amazon* **2007**, *37*, 511–516.

Rodríguez, J.C.; Gómez, D.; Pacetti, D.; Núnnez, O.; Gagliardi, R.; Frega, N.G.; Ojeda, M.L.; Loizzo, M.R.; Tundis, R.; Lucci, P. Effects of the Fruit Ripening Stage on Antioxidant Capacity, Total Phenolics, and Polyphenolic Composition of Crude Palm Oil from Interspecific Hybrid *Elaeis oleifera* × *Elaeis guineensis. Journal of Agricultural and Food Chemistry* **2016**, *64*, 852–859, doi:10.1021/acs.jafc.5b04990.

Rojas-Garbanzo, C.; Pérez, A.M.; Bustos-Carmona, J.; Vaillant, F. Identification and Quantification of Carotenoids by HPLC-DAD During the Process of Peach Palm (*Bactris gasipaes* H.B.K.) Flour. *Food Research International* **2011**, *44*, 2377–2384, doi:10.1016/J.FOODRES.2011.02.045.

Rudke, A.R.; Mazzutti, S.; Andrade, K.S.; Vitali, L.; Ferreira, S.R.S. Optimization of Green PLE Method Applied for the Recovery of Antioxidant Compounds from Buriti (*Mauritia flexuosa* L.) Shell. *Food Chemistry* **2019**, *298*, 125061, doi:10.1016/j.foodchem.2019.125061.

Sagrillo, M.R.; Garcia, L.F.M.; De Souza Filho, O.C.; Duarte, M.M.M.F.; Ribeiro, E.E.; Cadoná, F.C.; Da Cruz, I.B.M. Tucumã Fruit Extracts (*Astrocaryum aculeatum* Meyer) Decrease Cytotoxic Effects of Hydrogen Peroxide on Human Lymphocytes. *Food Chemistry* **2015**, *173*, 741–748, doi:10.1016/j.foodchem.2014.10.067.

Sampaio, K.A.; Ayala, J.V.; Silva, S.M.; Ceriani, R.; Verhé, R.; Meirelles, A.J.A. Thermal Degradation Kinetics of Carotenoids in Palm Oil. *JAOCS, Journal of the American Oil Chemists' Society* **2013**, *90*, 191–198, doi:10.1007/s11746-012-2156-1.

Santos, M.D.F.G.D.; Alves, R.E.; Brito, E.S.D.; Silva, S.D.M.; Silveira, M.R.S.D. Quality Characteristis of Fruits and Oils of Palms Native to the Brazilian Amazon. *Revista Brasileira de Fruticultura* **2017**, *39*, doi:10.1590/0100-29452017.

Santos, M.F.G.; Alves, R.E.; Roca, M. Carotenoid Composition in Oils Obtained from Palm Fruits from the Brazilian Amazon. *Grasas y Aceites* **2015**, *66*, doi:10.3989/gya.1062142.

Santos, P.D. de F.; Batista, P.S.; Torres, L.C.R.; Thomazini, M.; de Alencar, S.M.; Favaro-Trindade, C.S. Application of Spray Drying, Spray Chilling and the Combination of Both Methods to Produce Tucumã Oil Microparticles: Characterization, Stability, and β-Carotene Bioaccessibility. *Food Research International* **2023**, *172*, 113174, doi:10.1016/j.foodres.2023.113174.

Santos, R.C.V.; Cruz, I.B.M.; Sagrillo, M.R.; Ribeiro, E.E. The Tucumã of Amazonas-*Astrocaryum aculeatum*. In *Exotic Fruits*; Rodrigues, S.; Silva, E.O.; Brito, E.S., Eds.; Cambridge, MA: Academic Press, 2018; ISBN 9780128031384.

Santos, W.O. dos; Rodrigues, A.M. da C.; Silva, L.H.M. da Chemical Properties of the Pulp Oil of Tucumã-i-Da-Várzea (*Astrocaryum giganteum* Barb. Rodr.) Obtained by Enzymatic Aqueous Extraction. *Lwt* **2022**, *163*, 113534, doi:10.1016/j.lwt.2022.113534.

Serra, J.L.; Rodrigues, A.M. da C.; de Freitas, R.A.; Meirelles, A.J. de A.; Darnet, S.H.; Silva, L.H.M. da Alternative Sources of Oils and Fats from Amazonian Plants: Fatty Acids, Methyl Tocols, Total Carotenoids and Chemical Composition. *Food Research International* **2019**, *116*, 12–19, doi:10.1016/j.foodres.2018.12.028.

Shahidi, F.; Zhong, Y. Measurement of Antioxidant Activity. *Journal of Functional Foods* **2015**, *18*, 757–781, doi:10.1016/j.jff.2015.01.047.

Shanley, P.; Medina, G. *FrutÍferas e Plantas Úteis Na Vida Amazônica*; Belém: CIFOR, 2005; ISBN 8588808021.

Silva, A.G.M.; de Lima, S.C.G.; de Oliveira, P.D.; Moraes, M. dos S.; Guimarães, C.M.C.; da Silva, J.A.R.; Garcia, A.R.; Nahúm, B. de S.; Neres, L. de S.; Noronha, G.N.; et al. Production, Chemical Composition, and Fatty Acid Profile of Milk from Buffaloes Fed with Cupuaçu (*Theobroma grandiflorum*) Cake and Murumuru (*Astrocaryum murumuru*) Cake in the Eastern Amazon. *Animal Science Journal* **2021**, *92*, doi:10.1111/asj.13576.

Silva, D.S.N.; Silva, M. de S.; Coelho, T.L.S.; Dantas, C.; Lopes Júnior, C.A.; Caldas, N.M.; Vieira, E.C. Combining High Intensity Ultrasound and Experimental Design to Improve Carotenoid Extraction Efficiency from Buriti (*Mauritia flexuosa*). *Ultrason Sonochem* **2022**, *88*, 106076, doi:10.1016/j.ultsonch.2022.106076.

Silva, R.S.; de Santos, L.C.; Mar, J.M.; Kluczkovski, A.M.; de Figueiredo, A.J.; Borges, S.V.; Bakry, A.M.; Sanches, E.A.; Campelo, P.H. Physicochemical Properties of Tucumã (*Astrocaryum aculeatum*) Powders with Different Carbohydrate Biopolymers. *Lwt* **2018**, *94*, 79–86, doi:10.1016/j.lwt.2018.04.047.

Singh, R.V.; Sambyal, K. An Overview of β-Carotene Production: Current Status and Future Prospects. *Food Bioscience* **2022**, *47*.

Soares, S.D.; Santos, O.V. Dos; Nascimento, F.D.C.A.D.; Pena, R. da S. A Review of the Nutritional Properties of Different Varieties and Byproducts of Peach Palm (*Bactris gasipaes*) and Their Potential as Functional Foods. *International Journal of Food Properties* **2022**, *25*, 2146–2164, doi:10.1080/10942912.2022.2127761.

Speranza, P.; De Oliveira Falcão, A.; Alves Macedo, J.; Da Silva, L.H.M.; Da Rodrigues, C.A.M.; Alves Macedo, G. Amazonian Buriti Oil: Chemical Characterization and Antioxidant Potential. *Grasas y Aceites* **2016**, *67*, e135, doi:10.3989/gya.0622152.

Speranza, P.; Ribeiro, A.P.R.; Cunha, R.L.; Macedo, J.A.; Macedo, G.A. Influence of Emulsion Droplet Size on Antimicrobial Activity of Interesterified Amazonian Oils. *LWT – Food Science and Technology* **2015**, *60*, 207–212, doi:10.1016/j.lwt.2014.07.022.

Sun, T.; Rao, S.; Zhou, X.; Li, L. Plant Carotenoids: Recent Advances and Future Perspectives. *Molecular Horticulture* **2022**, *2*.

Szydłowska-Czerniak, A.; Trokowski, K.; Karlovits, G.; Szłyk, E. Effect of Refining Processes on Antioxidant Capacity, Total Contents of Phenolics and Carotenoids in Palm Oils. *Food Chemistry* **2011**, *129*, 1187–1192, doi:10.1016/j.foodchem.2011.05.101.

Teixeira, G.L.; Ibañez, E.; Block, J.M. Emerging Lipids from Arecaceae Palm Fruits in Brazil. *Molecules* **2022**, *27*.

Tezara, W.; Domínguez, T.S.T.; Loyaga, D.W.; Ortiz, R.N.; Chila, V.H.R.; Ortega, M.J.B. Photosynthetic Activity of Oil Palm (*Elaeis guineensis*) and Interspecific Hybrid Genotypes (*Elaeis oleifera* × *Elaeis guineensis*), and Response of Hybrids to Water Deficit. *Scientia Horticulturae* **2021**, *287*, doi:10.1016/j.scienta.2021.110263.

Torres-Montilla, S.; Rodriguez-Concepcion, M. Making extra room for carotenoids in plant cells: New opportunities for biofortification. *Progress in Lipid Research* **2021**, *84*, 101128, doi:10.1016/j.plipres.2021.101128.

Trujillo-Quijano, J.; Rodriguez-Amaya, D.B.; Esteves, W.; Plonis, G.F. Carotenoid Composition and Vitamin A Values of Oils from Four Brazilian Palm Fruits. *Lipid/Fett* **1990**, *92*, 222–226.

USDA Foreign Agricultural Service. *Palm Oil 2023 World Production.* https://fas.usda.gov/data/production/commodity/4243000

Viana, A.L.; Mady, F.T.M.; Carmo, M.A.; Guimarães, D.F.S. Pecíolo de Inajá (*Maximiliana maripa* [Aubl.] Drud) Como Fonte de Matéria Prima Para Produção de Papel Na Amazônia. *Revista Eletrônica em Gestão, Educação e Tecnologia Ambiental – REGET* **2014**, *18*, 1512–1520, doi:10.5902/2236117015256.

Vidal, T. do C.S.; Simão, M.O. de A.R.; Almeida, V.F. de A Sustentabilidade Da Produção de Óleos e Manteigas Vegetais Em Comunidade Amazônica-RESEX Médio Juruá. *Research, Society and Development* **2021**, *10*, e32710313478, doi:10.33448/rsd-v10i3.13478.

Vidoca, L.P.; Almeida, E.S. de; Cardoso, M.F.; Otavio, L.; Valadares, L.F.; Monteiro, S. Extraction of Carotene from Crude Hybrid Palm Oil Using Polymeric Resin. *Journal of Food Engineering* **2020**, *278*, doi:10.1016/j.jfoodeng.2020.109944.

Yap, S.C.; Choo, y m; Goh, S.H.; Ong, A.S.; Ooi, C.K. Quantitative Analysis of Carotenes in the Oil from Different Palm Species. *Elaeis* **1991**, *3*, 369–378.

Yara-Varón, E.; Li, Y.; Balcells, M.; Canela-Garayoa, R.; Fabiano-Tixier, A.S.; Chemat, F. Vegetable Oils as Alternative Solvents for Green Oleo-Extraction, Purification and Formulation of Food and Natural Products. *Molecules* **2017**, *22*.

Yuyama, L.K.O.; Aguiar, J.P.L.; Yuyama, K.; Clement, C.R.; Macedo, S.H.M.; Fávaro, D.I.T.; Afonso, C.; Vasconcellos, M.B.A.; Pimentel, S.A.; Badolato, E.S.G.; et al. Chemical Composition of the Fruit Mesocarp of Three Peach Palm (*Bactris gasipaes*) Populations Grown in Central Amazonia, Brazil. *International Journal of Food Sciences and Nutrition* **2003**, *54*, 49–56, doi:10.1080/0963748031000061994.

12 Oilseeds from a Brazilian Semi-Arid Region

Edible Potential Regarding the Mineral Composition

Ivone M. C. Almeida, M. Teresa Oliva-Teles, Rita C. Alves, Joana Santos, Roberta S. Pinho, Suzene I. Silva, Cristina Delerue-Matos, and M. Beatriz P. P. Oliveira

12.1 INTRODUCTION

The plant species that grow in the arid and semi-arid land regions of the world have attracted considerable attention in recent years. The Caatinga, the main ecosystem in the Northeastern Brazil (~800,000 km^2), is a seasonally dry tropical forest composed by a heterogeneous mix of plant species, primarily deciduous xerophytic spiny shrubs, and trees [1, 2]. Despite the huge diversity, only a few of the species of this semi-arid region are exploited for industrial purposes [2]. The rural populations living in this area use many of these plants for food, fuel, timber, medicines, and livestock feed. These products contribute to income generation and offer an alternative source of livelihood when crops fail under the erratic and low rainfall conditions. However, human intervention over the years, such as deforestation, livestock pasture, and cropping, has contributed to the loss of vegetation cover and biodiversity and the extensive soil degradation in this biome [3–9]. Although different studies have already identified some plant species with great economic potential [2, 3, 8–11], the lack of scientific knowledge for the majority remains. The evaluation of the hemical composition of non-conventional tropical plants is essential not only for their fundamental knowledge but also to support their use as raw material for food, chemical, and pharmaceutical industries [5]. For instance, in a recent study, Andrade et al. showed that native species from Caatinga used by the local population to treat inflammatory disorders have also a good photoprotective potential and could be used for pharmaceutical preparations [11]. In addition, nuts and seeds from the tropics have been shown to be rich in lipids and relatively good sources of inexpensive and renewable carbohydrates and proteins, and compounds with high added value, like lipid-soluble vitamins, phytosterols, and other phytochemicals, as well as notable quantities of minerals [12–14].

The insufficiency of minerals in humans is mainly caused by an unbalanced diet. In fact, Fe, Ca, Mg, and Cu are among the mineral elements most commonly lacking in human diets [15], and their deficiency could have a negative impact on health. Information about the mineral composition of unconventional oilseeds from Caatinga is very scarce. The aim of this work is, therefore, to analyze the mineral profile of oilseeds from five native species of this region using high-resolution continuum source atomic absorption spectrometry (HR–CS AAS), in order to evaluate their edible potential as a source of minerals. The seeds selected for this study are from Brazilian plants traditionally used in folk medicine, namely, *Croton adamantinus* Mull. Arg. (Euphorbiaceae), *Barnebya harleyi* W.R. Anderson and B. Gates, *Diplopterys pubipetala* (A. Juss.) W.R. Anderson and C. Davis (Malpighiaceae), *Hippocratea volubilis* L. (Celastraceae), and *Couroupita guianensis* Aubl.

 DOI: 10.1201/9781003462804-16

(Lecythidaceae). The present work is, as far as we know, the first published report on the mineral composition of these oilseeds.

12.2 MATERIALS AND METHODS

12.2.1 Chemicals and Standard Solutions

Ultrapure water from a Simplicity 185 system (resistivity 18.2 MΩ cm; Millipore, Belford, USA) and nitric acid (65%; Suprapure®, Merck, Darmstadt, Germany) were used for the preparation of samples and standards. All reagents were of analytical or suprapure reagent grade. Working standard solutions were prepared from the stock solutions of Na, K, Ca, Mg, Fe, Cu, Cr, and Al (1,000 mg/L; Panreac, Barcelona, Spain) by proper dilution with a 0.5% and 1% (*v*/*v*) nitric acid solution. Cesium chloride (0.1% *w*/*v*; Sigma-Aldrich, Steinheim, Germany) was used as flame ionization chemical suppressor. For Ca analyses, lanthanum chloride heptahydrate (1% *w*/*v*) was added to all solutions in order to minimize the oxy–salts interferences in FAAS analysis. A 0.1% (*w*/*v*) amount of magnesium nitrate (10 g/L; Panreac, Barcelona, Spain) was used as matrix modifier for the electrothermal atomic absorption spectrometry (ETAAS) quantification of Fe and Al, 0.05% (*w*/*v*) magnesium and 0.1% (*w*/*v*) palladium nitrates (10 g/L; Merck, Darmstadt, Germany) for Cu analysis.

Potassium dihydrogen phosphate (99.5%; Riedel-de Haën, Seelze, Germany), ammonium heptamolybdate tetrahydrate (99.0%; Merck, Darmstadt, Germany), and ammonium monovanadate (99.0%; Merck, Darmstadt, Germany) were the reagents used for P analysis. Working standards solutions were manually prepared for HR–CS FAAS and P analysis. Solutions for HR–CS ETAAS calibration curves were automatically prepared by the autosampler.

All glassware and plastic materials were cleaned by treatment with nitric acid (10%, *v*/*v*) for 24 h and then rinsed with water prior to use.

12.2.2 Samples

Mature fruits of five different indigenous plant species (*Diplopterys pubipetala*, *Barnebya harleyi*, *Croton adamantinus*, *Hippocratea volubilis*, and *Couroupita guianensis*) were obtained from five individuals of each species in Pernambuco State (Brazil).

The fruits were dehydrated at 60°C for 48 h until constant weight, and seeds were manually removed from the fruits. Seeds of each species were thinly ground in a mill and stored in plastic containers at 4°C until use.

12.2.3 Sample Preparation

Aliquots (1 g) of the dried ground seed specimens were dry-ashed in a muffle furnace at 500°C overnight. To the resulting white ash, 200 μL of nitric acid (65%) was added and ashed for another hour at 500°C. An amount of 5.0 mL of nitric acid (65%) was used to re-dissolve minerals and, afterward, evaporated to a final volume of 2 mL. After filtration, the solutions were diluted to 50 mL with ultrapure water. Blanks were also carried out. The mineralization procedure was done in triplicate for each composite sample [16].

12.2.4 Analytical Methodologies

An Analytik Jena contrAA 700 atomic absorption spectrometer (Analytik Jena, Germany), equipped with a high-intensity xenon short-arc lamp as continuum radiation source, and automatic MPE60 and AS52s autosamplers (Analytik, Jena, Germany), for electrothermal and flame atomization, respectively, were used throughout this work. Air-acetylene flame (high-purity grade) was used for K, Mg, Ca, and Na quantification at 766.4908 nm, 285.2125 nm, 422.6728 nm, and 588.9953 nm, respectively. Transversely heated graphite furnace with platform-pyrolytically coated graphite tubes (Analytik Jena, Germany) and high-purity argon were used for electrothermal determinations of Fe (248.3270 nm), Cu (324.7540 nm), Cr (357.8687 nm), and Al (309.2713 nm). Pyrolysis and atomization temperatures were optimized to maximize absorbance signals and to minimize backgrounds and matrix interferences. The optimized electrothermal programs used are shown in Table 12.1. Measurements were performed at the selected primary atomic lines at a wavelength-integrated absorbance equivalent to 3 pixels (central pixel ± 1).

The method accuracy was assessed through recovery experiments of the seed samples spiked with three different concentrations (50%, 100%, and 125% of the sample contents) of the aqueous standards. Precision was evaluated through intra- and inter-day measurements by performing three sample measurements on the same day (intra-day) and over three days (inter-day). The limits of quantification (LOQ) were based on the residual standard deviation (10-fold) of calibration curves.

P was determined by the vanadomolybdophosphoric acid colorimetric method according to the Standard Method 4500-P [16]. The color development reagent for P determination was prepared by the addition of ammonium heptamolybdate tetrahydrate and ammonium monovanadate. Phosphorus measurements were performed at 420 nm in a dual-beam UV/vis spectrophotometer (Evolution 300, Thermo Scientific, Boston, MA, USA).

12.2.5 Statistical Analysis

The results were reported as mean ± standard deviation. One-way analysis of variance (ANOVA) and Tukey's HSD post-hoc test were applied to evaluate the possible significant differences ($p < 0.05$) in mineral content between the different seeds. The mineral compositions of the seeds were also compared using multivariate statistical analysis. A linear discriminant analysis (LDA) was performed with forward stepwise analysis, considering a p-value of 0.01, using Wilk's lambda (λ) as the selection criterion, and an F-statistic value to determine the significance of the changes λ when a new variable is tested. Following this was canonical analysis to retrieve more information about the nature of the discrimination between the samples. Statistical analysis was carried out using STATISTICA software v.10 for Windows (Statsoft Inc., Tulsa, USA).

12.3 RESULTS AND DISCUSSION

In previous research, Pinho et al. already identified *D. pubipetala*, *B. harleyi*, *C. adamantinus*, and *H. volubilis* as promising species for cultivation, based on their oil content and fatty acids profile

TABLE 12.1
Temperature Programs for Fe, Cu, Cr, and Al Determination in Seeds by High-Resolution Continuum Source Electrothermal Atomic Absorption Spectrometry (HRCS ETAAS)

	Temperature (Ramp Time, Hold Time); °C (°C/s, s)			
	Fe	Cu	Cr	Al
Drying	90 (3, 23.3)	90 (3, 23.3)	90 (3, 23.3)	90 (3, 23.3)
Pyrolysis	950 (300, 12.0)	900 (300, 11.8)	1,300 (300, 13.2)	1,400 (300, 13.5)
Atomization	1,900 (1,500, 4.6)	1,800 (1,500, 4.6)	2,400 (1,500, 4.7)	2,450 (1,500, 3.7)
Cleaning	2,450	2,450	2,450	2,500

[13]. *C. adamantinus* is used in the treatment of wounds due to its anti-inflammatory and analgesic properties. Their seeds contain 37.1% of oil, with high concentrations of essential linoleic (44.2%) and linolenic (45.2%) fatty acids, and are rich in vitamin E—interesting characteristics for the human diet [13, 17]. *B. harleyi* and *D. pubipetala* seeds contain high amounts of protein and are promising sources of oil (46.4% and 45.3%, respectively), in particular, of α-linolenic acid (42.8% and 31.9%, respectively) [13]. *H. volubilis* is a traditional expectorant. The seeds contain edible oil (45–50%), but with a slightly bitter taste. *C. guianensis* leaves, flowers, and barks are known to contain many active ingredients that possess antibiotic, antifungal, antiseptic, antinociceptive, and immunostimulant activities [18–20]. They contain about ~30% of oil, consisting mainly in linoleic acid (>80%) [13, 21, 22]. The minerals analyzed in this work were selected based on their physiological importance (e.g., Ca, Mg, K, . . .) or relevance as eventual indicators of environmental contamination (e.g., Cr and Al), since the perception of their composition is essential to better understand their potential to be used in human and/or animal nutrition or, for instance, in the dietary supplement industry.

Flame (FAAS) and electrothermal atomic absorption spectrometry (ETAAS), inductively coupled plasma optical emission spectrometry (ICP–OES), and inductively coupled plasma mass spectrometry (ICP–MS) are the main methods employed in mineral analysis [23]. In this work, the more recent technique HR–CS AAS was employed to evaluate the mineral content of the referred seeds. This equipment presents main advantages over the traditional line source AAS, namely, the visualization of the spectral environment in the vicinity of the analytical line, the simultaneous background correction, and the improvement in precision and detection limits due to the higher signal-to-noise ratio [24, 25].

12.3.1 Method Validation

Method parameters, including linear range and correlation coefficient (*r*), LOQ values, repeatability, and mean recovery percentages, are presented in Table 12.2. Na, K, Ca, Mg, P, Fe, Cu, Cr, and Al were quantified in seeds by the external calibration method with aqueous standard solutions. Good linearity was achieved for all elements over the range of tested concentrations, with

TABLE 12.2
Methods Validation Parameters: Linear Range and Correlation Coefficients (*r*) of the Analytical Calibration Curves, Intra- and Inter-Day Precisions (RSD, %), and Limits of Quantification (LOQ) for the Elements Studied

Element	Wavelength	Detection	Calibration	*r*	LOQ	Precision RSD[a] (%)		Recovery[b]
–	(nm)	–	–	–	(μg/g)	Intra-Day	Inter-Day	(%)
Na	588.9953	HR-CS FAAS	0.1–1 (mg/L)	0.999	8.4	1.6	2.4	85
K	766.4908	HR-CS FAAS	0.1–1 (mg/L)	0.999	8.6	1.5	2.3	112
Ca	422.6728	HR-CS FAAS	0.1–1 (mg/L)	1	4.6	3.5	4.3	112
Mg	285.2125	HR-CS FAAS	0.1–1 (mg/L)	1	6.5	3.4	4.2	103
P	420	UV/Vis	0.1–3 (mg/L)	1	11	1.6	2.1	104
Fe	248.327	HR-CS ETAAS	3–15 (μg/L)	0.998	0.09	5.1	6.3	93
Cu	324.754	HR-CS ETAAS	3–15 (μg/L)	1	0.03	3.7	4.3	106
Cr	357.8687	HR-CS ETAAS	3–15 (μg/L)	1	0.04	5.2	7.9	110
Al	309.2713	HR-CS ETAAS	0.1–40 (μg/L)	1	0.11	3.4	4.8	114

Note:[a] Average RSD: inter-day ($n = 3$); intra-day ($n = 3$, three days). [b]Mean recovery of experiments with three levels of sample spiking (50%, 100%, and 125% of the sample contents).

correlation coefficients higher than 0.997 for all calibration plots. The LOQ ranged from 4.6 (Ca) to 11 μg/g (Ca) for major elements, and from 0.025 (Cu) to 0.11 μg/g (Al) for minor elements. Accuracy and precision of the method were good: Recoveries ranged from 85% to 114%, intra-day precisions from 1.5% to 5.2%, and inter-day relative standard deviation (RSD) between 2.1% and 7.9%. These data confirmed the suitability of the method for the designated application.

12.3.2 Seed Analysis

The potential of the selected oilseeds as sources of minerals (Na, K, Ca, Mg, P, Fe, Cu, Cr) was ascertained and is described in Table 12.3.

Physiologically, the macroelements Na, K, Ca, and Mg are essential for several human body functions, while P is the principal reservoir for metabolic energy and a co-factor of many enzymes [26]. K was the most prevalent element in the mineral composition of these seeds, followed, in decreasing order, by Mg, P, Ca, and Na. Among the studied samples, *B. harleyi* contained the highest K concentration (3,649 μg/g). K contents ranged from 1,636 to 3,649 μg/g, levels considerably higher than those reported by Onyeike and Acheru [27] for castor seeds, coconut seeds, dikanut seeds, groundnut seeds, melon seeds, oil bean seeds, and palm kernel seeds (134–168 μg/g). This is an interesting characteristic, taking into account the health importance of K.

The Mg contents in the samples (362 to 586 μg/g) were very modest compared to the levels found for some common culinary nuts and seeds (almonds, ~2,000 μg/g; pecans, ~1,500 μg/g; hazelnuts 2,000 μg/g) [28] and mustard oilseeds (~7,000 μg/g) [29].

The P content ranged from 224 to 499 μg/g. These results are lower than those reported for umbu (*Spondias tuberosa* Arr. Cam) seeds (7,720–8,250 μg/g) [10] and baru (*Dipteryx alata* Vog.) almonds (7,300 μg/g) [30], two Brazilian plant species.

Ca levels (134–227 μg/g) were lower than those reported for pumpkin seeds (390–420 μg/g), sunflower seeds, pistachios (~900 μg/g), and peanuts (520–530 μg/g) [28], but seeds from *D. pubipetala*, *B. harleyi*, and *H. volubilis* presented Ca values identical to that of pine nuts (130 μg/g) [28]. The richest source of Ca were the seeds of *C. guianensis* (227 μg/g), with levels slightly higher to those found in the nuts of *Cyperus esculentus* (188 μg/g) [31].

The Na content was low for all the samples (<106 μg/g), another good characteristic for edible seeds having in view its health effects.

Fe and Cu are essential trace elements. About two-thirds of the total body iron are involved in metabolic or enzymatic functions, mostly in erythrocytes (hemoglobin) and muscles (myoglobin), while the physiological functions of Cu arise directly from its role in a number of Cu-containing enzymes, essential for brain growth and maturation. Cr is also an essential element usually present in food in the trivalent form. In living organisms, Cr (III) is a component of enzymes that controls glucose metabolism and synthesis of fatty acids and cholesterol. Cr (VI) is toxic and carcinogenic to humans, but it is not usually present in food [32, 33].

From a food safety perspective, Al contents of the oilseeds were also evaluated. Most of the population is not at risk for Al toxicity, since humans have diverse mechanisms to prevent significant absorption and to aid its elimination. However, when protective gastrointestinal mechanisms are bypassed, renal function is impaired, or exposure is high, Al can accumulate in the body and originate impaired neurological development, Alzheimer's disease, metabolic bone disease, dyslipidemia, or even genotoxic activity [34].

The levels obtained for the minor minerals Cu, Cr, and Al were generally low for all samples. Fe was the most abundant of the analyzed oligoelements in all seeds, ranging from 2.34 μg/g in *C. adamantinus* to ten-fold more (25.6 μg/g) in *B. harleyi*. These values are lower than those described for mustard seeds (75–170 μg/g) [29], but the Fe level in *B. harleyi* is comparable to the values found in the literature for some well-known edible nuts, such as pecan nuts (24–26 μg/g) and peanuts (20 μg/g) [28], and to the tropical babassu (*Orbignya speciosa*) nuts (18–33 μg/g) and sapucaia (*Lecythis pisonis*) nuts (21–36 μg/g) [14]. *H. volubilis* and *C. guianensis* oilseeds contained similar Fe concentrations to those reported for cupuassu (*Theobroma grandiflorum*) seeds (7.5 μg/g) [14]. *D. pubipetala* and *H. volubilis* seeds contained less than quantifiable levels of Cu (<0.025 μg/g), and the remaining samples contained lower amounts than those found in mustard seeds (6–13 μg/g). Cr concentrations fell below its quantification limit (<0.039 μg/g) in all seeds. Aluminum concentrations ranged from 0.044 μg/g in *D. pubipetala* to 1.80 μg/g in *C. guianensis* seeds, less than the concentration indicated for *Allantoma lineata* seeds (8 μg/g), an Amazon tree from the same family [22].

In sum, *B. harley* seeds had the highest content of K, P, Na, Fe, and Cu, while *C. guianensis* presented the highest levels of Mg, Ca, and Al from the five analyzed seeds.

Although all the oilseeds analyzed showed, in general, a similar mineral profile (high concentration of K followed by Mg > P > Ca > Na > Fe > Al > Cu), their composition showed several significant ($p < 0.05$) differences between them (Table 12.3). A linear discriminant analysis (LDA) was performed to determine which minerals were able to discriminate between two or more naturally occurring groups (Table 12.4).

TABLE 12.3
Mineral Concentrations (μg/g Dry Weight (dw)) of the Analyzed Oilseeds

	Species				
Element	***Diplopterys pubipetala***	***Barnebya harleyi***	***Croton adamantinus***	***Hippocratea volubilis***	***Couroupita guianensis***
Na	38 ± 1 c	106 ± 6 a	90 ± 12 ab	74 ± 14 b	8 ± 1 d
K	1,636 ± 135 c	3,649 ± 215 a	1,719 ± 282 c	2,554 ± 89 b	1,732 ± 187 c
Ca	134 ± 6 c	134 ± 9 c	147 ± 14 cb	170 ± 2 b	227 ± 12 a
Mg	447 ± 13 b	494 ± 39 b	362 ± 0.4 c	384 ± 30 c	586 ± 12 a
P	288 ± 55 c	499 ± 30 a	397 ± 1 b	224 ± 36 c	419 ± 4 ab
Fe	3.69 ± 0.03 c	25.6 ± 2.5 a	2.34 ± 0.23 c	6.88 ± 0.19 b	7.50 ± 0.54 b
Cu	<LOQ	0.77 ± 0.15 a	0.046 ± 0.002 b	<LOQ	0.043 ± 0.005 b
Cr	<LOQ	<LOQ	<LOQ	<LOQ	<LOQ
Al	<LOQ	1.00 ± 0.14 b	<LOQ	0.192 ± 0.017 c	1.80 ± 0.08 a

Note: Values are presented as mean standard deviation of triplicate determinations. Within each line, different letters represent significant differences between species, at $p < 0.05$.

TABLE 12.4
Summary of Discriminant Function Analysis

–	Wilks's Lambda	Partial Lambda	F-Remove (4,4)	*p*-Level	Toler.
Na	0.000	0.163	3.840	0.149	0.173
K	0.000	0.302	1.737	0.339	0.333
Ca	0.000	0.205	2.910	0.203	0.290
Mg	0.000	0.557	0.597	0.692	0.289
P	0.000	0.391	1.168	0.468	0.732
Fe	0.000	0.094	7.270	0.067	0.069
Cu	0.000	0.125	5.265	0.102	0.106
Al	0.000	0.016	45.689	0.005	0.056

Wilks's Lambda: 0.00000 approx. F (28.15) = 56.326, $p < 0.0000$.

The model was generated through forward stepwise analysis, considering variables one by one, choosing those with the most significant contribution to the data discrimination. Cr content was not considered in this multivariate analysis, as it was always inferior to the LOQ of the analytical method. All the other minerals (K, Mg, P, Ca, Na, Fe, Cu, and Al) were selected for the model generated by the LDA, Al and Fe being the variables that showed the highest contribution to the model. A subsequent canonical analysis generated, in total, four canonical variates (CV). The minerals and the case samples were displayed in a canonical variate scatterplot (Figure 12.1) of the first two discriminant functions, which comprised 98.9% of the information contained in the generated model. The first dimension (CV 1) represents 70.0% of the data variance, being more related to their Al, Fe, and Cu content. The second canonical variate (CV 2) adds more than 28.9% of information to the model, separating the samples according to their Fe, K, Ca, and Mg content. The relations of the variables with the first two canonical variates are presented by the resulting equations:

$$\begin{aligned} CV\,1 = {} & -4.18\,[Al] + 3.34\,[Fe] - 0.97\,[Mg] - 0.048\,[P] + 1.29\,[Ca] - 2.65\,[Cu] \\ & - 1.73\,[K] - 0.81\,[Na] \end{aligned} \tag{12.1}$$

$$\begin{aligned} CV\,2 = {} & -0.15\,[Al] + 1.19\,[Fe] - 0.73\,[Mg] + 0.13\,[P] + 0.83\,[Ca] + 1.06\,[Cu] \\ & + 1.23\,[K] + 0.43\,[Na] \end{aligned} \tag{12.2}$$

From the graphical display of the studied cases (see Figure 12.1), there were two seed species, namely, *B. harleyi* and *C. guianensis*, visibly separated from the others. In the CV1, the Al content of the samples allowed to clearly distinguish the separation between the *C. guianensis* seeds and the others. The *B. harleyi* seeds were discriminated by their highest content of Fe, K, and Al. The seeds *D. pubipetala*, *C. adamantinus*, and *H. volubilis* were displayed more closely to the positive side of the CV1, showing that the mineral composition of these three seeds was more similar.

12.4 CONCLUSIONS

The mineral contents of oilseeds from five native plant species from Brazil were evaluated. All the samples analyzed seem to be good sources of K, Mg, and P and presenting also low levels of Na. Nevertheless, *B. harley* oilseeds presented significantly higher amounts ($p < 0.05$) of K and P, Fe, and Cu compared with the other species. *C. guianensis* seeds were the best source of Mg. These two samples were, indeed, clearly differentiated from the remaining ones through linear discriminant analysis. In general, Cr and Al contents were low for all samples. The results show that these oilseeds from the Brazilian Caatinga semi-arid region have the potential to be used as food products and as mineral sources (mainly K).

12.5 AUTHOR CONTRIBUTIONS

Conceptualization: R.S.P., S.I.S., and M.B.P.P.O. Formal analysis: I.M.C.A., R.C.A., and J.S. Funding acquisition: C.D.-M., M.B.P.P.O. Investigation: I.M.C.A. and M.T.O.-T. Methodology: M.T.O.-T. Project administration: S.I.S. and M.B.P.P.O. Resources: R.S.P., S.I.S., and C.D.-M. Supervision: M.T.O.-T., C.D.-M., and M.B.P.P.O. Validation: M.T.O.-T. Writing, original draft: I.M.C.A., R.C.A., and J.S. Writing, review and editing: M.T.O.-T. and R.C.A.

12.6 FUNDING

The work was supported by UIDB/50006/2020, with funding from FCT/MCTES through national funds.

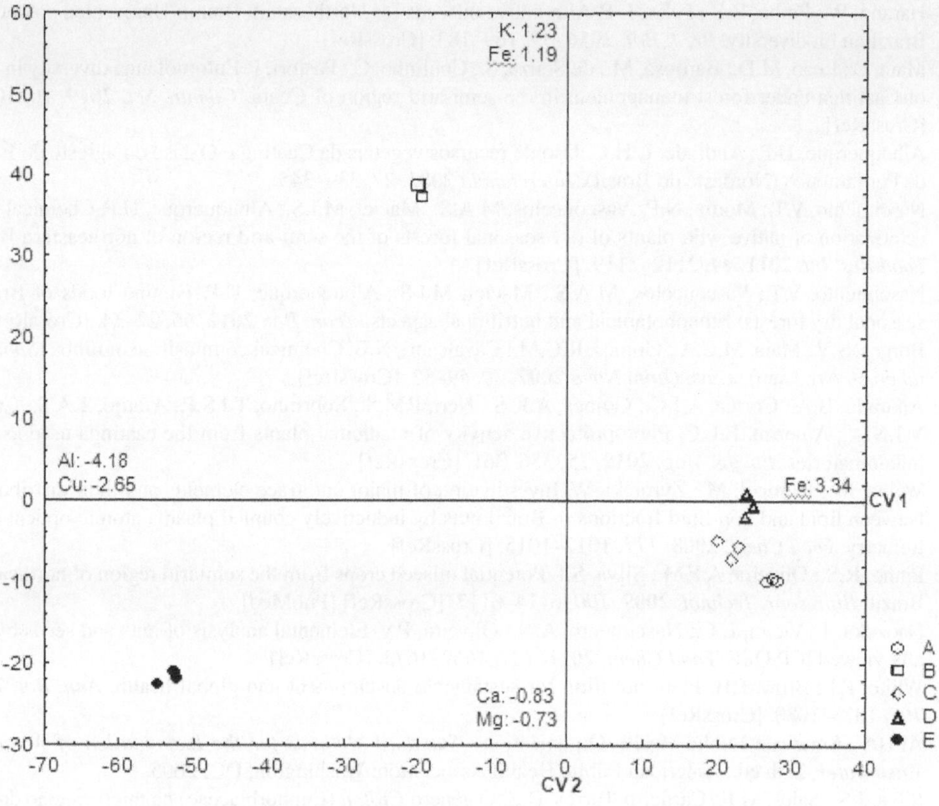

FIGURE 12.1 Scatterplot of all analyzed cases on the first canonical variate (CV1) versus the second canonical variate (CV2): (A) *Diplopterys pubipetala*; (B) *Barnebya harleyi*; (C) *Croton adamantinus*; (D) *Hippocratea volubilis*; (E) *Couroupita guianensis*.

ACKNOWLEDGMENTS

R. C. Alves is grateful to *Fundação para a Ciência e a Tecnologia* for the CEECIND/01120/2017 contract.

CONFLICTS OF INTEREST

The authors declare no conflicts of interest.

REFERENCES

1. Araújo, E.L.; Castro, C.C.; Albuquerque, U.P. Dynamics of Brazilian Caatinga—a review concerning the plants, environment and people. *Funct. Ecosystems. Commun.* **2007**, *1*, 15–28.
2. Silva, R.B.; Silva-Júnior, E.V.; Rodrigues, L.C.; Andrade, L.H.C.; Silva, S.I.; Harand, W.; Oliveira, A.F. A comparative study of nutritional composition and potential use of some underutilised tropical fruits of Arecaceae. *An. Acad. Bras. Ciênc.* **2015**, *87*, 1701–1709. [CrossRef] [PubMed]
3. Crepaldi, I.C.; Almeida-Muradian, L.B.D.; Rios, M.D.G.; Penteado, M.D.V.C.; Salatino, A. Composição nutricional do fruto de licuri (*Syagrus coronata* (Martius) Beccari). *Br. J. Bot.* **2001**, *24*, 155–159. [CrossRef]
4. Lucena, R.F.P.; Albuquerque, U.P.; Monteiro, J.M.; de Almeida, C.F.; Florentino, A.T.; Ferraz, J.S. Useful plants of the semi-Arid Northeastern region of Brazil—a look at their conservation and sustainable use. *Environ. Monit. Assess.* **2007**, *125*, 281–290. [CrossRef] [PubMed]

5. Harand, W.; Pinho, R.S.; Felix, L.P. Alternative oilseeds for Northeastern Brazil: Unrevealed potential of Brazilian biodiversity. *Br. J. Bot.* **2016**, *39*, 169–183. [CrossRef]
6. Maia, L.; Leão, M.D.; Barbosa, M.; de Souza, S.; Coutinho, C.; Pastori, P. Entomofauna diversity in areas of Caatinga under forest management in the semi-arid region of Ceará. *Comun. Sci.* **2019**, *10*, 10–20. [CrossRef]
7. Albuquerque, U.P.; Andrade, L.H.C. Uso de recursos vegetais da Caatinga: O caso do agreste do Estado de Pernambuco (Nordeste do Brasil). *Interciencia* **2002**, *27*, 336–345.
8. Nascimento, V.T.; Moura, N.P.; Vasconcelos, M.A.S.; Maciel, M.I.S.; Albuquerque, U.P. Chemical characterization of native wild plants of dry seasonal forests of the semi-arid region of northeastern Brazil. *Food Res. Int.* **2011**, *44*, 2112–2119. [CrossRef]
9. Nascimento, V.T.; Vasconcelos, M.A.S.; Maciel, M.I.S.; Albuquerque, U.P. Famine foods of Brazil's seasonal dry forests: Ethnobotanical and nutritional aspects. *Econ. Bot.* **2012**, *66*, 22–34. [CrossRef]
10. Borges, S.V.; Maia, M.C.A.; Gomes, R.C.M.; Cavalcanti, N.B. Chemical composition of umbu (*Spondias tuberosa* Arr. Cam) seeds. *Quím Nova* **2007**, *30*, 49–52. [CrossRef]
11. Andrade, B.A.; Corrêa, A.J.C.; Gomes, A.K.S.; Neri, P.M.S.; Sobrinho, T.J.S.P.; Araújo, T.A.S.; Castro, V.T.N.A.; Amorim, E.L.C. Photoprotective activity of medicinal plants from the caatinga used as anti-inflammatories. *Phcog. Mag.* **2019**, *15*, 356–361. [CrossRef]
12. Welna, M.; Klimpel, M.; Zyrnicki, W. Investigation of major and trace elements and their distributions between lipid and non-lipid fractions in Brazil nuts by inductively coupled plasma atomic optical spectrometry. *Food Chem.* **2008**, *111*, 1012–1015. [CrossRef]
13. Pinho, R.S.; Oliveira, A.F.M.; Silva, S.I. Potential oilseed crops from the semiarid region of northeastern Brazil. *Bioresour. Technol.* **2009**, *100*, 6114–6117. [CrossRef] [PubMed]
14. Naozuka, J.; Vieira, E.C.; Nascimento, A.N.; Oliveira, P.V. Elemental analysis of nuts and seeds by axially viewed ICP OES. *Food Chem.* **2011**, *124*, 1667–1672. [CrossRef]
15. White, P.J.; Brow, P.H. Plant nutrition for sustainable development and global health. *Ann. Bot.* **2010**, *105*, 1073–1080. [CrossRef]
16. APHA (American Public Health Organization). *Standard Methods for the Examination of Water and Wastewater*, 21th ed.; American Public Health Association: Washington, DC, 2005.
17. Silva, J.S.; Sales, M.F.; Carneiro-Torres, D.S. O gênero *Croton* (Euphorbiaceae) na microrregião do Vale do Ipanema, Pernambuco, Brasil. *Rodriguésia* **2009**, *60*, 879–901. [CrossRef]
18. Andrade, E.H.A.; Zoghbi, M.G.B.; Maia, J.G.S. The volatiles from flowers of *Couroupita guianensis* Aubl., *Lecythis usitata* Miers. var. paraensis (Ducke) R. Kunth. and *Eschweilera coriacea* (A. P. DC.) Mori (Lecythidaceae). *J. Essent. Oil Res.* **2000**, *12*, 163–166. [CrossRef]
19. Umachigi, P.S.; Jayaveera, K.N.; Kumar, C.K.A.; Kumar, S. Antimicrobial, wound healing and antioxidant potential of *Couroupita guianensis* in rats. *Pharmacologyonline* **2007**, *3*, 269–281.
20. Pinheiro, M.M.G.; Bessa, S.O.; Fingolo, C.E.; Kuster, R.M.; Matheus, M.E.; Menezes, F.S.; Fernandes, P.D. Antinociceptive activity of fractions from *Couroupita guianensis* Aubl. leaves. *J. Ethnopharmacol.* **2010**, *127*, 407–413. [CrossRef]
21. Lago, R.C.A.; Pereira, D.A.; Siqueira, F.A.R.; Szpiz, R.R.; Oliveira, J.P. Estudo preliminar das sementes e doóleo de cinco espécies da Amazônia. *Acta Amaz.* **1987**, *16*, 369–376. [CrossRef]
22. Andrade, E.H.A.; Maia, J.G.S.; Streich, R.; Marx, F. Seed composition of Amazonian Lecythidaceae species: Part 3 in the series "Studies of edible Amazonian plants". *J. Food Comp. Anal.* **1999**, *12*, 37–51. [CrossRef]
23. Taylor, A.; Branch, S.; Halls, D.; Patriarca, M.; White, M. Atomic spectrometry update. Clinical and biological materials, foods and beverages. *J. Anal. Atomic. Spectrom.* **2005**, *20*, 323–369. [CrossRef]
24. Welz, B. High-resolution continuum source AAS: The better way to perform atomic absorption spectrometry. *Anal. Bioanal. Chem.* **2005**, *381*, 69–71. [CrossRef] [PubMed]
25. Oliveira, S.R.; Raposo, J.L.; Gomes Neto, J.A. Fast sequential multi-element determination of Ca, Mg, K, Cu, Fe, Mn and Zn for foliar diagnosis using high-resolution continuum source flame atomic absorption spectrometry: Feasibility of secondary lines, side pixel registration and least-squares background correction. *Spectrochim. Acta Part B* **2009**, *64*, 593–596.
26. Stein, A.J. Global impacts of human mineral malnutrition. *Plant Soil* **2010**, *335*, 133–154. [CrossRef]
27. Onyeike, E.N.; Acheru, G.N. Chemical composition of selected Nigerian oil seeds and physicochemical properties of the oil extracts. *Food Chem.* **2002**, *77*, 431–437. [CrossRef]
28. Rodushkin, I.; Engström, E.; Sörlin, D.; Baxter, D. Levels of inorganic constituents in raw nuts and seeds on the Swedish market. *Sci. Total Environ.* **2008**, *392*, 290–304. [CrossRef]
29. Sharif, R.; Paul, R.; Bhattacharjya, D.; Ahmed, K. Physicochemical characters of oilseeds from selected mustard genotypes. *J. Bangladesh Agric. Univ.* **2017**, *15*, 27–40. [CrossRef]

30. Vera, R.; Junior, M.S.; Naves, R.V.; Souza, E.R.B.; Fernandes, E.P.; Caliari, M.; Leandro, W.M. Características químicas de amêndoas de barueiros (*Dipteryx alata* vog.) de ocorrência natural no cerrado do estado de Goiás, Brasil. *Rev. Br. Frutic.* **2009**, *31*, 112–118. [CrossRef]
31. Glew, R.H.; Glew, R.S.; Chuang, L.-T.; Huang, Y.-S.; Millson, M.; Constans, D.; Vanderjagt, D.J. Amino acid, mineral and fatty acid content of pumpkin seeds (*Cucurbita* spp.) and *Cyperus esculentus* nuts in the Republic of Niger. *Plant Foods Hum. Nutr.* **2006**, *61*, 49–54. [CrossRef]
32. Pang, Y.; MacIntosh, D.L.; Ryan, P.B. A longitudinal investigation of aggregate oral intake of copper. *J. Nutr.* **2001**, *131*, 2171–2176. [CrossRef] [PubMed]
33. Scheiber, I.F.; Mercer, J.F.B.; Dringen, R. Metabolism and functions of copper in brain. *Prog. Neurobiol.* **2014**, *116*, 33–57. [CrossRef] [PubMed]
34. Hernández-Sánchez, A.; Tejada-González, P.; Arteta-Jiménez, M. Aluminium in parenteral nutrition: A systematic review. *Eur. J. Clin. Nutr.* **2013**, *67*, 230–238. [CrossRef] [PubMed]

13 The Amazon's Andiroba Oil

Overview on Characterization, Pharmacological Properties, and Clinical Effects

Deiweson Souza-Monteiro, Cristian Kallahan Silva Chagas, Vinicius Ruan Neves dos Santos, and Rafael Rodrigues Lima

13.1 AN AMAZONIAN OIL CALLED ANDIROBA

The Amazon rainforest, with its vast array of plants, bioproducts, and natural wealth, showcases an immense and largely diverse biodiversity. Combined with the traditional knowledge of various Indigenous peoples from the region, the potential applications of these compounds for improving various aspects of society represent a significant area of research (Elisabetsky & Shanley, 1994).

One of the most famous and widely used oils by traditional Amazonian communities is andiroba oil (Lira et al., 2021). Andiroba oil emerges in traditional contexts due to its potential applications as an analgesic, anti-inflammatory, antipyretic, healing/cicatricial, bactericidal, fungicidal, insect repellent, anthelmintic, antiparasitic, and other reported biological activities (Brasil, 2021). Andiroba oil is used in its natural form to treat the health problems previously cited but is also commonly found as part of nanoemulsions, creams, and cosmetic products, such as soaps and shampoos (Dias et al., 2023). Andiroba oil is obtained through the processing of the species *Carapa guianensis* Aublet (Mendonça & Ferraz, 2007). It is popularly known as andiroba, crabwood (in English-speaking countries), andirova, carapa, cedro macho, krappa, mandiroba, iandiroba, and audirova (Brasil, 2021). *Andiroba* is a word that comes from the Tupi-Guarani language (Indigenous language) and means "bitter taste" (Santos et al., 2013).

Andiroba oil has an amber or light-yellow color, thick viscosity, transparency, prominent odor, and bitter taste. It solidifies at temperatures below 25ºC (de Sousa et al., 2021). Natives produce extracts based on various parts of the tree, such as the seeds, leaves, and stem. However, the oil obtained from the seeds is the most popular (Lira et al., 2021). Manual extraction of andiroba oil varies according to the culture and knowledge of each community. Usually, the traditional artisanal extraction process is complex and can take up to two months. The seeds are cooked for about three hours and stored for 20 days, after which they are mashed. Then, the mash is placed on an inclined surface for oil dripping, which can take up to 30 days (Mendonça & Ferraz, 2007). Part of this process as well as the final state of the oil can be seen in Figure 13.1.

13.2 ANDIROBA AROUND THE WORLD

Carapa guianensis Aublet is a species of native origin and endemic to South America (Inoue et al., 2013). It belongs to the Meliaceae family, and it is distributed from Central America to northern South America, including Guiana and French Guiana, Trinidad, the Caribbean coast, Venezuela, Ecuador, Colombia, Peru, and Brazil, as well as parts of the West Indies and southern Africa. It

DOI: 10.1201/9781003462804-17

FIGURE 13.1 Andiroba oil artisanal production: (left) andiroba oilseeds; (center) artisanal cooking of the seeds; (right) the final aspect of the oil, collected and bottled.

Source: Photographs by the authors.

was collected and cataloged in French Guiana by the French pharmacist and botanist Jean Baptiste Aublet in 1775 (Brasil, 2021).

Since its first description in 1775, approximately 40 names have been referenced to species belonging to the genus *Carapa* (Kenfack, 2011a). However, only four species are currently recognized: *Carapa akuri* Poncy, *C. megistocarpa*, *C. guianensis Aubl.*, and *C. procera* (Forget et al., 2009). There is a consensus within the scientific community that there exists significant variability, both morphologically and molecularly, among these species (Kenfack, 2011b).

Carapa procera is the most commonly found species in Africa, and local studies reveal that various communities also utilize its extract, whether oil or not, in a similar manner to Amazonian communities, primarily for its potential as a healing agent (Dembélé et al., 2015). A taxonomic study of the *Carapa* genus revealed that *Carapa guianensis* Aublet is the species found in Brazil, mainly in wetlands across the Amazon River basin (Kenfack, 2011a, 2011b).

In the Brazilian Amazon region, *Carapa guianensis* Aublet emerges as a prominent botanical entity, commonly referred to as andirobeira. This species epitomizes a towering presence, reaching heights of up to 30–50 m, characterized by a cylindrical and erect trunk extending from the roots to the branches. The bark exhibits substantial thickness and a bitter taste, displaying hues ranging from reddish to grayish tones. Its leaves are arranged alternately, exhibiting compound structures with entire margins, and are endowed with a lustrous dark-green hue. Inflorescences, harboring white to cream-colored flowers, are conspicuously clustered at the terminal ends of branches, emitting a delicate fragrance (Brasil, 2021). In Figure 13.2, the andiroba tree and seeds can be observed.

The fruit is a round burr which is coriaceous, hard, and brown in color. Upon ripening, the burr opens, releasing onto the ground seven to nine seeds resembling chestnuts. These seeds are polygonal, flat on the inner side and convex on the outer side, with a smooth and slightly spongy brown shell covering a compact but soft and oily white mass tinged with a faint pink hue. The seed comprises approximately 75% oily mass containing 43% oil (Tappin, 2007).

From another context, *Carapa guianensis* Aublet also shows economic importance in the wood industry. Its wood is prized for solid products, such as furniture making, building, veneers, and plywood (Firmino, 2019).

FIGURE 13.2 Andiroba tree and seed: (A) a *Carapa guianensis* Aublet tree; (B) an open chestnut with a seed inside.

Source: Photographs by the author.

13.3 PHARMACOLOGICAL PROPERTIES

Although other parts of the plant are quite used for many purposes, as previously mentioned, the main focus of this chapter are the pharmacological aspects of the seed oil. Current scientific studies are limited when considering the vast amount of needed information and compared to other herbal medicines or essential oils. Nevertheless, existing literature highlights and discusses several observed pharmacological effects of the oil. Numerous in vitro and in vivo studies indicate diverse characteristics and properties derived from the oil. Additionally, research involving traditional communities reveals a broad spectrum of applications for various pathologies by these populations. This chapter will focus on the effects with the most scientific evidence, emphasizing its anti-inflammatory, antibacterial, insecticidal, and wound healing properties.

13.3.1 Anti-Inflammatory Activity

One of the most notable activities of andiroba oil is its anti-inflammatory activity. Many people use it for conditions such as sore throat, joint pain, and arthritis (Lorenzi & Matos, 2008). Unsaturated fatty acids present in andiroba oil, such as oleic and linoleic acids, serve as precursors for eicosanoids, playing a pivotal role in anti-inflammatory mediation, reducing the risk of cardiovascular diseases, and elevating high-density lipoprotein (HDL) cholesterol levels (Silva, 2018).

Andiroba oil has demonstrated the ability to reduce inflammation levels by decreasing leukocyte infiltration, thereby dampening the inflammatory response and enhancing macrophage activity. This effect may be attributed to the chemical compounds present in its constitution, such as fatty acids, tetranortriterpenoids, limonoids, and polyphenols, which may act synergistically and are later discussed in this chapter. Additionally, andiroba oil has been observed to induce a production of nitric oxide, hydrogen peroxide, and TNF-α (da Fonseca et al., 2024). Assessing the efficacy of andiroba oil in a periodontitis model reveals its anti-inflammatory activity, resulting in a reduction in the number of inflammatory cells (Carmona et al., 2013).

A review examining the anti-inflammatory properties of andiroba oil determined that it possesses the capability to regulate the immune response during inflammation. This is achieved by diminishing leukocyte activity while enhancing the presence and function of both conventional and alternatively activated macrophages (da Fonseca et al., 2024).

13.3.2 Antibacterial, Antiprotozoal, and Antifungal Activity

Andiroba oil was able to inhibit the growth of the strains *Paenibacillus alginolyticus*, *Paenibacillus pabuli*, *Paenibacillus azotofixans*, *Paenibacillus borealis*, *Paenibacillus gluconolyticus*, *Paenibacillus validus*, *Paenibacillus thiaminolyticus*, and *P. larvae* (Santos et al., 2012). In a study evaluating the essential oil obtained by hydrodistillation and the methanolic extract of *Carapa guianensis* leaves, activity was assessed only against *Staphylococcus aureus* and *Enterococcus faecalis*, with positive results (Meccia et al., 2013). For a methanolic extract of *Carapa procera* bark, antimicrobial activity was observed against the strains of *Escherichia coli*, *Pseudomonas aeruginosa*, *Streptococcus pyrogenes*, *Staphylococcus aureus*, and *Candida albicans* (Owusu et al., 2021).

Aiming to evaluate the antifungal capacity of andiroba oil at 10 and 20 days against *Trametes versicolor* and *Postia placenta* strains, both pure oil and iodine-enriched oil managed to inhibit more than 50% of *Trametes versicolor* growth. However, for *Postia placenta*, pure oil could not inhibit more than 50% of growth, whereas the iodine-enriched oils achieved this only after 20 days of exposure (Sousa et al., 2018).

Assessing antitrypanosome activity, andiroba oil reduced viability at concentrations of 0.5% and 1.0%. After one hour of exposure at 0.5%, viability decreased by 94%, with no live parasites observed in the second and third hours compared to the control (Baldissera et al., 2013).

Antileishmanial effect of the oil was evaluated at 24 and 48 hours against *Leishmania infantum* and *Leishmania amazonensis* with IC_{50} values of 16 mg/mL (24 hours) and 18 mg/mL (48 hours) against *L. infantum*, and 18 mg/mL (24 hours) and 30 mg/mL (48 hours) against *L. amazonensis* (Moraes et al., 2018). In another study, assessing seed oil with limonoid-rich fractions against *Leishmania amazonensis*, contrary to other studies, the oil alone did not exhibit activity (IC_{50} greater than 500 µg/mL). However, its fractions showed excellent activity against both promastigote and amastigote forms, with higher IC_{50} values, reducing both the number of infected cells and amastigotes (Oliveira et al., 2018).

Regarding its antimalarial activity, using andiroba oil and its limonoid-rich fraction against W2 and Dd2 forms of *Plasmodium falciparum*, inhibition of 100% was observed for both the oil and the fraction against the W2 form. For the Dd2 form, in 24 hours, the oil caused 31% inhibition with an IC_{50} of 82, while the fraction caused 56% inhibition with an IC_{50} of 9.4 mg/mL. In 48 hours, the oil inhibited 71% with an IC_{50} of 9.4 mg/mL and the fraction 64% with an IC_{50} of 2.4 mg/mL. Lastly, in 72 hours, the oil inhibited 88% with an IC_{50} of 8.4 and the fraction 82% with an IC_{50} of 0.4 mg/mL (Miranda Júnior et al., 2012). Its antimalarial activity has also been evaluated using isolated compounds against *Plasmodium falciparum* with EC50 values using breast tumor cells as a host model, showing selectivity between 0.24 and 1.93 for the nine compounds studied (Tanaka et al., 2012).

13.3.3 Larvicide, Insecticide, and Repellent

Larvicidal effect of andiroba seed oil (Carapa guianensis) against F1 progeny of a field-colonized and laboratory colony of *Aedes albopictus* was assessed. Although the larval mortality rate was high for the laboratory colony (LC50 1,810 ppm for the third stage and 1,820 ppm for the fourth stage), a higher mortality rate was achieved for the field population (LC50 of 740 ppm for the third stage and 660 ppm for the fourth stage) (Silva et al., 2006). In another study evaluating the bark and seed extracts of *Carapa guianensis* against *Aedes aegypti*, the bark extract showed no larvicidal activity, while the seed extract managed to kill 100% of larvae at 24, 48, and 72 hours, with respective lethal concentrations of 141, 130, and 62 µg/mL (Correa de Oliveira et al., 2022).

Several studies focusing on larvicidal activity using andiroba as a base have been conducted. For instance, a nanoemulsion with 5% *of Carapa guianensis* seed oil at a concentration of 250 ppm showed no statistical differences in the first cycle compared to the second. However, in the third test cycle, it demonstrated a mortality rate of 53.33%, indicating a time-dependent action (Jesus et al.,

2017). Similarly, an ethanol emulsion with silk protein and andiroba oil exhibited larvicidal activities against *Aedes aegypti* at 24 hours (LC50: 212.333 μg/mL; LC90: 93.7 μg/mL) and 48 hours (LC50: 94.455 μg/mL; LC90: 94.275 μg/mL) (Sarquis et al., 2020).

Andiroba is also known for its insecticidal activity, protecting against *Cryptotermes brevis* by reducing wood damage and mass loss while causing insect mortality (Sousa et al., 2019). Furthermore, against *Anastrepha fraterculus*, andiroba oil managed to induce a high level of mortality (37.50% compared to the 70% ethanol control with 2.50%), and it also reduced the number of insects in the offspring by 100% (Wille et al., 2021). Against the maize weevil (*Sitophilus zeamais*), andiroba oil caused 100% mortality in the adult form of the insect, with LD50 of 60.0% and LD90 of 68.6% oil (Santos et al., 2015).

Andiroba oil is traditionally used as an insect repellent. Some studies have shown that the oil protects against *Aedes aegypti* bites, with an average protection time of 94.6 minutes compared to ethyl butylacetylaminopropionate or IR3535, a known repellent, which provides protection for 22.9 minutes (Fradin, 1998). Additionally, it also exhibits repellent activity against *Musca domestica* for 24 hours (Zortéa et al., 2017) and insecticidal effects at concentrations of 1% and 5% in a dose-dependent manner, albeit with a smaller area of effect (Klauck et al., 2014).

Against the fruit fly (*Anastrepha fraterculus* Wiedemann), andiroba oil was able to reduce the attractiveness of hydrolyzed protein in a dose-dependent manner. At 5% oil, it captured 32.8% of flies, while at 1% oil, it captured 9.1%, and at 2% oil, approximately 6.8%. In comparison, the control captured about 51.3% of the total number of flies. A lower number of captured flies indicates that the oil repels insects from the hydrolyzed protein (Machado da Rosa et al., 2013).

13.3.4 Antioxidant

Andiroba seed oil is rich in phenolic compounds (mainly coumarins, flavonoids), fatty acids, and limonoids. Through the DPPH method, radical scavenging activity was identified with a half-maximal effective concentration (EC50) of 89.07 ± 3.67 μL/mL (Araujo-Lima et al., 2018). Comparing andiroba oil collected from different locations using the DPPH method, different EC50 (half-maximal effective concentration) values were found for each municipality: for oil collected in Nova Ipixuna, EC50: 86.62 μL/mL; in Abaetetuba, EC50: 23.05 μL/mL; in Santarém, EC50: 43.56 μL/mL; in Curralinho, EC50: 468.9 μL/mL; in Santa Isabel do Pará, EC50: 326.3 μL/mL; and in Porto de Moz, EC50: 425.4 μL/mL (Milhomem-Paixão et al., 2016)

13.3.5 Healing Properties

Studies testing the healing properties of andiroba oil showed that it can increase the presence of polymorphonuclear and mononuclear macrophages compared to the control, along with causing greater collagen deposition, facilitating wound healing (Araujo et al., 2017). In murine models, the oil exhibited wound healing capacity similar to Vaseline (Fernandes et al., 2014) and facilitated the wound healing process with laser application, stimulating the synthesis of collagen type I and type III (Oliveira et al., 2021). In different wound healing models, an ethanol extract of *Carapa guianensis* leaves demonstrated once again that *C. guianensis* can stimulate faster healing. Additionally, it increased the breaking strength from 380.10 mg/100g in the control to 576.60 mg/100 g, indicating that the scar becomes stronger and firmer. Furthermore, it increased the hydroxyproline content to 58 mg compared to the control's 38.16 mg, suggesting a possible increase in collagen in the wound area, which facilitates and makes the scar more resistant (Nayak et al., 2011).

Topical application of an emulsion with andiroba oil was able to modulate the levels of TGFβ3, decrease the number of macrophages and myofibroblasts, stimulate angiogenesis and the deposition of collagen type II and III, leading to a shorter healing time, increased wound tensile strength, and a better appearance of the final scar (Chia et al., 2018).

13.3.6 Other Activities

Other activities of andiroba have been identified, such as anti-obesity (Matsumoto et al., 2019a; Matsumoto et al., 2019b), anticonvulsant (Rodrigues et al., 2020), antitumor, antiallergic, hepatoprotective (Ninomiya et al., 2016), analgesic, and antiplatelet (Penido et al., 2005), among others.

13.4 WHAT ARE THE MAJOR COMPONENTS OF ANDIROBA OIL?

The biological effects of essential oils are closely linked to their phytochemical composition. It is known that the concentration of these compounds can be affected by genetic, environmental, geographical, seasonal, and circadian factors (Ribeiro et al., 2018). Thus, the variability in the concentration of these bioactive compounds in essential oils can be related to significant changes in the magnitude of their biological effects, which is why the process of characterization is necessary to ensure the viability of their use for commercial, medicinal, or research purposes.

Andiroba oil may also be affected by this variability. However, the oil extracted from andiroba seeds has estimated proportions of its phytochemical content, with the majority being triacylglycerols with high levels of unsaturated fatty acids, such as oleic (51.81%), palmitic (25.76%), stearic (9.08%), and linoleic acid (8.3%) (Cabral et al., 2013). It also contains around 2–5% triterpenes, steroids, coumarins, flavonoids, and limonoids (Cabral et al., 2013).

In 2010, the Mediterranean diet was recognized by UNESCO, and in 2012, it was considered one of the most sustainable and health-beneficial diets. A key component of this diet is olive oil, whose primary health benefits are attributed to its minor components, such as polyphenols and oleic acid (Dernini & Berry, 2015). Oleic acid comprises about 70–80% of olive oil's composition. Oleic acid is produced both endogenously and through dietary intake. One of the most notable effects of this compound is its antioxidant capacity, which directly influences both the synthesis and the activity of antioxidant enzymes. This antioxidant property is potentially linked to its hypotensive effects, as it improves endothelial dysfunction. Additionally, oleic acid has demonstrated anti-inflammatory effects by reducing oxidative stress–promoting mediators, such as lipopolysaccharides, phorbol esters, and cytokines (Santa-María et al., 2023).

Palmitic acid is another prominent fatty acid found in andiroba oil. Palmitic acid is a saturated free fatty acid and is the most abundant endogenous fatty acid in the human body (Pascual et al., 2021). Its nutritional value is quite controversial due to its reported effects; some authors assert that it may be associated with toxic effects, particularly in the cardiovascular system (Lv et al., 2023). However, many other authors suggest that moderate consumption of foods containing this compound can be beneficial, maintaining cellular structure and supporting various cell-signaling mechanisms. These beneficial effects were linked to increased glycolytic energy capacity and non-mitochondrial respiration (Mthembu et al., 2024).

A study evaluating the effects of a diet enriched with stearic acid, compared to diets enriched with other fatty acids, revealed that it resulted in lower fecal serum amyloid A (SAA) concentrations. This suggests that the hypocholesterolemic effects of dietary stearic acid may be mediated through differential effects on bile acid metabolism. These findings indicate that the plasma low-density lipoprotein (LDL) cholesterol response to dietary stearic acid was not due to increased bile acid excretion. Instead, dietary stearic acid may have inhibited the synthesis of intestinal hydrophobic SAA, thereby reducing the efficiency of cholesterol solubilization and absorption (Meng et al., 2019). Andiroba oil is also rich in stearic acid, making it a potential compound of positive biological responses, as previously described.

Linoleic acid, like all fatty acids, can be used as a source of energy. As a crucial constituent of ceramides, linoleic acid plays a pivotal role in preserving the transdermal water barrier within the epidermis. Its inclusion is commonplace in enteral, parenteral, and infant formulas, where the fat composition may vary based on intended application. Moreover, topical administration of linoleic acid offers a viable therapeutic approach for addressing skin disorders associated with its deficiency (Whelan,

2013). It is important to highlight that all these fatty acids mentioned remain constituents of unsaturated fat, and therefore, excessive consumption can be harmful. The biological effects described here are associated with investigations conducted in models of consumption or utilization of low doses.

The occurrence of these bioactive compounds in andiroba oil is associated with various biological effects (Dias et al., 2023). In this context, tetranortriterpenoids (TNTP), another class of compounds found in andiroba, exhibit notable anti-inflammatory activity, as evidenced by their ability to diminish zymosan-induced arthritis in the knees of Swiss rats. This activity is possibly mediated by the suppression of inflammatory mediators TNF-α, IL-1b, and CXCL8/IL-8 through NFκβ pathway inhibition (Ferraris et al., 2011).

TNTP present in andiroba oil also has antiallergic activity, reducing the size of edema induced by ovalbumin at doses from 100 to 400 mg/kg, compared to promethazine (antihistamine control). They also decreased the size of edema stimulated by histamine, reducing exudative plasma even at a concentration of 12.5 mg/kg. This activity is mediated by blocking histamine, bradykinin, and platelet-activating factor signaling, which consequently stimulates prostaglandin E2 formation (Penido et al., 2005). In addition, tetranortriterpenoids can reduce leukocyte and eosinophil influx at the site of inflammation and inhibit the synthesis of CC chemokines that stimulate eosinophil chemotaxis, which are processes responsible for reducing inflammation (Penido et al., 2005). In addition to this activity, tetranortriterpenoids can modulate the action of T lymphocytes by decreasing their activation through the NFκβ pathway and directly affecting eosinophil transmigration without affecting endothelial cell adhesion (Ferraris et al., 2011).

Another interesting compound present in *Carapa guianensis* oil is limonoid. Limonoids are tetranortriterpenoids that belong to the triterpenoid class and generally have four highly oxidized rings. They are made up of compounds with fused tetracyclic carbon skeletons, and through cleavage processes and the rearrangement of these carbon skeletons, various structures are formed that have biological activities (Higuchi et al., 2017; Kikuchi et al., 2020). Among the main limonoids present in andiroba seed oil are 17b-hydroxyazadiradione, gedunin, 6a-acetoxy-gedunin, 7-deacetoxy-7-oxogedunin, 1,2-dihydro-3b-hydroxy-7-deacetoxy-7-oxogedunin, methyl angolensate, and xyloccensin k (Cabral et al., 2013).

The versatility of the use of andiroba may be largely due to the presence of limonoids. It has already been reported in the literature that the presence of these bioactives in the andiroba flower and andiroba seed oil has anti-infective actions, such as antifungal, antibacterial, antiparasitic, and antimalarial actions, as well as analgesic and anti-inflammatory effects (Miranda Júnior et al., 2012; Pereira et al., 2014; Penido et al., 2005; Dias et al., 2023). In addition to their efficacy against wounds and bruises, limonoids also have curious anticancer and antitumor effects (Tsukamoto et al., 2019; Dias et al., 2023). Other biological activities attributed to the presence of limonoids in andiroba oil include antipyretic and anti-irritant activity, as well as acting as an insecticide and repellent (Tsukamoto et al., 2019; Dias et al., 2023).

Regarding the insecticidal activity of isolated compounds from *Carapa guianensis* against *Spodoptera frugiperda*, the limonoid 7-deacetoxy-7-oxogedunin affected larval development by prolonging the larval stage and reducing weight. The limonoid β-photogedunin was the only compound that affected pupal development at concentrations of 10 and 50 mg, causing larval mortality of 53% and pupal mortality of 20% (Sarria et al., 2011).

The biological properties of andiroba oil are also linked to the presence of phenolic compounds. Among natural antioxidant compounds, phenolic compounds (including phenolic acids, phenolic alcohols, and flavonoids) play a crucial role in preventing auto-oxidation and are closely related to the oil's stability.

13.5 CLINICAL EVIDENCE AND BIOTECHNOLOGY APPLICATIONS

Clinical studies are valuable research endeavors that employ rigorous scientific methods and are crucial for decision-making. However, to conduct these studies, it is necessary to build and strengthen preliminary studies that minimize the chances of errors in these trials and provide support for better design by truly understanding the positive effects and possible side effects of the substance being tested. In this context, research needs to advance to better understand the potential of andiroba oil and to intensify the production of clinical trials. This may explain why the number of clinical studies currently available is so limited.

One clinical study evaluated the potential of an Orabase made with 3% andiroba oil in the treatment of oral mucositis in children undergoing leukemia treatment. The andiroba oil was able to reduce both the pain and healing time, likely due to its modulatory activity on the inflammatory mediators present in mucositis (Soares et al., 2021).

In light of the considerable potential exhibited by andiroba oil, it emerges as a promising candidate for exploration and application within the biotechnological domain. Biotechnological interventions hold promise in enhancing various facets of the oil's utilization, spanning from refining extraction methodologies to formulating derivative products. This chapter elucidates the expansive spectrum of applications for andiroba oil and underscores the versatility inherent in its usage. Furthermore, the identification of the compounds within andiroba oil underscores its relevance in the realm of nanotechnology, thus advocating for its comprehensive exploration within this domain. Many products are currently undergoing scrutiny and investigation by research groups, whether in the form of Orabase, nanoemulsion, emulgel, or pure oil, further accentuating the attractiveness of biotechnological applications.

13.6 FINAL CONSIDERATIONS

Andiroba oil holds significant appeal within traditional communities, rendering it one of the most widely utilized oils for various purposes, ranging from treating sore throats to serving as a mosquito repellent. Consequently, numerous studies have shifted their focus toward exploring diverse applications of the oil, resulting in a heterogeneous literature on the subject. Nonetheless, existing evidence positions the oil as a promising ethno-pharmaceutical agent, primarily attributed to its components possessing anti-inflammatory, healing, and insecticidal properties.

The broad spectrum of oil utilities also positions it as an intriguing product for both industrial and pharmaceutical sectors. Based on the exposed in this chapter, the extensive range of applications that the oil can offer for the development of new products and innovations becomes apparent. Thus, the direction of research and production of further studies is evident, aiming to deepen the understanding of all mechanisms and molecular processes involved in oil extraction and production. This includes exploring its industrial, pharmacological, and economic potential applications.

REFERENCES

Araujo, A. L.; Teixeira, F. A.; Lacerda, T. F.; Flecher, M. C.; Souza, V. R. C.; Coelho, C. S. Effects of Topical Application of Pure and Ozonized Andiroba Oil on Experimentally Induced Wounds in Horses. *Braz J Vet Res Anim Sci*, 2017, 54 (1), 66. https://doi.org/10.11606/issn.1678-4456.bjvras.2017.113776.

Araujo-Lima, C. F.; Fernandes, A. S.; Gomes, E. M.; Oliveira, L. L.; Macedo, A. F.; Antoniassi, R.; Wilhelm, A. E.; Aiub, C. A. F.; Felzenszwalb, I. Antioxidant Activity and Genotoxic Assessment of Crabwood (Andiroba, *Carapa guianensis* Aublet) Seed Oils. *Oxid Med Cell Longev*, 2018, 2018, 1–11. https://doi.org/10.1155/2018/3246719.

Baldissera, M. D.; Da Silva, A. S.; Oliveira, C. B.; Zimmermann, C. E. P.; Vaucher, R. A.; Santos, R. C. V.; Rech, V. C.; Tonin, A. A.; Giongo, J. L.; Mattos, C. B.; Koester, L.; Santurio, J. M.; Monteiro, S. G. Trypanocidal Activity of the Essential Oils in Their Conventional and Nanoemulsion Forms: In Vitro Tests. *Exp Parasitol*, 2013, 134 (3), 356–361. https://doi.org/10.1016/j.exppara.2013.03.035.

Brasil. Ministério da Saúde. Secretaria de Ciência, Tecnologia e Insumos Estratégicos. *Informações sistematizadas da Relação Nacional da Plantas Medicinais de interesse ao Sus Carapa guianensis Aublet. Meliaceae—Andiroba. Brasília, DF*, 2021, 1, 81.

Cabral, E. C.; da Cruz, G. F.; Simas, R. C.; Sanvido, G. B.; Gonçalves, L. D. V.; Leal, R. V.; da Silva, R. C.; da Silva, J. C.; Barata, L. E.; da Cunha, V. S.; de Franca, L.F. Typification and Quality Control of the Andiroba (*Carapa guianensis*) Oil via Mass Spectrometry Fingerprinting. *Anal Methods*, 2013, 5 (6), 1385–1391, 2013.

Carmona, G. B.; Teixeira, R. K. C.; Brito, M. V. H.; Pontes, F. S. C.; Andrade, E. H. A.; Fonseca, F. P.; Costa, R. M. B.; Carvalho, F. M. Effect of Andiroba Oil on Periodontitis in Wistar Rats. *Acta Cir Bras*, 2013, 28 (6), 430–434. https://doi.org/10.1590/S0102-86502013000600005.

Chia, C. Y.; Medeiros, A. D.; Corraes, A. de M. S.; Manso, J. E. F.; Silva, C. S. C. da; Takiya, C. M.; Vanz, R. L. Healing Effect of Andiroba-Based Emulsion in Cutaneous Wound Healing via Modulation of Inflammation and Transforming Growth Factor Beta 3. *Acta Cir Bras*, 2018, 33 (11), 1000–1015. https://doi.org/10.1590/s0102-865020180110000007.

Correa de Oliveira, P. M.; Barreto Sousa, J. P.; Albernaz, L. C.; Coelho-Ferreira, M.; Salmen Espindola, L. Bioprospection for New Larvicides Against *Aedes aegypti* Based on Ethnoknowledge from the Amazonian São Sebastião de Marinaú Riverside Community. *J Ethnopharmacol*, 2022, 293, 115284. https://doi.org/10.1016/j.jep.2022.115284.

da Fonseca, A. S. A.; de Souza Monteiro, I.; Dos Santos, C. R.; Carneiro, M. L. B.; Morais, S. S.; Araújo, P. L.; Santana, T. F.; Joanitti, G. A. Effects of Andiroba Oil (*Carapa guianensis* Aublet) on the Immune System in Inflammation and Wound Healing: A Scoping Review. *J Ethnopharmacol*, 2024, 118004.

de Sousa, R. L.; Silva, S. G.; Costa, J. M.; da Costa, W. A.; Maia, A. A. B.; de Oliveira, M. S.; de Aguiar Andrade, E. H. Chemical Profile of Manually Extracted Andiroba Oil (*Carapa guianensis* Aubl., Meliaceae) from Mamangal Community, Located in Igarapé-Miri, Pará, Brazil. *Sci Plena*, 2021, 17 (12).

Dembélé, U.; Lykke, A. M.; Koné, Y.; Témé, B.; Kouyaté, A. M. Use-Value and Importance of Socio-Cultural Knowledge on *Carapa procera* Trees in the Sudanian Zone in Mali. *J Ethnobiol Ethnomed*, 2015, 11, 1–10.

Dernini, S.; Berry, E. M. Mediterranean Diet: From a Healthy Diet to a Sustainable Dietary Pattern. *Frontiers in Nutrition*, 2015, 2, 15.

Dias, K. K. B.; Cardoso, A. L.; da Costa, A. A. F.; Passos, M. F.; da Costa, C. E. F.; da Rocha Filho, G. N.; de Aguiar Andrade, E. H.; Luque, R.; do Nascimento, L. A. S.; Noronha, R. C. R. Biological activities from andiroba (*Carapa guianensis* Aublet.) and its Biotechnological Applications: A Systematic Review. *Arab J Chem*, 2023, 16 (4), 104629.

Elisabetsky, E.; Shanley, P. Ethnopharmacology in the Brazilian Amazon. *Pharmacol Ther*, 1994, 64 (2), 201–214. https://doi.org/10.1016/0163-7258(94)90039-6

Fernandes, C. P. M.; Lima, C. S. de; Lopes, T. V.; Félix, S. R.; Schons, S. de V.; Fernandes, C. G.; Nobre, M. de O. The Use of the Andiroba Oil (Carapaguianensis) on Cutaneous Wounds in Wistar Rats. *Revista Brasileira de Higiene e Sanidade Animal*, 2014, 8 (3). https://doi.org/10.5935/1981-2965.20140124.

Ferraris, F. K.; Rodrigues, R.; da Silva, V. P.; Figueiredo, R.; Penido, C.; Henriques, M. das G. M. O. Modulation of T Lymphocyte and Eosinophil Functions in Vitro by Natural Tetranortriterpenoids Isolated from *Carapa guianensis* Aublet. *Int Immunopharmacol*, 2011, 11 (1), 1–11. https://doi.org/10.1016/j.intimp.2010.09.010.

Firmino, A. V.; Vidaurre, G. B.; Oliveira, J. T. D. S.; Guedes, M.; Almeida, M. N. F. D.; Silva, J. G. M. D.; Latorraca, J. V. D. F.; Zanuncio, J. C. Wood Properties of *Carapa guianensis* from Floodplain and Upland Forests in Eastern Amazonia, Brazil. *Sci Rep*, 2019, 9 (1), 10641.

Forget, P. M.; Poncy, O.; Thomas, R. S.; Hammond, D. S.; Kenfack, D. A New Species of Carapa (Meliaceae) from Central Guyana. *Brittonia*, 2009, 61, 366–374.

Fradin, M. S. Mosquitoes and Mosquito Repellents: A Clinician's Guide. *Ann Intern Med*, 1998, 128 (11), 931–940.

Higuchi, K.; Miyake, T.; Ohmori, S.; Tani, Y.; Minoura, K.; Kikuchi, T.; Yamada, T.; Tanaka, R. Carapanosins A–C from Seeds of Andiroba (*Carapa guianensis*, Meliaceae) and Their Effects on LPS-Activated NO Production. *Molecules*, 2017, 22 (3), 502.

Inoue, T.; Matsui, Y.; Kikuchi, T.; In, Y.; Yamada, T.; Muraoka, O.; Matsunaga, S.; Tanaka, R. Guianolides A and B, new carbon skeletal limonoids from the seeds of *Carapa guianensis*. *Org Lett*, 2013, 15 (12), 3018–3021. https://doi.org/10.1021/ol400924u

Jesus, F. L. M.; de Almeida, F. B.; Duarte, J. L.; Oliveira, A. E. M. F. M.; Cruz, R. A. S.; Souto, R. N. P.; Ferreira, R. M. A.; Kelmann, R. G.; Carvalho, J. C. T.; Lira-Guedes, A. C.; Guedes, M. Preparation of a Nanoemulsion with *Carapa guianensis* Aublet (Meliaceae) Oil by a Low-Energy/Solvent-Free Method and Evaluation of its Preliminary Residual Larvicidal Activity. *eCAM*, 2017, 2017, 1–8. https://doi.org/10.1155/2017/6756793.

Kenfack, D. A Synoptic Revision of Carapa (Meliaceae). *HPB*, 2011a, 16 (2), 171–231.

Kenfack, D. Resurrection in Carapa (Meliaceae): A Reassessment of Morphological Variation and Species Boundaries Using Multivariate Methods in a Phylogenetic Context. *Bot J Linn Soc*, 2011b, 165 (2), 186–221.

Kikuchi, T. Akita, K.; Koike, H. In, Y.; Yamada, T.; Tanaka, R. Carapanins A–C: New Limonoids from Andiroba (*Carapa guianensis*) Fruit Oil. *Org. Biomol. Chem.* 2020. https://doi.org/10.1039/d0ob01872d.

Klauck, V.; Pazinato, R.; Stefani, L. M.; Santos, R. C.; Vaucher, R. A.; Baldissera, M. D.; Raffin, R.; Boligon, A.; Athayde, M.; Barctta, D.; Machado, G.; Da Silva, A. S. Insecticidal and Repellent Effects of Tea Tree and Andiroba Oils on Flies Associated with Livestock. *Med Vet Entomol*, 2014, 28 (S1), 33–39. https://doi.org/10.1111/mve.12078.

Lira, G. B.; da Costa Lopes, A. S.; de Araújo Nascimento, F. C.; dos Santos Conceição, G.; Brasil, D. D. S. B. Processos de extração e usos industriais de óleos de andiroba e açaí: uma revisão. *Res Soc Dev*, 2021, 10 (12), e229101220227.

Lorenzi, H.; Matos, F. J. A. *Plantas medicinais no Brasil: Nativas e exóticas*. 2nd ed. Nova Odessa: Instituto Plantarum de Estudos da Flora Ltda, 2008.

Lv, L.; Wang, X.; Wu, H. Assessment of Palmitic Acid Toxicity to Animal Hearts and Other Major Organs Based on Acute Toxicity, Network Pharmacology, and Molecular Docking. *Comput Biol Med*, 2023, 158, 106899.

Machado da Rosa, J.; Carissimi Boff, M. I.; Gonçalves, P. A.; Boff, P.; Zanelato Nunes, M. Andiroba Oil (*Carapa guianensis* Aubl) in the Capture of the Fruit Fly (*Anastrepha fraterculus* Wiedemann) in Feijoa (*Acca sellowiana* (Berg) Burret). *Idesia (Arica)*, 2013, 31 (3), 97–101. https://doi.org/10.4067/S0718-34292013000300013.

Matsumoto, C.; Koike, A.; Tanaka, R.; Fujimori, K. A Limonoid, 7-Deacetoxy-7-Oxogedunin (CG-1) from Andiroba (*Carapa guianensis*, Meliaceae) Lowers the Accumulation of Intracellular Lipids in Adipocytes via Suppression of IRS-1/Akt-Mediated Glucose Uptake and a Decrease in GLUT4 Expression. *Molecules*, 2019b, 24 (9), 1668. https://doi.org/10.3390/molecules24091668.

Matsumoto, C.; Maehara, T.; Tanaka, R.; Fujimori, K. Limonoid 7-Deacetoxy-7-Oxogedunin from Andiroba, *Carapa guianensis*, Meliaceae, Decreased Body Weight Gain, Improved Insulin Sensitivity, and Activated Brown Adipose Tissue in High-Fat-Diet-Fed Mice. *J Agric Food Chem*, 2019a, 67 (36), 10107–10115. https://doi.org/10.1021/acs.jafc.9b04362.

Meccia, G.; Quintero, P.; Rojas, L. B.; Usubillaga, A.; Velasco, J.; Diaz, T.; Diaz, C.; Velásquez, J.; Toro, M. Chemical Composition of the Essential Oil from the Leaves of *Carapa guianensis* Collected from *Venezuelan guayana* and the Antimicrobial Activity of the Oil and Crude Extracts. *Nat Prod Commun*, 2013, 8 (11), 1934578X1300801. https://doi.org/10.1177/1934578X1300801135.

Mendonça, A. P.; Ferraz, I. D. K. Óleo de andiroba: processo tradicional da extração, uso e aspectos sociais no estado do Amazonas, Brasil. *Acta Amazon*, 2007, 37, 353–364.

Meng, H.; Matthan, N. R.; Wu, D.; Li, L.; Rodríguez-Morató, J.; Cohen, R.; Galluccio, J. M.; Dolnikowski, G. G.; Lichtenstein, A. H. Comparison of Diets Enriched in Stearic, Oleic, and Palmitic Acids on Inflammation, Immune Response, Cardiometabolic Risk Factors, and Fecal Bile Acid Concentrations in Mildly Hypercholesterolemic Postmenopausal Women—Randomized Crossover Trial. *Am J Clin Nut*, 2019, 110 (2), 305–315.

Milhomem-Paixão, S. S. R.; Fascineli, M. L.; Roll, M. M.; Longo, J. P. F.; Azevedo, R. B.; Pieczarka, J. C.; Salgado, H. L. C.; Santos, A. S.; Grisolia, C. K. The Lipidome, Genotoxicity, Hematotoxicity and Antioxidant Properties of Andiroba Oil from the Brazilian Amazon. *Genet Mol Biol*, 2016, 39 (2), 248–256. https://doi.org/10.1590/1678-4685-gmb-2015-0098.

Miranda Júnior, R. N. C.; Dolabela, M. F.; da Silva, M. N.; Póvoa, M. M.; Maia, J. G. S. Antiplasmodial Activity of the Andiroba (*Carapa guianensis* Aubl., Meliaceae) Oil and its Limonoid-Rich Fraction. *J Ethnopharmacol*, 2012, 142 (3), 679–683. https://doi.org/10.1016/j.jep.2012.05.037.

Moraes, A. R. D. P. de; Tavares, G. D.; Rocha, F. J. S.; de Paula, E.; Giorgio, S. Effects of Nanoemulsions Prepared with Essential Oils of Copaiba- and Andiroba Against *Leishmania infantum* and *Leishmania amazonensis* Infections. *Exp Parasitol*, 2018, 187, 12–21. https://doi.org/10.1016/j.exppara.2018.03.005.

Mthembu, S. X.; Mazibuko-Mbeje, S. E.; Silvestri, S.; Orlando, P.; Marcheggiani, F.; Cirilli, I.; Bongani, B.; Nkambule, E.; Christo, J. F.; Muller, A. F. G.; Luca Tiano Dludla, P. V. Low Levels and Partial Exposure to Palmitic Acid Improves Mitochondrial Function and the Oxidative Status of Cultured Cardiomyoblasts. *Toxicol Rep*, 2024, 12, 234–243.

Nayak, B. S.; Kanhai, J.; Milne, D. M.; Pereira, L. P.; Swanston, W. H. Experimental Evaluation of Ethanolic Extract of *Carapa guianensis* L. Leaf for its Wound Healing Activity Using Three Wound Models. *eCAM*, 2011, 2011, 1–6. https://doi.org/10.1093/ecam/nep160.

Ninomiya, K.; Miyazawa, S.; Ozeki, K.; Matsuo, N.; Muraoka, O.; Kikuchi, T.; Yamada, T.; Tanaka, R.; Morikawa, T. Hepatoprotective Limonoids from Andiroba (*Carapa guianensis*). *Int J Mol Sci*, 2016, 17 (4), 591. https://doi.org/10.3390/ijms17040591.

Oliveira, I. dos S. da S.; Moragas Tellis, C. J.; Chagas, M. do S. dos S.; Behrens, M. D.; Calabrese, K. da S.; Abreu-Silva, A. L.; Almeida-Souza, F. *Carapa guianensis* Aublet (Andiroba) Seed Oil: Chemical Composition and Antileishmanial Activity of Limonoid-Rich Fractions. *Biomed Res Int*, 2018, 2018, 1–10. https://doi.org/10.1155/2018/5032816.

Oliveira, R. S. de; Fernandes, M. M. P.; Mesquita, M. do N.; Cruz, A. C. L. da; Pelizzari, C.; Neves, E. C. das; Peruquetti, R. C.; Berón, M. M.; Viott, A. de M.; Souza, S. F. de. Therapeutic Laser with or without Andiroba Oil in Treating Cutaneous Wounds by Second Intention in Wistar Rats. *Acta Vet Bras*, 2021, 15 (4), 316–322. https://doi.org/10.21708/avb.2021.15.4.10027.

Owusu, D. A.; Afedzi, A. E. K.; Quansah, L. Phytochemical and Proximate Content of Carapa Procera Bark and its Antimicrobial Potential Against Selected Pathogens. *PLoS One*, 2021, 16 (12), e0261755. https://doi.org/10.1371/journal.pone.0261755.

Pascual, G.; Domínguez, D.; Elosúa-Bayes, M.; Beckedorff, F.; Laudanna, C.; Bigas, C.; Douillet, D.; Greco, C.; Symeonidi, A.; Hernández, I.; Gil, S. R.; Prats, N.; Bescós, C.; Shiekhattar, R.; Amit, M.; Heyn, H.; Shilatifard, A.; Benitah, S. A. Dietary Palmitic Acid Promotes a Prometastatic Memory via Schwann Cells. *Nature*, 2021, 599 (7885), 485–490.

Penido, C.; Costa, K. A.; Pennaforte, R. J.; Costa, M. F. S.; Pereira, J. F. G.; Siani, A. C.; Henriques, M. G. M. O. Anti-Allergic Effects of Natural Tetranortriterpenoids Isolated from Carapa Guianensis Aublet on Allergen-Induced Vascular Permeability and Hyperalgesia. *Inflamm Res*, 2005, 54 (7), 295–303. https://doi.org/10.1007/s00011-005-1357-6.

Pereira, T. B.; Rocha e Silva, L. F.; Amorim, R. C.; Melo, M. R.; Zacardi de Souza, R. C.; Eberlin, M. N.; Lima, E. S.; Vasconcellos, M. C.; Pohlit, A. M. In Vitro and In Vivo Anti-Malarial Activity of Limonoids Isolated from the Residual Seed Biomass from *Carapa guianensis* (Andiroba) Oil Production. *Malar J*, 2014, 13, 1–8.

Ribeiro, S. M.; Bonilla, O. H.; Lucena, E. M. P. Influência da sazonalidade e do ciclo circadiano no rendimento e composição química dos óleos essenciais de Croton spp. da Caatinga. *Iheringia, Série Botânica*, 2018, 73 (1), 31–38.

Rodrigues de Oliveira, F.; Eleuterio Rodrigues, K.; Hamoy, M.; Rodrigues Sarquis, Í.; Otake Hamoy, A.; Elena Crespo Lopez, M.; Maciel Ferreira, I.; de Matos Macchi, B.; Luiz Martins do Nascimento, J. Fatty Acid Amides Synthesized from Andiroba Oil (*Carapa guianensis* Aublet.) Exhibit Anticonvulsant Action with Modulation on GABA-A Receptor in Mice: A Putative Therapeutic Option. *Pharmaceuticals*, 2020, 13 (3), 43. https://doi.org/10.3390/ph13030043.

Santa-María, C.; López-Enríquez, S.; Montserrat-de la Paz, S.; Geniz, I.; Reyes-Quiroz, M. E.; Moreno, M.; Palomares, F.; Sobrino, F.; Alba, G. Update on Anti-Inflammatory Molecular Mechanisms Induced by Oleic Acid. *Nutrients*, 2023, 15 (1), 224.

Santos, A. C. V.; Fernandes, C. C.; Lopes, L. M.; Sousa, A. H. Use of Plant Oils from the Southwestern Amazon for the Control of Maize Weevil. *J Stored Prod Res*, 2015, 63, 67–70. https://doi.org/10.1016/j.jspr.2015.07.002.

Santos, O. J. D.; Malafaia, O.; Ribas-Filho, J. M.; Czeczko, N. G.; Santos, R. H. P.; Santos, R. A. P. Efeito de *Schinus terebinthifolius* Raddi (aroeira) e *Carapa guianensis* Aublet (andiroba) na cicatrização de gastrorrafias. *Arq Bras Cir Dig* (São Paulo), 2013, 26, 84–91.

Santos, R. C. V.; Alves, C. F. dos S.; Schneider, T.; Lopes, L. Q. S.; Aurich, C.; Giongo, J. L.; Brandelli, A.; de Almeida Vaucher, R. Antimicrobial Activity of Amazonian Oils Against Paenibacillus Species. *J Invertebr Pathol*, 2012, 109 (3), 265–268. https://doi.org/10.1016/j.jip.2011.12.002.

Sarquis, I. R.; Sarquis, R. S. F. R.; Marinho, V. H. S.; Neves, F. B.; Araújo, I. F.; Damasceno, L. F.; Ferreira, R. M. A.; Souto, R. N. P.; Carvalho, J. C. T.; Ferreira, I. M. *Carapa guianensis* Aubl. (Meliaceae) Oil Associated with Silk Fibroin, as Alternative to Traditional Surfactants, and Active against Larvae of the Vector *Aedes aegypti*. *Ind Crops Prod*, 2020, 157, 112931. https://doi.org/10.1016/j.indcrop.2020.112931.

Sarria, A. L. F.; Soares, M. S.; Matos, A. P.; Fernandes, J. B.; Vieira, P. C.; Silva, M. F. das G. F. da. Effect of Triterpenoids and Limonoids Isolated from *Cabralea canjerana* and *Carapa guianensis* (Meliaceae) Against *Spodoptera frugiperda* (J. E. Smith). *Zeitschrift für Naturforschung C*, 2011, 66 (5–6), 245–250. https://doi.org/10.1515/znc-2011-5-607.

Silva, L. R. Propriedades físico-químicas e perfil dos ácidos graxos do óleo da andiroba. *Nativa*, 2018, 6 (2), 147–152.

Silva, O. S.; Prophiro, J. S.; Nogared, J. C.; Kanis, L.; Emerick, S.; Blazius, R. D.; Romão, P. R. Larvicidal Effect of Andiroba Oil, *Carapa guianensis* (Meliaceae), against *Aedes aegypti*. *Journal of the American Mosquito Control Association*, 2006, 22 (4), 699–701.

Soares, A. dos S.; Wanzeler, A. M. V.; Cavalcante, G. H. S.; Barros, E. M. da S.; Carneiro, R. de C. M.; Tuji, F. M. Therapeutic Effects of Andiroba (*Carapa guianensis* Aubl) Oil, Compared to Low Power Laser, on Oral Mucositis in Children Underwent Chemotherapy: A Clinical Study. *J Ethnopharmacol*, 2021, 264, 113365. https://doi.org/10.1016/j.jep.2020.113365.

Sousa, S. F. de; Paes, J. B.; Arantes, M. D. C.; Lopes Junior, D. V.; Nicácio, M. A. Efficiency of Vegetable Oils in Wood Resistance to Cryptotermes Brevis Termites. *Floresta e Ambiente*, 2019, 26 (2). https://doi.org/10.1590/2179-8087.078017.

Sousa, S. F. de; Paes, J. B.; Arantes, M. D. C.; Martinez Lopez, Y.; Brocco, V. F. Análise Física e Avaliação do Efeito Antifúngico Dos Óleos d e Andiroba, Copaíba E Pinhão-Manso. *Floresta*, 2018, 48 (2), 153. https://doi.org/10.5380/rf.v48i2.52280.

Tanaka, Y.; Sakamoto, A.; Inoue, T.; Yamada, T.; Kikuchi, T.; Kajimoto, T.; Muraoka, O.; Sato, A.; Wataya, Y.; Kim, H.-S.; Tanaka, R. Andirolides H–P from the Flower of Andiroba (*Carapa guianensis*, Meliaceae). *Tetrahedron*, 2012, 68 (18), 3669–3677. https://doi.org/10.1016/j.tet.2011.12.076.

Tappin, M. R. R. *Desenvolvimento de metodologia analítica para a quantificação de tetranortriterpenóides em óleo de "andiroba"* [Dissertation]. Rio de Janeiro: Instituto Nacional de Controle de Qualidade em Saúde Fundação Oswaldo Cruz; 2007.

Tsukamoto, Y.; Oya, H.; Kikuchi, T.; Yamada, T.; Tanaka, R. Guianofruits C–I from Fruit Oil of Andiroba (*Carapa guianensis*, Meliaceae). *Tetrahedron*, 2019, 75 (9), 1149–1156.

Whelan, J.; Fritsche, K. Linoleic Acid. *Adv Nut*, 2013, 4 (3), 311.

Wille, C. L.; Rosa, J. M. da; Wille, P. E.; Brillinger, D.; Franco, C. R.; Boff, P.; Gonçalves, P. A. de S.; Boff, M. I. C. Use of Andiroba Oil to Control *Anastrepha fraterculus* (Diptera: Tephritidae) in Different Fruit Hosts. *Rev Colomb Entomol*, 2021, 47 (1), 2–4. https://doi.org/10.25100/socolen.v47i1.7698.

Zortéa, T.; Baretta, D.; Volpato, A.; Lorenzetti, W. R.; Segat, J. C.; Maccari, A. P.; Santos, R. C.; Vaucher, R. A.; Stefani, L. M.; Silva, A. S. da. Repellent Effects of Andiroba and Copaiba Oils against *Musca domestica* (Common House Fly) and Ecotoxicological Effects on the Environment. *Acta Sci Vet*, 2017, 45 (1), 8. https://doi.org/10.22456/1679-9216.79775.

Section 5

Oil Resins

14 Copaifera of the Neotropics
A Review of the Phytochemistry and Pharmacology

Rafaela da Trindade, Joyce Kelly da Silva, and William N. Setzer

14.1 INTRODUCTION TO THE GENUS *COPAIFERA*

The copaiba trees belong to the genus *Copaifera*, family Fabaceae, and subfamily Caesalpinoideae. The genus was described the first time in 1638 by Marcgraf and Piso, who employed the name "Copaiba" without designating the species [1]. In 1760, Nicolaus Joseph Von Jacquin described the species *Copaiva officinalis* in the work *Enumeratio Systematica Plantarum* [2]. Afterward, in the year 1764, Carl von Linnaeus did a more detailed study of the genus in the work *Species Plantarum*, in which he described the type species *Copaifera officinalis* (Jacq.) L. [3]. There are more than 70 *Copaifera* species distributed throughout the world, with widespread occurrence in Central and South America; there are also four species found in Africa and one species found on the island of Borneo, situated in the Pacific Ocean [4]. Brazil is the country with the greatest biodiversity of *Copaifera*, with 26 species and 8 varieties [5].

The vernacular name copaiba probably originated from the Tupi-Guarani and alludes to the names used by Indigenous peoples, copaíva and copahu (kupa'iwa and kupa'u, respectively), which refer to the tree exudate, in reference to the oil stored in its interior [6]. Sixteenth-century records produced by chroniclers during the Brazilian colonization report the widespread use of copaiba oil among the Natives as anti-inflammatory and healing agents, and also for esoteric purposes, such as aphrodisiac and contraceptive [4, 6, 7]. This natural product is known and valued to the present day, mainly in the Amazon region, where the rural population has little access to industrialized pharmaceutical products and public health care [6, 8].

The copaiba trees have shrub or arboreal habits, can reach up to 40 m in height and 4 m in diameter at breast height (dbh), have slow growth, and can live up to 400 years [6]. Their cylindrical trunks contain intercellular secretory channels arranged in bands of marginal axial parenchyma, the lumen from secretory cells is formed schizogenously, and the oleoresin is synthesized in parenchyma cells of the canal. The species have alternate leaves, which are pinnate, with 2–12 pairs of leaflets (opposite, alternate, or subopposite), usually glabrous, and may have translucent points and glands at the base of the marginal vein; they have small and interpetiolar stipules and are generally deciduous. The inflorescences are alternate panicles, and the flower buds are protected by small bracts; they have small flowers, numerous and sessile, which are monoclamids, with a tetramer chalice that forms short tubes and contains internally hirsute sepals. The androecium holds ten free stamens, glabrous fillets, and oblong and rimose anther, and the gynoecium presents a sessile ovary with two elongate ovules, filiform style, and globular and papillary stigma. The fruits are bivalved, dehiscent, laterally compressed, and monospermic. The seed is a pendulum, oblong-globose, covered by abundant white or yellow aril, and lacking endosperm [1, 9–11].

Reprinted from Intl J Molecular Sci **2018**, 19, 1511; doi 10.3390/ijms19051511.

DOI: 10.1201/9781003462804-19

Although the *Copaifera* genus has been extensively studied taxonomically, there are still difficulties in identifying some species, mainly due to their intricate floral morphology and the absence of reproductive structures in the samples studied. With regard to the Amazonian species, the scarcity of field information and illustrations of specimens comprises the main limitations for botanical descriptions of the group. These taxonomic problems have restricted the advance of chemical and pharmacological research, limited the industrial and rational uses of resin oils and wood, and also hampered the development of projects, plans for sustainable management, and conservation of commercially targeted species [9, 12]. The main economic contributions of *Copaifera* species have been wood and oleoresins.

Among *Copaifera* species that are used in the production of oleoresins, *C. reticulata* is the most frequent, representing 70% of the production [6]. Copaiba oleoresin is one of the most important renewable natural remedies for the Indigenous people from the Amazon region, and its use is widely diffused due its various pharmacological properties [13]. The oleoresin is a transparent, colored liquid with variable viscosity and is constituted by a non-volatile fraction composed of diterpenes and a volatile fraction composed of sesquiterpenes [14, 15]. Its chemical profile may vary according to species, seasonal and climatic characteristics of the environment, soil type and composition, and rainfall index. Biotic pressures, such as insect predation and pathogen infection, also cause differences in oleoresin composition [16, 17]. The extraction of copaiba oil is done through the perforation of the trunk with a punch, and the resin is collected with the help of a polyvinyl chloride (PVC) pipe, through which the oil flows and is then stored. This practice is mainly done by plant extraction; therefore, the product of several trees is often mixed, resulting in an additional obstacle to the botanical identity of the copaiba trees. In addition, the lack of parameters to characterize the oil and to perform quality control of the botanical drug also constitutes an obstacle for the registration and exportation of herbal products containing copaiba [18, 19].

14.2 ECOLOGY AND DISTRIBUTION OF *COPAIFERA*

The genus *Copaifera* is native to tropical regions of Latin America, an area of great species diversity [1]. Distributed widely in the Americas, stretching from Mexico to northern Argentina, the genus also occurs in West Africa and Asia [20]. The greatest richness of species occurs in Brazil, where they are distributed from the north to the south of the country. The most common species are *C. multijuga* Hayne, which is found in the Amazonas, Pará, and Rondônia States; *C. reticulata* Ducke, which occurs in Amapá, Pará, and Roraima; and *C. langsdorffii* Desf., which can occur from the northern to the southern regions of Brazil [5]. Other species have more restricted distribution, such as *C. guyanensis* Desf. (Amazonas), *C. majorina* Dwyer (Bahia), *C. cearensis* Huber ex Ducke (Ceará, Bahia, Piauí, and Rio de Janeiro), *C. elliptica* Mart. (Goias and Mato Grosso), *C. paupera* (Herzog) Dwyer (Acre), and *C. lucens* Dwyer (Bahia, Espírito Santo, Minas Gerais, Rio de Janeiro, São Paulo) [5]. Although many species of *Copaifera* have wide occurrence within the Brazilian territory and may occur in different phytogeographic domains (e.g., *C. langsdorffii*), some feature endemism, such as *C. trapezifolia* Hayne, which occurs in an extremely disturbed region of the Atlantic rainforest, of which only 11.6% of the natural vegetation cover remains [21]. Thus, morphological, physiological, and ecological studies are highly relevant for the preservation of species and their natural environment [11]. A study conducted in the Minas Gerais state on the geographical distribution and environmental characteristics of arboreal species showed that *C. langsdorffii* has wide occurrence throughout the whole state, where latosol-type soil predominates, but additionally has a preference for ustic soils (62%) [22].

In relation to the ecological group, copaiba are classified as long-living, late-secondary, and climax tree species, demanding of light but tolerant to shade [23]. They are considered generalists because they are adapted to a wide variety of environments. They can occur in floodplains, riparian forests, and the streams of the Amazon basin and the forests of the Cerrado in the center of Brazil [24]. *C. langsdorffii*, for example, has great ecological plasticity, occurring in several biomes, such as the Cerrado, Atlantic Forest, Caatinga, and Amazon rainforest [23]. *Copaifera* species have great plasticity in relation to edaphic conditions; they occur in areas with fertile soil and well-drained soil and in areas with very poor acidic soils, such as Cerrado fields. They grow well on sandy and clayey soils and generally occupy the forest canopy [25, 26].

Phenological studies on *Copaifera* are important for the rational use of the species and for the preparation of management plans [27]. The reproduction of copaibas occurs from the fifth year of growth after planting in a climax forest ecosystem [6]. *C. multijuga*, commonly found in the Amazon, blooms in the rainy season—between the months of December and April—and fructifies between April and July [27]. Blooming of *C. reticulata* occurs from January to March, with fruiting from March to August, lasting into October [26, 28]. *C. langsdorffii*, observed in the Tijuca Forest, Rio de Janeiro, blooms between March and April and fructifies between August and September. Another survey carried out near Campinas, São Paulo, showed that the flowering of *C. langsdorffii* occurs in the middle of the rainy season (December–February), with development of fruit during the dry season (April–September) [29, 30]. The phenophases of *C. officinalis* were monitored in the municipality of Boa Vista (Roraima State, Brazil) and showed that the flowering of the species occurs between the months of September and November, and the fruiting from November to March. Depending on the stage of fruit ripening, the dehiscence can begin in January, in which the seeds enveloped by the aril are exposed, allowing for their dispersal [28].

Copaifera is a hermaphrodite plant of mixed reproduction with a predominance of allogamy. The trees are generally bee-pollinated (melittophily), and *Apis mellifera* and *Trigona* spp. are its main pollinating agents [25]. *C. langsdorffii* has high fecundity, producing large quantities of fruits in a short period of time. Its seeds have low nutritional value, mainly composed of carbohydrates, but can attract a wide variety of animals with a general diet [30]. The dispersion of the copaiba seeds occurs mainly in zoocoric and barocoric forms [27, 28]. Some vertebrates, such as birds and mammals, have been observed visiting the fruits of *Copaifera* [31]. Its seeds have morphological characteristics that fit the ornithocoria syndrome, mainly because they are black with colored, fleshy arils which, after being swallowed, can be regurgitated intact and remain viable for germination [32]. A study revealed that ten species of *Passeriformes*, such as *Ramphastos toco*, *Cyanocorax cristatellus*, and *Turdus rufiventris*, visited the fruits of *C. langsdorffii*. Likewise, monkeys of the species *Eriodes arachnoides* and *Cebus paella* also eat the fruits of *C. langsdorffii* [31]. Copaiba seeds may also present hydrocoric dispersion due to their frequent occurrence near waterways [25]. Copaiba seeds are of conventional behavior and may be conserved in the long term ex situ, with dormancy due to the deposition of coumarin in the tegument, and its germination is of the epigene type [25]. A tree can produce from 2 to 3 kg of seeds [33].

The population density of copaiba trees in an area is usually very low. It is possible to find only one tree every 5 ha, but they may occur in densities of one to two trees per hectare. The production of oleoresin by species is fairly variable and can be influenced by genetic differences among species, habitat, soil, and intensity of exploitation [34]. The production of oleoresin per tree ranges from 100 mL to 60 L per year. In addition, not all trees produce oil [24]. Therefore, detailed investigations regarding extraction methods and equipment that do not harm the plant, correlation of genetic data to botanical identification of species, floristic inventory of copaiba populations, and ecological studies on its ecosystems are indispensable for the sustainable and rational use of this resource [35, 36].

14.3 TRADITIONAL USES OF *COPAIFERA*

14.3.1 Medicinal Uses

In Pará State (Amazon region, Brazil), people of all ages and social classes consider copaiba one of the most important natural remedies from the Amazon region. Several parts and preparations of the plant are used in folk medicine [24]. The oleoresin or bark decoction is used as an anti-inflammatory and contraceptive by the Native people from the Brazilian Amazon. The topical application of oil on the skin serves to heal wounds. It is used in massages on the head to cure paralysis, pains, and convulsions. In Amapá State, it is recommended to soak a cotton ball in oil and place it on tumors, ulcers, or hives. The daily intake of two drops of oil mixed with one tablespoon of honey is indicated for inflammation, syphilis, bronchitis, and cough [6, 37, 38]. In Venezuela, the oil is used to prepare a patch that is applied to heal ulcers and wounds, and the decoction of the bark in the form of a bath is used to combat rheumatism; to wash infected wounds, such as dog bites; and to use as an anti-tetanus [37, 38]. A tea from the seeds is also used as a purgative and for the treatment of asthma. In northern Brazil, the practice of "embrocation" (applying oil directly to the throat) is common to treat throat infections [39]. In Belém, the "garrafada"—an infusion of the bark sold in bottles—is currently used as a substitute for the oleoresin due to the difficulty in obtaining the oil in the city [38].

Copaiba has a wide range of ethnopharmacological indications, including for the treatment of cystitis, urinary incontinence, gonorrhea, and syphilis; respiratory ailments, including bronchitis, strep throat, hemoptysis, pneumonia, and sinusitis; infections in the skin and mucosa, such as dermatitis, eczema, psoriasis, and wounds; ulcers and lesions of the uterus; leishmaniasis and leucorrhea; anemia; headaches; and snake bites. It is also used for its aphrodisiac, stimulant, anti-inflammatory, antiseptic, antitetanus, antirheumatic, antiherpetic, anthelminthic, anticancer, antitumor (prostate tumors), and antiparalytic properties [4, 6, 26, 38, 40]. *Copaifera* species are used by people of Igarapé Miri (Pará State) for healing wounds [41].

Studies have shown that the ingestion of high doses of copaiba oil can cause adverse side effects, such as gastrointestinal irritation, sialorrhea, and central nervous system depression. A dose of 10 g may cause symptoms of intolerance, nausea, vomiting, colic and diarrhea, and exanthema. Prolonged use may cause kidney damage and topical reactions in susceptible individuals [39, 42]. Thus, the advance in pharmacological and quality control studies of copaiba formulations sold at herbal markets is indispensable for the safe use of this plant drug.

14.3.2 Human Nutrition

Copaiba oil was approved in the United States as a food additive and is used in small amounts as a flavoring agent in foods and beverages [43].

14.3.3 Cosmetic Uses

The species of *Copaifera* are intensively pursued for inclusion in the cosmetics market due to their therapeutic properties and the fragrant value of their oils [44]. Copaiba oil is currently used in the cosmetic industry as a fixative for perfumes and perfuming soaps [38]. As an emollient, bactericidal, and anti-inflammatory agent, copaiba oil is used in the production of soaps, lotions, creams and moisturizers, bath foams, shampoos, and hair conditioners [6, 24]. In addition, it aids in the treatment of dandruff and acne [38, 45]. Despite its fragrant value, little information regarding its odorant potential is available in the literature [44].

14.3.4 Fuel

As a renewable source of hydrocarbons, copaiba oil has been evaluated for use as an ecologically clean fuel. Experimental plantations were started in the early 1980s near Manaus, Brazil, to test its viability as an alternative energy source to fossil fuels [7]. For potential use as fuel, a combination with diesel oil in a ratio of 9:1 (diesel oil to copaiba) has been recommended [6]. Various reports indicate that the liquid can be poured directly into the fuel tank of a diesel-powered car and the vehicle will run normally, with a bluish exhaust smoke being the only noticeable difference [46]. Traditionally, the oil is used in lamps as fuel for lighting [24].

14.3.5 Wood

The copaiba trees are considered hardwoods, with high demand due to their properties of strength, as well as insect and xylophagous fungi repellency. The wood is saturated with oil and resin and has been used in both shipbuilding and civil construction, especially in the manufacture of steam caves, pool cues, and decorative and furniture coverings. It is also used in the preparation of lumbers, rafters, door- and window frames, and boards in general, including for agricultural implements, general carpentry, flooring, furniture, coatings, lamination, plywood sheets. The wood has a high content of lignin and is very good for the production of alcohol and charcoal. *C. langsdorffii* has traditionally been exploited extensively for charcoal in the Cariri Region, south of Ceará [24, 47].

14.3.6 Veterinary Uses

In southern Pará State, farmers have used copaiba oil to prevent foot-and-mouth infection in cattle. The oil is poured on the floor next to the salt lick so that when cattle approach to eat salt, they step in oil, soaking their feet [24]. When wounded, some animals lick and rub their bodies in the oil that flows from the trees [24].

14.3.7 Other Uses

Hunters often hunt under the copaiba tree during fruiting because the seeds and oil attract animals [24]. The oleoresin is used in the photographic industry to improve image clarity in areas of low contrast and resolution. The resin has also been used in paper making, as an additive for butadiene in the production of synthetic rubber, as a source of a chiral substrate in the synthesis of biomarkers of sediment and oil residues, and as fixative in the manufacture of varnish, perfume, and paints used in the painting of porcelain, fabrics, and for dying cotton yarn [6, 24, 38].

14.4 ESSENTIAL OIL CHEMISTRY OF *COPAIFERA*

The major components of the essential oils from *Copaifera* species are summarized in Table 14.1. In general, copaiba oils derived from *Copaifera* oleoresins are rich in sesquiterpene hydrocarbons and often dominated by β-caryophyllene [15]. Some copaiba oils, however, also show significant concentrations of diterpene acids, which are generally analyzed as their methyl esters [15]. A perusal of internet sources of copaiba oil suggests that the most important commercial sources of copaiba oil are *C. langsdorffii*, *C. officinalis*, and *C. reticulata*, and the most prized copaiba oils are rich in β-caryophyllene. The oleoresin essential oils from these three *Copaifera* species can have as much as 33% (*C. langsdorffii*), 87% (*C. officinalis*), and 68% (*C. reticulata*) β-caryophyllene (see Table 14.1).

14.5 NON-VOLATILE CHEMISTRY OF *COPAIFERA*

The oleoresins of several *Copaifera* species have been shown to be rich sources of clerodane, kaurane, and labdane triterpenoids (Figures 14.1–3, Table 14.2). In particular, *C. langsdorffii* resin is composed of biologically active copalic acid [70, 71] and kaurenoic acid [72–74]. *C. multijuga* [74] and *C. paupera* [75] resins are also good sources of copalic acid.

14.6 BIOLOGICAL ACTIVITIES OF *COPAIFERA*

Copaifera oleoresins have shown remarkable biological activities, many of which have been attributed to diterpenoid acids (see Table 14.3). Generally, *Copaifera* oleoresins and their diterpenoid constituents have shown antibacterial, anti-inflammatory, antileishmanial, antiproliferative, antitrypanosomal, and wound healing activities.

TABLE 14.1
Major Components of the Essential Oils of *Copaifera* Species

Copaifera spp.	EO Source	Major Components (>5%)	Ref.
C. cearensis Huber ex Ducke	Oleoresin	α-copaene (8.2%), β-caryophyllene (19.7%), δ-cadinene (7.2%), β-bisabolol (8.2%), hardwickiic acid (6.2%), clorechinic acid (11.3%)	[13]
C. cearensis Huber ex Ducke	Oleoresin	α-copaene (8.2%), β-caryophyllene (19.7%), hardwickiic acid (6.2%)	[48]
C. duckei Dwyer	Oleoresin	β-caryophyllene (0.7–6.2%), *trans*-α-bergamotene (3.4–7.9%), β-selinene (5.5–7.3%), β-bisabolene (8.9–12.1%), kaur-16-en-19-oic acid (19.8–24.5%), polyalthic acid (17.1–27.7%), hardwickiic acid (0–24.3%)	[8]
C. duckei Dwyer	Oleoresin	β-elemene (8.3–9.4%), β-caryophyllene (13.0–15.5%), *trans*-α-bergamotene (8.3–10.6%), β-selinene (13.8–15.4%), α-selinene (8.8–9.9%), β-bisabolene (15.7–17.6%)	[49]
C. duckei Dwyer	Oleoresin	β-caryophyllene (25.1–50.2%), *trans*-α-bergamotene (6.4–12.0%), (*E*)-β-farnesene (2.9–5.8%), β-selinene (1.8–6.7%), β-bisabolene (5.2–33.6%)	[49]
C. guianensis Desf.	Oleoresin	*trans*-α-bergamotene (7.2%), caryophyllene oxide (19.1%), kaur-16-en-19-oic acid (17.5%), polyalthic acid (10.6%), hardwickiic acid (11.0%)	[8]
C. langsdorffii Desf.	Oleoresin	β-caryophyllene (32.8%), kaurenoic acid (44.3%), copalic acid (5.6%), hardwickiic acid (8.2%)	[48]
C. langsdorffii Desf.	Oleoresin	cyclosativene (5.0%), β-elemene (5.1%), β-caryophyllene (5.5%), *trans*-α-bergamotene (48.4%), β-selinene (5.0%), α-himachalene (11.2%)	[50]

TABLE 14.1 (*Continued*)
Major Components of the Essential Oils of *Copaifera* Species

***Copaifera* spp.**	**EO Source**	**Major Components (>5%)**	**Ref.**
C. langsdorffii Desf.	Oleoresin	α-copaene (5.8%), γ-muurolene (22.7%), eremophilone (6.8%), kaurene (6.8%), methyl oleate (26.5%)	[51]
C. langsdorffii Desf.	Oleoresin	β-elemene (8.0%), β-caryophyllene (31.4%), *trans*-α-bergamotene (10.2%), γ-muurolene (16.1%)	[52]
C. langsdorffii Desf.	Oleoresin	β-caryophyllene (1.1–9.0%), germacrene D (4.0–18.0%), bicyclogermacrene (1.5–5.7%), spathulenol (12.6–35.7%), caryophyllene oxide (7.4–16.6%), α-cadinol (3.2–7.9%)	[53]
C. langsdorffii Desf.	Pericarp	α-copaene (3.2–14.4%), β-elemene (0–11.1%), β-caryophyllene (2.7–10.5%), germacrene D (1.9–10.7%), bicyclogermacrene (0–6.3%), spathulenol (2.2–16.2%), caryophyllene oxide (4.0–5.1%), *iso*-spathulenol (5.6–21.6%)	[54]
C. langsdorffii Desf.	Leaf	α-copaene (1.8–6.9%), β-elemene (0–8.4%), β-caryophyllene (5.7–17.5%), germacrene D (0–17.3%), bicyclogermacrene (0–11.5%), δ-cadinene (1.6–6.1%), spathulenol (3.8–12.4%), caryophyllene oxide (0–15.6%), α-muurolol (4.8–6.2%), α-cadinol (4.9–6.8%)	[54]
C. langsdorffii Desf.	Leaf	β-caryophyllene (10.1–16.8%), germacrene D (9.1–45.2%), bicyclogermacrene (4.8–21.1%), spathulenol (4.9–29.4%), caryophyllene oxide (3.8–18.8%)	[55]
C. langsdorffii Desf.	Seed	coumarin (0–12.6%), spathulenol (19.4–38.9%), caryophyllene oxide (0–21.8%), humulene epoxide II (0–5.1%), *iso*-spathulenol (6.9–25.8%), τ-muurolol (1.3–5.0%), α-cadinol (2.0–10.4%)	[54]

(*Continued*)

TABLE 14.1 (*Continued*)
Major Components of the Essential Oils of *Copaifera* Species

Copaifera spp.	EO Source	Major Components (>5%)	Ref.
C. langsdorffii Desf.	Stem	β-caryophyllene (2.4–13.9%), germacrene D (0–19.1%), bicyclogermacrene (0–8.0%), δ-cadinene (0–5.7%), spathulenol (3.6–13.7%), caryophyllene oxide (4.9–13.3%), *iso*-spathulenol (0–7.9%), τ-muurolol (3.4–7.9%), α-cadinol (4.9–11.5%)	[54]
C. lucens Dwyer	Oleoresin	polyalthic acid (69.8%), copalic acid (11.1%)	[48]
C. martii Hayne	Oleoresin	α-copaene (36.4–51.2%), β-elemene (4.1–6.2%), *allo*-aromadendrene (4.2–5.0%), δ-cadinene (13.7–17.2%)	[56]
C. martii Hayne	Oleoresin	β-bisabolene (10.7%), zingiberene (7.2%), kaurenoic acid (7.9%), kovalenic acid (29.0%)	[48]
C. multijuga Hayne	Oleoresin	α-copaene (2.1–5.2%), β-caryophyllene (42.9–60.3%), *trans*-β-bergamotene (2.0–7.0%), caryophyllene oxide (tr–8.8%), copalic acid (1.9–11.0%), 3-acetoxycopalic acid (0.8–6.2%)	[8]
C. multijuga Hayne	Oleoresin	β-caryophyllene (57.5%), α-humulene (8.3%), copalic acid (6.2%)	[57]
C. multijuga Hayne	Oleoresin	β-caryophyllene (57.5%), α-humulene (8.3%), copalic acid (6.2%)	[13]
C. multijuga Hayne	Oleoresin	β-caryophyllene (60.2%), *trans*-α-bergamotene (6.4%), α-humulene (8.6%), copalic acid (9.5%)	[44]
C. multijuga Hayne	Oleoresin	β-caryophyllene (57.5%), copalic acid (6.2%)	[48]
C. multijuga Hayne	Oleoresin	α-copaene (18.8%), β-caryophyllene (36.0%), *trans*-α-bergamotene (7.0%), β-bisabolene (8.5%), δ-cadinene (6.1%)	[58]
C. multijuga Hayne	Oleoresin	α-copaene (2.0–15.0%), β-caryophyllene (5.1–64.0%), α-humulene (0–8.9%), germacrene D (0–16.7%), δ-cadinene (0–5.4%), caryophyllene oxide (0.2–31.5%), copalic acid (1.7–7.1%)	[17]

TABLE 14.1 (*Continued*)
Major Components of the Essential Oils of *Copaifera* Species

Copaifera spp.	EO Source	Major Components (>5%)	Ref.
C. multijuga Hayne	Oleoresin	β-caryophyllene (57.1%), α-humulene (10.2%), β-sesquiphellandrene (9.9%)	[59]
C. multijuga Hayne	Oleoresin	α-copaene (2.5–14.9%), β-caryophyllene (10.6–62.7%), α-humulene (2.4–8.7%), germacrene D (0–18.9%), caryophyllene oxide (0.2–32.5%), copalic acid (1.1–5.2%)	[60]
C. multijuga Hayne	Oleoresin	α-copaene (5.0%), β-gurjunene (5.3%), β-caryophyllene (29.6%), α-humulene (5.7%), caryophyllene alcohol (5.8%), caryophyllene oxide (13.0%)	[61]
C. multijuga Hayne	Oleoresin	β-caryophyllene (58.4%), α-humulene (8.4%), copalic acid (6.3%)	[61]
C. officinalis (Jacq.) L.	Oleoresin	β-caryophyllene (8.5%), copalic acid (13.9%), hardwickiic acid (30.7%)	[48]
C. officinalis (Jacq.) L.	Oleoresin	β-caryophyllene (24.9%), ***allo***-aromadendrene (7.5%), germacrene B (5.1%), β-bisabolene (6.3%), δ-cadinene (15.3%), α-cadinene (5.6%)	[62]
C. officinalis (Jacq.) L.	Oleoresin	β-caryophyllene (62.7%), ***trans***-α-bergamotene (7.6%), α-humulene (8.1%)	Setzer a
C. officinalis (Jacq.) L.	Oleoresin	β-caryophyllene (87.3%)	Setzer b
C. paupera (Herzog) Dwyer	Oleoresin	β-bisabolene (20.2%), zingiberene (19.4%), kaurenoic acid (13.3%), copalic acid (6.1%)	[48]
C. paupera (Herzog) Dwyer	Oleoresin	α-cubebene (5.5%), α-copaene (42.5%), β-caryophyllene (14.1%), δ-cadinene (10.4%)	[63]
C. piresii Ducke	Oleoresin	α-copaene (45.5%), β-caryophyllene (10.3%), δ-cadinene (13.7%)	[63]
C. publflora Benth.	Oleoresin	β-caryophyllene (65.9%), α-humulene (7.3%), β-selinene (10.2%), α-selinene (5.5%)	[63]

(*Continued*)

TABLE 14.1 (*Continued*)
Major Components of the Essential Oils of *Copaifera* Species

Copaifera spp.	EO Source	Major Components (>5%)	Ref.
C. reticulata Ducke	Oleoresin	β-caryophyllene (40.9%), α-humulene (6.0%), germacrene D (5.0%)	[13]
C. reticulata Ducke	Oleoresin (Pará)	β-caryophyllene (40.9%)	[48]
C. reticulata Ducke	Oleoresin (Acre)	α-copaene (25.1%), β-caryophyllene (13.1%), copalic acid (7.7%), kaurenoic acid (7.5%), hardwickiic acid (6.9%)	[48]
C. reticulata Ducke	Oleoresin	β-elemene (0.5–5.6%), β-caryophyllene (1.4–68.0%), ***trans***-α-bergamotene (2.4–29.6%), α-humulene (1.1–9.7%), β-selinene (0–20.6%), α-selinene (0–13.2%), β-bisabolene (3.7–42.4%), caryophyllene oxide (0.1–15.2%)	[64]
C. reticulata Ducke	Oleoresin	β-elemene (0–6.0%), β-caryophyllene (0–43.4%), ***trans***-α-bergamotene (12.0–32.8%), α-guaiene (0–9.5%), α-humulene (0–7.0%), β-selinene (0–17.1%), α-selinene (0–10.4%), ***trans***-β-guaiene (0–5.8%), β-bisabolene (24.2–50.3%)	[65]
C. reticulata Ducke	Oleoresin	β-caryophyllene (25.1–50.2%), ***trans***-α-bergamotene (6.4–12.0%), α-humulene (4.1–5.8%), β-selinene (1.8–6.7%), β-bisabolene (5.2–17.4%)	[66]
C. reticulata Ducke	Oleoresin	β-caryophyllene (37.3%), ***trans***-α-bergamotene (9.0%), α-humulene (5.4%), β-bisabolene (14.5%)	[67]
C. reticulata Ducke	Oleoresin	β-caryophyllene (7.7%), ***trans***-α-bergamotene (22.0%), β-selinene (12.2%), α-selinene (11.4%), β-bisabolene (24.9%)	[68]
C. trapezifolia Hayne	Leaf	β-caryophyllene (33.5%), α-humulene (6.2%), germacrene D (11.0%), spathulenol (7.6%)	[69]

Note: Unpublished analysis of a commercial essential oil from New Directions Aromatics (Sydney, Australia). Unpublished analysis of a commercial essential oil from Améo Essential Oils/Zija International (Lehi, Utah).

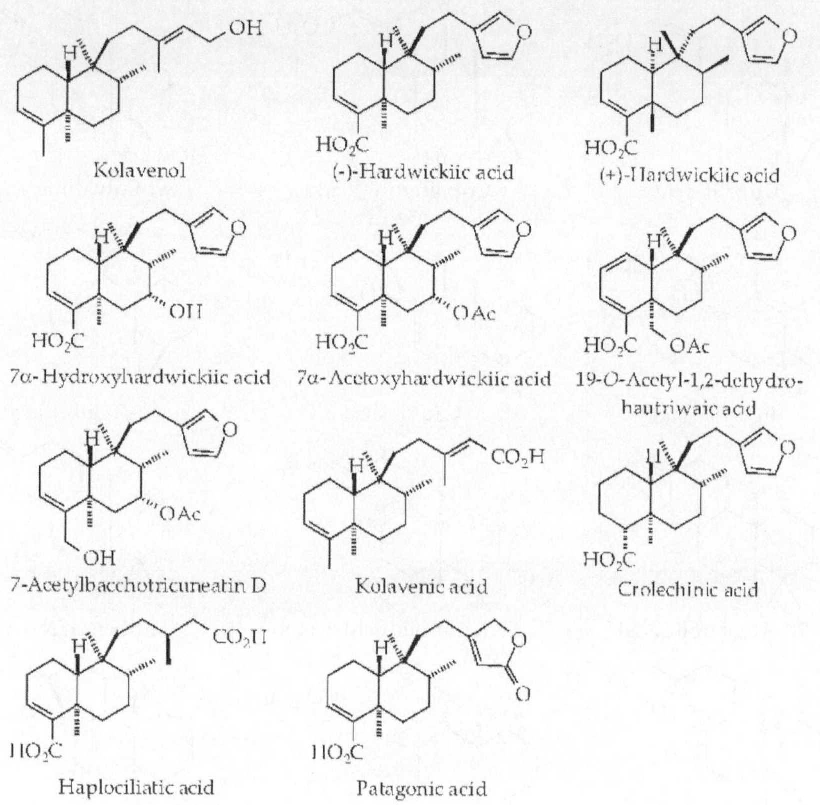

FIGURE 14.1 Clerodane diterpenoids found in *Copaifera* species.

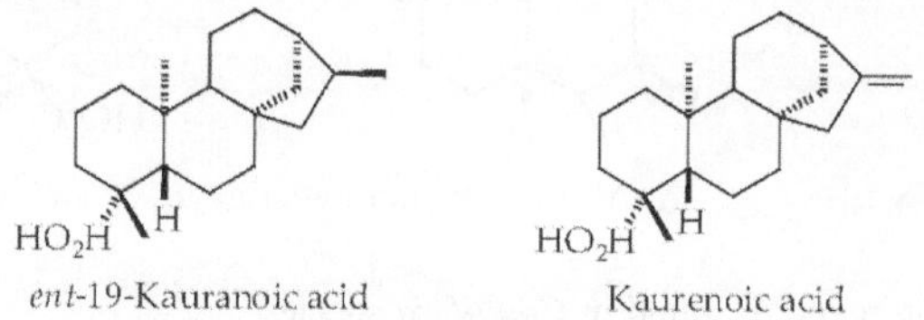

FIGURE 14.2 Kaurane diterpenoids found in Copaifera species.

14.6.1 Antiparasitic Activity of Copaiba

Several *Copaifera* oleoresin oils have shown in vitro antiparasitic activity against *Leishmania amazonensis* promastigotes, including *C. cearensis*, *C. langsdorffii*, *C. lucens*, *C. martii*, *C. multijuga*, *C. officinalis*, *C. paupera*, and *C. reticulata* [48]. The resin oil of *C. martii* showed in vivo antileishmanial activity in a mouse model [97], and *C. reticulata* resin oil showed activity against *L. amazonensis* axenic amastigotes (IC_{50} = 15.0 μg/mL) and intracellular amastigotes (IC_{50} = 20 μg/mL) [109]. Diterpenoids isolated from *C. officinalis*—agathic acid, alepterolic acid, kaurenoic acid, methyl copalate, pinifolic acid, and *ent*-polyalthic acid—showed antileishmanial activity against *L. amazonensis* promastigotes [104].

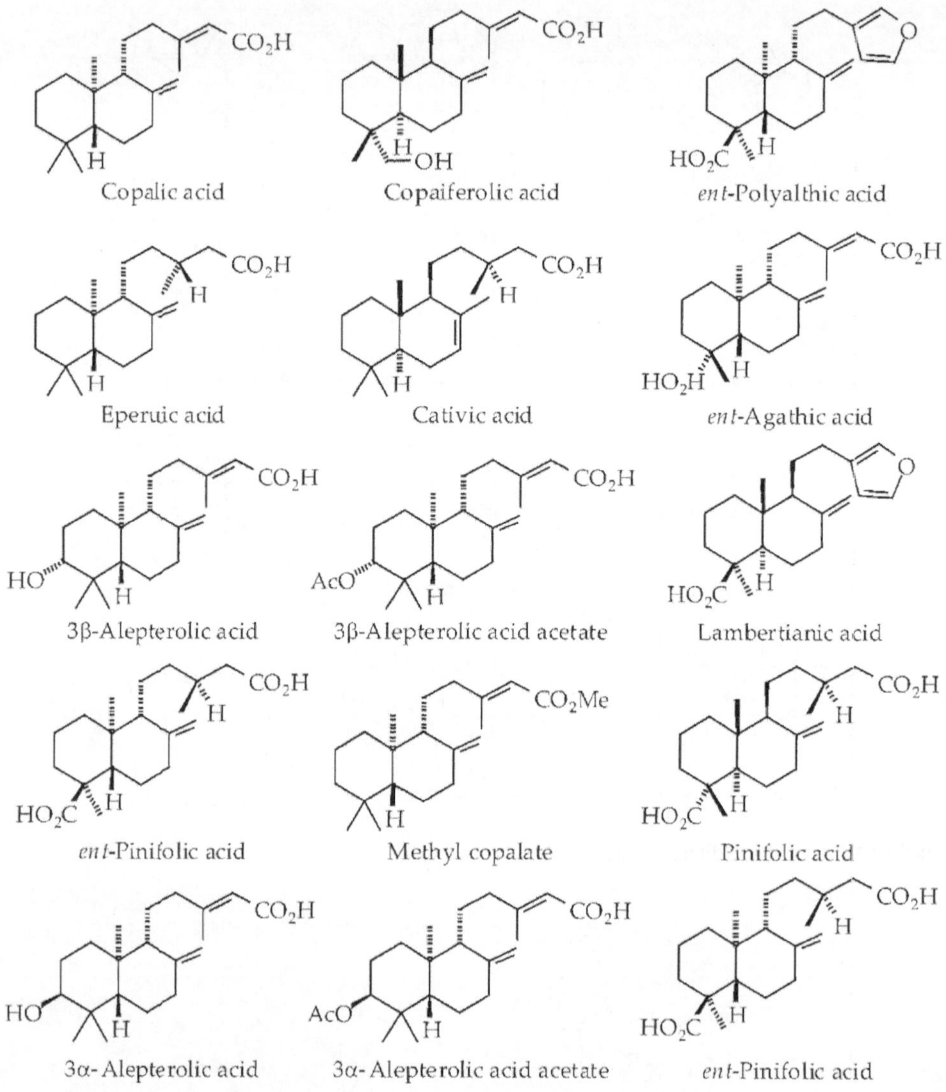

FIGURE 14.3 Labdane diterpenoids found in *Copaifera* species.

Copaifera oleoresins and diterpene acids have also shown antitrypanosomal activities. *C. duckei* and *C. reticulata* resins showed in vitro activity against *T. evansi* trypomastigotes [86]. The diterpene acids—agathic acid, copalic acid, alepterolic acid, kaurenoic acid, methyl copalate, pinifolic acid, and *ent*-polyalthic acid—all showed antitrypanosomal activity against *T. cruzi*, including in the epimastigote, trypomastigote, and amastigote forms of the protozoan [111].

A number of parasitic protozoal proteins have been identified as potential targets for antiparasitic chemotherapy [113]. In conjunction with this review, we have examined the potential parasitic targets of *Copaifera* diterpenoids using molecular docking. It is currently not known what biomolecular targets from *Leishmania* or *Trypanosoma* may be responsible for the antiprotozoal activities of copaiba. The *Copaifera* diterpenoids (Figures 14.1–3) were screened, in silico, against *Leishmania* drug targets [114–116] and *Trypanosoma cruzi* protein targets [117] using Molegro Virtual Docker v. 6.0.1, as previously described [114–117]. The docking energies are summarized in Tables 14.4 and 14.5.

TABLE 14.2
Non-Volatile Components Isolated and/or Identified from *Copaifera* Species

Copaifera spp.	Plant Part	Compounds Isolated and/or Identified	Ref.
C. cearensis Huber ex Ducke	Oleoresin	Epuric acid, cativic acid, copalic acid, kolavenic acid, crolechinic acid, hardwickiic acid, haplociliatic acid, labdanolic acid, patagonic acid.	[76]
C. guianensis Desf.	Oleoresin	Methyl copalate	[74]
C. langsdorffii Desf.	Oleoresin	Copalic acid, kaurenoic acid, acetoxycopalic acid (=3α-alepterolic acid acetate), ***ent***-agathic acid, hydroxycopalic acid (=3α-alepterolic acid)	[71]
C. langsdorffii Desf.	Oleoresin	Copalic acid, acetoxycopalic acid (=3α-alepterolic acid acetate), 3-hydroxy-14,15-dinorlabd-8(17)-en-13-one, ***ent***-agathic acid, hydroxycopalic acid (=3α-alepterolic acid)	[70]
C. langsdorffii Desf.	Oleoresin	Kaurenoic acid	[72]
C. langsdorffii Desf.	Oleoresin	Kaurenoic acid	[73]
C. langsdorffii Desf.	Oleoresin	Kaurenoic acid	[74]
C. langsdorffii Desf.	Oleoresin	Sclareol, manool, copalic acid, acetoxycopalic acid (=3α-alepterolic acid acetate), hydroxycopalic acid (=3α-alepterolic acid), ***ent***-agathic acid	[77]
C. langsdorffii Desf.	Leaves	Kaurenoic acid, quercitrin, afzelin	[78]
C. langsdorffii Desf.	Leaves	Rutin, quercetin-3-***O***-α-l-rhamnopyranoside, kaempferol-3-***O***-α-l-rhamnopyranoside, quercetin, kaempferol	[79]
C. langsdorffii Desf.	Fruit	Gallic acid, epicatechin gallate, catechin, epicatechin, isoquercitrin	[80]
C. langsdorffii Desf.	Leaves	Kaurenoic acid, quercitrin, afzelin, eupatorin, galloyl quinic acid, gallic acid 4-***O***-glucoside	[81]
C. multijuga Hayne	Oleoresin	Copalic acid, 3-hydroxycopalic acid, (=3α-alepterolic acid), 3-acetoxycopalic acid, (=3α-alepterolic acid acetate)	[74]
C. paupera (Herzog) Dwyer	Oleoresin	Copalic acid, methyl copalate, agathic acid 15-methyl ester, agathic acid 15,19-dimethyl ester, ***ent***-polyalthic acid, methyl ***ent***-polyalthate, ***ent***-pinifolic acid, methyl 3β-hydroxy-labda-8(17),13-dien-15-ate, methyl 18-hydroxy-copaiferolate, 14,15-bisnorlabd-8(17)-en-13-one, ***ent***-kaurenic acid, 16β-kauran-19-oic acid, 3-methyl-5-(2,2,6-trimethyl-6-hydroxycyclohexyl)-pentanoic acid, pauperol	[75]
C. reticulata Ducke	Oleoresin	3β-alepterolic acid, 3β-alepterolic acid acetate, 3β-hydroxylabdan-8(17)-en-15-oic acid, ***ent***-agathic acid	[82]

The *Leishmania* protein target with the best overall docking properties with *Copaifera* diterpenoids was *L. major* dihydroorotate dehydrogenase (average E_{dock} =−109.2 kJ/mol). These docking energies were better than the docking energy for the normal substrate, dihydroorotate (E_{dock} = −72.1 kJ/mol) and comparable to the co-crystallized ligand for this protein, nitroorotate (E_{dock}= −104.2 kJ/mol). Docking energies for *Copaifera* diterpenoids with TcDHODH (average −92.5 kJ/mol) were also better than the normal substrate (dihydroorotate, E_{dock} = −64.2 kJ/mol), but worse than the synthetic TcDHODH inhibitor, 5-[2-(5-carboxynaphthalen-2-yl)ethyl]-2,6-dioxo-1,2,3,6-tetrahydropyrimidine-4-carboxylic acid (TT2-2-199, E_{dock} = −140.7 kJ/mol). Similarly, *Copaifera* diterpenoids docked with *L. donovani* DHODH (average E_{dock} = −89.9 kJ/mol) better than dihydroorotate (E_{dock} = −60.9 kJ/mol). Based on these docking energies, protozoal dihydroorotate dehydrogenases are likely targets for *Copaifera* diterpenoids.

Leishmania major methionyl–tRNA synthetase was another *Leishmania* protein target with good docking energies. Although the docking energies with this protein were excellent (average E_{dock} = −106.9 kJ/mol), they are much poorer than the docking energy of the normal substrate, methionyl adenylate (E_{dock} = −168.1 kJ/mol). Similarly, the *T. cruzi* target protein with the best

TABLE 14.3
Biological Activities of *Copaifera* Oleoresins, Essential Oils, and Isolated Components

Copaifera spp.	Material	Biological Activity	Ref.
C. cearensis Huber ex Ducke	Oleoresin	Anti-inflammatory: At a concentration of 50 μg/mL, in vitro NO production in mouse peritoneal macrophages was significantly reduced by ***C. cearensis*** oil.	[13]
C. cearensis Huber ex Ducke	Oleoresin	Antileishmanial: ***L. amazonensis*** promastigotes (IC_{50} = 18.0 μg/mL).	[48]
C. cearensis Huber ex Ducke	Oleoresin	Antibacterial: Methicillin-resistant ***Staphylococcus aureus*** (MRSA, MIC = 125 μg/mL), ***B. subtilis*** (MIC = 62.5 μg/mL), ***Enterococcus faecalis*** (MIC = 62.5 μg/mL).	[83]
C. duckei Dwyer	Oleoresin	Anti-inflammatory: Carrageenin-induced paw edema test (rats, 18% edema inhibition with dose of 1,802 mg/kg; granuloma test, 42% inhibition with dose of 1,802 mg/kg); croton oil–induced dermatitis test (mice, IC_{50} = 663 mg/kg)	[84]
C. duckei Dwyer	Oleoresin	Antinociceptive: Mouse model (acetic acid–induced writhing test, IC_{50} = 704 mg/kg).	[84]
C. duckei Dwyer	Oleoresin	Antiproliferative: In vivo hepatocellular proliferation, partial hepatectomy, rats. Hepatocellular proliferation and liver mitochondrial respiration were significantly lower in ***C. duckei***–treated rats compared to control (saline solution).	[85]
C. duckei Dwyer	Oleoresin	Antitrypanosomal: ***T. evansi***, in vivo mouse model, doses of 0.63 mL/kg/day over five days, showed no curative effects. ***T. evansi***, in vitro trypomastigotes, 0.5% solution of ***C. duckei*** oil, showed 100% killing after 6 h.	[86]
C. langsdorffii Desf.	Oleoresin	Anti-inflammatory: Preincubation of LPS-stimulated human THP-1 monocytes with diterpenoid-rich oleoresin reduced the release of proinflammatory cytokines (IL-1β, IL-6, TNFα).	[50]
C. langsdorffii Desf.	Oleoresin	Antifungal: ***Tricophyton mentagrophytes*** (MIC = 170 μg/mL). Scanning electron microscopic (SEM) analysis revealed physical damage and morphological alterations of the fungi upon exposure to copaiba oleoresin.	[52]
C. langsdorffii Desf.	Oleoresin	Antipsoriatic: Human clinical trial, topical application of copaiba resin exhibited a significant improvement of the typical signs of psoriasis.	[50]
C. langsdorffii Desf.	Oleoresin	Gastroprotective: Ethanol- or indomethacin-induced ulcer in rats, oral administration of copaiba resin at doses of 400 mg/kg provided dose-dependent significant protection against gastric damage caused by ethanol or indomethacin.	[87]
C. langsdorffii Desf.	Oleoresin	Gastroprotective: Mesenteric ischemia/reperfusion (I/R) in rats, copaiba resin treatment caused significant attenuations in I/R-associated increases of myeloperoxidase, malondialdehyde, and catalase and effectively prevented the I/R-associated depletion of glutathione.	[88]
C. langsdorffii Desf.	Oleoresin	Wound healing: Rat incision wounds, 4% oleoresin topically applied showed significant wound contraction and tensile strength compared to controls.	[89]
C. langsdorffii Desf.	Oleoresin	Wound healing: Rat dorsal skin flaps, oral administration of copaiba oleoresin (400 mg/kg), copaiba treatment presented discrete anti-lipoperoxidation action, intense antioxidant action, and anti-inflammatory activity during the ischemia and reperfusion of randomized cutaneous flaps.	[90]

TABLE 14.3 (*Continued*)
Biological Activities of *Copaifera* Oleoresins, Essential Oils, and Isolated Components

Copaifera spp.	Material	Biological Activity	Ref.
C. langsdorffii Desf.	Oleoresin	Anti-inflammatory: Rat model of experimental endometriosis. Copaiba oil caused a marked reduction in endometrial growth.	[91]
C. langsdorffii Desf.	Oleoresin	Antileishmanial: ***L. amazonensis*** promastigotes (IC_{50} = 20.0 μg/mL).	[48]
C. langsdorffii Desf.	Oleoresin	Antibacterial: ***B. subtilis*** (MIC = 62.5 μg/mL).	[83]
C. langsdorffii Desf.	10% copaiba oil ointment	Wound healing: Rat dorsal skin flaps, topical copaiba oil ointment favors angiogenesis and accelerates the viability of random skin flaps in rats.	[51]
C. langsdorffii Desf.	10% oleoresin cream	Antibacterial: Open wounds on rats inoculated with ***Streptococcus pyogenes*** or ***Staphylococcus aureus***. Treatment with 10% copaiba cream reduced bacterial populations to 0.02% (***S. pyogenes***) and 0.3% (***S. aureus***) after 14 days.	[92]
C. langsdorffii Desf.	10% oleoresin cream	Wound healing: Rabbit ears, 10% oleoresin cream–treated wounds presented better clinical outcomes, confirmed by histology with evidence of fibroblastic activity by day 7 and organized collagen fibers observed from day 14.	[93]
C. langsdorffii Desf.	10% oleoresin cream	Wound healing: Rat skin biopsy punch on dorsal surface, 10% oleoresin cream–treated wounds showed a faster wound healing rate compared to saline or cream-only controls, by regulating matrix metalloproteinase, (MMP)-2, and MMP-9 activities, stimulating collagen synthesis, and promoting tissue remodeling and re-epithelialization.	[94]
C. langsdorffii Desf.	3α-alepterolic acid acetate	Antibacterial: Cariogenic ***Streptococcus*** spp.; MIC range 12.0–60.0 μg/mL.	[70]
C. langsdorffii Desf.	Aqueous leaf extract	Insecticidal: 5% extract inhibited ***Bemisia tabaci*** infestation of tomato plants in the field.	[95]
C. langsdorffii Desf.	Copalic acid	Antibacterial: Gram-positive bacteria (MIC range 0.5 μg/mL to 15.0 μg/mL).	[71]
C. langsdorffii Desf.	Copalic acid	Antibacterial: Cariogenic ***Streptococcus*** spp. (MIC range 2.0–6.0 μg/mL)	[70]
C. langsdorffii Desf.	Copalic acid	Antibacterial: Periodontal anaerobic bacteria: ***Actinomyces naeslundii*** (MIC 6.2 μg/mL), ***Bacteroides fragilis*** (MIC 25.0 μg/mL), ***Peptostreptococcus anaerobius*** (MIC 3.1 μg/mL), ***Porphyromonas gingivalis*** (MIC 3.1 μg/mL).	[77]
C. langsdorffii Desf.	Copalic acid	Antiproliferative: In vitro cytotoxicity on MO59J (human glioblastoma cells, IC_{50} = 68.3 μg/mL) and HeLa (human cervical adenocarcinoma cells, IC_{50} = 44.0 μg/mL).	[71]
C. langsdorffii Desf.	EtOH/H2O leaf extract	Gastroprotective: Ethanol/HCl-induced ulcer in mice, the extract (500 mg/kg) showed a significant decrease in the total gastric juice acidity and an increase in mucus production; isolated compounds (30 mg/kg) α-humulene, β-caryophyllene, and caryophyllene oxide showed greater gastroprotective activity in the ethanol/HCl-induced ulcer model.	[78]
C. langsdorffii Desf.	Kaurenoic acid	Anti-inflammatory: Rat model of acetic acid–induced colitis. A marked reduction in gross damage score (52% and 42%) and wet weight of damaged colon tissue (39% and 32%) was observed in rats that received 100 mg/kg kaurenoic acid, respectively, by rectal and oral routes.	[72]
C. langsdorffii Desf.	Kaurenoic acid	Antibacterial: Gram-positive bacteria (MIC range 5.0 μg/mL to 100.0 μg/mL).	[71]

(*Continued*)

TABLE 14.3 (*Continued*)
Biological Activities of *Copaifera* Oleoresins, Essential Oils, and Isolated Components

Copaifera spp.	Material	Biological Activity	Ref.
C. langsdorffii Desf.	Kaurenoic acid	Antiproliferative: In vitro cytotoxicity, 78 μM concentration, on CEM (human leukemia, 95% growth inhibition), MCF-7 (human breast tumor, 45% growth inhibition), and HCT-8 (human colon tumor, 45% growth inhibition) cells.	[73]
C. langsdorffii Desf.	Kaurenoic acid	Antiproliferative: In vitro cytotoxicity, 20 μM concentration, on AGP01 (human gastric cancer, 28% growth inhibition) and SF-295 (human glioblastoma, 28% growth inhibition) cells.	[74]
C. langsdorffii Desf.	Kaurenoic acid	Smooth muscle relaxant: Rat uterine muscle ex vivo, kaurenoic acid exerts a uterine relaxant effect acting principally through calcium blockade and, in part, by the opening of ATP-sensitive potassium channels.	[96]
C. lucens Dwyer	Oleoresin	Antileishmanial: ***L. amazonensis*** promastigotes (IC_{50} = 20.0 μg/mL).	[48]
C. lucens Dwyer	Oleoresin	Antibacterial: ***S. aureus*** (MIC = 125 μg/mL), ***B. subtilis*** (MIC = 125 μg/mL)	[83]
C. martii Hayne	Oleoresin	Antileishmanial: ***L. amazonensis*** promastigotes (IC_{50} = 14.0 μg/mL).	[48]
C. martii Hayne	Oleoresin	Antileishmanial: In vivo mouse model, copaiba oil oral treatment (100 mg/kg/day) caused a significant reduction in the average lesion size (1.1 ± 0.4 mm) against ***Leishmania amazonensis*** lesions compared with untreated mice (4.4 ± 1.3 mm).	[97]
C. martii Hayne	Oleoresin	Antibacterial: ***S. aureus*** (MIC = 62.5 μg/mL), MRSA (MIC = 62.5 μg/mL), ***B. subtilis*** (MIC = 15.6 μg/mL), ***S. epidermidis*** (MIC = 62.5 μg/mL), ***Enterococcus faecalis*** (MIC = 62.5 μg/mL).	[83]
C. multijuga Hayne	Oleoresin	Antiproliferative: In vitro cytotoxicity, B16F10 (murine melanoma cells, IC_{50} = 457 μg/mL).	[57]
C. multijuga Hayne	Oleoresin	Antiproliferative: In vivo lung metastasis and tumor growth, in mouse model, oral administration of ***C. multijuga*** oleoresin (at 2 g/Kg in the days 3, 5, 7, 10, 12, and 14 after inoculation of tumoral cells) reduced tumor growth by 58% and tumor weight by 76% and reduced the number of lung nodules by 47.1%.	[57]
C. multijuga Hayne	Oleoresin	Antiproliferative: In vivo Ehrlich ascitic and solid tumor, mouse model: ***C. multijuga*** oleoresin (doses varying between 100 and 200 mg/kg) showed antineoplastic properties against Ehrlich ascitic tumor (EAT) and solid tumor during ten consecutive days of treatment.	[98]
C. multijuga Hayne	Oleoresin	Insecticidal: Mosquito larvicidal activity (***Anopheles darlingi***, LC50 = 31 μg/mL; ***Aedes aegypti***, LC50 = 93 μg/mL)	[59]
C. multijuga Hayne	Oleoresin	Anti-inflammatory: The β-caryophyllene-rich oleoresin oil of ***C. multijuga***, at a dose of 100 mg/kg, inhibited zymosan-induced pleurisy in a mouse model, reducing total leukocytes by 45% and neutrophil accumulation by 73%. ***C. multijuga*** oil also showed in vitro reduction of NO production in mouse peritoneal macrophages at a concentration of 5 μg/mL.	[13]
C. multijuga Hayne	Oleoresin	Anti-inflammatory: Rat pleurisy model, doses of 100 mg/kg and 200 mg/kg, presented in vivo anti-inflammatory effects.	[58]
C. multijuga Hayne	Oleoresin	Antileishmanial: ***L. amazonensis*** promastigotes (IC_{50} = 10.0 μg/mL).	[48]
C. multijuga Hayne	Oleoresin	Antibacterial: MRSA (MIC = 125 μg/mL), ***B. subtilis*** (MIC = 125 μg/mL).	[83]

TABLE 14.3 (*Continued*)
Biological Activities of *Copaifera* Oleoresins, Essential Oils, and Isolated Components

Copaifera spp.	Material	Biological Activity	Ref.
C. multijuga Hayne	Oleoresin	Antinociceptive: Mouse model (acetic acid–induced writhing, tail flick, hot plate), oral administration of ***C. multijuga*** oil with doses of 30–150 mg/kg significantly showed antinociception, which was reversed with naloxone.	[99]
C. multijuga Hayne	Oleoresin	Insecticidal: Mosquito larvicidal activity (***Anopheles darlingi***, LC50 = 128 μg/mL; ***Aedes aegypti***, LC50 = 18 μg/mL).	[59]
C. multijuga Hayne	Oleoresin extracts	Anti-inflammatory: The CH2Cl2 and MeOH fractions obtained from ***C. multijuga*** oleoresin, given by the intraperitoneal route, caused a significant inhibition of carrageenan-induced rat paw edema with inhibition of 49 ± 13% and 64 ± 9%, respectively.	[61]
C. multijuga Hayne	Oleoresin extracts	Anti-inflammatory: The hexane, chloroform, and methanol fractions obtained from ***C. multijuga*** oleoresin, given by the oral gavage, caused a significant inhibition of carrageenan-induced rat paw edema.	[100]
C. multijuga Hayne	Oleoresin extracts	Antinociceptive: The hexane, chloroform, and methanol fractions obtained from ***C. multijuga*** oleoresin, given by oral gavage, caused a significant inhibition (in a concentration-dependent way) of the number of contortions induced by acetic acid and the second phase of formalin-induced licking response. Similar results were observed in the tail flick model; administration of naloxone inhibited the antinociceptive effect of the fractions, indicating that copaiba resin may be acting on opioid receptors.	[100]
C. multijuga Hayne	3β-alepterolic acid	Antiproliferative: In vitro cytotoxicity, 20 μM concentration, on AGP01 (human gastric cancer, 8.5% growth inhibition) and SF-295 (human glioblastoma, 21% growth inhibition) cells.	[74]
C. multijuga Hayne	3β-alepterolic acid acetate	Antiproliferative: In vitro cytotoxicity, 20 μM concentration, on AGP01 (human gastric cancer, 13% growth inhibition), and SF-295 (human glioblastoma, 18% growth inhibition) cells.	[74]
C. officinalis (Jacq.) L.	Oleoresin	Antiproliferative: In vivo Walker 256 carcinoma inoculated into the vagina and uterine cervix of female rats, ***C. officinalis*** oleoresin stimulated tumor growth by 70%.	[101]
C. officinalis (Jacq.) L.	Oleoresin EO	Anti-inflammatory: Dias and coworkers investigated the immunomodulatory effects of ***C. officinalis*** oleoresin essential oil on inflammatory cytokines (NO, H2O2, TNF-α, IFN-γ, and IL-17) in a murine model of experimental autoimmune encephalomyelitis. At a concentration of 100 μg/mL, ***C. officinalis*** oil inhibited the in vitro production of the inflammatory cytokines, modulating the immune response.	[62]
C. officinalis (Jacq.) L.	Oleoresin	Antileishmanial: ***L. amazonensis*** promastigotes (IC_{50} = 20.0 μg/mL).	[48]
C. officinalis (Jacq.) L.	Oleoresin	Antibacterial: ***S. aureus*** (MIC = 62.5 μg/mL), MRSA (MIC = 125 μg/mL), ***B. subtilis*** (MIC = 31.3 μg/mL), ***S. epidermidis*** (MIC = 31.3 μg/mL), ***Enterococcus faecalis*** (MIC = 31.3 μg/mL).	[83]
C. officinalis (Jacq.) L.	Oleoresin	Antibacterial: ***Streptococcus mutans*** (MIC = 780 μg/mL).	[102]
C. officinalis (Jacq.) L.	Oleoresin	Antibacterial: ***Staphylococcus aureus*** (MIC = 312.5 μg/mL).	[103]

(*Continued*)

TABLE 14.3 (*Continued*)
Biological Activities of *Copaifera* Oleoresins, Essential Oils, and Isolated Components

Copaifera spp.	Material	Biological Activity	Ref.
C. officinalis (Jacq.) L.	Agathic acid	Antileishmanial: ***L. amazonensis*** promastigotes (IC_{50} = 28.0 μg/mL), amastigotes (IC_{50} = 17.0 μg/mL).	[104]
C. officinalis (Jacq.) L.	Alepterolic acid (=hydroxycopalic acid)	Antileishmanial: ***L. amazonensis*** promastigotes (IC_{50} = 2.5 μg/mL), amastigotes (IC_{50} = 18.0 μg/mL)	[104]
C. officinalis (Jacq.) L.	Kaurenoic acid	Antileishmanial: ***L. amazonensis*** promastigotes (IC_{50} = 28.0 μg/mL), amastigotes (IC_{50} = 3.5 μg/mL).	[104]
C. officinalis (Jacq.) L.	Methyl copalate	Antileishmanial: ***L. amazonensis*** promastigotes (IC_{50} = 6.0 μg/mL), amastigotes (IC_{50} = 14.0 μg/mL).	[104]
C. officinalis (Jacq.) L.	Pinifolic acid	Antileishmanial: ***L. amazonensis*** promastigotes (IC_{50} = 70.0 μg/mL), amastigotes (IC_{50} = 4.0 μg/mL).	[104]
C. officinalis (Jacq.) L.	***Ent***-polyalthic acid	Antileishmanial: ***L. amazonensis*** promastigotes (IC_{50} = 35.0 μg/mL), amastigotes (IC_{50} = 15.0 μg/mL).	[104]
C. paupera (Herzog) Dwyer	Oleoresin	Antileishmanial: ***L. amazonensis*** promastigotes (IC_{50} = 11.0 μg/mL).	[48]
C. paupera (Herzog) Dwyer	Oleoresin	Antibacterial: ***B. subtilis*** (MIC = 62.5 μg/mL), ***Enterococcus faecalis*** (MIC = 62.5 μg/mL)	[83]
C. paupera (Herzog) Dwyer	Copalic acid	Antibacterial: ***Bacillus subtilis*** (MIC = 3.1–6.3 μg/mL), ***Staphylococcus aureus*** (MIC = 8–10 μg/mL), ***Staphylococcus epidermidis*** (MIC = 4–5 μg/mL).	[75]
C. paupera (Herzog) Dwyer	***Ent***-polyalthic acid	Antibacterial: ***Bacillus subtilis*** (MIC = 20–30 μg/mL), ***Staphylococcus aureus*** (MIC = 40–50 μg/mL), ***Staphylococcus epidermidis*** (MIC = 40 μg/mL).	[75]
C. paupera (Herzog) Dwyer	Kaurenoic acid	Antibacterial: ***Bacillus subtilis*** (MIC = 2.5–5 μg/mL), ***Staphylococcus aureus*** (MIC = 6–8 μg/mL), ***Staphylococcus epidermidis*** (MIC = 4–6 μg/mL).	[75]
C. paupera (Herzog) Dwyer	Methyl copalate	Antiproliferative: In vitro cytotoxicity, P-388 (murine lymphoma, IC_{50} = 2.5 μg/mL), A-549 (human lung carcinoma, IC_{50} = 5 μg/mL), HT-29 (human colon carcinoma, IC_{50} = 5 μg/mL), MEL-28 (human melanoma, IC_{50} = 10 μg/mL).	[75]
C. reticulata Ducke	Oleoresin	Antibacterial: ***Porphyromonas gingivalis*** (MIC = 6.25 μg/mL), ***Streptococcus*** spp. (MIC 25–50 μg/mL).	[68]
C. reticulata Ducke	Oleoresin	Antiproliferative: In vitro cytotoxicity, GM07492-A (human lung fibroblast cells, IC_{50} = 51.85 μg/mL).	[68]
C. reticulata Ducke	Oleoresin	Anxiolytic: elevated plus-maze test with rats—oral doses of 100, 400, and 800 mg/kg produced a dose-dependent anxiolytic-like effect over the dose range tested.	[105]
C. reticulata Ducke	Oleoresin	Insecticidal: Mosquito larvicidal activity (***Culex quinquefasciatus***, LC50 = 0.4, 0.9, 39, and 90 μg/mL against the 1st, 2nd, 3rd, and 4th larval instars, respectively).	[106]
C. reticulata Ducke	Oleoresin	Insecticidal: Mosquito larvicidal activity (***Aedes aegypti***, LC50 = 8.9 μg/mL against the 3rd larval instar).	[107]
C. reticulata Ducke	Oleoresin	Neuroprotective: Rat model of motor cortex excitotoxic injury, ***C. reticulata*** resin treatment induces neuroprotection by modulating inflammatory response following an acute damage to the central nervous system.	[67]
C. reticulata Ducke	Oleoresin	Acaricidal: ***Rhipicephalus (Boophilus) microplus*** (LC50 = 1,579 μg/mL).	[108]

TABLE 14.3 (*Continued*)
Biological Activities of *Copaifera* Oleoresins, Essential Oils, and Isolated Components

Copaifera spp.	Material	Biological Activity	Ref.
C. reticulata Ducke	Oleoresin	Anti-inflammatory: At a concentration of 500 μg/mL, ***C. reticulata*** oleoresin oil showed 85% inhibition of NO production in mouse peritoneal macrophages in vitro.	[13]
C. reticulata Ducke	Oleoresin	Antileishmanial: A β-caryophyllene-rich ***C. reticulata*** (from Pará State) oleoresin oil showed remarkable activity against ***L. amazonensis*** promastigotes with IC_{50} of 5.0 μg/mL. Another sample of ***C. reticulata*** oil (from Acre) with lower concentration of β-caryophyllene was less active (IC_{50} = 22.0 μg/mL).	[48]
C. reticulata Ducke	Oleoresin EO	Antileishmanial: ***C. reticulata*** oleoresin essential oil showed significant antileishmanial activity against axenic amastigotes (IC_{50} = 15.0 μg/mL) and intracellular amastigotes (IC_{50} = 20 μg/mL) of ***L. amazonensis***. Note that the major component of the oil, β-caryophyllene, was inactive against the amastigotes. Interference with the mitochondrial membrane was suggested as the mechanism for antileishmanial activity.	[109]
C. reticulata Ducke	Oleoresin	Antinociceptive: Mouse model (acetic acid–induced writhing, tail flick, hot plate), oral administration of ***C. reticulata*** oil with doses of 30–150 mg/kg significantly showed antinociception, which was reversed with naloxone.	[99]
C. reticulata Ducke	Oleoresin	Antitrypanosomal: ***T. evansi***, in vivo mouse model, doses of 0.63 mL/kg/day over five days showed no curative effects. ***T. evansi***, in vitro trypomastigotes, 0.5% solution of ***C. reticulata*** oil showed 100% killing after 6 h.	[86]
C. reticulata Ducke	Oleoresin EO (Acre)	Antibacterial: ***S. aureus*** (MIC = 62.5 μg/mL), MRSA (MIC = 125 μg/mL), ***B. subtilis*** (MIC = 31.3 μg/mL), ***S. epidermidis*** (MIC = 62.5 μg/mL), ***Enterococcus faecalis*** (MIC = 62.5 μg/mL).	[83]
C. reticulata Ducke	3β-alepterolic acid	Insecticidal: Mosquito larvicidal activity (***Aedes aegypti***, LC50 = 87.3 μg/mL against the 3rd larval instar).	[82]
C. reticulata Ducke	3β-alepterolic acid acetate	Insecticidal: Mosquito larvicidal activity (***Aedes aegypti***, LC50 = 0.8 μg/mL against the 3rd larval instar).	[82]
Copaifera spp.	Oleoresin	Antibacterial: Oleoresin oils from unidentified species of ***Copaifera*** showed remarkable antibacterial activity against Gram-positive ***Bacillus subtilis*** and ***Staphylococcus aureus*** (MIC = 5 μg/mL). The oils were inactive against Gram-negative organisms.	[110]
Copaifera spp.	Agathic acid	Antitrypanosomal: ***T. cruzi*** epimastigotes (IC_{50} = 86.8 μM), trypomastigotes (IC_{50} = 823 μM), amastigotes (IC_{50} = 14.9 μM).	[111]
Copaifera spp.	Copalic acid	Antitrypanosomal: ***T. cruzi*** epimastigotes (IC_{50} = 47.2 μM), trypomastigotes (IC_{50} = 444 μM), amastigotes (IC_{50} = 1.3 μM). Note: β-caryophyllene + copalic acid showed a significant synergistic effect against ***T. cruzi*** trypomastigotes.	[111]
Copaifera spp.	Alepterolic acid (=hydroxycopalic acid)	Antitrypanosomal: ***T. cruzi*** epimastigotes (IC_{50} = 41.2 μM), trypomastigotes (IC_{50} = 453 μM), amastigotes (IC_{50} = 1.8 μM).	[111]
Copaifera spp.	Kaurenoic acid	Antitrypanosomal: ***T. cruzi*** epimastigotes (IC_{50} = 167 μM), trypomastigotes (IC_{50} = 596 μM), amastigotes (IC_{50} = 16.5 μM).	[111]
Copaifera spp.	Methyl copalate	Antitrypanosomal: ***T. cruzi*** epimastigotes (IC_{50} = 83.3 μM), trypomastigotes (IC_{50} = 377 μM), amastigotes (IC_{50} = 2.5 μM).	[111]

(*Continued*)

TABLE 14.3 (*Continued*)
Biological Activities of *Copaifera* Oleoresins, Essential Oils, and Isolated Components

Copaifera spp.	Material	Biological Activity	Ref.
Copaifera spp.	Pinifolic acid	Antitrypanosomal: *T. cruzi* epimastigotes (IC_{50} = 854 μM), trypomastigotes (IC_{50} = 1,630 μM), amastigotes (IC_{50} = 18.6 μM).	[111]
Copaifera spp.	*Ent*-polyalthic acid	Antitrypanosomal: *T. cruzi* epimastigotes (IC_{50} = 168 μM), trypomastigotes (IC_{50} = 965 μM), amastigotes (IC_{50} = 28.4 μM).	[111]
Copaifera spp.	β-caryophyllene	Antileishmanial: *L. amazonensis* amastigotes (IC_{50} = 1.3 μg/mL).	[112]
Copaifera spp.	β-caryophyllene	Antitrypanosomal: *T. cruzi* epimastigotes (IC_{50} = 78.4 μM), trypomastigotes (IC_{50} = 1,593 μM), amastigotes (IC_{50} = 63.7 μM).	[111]

TABLE 14.4
MolDock Docking Energies (kJ/mol) of *Copaifera* Diterpenoids with *Leishmania* Protein Targets

Leishmania Targets	PDB[a]	E_{dock} (ave)	E_{dock} (min)	Best-Docking Diterpenoid Ligand
Cathepsin B (LdonCatB)	homology	−84.6	−100.6	3α-alepterolic acid acetate
Cathepsin B (LmajCatB)	homology	−80.8	−93.7	7α-acetoxyhardwickiic acid
Cyclophilin A (LdonCypA)	3EOV	−83.3	−102.6	*ent*-pinifolic acid
Deoxyuridine triphosphate nucleotidohydrolase (LmajdUTPase)	2YAY	−85.3	−103.8	19-*O*-acetyl-1,2-dehydrokautriwaic acid
Dihydroorotate dehydrogenase (LdonDHODH)	3GYE	−89.9	−102.7	7α-acetoxyhardwickiic acid
Dihydroorotate dehydrogenase (LmajDHODH)	3MHU	−109.2	−126.7	7α-acetoxyhardwickiic acid
Glucose-6-phosphate isomerase (LmexGPI)	1Q50	−73.0	−85.3	19-*O*-acetyl-1,2-dehydrokautriwaic acid
Glyceraldehyde-3-phosphate dehydrogenase (LmexGAPDH)	1A7K	−74.2	−83.0	19-*O*-acetyl-1,2-dehydrokautriwaic acid
Glycerol-3-phosphate dehydrogenase (LmexGPDH)	1N1E	−100.4	−114.3	3α-alepterolic acid acetate
Methionyl-tRNA synthetase (LmajMetRS)	3KFL	−106.9	−123.0	3α-alepterolic acid acetate
Nicotinamidase (LinfPnC1)	3R2J	−61.3	−75.9	3β-alepterolic acid
N-myristoyl transferase (LmajNMT)	4A30	−92.3	−104.0	19-*O*-acetyl-1,2-dehydrokautriwaic acid
Nucleoside diphosphate kinase b (LmajNDKb)	3NGS	−83.9	−105.8	7α-acetoxyhardwickiic acid
Nucleoside hydrolase (LmajNH)	1EZR	−80.5	−90.5	7-acetylbacchotricuneatin D
Oligopeptidase B (LmajOPB)	2XE4	−97.8	−106.1	7α-acetoxyhardwickiic acid
Phosphodiesterase B1 (LmajPDEB1)	2R8Q	−89.5	−105.8	3β-alepterolic acid acetate
Phosphomannumutase (LmexPMM)	2I55	−94.2	−117.5	19-*O*-acetyl-1,2-dehydrokautriwaic acid
Pteridine reductase 1 (LmajPTR1)	1E7W	−93.8	−110.7	Copaiferolic acid
Pyruvate kinase (LmexPYK)	1PKL	−103.4	−113.5	7α-acetoxyhardwickiic acid
Sterol 14α-demethylase (LinfCYP51)	3L4D	−90.2	−111.3	19-*O*-acetyl-1,2-dehydrokautriwaic acid
Thiol-dependent reductase I (LinfTDR1)	4AGS	−78.7	−88.8	19-*O*-acetyl-1,2-dehydrokautriwaic acid
Triosephosphate isomerase (LmexTIM)	2VXN	−90.7	−101.5	3α-alepterolic acid acetate
Trypanothione reductase (LinfTR)	4APN	−92.5	−109.0	19-*O*-acetyl-1,2-dehydrokautriwaic acid
Tyrosyl–tRNA synthetase (LmajTyrRS)	3P0J	−92.4	−102.9	Patagonic acid
Uridine diphosphate–glucose pyrophosphorylase (LmajUGPase)	2OEG	−99.9	−113.9	Patagonic acid

Note: [a]*PDB* = protein data bank.

TABLE 14.5
MolDock Docking Energies (kJ/mol) of *Copaifera* Diterpenoids with *Trypanosoma cruzi* Protein Targets

***Trypanosoma cruzi* Targets**	**PDB**	**E_{dock} (ave)**	**E_{dock} (min)**	**Best-Docking Diterpenoid Ligand**
Cruzain	3IUT	−80.2	−92.6	Patagonic acid
Cyclophilin (TcCyp19)	1XQ7	−78.9	−92.0	3β-alepterolic acid acetate
Deoxyuridine triphosphatase (TcUTPase)	1OGK	−83.4	−101.0	3β-alepterolic acid acetate
Dihydrofolate reductase–thymidylate synthase (TcDHFR-TS)	3IRN	−93.2	−110.7	7α-acetoxyhardwickiic acid
Dihydroorotate dehydrogenase (TcDHODH)	3W6Y	−92.5	−109.7	7α-acetoxyhardwickiic acid
Farnesyl diphosphate synthase (TcFPPS)	3ICZ	−96.2	−109.8	7α-acetoxyhardwickiic acid
Glyceraldehyde-3-phosphate dehydrogenase (TcGAPDH)	1QXS	−70.3	−85.3	Copaiferolic acid
Hypoxanthine phosphoribosyltransferase (TcHPRT)	1P19	−82.1	−94.4	7α-hydroxyhardwickiic acid
Nucleoside diphosphate kinase B (TcNDKb)	3PRV	−71.6	−88.4	Crolechinic acid
Old yellow enzyme (=Prostaglandin F2α synthase) (TcPGFS)	3ATY	−85.6	−97.3	Patagonic acid
Pteridine reductase 2 (TcPTR2)	1MXH	−96.8	−118.4	(+)-hardwickiic acid
Pyruvate kinase (TcPYK)	3QV9	−80.3	−87.4	(−)-hardwickiic acid
Spermidine synthase (TcSpdSyn)	3BWC	−96.8	−106.8	19-*O*-acetyl-1,2-dehydrokautriwaic acid
Sterol 14α-demethylase (TcCYP51)	3KLO	−89.5	−101.8	19-*O*-acetyl-1,2-dehydrokautriwaic acid
Triosephosphate isomerase (TcTIM)	1SUX	−88.2	−100.7	*ent*-polyalthic acid
Trypanothione reductase (TcTR)	1BZL	−81.9	−95.8	Copaiferolic acid
UDP–galactose mutase (TcUGM)	4DSH	−104.5	−115.7	Copaiferolic acid

docking was UDP–galactose mutase (average E_{dock} = −104.5 kJ/mol), but the normal substrate and co-crystallized ligand, uridine diphosphate (UDP), had a much superior docking energy (E_{dock} = −232.8 kJ/mol). Likewise, *L. major* UDP–glucose pyrophosphorylase showed an average docking energy of −99.9 kJ/mol, which was much worse than UDP itself (E_{dock} = −145.9 kJ/mol). The diterpenoids showed good docking to *T. cruzi* spermidine synthase, with an average docking energy of −96.8 kJ/mol; however, these are much worse than the docking energy of the co-crystallized ligand, *S*-adenosyl methionine, with a docking energy of −133.0 kJ/mol. Thus, although they exhibited good docking properties, it is unlikely that *Copaifera* diterpenoids can compete with the normal substrate ligands for these proteins.

Copaifera diterpenoids showed excellent docking to *L. mexicana* pyruvate kinase (average E_{dock} = −103.4 kJ/mol), much better than the normal substrate, phosphoenolpyruvate (E_{dock} = −59.8 kJ/mol). Docking energies with *T. cruzi* pyruvate kinase were not as impressive (average −80.3 kJ/mol), but still better than phosphoenolpyruvate (E_{dock} = −48.6 kJ/mol) and comparable to the TcPYK inhibitor, ponceau S (E_{dock} = −83.6 kJ/mol). Parasite pyruvate kinases can be expected to be target proteins for *Copaifera* diterpenoids.

Protozoal triosephosphate isomerases (LmexTIM and TcTIM) are expected to be targeted by *Copaifera* diterpenoids. The average docking energy with LmexTIM (−90.7 kJ/mol) was much better than either the normal substrate (dihydroxyacetone phosphate, E_{dock} = −52.4 kJ/mol) or the co-crystallized ligand, phosphoglycolohydroxamic acid (E_{dock} = −61.1 kJ/mol). Likewise, docking energies with TcTIM (average −88.2 kJ/mol) were better than the dihydroxyacetone phosphate

(E_{dock} = −59.7 kJ/mol) and comparable to the TcTIM inhibitor, 3-(2-benzothiazolylthio)-1-propanesulfonic acid (E_{dock} = −85.5 kJ/mol).

Both *L. major* pteridine reductase and *T. cruzi* pteridine reductase had docking properties with *Copaifera* diterpenoids with comparable energies (average E_{dock} = −93.8 and −96.8 kJ/mol, respectively) with the normal substrate dihydrobiopterin (E_{dock} = −96.9 and −100.1 kJ/mol, respectively). Thus, *Copaifera* diterpenoids may compete with dihydrobiopterin for teridine reductase. Sterols are the normal substrates for sterol 14α-demethylase (CYP51), and triterpenoids are expected to also target this protein as inhibitors [118]. Nevertheless, *Copaifera* diterpenoids showed docking energies that may compete with normal sterols for these protein targets. *L. infantum* CYP51 had an average docking energy with the diterpenoids of −90.2 kJ/mol, which was generally not as good as a normal sterol substrate (obtusifoliol, E_{dock} = −104.4 kJ/mol), but comparable to the known LinfCYP51 inhibitor fluconazole (E_{dock} = −87.5 kJ/mol). Likewise, *T. cruzi* CYP51 had an average diterpenoid docking energy of −89.5 kJ/mol, but substrate (obtusifoliol) docking of −105.6 kJ/mol, and fluconazole docking energy of −90.9 kJ/mol.

Copaifera diterpenoids generally showed weak docking energies against the parasite cysteine proteases, *L. donovani* cathepsin B, *L. major* cathepsin B, or cruzain. This docking behavior of diterpenoids with *Leishmania* cathepsin B [114] and cruzain [117] was previously observed. *Leishmania donovani* and *T. cruzi* cyclophilins also showed weak docking energies.

Although *Copaifera* diterpenoids showed only weak docking to parasite glyceraldehyde-3-phosphate dehydrogenases, they may still target these proteins. LmexGAPDH had an average E_{dock} of −73.0 kJ/mol, and TcGAPDH had an average E_{dock} of −70.3 kJ/mol, but these docking energies are better than the docking energies of the normal substrate, glyceraldehyde-3-phosphate (E_{dock} = −58.9 and −52.6 kJ/mol, respectively).

Additional *Leishmania* proteins expected to be targeted by *Copaifera* diterpenoids include glycerol-3-phosphate dehydrogenase, which showed excellent docking energies (average −100.4 kJ/mol) to LmexGPDH, better than the normal substrate, glycerol-3-phosphate (E_{dock} = −62.5 kJ/mol). Also targeted with a weak docking energy are glucose-6-phosphate isomerase (Lmex GPI E_{dock} average = −73.0 kJ/mol), though better than the docking energy of the normal substrate (glucose-6-phosphate, E_{dock} = −62.0 kJ/mol), and phosphomannomutase (LmexPMM E_{dock} average = −94.2 kJ/mol), which is better compared to the normal substrate (mannose-6-phosphate, Edock = −72.5 kJ/mol).

Additional *T. cruzi* protein targets may be dihydrofolate reductase–thymidylate synthase (TcDHFR–TS), which showed an average docking energy with *Copaifera* diterpenoids of −93.2 kJ/mol, comparable to the docking energy of the normal substrate (dihydrofolate, −99.3 kJ/mol), as well as the TcDHFR–TS inhibitor cycloguanil (E_{dock} = −83.1 kJ/mol); farnesyl diphosphate synthase (TcFPPS), with docking energies that averaged −96.2 kJ/mol, which is comparable to the docking energy of the normal substrate, isopentenyl diphosphate (E_{dock} = −98.9 kJ/mol); and hypoxanthine phosphoribosyltransferase (TcHPRT), having an average E_{dock} = −82.1 kJ/mol, compared to the normal substrate, hypoxanthine, with E_{dock} of −65.9 kJ/mol.

14.6.2 Antibacterial Activity of Copaiba

Copaiba oleoresin has shown antibacterial activity against several strains, in particular, Gram-positive *Bacillus subtilis* and *Staphylococcus aureus*, with minimum inhibitory concentration (MIC) values of 5 μg/mL for both organisms [110]. Copalic acid, isolated from *C. langsdorffii*, showed excellent antibacterial activity against *Bacillus cereus* (MIC 8.0 μg/mL), *B. subtilis* (MIC 5.0 μg/mL), *Kocuria rhizophila* (MIC 5.0 μg/mL), *Streptococcus pyogenes* (MIC 3.0 μg/mL), *S. pneumoniae* (MIC 3.0 μg/mL), *S. agalactiae* (MIC 2.0 μg/mL), *S. dysgalactiae* (MIC 1.0 μg/mL), *S. epidermidis* (MIC 0.5 μg/mL) [71], *S. salivarius* (MIC 2.0 μg/mL), *S. mutans* (MIC 3.0 μg/mL), *S. mitis* (MIC 5.0 μg/mL), *S. sobrinus* (MIC 3.0 μg/mL), and *S. sanguinis* (MIC 6.0 μg/mL) [70]. Likewise, kaurenoic acid showed remarkable activity against *S. pyogenes* (MIC 5.0 μg/mL),

TABLE 14.6
MolDock Docking Energies (kJ/mol) of *Copaifera* Diterpenoids with Bacterial Protein Targets

Bacterial Protein Targets	PDB	E_{dock} (ave)	E_{dock} (min)	Best Docking Diterpenoid Ligand
Pseudomonas aeruginosa peptide deformylase (PaPDF)	1LRY	−96.3	−113.5	(+)-hardwickiic acid
Streptococcus pneumoniae peptide deformylase (SpPDF)	2AIE	−100.2	−115.4	7α-acetoxyhardwickiic acid
Mycobacterium tuberculosis peptide deformylase (MtPDF)	3E3U	−95.7	−107.3	(+)-hardwickiic acid
Escherichia coli topoisomerase IV (EcTopoIV)	1S16	−100.5	−118.8	7α-acetoxyhardwickiic acid
Mycobacterium tuberculosis DNA gyrase B (MtDNAGyrB)	3ZKD	−101.3	−118.3	3α-alepterolic acid acetate
Mycobacterium tuberculosis protein tyrosine phosphatase (MtPTPB)	2OZ5	−89.2	−107.2	***Ent***-polyalthic acid
Mycobacterium tuberculosis UDP-galactopyranose mutase (MtUGM)	4RPL	−92.2	−104.4	19-***O***-acetyl-1,2-dehydrokautriwaic acid
Mycobacterium tuberculosis mycocyclosin synthase (MtCYP121)	5IBE	−87.0	−108.2	19-***O***-acetyl-1,2-dehydrokautriwaic acid
Escherichia coli DNA ligase (EcLigA)	2OWO	−97.8	−108.5	19-***O***-acetyl-1,2-dehydrokautriwaic acid
Mycobacterium tuberculosis DNA ligase (MtLigA)	1ZAU	−89.0	−107.2	19-***O***-acetyl-1,2-dehydrokautriwaic acid
Staphylococcus aureus DNA ligase (SaLigA)	4CC6	−85.2	−97.8	Methyl copalate
Streptococcus pneumoniae DNA ligase (SpLigA)	4GLW	−88.1	−109.5	Cativic acid

S. pneumoniae (MIC 5.0 μg/mL), *S. dysgalactiae* (MIC 8.0 μg/mL) [71], *S. epidermidis* (MIC 4–5 μg/mL), *B. subtilis* (MIC 3.1–6.3 μg/mL), and *S. aureus* (MIC 8–10 μg/mL) [75]. 3α-alepterolic acid acetate (acetoxycopalic acid) showed moderate antibacterial activity against cariogenic *Streptococcus* bacteria, with MIC values ranging from 12.0 to 60.0 μg/mL [70]. *Ent*-polyalthic acid also showed moderate antibacterial activity against *B. subtilis* (MIC 20–30 μg/mL), *S. aureus* (MIC 40–50 μg/mL), and *S. epidermidis* (MIC 40 μg/mL) [75].

In order to provide some insight into the mechanisms of activity, a virtual screening of copaiba diterpenoids has been carried out against several bacterial protein targets, including peptide deformylase, DNA gyrase, topoisomerase IV, UDP-galactopyranose mutase, protein tyrosine phosphatase, cytochrome P450 CYP 121, and nicotinamide adenine dinucleotide (NAD+)–dependent DNA ligase [119] (see Table 14.6). The best bacterial target for copalic acid was *Mycobacterium tuberculosis* DNA gyrase B (PDB 3ZKD), with a docking energy (E_{dock}) of −105.7 kJ/mol. The protein with the best docking energy with kaurenoic acid was *S. pneumoniae* peptide deformylase (PDB 2AIE, E_{dock} = −89.7 kJ/mol). 3α-alepterolic acid acetate was the best docking ligand to *Escherichia coli* topoisomerase IV (PDB 1S16) and *M. tuberculosis* DNAGyrB (PDB 3ZKD), with docking energies of −118.8 and −118.3 kJ/mol, respectively. 3β-alepterolic acid acetate also showed excellent docking to these two proteins, with docking energies of −117.1 and −117.3 kJ/mol, respectively. The best bacterial target for *ent*-polyalthic acid was *M. tuberculosis* protein tyrosine phosphatase (PDB 2OZ5, E_{dock} = −107.2 kJ/mol). The copaiba diterpenoid ligand with the best docking properties was 7α-acetoxyhardwickiic acid with *S. aureus* peptide deformylase (PDB 3U7M, E_{dock} = −120.6 kJ/mol).

14.6.3 Antiproliferative Activity of Copaiba

Copaiba oleoresins have exhibited both in vitro and in vivo antiproliferative activities (Table 14.3). *Copaifera reticulata* oleoresin, for example, has shown in vitro cytotoxic activity against GM07492-A human lung fibroblast cells with an IC_{50} of 51.85 μg/mL [68]. The oleoresin of *C. multijuga* has shown in vitro cytotoxic activity against B16F10 murine melanoma cells with an IC_{50} of 457 μg/mL [57]. Furthermore, in a mouse model of lung metastasis and tumor growth, oral administration of *C. multijuga* oleoresin reduced tumor growth, tumor mass, and number of lung nodules after inoculation of B16F10 tumor cells [57]. Likewise, *C. multijuga* oleoresin, in doses varying between 100 and 200 mg/kg, showed antineoplastic properties against Ehrlich ascetic tumors and solid tumors in an in vivo mouse model [98]. On the other hand, *C. officinalis* oleoresin actually stimulated growth of Walker 256 carcinoma by 70% in an in vivo rat model [101].

Diterpenoids isolated from *Copaifera* species have shown cytotoxic activities (Table 14.3). Copalic acid, isolated from *C. langsdorffii*, showed in vitro cytotoxicity on MO59J human glioblastoma cells and HeLa human cervical adenocarcinoma cells with IC_{50} of 68.3 and 44.0 μg/mL, respectively [71]. Kaurenoic acid has demonstrated cytotoxicity against several human tumor cell lines, including CEM leukemia, MCF-7 breast tumor, HCT-8 colon tumor [73], AGP01 gastric tumor, and SF-295 glioblastoma [74]. Growth inhibition of AGP01 and SF-295 cells was also demonstrated by 3β-alepterolic acid and 3β-alepterolic acid acetate [73]. Methyl copalate showed remarkable cytotoxic activity on P-388 murine lymphoma (IC_{50} = 2.5 μg/mL), A-549 human lung carcinoma (IC_{50} = 5 μg/mL), HT-29 human colon carcinoma (IC_{50} = 5 μg/mL), and MEL-28 human melanoma (IC_{50} = 10 μg/mL) cells [75].

Molecular docking (Molegro Virtual Docker, Aarhus, Denmark) has been carried out with *Copaifera* diterpenoids on cancer molecular targets, including androgen receptor; aromatase; caseine kinase II; cyclin-dependent kinases 2, 4, and 6; cyclooxygenase 2; DNA (cytosine-5)-methyltransferase-1 and -3A; epidermal growth factor receptor; estrogen receptor α; estrogen receptor β; heat shock protein 90; insulin-like growth factor 1 receptor; 5-lipoxygenase; mitogen-activated protein kinase 1; NF-κB; p90 ribosomal protein S6 kinase; P-glycoprotein; phosphatidylinositol-4,5-bisphosphate 3-kinase; topoisomerase I; topoisomerase IIα; topoisomerase IIβ; tubulin; and vascular endothelial growth factor receptor (Table 14.7). The best overall cancer targets for the copaiba diterpenoids were human DNA (cytosine-5)-methyltransferase-1 (HsDNMT1), human estrogen receptor β (HsERβ), and human mitogen-activated protein kinase 1 (HsMEK1), with average MolDock docking energies of −102.7, −99.2, and −101.5 kJ/mol, respectively. DNA (cytosine-5)-methyltransferase-1 (DNMT1) is the enzyme responsible for DNA methylation of carbon-5 of cytosine within CpG dinucleotides [120]. The enzyme is required for embryonic development [121] but is overexpressed in lung, liver, colorectal, gastric, breast, and lung tumors [122]. Thus, DNMT1 has emerged as an attractive target for cancer chemotherapy [123, 124]. The mitogen-activated protein kinase (MAPK) signaling cascade is one of the most important pathways involved in cellular proliferation and differentiation [125], and therefore, inhibition of components of this pathway, such as MEK1, can potentially target tumors that depend on MAPK signaling [126]. Agonism of estrogen receptor α (ERα) stimulates proliferation of breast, uterus, and prostate tissues, whereas ERβ agonism inhibits proliferation of these tissues [127]. Thus, compounds that can selectively bind and activate ERβ, but not ERα, could represent effective antitumor agents for treatment of prostate and breast cancer [128]. Copalic acid and methyl copalate both targeted HsMEK1, with docking energies of −108.2 and −111.0 kJ/mol, respectively, while 3β-alepterolic acid and 3β-alepterolic acid acetate showed excellent docking with HsDNMT1 (E_{dock}= −107.2 and −121.7 kJ/mol, respectively). Kaurenoic acid was a relatively weakly docking ligand but did show selective docking to aromatase (E_{dock} = −93.7 kJ/mol). The best-docking ligand was patagonic acid, which had a docking energy of −121.8 kJ/mol with HsDNMT1.

TABLE 14.7
MolDock Docking Energies (kJ/mol) of *Copaifera* Diterpenoids with Cancer-Relevant Protein Targets

Cancer-Relevant Protein Targets	PDB	E_{dock} (ave)	E_{dock} (min)	Best Docking Ligand
Human androgen receptor (HsAR)	5VO4	−67.6	−103.2	Cativic acid
Human aromatase (HsCYP19A1)	5JKW	−97.1	−112.7	*Ent*-pinifolic acid
Human casein kinase II (HsCK2)	5N9K	−83.3	−101.0	3α-alepterolic acid acetate
Human cyclin-dependent kinase 2 (HsCDK2)	5JQ8	−91.1	−104.1	7α-acetoxyhardwickiic acid
Human cyclin-dependent kinase 4 (HsCDK4)	2W96	−95.2	−117.6	3α-alepterolic acid acetate
Human cyclin-dependent kinase 6 (HsCDK6)	5L2S	−87.3	−99.0	(+)-hardwickiic acid
Murine cyclooxygenase 2 (MmCOX-2)	6COX	−93.7	−106.9	Kolavenic acid
Human DNA (cytosine-5)-methyltransferase 1 (HsDNMT1)	3SWR	−102.7	−121.8	Patagonic acid
Human DNA (cytosine-5)-methyltransferase 3A (HsDNMT3A)	2QRV	−94.3	−113.4	7α-hydroxyhardwickiic acid
Human epidermal growth factor receptor (HsEGFR)	1XKK	−81.2	−98.8	19-*O*-acetyl-1,2-dehydrokautriwaic acid
Human estrogen receptor α (HsERα)	1X7E	−96.5	−107.5	19-*O*-acetyl-1,2-dehydrokautriwaic acid
Human estrogen receptor β (HsERβ)	1U3S	−99.6	−120.1	19-*O*-acetyl-1,2-dehydrokautriwaic acid
Human heat shock protein HSP 90-α	5J2X	−83.8	−92.6	Cativic acid
Human insulin-like growth factor 1 receptor (HsIGF1R)	3LW0	−86.1	−94.2	Copaiferolic acid
Human 5-lipoxygenase (Hs5-LOX)	3V99	−89.9	−106.2	3α-alepterolic acid acetate
Human mitogen-activated protein kinase 1 (HsMEK1)	3OS3	−101.5	−113.5	3β-alepterolic acid acetate
Murine nuclear factor κ-light-chain-enhancer of activated B cells (MmNF-κB)	1VKX	−74.2	−87.7	3α-alepterolic acid acetate
Human p90 ribosomal protein S6 kinase (HsRSK2) *C*-terminal domain	4D9U	−75.3	−90.7	3β-alepterolic acid acetate
Human p90 ribosomal protein S6 kinase (HsRSK2) *N*-terminal domain	4NW6	−86.4	−99.7	7α-acetoxyhardwickiic acid
Murine P-glycoprotein	3G60	−97.6	−116.1	Patagonic acid
Human phosphatidylinositol-4,5-bisphosphate 3-kinase (HsPI3K)	2A5U	−84.5	−97.6	7α-acetoxyhardwickiic acid
Human topoisomerase I (HsTOPO-I)	1NH3	−83.7	−99.6	Copaiferolic acid
Human topoisomerase IIα (HsTOPO-IIα)	4FM9	−94.2	−105.2	19-*O*-acetyl-1,2-dehydrokautriwaic acid
Human topoisomerase IIβ (HsTOPO-IIβ)	4J3N	−85.6	−98.9	3α-alepterolic acid acetate
Bovine tubulin (colchicine binding site)	1SA1	−94.2	−103.8	3α-alepterolic acid acetate
Bovine tubulin (paclitaxel binding site)	1JFF	−79.1	−88.6	7α-acetoxyhardwickiic acid
Bovine tubulin (vinblastine binding site)	1Z2B	−89.1	−101.0	(+)-hardwickiic acid
Human vascular endothelial growth factor receptor (HsVEGFR)	4ASE	−92.9	−105.5	7-acetylbacchotricuneatin D

14.6.4 Anti-Inflammatory Activity of Copaiba

Inflammation is the biological response of body tissues to detrimental stimuli, such as pathogenic microorganisms, chemical or physical irritants, or injury. Inflammation is manifested by redness, swelling, heat, and sometimes pain. While acute inflammation is a normal part of the healing process, chronic inflammation often plays a role in chronic diseases, such as osteoarthritis, lupus, and inflammatory bowel disease, and can be problematic. Several copaiba oleoresins have shown anti-inflammatory activity, including *C. cearensis* [13], *C. duckei* [84], *C. langsdorffii* [50, 91], *C. multijuga* [13, 58, 61, 100], *C. officinalis* [62], and *C. reticulata* [13] (Table 14.3).

The immune response is a complex cascade of interacting cytokines and reactions, and there are several biomolecular targets important in treating chronic inflammation. We have carried out virtual screening of copaiba diterpenoids against soluble epoxide hydrolase (EPHX2), fibroblast collagenase, phospholipase A2 (PLA2), 5-lipoxygenase, inducible nitric oxide synthase, phosphoinositide 3-kinase, interleukin-1 receptor-associated kinase 4, glutathione *S*-transferase ω-1, cyclooxygenase-1, cyclooxygenase-2, c-Jun *N*-terminal kinase, nuclear factor κ-light-chain-enhancer of activated B cells (NF-κB), inhibitor of κB kinase β, NF-κB essential modulator, lipid binding protein MD-2, myeloperoxidase, p38 mitogen-activated protein kinase, peroxisome proliferator-activated receptor γ, and cAMP-specific 3′,5′-cyclic phosphodiesterase 4D (Table 14.8). The overall best target proteins were murine soluble epoxide hydrolase and murine phospholipase A2, with average docking energies of −108.3 and −100.0 kJ/mol. Secretory phospholipase A2 and cytosolic phospholipase

TABLE 14.8
MolDock Docking Energies (kJ/mol) of *Copaifera* Diterpenoids with Cancer-Relevant Protein Targets

Inflammation-Relevant Protein Targets	PDB	E_{dock} (ave)	E_{dock} (min)	Best-Docking Ligand
Murine soluble epoxide hydrolase (MmEPHX2)	1CR6	−108.3	−125.6	19-***O***-acetyl-1,2-dehydrokautriwaic acid
Human soluble epoxide hydrolase (HsEPHX2)	4HAI	−95.1	−104.5	***ent***-agathic acid
Human fibroblast collagenase (HsMMP-1)	1CGL	−96.5	−109.0	19-***O***-acetyl-1,2-dehydrokautriwaic acid
Porcine phospholipase A2 (SsPLA2)	2B03	−100.0	−112.4	(+)-hardwickiic acid
Human phospholipase A2 (HsPLA2)	1J1A	−95.4	−109.5	7α-acetoxyhardwickiic acid
Human 5-lipoxygenase (Hs5-LOX)	3V99	−89.9	−106.2	3α-alepterolic acid acetate
Murine inducible nitric oxide synthase (MmiNOS)	1M8D	−87.5	−110.4	Copaiferolic acid
Human phosphatidylinositol-4,5-bisphosphate 3-kinase γ (HsPI3Kγ)	2A5U	−84.5	−97.6	7α-acetoxyhardwickiic acid
Human interleukin-1 receptor-associated kinase 4 (HsIRAK4)	5T1S	−90.7	−100.6	7α-acetoxyhardwickiic acid
Human glutathione ***S***-transferase ω-1 (HsGSTO1)	5V3Q	−82.7	−94.2	19-***O***-acetyl-1,2-dehydrokautriwaic acid
Ovine cyclooxygenase-1 (OaCOX-1)	3N8Z	−62.1	−92.0	Crolechinic acid
Murine cyclooxygenase-2 (MmCOX-2)	6COX	−93.7	−106.9	Kolavenic acid
Human c-Jun ***N***-terminal kinase (HsJNK)	4Y46	−83.7	−97.7	19-***O***-acetyl-1,2-dehydrokautriwaic acid
Xenopus laevis inhibitor of κB kinase β (XlIKKB)	3RZF	−88.1	−100.9	Patagonic acid
Human NF-κB essential modulator (HsNEMO)	3BRT	−85.9	−105.3	3α-alepterolic acid acetate
Human lipid binding protein MD-2 (HsMD-2)	2E59	−71.2	−84.3	19-***O***-acetyl-1,2-dehydrokautriwaic acid
Human myeloperoxidase (HsMPO)	4C1M	−84.3	−98.8	(+)-hardwickiic acid
Murine nuclear factor κ-light-chain-enhancer of activated B cells (MmNF-κB)	3DO7	−72.7	−85.9	(+)-hardwickiic acid
Human p38 mitogen-activated protein kinase (Hsp38MAPK)	1OZ1	−91.3	−116.4	19-***O***-acetyl-1,2-dehydrokautriwaic acid
Human peroxisome proliferator-activated receptor γ (HsPPARγ)	3ADV	−91.9	−107.6	Copaiferolic acid
Human cAMP-specific 3′,5′-cyclic phosphodiesterase 4D (HsPDE4D)	5K32	−90.2	−107.3	7α-acetoxyhardwickiic acid

A2 are both targets for anti-inflammatory drug development [129]. Soluble epoxide hydrolase has been identified as a molecular target not only for inflammatory diseases but also as a target for neurodegenerative diseases and for treatment of pain [130]. Thus, targeting EPHX2 and/or PLA2 by copaiba diterpenoids may explain the anti-inflammatory activities of copaiba oleoresins.

14.7 COMPUTATIONAL METHODS—MOLECULAR DOCKING

Molecular docking analyses were carried out using Molegro Virtual Docker (v. 6.0.1, Molegro ApS, Aarhus, Denmark) against known bacterial [119], *Leishmania* [114–116], *Trypanosoma cruzi* [117], and cancer-relevant and inflammation-relevant protein targets [131], as previously described [114–117, 119, 132].

14.8 CONCLUSIONS

The oleoresins from *Copaifera* species (copaiba) have been used by the Native peoples of the Amazon region for thousands of years. These materials have shown remarkable biological activities, including antibacterial, antiparasitic, antineoplastic, and anti-inflammatory activities. Copaiba resins have been distilled to give essential oils that are largely composed of sesquiterpenoids, particularly β-caryophyllene. The resins are also composed of diterpene acids, which are responsible for many of the observed biological activities. Molecular docking of copaiba diterpene acids with documented protein targets has revealed potential mechanisms of activity for these bioactive constituents. Future research to validate the molecular mechanisms of copaiba diterpenoids is encouraged.

14.9 AUTHOR CONTRIBUTIONS

R.d.T., J.K.d.S., and W.N.S., authors, contributed equally to literature searching and manuscript writing, review, and editing. W.N.S. carried out the molecular docking.

14.10 FUNDING

W.N.S. is grateful to do TERRA International (available online: https://www.doterra.com/US/en) for financial support of the Aromatic Plant Research Centers (APRC; available online: https://aromaticplant.org/).

CONFLICTS OF INTEREST

The authors declare no conflict of interest.

REFERENCES

1. Dwyer, J.D. The Central American, West Indian, and South American species of *Copaifera* (Caesalpiniaceae). *Brittonia* **1951**, *7*, 143–172. [CrossRef]
2. Jacquin, N.J. *Enumeratio Systematica Plantarum*; Apud Theodorum Haak: Leiden, The Netherlands, 1760.
3. Linné, C. *Species Plantarum*; Joannis Thomae de Trattner: Vienna, Austria, 1764.
4. Joyce, B.L.; Al-Ahmad, H.; Chen, F.; Stewart, C.N. Diesel trees. In *Handbook of Bioenergy Crop Plants*; Kole, C.; Joshi, C.P.; Shonnard, D.R., Eds.; CRC Press: Boca Raton, FL, 2012; pp. 619–629.
5. Costa, J.A.S.; Copaifera. In Flora do Brasil 2020, em Construção. Available online: http://floradobrasil.jbrj. gov.br/reflora/floradobrasil/FB22895 (accessed on 6 March 2018).
6. Veiga Junior, V.F.; Pinto, A.C. O gênero *Copaifera* L. *Quim. Nova* **2002**, *25*, 273–286. [CrossRef]
7. Plowden, C. The ethnobotany of copaiba (*Copaifera*) oleoresin in the Amazon. *Econ. Bot.* **2004**, *58*, 729–733. [CrossRef]

8. Cascon, V.; Gilbert, B. Characterization of the chemical composition of oleoresins of *Copaifera guianensis* Desf., *Copaifera duckei* Dwyer and *Copaifera multijuga* Hayne. *Phytochemistry* **2000**, *55*, 773–778. [CrossRef]
9. Martins-da-Silva, R.C.V.; Pereira, J.F.; de Lima, H.C. O gênero *Copaifera* (Leguminosae—Caesalpinioideae) na Amazônia Brasileira. *Rodriguésia* **2008**, *59*, 455–476.
10. Souza, V.C.; Lorenzi, H. *Botânica Sistemática: Guia Ilustrado para Identificação das Famílias de Angiospermas da Flora Brasileira, Baseado na APG III*, 3rd ed.; Instituto Plantarum: Nova Odessa, Brazil, 2012.
11. Milani, J.F.; Rocha, J.F.; de Pádua Teixeira, S. Oleoresin glands in copaíba (*Copaifera trapezifolia* Hayne: Leguminosae), a Brazilian rainforest tree. *Trees* **2012**, *26*, 769–775. [CrossRef]
12. Soares, N.S.; Goncalves, C.A.; Araujo, G.M.; Lomonaco, C. Floristic composition and abundance in forest fragments: A case study from southern Goias, Brazil. *Biosci. J.* **2015**, *31*, 1238–1252. [CrossRef]
13. Veiga Junior, V.F.; Rosas, E.C.; Carvalho, M.V.; Henriques, M.G.M.O.; Pinto, A.C. Chemical composition and anti-inflammatory activity of copaiba oils from *Copaifera cearensis* Huber ex Ducke, *Copaifera reticulata* Ducke and *Copaifera multijuga* Hayne—a comparative study. *J. Ethnopharmacol.* **2007**, *112*, 248–254. [CrossRef] [PubMed]
14. Langenheim, J.H. *Plant Resins: Chemistry, Evolution, Ecology, and Ethnobotany*; Timber Press, Inc.: Portland, OR, 2003.
15. Leandro, L.M.; de Sousa Vargas, F.; Barbosa, P.C.S.; Neves, J.K.O.; da Silva, J.A.; da Veiga-Junior, V.F. Chemistry and biological activities of terpenoids from copaiba (*Copaifera* spp.) oleoresins. *Molecules* **2012**, *17*, 3866–3889. [CrossRef] [PubMed]
16. Langenheim, J.H. Higher plant terpenoids: A phytcentric overview of their ecological roles. *J. Chem. Ecol.* **1994**, *20*, 1223–1280. [CrossRef] [PubMed]
17. Barbosa, P.C.S.; Medeiros, R.S.; Sampaio, P.T.B.; Vieira, G.; Wiedemann, L.S.M.; Veiga-Junior, V.F. Influence of abiotic factors on the chemical composition of copaiba oil (*Copaifera multijuga* Hayne): Soil composition, seasonality and diameter at breast height. *J. Braz. Chem. Soc.* **2012**, *23*, 1823–1833. [CrossRef]
18. Biavatti, M.W.; Dossin, D.; Deschamps, F.C.; Lima, M.D.P. Análise de óleos-resinas de copaíba: Contribuição para o seu controle de qualidade. *Rev. Bras. Farmacogn.* **2006**, *16*, 230–235. [CrossRef]
19. Pieri, F.A.; Mussi, M.C.; Moreira, M.A.S. Óleo de copaíba (*Copaifera* sp.): Histórico, extração, aplicações industriais e propriedades medicinais. *Rev. Bras. Plantas Med.* **2009**, *11*, 465–472. [CrossRef]
20. Rosa, J.C.; da Silva Gomes, A.M. Os aspectos etnobotânicos da copaíba. *Rev. Geogr.* **2009**, *4*, 59–77. [CrossRef]
21. Ribeiro, R.D.; de Lima, H.C. Riqueza e distribuição geográfica de espécies arbóreas da família Leguminosae e implicações para conservação no Centro de Diversidade Vegetal de Cabo Frio, Rio de Janeiro, Brasil. *Rodriguésia* **2009**, *60*, 111–127. [CrossRef]
22. Carvalho, M.C.; Gomide, L.R.; Santos, R.M.D.; Scolforo, J.R.S.; Carvalho, L.M.T.D.; Mello, J.M.D. Modeling ecological niche of tree species in Brazilian tropical area. *Cerne* **2017**, *23*, 229–240. [CrossRef]
23. Carvalho, P.E.R. *Espécies Arbóreas Brasileiras, Vol. 1*; EMBRAPA Informação Tecnológica: Brasilia, Brazil, 2003.
24. Shanley, P.; Leite, A.; Alechandre, A.; Azevedo, C. Copaíba: *Copaifera* spp. In *Frutíferas e Plantas Úteis na Vida Amazônica*; Shanley, P.; Medina, G., Eds.; CIFOR and IMAZON: Belém, Brazil, 2005; pp. 85–94.
25. Carvalho, P.E.R. *Espécies Florestais Brasileiras: Recomendações Silviculturais, Potencialidades e Uso da Madeira*; EMBRAPA-CNPF: Colombo, Brazil, 1994.
26. Rigamonte-Azevedo, O.C.; Wadt, P.G.S.; Wadt, L.D.O. *Copaíba: Ecologia e Produção de Óleo-Resina*; Embrapa Acre: Rio Branco, Brazil, 2004.
27. Alencar, J.D.C. Estudos silviculturais de uma população natural de *Copaifera multijuga* Hayne—Leguminosae, na Amazônia central. IV. Interpretação de dados fenológicos em relação a elementos climáticos. *Acta Amaz.* **1988**, *18*, 199–209. [CrossRef]
28. Muniz, F.H. Padrões de floração e frutificação de árvores da Amazônia Maranhense. *Acta Amaz.* **2008**, *38*, 617–626. [CrossRef]
29. Dos Santos, N. Fenologia. *Rodriguésia* **1979**, *31*, 223–226.
30. Pedroni, F.; Sanchez, M.; Santos, F.A.M. Fenologia da copaíba (*Copaifera langsdorffii* Desf.—Leguminosae, Caesalpinioideae) em uma floresta semidecídua no sudeste do Brasil. *Rev. Bras. Botânica* **2002**, *25*, 183–194. [CrossRef]
31. Pedroni, F. *Ecologia da Copaíba (Copaifera langsdorffii Desf. Caelsapiniaceae) na Reserva Municipal de Santa Genebra, Campinas, SP*; Universidade Estadual de Campinas: Barão Geraldo, Brazil, 1993.

32. Motta Junior, C.J.; Lombardi, J.A. Aves como agentes dispersores da copaiba (*Copaifera langsdorffii*, Caesalpiniaceae) em São Carlos, estado de São Paulo. *Rev. Bras. Ornitol.* **1990**, *1*, 105–106.
33. Brum, H.D.; Camargo, J.L.C.; Ferraz, I.D.K. Copaíba-roxa, *Copaifera mutlijuga* Hayne. In *Manual de Sementes da Amazônia*; Ferraz, I.D.K.; Camargo, J.L.C., Eds.; INPA: Manaus, Brazil, 2009; p. 12.
34. Pinto, A.; Amaral, P.; Gaia, C.; de Oliveira, W. *Boas Práticas para Manejo Florestal e Agroindustrial de Produtos Florestais Não Madeireiros: Açaí, Andiroba, Babaçu, Castanha-do-Brasil, Copaíba e Unha-de-Gato*; Sebrae-AM: Belém, Brazil, 2010.
35. Dos Santos, J.C.; Leite, A.C.P.; Wadt, L.H.O.; Borges, K.H.; Andrade, F.G.; Menezes, R.S.; Muniz, P.S.B. *Demandas Tecnologicas para o Sistema Produtivo de Óleo de Copaíba (Copaifera spp.) no Estado do Acre*; Embrapa Acre: Rio Branco, Brazil, 2001.
36. Newton, P.; Watkinson, A.R.; Peres, C.A. Determinants of yield in a non-timber forest product: *Copaifera* oleoresin in Amazonian extractive reserves. *For. Ecol. Manag.* **2011**, *261*, 255–264. [CrossRef]
37. Estrella, E. *Plantas Medicinales Amazónicas: Realidad y Perspectivas*; Tratado de Cooperación Amazónica: Lima, Peru, 1995.
38. Rios, M.N.D.S.; Pastore Junior, F. *Plantas da Amazônia: 450 Espécies de Uso Geral*; Universidade de Brasília: Brasilia, Brazil, 2011.
39. Maciel, M.A.M.; Pinto, A.C.; Veiga Junior, V.F. Plantas medicinais: A necessidade de estudos multidisciplinares. *Quim. Nova* **2002**, *25*, 429–438. [CrossRef]
40. Luz, F.J.F. Plantas medicinais de uso popular em Boa Vista, Roraima, Brasil. *Hortic. Bras.* **2001**, *19*, 88–96. [CrossRef]
41. De Albuquerque, K.C.O.; da Veiga, A.D.S.S.; da Silva e Silva, J.V.; Brigido, H.P.C.; Ferreira, E.P.D.R.; Costa, E.V.S.; Marinho, A.M.D.R.; Percário, S.; Dolabela, M.F. Brazilian Amazon traditional medicine and the treatment of difficult to heal leishmaniasis wounds with *Copaifera*. *Evid.-Based Complement. Altern. Med.* **2017**, *2017*. [CrossRef] [PubMed]
42. Basile, A.C.; Sertié, J.A.; Freitas, P.C.; Zanini, A.C. Anti-inflammatory activity of oleoresin from Brazilian *Copaifera*. *J. Ethnopharmacol.* **1988**, *22*, 101–109. [CrossRef]
43. Taylor, L. *The Healing Power of Rainforest Herbs: A Guide to Understanding and Using Herbal Medicinals*; Square One Publishers: Garden City Park, NY, 2005.
44. Sant'Anna, B.M.P.; Fontes, S.P.; Pinto, A.C.; Rezende, C.M. Characterization of woody odorant contributors in copaiba oil (*Copaifera multijuga* Hayne). *J. Braz. Chem. Soc.* **2007**, *18*, 984–989. [CrossRef]
45. Revilla, J. *Apontamentos para a Cosmética Amazônica*; SEBRAE-INPA: Manaus, Brazil, 2002.
46. Shanley, P.; Pierce, A.R.; Laird, S.A.; Guillén, A. *Tapping the Green Market: Certification & Management of Non-Timber Forest Products*; Earthscan: London, UK, 2002.
47. Gonzaga, A.L. *Madeira: Uso e Conservação*; IPHAN/MONUMENTA: Brasilia, Brazil, 2006.
48. Santos, A.O.; Ueda-Nakamura, T.; Dias Filho, B.P.; Veiga Junior, V.F.; Pinto, A.C.; Nakamura, C.V. Effect of Brazilian copaiba oils on *Leishmania amazonensis*. *J. Ethnopharmacol.* **2008**, *120*, 204–208. [CrossRef] [PubMed]
49. Lameira, O.A.; Martins-da-Silva, R.C.V.; Zoghbi, M.D.G.B.; Oliveira, E.C.P. Seasonal variation in the volatiles of *Copaifera duckei* Dwyer growing wild in the state of Pará—Brazil. *J. Essent. Oil Res.* **2009**, *21*, 105–108. [CrossRef]
50. Gelmini, F.; Beretta, G.; Anselmi, C.; Centini, M.; Magni, P.; Ruscica, M.; Cavalchini, A.; Maffei Facino, R. GC-MS profiling of the phytochemical constituents of the oleoresin from *Copaifera langsdorffii* Desf. and a preliminary in vivo evaluation of its antipsoriatic effect. *Int. J. Pharm.* **2013**, *440*, 170–178. [CrossRef] [PubMed]
51. Estevão, L.R.M.; de Medeiros, J.P.; Baratella-Evêncio, L.; Simões, R.S.; Mendonça, F.D.S.; Evêncio-Neto, J. Effects of the topical administration of copaiba oil ointment (*Copaifera langsdorffii*) in skin flaps viability of rats. *Acta Cirúrgica Bras.* **2013**, *28*, 863–869. [CrossRef]
52. Zimmermam-Franco, D.C.; Bolutari, E.B.; Polonini, H.C.; do Carmo, A.M.R.; Chaves, M.D.G.A.M.; Raposo, N.R.B. Antifungal activity of *Copaifera langsdorffii* Desf oleoresin against dermatophytes. *Molecules* **2013**, *18*, 12561–12570. [CrossRef] [PubMed]
53. De Almeida, L.F.R.; Portella, R.D.O.; Bufalo, J.; Marques, M.O.M.; Facanali, R.; Frei, F. Non-oxygenated sesquiterpenes in the essential oil of *Copaifera langsdorffii* Desf. increase during the day in the dry season. *PLoS ONE* **2016**, *11*, e0149332. [CrossRef] [PubMed]
54. Do Nascimento, M.E.; Zoghbi, M.D.G.B.; Pinto, J.E.B.P.; Bertolucci, S.K.V. Chemical variability of the volatiles of *Copaifera langsdorffii* growing wild in the southeastern part of Brazil. *Biochem. Syst. Ecol.* **2012**, *43*, 1–6. [CrossRef]
55. De Almeida, L.F.R.; Portella, R.D.O.; Facanali, R.; Marques, M.O.M.; Frei, F. Dry and wet seasons set the phytochemical profile of the *Copaifera langsdorffii* Desf. essential oils. *J. Essent. Oil Res.* **2014**, *26*, 292–300. [CrossRef]

56. Zoghbi, M.D.G.B.; Lameira, O.A.; Oliveira, E.C.P. Seasonal variation of oleoresin and volatiles from *Copaifera martii* Hayne growing wild in the State of Pará, Brazil. *J. Essent. Oil Res.* **2007**, *19*, 504–506. [CrossRef]
57. Lima, S.R.M.; Veiga Junior, V.F.; Christo, H.B.; Pinto, A.C.; Fernandes, P.D. In vivo and in vitro studies on the anticancer activity of *Copaifera multijuga* Hayne and its fractions. *Phyther. Res.* **2003**, *17*, 1048–1053. [CrossRef] [PubMed]
58. Kobayashi, C.; Fontanive, T.O.; Enzweiler, B.G.; de Bona, L.R.; Massoni, T.; Apel, M.A.; Henriques, A.T.; Richter, M.F.; Ardenghi, P.; Suyenaga, E.S. Pharmacological evaluation of *Copaifera multijuga* oil in rats. *Pharm. Biol.* **2011**, *49*, 306–313. [CrossRef] [PubMed]
59. Trindade, F.T.T.; Stabeli, R.G.; Pereira, A.A.; Facundo, V.A.; Silva, A.D.A. *Copaifera multijuga* ethanolic extracts, oil-resin, and its derivatives display larvicidal activity against *Anopheles darlingi* and *Aedes aegypti* (Diptera: Culicidae). *Braz. J. Pharmacogn.* **2013**, *23*, 464–470. [CrossRef]
60. Souza Barbosa, P.C.; Moreira Wiedemann, L.S.; da Silva Medeiros, R.; Barbosa Sampaio, P.D.T.; Vieira, G.; da Veiga-Junior, V.F. Phytochemical fingerprints of copaiba oils (*Copaifera multijuga* Hayne) determined by multivariate analysis. *Chem. Biodivers.* **2013**, *10*, 1350–1360. [CrossRef] [PubMed]
61. Veiga Junior, V.F.; Luciano, Z.; Patitucci, M.L.; Pinto, A.C.; Calixto, J.B. The inhibition of paw oedema formation caused by the oil of *Copaifera multijuga* Hayne and its fractions. *J. Pharm. Pharmacol.* **2006**, *58*, 1405–1410. [CrossRef] [PubMed]
62. Dias, D.S.; Fontes, L.B.A.; Crotti, A.E.M.; Aarestrup, B.J.V.; Aarestrup, F.M.; da Silva Filho, A.A.; Corrêa, J.O.A. Copaiba oil suppresses inflammatory cytokines in splenocytes of C57Bl/6 mice induced with experimental autoimmune encephalomyelitis (EAE). *Molecules* **2014**, *19*, 12814–12826. [CrossRef] [PubMed]
63. Zoghbi, M.D.G.B.; Martins-da-Silva, R.C.V.; Trigo, J.R. Volatiles of oleoresins of *Copaifera paupera* (Herzog) Dwyer, *C. piresii* Dwyer and *C. pubiflora* Benth. (Leguminosae). *J. Essent. Oil Res.* **2009**, *21*, 403–405. [CrossRef]
64. Zoghbi, M.D.G.B.; Andrade, E.H.A.; Martins-Da-Silva, R.C.V.; Trigo, J.R. Chemical variation in the volatiles of *Copaifera reticulata* Ducke (Leguminosae) growing wild in the states of Pará and Amapá, Brazil. *J. Essent. Oil Res.* **2009**, *21*, 501–503. [CrossRef]
65. Herrero-Jáuregui, C.; Casado, M.A.; Zoghbi, M.D.G.B.; Martins-Da-Silva, R.C. Chemical variability of *Copaifera reticulata* Ducke oleoresin. *Chem. Biodivers.* **2011**, *8*, 674–685. [CrossRef] [PubMed]
66. Sachetti, C.G.; de Carvalho, R.R.; Paumgartten, F.J.R.; Lameira, O.A.; Caldas, E.D. Developmental toxicity of copaiba tree (*Copaifera reticulata* Ducke, Fabaceae) oleoresin in rat. *Food Chem. Toxicol.* **2011**, *49*, 1080–1085. [CrossRef] [PubMed]
67. Guimarães-Santos, A.; Santos, D.S.; Santos, I.R.; Lima, R.R.; Pereira, A.; de Moura, L.S.; Carvalho, R.N.; Lameira, O.; Gomes-Leal, W. Copaiba oil-resin treatment is neuroprotective and reduces neutrophil recruitment and microglia activation after motor cortex excitotoxic injury. *Evid.-Based Complement. Altern. Med.* **2012**, *2012*. [CrossRef] [PubMed]
68. Bardají, D.K.R.; da Silva, J.J.M.; Bianchi, T.C.; de Souza Eugênio, D.; de Oliveira, P.F.; Leandro, L.F.; Rogez, H.L.G.; Venezianni, R.C.S.; Ambrosio, S.R.; Tavares, D.C.; et al. *Copaifera reticulata* oleoresin: Chemical characterization and antibacterial properties against oral pathogens. *Anaerobe* **2016**, *40*, 18–27. [CrossRef] [PubMed]
69. Veiga, V.F.; Pinto, A.C.; de Lima, H.C. The essential oil composition of *Copaifera trapezifolia* Hayne leaves. *J. Essent. Oil Res.* **2006**, *18*, 430–431. [CrossRef]
70. Souza, A.B.; Martins, C.H.G.; Souza, M.G.M.; Furtado, N.A.J.C.; Heleno, V.C.G.; de Sousa, J.P.B.; Rocha, E.M.P.; Bastos, J.K.; Cunha, W.R.; Veneziani, R.C.S.; et al. Antimicrobial activity of terpenoids from *Copaifera langsdorffii* Desf. against cariogenic bacteria. *Phyther. Res.* **2011**, *25*, 215–220. [CrossRef] [PubMed]
71. Abrão, F.; de Araújo Costa, L.D.; Alves, J.M.; Senedese, J.M.; de Castro, P.T.; Ambrósio, S.R.; Veneziani, R.C.S.; Bastos, J.K.; Tavares, D.C.; Martins, C.H.G. *Copaifera langsdorffii* oleoresin and its isolated compounds: Antibacterial effect and antiproliferative activity in cancer cell lines. *BMC Complement. Altern. Med.* **2015**, *15*, 1–10. [CrossRef] [PubMed]
72. Paiva, L.A.F.; Gurgel, L.A.; Silva, R.M.; Tomé, A.R.; Gramosa, N.V.; Silveira, E.R.; Santos, F.A.; Rao, V.S.N. Anti-inflammatory effect of kaurenoic acid, a diterpene from *Copaifera langsdorffii* on acetic acid-induced colitis in rats. *Vascul. Pharmacol.* **2003**, *39*, 303–307. [CrossRef]
73. Costa-Lotufo, L.V.; Cunha, G.M.A.; Farias, P.A.M.; Viana, G.S.B.; Cunha, K.M.A.; Pessoa, C.; Moraes, M.O.; Silveira, E.R.; Gramosa, N.V.; Rao, V.S.N. The cytotoxic and embryotoxic effects of kaurenoic acid, a diterpene isolated from *Copaifera langsdorffii* oleo-resin. *Toxicon* **2002**, *40*, 1231–1234. [CrossRef]

74. Vargas, F.D.S.; de Almeida, P.D.O.; Aranha, E.S.P.; Boleti, A.P.D.A.; Newton, P.; de Vasconcellos, M.C.; Veiga Junior, V.F.; Lima, E.S. Biological activities and cytotoxicity of diterpenes from *Copaifera* spp. oleoresins. *Molecules* **2015**, *20*, 6194– 6210. [CrossRef] [PubMed]
75. Tincusi, B.M.; Jiménez, I.A.; Bazzocchi, I.L.; Moujir, L.M.; Mamani, Z.A.; Barroso, J.P.; Ravelo, A.G.; Hernández, B.V. Antimicrobial terpenoids from the oleoresin of the Peruvian medicinal plant *Copaifera paupera*. *Planta Med.* **2002**, *68*, 808–812. [CrossRef] [PubMed]
76. Pinto, A.C.; Braga, W.F.; Rezende, C.M.; Garrido, F.M.S.; Veiga, V.F.; Bergter, L.; Patitucci, M.L.; Antunes, O.A.C. Separation of acid diterpenes of *Copaifera cearensis* Huber ex Ducke by flash chromatography using potassium hydroxide impregnated silica gel. *J. Braz. Chem. Soc.* **2000**, *11*, 355–360. [CrossRef]
77. Souza, A.B.; de Souza, M.G.M.; Moreira, M.A.; Moreira, M.R.; Furtado, N.A.J.C.; Martins, C.H.G.; Bastos, J.K.; dos Santos, R.A.; Heleno, V.C.G.; Ambrosio, S.R.; et al. Antimicrobial evaluation of diterpenes from *Copaifera langsdorffii* oleoresin against periodontal anaerobic bacteria. *Molecules* **2011**, *16*, 9611–9619. [CrossRef] [PubMed]
78. Lemos, M.; Santin, J.R.; Mizuno, C.S.; Boeing, T.; de Sousa, J.P.B.; Nanayakkara, D.; Bastos, J.K.; de Andrade, S.F. *Copaifera langsdorffii*: Evaluation of potential gastroprotective of extract and isolated compounds obtained from leaves. *Braz. J. Pharmacogn.* **2015**, *25*, 238–245. [CrossRef]
79. De Sousa, J.P.B.; Brancalion, A.P.S.; Junior, M.G.; Bastos, J.K. A validated chromatographic method for the determination of flavonoids in *Copaifera langsdorffii* by HPLC. *Nat. Prod. Commun.* **2012**, *7*, 25–28. [PubMed]
80. Batista, Â.G.; Ferrari, A.S.; da Cunha, D.C.; da Silva, J.K.; Cazarin, C.B.B.; Correa, L.C.; Prado, M.A.; de Carvalho-Silva, L.B.; Esteves, E.A.; Maróstica Júnior, M.R. Polyphenols, antioxidants, and antimutagenic effects of *Copaifera langsdorffii* fruit. *Food Chem.* **2016**, *197*, 1153–1159. [CrossRef] [PubMed]
81. Costa, A.R.M.; Freitas, L.A.P.; Mendiola, J.; Ibáñez, E. *Copaifera langsdorffii* supercritical fluid extraction: Chemical and functional characterization by LC/MS and in vitro assays. *J. Supercrit. Fluids* **2015**, *100*, 86–96. [CrossRef]
82. Geris, R.; da Silva, I.G.; da Silva, H.H.G.; Barison, A.; Rodrigues-Filho, E.; Ferreira, A.G. Diterpenoids from *Copaifera reticulata* Ducke with larvicidal activity against *Aedes aegypti* (L.) (Diptera, Culicidae). *Rev. Inst. Med. Trop. Sao Paulo* **2008**, *50*, 25–28. [PubMed]
83. Dos Santos, A.O.; Ueda-Nakamura, T.; Prado Dias Filho, B.; da Veiga Junior, V.F.; Pinto, A.C.; Nakamura, C.V. Antimicrobial activity of Brazilian copaiba oils obtained from different species of *Copaifera*. *Mem. Inst. Oswaldo Cruz* **2008**, *103*, 277–281. [CrossRef]
84. Carvalho, J.C.T.; Cascon, V.; Possebon, L.S.; Morimoto, M.S.S.; Cardoso, L.G.V.; Kaplan, M.A.C.; Gilbert, B. Topical antiinflammatory and analgesic activities of *Copaifera duckei* Dwyer. *Phyther. Res.* **2005**, *19*, 946–950. [CrossRef] [PubMed]
85. Castro-e-Silva, O., Jr.; Zucoloto, S.; Ramalho, F.S.; Ramalho, L.N.Z.; Reis, J.M.C.; Bastos, A.A.C.; Brito, M.V.H. Anti-proliferative activity of oleoresin from Brazilian *Copaifera duckei* oleoresin on liver regeneration in rats. *Phyther. Res.* **2004**, *18*, 92–94. [CrossRef] [PubMed]
86. Dorneles, F.D.S.; Da Silva, A.S.; Oliveira, C.B.; Zimmermann, C.E.P.; Rosa, L.D.; Tonin, A.A.; de Oliveira, E.C.P.; Santurio, J.M.; Monteiro, S.G. Susceptibility of *Trypanosoma evansi* in the Copaiba oil: In vitro test and in mice experimentally infected with the parasite. *Acta Sci. Vet.* **2013**, *41*, 1136.
87. Paiva, L.A.F.; Rao, V.S.N.; Gramosa, N.V.; Silveira, E.R. Gastroprotective effect of *Copaifera langsdorffii* oleo-resin on experimental gastric ulcer models in rats. *J. Ethnopharmacol.* **1998**, *62*, 73–78. [CrossRef]
88. Paiva, L.A.F.; Gurgel, L.A.; Campos, A.R.; Silveira, E.R.; Rao, V.S.N. Attenuation of ischemia/reperfusion-induced intestinal injury by oleo-resin from *Copaifera langsdorffii* in rats. *Life Sci.* **2004**, *75*, 1979–1987. [CrossRef] [PubMed]
89. Paiva, L.A.F.; de Alencar Cunha, K.M.; Santos, F.A.; Gramosa, N.V.; Silveira, E.R.; Rao, V.S.N. Investigation on the wound healing activity of oleo-resin from *Copaifera langsdorffii* in rats. *Phyther. Res.* **2002**, *16*, 737–739. [CrossRef] [PubMed]
90. De Lima Silva, J.J.; Guimarães, S.B.; da Silveira, E.R.; de Vasconcelos, P.R.L.; Lima, G.G.; Torres, S.M.; de Vasconcelos, R.C. Effects of *Copaifera langsdorffii* Desf. on ischemia-reperfusion of randomized skin flaps in rats. *Aesthet. Plast. Surg.* **2009**, *33*, 104–109. [CrossRef] [PubMed]
91. Neto, J.N.; Lindoso, M.J.D.S.; Coelho, L.F.; Antonio, R.; Carvalho, R.A.F.; Rodrigues, T.G.P.D.M.; de Araújo, A.G.P.; Girão, M.J.B.C.; Schor, E. Changes in the volume and histology of endometriosis foci in rats treated with copaiba oil (*Copaifera langsdorffii*). *Acta Cirúrgica Bras.* **2011**, *26*, 20–24. [CrossRef]
92. Masson-Meyers, D.S.; Enwemeka, C.S.; Bumah, V.; Andrade, T.A.M.; Cashin, S.E.; Frade, M.A.C. Antimicrobial effects of *Copaifera langsdorffii* oleoresin in infected rat wounds. *Int. J. Appl. Microbiol. Sci.* **2013**, *2*, 9–20.

93. Masson-Meyers, D.S.; Andrade, T.A.M.; Leite, S.N.; Frade, M.A.C. Cytotoxicity and wound healing properties of *Copaifera langsdorffii* oleoresin in rabbits. *Int. J. Nat. Prod. Sci.* **2013**, *3*, 10–20.
94. Masson-Meyers, D.; Enweme, C.S.; Bumah, V.; Andrade, T.; Frade, M.A. Topical treatment with *Copaifera langsdorffii* oleoresin improves wound healing in rats. *Int. J. Phytomed.* **2013**, *5*, 378–386.
95. Barbosa, F.S.; Leite, G.L.D.; Alves, S.M.; Nascimento, A.F.; D'Ávila, V.D.A.; da Costa, C.A. Insecticide effects of *Ruta graveolens*, *Copaifera langsdorffii* and *Chenopodium ambrosioides* against pests and natural enemies in commercial tomato plantation. *Acta Sci. Agron.* **2011**, *33*, 37–43.
96. De Alencar Cunha, K.M.; Paiva, L.A.; Santos, F.A.; Gramosa, N.V.; Silveira, E.R.; Rao, V.S. Smooth muscle relaxant effect of kaurenoic acid, a diterpene from *Copaifera langsdorffii* on rat uterus in vitro. *Phyther. Res.* **2003**, *17*, 320–324. [CrossRef] [PubMed]
97. Dos Santos, A.O.; Costa, M.A.; Ueda-Nakamura, T.; Dias-Filho, B.P.; da Veiga-Júnior, V.F.; de Souza Lima, M.M.; Nakamura, C.V. *Leishmania amazonensis*: Effects of oral treatment with copaiba oil in mice. *Exp. Parasitol.* **2011**, *129*, 145–151. [CrossRef] [PubMed]
98. Gomes, N.D.M.; Rezende, C.D.M.; Fontes, S.P.; Hovell, A.M.C.; Landgraf, R.G.; Matheus, M.E.; Pinto, A.D.C.; Fernandes, P.D. Antineoplasic activity of *Copaifera multijuga* oil and fractions against ascitic and solid Ehrlich tumor. *J. Ethnopharmacol.* **2008**, *119*, 179–184. [CrossRef] [PubMed]
99. Gomes, N.M.; Rezende, C.M.; Fontes, S.P.; Matheus, M.E.; Fernandes, P.D. Antinociceptive activity of Amazonian copaiba oils. *J. Ethnopharmacol.* **2007**, *109*, 486–492. [CrossRef] [PubMed]
100. Gomes, N.D.M.; de Rezende, C.M.; Fontes, S.P.; Matheus, M.E.; Pinto, A.D.C.; Fernandes, P.D. Characterization of the antinociceptive and anti-inflammatory activities of fractions obtained from *Copaifera multijuga* Hayne. *J. Ethnopharmacol.* **2010**, *128*, 177–183. [CrossRef] [PubMed]
101. Brito, N.M.B.; Brito, M.V.H.; Carvalho, R.D.K.V.; Matos, L.T.D.M.B.; Lobato, R.C.; Correa, S.C.; Brito, R.B. The effect of copaiba balsam on Walker 256 carcinoma inoculated into the vagina and uterine cervix of female rats. *Acta Cir. Bras.* **2010**, *25*, 176–180. [CrossRef] [PubMed]
102. Pieri, F.A.; Mussi, M.C.M.; Fiorini, J.E.; Moreira, M.A.S.; Schneedorf, J.M. Bacteriostatic effect of copaiba oil (*Copaifera officinalis*) against *Streptococcus mutans*. *Braz. Dent. J.* **2012**, *23*, 36–38. [CrossRef] [PubMed]
103. Guimarães, A.L.; Cunha, E.A.; Matias, F.O.; Garcia, P.G.; Danopoulos, P.; Swikidisa, R.; Pinheiro, V.A.; Nogueira, R.J. Antimicrobial activity of copaiba (*Copaifera officinalis*) and pracaxi (*Pentaclethra macroloba*) oils against *Staphylococcus aureus*: Importance in compounding for wound care. *Int. J. Pharm. Compd.* **2016**, *20*, 58–62. [PubMed]
104. Dos Santos, A.O.; Izumi, E.; Ueda-Nakamura, T.; Dias-Filho, B.P.; da Veiga-Júnior, V.F.; Vataru Nakamura, C. Antileishmanial activity of diterpene acids in copaiba oil. *Mem. Inst. Oswaldo Cruz* **2013**, *108*, 59–64. [CrossRef]
105. Curio, M.; Jacone, H.; Perrut, J.; Pinto, Â.C.; Veiga Filho, V.F.; Silva, R.C.B. Acute effect of *Copaifera reticulata* Ducke copaiba oil in rats tested in the elevated plus-maze: An ethological analysis. *J. Pharm. Pharmacol.* **2009**, *61*, 1105–1110. [CrossRef] [PubMed]
106. Silva, I.G.; Zanon, V.O.M.; Silva, H.H.G. Larvicidal activity of *Copaifera reticulata* Ducke oil-resin against *Culex quinquefasciatus* Say (Diptera: Culicidae). *Neotrop. Entomol.* **2003**, *32*, 729–732. [CrossRef]
107. Da Silva, H.H.G.; Geris, R.; Rodrigues Filho, E.; Rocha, C.; da Silva, I.G. Larvicidal activity of oil-resin fractions from the Brazilian medicinal plant *Copaifera reticulata* Ducke (Leguminosae-Caesalpinoideae) against *Aedes aegypti* (Diptera, Culicidae). *Rev. Soc. Bras. Med. Trop.* **2007**, *40*, 264–267. [CrossRef] [PubMed]
108. Fernandes, F.D.F.; Freitas, E.D.P.S. Acaricidal activity of an oleoresinous extract from *Copaifera reticulata* (Leguminosae: Caesalpinioideae) against larvae of the southern cattle tick, *Rhipicephalus* (*Boophilus*) *microplus* (Acari: Ixodidae). *Vet. Parasitol.* **2007**, *147*, 150–154. [CrossRef] [PubMed]
109. Dos Santos, A.O.; Ueda-Nakamura, T.; Dias Filho, B.P.; da Veiga Junior, V.F.; Nakamura, C.V. Copaiba oil: An alternative to development of new drugs against leishmaniasis. *Evid.-Based Complement. Altern. Med.* **2012**, *2012*. [CrossRef] [PubMed]
110. Pacheco, T.A.R.C.; Barata, L.E.S.; Duarte, M.C.T. A ntimicrobial activity of copaiba (*Copaifera* spp.) balsams. *Rev. Bras. Plantas Med.* **2006**, *8*, 123–124.
111. Izumi, E.; Ueda-Nakamura, T.; Veiga, V.F.; Pinto, A.C.; Nakamura, C.V. Terpenes from *Copaifera* demonstrated in vitro antiparasitic and synergic activity. *J. Med. Chem.* **2012**, *55*, 2994–3001. [CrossRef] [PubMed]
112. Soares, D.C.; Portella, N.A.; Ramos, M.F.D.S.; Siani, A.C.; Saraiva, E.M. *trans*-β-Caryophyllene: An effective antileishmanial compound found in commercial copaiba oil (*Copaifera* spp.). *Evid.-Based Complement. Altern. Med.* **2013**, *2013*. [CrossRef] [PubMed]

113. Ogungbe, I.V.; Setzer, W.N. The potential of secondary metabolites from plants as drugs or leads against protozoan neglected diseases-Part III: In-silico molecular docking investigations. *Molecules* **2016**, *21*, 1389. [CrossRef] [PubMed]
114. Ogungbe, I.V.; Setzer, W.N. In-silico *Leishmania* target selectivity of antiparasitic terpenoids. *Molecules* **2013**, *18*, 7761–7847. [CrossRef] [PubMed]
115. Ogungbe, I.V.; Ng, J.D.; Setzer, W.N. Interactions of antiparasitic alkaloids with *Leishmania* protein targets: A molecular docking analysis. *Future Med. Chem.* **2013**, *5*, 1777–1799. [CrossRef] [PubMed]
116. Ogungbe, I.V.; Erwin, W.R.; Setzer, W.N. Antileishmanial phytochemical phenolics: Molecular docking to potential protein targets. *J. Mol. Graph. Model.* **2014**, *48*, 105–117. [CrossRef] [PubMed]
117. Mcculley, S.F.; Setzer, W.N. An in-silico investigation of anti-Chagas phytochemicals. *Curr. Clin. Pharmacol.* **2014**, *9*, 205–257. [CrossRef] [PubMed]
118. Warfield, J.; Setzer, W.N.; Ogungbe, I.V. Interactions of antiparasitic sterols with sterol 14α-demethylase (CYP51) of human pathogens. *SpringerPlus* **2014**, *3*, 679. [CrossRef] [PubMed]
119. Setzer, M.S.; Sharifi-Rad, J.; Setzer, W.N. The search for herbal antibiotics: An in-silico investigation of antibacterial phytochemicals. *Antibiotics* **2016**, *5*, 30. [CrossRef] [PubMed]
120. Veeck, J.; Esteller, M. Breast cancer epigenetics: From DNA methylation to microRNAs. *J. Mammary Gland Biol. Neoplasia* **2010**, *15*, 5–17. [CrossRef] [PubMed]
121. Chen, T.; Hevi, S.; Gay, F.; Tsujimoto, N.; He, T.; Zhang, B.; Ueda, Y.; Li, E. Complete inactivation of DNMT1 leads to mitotic catastrophe in human cancer cells. *Nat. Genet.* **2007**, *39*, 391–396. [CrossRef] [PubMed]
122. Lin, R.K.; Wu, C.Y.; Chang, J.W.; Juan, L.J.; Hsu, H.S.; Chen, C.Y.; Lu, Y.Y.; Tang, Y.A.; Yang, Y.C.; Yang, P.C.; et al. Dysregulation of p53/Sp1 control leads to DNA methyltransferase-1 overexpression in lung cancer. *Cancer Res.* **2010**, *70*, 5807–5817. [CrossRef] [PubMed]
123. Medina-Franco, J.L.; Caulfield, T. Advances in the computational development of DNA methyltransferase inhibitors. *Drug Discov. Today* **2011**, *16*, 418–425. [CrossRef] [PubMed]
124. Maldonado-Rojas, W.; Olivero-Verbel, J.; Marrero-Ponce, Y. Computational fishing of new DNA methyltransferase inhibitors from natural products. *J. Mol. Graph. Model.* **2015**, *60*, 43–54. [CrossRef] [PubMed]
125. Isshiki, Y.; Kohchi, Y.; Iikura, H.; Matsubara, Y.; Asoh, K.; Murata, T.; Kohchi, M.; Mizuguchi, E.; Tsujii, S.; Hattori, K.; et al. Design and synthesis of novel allosteric MEK inhibitor CH4987655 as an orally available anticancer agent. *Bioorg. Med. Chem. Lett.* **2011**, *21*, 1795–1801. [CrossRef] [PubMed]
126. Weekes, C.D.; Von Hoff, D.D.; Adjei, A.A.; Leffingwell, D.P.; Eckhardt, S.G.; Gore, L.; Lewis, K.D.; Weiss, G.J.; Ramanathan, R.K.; Dy, G.K.; et al. Multicenter phase I trial of the mitogen-activated protein kinase 1/2 inhibitor BAY 86-9766 in patients with advanced cancer. *Clin. Cancer Res.* **2013**, *19*, 1232–1243. [CrossRef] [PubMed]
127. Warner, M.; Gustafsson, J.Å. The role of estrogen receptor β (ERβ) in malignant diseases—a new potential target for antiproliferative drugs in prevention and treatment of cancer. *Biochem. Biophys. Res. Commun.* **2010**, *396*, 63–66. [CrossRef] [PubMed]
128. Warner, M.; Huang, B.; Gustafsson, J.A. Estrogen receptor β as a pharmaceutical target. *Trends Pharmacol. Sci.* **2017**, *38*, 92–99. [CrossRef] [PubMed]
129. Tibes, U. Phospholipase A2 inhibitors in development. *Expert Opin. Investig. Drugs* **1997**, *6*, 279–298. [CrossRef] [PubMed]
130. Wagner, K.M.; McReynolds, C.B.; Schmidt, W.K.; Hammock, B.D. Soluble epoxide hydrolase as a therapeutic target for pain, inflammatory and neurodegenerative diseases. *Pharmacol. Ther.* **2017**, *180*, 62–76. [CrossRef] [PubMed]
131. Setzer, W.N.; Byler, K.G. In-silico approaches to natural products drug discovery: A review of the recent literature. In *Natural Products and Drug Discovery: From Pharmacochemistry to Pharmacological Approaches*; Diniz, M.F.F.M., Scotti, M.T., Scotti, L., Alves, M.F.E., Eds.; Editora UFPB: João Pessoa, Brazil, 2018.
132. Setzer, M.S.; Byler, K.G.; Ogungbe, I.V.; Setzer, W.N. Natural products as new treatment options for trichomoniasis: A molecular docking investigation. *Sci. Pharm.* **2017**, *85*, 5. [CrossRef] [PubMed]

15 Oil Resin
Characterization, Chemical Compositions, Food Applications, and Biological Activities

Karyme do Socorro de Souza Vilhena, Leonardo Souza da Costa, Anderson de Santana Botelho, Oberdan Oliveira Ferreira, Fábio José Bonfim Cardoso, Marcilene Paiva da Silva, Mozaniel Santana de Oliveira, and Eloisa Helena Aguiar Andrade

15.1 INTRODUCTION

Oil resins are viscous substances produced by some plant species, and they are known to be composed of complex mixtures containing both volatile and non-volatile compounds (Ali et al. 2023; Oriani et al. 2018). The growing interest in oil resins can be mainly attributed to the non-volatile constituents present in the resin fraction, which include fatty acids, pigments, and phenolic compounds. These species are responsible for imparting different, and often more attractive, properties than those observed in essential oils (Procopio et al. 2022; Song et al. 2020).

Oil resins are widely used in traditional medicine to treat a large variety of illnesses, especially in developing countries. However, these substances have become more popular in recent decades due to the trend toward valuing natural products as alternative treatments to synthetic pharmaceutical products (Becker et al. 2020; do Carmo Silva et al. 2020; Peivastegan et al. 2020). Ethnopharmacological studies support some of these uses, relating the beneficial properties of oil resins to their active components, which can promote healing, in addition to exhibiting various other biological properties, such as anti-inflammatory and antibacterial activities (Diefenbach et al. 2018; Menezes et al. 2022).

Unlike essential oils, the more complex profile of this type of extract allows the use of smaller quantities in the formulation of food products, wherein they are widely employed because of their desirable flavor, aroma, and color characteristics (Procopio et al. 2022). The application of oil resins in the food industry is also desirable due to their antifungal, antioxidant, and antibacterial activities, which are observed in oil resins extracted from the *Copaifera*, *Zingiber*, and *Capsicum* genera (do Carmo Silva et al. 2020; Melgar-Lalanne et al. 2017; P.S. Murthy, Gautam, and Pura 2015).

To date, various methods have been reported for the extraction of oil resins, including hydrodistillation, organic solvent extraction, microwave-assisted extraction, and supercritical fluid extraction (Melgar-Lalanne et al. 2017). The first two methods continue to be the most widely used in industry, mainly due to their lower implementation and operating costs. Despite this, these techniques are environmentally unfriendly due to the requirement for large solvent volumes. Furthermore, they tend to be low-yielding, and the resulting extracts can be of poor quality due to the degradation of some temperature-sensitive compounds (Kamaruddin et al. 2023).

 DOI: 10.1201/9781003462804-20

To discover the potential applications of such natural products, it is necessary to investigate and understand their characteristics and properties. With this in mind, the current chapter aims to provide an overview of oil resins, including their chemical compositions and biological properties, as reported in the literature to date.

15.2 OIL RESINS: CHARACTERIZATION AND CHEMICAL COMPOSITIONS

Oil resin is a viscous liquid ranging in color from yellow to light brown, and it is collected directly through artificial exudation by making a cut in the tree trunk (de Lima et al. 2022; de Oliveira Neves et al. 2018). Generally, oil resins are composed of various phases. Firstly, the solid phase, also called resin, is non-volatile in nature and consists mainly of compounds such as acidic diterpenes. Secondly, an oily and volatile phase that is composed of lower-molecular-weight substances, such as sesquiterpenes, is also present (Arruda et al. 2019).

Seven genera belonging to the Calophyllaceae family have been recognized and catalogued in Brazil, including the *Calophyllum*, *Caraipa*, *Copaifera*, *Protium*, *Clusiella*, *Boswellia*, and *Haploclathra* genera. The species belonging to this family are known to produce exudates with color variations ranging from colorless to white, yellow, and light brown (Cabral et al. 2021). The following subsections describe the characteristics and chemical compositions of oil resins obtained from species belonging to several of these genera, with emphasis on those obtained from the Calophyllaceae families.

15.3 *CALOPHYLLUM* GENUS

Calophyllum inophyllum L. is a large, broad-leaved evergreen tree. This species is considered pantropical and is found as a native or cultivated species in East Africa, India, Sri Lanka, Malaysia, Southeast Asia, Northern Australia, and the Pacific Islands. For example, in Vietnam, it is habitually found in coastal areas and is popularly known as tamanu (Pribowo et al. 2021; Hapsari et al. 2023). The oils extracted from these species, which are known as tamanu oils (TOs), are commonly employed to treat rheumatism and skin diseases, such as mycoses and dermatoses (Hien, Viet Hung, and Heng 2023). Furthermore, various studies have reported that the oil resin obtained from the seeds of *C. inophyllums* can be used as a biofuel (Karthik et al. 2022; Abdul Sani et al. 2023; Hapsari et al. 2023). Moreover, medicinal *in vitro* studies have demonstrated that the bioactive pyranocoumarins (e.g., calanolide A) extracted from the inophyllum plant act as inhibitors of the human immunodeficiency virus (HIV) reverse transcriptase enzyme (Hien, Viet Hung, and Heng 2023; Spino, Dodier, and Sotheeswaran 1998). It has also been reported that the fruits of this species demonstrate antimicrobial properties against *Staphylococcus aureus* and *Mycobacterium smegmatis* (Yimdjo et al. 2004; Hien, Viet Hung, and Heng 2023). Additionally, Hien, Viet Hung, and Heng (2023) studied the oil resin obtained by cold-pressing the fruit seeds of *C. inophyllum* grown in Vietnam. The resulting oil resin was extracted with methanol/water (9:1, v/v) to produce two fractions, one resinous (14.5%) and the other consisting of a refined oil (36.5%). Carrying out the same extraction using less-polar solvents (i.e., *n*-hexane/ethyl acetate, 96:4) followed by recrystallization yielded two substances (Figure 15.1), which were identified as inocalophylline C and calophyllolide.

Another study analyzed the TO resin obtained by cold-pressing the seeds of a *C. inophyllum* species from Huahine Island, French Polynesia. It was demonstrated that extracting the resin using ethanol produced two fractions, one neutral and the other acidic, which were rich in phenolics, flavonoids, and pyranocoumarins. In particular, the key constituents were identified as inophyllum P, 12-oxocalanolide A, calanolide D, and inocalophyllin B (Figure 15.2) (Cassien et al. 2021).

The *C. brasiliense* species is common in countries ranging from Mexico to Brazil and is usually found in flooded areas. In Brazil, it is popularly known as "guanandi" or "jacareúba," and the oil

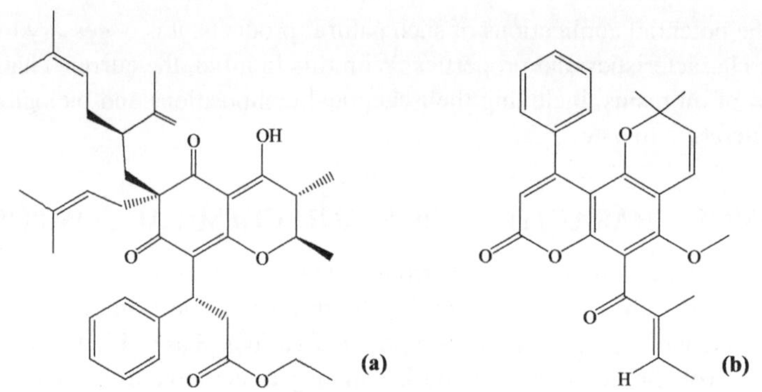

FIGURE 15.1 Chemical structures of (a) inocalophylline C and (b) calophyllolide, which were isolated from the oil resin of *C. inophyllum* fruit seeds collected in Vietnam.

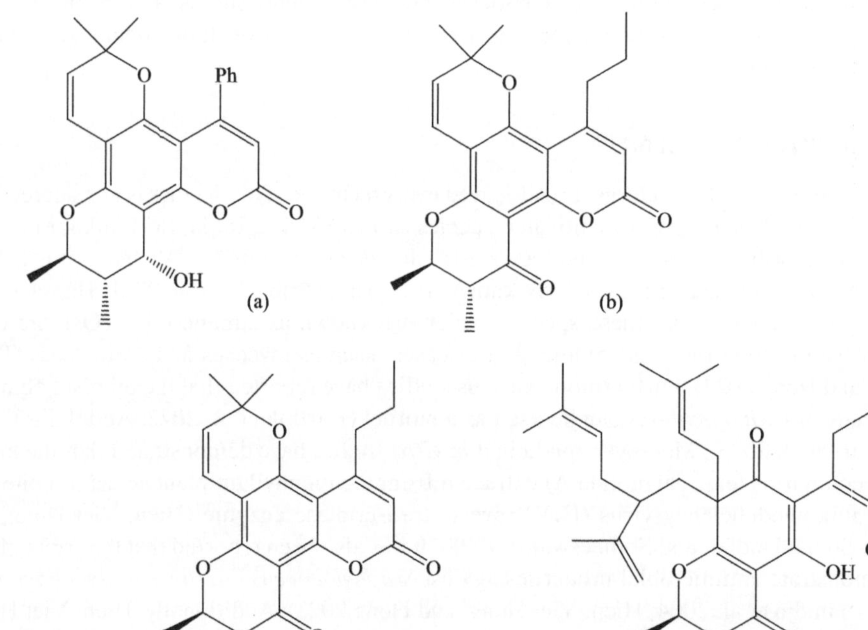

FIGURE 15.2 Chemical structures of (a) inophyllum P, (b) 12-oxocalanolide A, (c) calanolide D, and (d) inocalophyllin B isolated from the ethanol extract of *C. inophyllum* seeds collected in French Polynesia.

resin obtained from the stem is used in folk medicine to treat inflammation, pain, rheumatism, and ulcers (Pessoa et al. 2021). Using a specimen collected in the municipality of Cuiabá, Mato Grosso, Brazil, the stem bark was macerated with hexane in a 1:5 w/v ratio, and the chemical composition of the extract was determined. Consequently, two chromanones were identified, namely, (a) brasiliensic acid and (b) isobrasiliensic acid, as shown in Figure 15.3 (Lemos et al. 2016).

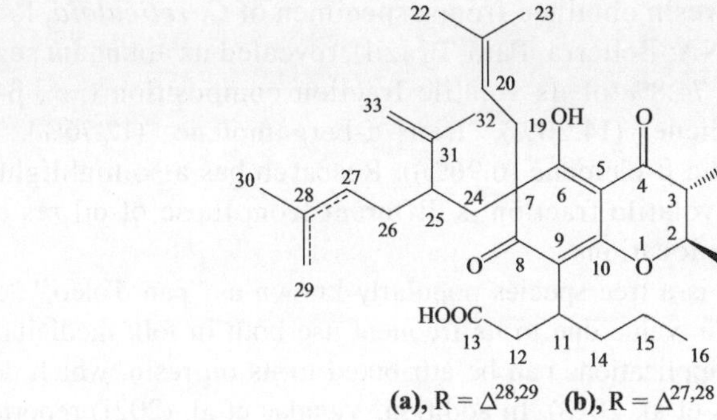

FIGURE 15.3 Chemical structures of brasiliensic acid (a) and isobrasiliensic acid (b).

15.4 *COPAIFERA* GENUS

The *Copaifera* genus, which belongs to the Leguminosae family, is composed of large trees found in the Brazilian region, including in the Amazon, Cerrado, and Atlantic rainforests (Lopes de Mesquita Vieira et al. 2024). Species of this genus are also found on the African continent, mainly in countries such as the Democratic Republic of the Congo, Cameroon, Guinea, and Angola (Arruda et al. 2019). More than 60 species of this genus have been cataloged worldwide, out of which 16 are found in Brazil, and 9 are only found in the Brazilian Amazon (Gurgel et al. 2019). Some of the most commonly occurring species in Brazil are *Copaifera officinalis* L., *C. reticulata* Ducke, *C. multijuga* Hayne, *C. confertiflora* Bth, *C. langsdorffii* Desf., *C. coriaceae* Mart., and *C. cearenses* Huber ex Ducke (Singh et al. 2023).

The oil resin produced by the *Copaifera* genus is an important natural product that highlights the biodiversity of the Amazon. The copaiba oil resin can be extracted from trees of this genus and is one of the oldest and most important natural medicines used by the people of the Amazon region (Teixeira et al. 2017). For example, it has been used to treat various illnesses and diseases, including respiratory, urinary, and skin conditions. Other therapeutic benefits associated with copaiba oil include its neuroprotective, anti-inflammatory, antinociceptive, antimicrobial, and healing effects (Pulcinelli et al. 2022; Arruda et al. 2019; Teixeira et al. 2017). The oil resin from species of this genus is rich in sesquiterpene hydrocarbons, which account for ~90% of its composition; acid diterpenes are also found in smaller amounts (Pasquel-Reátegui et al. 2022).

The *Copaifera multijuga* Hayne species is an Amazonian tree popularly known as "mari mari" or "copaíba roxa" (Oliveira et al. 2020). Silva et al. (2017) investigated the compositions of oil resins obtained from 11 specimens of *C. multijuga* found in distinct locations in the municipality of Oriximiná (Pará, Brazil). These oil resins were extracted by hydrodistillation, and chemical characterization of the volatile fraction indicated the presence of abundant hydrocarbon sesquiterpenes (93.48–99.67%). All 11 samples contained β-caryophyllene (36.20–85.4%) as the major constituent, followed by α-humulene (4.15–11.26%), δ-elemene (0.08–8.52%), β-elemene (0.68–2.86%), and ciperene (0.22–1.11%) (Silva et al. 2017). Viscosity analyses of the oil resins obtained from the 11 specimens of *C. multijuga* provided a range of 9.46–92.62 mPa.s. Upon correlating the observed viscosities with the chemical constituents, it was found that a higher content of β-caryophyllene was associated with a lower viscosity (Silva et al. 2017).

Another species of this genus found in the Amazon region is *Copaifera reticulata* Ducke, which is the most common and widely used species in this region. The oil resin obtained from *C. reticulata* is largely utilized in traditional medicine due to its anti-inflammatory properties (de Almeida Júnior et al. 2021; Teixeira et al. 2017). Characterization of the volatile constituents of the oil resin obtained from a specimen of *C. reticulata*, located in the Tapajós National Forest (FLONA; Belterra, Pará, Brazil), revealed its abundant sesquiterpene content, which accounted for 74.8% of its volatile fraction composition (i.e., β-bisabolene (25.15%), *cis*-eudesma-6,11-diene (14.20%), trans-α-bergamotene (12.76%), β-selinene (8.70%), α-selinene (7.03%), and β-elemene (6.96%)). Research has also highlighted that the chemical composition of this volatile fraction is different from those of oil resins obtained from the same species in other locations.

Copaifera langsdorffii Desf. is a tree species popularly known as "pau d'óleo," "cupiúva," or "cupaúba." It has great economic value due to its frequent use both in folk medicine and in the pharmaceutical industry. Such applications can be attributed to its oil resin, which demonstrates phytotherapeutic effects (Santos et al. 2023). In addition, Valadas et al. (2021) reported the antibacterial activity of the *C. langsdorffii* oil resin against Gram-positive and Gram-negative bacteria. Furthermore, the lacquer produced from the oil resin was found to demonstrate antimicrobial activity against *Streptococcus mutans*. Compositional analysis of the volatile constituents present in the oil resin showed that the main constituents are caryophylene oxide (54.2%), β-caryophylene (6.08%), β-element (4.43%), α-*cis*-bergamotene (4.56%), and ar-curcumene (4.63%), accounting for 84.69% of the total bioactive content.

The *Copaifera officinalis* L. species also stands out for its oil resin production. Following the separation of the *C. officinalis* oil resin components by means of supercritical adsorption with CO_2 and using diverse adsorbents (e.g., zeolites, aluminum oxide, and silicon oxide), the volatile compounds were characterized. Notably, the volatile fraction represented 78.57% of the analyzed oil resin and is composed mainly of sesquiterpenes and a smaller fraction (21.43%) of diterpenic acids. The main volatile constituents identified were β-caryophyllene (39.61%), α-bergamotene (14.07%), germacrene D (7.75%), α-humulene (4.35%), cycloisosativene (2.85%), γ-muurolene (2.57%), β-bisabolene (2.04%), δ-cadinene (2.10%), and β-elemene (1.66%). The most abundant diterpenic acids identified were copalic acid (7.82%), polyalthic acid (3.51%), agathic acid (2.00%), and hardwickii acid (1.49%) (Pasquel-Reátegui et al. 2022).

15.5 *CARAPA* GENUS

One of the most relevant species of this genus is the *Carapa guianensis* Aubl. (Meliaceae) species, which is found in various regions, ranging from Central America to the north of South America, such as Venezuela, Ecuador, Colombia, Brazil, and Peru. The *C. guianensis* tree is extremely large, reaching heights of up to 30 m. This species is popularly known as "andiroba," and its fruits produce an oil with medicinal (e.g., anti-inflammatory) properties. Consequently, it is commonly used to treat various ailments and in the production of cosmetics (Nascimento et al. 2019). The annual consumption of andiroba oil in Brazil is approximately 30,000 L, and its exports reach 450,000 L per year (Ibiapina et al. 2022).

Nascimento et al. (2019) analyzed the lipid profiles of the oils produced from the fruits of both *C. guianensis* and *C. vasquezii*, wherein the seeds were collected from ten species within the Adolpho Ducke Forest Reserve (Manaus, Amazonas). The seeds were processed by means of a Soxhlet extraction using *n*-hexane. The main substances identified in the vegetable oil of *C. guianensis* were methyl oleate (42.3%), methyl palmitate (26.0%), and oleic acid (12.5%). The vegetable oil obtained from the seeds of *C. vasquezii* possessed a similar composition, with methyl oleate, methyl palmitate, and oleic acid contents of 38.6, 20.5, and 14.5%, respectively. It was also revealed that the oil from *C. vasquezii* exhibited a higher inhibition potential against *A. flavus*, *A. niger*, and *F. oxysporum* when compared to *C. guianensis*.

According to Kikuchi et al. (2020), the oil obtained from the seeds of *C. guianensis* is widely used in popular and Indigenous medicine due to its insecticidal, antimalarial, antiparasitic, anti-inflammatory, antipyretic, and leprostatic properties. Research carried out on the oil obtained from the fruit of a species of *C. guianensis* from the Amazon rainforest

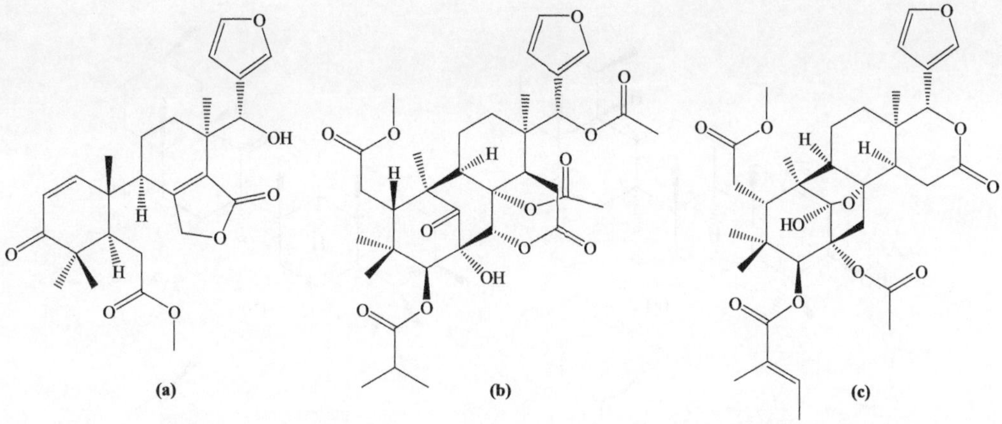

FIGURE 15.4 Chemical structures of limonoid compounds (a) Carapanins A, (b) Carapanins B (**10**), and (c) Carapanins C obtained from *C. guianensis*.

identified three compounds belonging to the limonoid class, namely, Carapanins A, B, and C (Figure 15.4). This research also demonstrated that the compounds Carapanins B and C are strong nitric oxide inhibitors.

15.6 *PROTIUM* GENUS

Protium is known as the main genus of the *Bursaceae* family and is commonly found in South American countries. Approximately 135 species of *Protium* are known to grow in the Amazon region, especially in Brazil, where they are known by the terminology "breu." The gums and oil resins produced by the species of this genus are commonly used in folk medicine as a tonic and stimulant, although they are also employed to treat ulcers and inflammation (de Almeida et al. 2015; Asteggiano et al. 2021).

In one study, de Almeida et al. (2015) investigated the chemical composition of the oil resin obtained from *Protium paniculatum* var. *modestum* grown in the Adolpho Ducke Reserve (Manaus, Amazonas, Brazil), and evaluated the anti-inflammatory activities of its chemical constituents. Upon washing with hexane, the lighter and more non-polar oily fraction, consisting of monoterpenes and sesquiterpenes, was separated from the fraction composed mainly of monohydroxylated (amyrins), dehydroxylated, and cabonilic triterpenes. This second triterpene fraction was analyzed, and the compounds (a) α-amyrin, (b) β-amyrin, (c) brein, and (d) maniladiol were identified. The chemical structures of these constituents can be seen in Figure 15.5.

Protium heptaphyllum (Aubl.) Marchand is an endemic species found in tropical regions. The oil resin produced by this species is popularly known by different names, including Breu, Breu Branco, Brazilian Olíbano, Almecéga Resin, Amazonic Breu Branco (Asteggiano et al. 2021; Faustino et al. 2020). A recent study carried out by Asteggiano et al. (2021) investigated the chemical composition of the volatile fraction and of the extracts rich in acids and neutral triterpenes, which were obtained from the oil resin of *P. heptaphyllum*. Their results indicated that the less-polar volatile fraction was composed predominantly of sesquiterpenes, such as α-pinene (12.64%), *p*-cymene (11.11%), δ-3-carene (7.04%), β-phellandrene (6.73%), α-phellandrene (4.21%), β-pinene (1.33%), and *trans*-verbenol (1.29%). In contrast, the extracts were found to be rich in neutral triterpenes, including α-amyrin and β-amyrin (Figure 15.5a and b), as well as smaller proportions of α-amyron, β-amyron (Figure 15.6a and b), brein, and maniladiol (Figure 15.5c and d). This study demonstrated that neutral triterpenes (mainly α-amyrin e β-amyrin) account for 48% (w/w) of the analyzed oil resin.

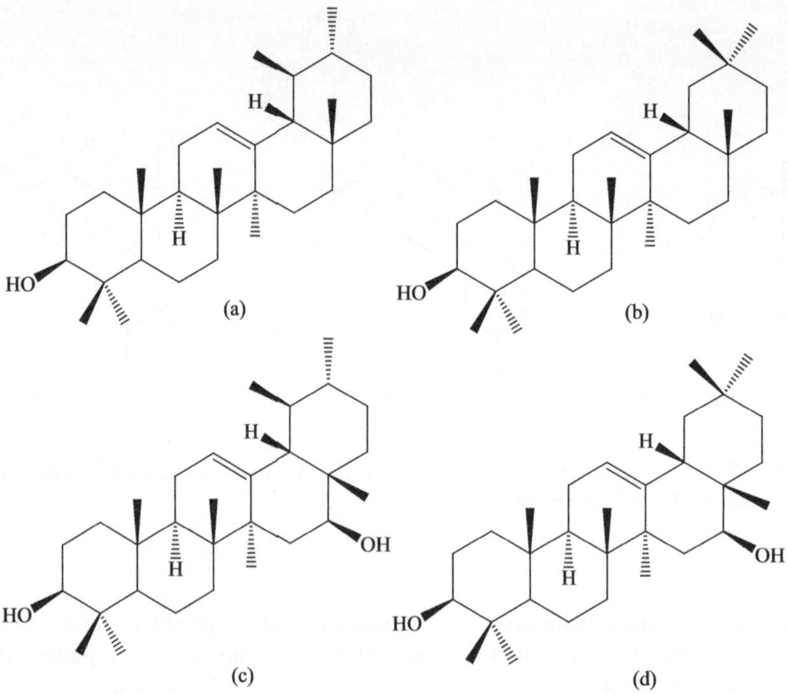

FIGURE 15.5 Chemical structures of (a) α-amyrin, (b) β-amyrin, (c) brein, and (d) maniladiol isolated from the oil resin of *P. paniculatum*.

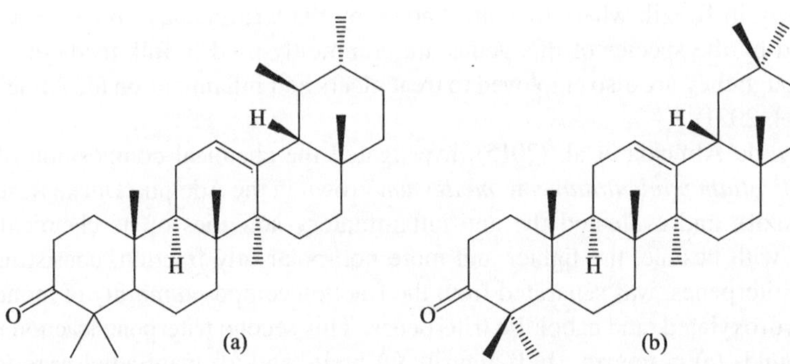

FIGURE 15.6 Chemical structures of the (a) α-amyrone and (b) β-amyrone neutral triterpenes isolated from the oil resin of *P. heptaphyllum*.

Following extraction of the essential oil from the oil resin of *P. heptaphyllum*, Faustino et al. (2020) evaluated the biocidal potential of a nanoemulsion of this essential oil against the various development stages of the *Aedes aegypti* mosquito. The oil resin was produced from a species grown in the municipality of Porto (Amapá, Brazil), and the resinous essential oil was extracted using a Clevenger-type apparatus. The essential oil was transparent and possessed a strong aroma. Its composition was predominantly rich in hydrocarbon sesquiterpenes, among which *p*-cimene (27.70%) and α-pinene (22.31%) were the main components. In addition, the prepared nanoemulsion exhibited a pupicidal, adulticidal, and repellent activity against the *Aedes aegypti* mosquito. These results are of particular interest since this mosquito is the main vector for the transmission of the dengue, Zika, and chikungunya viruses, which are considered endemic in tropical and subtropical countries (e.g., Central America, Venezuela, Colombia, and Brazil) (Zardini et al. 2024).

15.7 *BOSWELLIA* GENUS

Frankincense is a terpenoid oil–gum resin obtained from trees of the genus *Boswellia* (Burseraceae). This genus comprises around 20 species distributed over various countries, including India, Ethiopia, Somalia, and the Arabian Peninsula. Its production is associated with defense mechanisms against insects, herbivores, infections, and other damage to the plants of this genus (DeCarlo et al. 2018; Johnson et al. 2019). The oil–gum resin can be obtained from the *B. carteri* Birdw. (sin. *B. sacra*), *B. serrata* Roxb., *B. papyrifera* (Delile ex Caill.), Hochst. E *B. frereana* Birdw. species (Massei et al. 2023).

More specifically, *Boswellia frereana* Birdw. is a native species of Somalia and Somaliland and is widely known for the production of large quantities of oil–gum resin, which has recently become popular for its yellow-colored essential oil. In this context, Johnson et al. (2021) obtained essential oils from *B. frereana* via the hydrodistillation of oil–gum resins collected from 12 species in the Somaliland region. Subsequently, they identified the chemotypes of *B. frereana* based on characterization of these essential oils and identified the main constituents of the samples as α-thujene (14.5–43.9%), α-pinene (3.0–63.0%), sabinene (3.3–8.0%), β-pinene (0.3–4.1%), and *p*-cymene (1.7–13.0%). One of the investigated samples also contained significant levels of α-phellandrene (6.0%) and its dimer (10.0%) (Johnson et al. 2021). In contrast, the resinous fractions extracted using dichloromethane contained high levels of lupeol (14.7–32.5%; Figure 15.7a), α-amyrin (13.0–5.2%; Figure 15.5a), 3-epi-lupeol (6.4–4.2%; Figure 15.7b), and β-amyrin (5.3–8.0%, Figure 15.5b) as the major constituents.

The *Boswellia serrata* Roxb. species is found in the dry areas of Pakistan, Yemen, Oman, India, Nigeria, and Saudi Arabia. This species is of great scientific interest and commercial importance as it produces an oil–gum resin with already-identified pharmacological properties (Ayub, Hanif, et al. 2023; Gupta, Singh, et al. 2022). Ayub, Hanif, et al. (2023) investigated the variation in the chemical composition of the oil–gum resin collected from a specimen growing in the Balochitan District of Pakistan. In addition, following extraction of essential oil from

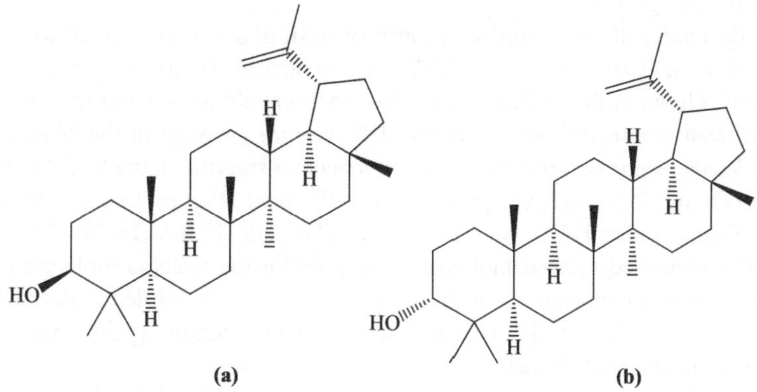

FIGURE 15.7 Chemical structures of the (a) lupeol and (b) 3-epi-lupeol triterpenes isolated from the oil resin of *B. frereana*.

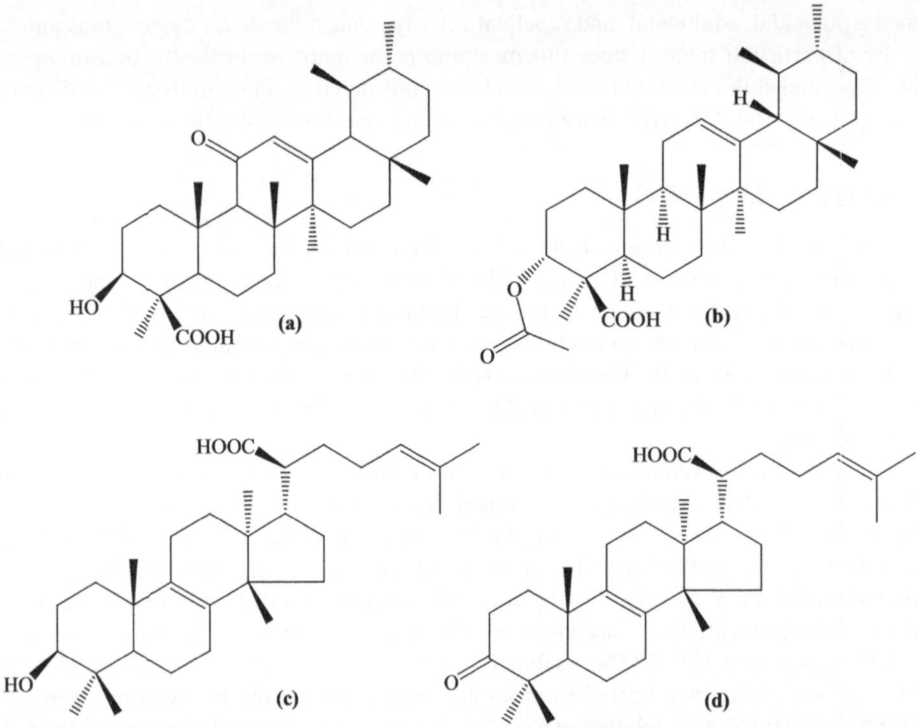

FIGURE 15.8 Chemical structure of (a) 3-hydroxy-11-keto-β-boswellic acid, (b) acetyl-β-boswellic acid, (c) 3-hydroxytirucallic acid, and (d) 3-ketotirucallic acid, isolated from the oil–gum resin of *B. serrata*.

the resin using superheated steam at various temperatures, they evaluated the antimicrobial activity of the oil. It was found that the essential oil yield varied according to the extraction temperature, with an optimal yield being obtained at 210°C. A total of 21 compounds were identified, including α-pinene (89.07–77.95%), α-thujene (4.49–3.72%), *trans*-verbenol (2.48–0.68%), β-thujone (2.21–0.86%), *p*-cymene (2.11–0.93%), *m*-cymene (1.89–0.89%), and sabinene (1.66–0.93%). This study also highlighted the antifungal and antibacterial activities of the essential oil obtained from the oil–gum resin.

Another study analyzed the chemical profile of a similar oil–gum resin to detect the presence of pentacyclic acid triterpenoids, which are considered chemical markers for the species. More specifically, Gupta et al. (2022) studied the fractions obtained from the methanolic extract of the oil–gum resin obtained from a species of *B. serrata* growing in the Mandsaur District of India. Four molecules were isolated from the ethyl acetate fraction, namely, 3-hydroxy-11-keto-β-boswellic acid (Figure 15.8a; KBA), acetyl-β-boswellic acid (Figure 15.8b; ABBA), 3-hydroxytirucallic acid (Figure 15.8c; HTA), and 3-ketotirucallic acid (Figure 15.8d; HTA). Interestingly, this research also identified various biological activities for the isolated molecules. More specifically, KBA significantly decreased the proliferation of cancer cells, while the isolated fractions, the methanolic extract, and KBA itself exhibited antibacterial potential against methicillin-resistant *S. aureus* (Gupta, Verma, et al. 2022).

As a summary, Table 15.1 shows the chemical compositions of the fractions and oil resins obtained from the fruits, seeds, and stems of the different plant species described in the preceding text.

TABLE 15.1
Major Components of the Oil Resins (Volatile and Fixed Constituents) Obtained from the Aforementioned Plant Species

Species	Collection Location	Plant Section	Identified Compounds	Reference
***Calophyllum inophyllum* L.**	Ben Tre Province, Vietnam	Fruit oil resin	Calophyllolide and inocalophylline C	Hien, Viet Hung, and Heng (2023)
***Calophyllum inophyllum* L.**	Huahine Island, French Polynesia	Mature seeds	Inophyllum P, 12-oxocalanolide A, calanolide D, and inocalophyllin B	Cassien et al. (2021)
***Calophyllum brasiliense* Cambess**	Cuiabá, Mato Grosso, Brazil	*n*-hexane extract from the stem bark	Brasiliensic acid, isobrasiliensic acid	Lemos et al. (2016)
***Copaifera multijuga* Hayne**	Oriximiná, Pará, Brazil	Volatile fraction obtained from the oil resin (11 samples)	β-caryophyllene (36.20–85.4%), α-humulene (4.15–11.26%), δ-elemene (0.08–8.52%), β-elemene (0.68–2.86%), and ciperene (0.22–1.11%)	M.T. Silva et al. (2017)
***Copaifera reticulata* Ducke**	Belterra, Pará, Brazil	Volatile fraction obtained from the oil resin of the trunk exudate	β-bisabolene (25.15%), cis-eudesma-6,11-diene (14.20%), trans-α-bergamotene (12.76%), β-selinene (8.70%), α-selinene (7.03%), and β-elemene (6.96%)	Couto et al. (2020)Guimarães-Santos et al. (2012)
***Copaifera langsdorffii* Desf.**	Juruena Valley, Mato Grosso, Brazil	Volatile fraction obtained from the oil resin of the trunk exudate	Caryophyllene oxide (54.2%), β-caryophyllene (6.08%), β-element (4.43%), α-cis-bergamotene (4.56%), and ar-curcumene (4.63%)	Valadas et al. (2021)
***Copaifera officinalis* L.**	Industrial (Ferquima Industry and Commerce LTDA)	Volatile fraction obtained from the oil resin	Sesquiterpenes: β-caryophyllene (39.61%), α-bergamotene (14.07%), germacrene D (7.75%), α-humulene (4.35%), cycloisosativene (2.85%), γ-muurolene (2.57%), β-bisabolene (2.04%), δ-cadinene (2.10%), and β-elemene (1.66%)	Pasquel-Reátegui et al. (2022)
		Acid fraction obtained from the oil resin via adsorption separation	Diterpenes acids: copalic acid (7.82%), polyalthic acid (3.51%), agathic acid (2.00%), and Hardwickii acid (1.49%)	
***C. guianensis* Aubl.**	Manaus, Amazônia, Brazil	Vegetable oils extracted from the seeds using hexane	Methyl oleate (42.3%), methyl palmitate (26.0%), and oleic acid (12.5%)	Nascimento et al. (2019)
***C. guianensis* Aubl.**	Amazon Forest, Brazil	Fruit oils	Limonoids: carapanin A, carapanin B, and carapanin C	Kikuchi et al. (2020)
***C. vasquezii* Kenfack**	Manaus, Amazônia, Brazil	Vegetable oils extracted from the seeds using hexane	Methyl oleate (38.6%), methyl palmitate (20.5%), and oleic acid (14.5%)	Nascimento et al. (2019)

(*Continued*)

TABLE 15.1 *(Continued)*
Major Components of the Oil Resins (Volatile and Fixed Constituents) Obtained from the Aforementioned Plant Species

Species	Collection Location	Plant Section	Identified Compounds	Reference
Protium paniculatum* var. *modestum	Manaus, Amazônia, Brazil	Insoluble material obtained from hexane extraction of the oil resin	α-amyrin, β-amyrin, brein, and maniladiol	de Almeida et al. (2015)
***Protium heptaphyllum* (Aubl.)**	Industrial (Abel Nutraceuticals, Turin, Italy)	Volatile fraction obtained from the oil resin	α-pinene (12.64%), ***p***-cymene (11.11%), δ-3-carene (7.04%), β-phellandrene (6.73%), α-phellandrene (4.21%), β-pinene (1.33%), and ***trans***-verbenol (1.29%)	Asteggiano et al. (2021)
		Extract containing neutral triterpenes	α-amyrin, β-amyrin, α-amyron, β-amyron, maniladiol, and brein	
***Protium heptaphyllum* (Aubl.)**	Porto, Amapá, Brazil	Resinous essential oil obtained from the oil resin	***p***-cimene (27.70%) and α-pinene (22.31%)	Faustino et al. (2020)
***Boswellia frereana* Birdw.**	Somaliland, Africa	Essential oil obtained from the oil–gum resin	α-thujene (14.5–43.9%), α-pinene (3.0–63.0%), sabinene (3.3–8.0%), β-pinene (0.3–4.1%), and ***p***-cymene (1.7–13.0%)	Johnson et al. (2021)
		Resinous fraction extracted from the oil–gum resin using dichloromethane	Lupeol (14.7–32.5%), α-amyrin (13.0–25.2%), 3-epi-lupeol (6.4–14.2%), and β-amyrin (5.3–8.0%)	
***Boswellia serrata* Roxb.**	Balochitan, Pakistan	Essential oil obtained from the oil–gum resin	α-pinene (89.07–77.95%), α-thujene (4.49–3.72%), trans-verbenol (2.48–0.68%), β-thujone (2.21–0.86%), ***p***-cymene (2.11–0.93%), ***m***-cymene (1.89–0.89%), and sabinene (1.66–0.93%)	Ayub, Hanif, et al. (2023)
***Boswellia serrata* Roxb.**	Mandsaur, India	Ethyl acetate fraction obtained from the methanol extract	3-hydroxy-11-keto-β-boswellic acid, acetyl-β-boswellic acid, 3-hydroxytirucallic acid, and 3-ketotirucallic acid	Gupta, Verma, et al. (2022)

Note: Only the percentages of the constituents derived from essential oils are shown in parentheses.

15.8 USE OF THE OIL RESINS IN FOOD APPLICATIONS AND TECHNOLOGIES

15.8.1 *Rosmarinus officinalis* (Rosemary)

Rosemary (*Rosmarinus officinalis*) oil resin has a range of applications in the food industry, offering benefits in terms of both food quality and safety (Cheng et al. 2022; Teoh, Ting, and Thoo 2023; Upadhyay, Sehwag, and Niwas Mishra 2017). Due to its antioxidant properties, rosemary oil has been widely studied as an ingredient for a variety of food products, especially in the context of a natural preservative (Che Man et al. 2003; Murphy et al. 1998; Rohlik, Pipek, and Panek 2010; Upadhyay and Mishra 2015b; Vergara, Cézar, and Rubio 2021). Its properties can be mainly attributed to the carnosic acid component present in the oil, which is able to neutralize free radicals and inhibit both lipid and protein oxidation. Notably, these are the main causes of food spoilage, which result in undesirable sensory changes (e.g., a rancid odor or a foul taste) and reduce the shelf lives of the corresponding products (Vergara, Cézar, and Rubio 2021; Visentín, Cismondi, and Maestri 2011).

Due to its antioxidant potential, rosemary oil resin has been widely employed for decades in products of animal origin, such as beef steaks, pork steaks, turkey sausages, and roast beef, with the aim of reducing oxidation and extending shelf life (Barbut, Josephson, and Maurer 1985; Liu East Lansing Mi et al. 1992; Stoick East Lansing Mi et al. 1991). Importantly, the application of oil resin to these foods has had a positive effect on their preservation without affecting their sensory characteristics, such as color, taste, odor, texture, and general acceptability. A study by Kalaikannan et al. (2022) found that the addition of 0.1% oil to chicken burgers extended the shelf life by up to one day at room temperature without affecting the sensory properties of the product. Thus, the addition of this oil could reduce refrigeration costs. Similarly, the use of rosemary oil extended the shelf life of lamb burgers from 10 to 14 days when added at a concentration of 0.08% (Vergara, Cézar, and Rubio 2021).

Besides conservation, rosemary oil can also enhance the color stability of animal products, as observed in sausages, wherein its addition reduced oxidation of the heme pigment and maintained the redness of products in contact with air and light. Importantly, this maintained the characteristic and attractive color of the sausages for a longer period of time (Rohlik, Pipek, and Panek 2010).

Other studies have also highlighted the potential of applying rosemary oil resin to preserve aquatic products. For example, Chakraborty, Chakkalakal, and Joseph (2016) demonstrated that the addition of oil resin to a freeze-dried concentrate of green mussels (*Perna viridis*) protected the unsaturated fatty acids from oxidation at an optimal concentration of 0.4 g/100 g, thereby extending the shelf life of the product and maintaining the nutritional composition at the desired levels. In addition, technologies such as the application of nano-liposomal oil coatings have been developed to exploit its antioxidant activity in the preservation of dried oysters, wherein encapsulation was observed to improve thermal stability compared to that of the isolated oil. As a result, its antioxidant action was potentiated, thereby prolonging the product shelf life (Cheng et al. 2022).

Rosemary oil resin has also been reported to enhance the oxidative stabilities of vegetable oils used in frying. For example, the addition of this oil resin to palm oil improves the stability of the oil when frying banana chips for five days, by reducing various oxidative parameters (e.g., the peroxide and anisidine contents), without generating significant differences in the product crispiness, color, taste, odor, texture, or general acceptability (Che Man et al. 2003). Similarly, this oil increased the shelf life of sunflower oil to 194 days by reducing various oxidation-related physicochemical parameters, even when added in combination with ascorbyl palmitate (Upadhyay and Mishra 2015a, 2015b; Upadhyay, Sehwag, and Niwas Mishra 2017). Another study, carried out by Visentín, Cismondi, and Maestri (2011), revealed the antioxidant action of rosemary oil when added to soybean oil. Notably, this property, which was enhanced following supercritical fluid extraction, wherein the main antioxidant compound (i.e., carnosic acid) was enriched, leading to a higher activity than that of di-*tert*-butyl methyl phenol (BHT), one of the most widely used antioxidants in the food industry.

The antioxidant properties of rosemary resin oil also render it applicable in special packaging technologies, with the aim of extending the shelf lives of various products. For example, Teoh, Ting, and Thoo (2023) evaluated an active film based on zein/gellan gum containing 20% rosemary oil resin and found that this film significantly reduced oxygen transfer compared to the control. In addition, the hydrophobic nature of the oil led to an enhanced water resistance for the film, and it was also found to exhibit an excellent antioxidant activity, demonstrating its potential for use in food packaging. With the same purpose, Song et al. (2020) evaluated active films based on 8% rosemary oil resin and concluded that these films are capable of eliminating free radicals and preventing lipid/protein oxidation in minced pork. Their results, therefore, corroborated other studies regarding the potential application of this oil resin in the production of antioxidant packaging.

Overall, rosemary oil resin has emerged as a promising agent in the food industry, standing out for its antioxidant properties and its ability to extend the shelf lives of various foods without interfering with the sensory characteristics of the products. Furthermore, its use in active packaging reveals a new application in product preservation. Thus, rosemary oil resin is beneficial not only in terms of its ability to improve food quality and safety but also as an alternative for food preservation, therefore contributing to a more sustainable society.

15.8.2 *Zingiber officinale* (Ginger)

Ginger is a perennial herb belonging to the Zingiberaceae family and measuring approximately 30–100 cm in height. This herb has been widely used as a spice, flavoring agent, and nutritional supplement for over 3,000 years in countries such as China, Germany, India, Greece, the United States, and Indonesia. Notably, ginger contains a variety of nutritional components, such as carbohydrates, proteins, amino acids, lipids, fatty acids, and minerals, which favor its use and consumption as a nutritional additive in food products (Yang et al. 2024).

Additionally, this herb produces an oil resin that exhibits a strong, acidic, and pungent flavor (Beristain et al. 2019). This oil resin is isolated as a brown-colored liquid which contains ~15–35% of the essential oil, along with other components that provide ginger with its characteristic flavor (i.e., gingerol, shogaol, and zingerone). This liquid is utilized in food processing systems and is obtained from the extraction of fresh, dried, or powdered rhizomes (Yulianto et al. 2022).

In terms of its more specific applications, one study has demonstrated the possible application of ginger oil resin as a bio-conservative in sugarcane juice (*Saccharum officinarum*) (Murthy, Gautam, and Naik Pura 2015). In this study, the oil resin was dissolved in ethanol, and a measured concentration of the resin (20–200 μL/mL) was added to 10 mL of fresh sugarcane juice, taking into account taste and general acceptability; a concentration of 100 μL/10 mL was found to be ideal. In addition, the antimicrobial profile of the oil resin was studied due to its ability to inhibit fermentation caused by bacteria present in the sugarcane juice. According to the authors, the addition of ginger oil resin achieved microbial decontamination and extended the shelf life of the fresh sugarcane juice.

In another study, Bellik (2014) evaluated the antioxidant and antimicrobial potentials of the oil resin of *Zingiber officinale* Roscoe to determine its future perspectives for application in food products. Consequently, the oil resin was found to exhibit a higher antioxidant activity (IC_{50} = 1.820 ± 0.034 mg/mL) than the essential oil (IC_{50} = 110.14 ± 8.44 mg/mL). Moreover, in terms of its antimicrobial activity, the oil resin was found to be most effective against *Escherichia coli, Bacillus subtilis*, and *S. aureus*, but less effective against *Bacillus cereus, Aspergillus niger*, and *Penicillium* spp.

One of the techniques associated with the application of ginger oil resin in foods is known as encapsulation, which preserves the biological functions of the bioactive compounds present in the samples (Majeed et al. 2015). This was also evidenced by Krisanti, Astuty, and Mulia (2017), who encapsulated the oil resin extracted from red ginger (*Z. officinale*) into chitosan–alginate microparticles and evaluated its impact on the preservation of milk freshness. This evaluation was based mainly on organoleptic evaluations and on the total plate count to determine the inhibition of bacterial growth. They reported that the chitosan–alginate microparticles loaded with the oil resin were effective preservatives for fresh milk.

Importantly, the antimicrobial potential of ginger oil resin depends on a number of factors, including its chemical composition, the type of solvent used for extraction, the extraction method, and the ginger processing procedure (e.g., the drying time). It has also been reported that the phenolic compounds present in the oil resin (e.g., eugenol, shogaols, zingerone, gingerdiols, and gingerols) exhibit synergistic relationships with other resin components (e.g., β-sesquiphellandrene, *cis*-caryophillene, zingiberene, α-farnesene, α-bisabolene, and β-bisabolene), and these relationships are mainly responsible for the observed biological activity (Beristain et al. 2019).

In terms of the different techniques used to obtain the oil resin, Subtain et al. (2024) evaluated five different extraction methods and compared the yields of the powdered ginger oil resins obtained using these approaches (i.e., organic solvent extraction, steam distillation, supercritical fluid extraction (CO_2, SCFE), extraction through three-phase partitioning, and extraction after cryogenic grinding). Their results showed that the oil resin obtained by SCFE exhibited the highest total phenolic content (TPC), total flavonoid content (TFC), total volatiles, and antioxidant activity. Although the oil resin extracted using SCFE had a yield of 3.98 ± 0.05%, it possessed a higher purity and good phytochemical properties. Ultimately, the yield and quality could be further improved using alternative approaches based on future developments.

15.8.3 *Capsicum* Genus (Paprika)

The *Capsicum* genus belongs to the Solanaceae family, which comprises >30 species and grows mainly in tropical and subtropical regions. The species of this genus offer many benefits to human health due to the presence of numerous key compounds in its fruits (i.e., capsaicinoides, carotenoids, phenolic compounds, flavonoids, vitamins, minerals, essential oils, and aromatic substances) (Taiti et al. 2024). Among these compounds, which are classed as secondary metabolites, the most important are the carotenoids and capsaicinoides, which are present in the oil resin and demonstrate various biological activities.

To date, a number of studies have analyzed the antimicrobial properties of the *Capsicum* oil resin. For example, Eid et al. (2024) investigated the potential of the *Capsicum annuum* oil resin to act as an antifungal and antibacterial compound against organisms such as *Candida albicans*, *S. aureus*, *E. coli*, and *Klebsiella pneumoniae*. They reported high inhibition potentials against methicillin-resistant *S. aureus* (MRSA) (inhibition zone: 29 ± 1.9 mm), *E. coli* (33 ± 0.9 mm), *K. pneumonia* (30 ± 1.4 mm), and *C. albicans* (21 ± 1.5 mm) compared to those obtained for the reference antibiotic ampicillin (18 ± 1.4 mm against *K. pneumonia*) and the antifungal agent fluconazole (12 ± 0.1 mm against *C. albicans*).

In another study, Yilmaz et al. (2023) revealed the dietetic effects of the *C. annuum* oil (CAO) resin in the growth, body composition, sensory characteristics, intestinal microbiome, and intestinal histomorphometry of rainbow trout. During this study, the fish were fed with various levels of CAO for 30 days (i.e., 0–28 g kg^{-1}), and it was found that the cooked trout fillets exhibited a strong flavor and taste originating from the CAO. This was most evident in the group treated with 14 g kg^{-1} CAO. Ultimately, this study revealed for the first time that dietetic CAO (7–14 g kg^{-1}) can favor the growth of rainbow trout with no negative effects on the sensorial characteristics or intestinal health.

The oil resin of *C. annuum* L. is considered a highly valuable natural product, offering many application opportunities as a spice, medicine, and source of antioxidants. Research has shown that capsaicin, the compound responsible for the flavor of the *C. annuum* L. oil resin, has significant application potential in appetite control, pain relief, and even cancer prevention. However, data on capsaicin and the other important biochemical components of this oil resin are often lacking. Thus, breeding efforts are required to improve crop quality to enhance its health and economic benefits (Hulagannavar et al. 2024).

Moreover, *capsicum* (CEC) oil resin has been added to lamb meat to investigate its effects on the growth performance, meat quality, muscle fatty acid and amino acid compositions, fiber characteristics, and antioxidant capacity (Wang et al. 2023). It was revealed that lambs fed with the CEC diet exhibited a higher average daily gain ($P < 0.05$) compared to the control group. In addition, the 24-hour pH levels of the lambs fed with the CEC diet were higher than that of the control. The authors therefore considered that CEC is an effective feed additive with great prospects for improving the meat quality, increasing the antioxidant capacity, and regulating the muscle fiber composition.

15.9 BIOLOGICAL ACTIVITY

The use of oil resins extracted from the *Copaifera* genus and from the nuts or seeds of the *Carapa guaianensis* species has been reported since the beginning of colonization by Native tribes. The biological properties of these oils were already known to this generation, and several studies have confirmed and demonstrated that these oil resins exhibit anti-inflammatory, analgesic, healing, antitumor, antimicrobial, antiallergic, insecticidal, and larvicidal properties. Consequently, these plant products have attracted great interest from the pharmaceutical and cosmetic industries (de Almeida Vaucher et al. 2015; Lopes de Mesquita Vieira et al. 2024; Silva et al. 2021).

The majority of oil resins are composed of diterpenes, triterpenes, and sesquiterpenes, whose concentrations vary considerably, depending on the species. In addition, depending on the composition and concentration, the oil resins can impart varying bioactivities in different applications (Dias et al. 2023; Diefenbach et al. 2018). For example, *Carapa guianensis* is known to demonstrate leishmanicidal activity against the promastigotes and amastigotes of *Leishmania amazonensis*. This activity has been attributed to its high limonoid content, including andirobin and 17-glycolyldesoxygedunin. The limonoids are a structurally diverse series of tetranortriterpenoids that are found mainly in the *Meliaceae* family (Oliveira et al. 2018). Carapanosin-type limonoids have also been isolated from the seeds of *Carapa guianensis* (Higuchi et al. 2017), and their inhibitory effects against nitric oxide production have been evaluated. Overall, it was found that these compounds were non-cytotoxic, and that the oil resin had potential to act as an anti-inflammatory agent. In another study, Abrão et al. (2015) evaluated the antibacterial and antiproliferative activities of *Copaifera langsdorffii* oil resin and concluded that a diterpene, namely, copalic acid, was responsible for the antimicrobial activity. The potential for developing novel antimicrobial agents to treat infections caused by multidrug-resistant microorganisms was clearly demonstrated, and it was proposed that this diterpene could be a starting point for the development of novel anticancer drugs. Moreover, research by Becker et al. (2020) focused on the antinociceptive and anti-inflammatory effects of a topical formulation containing *Copaifera* L. (3%) oil resin. They found that the effects could be attributed to the presence of β-caryophyllene and other sesquiterpenes that were identified by gas chromatography–mass spectrometry. Additionally, the topical formulation containing copaiba oil resin exhibited antinociceptive and anti-inflammatory effects, which could render it suitable for the treatment of inflammatory pain associated with sunburn. As a summary, Table 15.2 highlights some common biological activities associated with the resinous oils of selected plant species.

Overall, oil resins are particularly important matrices of plant origin since they are the source of various molecules that have applications in the pharmaceutical, cosmetics, fuel, and food industries, among others. Numerous studies have reported their desirable properties, including anti-inflammatory, antibiotic, anticancer, and anti-HIV activities. Thus, oil resins can be considered to be sources of biomolecules that are of significant importance to society. The study of their chemical compositions and properties is therefore fundamental for the development of novel bioactive molecules.

TABLE 15.2
Bioactive Effects and Bioactive Compounds of Various Oil Resins

Species	Bioactivity	Bioactive Compounds and Oil Applications	Reference
Carapa guianensis Aublet.	Anticancer	Limonoids (gedunin, 7-deacetoxy-7-hydroxylgedunin, and 7-deacetoxy-7-oxogedunin) reduce triglyceride levels in human hepatocellular carcinoma cells.	Inoue et al. (2015)
		Triterpene-rich andiroba oils (epoxydeacetylgedunin and deacetilgedunin) exhibit cytotoxic and mutagenic effects in a gastric cancer cell line ***in vitro***.	Porfírio-Dias et al. (2020)
	Larvicide	Emulsions (oil and derivatives associated with silk fibroin) act against the ***Aedes aegypti*** vector larvae.	Sarquis et al. (2020)
	Anti-inflammatory and analgesic	Andiroba oil Orabase is effective in the treatment of oral mucositis.	Soares et al. (2021)
	Antifungal	The major lipid components exhibit a synergistic activity against phytopathogenic fungi.	Nascimento et al. (2019)
	Antimicrobial	Nanoemulsions of andiroba oil at 10% show a 26% toxic effect against the larvae tested and prove to be a potential candidate for the treatment and prevention of American foulbrood disease (AFD).	de Almeida Vaucher et al. (2015)
	Anticonvulsant	The biocatalytic production of fatty acids from andiroba oil using ethanolamine gives compounds for use in seizure control.	Rodrigues de Oliveira et al. (2020)
	Antimalarial	Limonoid (6α-acetoxygedunin) exhibits a significant antimalarial activity ***in vivo*** against ***Plasmodium falciparum***.	Pereira et al. (2014)
	Antileishmaniasis	The limonoid-rich fractions of ***C. guianensis*** (11β-hydroxygedunin and 6α,11β-diacetoxygedunin) seed oil exhibit antileishmaniasis activity.	Oliveira et al. (2018)
Copaifera langsdorffii	Antifungal	A nanoemulsion of copaiba oil resin is used as an antifungal agent against ***Paracoccidioides***.	do Carmo Silva et al. (2020)
		Copaiba oil resin (β-caryophyllene) exhibits fungicidal activity against ***T. mentagrophytes*** and ***T. rubrum***.	Zimmermam-Franco et al. (2013)
Copaifera multijuga		Oil resin–based microemulsions display activity against ***Cryptococcus neoformans*** due to the release of β-caryophyllene.	de Oliveira Neves et al. (2018)
Copaifera reticulata		The crude oil resin (containing ent-polyalthic acid) exhibits antifungal activity against the dermatophyte fungus of the ***Trichophyton*** genus.	Pfeifer Barbosa et al. (2019)

(Continued)

TABLE 15.2 (*Continued*)
Bioactive Effects and Bioactive Compounds of Various Oil Resins

Species	Bioactivity	Bioactive Compounds and Oil Applications	Reference
Copaifera reticulata	Antibacterial	The major constituents of this oil resin (β-bisabolene, ***trans***-α-bergamotene, β-selinene, α-selinene, terpenic acids, ent-agathic-15-methyl ester, ent-copalic acid, and ent-polyhalic acid) exhibit an antibacterial activity against some microorganisms that cause dental caries and periodontitis.	Bardají et al. (2016)
Copaifera multijuga		The major sesquiterpene constituent (β-caryophyllene) present in copaiba oil gel shows antibacterial activity against ***Streptococcus*** sp.	Simões et al. (2016)
Copaifera officinalis		Application of copaiba oil through microdilution of the minimum bactericidal concentration leads to antibacterial activity against ***S. aureus***.	Guimarães et al. (2016)
Copaifera langsdorffii		Copaiba oil nanostructured emulsions (containing palmitic and oleic acid esters, α-curcumene, α-himachalene, isotujol, and α-fenchene) exhibit an antimicrobial activity against ***Staphylococcus*** and ***Candida*** strains.	Alencar et al. (2015)
Copaifera multijuga	Anti-inflammatory	The topical use of a copaiba oil nanoemulsion incorporated into a hydrogel rich in β-caryophyllene demonstrates a potential anti-inflammatory effect during ***in vivo*** tests.	Lucca et al. (2018)
Copaifera reticulata	Healing	The oil resin modulates the inflammatory infiltrate and promotes faster resolution and closure of tongue wounds with the deposition of collagen fibers.	Alvarenga et al. (2020)
Copaifera langsdorffii	Anticancer	The diterpene kaurenoic acid (ent-kaur-16-en-19-oic acid) exhibits anticancer effects on gastric cancer cell lines ***in vitro***.	Cardoso et al. (2017)
Copaifera officinalis	Antiviral	A nanoemulsion of copaiba oil (derived from copalic acid, the diterpene labdane, and caryophyllene) exhibits antiviral potential against the Zika virus.	Carvalho et al. (2023)
Protium heptaphyllum	Larvicide	A nanoemulsion of the resin essential oil (containing ***p***-cimene and α-cinene) is effective against ***Aedes aegypti***.	Faustino et al. (2020)
Boswellia dalzielli	Anti-inflammatory	The main volatile constituents (α-pinene, α-thujene, and α-phellandren-8-ol) exhibit anti-inflammatory activities.	Massei et al. (2023)
Pinus roxburghii	Antimicrobial	Oil resin (containing 3-carene, α-pinene, and β-pinene) extracted using supercritical CO_2 is evaluated via a diffusion test based on the minimum inhibitory concentration for antibacterial activity. The extraction technique applied improves the yield of essential oil obtained from resin oils, and also the biological activities.	Ayub, Hanif, et al. (2023)
Boswellia serrata	Antibacterial	The ***B. serrata*** essential oil constituents (camphene, sabinene, β-pinene, and ***m***-cymene) exhibit antibacterial activity, as determined by disk diffusion assays.	Ayub, Choobkar, et al. (2023)
Boswellia carterii	Antimicrobial	The resin essential oil, rich in β-pinene (61.04%), demonstrates its potential as a natural antimicrobial agent to prevent diseases of food origin and to inhibit plant pathogenic fungi.	Perveen and Alsayed (2023)

REFERENCES

Abdul Sani, Amiril Sahab, Zubaidah Zamri, Puteri Humairah Megat Ahmad Radzi, Ainaa Mardhiah Sabri, and Norfazillah Talib. 2023. "Modified Tamanu Plant-Based Oil From Pahang Malaysia as Biodegradable Metalworking Fluids." *Materials Today: Proceedings* 75: 39–45. https://doi.org/10.1016/j.matpr.2022.10.015. https://www.sciencedirect.com/science/article/pii/S2214785322065087.

Abrão, Fariza, Luciana Costa, Jacqueline Alves, Juliana Senedese, Pâmela Castro, Sergio Ambrosio, Rodrigo Veneziani, Denise Tavares, and Carlos Henrique Martins. 2015. "*Copaifera langsdorffii* Oleoresin and its Isolated Compounds: Antibacterial Effect and Antiproliferative Activity in Cancer Cell Lines." *BMC Complementary and Alternative Medicine* 15. https://doi.org/10.1186/s12906-015-0961-4.

Alencar, Éverton N., Francisco H. Xavier-Júnior, Andreza R. V. Morais, Teresa R. F. Dantas, Nednaldo Dantas-Santos, Lourena M. Verissimo, Vera L. G. Rehder, Guilherme M. Chaves, Anselmo G. Oliveira, and E. Sócrates T. Egito. 2015. "Chemical Characterization and Antimicrobial Activity Evaluation of Natural Oil Nanostructured Emulsions." *Journal of Nanoscience and Nanotechnology* 15 (1): 880–888. https://doi.org/10.1166/jnn.2015.9187. http://openurl.ingenta.com/content/xref?genre=article&issn=1533-4880&volume=15&issue=1&spage=880.

Ali, Dalia E., Marwa M. Abd el-Aziz, Sherihan Salaheldin Abdelhamid Ibrahim, Eman Sheta, and Essam Abdel-Sattar. 2023. "Gastroprotective and Anti-Helicobacter Pylori Potentials of Essential Oils from the Oleoresins of *Araucaria bidwillii* and *Araucaria heterophylla*." *Inflammopharmacology* 31 (1): 465–483. https://doi.org/10.1007/s10787-022-01112-w.

Alvarenga, María Olimpia Paz, Leonardo Oliveira Bittencourt, Paulo Fernando Santos Mendes, Julia Turra Ribeiro, Osmar Alves Lameira, Marta Chagas Monteiro, Carlos Augusto Galvцëo Barboza, Manoela Domingues Martins, and Rafael Rodrigues Lima. 2020. "Safety and Effectiveness of Copaiba Oleoresin (*C. reticulata* Ducke) on Inflammation and Tissue Repair of Oral Wounds in Rats." *International Journal of Molecular Sciences* 21 (10): 3568. https://doi.org/10.3390/ijms21103568. https://www.mdpi.com/1422-0067/21/10/3568.

Arruda, Caroline, Jennyfer Andrea Aldana Mejía, Victor Pena Ribeiro, Carly Henrique Gambeta Borges, Carlos Henrique Gomes Martins, Rodrigo Cássio Sola Veneziani, Sérgio Ricardo Ambrósio, and Jairo Kenupp Bastos. 2019. "Occurrence, Chemical Composition, Biological Activities and Analytical Methods on *Copaifera genus*—A Review." *Biomedicine & Pharmacotherapy* 109: 1–20. https://doi.org/10.1016/j.biopha.2018.10.030. https://www.sciencedirect.com/science/article/pii/S0753332218348546.

Asteggiano, Alberto, Andrea Occhipinti, Andrea Capuzzo, Enrica Mecarelli, Riccardo Aigotti, and Claudio Medana. 2021. Quali—Quantitative Characterization of Volatile and Non-Volatile Compounds in *Protium heptaphyllum* (Aubl.) Marchand Resin by GC—MS Validated Method, GC—FID and HPLC—HRMS2. *Molecules* 26 (5). https://doi.org/10.3390/molecules26051447.

Ayub, Muhammad Adnan, Muhammad Asif Hanif, Joanne Blanchfield, Muhammad Zubair, Muhammad Amin Abid, and Muhammad Tahir Saleh. 2023. "Chemical Composition and Antimicrobial Activity of *Boswellia serrata* Oleo-Gum-Resin Essential Oil Extracted by Superheated Steam." *Natural Product Research* 37 (14): 2451–2456. https://doi.org/10.1080/14786419.2022.2044327. https://www.tandfonline.com/doi/full/10.1080/14786419.2022.2044327.

Ayub, Muhammad Adnan, Nasrin Choobkar, Muhammad Asif Hanif, Mazhar Abbas, Qurat Ul Ain, Muhammad Riaz, and Amir Daraei Garmakhany. 2023. "Chemical Composition, Antioxidant, and Antimicrobial Activities of *P. roxburghii* Oleoresin Essential Oils Extracted by Steam Distillation, Superheated Steam, and Supercritical Fluid CO2 Extraction." *Journal of Food Science* 88 (6): 2425–2438. https://doi.org/10.1111/1750-3841.16597. https://ift.onlinelibrary.wiley.com.

Barbut, S., D. B. Josephson, and A. J. Maurer. 1985. "Antioxidant Properties of Rosemary Oleoresin in Turkey Sausage." *Journal of Food Science* 50 (5): 1356–1359. https://doi.org/10.1111/j.1365-2621.1985.tb10476.x.

Bardají, Danae Kala Rodríguez, Jonas Joaquim Mangabeira da Silva, Thamires Chiquini Bianchi, Daniele de Souza Eugênio, Pollyanna Francielli de Oliveira, Luís Fernando Leandro, Hervê Louis Ghislain Rogez, Rodrigo Cassio Sola Venezianni, Sergio Ricardo Ambrosio, Denise Crispim Tavares, Jairo Kenupp Bastos, and Carlos Henrique G. Martins. 2016. "*Copaifera reticulata* Oleoresin: Chemical Characterization and Antibacterial Properties Against Oral Pathogens." *Anaerobe* 40: 18–27. https://doi.org/10.1016/j.anaerobe.2016.04.017. https://linkinghub.elsevier.com/retrieve/pii/S1075996416300440.

Becker, Gabriela, Indiara Brusco, Rosana Casoti, Marila Crivellaro Lay Marchiori, Letícia Cruz, Gabriela Trevisan, and Sara Marchesan Oliveira. 2020. "*Copaiba oleoresin* Has Topical Antinociceptive Activity in a UVB Radiation-Induced Skin-Burn Model in Mice." *Journal of Ethnopharmacology* 250:

112476. https://doi.org/10.1016/j.jep.2019.112476. https://www.sciencedirect.com/science/article/pii/S0378874119313583.

Bellik, Yuva. 2014. "Total Antioxidant Activity and Antimicrobial Potency of the Essential Oil and Oleoresin of *Zingiber officinale* Roscoe." *Asian Pacific Journal of Tropical Disease* 4 (1): 40–44. https://doi.org/10.1016/S2222-1808(14)60311-X.

Beristain, Silvia, Paola Hernandez Carranza, Teresa Cid-Pérez, Raúl Avila-Sosa, Irving Ruiz-López, and C. E. Ochoa Velasco. 2019. "Antimicrobial Activity of Ginger (*Zingiber Officinale*) and its Application in Food Products." *Food Reviews International* 35: 1–20. https://doi.org/10.1080/87559129.2019.1573829.

Cabral, Fernanda Nunes, Rafaela Jorge Trad, Bruno Sampaio Amorim, Cleiton Fantin, Patrícia Melchionna Albuquerque, and Nállarett Marina Dávila Cardozo. 2021. "Flora da Reserva Ducke, Amazonas, Brasil: Calophyllaceae." *Rodriguésia* 72. https://doi.org/10.1590/2175-7860202172002.

Cardoso, Plínio Cerqueira dos Santos, Carlos Alberto Machado da Rocha, Mariana Ferreira Leal, Marcelo de Oliveira Bahia, Diego Di Felipe D'ávila Alcântara, Raquel Alves dos Santos, Natália dos Santos Gonçalves, Sérgio Ricardo Ambrósio, Bruno Coêlho Cavalcanti, Caroline Aquino Moreira-Nunes, Claudia do ц ⌠ Pessoa, and Rommel Mário Rodríguez Burbano. 2017. "Effect of Diterpenoid Kaurenoic Acid on Genotoxicity and Cell Cycle Progression in Gastric Cancer Cell Lines." *Biomedicine & Pharmacotherapy* 89: 772–780. https://doi.org/10.1016/j.biopha.2017.02.085. https://linkinghub.elsevier.com/retrieve/pii/S0753332216324180.

Carvalho, Tamara, Marcela Guimarães Landim, Maria Letícia Duarte Lima, Cíntia Bittar, Beatriz Carvalho de Araújo Oliveira Faria, Paula Rahal, Milena Campelo Freitas de Lima, Valdir Florêncio da Veiga Junior, Graziella Anselmo Joanitti, and Marilia Freitas Calmon. 2023. "Synthesis of Copaiba (*Copaifera officinalis*) Oil Nanoemulsion and the Potential Against Zika Virus: An In Vitro Study." *PLoS ONE* 18 (9): e0283817. https://doi.org/10.1371/journal.pone.0283817. https://dx.plos.org/10.1371/journal.pone.0283817.

Cassien, Mathieu, Anne Mercier, Sophie Thétiot-Laurent, Marcel Culcasi, Emilie Ricquebourg, Alice Asteian, Gaëtan Herbette, Jean-Pierre Bianchini, Phila Raharivelomanana, and Sylvia Pietri. 2021. "Improving the Antioxidant Properties of *Calophyllum inophyllum* Seed Oil from French Polynesia: Development and Biological Applications of Resinous Ethanol-Soluble Extracts." *Antioxidants* 10 (2). https://doi.org/10.3390/antiox10020199.

Chakraborty, Kajal, Selsa J. Chakkalakal, and Deepu Joseph. 2016. "Antioxidative Status of a Nutrient Enriched Formulation of *Perna viridis* in a Time Series Shelf-Life Study." *Journal of Aquatic Food Product Technology* 25 (5): 666–683. https://doi.org/10.1080/10498850.2014.915909.

Che Man, Yaakob B., Wanna Ammawath, Russly A. Rahman, and Salmah Yusof. 2003. "Quality Characteristics of Refined, Bleached and Deodorized Palm Olein and Banana Chips After Deep-Fat Frying." *Journal of the Science of Food and Agriculture* 83 (5): 395–401. https://doi.org/10.1002/jsfa.1391.

Cheng, Xiaoyu, Mingwu Zang, Shouwei Wang, Xin Zhao, Guozhen Zhai, Le Wang, Xiang Li, Yan Zhao, and Yijing Yue. 2022. "Physicochemical and Antioxidant Properties of Nanoliposomes Loaded with Rosemary Oleoresin and Their Oxidative Stability Application in Dried Oysters." *Bioengineering* 9 (12). https://doi.org/10.3390/bioengineering9120818. https://www.mdpi.com/2306-5354/9/12/818.

Couto, Roberta S., Maria F. Rodrigues, Leila S. Ferreira, Ivana M. Diniz, Fernando D. Silva, Talita C. Lopez, Rafael R. Lima, and Márcia M. Marques. 2020. "Evaluation of Resin-Based Material Containing Copaiba Oleoresin (*Copaifera Reticulata* Ducke): Biological Effects on the Human Dental Pulp Stem Cells." *Biomolecules* 10 (7). https://doi.org/10.3390/biom10070972.

de Almeida Júnior, José S., Éden B. da Silva, Tânia M. Moraes, Aline A. Kasper, Adilson Sartoratto, Leopoldo C. Baratto, Elaine C. de Oliveira, Euzebio Oliveira, Lauro E. Barata, Antonio H. Minervino, and Waldiney P. Moraes. 2021. "Anti-Inflammatory Potential of the Oleoresin from the Amazonian Tree *Copaifera reticulata* with an Unusual Chemical Composition in Rats." *Veterinary Sciences* 8 (12). https://doi.org/10.3390/vetsci8120320.

de Almeida, Patrícia D. O., Ana Paula de A. Boleti, André Luis Rüdiger, Geane A. Lourenço, Valdir Florêncio da Veiga Junior, and Emerson S. Lima. 2015. "Anti-Inflammatory Activity of Triterpenes Isolated from *Protium paniculatum* Oil-Resins." *Evidence-Based Complementary and Alternative Medicine* 2015: 293768. https://doi.org/10.1155/2015/293768.

de Almeida Vaucher, Rodrigo, Janice Luehring Giongo, Leandro Perger Bolzan, Marcos Saldanha Côrrea, Viviane Pedroso Fausto, Camilla Filippi dos Santos Alves, Leonardo Quintana Soares Lopes, Aline Augusti Boligon, Margareth Linde Athayde, Attus Pereira Moreira, Adriano Brandelli, Renata Platchek Raffin, and Roberto Christ Vianna Santos. 2015. "Antimicrobial Activity of Nanostructured Amazonian Oils Against Paenibacillus Species and Their Toxicity on Larvae and Adult Worker Bees." *Journal of Asia-Pacific Entomology* 18 (2): 205–210. https://doi.org/10.1016/j.aspen.2015.01.004. https://linkinghub.elsevier.com/retrieve/pii/S1226861515000187.

de Lima, Milena Campelo Freitas, Rayssa Ribeiro, Josiane Elizabeth Almeida e Silva, Sthephanie Silva dos Santos Tavares, Yuri Campello Dias de Araujo, and Valdir F. da Veiga-Junior. 2022. "Chemistry, Biological Activities, and Uses of Copaiba Oil Resins." In *Gums, Resins and Latexes of Plant Origin: Chemistry, Biological Activities and Uses*, edited by Hosakatte Niranjana Murthy, 377–397. Cham: Springer International Publishing.

de Oliveira Neves, Jamilly Kelly, Alexsandra Conceição Apolinário, Karina Lidiane Alcantara Saraiva, Dayanne Tomaz Casimiro da Silva, Mysrayn Yargo de Freitas Araújo Reis, Bolivar Ponciano Goulart de Lima Damasceno, Adalberto Pessoa, Marcos Aurélio Moraes Galvão, Luiz Alberto Lira Soares, Valdir Florêncio da Veiga Júnior, José Alexsandro da Silva, and Attilio Converti. 2018. "Microemulsions Containing *Copaifera multijuga* Hayne Oil-Resin: Challenges to Achieve an Efficient System for a-Caryophyllene Delivery." *Industrial Crops and Products* 111: 185–192. https://doi.org/10.1016/j.indcrop.2017.10.025. https://linkinghub.elsevier.com/retrieve/pii/S0926669017307100.

DeCarlo, Anjanette, Stephen Johnson, Ambika Poudel, Prabodh Satyal, Loren Bangerter, and William N. Setzer. 2018. "Chemical Variation in Essential Oils from the Oleo-gum Resin of *Boswellia carteri*: A Preliminary Investigation." *Chemistry & Biodiversity* 15 (6): e1800047. https://doi.org/10.1002/cbdv.201800047.

Dias, Kaio Kelvin Barros, Adauto Lima Cardoso, Ana Alice Farias da Costa, Marcele Fonseca Passos, Carlos Emmerson Ferreira da Costa, Geraldo Narciso da Rocha Filho, Eloísa Helena de Aguiar Andrade, Rafael Luque, Luís Adriano Santos do Nascimento, and Renata Coelho Rodrigues Noronha. 2023. "Biological Activities from Andiroba (*Carapa guianensis* Aublet.) and its Biotechnological Applications: A Systematic Review." *Arabian Journal of Chemistry* 16 (4): 104629. https://doi.org/10.1016/j.arabjc.2023.104629. https://linkinghub.elsevier.com/retrieve/pii/S1878535223000904.

Diefenbach, Ana Lúcia, Francisco Wilker Mustafa Gomes Muniz, Harry Juan Rivera Oballe, and Cassiano Kuchenbecker Rosing. 2018. "Antimicrobial Activity of Copaiba Oil (*Copaifera* ssp.) on Oral Pathogens: Systematic Review." *Phytotherapy Research* 32 (4): 586–596. https://doi.org/10.1002/ptr.5992.

do Carmo Silva, Lívia, Meire Ane Costa Miguel Miranda, Julianna Veiga de Freitas, Sarah Fernanda Araújo Ferreira, Emília Celma de Oliveira Lima, Cecilia Maria Alves de Oliveira, Lucilia Kato, Ana Paula Terezan, Anselmo Fortunato Ruiz Rodriguez, Fernando Sergio Escócio Drumond Viana Faria, Célia Maria de Almeida Soares, and Maristela Pereira. 2020. "Antifungal Activity of Copaíba Resin Oil in Solution and Nanoemulsion Against *Paracoccidioides* spp." *Brazilian Journal of Microbiology* 51 (1): 125–134. https://doi.org/10.1007/s42770-019-00201-3.

Eid, M. Ahmad, Hiba Natsheh, Linda Issa, Majdulin Zoabi, Monia Amer, Effat Mahamid, and Ahmed Mousa. 2024. "*Capsicum annuum* Oleoresin Nanoemulgel—Design Characterization and *In Vitro* Investigation of Anticancer and Antimicrobial Activities." *Current Pharmaceutical Design* 30 (2): 151–160. https://doi.org/10.2174/0113816128283684231220062019. http://benthamscience.com/article/137387.

Faustino, Cleidjane Gomes, Fernando Antônio de Medeiros, Allan Kardec Ribeiro Galardo, Alex Bruno Lobato Rodrigues, Rosany Lopes Martins, Yuri de Medeiros Souza Lima, Josean Fechine Tavares, Marcos Antônio Alves de Medeiros, Jader dos Santos Cruz, and Sheylla Susan Moreira da Silva de Almeida. 2020. "Larvicide Activity on *Aedes aegypti* of Essential Oil Nanoemulsion from the *Protium heptaphyllum* Resin." *Molecules* 25 (22): 5333. https://doi.org/10.3390/molecules25225333. https://www.mdpi.com/1420-3049/25/22/5333.

Guimarães, Anna Luísa Aguijar, Elisa Alves Cunha, Fernanda Oliveira Matias, Patrícia Guedes Garcia, Panagiota Danopoulos, Rosita Swikidisa, Vanessa Alves Pinheiro, and Rodrigo José Lupatini Nogueira. 2016. "Antimicrobial Activity of Copaiba (*Copaifera officinalis*) and Pracaxi (*Pentaclethra macroloba*) Oils Against *Staphylococcus Aureus*: Importance in Compounding for Wound Care." *International Journal of Pharmaceutical Compounding* 20 (1): 58–62. http://www.ncbi.nlm.nih.gov/pubmed/27125055.

Guimarães-Santos, Adriano, Diego Siqueira Santos, Ijair Rogério Santos, Rafael Rodrigues Lima, Antonio Pereira, Lucinewton Silva de Moura, Raul Nunes Carvalho, Osmar Lameira, and Walace Gomes-Leal. 2012. "Copaiba Oil-Resin Treatment Is Neuroprotective and Reduces Neutrophil Recruitment and Microglia Activation After Motor Cortex Excitotoxic Injury." *Evidence-Based Complementary and Alternative Medicine* 2012: 918174. https://doi.org/10.1155/2012/918174.

Gupta, Madhuri, Sajendra Kumar Verma, Shilpi Singh, Laxmikant Trivedi, Prasant Kumar Rout, Prema G. Vasudev, Suaib Luqman, Mahendra P. Darokar, Rajendra S. Bhakuni, and Laxminarain Misra. 2022. "Anti-Proliferative and Antibacterial Activity of Oleo-Gum-Resin of *Boswellia serrata* Extract and its Isolate 3-Hydroxy-11-Keto-b-Boswellic Acid." *Journal of Herbal Medicine* 32: 100546. https://doi.org/10.1016/j.hermed.2022.100546. https://www.sciencedirect.com/science/article/pii/S2210803322000148.

Gupta, Madhuri, Shilpi Singh, Alka kurmi, Suaib Luqman, Dharmendra Saikia, Moni Thomas, and Prasant Kumar Rout. 2022. "Correlation of Boswellic Acids with Antiproliferative, Antioxidant and Antimicrobial Activities of Topographically Collected *Boswellia serrata* Oleo-Gum-Resin." *Phytomedicine Plus* 2 (3):

100289. https://doi.org/10.1016/j.phyplu.2022.100289. https://www.sciencedirect.com/science/article/pii/S2667031322000720.

Gurgel, Ely Simone Cajueiro, Mozaniel Santana de Oliveira, Marília Caldas Souza, Sebastiцëo Gomes da Silva, Maria Silvia de Mendonça, and Antônio Pedro da Silva Souza Filho. 2019. "Chemical Compositions and Herbicidal (Phytotoxic) Activity of Essential Oils of Three Copaifera Species (Leguminosae-Caesalpinoideae) from Amazon-Brazil." *Industrial Crops and Products* 142: 111850. https://doi.org/10.1016/j.indcrop.2019.111850. https://www.sciencedirect.com/science/article/pii/S092666901930860X.

Hapsari, Safrina, Nurul Jadid, Hakun Wirawasista Aparamarta, and Setiyo Gunawan. 2023. "Impact of Solvent Type, Solvent-Water Concentration, and Number of Stages on the Extraction of Coumarin Mixture from Tamanu (*Calophyllum inophyllum*) Oil and its Antioxidant Activity." *Arabian Journal of Chemistry* 16 (2): 104449. https://doi.org/10.1016/j.arabjc.2022.104449. https://www.sciencedirect.com/science/article/pii/S1878535222007651.

Hien, Ha Minh, Tran Viet Hung, and Paul W. S. Heng. 2023. "A New Chromanone Acid Derivative from the Nut Oil Resin of *Calophyllum inophyllum*." *Journal of Asian Natural Products Research* 25 (10): 1021–1028. https://doi.org/10.1080/10286020.2023.2208530.

Higuchi, Keiichiro, Teppei Miyake, Shoko Ohmori, Yoshimi Tani, Katsuhiko Minoura, Takashi Kikuchi, Takeshi Yamada, and Reiko Tanaka. 2017. "Carapanosins A–C From Seeds of Andiroba (Carapa Guianensis, Meliaceae) and Their Effects on LPS-Activated NO Production." *Molecules* 22 (3): 502–502. https://doi.org/10.3390/molecules22030502. http://www.mdpi.com/1420-3049/22/3/502.

Hulagannavar, Pushpa, Dileepkumar A. Masuthi, Lakshmidevamma TN, Abdul Kareem M, Shivayogi Ryavalad, Ratnakar Shet, and MH Tatagar. 2024. "Quantitative Determination of Capsaicin, Oleoresin and Ascorbic Acid Content in Chilli (*Capsicum annuum* L.)." *International Journal of Advanced Biochemistry Research* 8 (1): 634–637. https://doi.org/10.33545/26174693.2024.v8.i1i.471.

Ibiapina, Andréia, Larissa da Silva Gualberto, Bianca Barros Dias, Bárbara Catarina Bastos Freitas, Glêndara Aparecida de Souza Martins, and Antonio Alves Melo Filho. 2022. "Essential and Fixed Oils from Amazonian Fruits: Proprieties and Applications." *Critical Reviews in Food Science and Nutrition* 62 (32): 8842–8854. https://doi.org/10.1080/10408398.2021.1935702.

Inoue, Takanobu, Yuuki Matsui, Takashi Kikuchi, Takeshi Yamada, Yasuko In, Osamu Muraoka, Chie Sakai, Kiyofumi Ninomiya, Toshio Morikawa, and Reiko Tanaka. 2015. "Carapanolides M–S From Seeds of Andiroba (Carapa guianensis, Meliaceae) and Triglyceride Metabolism-Promoting Activity in High Glucose-Pretreated HepG2 Cells." *Tetrahedron* 71 (18): 2753–2760. https://doi.org/10.1016/j.tet.2015.03.017. https://linkinghub.elsevier.com/retrieve/pii/S0040402015003154.

Johnson, Stephen, Anjanette DeCarlo, Prabodh Satyal, Noura S. Dosoky, Aaron Sorensen, and William N. Setzer. 2019. "The Chemical Composition of *Boswellia occulta* Oleogum Resin Essential Oils." *Natural Product Communications* 14 (7): 1934578X19866307. https://doi.org/10.1177/1934578X19866307.

Johnson, Stephen, Anjanette DeCarlo, Prabodh Satyal, Noura S. Dosoky, Aaron Sorensen, and William N. Setzer. 2021. "The Chemical Composition of Single-Tree *Boswellia frereana* Resin Samples." *Natural Product Communications* 16 (9): 1934578X211043727. https://doi.org/10.1177/1934578X211043727.

Kalaikannan, A., A. Elango, J. Rajeshwar, and Shanthi Dhanasekar. 2022. "Evaluation of Essential Oils as a Hurdle for Ambient Temperature Storage of Chicken Patties." *Asian Journal of Dairy and Food Research*. https://doi.org/10.18805/ajdfr.DR-1844.

Kamaruddin, Muhamad Syafiq Hakimi, Gun Hean Chong, Nurizzati Mohd Daud, Nicky Rahmana Putra, Liza Md Salleh, and Norhidayah Suleiman. 2023. "Bioactivities and Green Advanced Extraction Technologies of Ginger Oleoresin Extracts: A Review." *Food Research International* 164: 112283. https://doi.org/10.1016/j.foodres.2022.112283. https://www.sciencedirect.com/science/article/pii/S0963996922013412.

Karthik, K., T. Ramesh, P. Sankarganesh, and I. Vinoth Kanna. 2022. "Performance and Emission Analysis of Tamanu Oil—Diesel Blends in CI Engine." *International Journal of Ambient Energy* 43 (1): 3117–3123. https://doi.org/10.1080/01430750.2020.1783353. https://doi.org/10.1080/01430750.2020.1783353.

Kikuchi, Takashi, Keiko Akita, Hiroki Koike, Yasuko In, Takeshi Yamada, and Reiko Tanaka. 2020. "Carapanins A–C: New Limonoids from Andiroba (*Carapa guianensis*) Fruit Oil." *Organic & Biomolecular Chemistry* 18 (45): 9268–9274. https://doi.org/10.1039/D0OB01872D.

Krisanti, Elsa, Rizka Astuty, and Kamarza Mulia. 2017. Microencapsulation of Oleoresin From Red Ginger (Zingiber Officinale var. Rubrum) in Chitosan and Alginate for Fresh Milk Preservatives. In *AIP Conference Proceedings* (Vol. 1817, No. 1). AIP Publishing.

Lemos, Larissa M. S., Rejane B. Oliveira, Bruno L. Sampaio, Gari V. Ccana-Ccapatinta, Fernando B. Da Costa, and Domingos T. O. Martins. 2016. "Brasiliensic and Isobrasiliensic Acids: Isolation from *Calophyllum brasiliense* Cambess. and Anti-Helicobacter Pylori Activity." *Natural Product Research* 30 (23): 2720–2725. https://doi.org/10.1080/14786419.2015.1137568.

Liu East Lansing Mi, H. F., A. M. Booren, J. I. Gray, and R. L. Crackel. 1992. "Antioxidant Efficacy of Oleoresin Rosemary and Sodium Tripolyphosphate in Restructured Pork Steaks." *Journal of Food Science* 57 (4): 803–806. https://doi.org/10.1111/j.1365-2621.1992.tb14298.x.

Lopes de Mesquita Vieira, Fernanda, Luiz Eduardo Celino Benedito, Alessandro Cézar de Oliveira Moreira, Jez William Batista Braga, and Aline Lima de Oliveira. 2024. "Applicability of NMR in Combination with Chemometrics for the Characterization and Differentiation of Oil-Resin Extracted from *Copaifera langsdorffii* and *Copaifera* spp." *Microchemical Journal* 197: 109850. https://doi.org/10.1016/j.microc.2023.109850. https://linkinghub.elsevier.com/retrieve/pii/S0026265X23014698.

Lucca, Letícia G., Sheila P. de Matos, Tainá Kreutz, Helder F. Teixeira, Valdir F. Veiga, Bibiana V. de Araújo, Renata P. Limberger, and Letícia S. Koester. 2018. "Anti-Inflammatory Effect from a Hydrogel Containing Nanoemulsified Copaiba oil (*Copaifera multijuga* Hayne)." *AAPS PharmSciTech* 19 (2): 522–530. https://doi.org/10.1208/s12249-017-0862-6. http://link.springer.com/10.1208/s12249-017-0862-6.

Majeed, Hamid, Yuan-Yuan Bian, Barkat Ali, Anjum Jamil, Usman Majeed, Qaiser Farid Khan, Khalid Javed Iqbal, Charles F. Shoemaker, and Zhong Fang. 2015. "Essential Oil Encapsulations: Uses, Procedures, and Trends." *RSC Advances* 5 (72): 58449–58463. https://doi.org/10.1039/C5RA06556A.

Massei, Kimberley, Thomas Michel, Girma Ilayas Obersat, Ahmed Al-Harrasi, and Nicolas Baldovini. 2023. "Phytochemical Study of *Boswellia dalzielii* Oleo-Gum Resin and Evaluation of its Biological Properties." *Phytochemistry* 213: 113751. https://doi.org/10.1016/j.phytochem.2023.113751. https://linkinghub.elsevier.com/retrieve/pii/S003194222300167X.

Melgar-Lalanne, Guiomar, Alan Javier Hernández-Álvarez, Maribel Jiménez-Fernández, and Ebner Azuara. 2017. "Oleoresins from Capsicum spp.: Extraction Methods and Bioactivity." *Food and Bioprocess Technology* 10 (1): 51–76. https://doi.org/10.1007/s11947-016-1793-z.

Menezes, Ana Carolina dos Santos, Lúsia Daltro Borges Alves, Daniel Cohen Goldemberg, Andréia Cristina de Melo, and Héliton Spindola Antunes. 2022. "Anti-Inflammatory and Wound Healing Effect of *Copaiba oleoresin* on the Oral Cavity: A Systematic Review." *Heliyon* 8 (2): e08993. https://doi.org/10.1016/j.heliyon.2022.e08993. https://www.sciencedirect.com/science/article/pii/S240584402200281X.

Murphy, Anna, Joe P. Kerry, Joe Buckley, and Ian Gray. 1998. "The Antioxidative Properties of Rosemary Oleoresin and Inhibition of Off-Flavours in Precooked Roast Beef Slices." *Journal of the Science of Food and Agriculture* 77 (2): 235–243. https://doi.org/10.1002/(SICI)1097-0010(199806)77:2<235::AID-JSFA30>3.0.CO;2-L.

Murthy, Pushpa S., Ranju Gautam, and J. Naik Pura. 2015. "Ginger Oleoresin Chemical Composition, Bioactivity and Application as Bio-Preservatives." *Journal of Food Processing and Preservation* 39 (6): 1905–1912. https://doi.org/10.1111/jfpp.12428.

Nascimento, Gleisson O., Diego P. Souza, Alberdan S. Santos, Josimar F. Batista, Bala Rathinasabapathi, Paulo Roberto Gagliardi, and José F. C. Gonçalves. 2019. "Lipidomic Profiles from Seed Oil of *Carapa guianensis* Aubl. and *Carapa vasquezii* Kenfack and Implications for the Control of Phytopathogenic Fungi." *Industrial Crops and Products* 129: 67–73. https://doi.org/10.1016/j.indcrop.2018.11.069. https://www.sciencedirect.com/science/article/pii/S0926669018310446.

Oliveira, Davi S., Leandro S. Lima, Ananda S. Antonio, Larissa S. M. Wiedemann, and Valdir F. Veiga-Junior. 2020. "Perfil metabólico dos extratos polares das folhas, galhos e cascas de *Copaifera multijuga* Hayne (Copaíba mair-mari da Amazônia) por ESI-EM." *Química Nova* 43.

Oliveira, Iara dos Santos da Silva, Carla Junqueira Moragas Tellis, Maria do Socorro dos Santos Chagas, Maria Dutra Behrens, Kátia da Silva Calabrese, Ana Lucia Abreu-Silva, and Fernando Almeida-Souza. 2018. "*Carapa guianensis* Aublet (Andiroba) Seed Oil: Chemical Composition and Antileishmanial Activity of Limonoid-Rich Fractions." *BioMed Research International* 2018: 1–10. https://doi.org/10.1155/2018/5032816. https://www.hindawi.com/journals/bmri/2018/5032816/.

Oriani, Vivian Boesso, Izabela Dutra Alvim, Bruno Nicolau Paulino, Fernanda Ramalho Procópio, Glaucia Maria Pastore, and Míriam Dupas Hubinger. 2018. "The Influence of the Storage Temperature on the Stability of Lipid Microparticles Containing *Ginger oleoresin*." *Food Research International* 109: 472–480. https://doi.org/10.1016/j.foodres.2018.04.066. https://www.sciencedirect.com/science/article/pii/S096399691830351X.

Pasquel-Reátegui, José Luis, Luana Cristina dos Santos, Francisco Manuel Barrales, Vanessa Laís Grober, Marcus Bruno Soares Forte, Adilson Sartoratto, Carmen Lucia Queiroga, and Julian Martínez. 2022. "Fractionation of Sesquiterpenes and Diterpenic Acids from Copaiba (*Copaifera officinalis*) Oleoresin Using Supercritical Adsorption." *The Journal of Supercritical Fluids* 184: 105565. https://doi.org/10.1016/j.supflu.2022.105565. https://www.sciencedirect.com/science/article/pii/S089684462200050X.

Peivastegan, M., M. Rajabi, H. Zaferani Arani, M. Olya, H. A. Atashi, and S. Abolghasemi. 2020. "Comparing the Effects of Oleoresin of Pistacia atlantica Tree and Diclofenac Gel on the Knee Osteoarthritis Improvement." *Shiraz E-Medical Journal* 21 (10): e98293. https://doi.org/10.5812/semj.98293. https://brieflands.com/articles/semj-98293.

Pereira, Tiago B., Luiz F. Rocha e Silva, Rodrigo C. N. Amorim, Márcia R. S. Melo, Rita C. Zacardi de Souza, Marcos N. Eberlin, Emerson S. Lima, Marne C. Vasconcellos, and Adrian M. Pohlit. 2014. "In Vitro and In Vivo Anti-Malarial Activity of Limonoids Isolated from the Residual Seed Biomass from *Carapa guianensis* (Andiroba) Oil Production." *Malaria Journal* 13 (1): 317–317. https://doi.org/10.1186/1475-2875-13-317. https://malariajournal.biomedcentral.com/articles/10.1186/1475-2875-13-317.

Perveen, Kahkashan, and Mashail Fahad Alsayed. 2023. "Antimicrobial Activity of Essential Oil of *Boswellia carterii* Birdw. Oleo Gum Resin and its Chemical Composition." *Journal of Essential Oil Bearing Plants* 26 (2): 386–395. https://doi.org/10.1080/0972060X.2023.2216733. https://www.tandfonline.com/doi/full/10.1080/0972060X.2023.2216733.

Pessoa, M. J. G., S. Pireda, P. Simioni, N. Bautz, and M. Da Cunha. 2021. "Structural and Histochemical Attributes of Secretory Ducts and Cavities in Leaves of Four Species of Calophyllaceae J. Agardh in *Amazonian savannas*." *Plant Biology* 23 (6): 1128–1140. https://doi.org/10.1111/plb.13321.

Pfeifer Barbosa, Anna Laís, Arlette Wenzel-Storjohann, José Diomedes Barbosa, Christian Zidorn, Christian Peifer, Deniz Tasdemir, and Serhat Sezai Çiçek. 2019. "Antimicrobial and Cytotoxic Effects of the *Copaifera reticulata* Oleoresin and its Main Diterpene Acids." *Journal of Ethnopharmacology* 233: 94–100. https://doi.org/10.1016/j.jep.2018.11.029. https://linkinghub.elsevier.com/retrieve/pii/S0378874118331957.

Porfírio-Dias, Clara L., Karina M. Melo, Carlos E. M. C. Bastos, Tássia A. A. Ferreira, Luana F. C. Azevedo, Hugo L. Salgado, Alberdan S. Santos, Jorge D. Rissino, Cleusa Y. Nagamachi, and Julio C. Pieczarka. 2020. "Andiroba Oil (*Carapa guianensis* Aubl) Shows Cytotoxicity but No Mutagenicity in the ACPP02 Gastric Cancer Cell Line." *Journal of Applied Toxicology* 40 (8): 1060–1066. https://doi.org/10.1002/jat.3966. https://analyticalsciencejournals.onlinelibrary.wiley.com.

Pribowo, Amadeus, Jyothsna Girish, Marsia Gustiananda, Rakrya Galih Nandhira, and Pietradewi Hartrianti. 2021. "Potential of Tamanu (*Calophyllum inophyllum*) Oil for Atopic Dermatitis Treatment." *Evidence-Based Complementary and Alternative Medicine* 2021: 6332867. https://doi.org/10.1155/2021/6332867.

Procopio, Fernanda Ramalho, Mariana Costa Ferraz, Bruno Nicolau Paulino, Paulo José do Amaral Sobral, and Miriam Dupas Hubinger. 2022. "Spice Oleoresins as Value-Added Ingredient for Food Industry: Recent Advances and Perspectives." *Trends in Food Science & Technology* 122: 123–139. https://doi.org/10.1016/j.tifs.2022.02.010. https://www.sciencedirect.com/science/article/pii/S0924224422000589.

Pulcinelli, Rianne Remus, Clara Lindenmayer Della Giustina, Solange Bandiera, Felipe Borges Almeida, Lucas dos Reis Izolan, Mauricio Schuler Nin, Mirna Bainy Leal, and Rosane Gomez. 2022. "Copaiba (*Copaifera reticulata*) Oleoresin Reduces Voluntary Alcohol Intake in Rats." *Acta Amazonica* 52. https://doi.org/10.1590/1809-4392202102342. https://www.scielo.br/j/aa/a/BpMgKFGg7YvNdbvPLxN5qkF/.

Rodrigues de Oliveira, Fábio, Keuri Eleuterio Rodrigues, Moisés Hamoy, Ícaro Rodrigues Sarquis, Akira Otake Hamoy, Maria Elena Crespo Lopez, Irlon Maciel Ferreira, Barbarella de Matos Macchi, and José Luiz Martins do Nascimento. 2020. "Fatty Acid Amides Synthesized from Andiroba Oil (*Carapa guianensis* Aublet.) Exhibit Anticonvulsant Action with Modulation on GABA-A Receptor in Mice: A Putative Therapeutic Option." *Pharmaceuticals* 13 (3): 43–43. https://doi.org/10.3390/ph13030043. https://www.mdpi.com/1424-8247/13/3/43.

Rohlik, Bo-Anne, Petr Pipek, and Jan Panek. 2010. "The Effect of Natural Antioxidants on the Colour of Dried/Cooked Sausages." *Czech Journal of Food Sciences* 28 (4): 249–257. http://doi.org/10.17221/345/2009-CJFS. https://cjfs.agriculturejournals.cz/artkey/cjf-201004-0001.php

Santos, Maria de Oliveira, Cicera Janaine Camilo, Daiany Alves Ribeiro, Julimery Gonçalves Ferreira Macedo, Carla de Fatima Alves Nonato, Fábio Fernandes Galvão Rodrigues, José Galberto Martins da Costa, and Marta Maria de Almeida Souza. 2023. "Chemical Composition Variation of Essential Oils of *Copaifera langsdorffii* Desf. from Different Vegetational Formations." *Natural Product Research* 37 (20): 3525–3530. https://doi.org/10.1080/14786419.2022.2081849.

Sarquis, Icaro R., Rosângela S. F. R. Sarquis, Victor H. S. Marinho, Fernando B. Neves, Inana F. Araújo, Leandro F. Damasceno, Ricardo M. A. Ferreira, Raiumundo N. P. Souto, José C. T. Carvalho, and Irlon M. Ferreira. 2020. "*Carapa guianensis* Aubl. (Meliaceae) Oil Associated with Silk Fibroin, as Alternative to Traditional Surfactants, and Active Against Larvae of the Vector *Aedes aegypti*." *Industrial Crops and Products* 157: 112931. https://doi.org/10.1016/j.indcrop.2020.112931. https://linkinghub.elsevier.com/retrieve/pii/S0926669020308487.

Silva, Debora F., Klinsmann T. Lima, Gilmara N. T. Bastos, Johnatt Allan R. Oliveira, Luís Adriano S. do Nascimento, Carlos Emmerson F. Costa, Geraldo N. R. Filho, Viktor O. C. Concha, and Marcele F. Passos. 2021. "PCL/Andiroba Oil (*Carapa guianensis* Aubl.) Hybrid Film for Wound Healing Applications." *Polymers* 13 (10): 1591. https://doi.org/10.3390/polym13101591. https://www.mdpi.com/2073-4360/13/10/1591.

Silva, Marcos T., Leonardo L. Borges, Tatiana de S. Fiuza, Leonice M. F. Tresvenzol, Edemilson da Conceição, C., Celyane R. Batista, Caio Matos, B., Valdir F. da Veiga Júnior, Rosa H. V. Mourão, Pedro H. Ferri, and José R. de Paula. 2017. "Viscosity of the Oil-Resins and Chemical Composition of the Essential Oils from Oils-Resins of *Copaifera multijuga* Hayne Growing in the National Forest Saracá-Taquera Brazil." *Journal of Essential Oil Bearing Plants* 20 (5). https://doi.org/10.1080/0972060X.2017.1358673.

Simões, Cláudia A. C. G., Nikeila C. de Oliveira Conde, Gisely N. Venâncio, Patrícia S. L. L. Milêrio, Maria F. C. L. Bandeira, and Valdir F. da Veiga Júnior. 2016. "Antibacterial Activity of Copaiba Oil Gel on Dental Biofilm." *The Open Dentistry Journal* 10 (Suppl-1, M6): 188–195. https://doi.org/10.2174/1874210601610010188. https://opendentistryjournal.com/VOLUME/10/PAGE/188/.

Singh, S., P. K. Chaurasia, and S. L. Bharati (Eds.). 2023. *Pharmacological Aspects of Essential Oils: Current and Future Trends*. 1st ed. CRC Press.

Soares, Artur dos Santos, Ana Márcia Viana Wanzeler, Gabriel Henrique Soares Cavalcante, Elsa Maria da Silva Barros, Rita de Cássia Matos Carneiro, and Fabrício Mesquita Tuji. 2021. "Therapeutic Effects of Andiroba (*Carapa guianensis* Aubl) Oil, Compared to Low Power Laser, on Oral Mucositis in Children Underwent Chemotherapy: A Clinical Study." *Journal of Ethnopharmacology* 264: 113365. https://doi.org/10.1016/j.jep.2020.113365. https://linkinghub.elsevier.com/retrieve/pii/S0378874120332505.

Song, Xue-Chao, Elena Canellas, Magdalena Wrona, Raquel Becerril, and Cristina Nerin. 2020. "Comparison of Two Antioxidant Packaging Based on Rosemary Oleoresin and Green Tea Extract Coated on Polyethylene Terephthalate for Extending the Shelf Life of Minced Pork Meat." *Food Packaging and Shelf Life* 26: 100588. https://doi.org/10.1016/j.fpsl.2020.100588. https://www.sciencedirect.com/science/article/pii/S2214289420306062.

Spino, Claude, Marco Dodier, and Subramaniam Sotheeswaran. 1998. "Anti-HIV Coumarins from Calophyllum Seed Oil." *Bioorganic & Medicinal Chemistry Letters* 8 (24): 3475–3478. https://doi.org/https://doi.org/10.1016/S0960-894X(98)00628-3. https://www.sciencedirect.com/science/article/pii/S0960894X98006283.

Stoick East Lansing Mi, S. M., J. I. Gray, A. M. Booren, and D. J. Buckley. 1991. "Oxidative Stability of Restructured Beef Steaks Processed with Oleoresin Rosemary, Tertiary Butylhydroquinone, and Sodium Tripolyphosphate." *Journal of Food Science* 56 (3): 597–600. https://doi.org/10.1111/j.1365-2621.1991.tb05337.x.

Subtain, Muhammad, Imran Pasha, Allah Rakha, and Amer Jamil. 2024. "Extraction Evaluation and Chemical Characterization of *Ginger oleoresins*; a Functional Food Ingredient." *Journal of Food Measurement and Characterization*. https://doi.org/10.1007/s11694-024-02518-6.

Taiti, Cosimo, Diego Comparini, Lavinia Moscovini, Simona Violino, Corrado Costa, and Stefano Mancuso. 2024. "Influence of the Drying Process on the Volatile Profile of Different *Capsicum* Species." *Plants* 13: 1131. https://doi.org/10.3390/plants13081131.

Teixeira, Francisco Bruno, Raíra de Brito Silva, Osmar Alves Lameira, Liana Preto Webber, Roberta Souza D'Almeida Couto, Manoela Domingues Martins, and Rafael Rodrigues Lima. 2017. "Copaiba Oil-Resin (*Copaifera reticulata* Ducke) Modulates the Inflammation in a Model of Injury to Rats' Tongues." *BMC Complementary and Alternative Medicine* 17 (1): 313. https://doi.org/10.1186/s12906-017-1820–2.

Teoh, Ru Wei, Adeline Su Yien Ting, and Yin Yin Thoo. 2023. "Characterization and Modeling of Diffusion Kinetics of Rosemary Oleoresin Extract from Gellan Gum-Based Film." *Journal of Food Science and Technology* 60 (12): 2978–2989. https://doi.org/10.1007/s13197-023-05826-9.

Upadhyay, Rohit, and Hari Niwas Mishra. 2015a. "Classification of Sunflower Oil Blends Stabilized by Oleoresin Rosemary (*Rosmarinus officinalis* L.) Using Multivariate Kinetic Approach." *Journal of Food Science* 80 (8): E1746–E1754. https://doi.org/10.1111/1750-3841.12966.

Upadhyay, Rohit, and Hari Niwas Mishra. 2015b. "A Multivariate Approach to Optimise the Synergistic Blend of Oleoresin Rosemary (*Rosmarinus officinalis* L.) and Ascorbyl Palmitate Added into Sunflower Oil." *International Journal of Food Science & Technology* 50 (4): 974–981. https://doi.org/10.1111/ijfs.12738.

Upadhyay, Rohit, Sneha Sehwag, and Hari Niwas Mishra. 2017. "Chemometric Approach to Develop Frying Stable Sunflower Oil Blends Stabilized with Oleoresin Rosemary and Ascorbyl Palmitate." *Food Chemistry* 218: 496–504. https://doi.org/10.1016/j.foodchem.2016.09.105. https://www.sciencedirect.com/science/article/pii/S030881461631500X.

Valadas, Lídia Audrey Rocha, Patrícia Leal Dantas Lobo, Said Gonçalves da Cruz Fonseca, Francisco Vagnaldo Fechine, Edilson Martins Rodrigues Neto, Marta Maria de França Fonteles, Lorena Ribeiro de Aguiar

Trévia, Hilda Lara Prado Vasconcelos, Sandra Mara da Silva Lima, Mara Assef Leitao Lotif, Analice Mendes Barreto Fernandes, and Mary Anne Medeiros Bandeira. 2021. "Clinical and Antimicrobial Evaluation of *Copaifera langsdorffii* Desf. Dental Varnish in Children: A Clinical Study." *Evidence-Based Complementary and Alternative Medicine* 2021: 6647849. https://doi.org/10.1155/2021/6647849.

Vergara, Herminia, Almudena Cézar, and Noemí Rubio. 2021. "Lamb Meat Burgers Shelf Life: Effect of the Addition of Different Forms of Rosemary (*Rosmarinus Officinalis* L.)." *CyTA—Journal of Food* 19 (1): 606–613. https://doi.org/10.1080/19476337.2021.1938238.

Visentín, Alexis, Martín Cismondi, and Damián Maestri. 2011. "Supercritical CO2 Fractionation of Rosemary Ethanolic Oleoresins as a Method to Improve Carnosic Acid Recovery." *Innovative Food Science & Emerging Technologies* 12 (2): 142–145. https://doi.org/10.1016/j.ifset.2011.01.004. https://www.sciencedirect.com/science/article/pii/S1466856411000075.

Wang, Wenwen, Yuan Wang, Tao Guo, Chang Gao, Yi Yang, Lei Yang, Zhiwei Cui, Jinju Mao, Na Liu, Xiaoping An, and Jingwei Qi. 2023. "Blend of Cinnamaldehyde, Eugenol, and Capsicum Oleoresin Improved Rumen Health of Lambs Fed High-Concentrate Diet as Revealed by Fermentation Characteristics, Epithelial Gene Expression, and Bacterial Community." *Animals* 13 (10). https://doi.org/10.3390/ani13101663.

Yang, Zhijun, Zhenghao Guo, Jun Yan, and Jing Xie. 2024. "Nutritional Components, Phytochemical Compositions, Biological Properties, and Potential Food Applications of Ginger (*Zingiber officinale*): A Comprehensive Review." *Journal of Food Composition and Analysis* 128: 106057. https://doi.org/10.1016/j.jfca.2024.106057. https://www.sciencedirect.com/science/article/pii/S0889157524000917.

Yilmaz, Sevdan, ekrem sanver Celik, Sebahattin Ergün, Mert Gürkan, Fevziye Kesbiç, and Hany Abdel—Latif. 2023. "The Effects of *Capsicum annuum* Oleoresin, as a Dietary Carotenoid, on Growth, Gut Microbiome, Intestinal Histomorphometry, and Sensory Characteristics of *Oncorhynchus* Mykiss." *Journal of the World Aquaculture Society* 55. https://doi.org/10.1111/jwas.13039.

Yimdjo, Marie C., Anatole G. Azebaze, Augustin E. Nkengfack, A. Michele Meyer, Bernard Bodo, and Zacharias T. Fomum. 2004. "Antimicrobial and Cytotoxic Agents from *Calophyllum inophyllum*." *Phytochemistry* 65 (20): 2789–2795. https://doi.org/10.1016/j.phytochem.2004.08.024. https://www.sciencedirect.com/science/article/pii/S0031942204004054.

Yulianto, Mohamad, Susiana Purwanitisari, Indah Hartati, Qurrotun Nisa, and Retno Nyamiati. 2022. "Subcritical Reactive Extraction of Shogaol and Gingerol: Effect of Time and Temperature." *International Food Research Journal* 29: 857–863. https://doi.org/10.47836/ifrj.29.4.12.

Zardini, Agnese, Francesco Menegale, Andrea Gobbi, Mattia Manica, Giorgio Guzzetta, Valeria d'Andrea, Valentina Marziano, Filippo Trentini, Fabrizio Montarsi, Beniamino Caputo, Angelo Solimini, Cecilia Marques-Toledo, André B. B. Wilke, Roberto Rosà, Giovanni Marini, Daniele Arnoldi, Ana Pastore y Piontti, Andrea Pugliese, Gioia Capelli, Alessandra della Torre, Mauro M. Teixeira, John C. Beier, Annapaola Rizzoli, Alessandro Vespignani, Marco Ajelli, Stefano Merler, and Piero Poletti. 2024. "Estimating the Potential Risk of Transmission of Arboviruses in the Americas and Europe: A Modelling Study." *The Lancet Planetary Health* 8 (1): e30–e40. https://doi.org/10.1016/S2542-5196(23)00252-8.

Zimmermam-Franco, Danielle, Edilene Bolutari, Hudson Polonini, Antônio do Carmo, Maria das Graças A. M. Chaves, and Nádia Raposo. 2013. "Antifungal Activity of *Copaifera langsdorffii* Desf Oleoresin Against Dermatophytes." *Molecules* 18 (10): 12561–12570. https://doi.org/10.3390/molecules181012561. http://www.mdpi.com/1420-3049/18/10/12561.

Section 6

Polar Compounds and Functional Foods

16 Advances in Pharmacological Activities of Terpenoids

Wenqiang Yang, Xu Chen, Yanli Li, Shaofen Guo, Zhen Wang, and Xiuling Yu

16.1 INTRODUCTION

Terpenoids constitute one of the largest and most structurally diverse groups of naturally occurring compounds. Terpenoids are a class of natural products derived from mevalonic acid (MVA) which are composed of a plurality of isoprene (C5) structural units. Terpenoids are widely found in nature, with various structures and a wide variety. To date, more than 50,000 terpenoids have been found in nature,[1] and most of them are isolated from plants. Some terpenoids play an important role in plant growth and development, such as gibberellin, as plant hormones regulate plant development and carotenoids participate in photosynthesis; some terpenoids play a role in the interaction between plants and the environment, such as participating in plant defense systems in the form of phytoalexins and interspecies competition as interspecific sensing compounds.[2] Many volatile terpenoids, such as menthol and perillyl alcohol, are used as raw materials for spices, flavorings, and cosmetics.[3] There are also some terpenoids with important economic value. They are used as pesticides, industrial raw materials, etc., such as pyrethrin and limonoids, which are often used as insecticides. Sesquiterpenes, farnesene and bisabolene, and monoterpenes, pinene, and limonene are recognized as precursor compounds for fuels.[4]

In the 1960s, the biological activity of terpenoids began to be noticed by humans, but only limited to certain terpenoids. By the 1970s, the pace of research on the biological activity of terpenoids was gradually accelerated, and the first small peak appeared. The research on artemisinin against malaria was completed at this stage. It was not until the late 1990s that the study on terpenoids attracted attention again, and the research focused on the separation and purification of active components, chemical composition analysis, and chemical structure research. Since then, the biological activity of terpenoids has increased year by year.

In recent years, with the deepening of research on terpenoids (especially terpenoids in medicinal plants), it has been found that such compounds play an increasingly prominent role in the field of medicine and have various biological activities, such as exhibiting antitumor, anti-inflammatory, antibacterial, antiviral, and antimalarial potential; promoting transdermal absorption; preventing and treating cardiovascular diseases; lowering blood sugar; and other effects. In addition, some terpenoids also have insecticidal, immunomodulatory, antioxidant, antiaging, and neuroprotective effects; the terpenoids paclitaxel and artemisinin have been widely used in clinical practice. Therefore, research on the biological activity of the terpenoids will contribute to the selection of drugs and the improvement of treatment methods and provide a theoretical basis for the development of new drugs, which are paid much attention by scholars.

16.2 OVERVIEW OF TERPENOIDS

Terpenoids are a general term for a class of compounds consisting of several isoprene structural units. Based upon the number of isoprene units, these have mainly been classified into monoterpene

DOI: 10.1201/9781003462804-22

(C10), sesquiterpene (C15), diterpene (C20), triterpene (C30), tetraterpene (C40), polyterpene (C > 40), etc. In addition to being in the form of terpene hydrocarbons, terpenoids are mostly present in the form of various oxygen-containing derivatives, including alcohols, aldehydes, carboxylic acids, ketones, esters, and glycosides. The synthetic pathways for terpenoids include the MVA pathway and the 1-deoxy-D-xylulose-5-phosphate (DXP) pathway. Isopentenyl diphosphate is the main metabolic intermediate for both ways. The MVA pathway exists in the cytoplasm, and the secondary metabolites, such as sesquiterpenes, sterols, and triterpenes, are mainly synthesized through this way; the DXP pathway is mainly present in plastids, and monoterpenes, diterpenes, and tetraterpenes are synthesized by this way.[5]

Terpenoids are mostly present in the form of volatile oils in higher medicinal plants and mainly exist in the following medicinal plant groups: Compositae, Ranunculaceae, Araliaceae, Oleaceae, Magnoliaceae, Lauraceae, Aristolochiaceae, Rutaceae, Labiatae, Pinaceae, Umbelliferae, Celastraceae, Acanthaceae, Taxaceae, etc. Most of the terpenes with bioactivities have been isolated in medicinal plants. Monoterpenes and sesquiterpenes are mainly found in essential oils of the medicinal plant; larger-molecular-weight terpenes, such as triterpene, are mainly found in balsam and resin.

16.3 PHARMACOLOGICAL ACTIVITIES OF TERPENOIDS

16.3.1 Antitumor Activity

Tumors are the most serious diseases that threaten human health, and the incidence of cancer in China is increasing year by year. The research on new high-efficiency antitumor drugs is imminent. Terpenoids have attracted the interest of many medicinal chemists with their distinctive structural features and good antitumor activity and have the potential to be the leading compounds in developing efficient and safe antitumor drugs. In the study of the antitumor effects of terpenoids (Figure 16.1), it was found that perillyl alcohol, geraniol, and paclitaxel are the terpenoids that have antitumor activities (Table 16.1).

Perillyl alcohol, a monocyclic monoterpene mainly present in the essential oils of medicinal plants, has attracted significant interest due to its potent antitumor activity. It has broad-spectrum, high-efficiency, low-toxicity antitumor properties. The growth of tumor cells was significantly inhibited after perillyl alcohol was added in the culture of tumor cells in many kinds of animals.[6] And perillyl alcohol plays a preventive and therapeutic role in cancer. The results showed that the administration of perillyl alcohol at a dose of 1–2 g/kg in rats can significantly reduce the incidence and multiplicity of colonic invasive adenocarcinoma caused by the injection of carcinogen azomethane.[6] Intraperitoneal injection of perillyl alcohol at a dose of 75 mg/kg three times per week significantly inhibited lung tumor formation caused by the simultaneous injection of the carcinogen 4-(methyl-nitrosamino)-1-(3-pyridyl)-1-butanone.[6] It was found that perillyl alcohol also exhibited high cytotoxic activity against OVCAR-8, HCT-116, and SF-295 human tumor cell lines, using the MTT assay. The cell growth inhibition percentage values were 90.92–95.82%.[32] Recent studies have revealed that the tumor growth inhibition rate of perillyl alcohol evaluated in mice was 35.3% and 45.4% at doses of 100 and 200 mg/kg/day, respectively. No toxicologically significant effect was found in liver and kidney parameters.[33] In in vitro anticancer studies, perillyl alcohol was found to exert cytotoxicity against HepG2 cell line with a half-maximal inhibitory concentration (IC_{50}) value of 409.2 μg/mL. However, this effect was not found to be selective.[34]

Geraniol is widely found in essential oils of aromatic medicinal plants. Current experimental evidence indicates that geraniol has therapeutic or prophylactic effects on different types of cancer (e.g., lung cancer,[7] colon cancer,[8] prostate cancer,[9] pancreatic cancer,[10] and liver cancer).[11] It has been demonstrated that geraniol can regulate a variety of signaling molecules and participate in a variety of life processes, such as cell cycle, cell proliferation, apoptosis, autophagy, and metabolism. It can be used as a multitarget drug for the treatment of cancer; the effect is significant

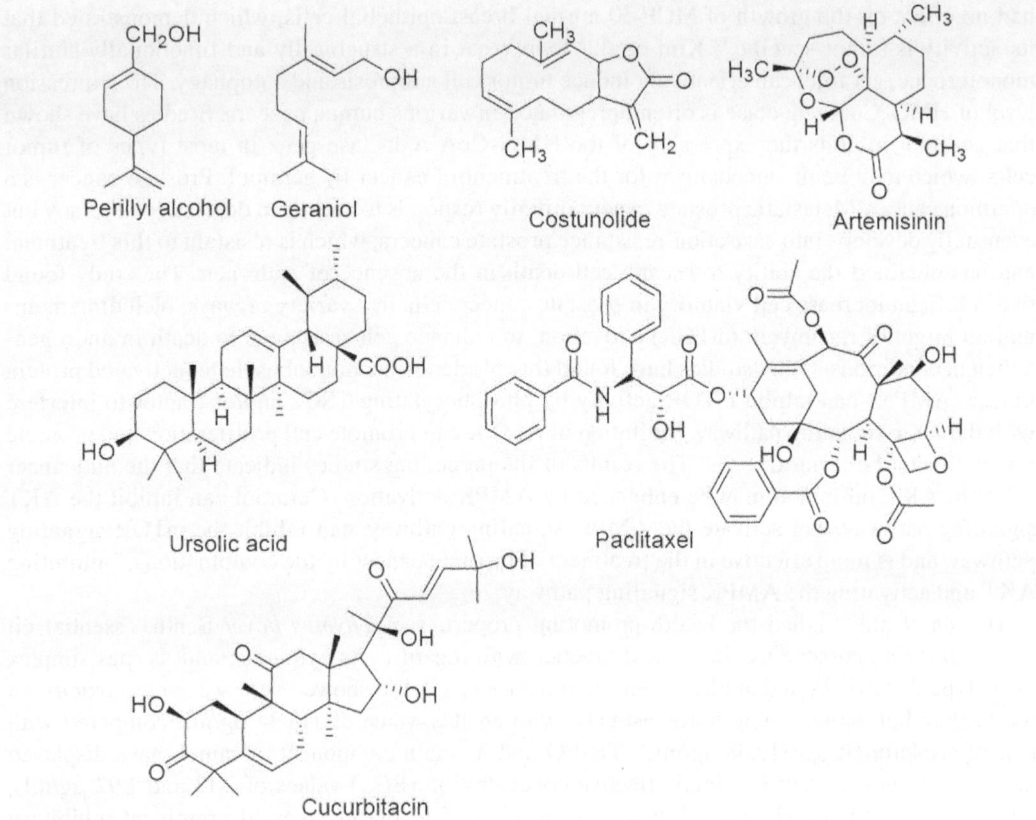

FIGURE 16.1 The structures of common terpenoids with antitumor activities.

TABLE 16.1
The Antitumor Activity of Terpenoids

Classification	Compound	Function	Reference
Monoterpene	Perillyl alcohol	Broad-spectrum anticancer	6
	Geraniol	Anti–lung cancer, colon cancer, prostate cancer, pancreatic cancer, and liver cancer	7–13
Sesquiterpene	Costunolide	Anti–bladder cancer, ovarian cancer, leukemia cells, prostate cancer, non-small-cell lung cancer, esophageal cancer	14–18
	Artemisinin and its derivatives	Antileukemia, melanoma, colon cancer, non-small-cell lung cancer, lung cancer, prostate cancer, breast cancer, and ovarian cancer	19–22
Diterpene	Paclitaxel	Anti–ovarian and breast cancer	23, 24
Triterpene	Ursolic acid	Anti–liver cancer, breast cancer, osteosarcoma, prostate cancer, and cervical cancer	25–30
	Cucurbitacin	Anti–bladder cancer, liver cancer, pancreatic cancer, breast cancer, and leukemia	31

and is not affected by adaptive resistance. For example, geraniol inhibits tumor cell growth by blocking the G1 phase of the Michigan Cancer Foundation (MCF)-7 breast cancer cell cycle. The levels of cyclin D1, cyclin-dependent kinase 4, cyclin E, and cyclin A were decreased in geraniol-treated MCF-7 cells, while the level of cyclin p27Kip1 was increased. At the same time, geraniol had no effect on the growth of MCF-10 normal breast epithelial cells, which demonstrated that its activity is tumor-specific.[12] Kim et al.[13] found that in a structurally and functionally similar monoterpene, geraniol can effectively induce tumor cell apoptosis and autophagy. The expression level of HMG-CoA reductase is often upregulated in various human cancers. Studies have shown that geraniol inhibits the expression of the HMG-CoA reductase gene in most types of tumor cells, which may be the mechanism for the treatment of cancer by geraniol. Prostate cancer is a common cancer. Metastatic prostate cancer initially responds to androgen deprivation therapy but eventually develops into castration-resistance prostate cancers, which is resistant to this treatment and has obtained the ability to escape cell death in the absence of androgen. The study found that AKT can increase cell viability in prostate cancer cells in a variety of ways, including mammalian target of rapamycin (mTOR) activation, to enhance cell resistance to death in androgen-deficient conditions. Other studies have found that 5′ adenosine monophosphate–activated protein kinase (AMPK) can inhibit mTOR activity by phosphorylating TSC2 and/or Raptor to interfere with the AKT signaling pathway. Inhibition of AMPK can promote cell proliferation and promote the malignant behavior of cells. The results of the preceding studies indicate that the anticancer effect of AKT inhibition may be enhanced by AMPK activation. Geraniol can inhibit the AKT signaling pathway, can activate the AMPK signaling pathway, can inhibit the mTOR signaling pathway, and is more effective in the treatment of prostate cancer by the combination of inhibiting AKT and activating the AMPK signaling pathway.

Hassan et al.[14] studied the health-promoting properties of *Thymus bovei* Benth. essential oil (TB-EO) for its possible use in clinical practice with regard to its cytotoxic, anti–herpes simplex virus type 2 (HSV-2), and antihypertensive properties. TB-EO showed very weak cytotoxicity on the healthy human fetal lung fibroblast cells with an IC_{50} value of 118.34 μg/mL compared with that of cisplatin (IC_{50} = 10.08 μg/mL). TB-EO and its main component, geraniol, have displayed anti-HSV-2 with the half-maximal effective concentration (EC_{50}) values of 2.13 and 1.92 μg/mL, respectively. TB-EO and geraniol at a concentration of 15 μg/mL showed prominent inhibitory activities against angiotensin-converting enzyme with the percentage of inhibition at 95.4% and 92.2%, respectively, compared with that of standard inhibitor captopril (99.8%; 15 μg/mL). The results may be a novel findings in the course of a search for new and active anticancer, anti-HSV-2, and antihypertensive agents.

Costunolide (CT) is a sesquiterpene lactone compound and is one of the main chemical constituents of the medicinal plant *Aucklandia lappa* Decne. Studies have shown that CT has effects of anti–bladder cancer,[15] ovarian cancer,[16] leukemia,[17] and prostate cancer,[18, 19] mainly by inhibiting cancer cell proliferation, inducing apoptosis and differentiation of cancer cells, inhibiting metastasis and invasion of cancer cells, reversing multidrug resistance, and inhibiting angiogenesis.

Artemisinin and its derivatives, well-known as antimalarial drugs, in recent years have been reported to have specific inhibitory effects on tumors in vitro and in vivo, with small side effects, obvious effects, and low costs.[20, 21] For example, artesunate is a semisynthetic derivative of artemisinin that exhibits anticancer activity against various types of tumors, such as leukemia, melanoma, colon cancer, non-small-cell lung cancer, lung cancer, prostate cancer, breast cancer, and ovarian cancer.[22] The present studies have shown that semisynthetic artemisinin derivatives can exert anticancer effects by causing cell cycle G0/G1 arrest, inducing cell apoptosis, and participating in oxidative stress, which result in more efficient antitumor activity than monomeric compounds, but this mechanism of action has not been elucidated.[23]

Paclitaxel is a kind of tetracyclic diterpenoid which is isolated from taxus plants and has a good therapeutic effect on cancers, such as ovarian cancer and breast cancer.[24] Paclitaxel was cytotoxic against SKOV3 cells in vitro, with a viability rate of 38.2 ± 1.3% for 1 μmol/L of paclitaxel, and the apoptotic rate of SKOV3 cells was 15.7 ± 1.7 for paclitaxel. Sun et al.[25] found that paclitaxel can activate the toll-like receptor 4-nuclear factor-kappa B (NF-κB) pathway and induce the expression of the *ABCB1* gene, which is of great significance for studying the drug resistance of paclitaxel in the treatment of ovarian cancer.

Ursolic acid can induce tumor cell apoptosis and has obvious antitumor effects, but its proapoptotic mechanism has not yet been elucidated. Recently, the mechanism of action of ursolic acid in the treatment of various types of tumors has been studied in depth.[26–29] The current investigation demonstrated that ursolic acid could protect hepatoma cells and reduce HBx-mediated autophagy through modulation of Ras homolog gene family member A (RhoA).[26] In the present study, it was found that ursolic acid at the low micromolar range may promote its anticancer action by targeting glycolysis in phenotypically distinct breast cancer cells.[27] Ursolic acid can also inhibit proliferation and induce apoptosis of human osteosarcoma 143B cells by inactivating Wnt/β-catenin signaling.[28] Research has shown that ursolic acid activates cell apoptosis in prostate cancer through rho-associated protein kinase/phosphatase and tensin homolog–mediated mitochondrial translocation of cofilin-1.[29] In addition, ursolic acid is made into nanoparticles which can improve anticancer efficacy and bioavailability.[30, 31] Cucurbitacin is a class of tetracyclic triterpenoids that also have antitumor activity.[35]

16.3.2 Anti-Inflammatory Activity

Inflammation is a common and very important pathological process, often manifesting as "red, swollen, hot, painful." It is a defensive response of living tissues with a vascular system to various damage factors, and it is closely related to various diseases, such as asthma, rhinitis, arthritis, and arteriosclerosis. If inflammation is allowed to develop and is not controlled, it can cause serious illness. The structures of terpenoids with anti-inflammatory activities are shown in Figure 16.2.

Paeoniflorin is a monoterpene glycoside compound isolated from the root of *Paeonia lactiflora* Pall. Bi et al.[36] studied the anti-inflammatory activity and mechanism of action of paeoniflorin, paeoniflorin derivatives, 4-*O*-methyl paeoniflorin, 4-*O*-methylbenzoyl paeoniflorin, and other monoterpenoids in peony. The results showed that most of the monoterpenes could inhibit the production of inflammatory factor nitric oxide (NO), interleukin-6 (IL-6), and tumor necrosis factor-alpha (TNF-α) induced by lipopolysaccharides (LPS). And it has a significant correlation with the dose.

Inula flower (*Inula japonica* Thunb.) is a traditional Chinese herbal medicine of the genus Asteraceae. Chen et al.[37] found that the sesquiterpene lactone compound IVSE in the inula flower can inhibit the production of NO induced by LPS, thereby exhibiting the anti-inflammatory activity. Wang et al.[38] isolated a novel sesquiterpene lactone compound, JEUD-38, in the *Inula* flower; this compound significantly attenuated LPS-induced NO production and had the effect of preventing and treating inflammatory diseases. In this research, the results showed that under the stimulation of LPS, the amount of NO increased by about 11 times compared with the blank group. After the addition of JEUD-38 (the cells were treated with different concentrations of JEUD-38), the production of NO was significantly inhibited, and the inhibitory effect was significantly correlated with the dose.

The Chinese traditional medicine *Tripterygium wilfordii* Hook. f. has been used to treat immune system diseases and inflammatory diseases for hundreds of years. Triptolidenol is the main bioactive component of *Tripterygium wilfordii.* It is also one of the most effective natural products of inflammation and immune regulation that has been found to treat various autoimmune and inflammation-related disorders.[39] Its main mechanism is to inhibit the production of inflammatory cytokines.[40] Research has also shown that triptolide, tripterine, and triptonide all have obvious anti-inflammatory effects.[41–43] In one study, it was found that both triptolide and triptonide significantly relieved the symptoms of acute lung injury (ALI) in mice. Triptolide can improve the pathological state of lung tissue in ALI mice. Triptonide can significantly reduce the expression of chemokines MIP-1α, MIP-1β, RANTES, MCP-1, and IP-10 induced by LPS.

Ginseng is a traditional Chinese medicine; its main component is ginsenoside. With the deepening of research, it has been found that ginsenoside-Rb1 (G-Rb1) is a potential anti-inflammatory agent. G-Rb1 can significantly inhibit the activation of NF-κB (NF-κB is a key factor in inflammation and

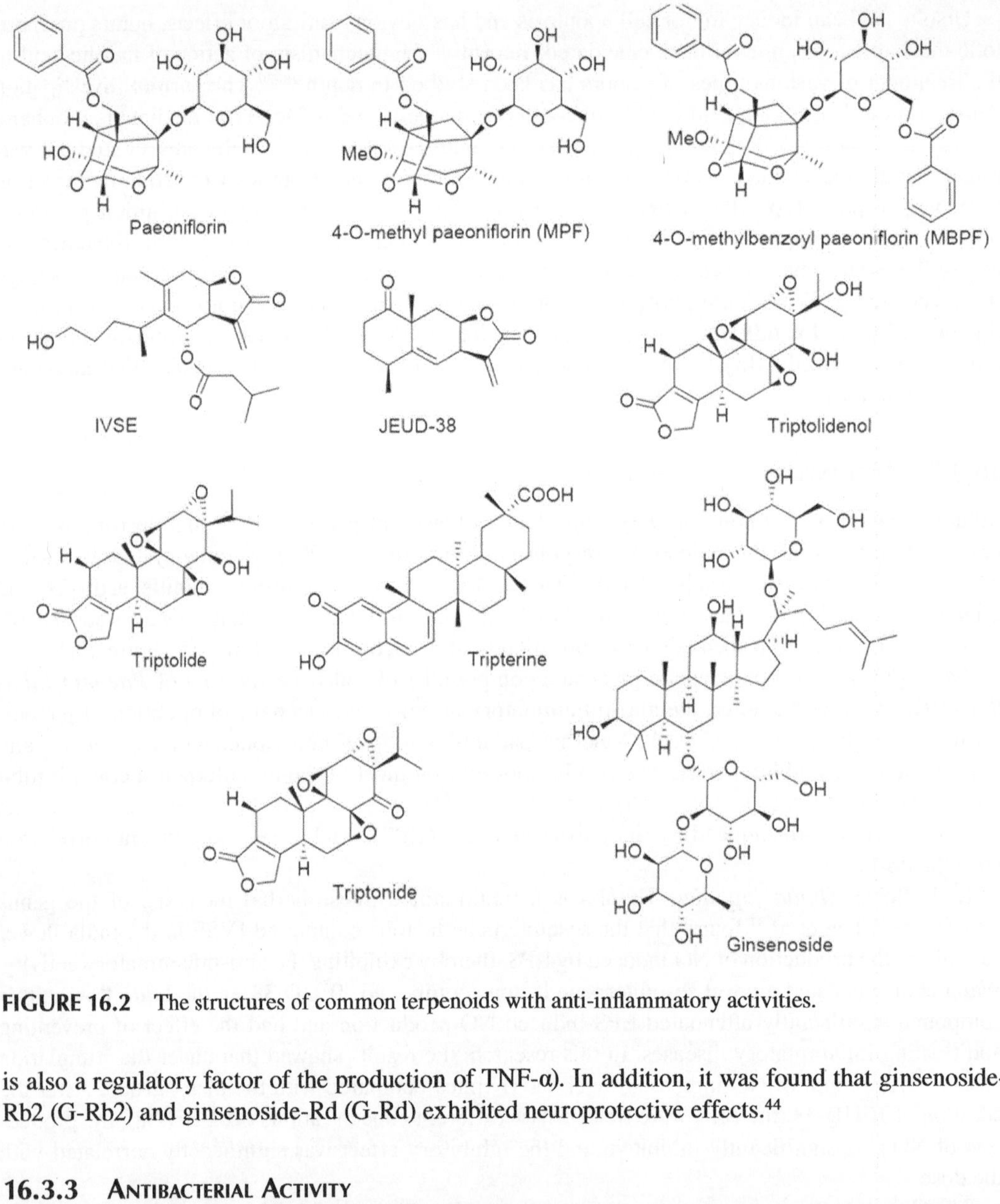

FIGURE 16.2 The structures of common terpenoids with anti-inflammatory activities.

is also a regulatory factor of the production of TNF-α). In addition, it was found that ginsenoside-Rb2 (G-Rb2) and ginsenoside-Rd (G-Rd) exhibited neuroprotective effects.[44]

16.3.3 Antibacterial Activity

Terpenoids (Figure 16.3) also have strong antibacterial effects[45–49] (Table 16.2). The monoterpenoids in terpenoids are mainly found in the genus *Mentha*, and previous studies have found that most of the compounds extracted from plants of the genus *Mentha* have strong antimicrobial activity.[50] *Menthol* is a cyclic monoterpene which has been shown to have antibacterial activity.[51–54] Raut et al.[55] found that menthol showed significant inhibitory activity of biofilm when studying the effects of plant-derived terpenoids on *Candida albicans*.

Patchouli alcohol (PA) is a tricyclic sesquiterpenoid compound found in *Pogostemon cablin* (Blanco) Benth. Xu et al.[58] found that it had anti–*Helicobacter pylori* activity in vitro and in vivo. The experimental data show that the bactericidal effect of PA is dependent on time and dose under different pH conditions, and the minimal bactericidal concentrations (MBCs) of PA were 25–75 µg/mL. In addition, PA has a significant inhibitory effect on the movement of *H. pylori* and the formation of flagella.

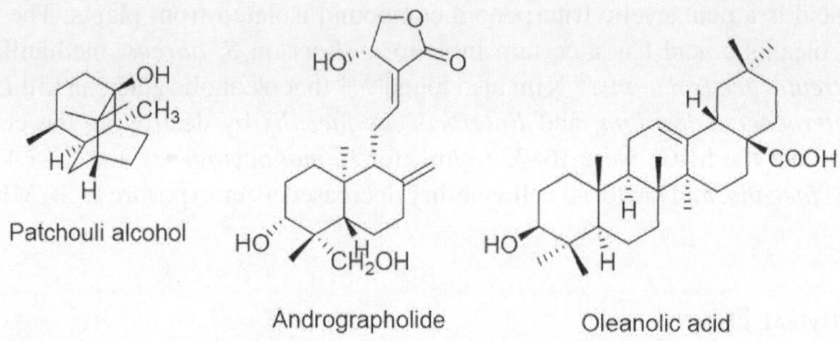

FIGURE 16.3 The structures of common terpenoids with antibacterial activities.

TABLE 16.2
The Antibacterial Activity of Terpenoids

Classification	Compound	Antibacterial Range	Reference
Monoterpene	1,8-eucalyptus	*Staphylococcus aureus, Bacillus subtilis, Escherichia coli,* and *Streptococcus, Listeria, Bacillus cereus*	45, 46
	Limonene, geranialdehyde	*B. subtilis, S. aureus, Streptococcus mutans, E. coli, Candida albicans,* methicillin-resistant *S. aureus,* and *Saccharomyces cerevisiae*	56
	Sabinene	*S. aureus* (Gram-positive) and *E. coli* (Gram-negative)	57
	Menthol	*S. aureus, Streptococcus pneumoniae, Streptococcus pyogenes, Haemophilus influenzae*	50–55
	Sabinol	Oral bacteria	47
	Carvone	*S. aureus, E. coli, B. subtilis, Salmonella typhimurium, Listeria*	48, 49
Sesquiterpene	Patchouli Alcohol	*Helicobacter pylori*	58
Diterpene	Artemisinin	Various pathogens, such as *B. subtilis, E. coli, Pseudomonas aeruginosa, S. cerevisiae, S. aureus, Mycobacterium tuberculosis*	59–61
	Andrographolide	*P. aeruginosa*	62, 63
Triterpene	Oleanolic acid	*S. aureus,* methicillin-resistant *S. aureus, S. mutans* and *Listeria monocytogenes, Enterococcus faecium, Enterococcus faecalis*	64–66

Studies have found that artemisinin drugs have different antibacterial activities against anaerobic bacteria, facultative anaerobic bacteria, microaerophilic bacteria, and aerobic bacteria. This antibacterial activity has specificity and concentration dependence, which is reflected in different bacteria with different antibacterial activities.[59–61]

Pseudomonas aeruginosa is a clinically important conditional pathogen with strong pathogenicity and high mortality. Due to the widespread use of antibiotics, its sensitivity to most antibiotics is reduced. It is a common pathogen. Andrographolide is a diterpene lactone compound in the Chinese medicine *Andrographis paniculata* Nees. Cheng et al.[62] found that andrographolide had a significant inhibitory effect on the biofilm of *P. aeruginosa* and synergistic antibacterial effect with azithromycin. Banerjee et al.[63] found that andrographolide exhibited potential antibacterial activity against most of the tested Gram-positive bacteria, among which it was most sensitive to *Staphylococcus aureus*, with a minimal inhibitory concentration (MIC) value of 100 μg/mL, and was found to have an inhibitory effect on the formation of the biofilm of *S. aureus*.

Oleanolic acid is a pentacyclic triterpenoid compound isolated from plants. The results of one study showed oleanolic acid has a certain inhibitory effect on *S. aureus*, methicillin-resistant *S. aureus*, and *Streptococcus mutans*.[64] Kim also found[65, 66] that oleanolic acid can kill *Listeria monocytogenes*, *Enterococcus faecium*, and *Enterococcus faecalis* by destroying the cell membranes of the bacteria, and the MICs were 16–32 μg/mL for *L. monocytogenes* and 32–64 μg/mL for *E. faecium* and *E. faecalis*, and bacterial cell viability decreased after exposure to 2× MIC of oleanolic acid.

16.3.4 Antiviral Effect

Terpenes are among the very promising source of new antimicrobial agents that have been shown to have activity against viruses, bacteria, fungi, and protozoa.[67] The structures of terpenoids with antiviral effects are shown in Figure 16.4. Monoterpenoids, isoborneol, and borneol, have strong activities against herpes simplex virus 1 (HSV-1). Isoborneol is a monoterpene found in a wide range of plant essential oils. This compound showed total inhibition of HSV-1 replication at a concentration of 0.06%.[68] Other monoterpene compounds, such as thymol, α-terpinene, γ-terpinene, 1,8-cineole, α-terpineol, and citral, isolated from tea tree, thyme, and eucalyptus, demonstrated in vitro antiviral activity against HSV-1 by inhibition of virus greater than 80%.[69]

Hassan et al.[70] isolated 12 pure compounds from the leaves and twigs of *Eucalyptus globulus*, and all of them were tested for antiherpetic activity against the replication of antigen types HSV-1 and HSV-2. The results demonstrated that the compound tereticornate A (IC_{50}: 0.96 μg/mL; selectivity index CC_{50}/IC_{50}: 218.8) showed the strongest activity in the anti-HSV-1 assay, even greater than acyclovir (IC_{50}: 1.92 μg/mL; selectivity index CC_{50}/IC_{50}: 109.4), a standard antiviral drug. And the compound also displayed moderate antibacterial effects and anti-inflammatory activity.

Putranjivain A, a diterpene isolated from *Euphorbia jolkini*, demonstrated antiviral effect against HSV-2 in Vero cells, with an IC_{50} value of 6.3 μM.[71] The two triterpenes compounds, moronic acid and betulonic acid, extracted from the plant *Rhus javanica* showed in vitro potent inhibitory effect on HSV-1 with EC_{50} values of 3.9 and 2.6 μg/mL, respectively.[72] Notoginsenoside ST-4i, isolated from the Chinese herb *Panax notoginseng*, demonstrated in vitro remarkable inhibitory activities against HSV-1 and HSV-2, with EC_{50} values of 16.4 and 19.44 μM, respectively.[73] The results of inhibitory activity against HSV-1 of glycyrrhizin (GR) indicated that GR eliminated the increased susceptibility of thermally injured mice to HSV infectivity during the induction of CD4+ contrasuppressor T cells.[74]

Artemisinin and its derivatives showed significant inhibitory activity on the hepatitis B virus (HBV) and hepatitis C virus (HCV).[75] Artesunate is the semisynthetic derivative of artemisinin, which has a more pronounced inhibitory effect on hepatitis B surface antigen (HBsAg). A synergistic effect was observed by combining artesunate with lamivudine (a drug for treating HBV).

Recent studies have found that andrographolide has a significant inhibitory effect on the replication process of the chikungunya virus (CHIKV) (a mosquito-borne alphavirus).[76] Andrographolide showed good inhibition of CHIKV infection and reduced virus production by approximately $3\log_{10}$, with a 50% EC_{50} of 77 μM without cytotoxicity. Time-of-addition and RNA transfection studies showed that andrographolide affected CHIKV replication, and the activity of andrographolide was shown to be cell type–independent. Dengue fever is the most common viral disease transmitted by arthropods in humans, but there are currently no drugs against the dengue virus (DENV). Panraksa showed that andrographolide exhibited significant anti-DENV activity in both cell lines (HepG2 and HeLa), significantly reducing both the levels of cellular infection and virus output, with 50% EC_{50} for DENV of 21.304 μM and 22.739 μM for HepG2 and HeLa, respectively.[77]

Harada et al.[78] found that GR (a triterpenoid saponin) can inhibit the replication of severe acute respiratory syndrome–associated virus and regulate the mobility of the envelope of the human immunodeficiency virus (HIV).

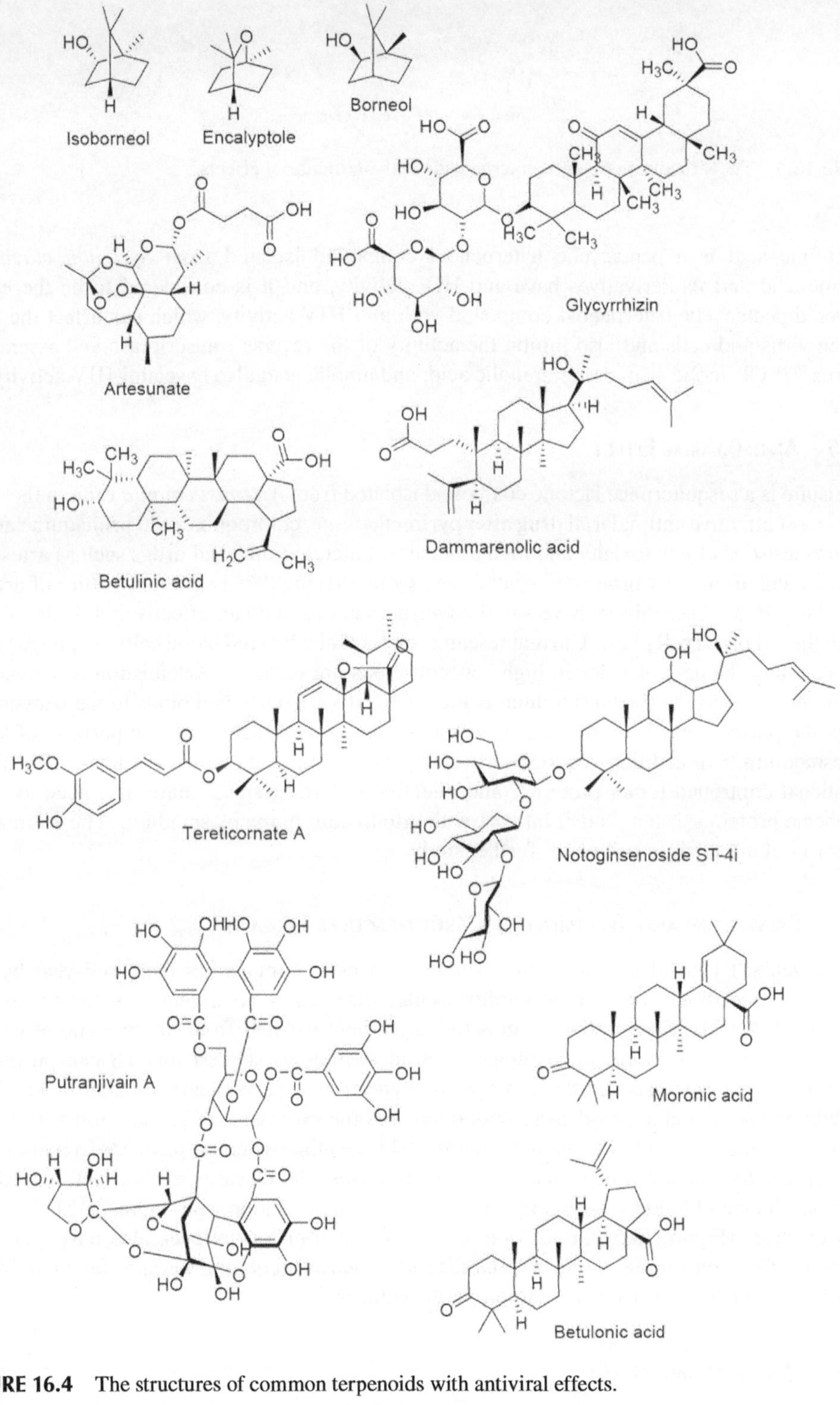

FIGURE 16.4 The structures of common terpenoids with antiviral effects.

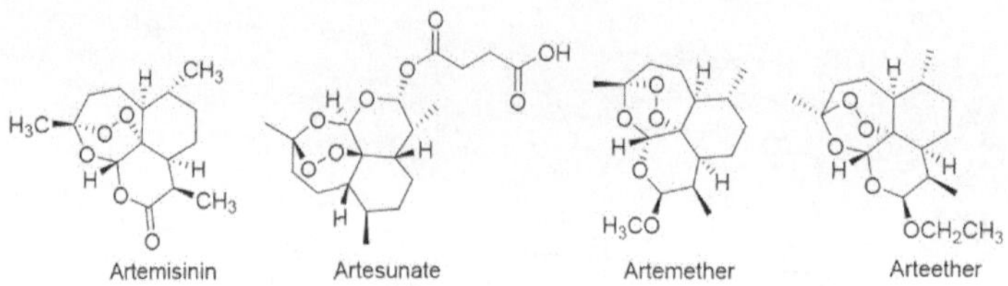

FIGURE 16.5 The structures of common terpenoids with antimalarial effects.

Betulinic acid is a pentacyclic triterpenoid compound isolated from *Syzigium claviforum*. Betulinic acid and its derivatives have anti-HIV activity, and it is considered to be the earliest discovered pentacyclic triterpenoid compound with anti-HIV activity, which can affect the fusion between virus and cells and also inhibit the activity of the reverse transcriptase and assembly of the virus.[79, 80] Oleanolic acid, dammarenolic acid, and ursolic acid also have anti-HIV activity.[81–83]

16.3.5 Antimalarial Effect

Artemisinin is a sesquiterpene lactone compound isolated from *Artemisia annua* Linn in the 1970s. It is the most effective antimalarial drug after pyrimethamine, chloroquine, and primaquine and has the characteristics of low toxicity and high efficiency. Later, antimalarial drugs such as artesunate, Arteether, and artemether have been synthesized by modifying the chemical structure of artemisinin (Figure 16.5). These drugs have small adverse reactions and can effectively kill the plasmodium in the red blood cell phase. Current research shows that when red blood cells are phagocytosed by plasmodium, heme molecules in high concentrations are released. Artemisinin is activated by heme in the place where the plasmodium is metabolized vigorously and binds to the parasite proteins in the plasmodium body to make the proteins inactivated, achieving the purpose of killing the plasmodium.[84] In addition, sarcoplasmic endoplasmic reticulum calcium ATPase (PfATP6),[85] translational controlled tumor protein,[86] and glutathione *S* transferase[87] have also been identified as nonheme protein substances that interact with artemisinin in the plasmodium. The antimalarial mechanism of artemisinin still needs further study.

16.3.6 Prevention and Treatment of Cardiovascular Diseases

In recent years, the incidence of cardiovascular diseases in humans has increased year by year. Finding effective drugs for treating cardiovascular diseases is an urgent task for researchers. Tanshinone IIA (TS) (Figure 16.6) is an active ingredient isolated from the rhizome of Chinese herbal medicine *Salvia miltiorrhiza* Bunge. The latest findings suggest that TS can prevent the formation of atherosclerosis and the damage and hypertrophy of the heart. In atherosclerosis, TS can inhibit the oxidation of low-density lipoprotein and the expression of proinflammatory factors, and TS also has certain activity and potential to stabilize atherosclerotic plaque.[88] In addition, TS has significant therapeutic effects on a variety of cardiovascular diseases, such as anti–myocardial infarction, dilation of blood vessels, improvement of microcirculation, anti-angina, etc.[89–91]

Ginsenosides (Figure 16.6), in addition to its anti-inflammatory effects, also have significant therapeutic effects on various cardiovascular diseases, such as regulating vascular function, inhibiting cardiomyocyte hypertrophy, and inhibiting thrombosis.[92]

16.3.7 Hypoglycemic Effect

Diabetes is a metabolic disease characterized by high blood sugar. The long-standing high blood sugar in diabetics can cause chronic damage and dysfunction in various tissues. Diterpenoid

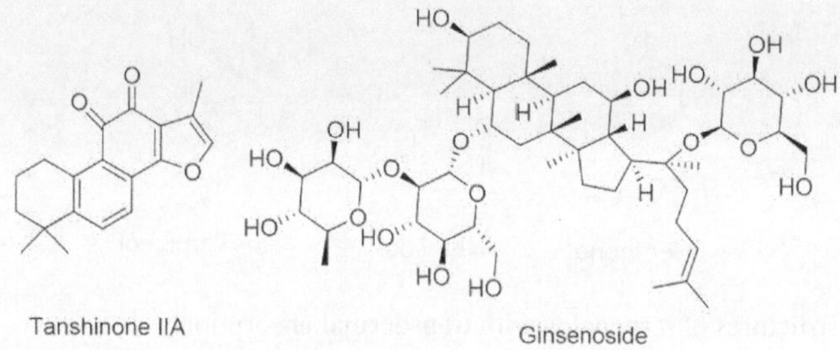

FIGURE 16.6 The structures of terpenoids with prevention and treatment of cardiovascular diseases.

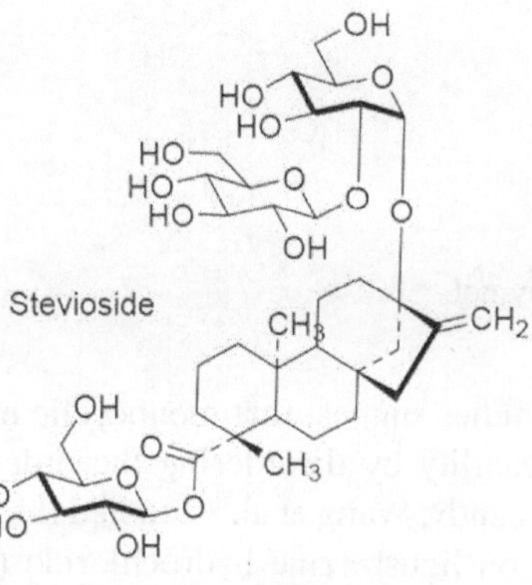

FIGURE 16.7 The structure of terpenoids with hypoglycemic effect.

stevioside (SVS) (Figure 16.7) is an active ingredient isolated from plant stevia (*Stevia rebaudiana* (Bertoni) Hemsl.). In the study by Jeppesen et al.,[93] it was found that SVS can significantly improve the hyperglycemia of rats. Its antihyperglycemic effect may be related to the expression of glycolytic-related genes, phosphorylation of liver mitochondrial ATP, and inhibition of nicotinamide adenine dinucleotide-oxidase activity, thereby achieving the effect of lowering blood sugar.

In recent years, artemisinin has been found to be a potential drug to improve type 1 diabetes. Ginsenosides have also made great progress in the research on the prevention of diabetes and hypoglycemic activity.

16.3.8 Promotion of Transdermal Absorption

The presence of the stratum corneum of the skin is a major obstacle in the percutaneous administration process.[94] In order to enhance the permeability of drugs through the skin, researchers have done a lot of research. The most popular method currently is the application of penetration enhancers, including natural terpenes, the most commonly used terpenes being menthol, menthone, 1,8-cineole, cinene, and nerolidol (Figure 16.8).[95] Natural terpene has high activity, little skin irritation, and low toxicity and can safely and effectively promote transdermal absorption of drugs. In addition, several terpenes (such as 1,8-cineole, menthol, and menthone) have been included in the list of "generally recognized as safe" issued by the US Food and Drug Administration. Hydrocarbon terpenes, such as D-cinene, have been approved as steroid activity enhancers.[96] Another study reported that α-terpineol is used as a transdermal enhancer for zidovudine and buspirone hydrochloride,[97, 98] and limonene has the effect of enhancing the skin penetration of ketoprofen and aceclofenac.[99, 100]

Terpenes interfere with the lipid arrangement in the intercellular regions of the stratum corneum (SC),[101] which leads to increased permeability of the skin and thus aids in drug absorption.

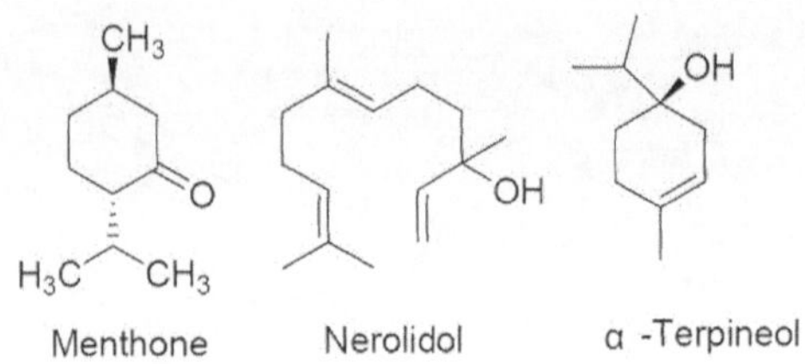

FIGURE 16.8 The structures of terpenoids with transdermal absorption.

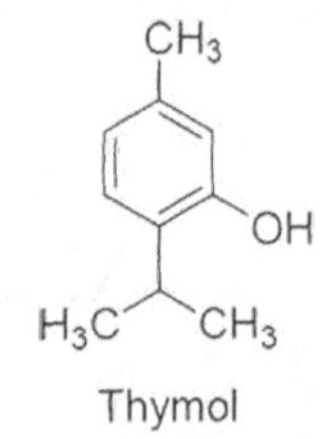

FIGURE 16.9 The structure of thymol.

The molecular mechanism studies suggest that monocyclic monoterpenes, such as menthol and menthone, enhance skin permeability by disordering the ordered organization of SC lipids and extracted part of SC lipids.[102] Recently, Wang et al.[103] studied the mechanism of transdermal absorption of menthol and menthone on ligustrazine hydrochloride (LH). The results suggest that the mechanism of action may involve hydrogen bonding and interference with the SC lipid arrangement. In addition, the polarity of terpenes is a key factor affecting drug absorption. Terpenes, such as limonene, have a strong promoting effect on the penetration of lipophilic drug molecules,[104] while terpenes with polar groups, such as menthol, have a better penetration-enhancing effect on hydrophilic drug molecules.[105]

16.3.9 Other Biological Activities

In addition to the aforementioned antitumor activity, anti-inflammatory activity, antibacterial activity, antiviral activity, antimalarial activity, prevention and treatment of cardiovascular diseases, and hypoglycemic activity, terpenoids also have other biological and pharmacological activities, such as insect resistance, immunoregulation, antioxidation, antiaging, and neuroprotection.

The results showed that monoterpenoid thymol (Figure 16.9) and its derivatives, menthol derivatives, have certain insecticidal activity.[106, 107] Through a series of screening and determination, it was found that 3,7-dimethyl-1-octanol (dihydrocitronellol) is the most effective terpene with antischistosomal activity (IC_{50} values of 13–52 μM)[108]; confocal laser scanning microscopy revealed that the compound induced severe tegumental damage in adult schistosomes, and a correlation between viability and tegumental changes was observed.

Oleanolic acid, ursolic acid, and triterpene acids of loquat (TAL) have good immunomodulatory effects.[109, 110] They are potent immunomodulators of macrophages infected with virulent *Mycobacterium tuberculosis* and their immunomodulatory activity induced through the activation of macrophages.

The ginsenoside Re isolated from American ginseng has an antioxidation effect and can protect cardiomyocytes from oxidative damage of internal and external oxidants.[111] Ginsenoside Rg1 regulates the expression of regulatory factors in the cell cycle and then exerts its antiaging effects. Currently, ginsenoside Rd is widely used for neuroprotection. In addition, Wan et al.[112] found that ginsenoside Rd can alleviate cognitive dysfunction.

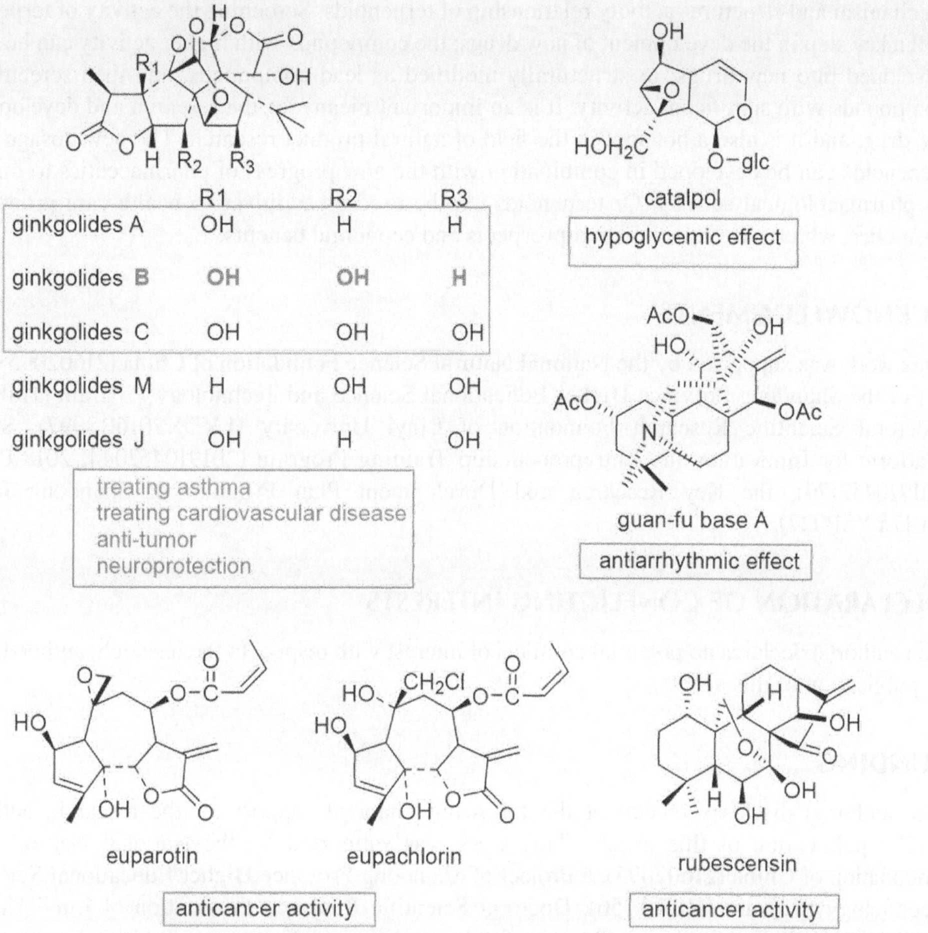

FIGURE 16.10 Structures of the most promising terpenoids.

16.4 CONCLUSIONS AND OUTLOOKS

The structures of the most promising terpenoids that could be used as templates for further investigations are listed in Figure 16.10. The current research results show that various terpenoids have been shown to have significant disease prevention and treatment effects and have antitumor, anti-inflammatory, antibacterial, antiviral, antimalarial effects and also show potential functions in immune regulation, neuroprotection, and antiallergy. The activity exhibited by terpenoids has a considerable role in the development of new drugs and improvements in existing treatment options. At the same time, the biological activities of many terpenoids still require in-depth and meticulous research.

Terpenoids are abundant in medicinal plants and are widely used, have enormous application potential, and have broad development prospects. Terpenoids can not only be extracted and isolated from plants but can also be obtained based on metabolic engineering, synthetic biology, and biotransformation, which are new sources of the synthesis of terpenoid and also solve the shortage of terpenoids. Structurally optimized terpenoids can be prepared in large quantities by synthetic biology to meet the needs of drug research and development.

At present, the mechanism of action of many terpenoids has not yet been elucidated. The combination of "omics" technology and molecular network pharmacology can be used to further study the mechanism and structure–activity relationship of terpenoids. Screening the activity of terpenoids is still a key step in the development of new drugs; the compounds with higher activity can be directly developed into new drugs, or structurally modified as lead compounds, and then screening new compounds with significant activity. It is an important means for the research and development of the drug, and it is also a hot spot in the field of natural product research. The new dosage form of terpenoids can be developed in combination with the new progress of pharmaceutics to maximize its pharmacological activity. Or terpenoids can be used as additives to health-care products and cosmetics, which have broad market prospects and economic benefits.

ACKNOWLEDGMENTS

This work was supported by the National Natural Science Foundation of China (21602095), a project of the Shandong Province Higher Educational Science and Technology Program (J18KA156), Doctoral Scientific Research Foundation of Linyi University (LYDX2016BS097), Students' Platform for Innovation and Entrepreneurship Training Program (201910452044, 201810452010, 201710452179), the Key Research and Development Plan Program of Shandong Province (2017YYSP027).

DECLARATION OF CONFLICTING INTERESTS

The author(s) declares no potential conflicts of interest with respect to the research, authorship, and/or publication of this article.

FUNDING

The author(s) discloses receipt of the following financial support for the research, authorship, and/or publication of this article: This work was supported by the National Natural Science Foundation of China (21602095), a Project of Shandong Province Higher Educational Science and Technology Program (J18KA156), Doctoral Scientific Research Foundation of Linyi University (LYDX2016BS097), Students' Platform for Innovation and Entrepreneurship Training Program (201910452044, 201810452010, 201710452179), the Key Research and Development Plan Program of Shandong Province (2017YYSP027).

REFERENCES

1. Sun LC., Li SY., Wang FZ., Xin FG. Research progresses in the synthetic biology of terpenoids. *Biotechnol Bull*. 2017;33(1):64–75.
2. Arimura G., Ozawa R., Shimoda T., Nishioka T., Boland W., Takabayashi J. Herbivory-induced volatiles elicit defence genes in lima bean leaves. *Nature*. 2000;406(6795):512–515. Crossref PubMed.
3. Martin VJJ., Pitera DJ., Withers ST., Newman JD., Keasling JD. Engineering a mevalonate pathway in *Escherichia coli* for production of terpenoids. *Nat Biotechnol*. 2003;21(7):796–802. Crossref PubMed.
4. George KW., Alonso-Gutierrez J., Keasling JD., Lee TS. Isoprenoid drugs, biofuels, and chemicals-artemisinin, farnesene, and beyond. *Adv Biochem Eng Biotechnol*. 2015;148:355–389. Crossref PubMed.
5. Liao P., Hemmerlin A., Bach TJ., Chye M-L. The potential of the mevalonate pathway for enhanced isoprenoid production. *Biotechnol Adv*. 2016;34(5):697–713. Crossref PubMed.
6. Chen TC., Fonseca COD., Schönthal AH. Preclinical development and clinical use of perillyl alcohol for chemoprevention and cancer therapy. *Am J Cancer Res*. 2015;5(5):1580–1593. PubMed.
7. Galle M., Crespo R., Kladniew BR., Villegas SM., Polo M., de Bravo MG. Suppression by geraniol of the growth of A549 human lung adenocarcinoma cells and inhibition of the mevalonate pathway in culture and in vivo: potential use in cancer chemotherapy. *Nutr Cancer*. 2014;66(5):888–895. Crossref PubMed.
8. Carnesecchi S., Schneider Y., Ceraline J., et al. Geraniol, a component of plant essential oils, inhibits growth and polyamine biosynthesis in human colon cancer cells. *J Pharmacol Exp Ther*. 2001;298(1):197–200. PubMed.

9. Kim S-H., Bae HC., Park E-J., et al. Geraniol inhibits prostate cancer growth by targeting cell cycle and apoptosis pathways. *Biochem Biophys Res Commun.* 2011;407(1):129–134. Crossref PubMed.
10. Burke YD., Stark MJ., Roach SL., Sen SE., Crowell PL. Inhibition of pancreatic cancer growth by the dietary isoprenoids farnesol and geraniol. *Lipids.* 1997;32(2):151–156. Crossref PubMed.
11. Ong TP., Heidor R., de Conti A., Dagli MLZ., Moreno FS. Farnesol and geraniol chemopreventive activities during the initial phases of hepatocarcinogenesis involve similar actions on cell proliferation and DNA damage, but distinct actions on apoptosis, plasma cholesterol and HMGCoA reductase. *Carcinogenesis.* 2006;27(6):1194–1203. Crossref PubMed.
12. Cho M., So I., Chun JN., Jeon J-H. The antitumor effects of geraniol: Modulation of cancer hallmark pathways review. *Int J Oncol.* 2016;48(5):1772–1782. Crossref PubMed.
13. Kim S-H., Park E-J., Lee CR., et al. Geraniol induces cooperative interaction of apoptosis and autophagy to elicit cell death in PC-3 prostate cancer cells. *Int J Oncol.* 2012;40(5):1683–1690. Crossref PubMed.
14. Hassan STS., Berchová-Bímová K., Šudomová M., Malaník M., Šmejkal K., Rengasamy KRR. In vitro study of multi-therapeutic properties of *Thymus bovei* Benth. essential oil and its main component for promoting their use in clinical practice. *J Clin Med.* 2018;7(9):283. Crossref.
15. Rasul A., Bao R., Malhi M., et al. Induction of apoptosis by costunolide in bladder cancer cells is mediated through ROS generation and mitochondrial dysfunction. *Molecules.* 2013;18(2):1418–1433. Crossref PubMed.
16. Yang Y-I., Kim J-H., Lee K-T., Choi J-H. Costunolide induces apoptosis in platinum-resistant human ovarian cancer cells by generating reactive oxygen species. *Gynecol Oncol.* 2011;123(3):588–596. Crossref PubMed.
17. Choi J-H., Lee K-T. Costunolide-induced apoptosis in human leukemia cells: involvement of c-Jun N-terminal kinase activation. *Biol Pharm Bull.* 2009;32(10):1803–1808. Crossref PubMed.
18. Hsu J-L., Pan S-L., Ho Y-F., Hwang T-L., Kung F-L., Guh J-H. Costunolide induces apoptosis through nuclear calcium2+ overload and DNA damage response in human prostate cancer. *J Urol.* 2011;185(5):1967–1974. Crossref PubMed.
19. Zhang JH., Liu WJ., Luo HM. The research of progress of the medicinal plant terpenoids. *World Science and Technology/Modernization of Traditional Chinese Medicine and Materia Medica.* 2018;20(3):419–430.
20. Efferth T., Kaina B. Toxicity of the antimalarial artemisinin and its dervatives. *Crit Rev Toxicol.* 2010;40(5):405–421. Crossref PubMed.
21. Crespo-Ortiz MP., Wei MQ. Antitumor activity of artemisinin and its derivatives: from a well-known antimalarial agent to a potential anticancer drug. *J Biomed Biotechnol.* 2012;2012:247597. Crossref PubMed.
22. Efferth T., Dunstan H., Sauerbrey A., Miyachi H., Chitambar CR. The anti-malarial artesunate is also active against cancer. *Int J Oncol.* 2001;18(4):767–773. Crossref PubMed.
23. Das AK. Anticancer effect of antimalarial artemisinin compounds. *Ann Med Health Sci Res.* 2015;5(2):93–102. Crossref PubMed.
24. Song Y., Xin X., Zhai X., Xia Z., Shen K. Sequential combination of flavopiridol with taxol synergistically suppresses human ovarian carcinoma growth. *Arch Gynecol Obstet.* 2015;291(1):143–150. Crossref PubMed.
25. Sun N-K., Huang S-L., Chang T-C., Chao CC-K. Tlr4 and NFκB signaling is critical for taxol resistance in ovarian carcinoma cells. *J Cell Physiol.* 2018;233(3):2489–2501. Crossref PubMed.
26. Chang C-D., Lin P-Y., Hsu J-L., Shih W-L. Ursolic acid suppresses hepatitis B virus X protein-mediated autophagy and chemotherapeutic drug resistance. *Anticancer Res.* 2016;36(10):5097–5108. Crossref PubMed.
27. Lewinska A., Adamczyk-Grochala J., Kwasniewicz E., Deregowska A., Wnuk M. Ursolic acid-mediated changes in glycolytic pathway promote cytotoxic autophagy and apoptosis in phenotypically different breast cancer cells. *Apoptosis.* 2017;22(6):800–815. Crossref PubMed.
28. Zhang R-X., Li Y., Tian D-D., et al. Ursolic acid inhibits proliferation and induces apoptosis by inactivating Wnt/β-catenin signaling in human osteosarcoma cells. *Int J Oncol.* 2016;49(5):1973–1982. Crossref PubMed.
29. Gai W-T., Yu D-P., Wang X-S., Wang P-T. Anti-cancer effect of ursolic acid activates apoptosis through ROCK/PTEN mediated mitochondrial translocation of cofilin-1 in prostate cancer. *Oncol Lett.* 2016;12(4):2880–2885. Crossref PubMed.
30. Li T., Chen X., Liu Y., et al. pH-Sensitive mesoporous silica nanoparticles anticancer prodrugs for sustained release of ursolic acid and the enhanced anti-cancer efficacy for hepatocellular carcinoma cancer. *Eur J Pharm Sci.* 2017;96:456–463. Crossref PubMed.

31. Wang S., Meng X., Dong Y. Ursolic acid nanoparticles inhibit cervical cancer growth in vitro and in vivo via apoptosis induction. *Int J Oncol*. 2017;50(4):1330–1340. Crossref PubMed.
32. Andrade LN., Lima TC., Amaral RG., et al. Evaluation of the cytotoxicity of structurally correlated p-menthane derivatives. *Molecules*. 2015;20(7):13264–13280. Crossref PubMed.
33. Andrade L., Amaral R., Dória G., et al. In vivo anti-tumor activity and toxicological evaluations of perillaldehyde 8,9-epoxide, a derivative of perillyl alcohol. *Int J Mol Sci*. 2016;17(1):32–11. Crossref.
34. Oturanel CE., Kıran İsmail., Özşen Özge., Çiftçi GA., Atlı Özlem., Kiran I., Cytotoxic AO. Cytotoxic, antiproliferative and apoptotic effects of perillyl alcohol and its biotransformation metabolite on A549 and HepG2 cancer cell lines. *Anticancer Agents Med Chem*. 2017;17(9):1243–1250. Crossref PubMed.
35. Duangmano S., Dakeng S., Jiratchariyakul W., Suksamrarn A., Smith DR., Patmasiriwat P. Antiproliferative effects of cucurbitacin B in breast cancer cells: down-regulation of the c-Myc/hTERT/Telomerase pathway and obstruction of the cell cycle. *Int J Mol Sci*. 2010;11(12):5323–5338. Crossref PubMed.
36. Bi X., Han L., Qu T., et al. Anti-inflammatory effects, SAR, and action mechanism of monoterpenoids from Radix Paeoniae alba on LPS-stimulated RAW 264.7 cells. *Molecules*. 2017;22(5):715. Crossref.
37. Chen X., Tang S-A., Lee E., et al. IVSE, isolated from Inula japonica, suppresses LPS-induced NO production via NF-κB and MAPK inactivation in RAW264.7 cells. *Life Sci*. 2015;124:8–15. Crossref PubMed.
38. Wang X., Tang S-A., Wang R., Qiu Y., Jin M., Kong D. Inhibitory effects of JEUD-38, a new sesquiterpene lactone from Inula japonica Thunb, on LPS-induced iNOS expression in RAW264.7 cells. *Inflammation*. 2015;38(3):941–948. Crossref PubMed.
39. Graziose R., Lila MA., Raskin I. Merging traditional Chinese medicine with modern drug discovery technologies to find novel drugs and functional foods. *Curr Drug Discov Technol*. 2010;7(1):2–12. Crossref PubMed.
40. Brinker AM., Ma J., Lipsky PE., Raskin I. Medicinal chemistry and pharmacology of genus Tripterygium Celastraceae. *Phytochemistry*. 2007;68(6):732–766. Crossref PubMed.
41. Fan DP., Guo QQ., Zheng K., Zhao YK., Aiping L., Xiaojuan H. Effects of triptolide on the expression of MIP–1α, eotaxin and MCP–1 in collagen–induced arthritis rats. *World Science and Technology–Modernization of Traditional Chinese Medicine*. 2016;18(9):1027–1032.
42. Li H., Zhang Y-Y., Tan H-W., Jia Y-F., Li D. Therapeutic effect of tripterine on adjuvant arthritis in rats. *J Ethnopharmacol*. 2008;118(3):479–484. Crossref PubMed.
43. Lv LP., Zhang YZ. Effects of triptolide on rheumatoid arthritis in rats. *Chinese Patent Medicine*. 2007;29:966–968.
44. Kim JH., Yi Y-S., Kim M-Y., Cho JY. Role of ginsenosides, the main active components of *Panax ginseng*, in inflammatory responses and diseases. *J Ginseng Res*. 2017;41(4):435–443. Crossref PubMed.
45. Sebei K., Sakouhi F., Herchi W., Khouja ML., Boukhchina S. Chemical composition and antibacterial activities of seven Eucalyptus species essential oils leaves. *Biol Res*. 2015;48(1):7–5. Crossref PubMed.
46. Dogan G., Kara N., Bagci E., Gur S. Chemical composition and biological activities of leaf and fruit essential oils from *Eucalyptus camaldulensis*. *Zeitschrift für Naturforschung C*. 2017;72(11–12):483–489. Crossref PubMed.
47. Wang T-H., Hsia S-M., Wu C-H., et al. Evaluation of the antibacterial potential of liquid and vapor phase phenolic essential oil compounds against oral microorganisms. *PLoS One*. 2016;11(9):e0163147. Crossref PubMed.
48. Chan YW., Siow KS., Ng PY., Gires U., Yeop Majlis B. Plasma polymerized carvone as an antibacterial and biocompatible coating. *Materials Science and Engineering: C*. 2016;68:861–871. Crossref PubMed.
49. Shahbazi Y. Chemical composition and in vitro antibacterial activity of *Mentha spicata* essential oil against common food-borne pathogenic bacteria. *J Pathog*. 2015;2015(2):916305. Crossref PubMed.
50. Park YJ., Baskar TB., Yeo SK., et al. Composition of volatile compounds and in vitro antimicrobial activity of nine *Mentha spp*. *Springerplus*. 2016;5(1):1628. Crossref PubMed.
51. Pattnaik S., Subramanyam VR., Bapaji M., Kole CR. Antibacterial and antifungal activity of aromatic constituents of essential oils. *Microbios*. 1997;89(358):39–46. PubMed.
52. Osawa K., Saeki T., Yasuda H., Hamashima H., Sasatsu M., Arai T. The antibacterial activities of peppermint oil and green tea polyphenols, alone and in combination, against enterohemorrhagic *Escherichia coli*. *Biocontrol Sci*. 1999;4(1):1–7. Crossref.
53. Inouye S., Takizawa T., Yamaguchi H. Antibacterial activity of essential oils and their major constituents against respiratory tract pathogens by gaseous contact. *J Antimicrob Chemother*. 2001;47(5):565–573. Crossref PubMed.

54. Trombetta D., Castelli F., Sarpietro MG., et al. Mechanisms of antibacterial action of three monoterpenes. *Antimicrob Agents Chemother*. 2005;49(6):2474–2478. Crossref PubMed.
55. Raut JS., Shinde RB., Chauhan NM., Karuppayil SM. Terpenoids of plant origin inhibit morphogenesis, adhesion, and biofilm formation by *Candida albicans*. *Biofouling*. 2013;29(1):87–96. Crossref PubMed.
56. Lee SB., Cha KH., Kim SN., et al. The antimicrobial activity of essential oil from dracocephalum foetidum against pathogenic microorganisms. *J Microbiol*. 2007;45(1):53–57. Crossref PubMed.
57. Wagner CS., De Gezelle J., Robertson M., Robertson K., Wilson M., Komarnytsky S. Antibacterial activity of medicinal plants from the physicians of Myddvai, a 14th century Welsh medical manuscript. *J Ethnopharmacol*. 2017;203:171–181. Crossref PubMed.
58. Xu YF., Lian DW., Chen YQ., et al. In vitro and in vivo antibacterial activities of patchouli alcohol, a naturally occurring tricyclic sesquiterpene, against *Helicobacter pylori* infection. *Antimicrob Agents Chemother*. 2017;61(6):10-1128. Crossref PubMed.
59. Appalasamy S., Lo KY., Ch'ng SJ., Nornadia K., Othman AS., Chan L-K. Antimicrobial activity of artemisinin and precursor derived from in vitro plantlets of *Artemisia annua* L. *Biomed Res Int*. 2014;2014(31):215872. Crossref PubMed.
60. Juteau F., Masotti V., Bessière JM., Dherbomez M., Viano J. Antibacterial and antioxidant activities of *Artemisia annua* essential oil. *Fitoterapia*. 2002;73(6):532–535. Crossref PubMed.
61. Kim W-S., Choi WJ., Lee S., et al. Anti-inflammatory, antioxidant and antimicrobial effects of artemisinin extracts from *Artemisia annua* L. *Korean J Physiol Pharmacol*. 2015;19(1):21–27. Crossref PubMed.
62. Cheng HJ., Liu J., Zhang G. The anti–bacterial effect of andrographolide against *Pseudomonas aeruginosa* biofilm and azithromycin. *Chinese J Microecology*. 2012;24:120–123.
63. Banerjee M., Parai D., Chattopadhyay S., Mukherjee SK. Andrographolide: antibacterial activity against common bacteria of human health concern and possible mechanism of action. *Folia Microbiol*. 2017;62(3):237–244. Crossref PubMed.
64. Zhao LX., Wang ZW., Zheng CJ., et al. Synthesis, characterization and antibacterial activity of oleanolic acid C–28 modification. *J Liaoning Normal University, Natural Science Edition*. 2013;36:529–533.
65. Kim S., Lee H., Lee S., Yoon Y., Choi K-H. Antimicrobial action of oleanolic acid on *Listeria monocytogenes*, *Enterococcus faecium*, and *Enterococcus faecalis*. *PLoS One*. 2015;10(3):e0118800. Crossref PubMed.
66. Kurek A., Nadkowska P., Pliszka S., Wolska KI. Modulation of antibiotic resistance in bacterial pathogens by oleanolic acid and ursolic acid. *Phytomedicine*. 2012;19(6):515–519. Crossref PubMed.
67. Hassan STS., Masarčíková R., Berchová K. Bioactive natural products with anti-herpes simplex virus properties. *J Pharm Pharmacol*. 2015;67(10):1325–1336. Crossref PubMed.
68. Armaka M., Papanikolaou E., Sivropoulou A., Arsenakis M. Antiviral properties of isoborneol, a potent inhibitor of herpes simplex virus type 1. *Antiviral Res*. 1999;43(2):79–92. Crossref PubMed.
69. Astani A., Reichling J., Schnitzler P. Comparative study on the antiviral activity of selected monoterpenes derived from essential oils. *Phytotherapy Res*. 2010;24(5):673–679. Crossref PubMed.
70. Brezáni V., Leláková V., Hassan S., et al. Anti-infectivity against herpes simplex virus and selected microbes and anti-inflammatory activities of compounds isolated from *Eucalyptus globulus* Labill. *Viruses*. 2018;10(7):360. Crossref.
71. Cheng H-Y., Lin T-C., Yang C-M., Wang K-C., Lin L-T., Lin C-C. Putranjivain A from *Euphorbia jolkini* inhibits both virus entry and late stage replication of herpes simplex virus type 2 in vitro. *J Antimicrob Chemother*. 2004;53(4):577–583. Crossref PubMed.
72. Kurokawa M., Basnet P., Ohsugi M., et al. Anti-herpes simplex virus activity of moronic acid purified from *Rhus javanica* in vitro and in vivo. *J Pharmacol Exp Ther*. 1999;289(1):72–78. PubMed.
73. Pei Y., Du Q., Liao P-Y., et al. Notoginsenoside ST-4 inhibits virus penetration of herpes simplex virus in vitro. *J Asian Nat Prod Res*. 2011;13(6):498–504. Crossref PubMed.
74. Utsunomiya T., Kobayashi M., Herndon DN., Pollard RB., Suzuki F., Glycyrrhizin SF. Glycyrrhizin (20 beta-carboxy-11-oxo-30-norolean-12-en-3 beta-yl-2-O-beta-D-glucopyranuronosyl-alpha-D-glucopyranosiduronic acid) improves the resistance of thermally injured mice to opportunistic infection of herpes simplex virus type 1. *Immunol Lett*. 1995;44(1):59–66. Crossref PubMed.
75. Wohlfarth C., Efferth T. Natural products as promising drug candidates for the treatment of hepatitis B and C. *Acta Pharmacol Sin*. 2009;30(1):25–30. Crossref PubMed.
76. Wintachai P., Kaur P., Lee RCH., et al. Activity of andrographolide against Chikungunya virus infection. *Sci Rep*. 2015;5(1):14179. Crossref PubMed.
77. Panraksa P., Ramphan S., Khongwichit S., Smith DR. Activity of andrographolide against dengue virus. *Antiviral Res*. 2017;139:69–78. Crossref PubMed.

78. Harada S. The broad anti-viral agent glycyrrhizin directly modulates the fluidity of plasma membrane and HIV-1 envelope. *Biochem J.* 2005;392(Pt 1):191–199. Crossref PubMed.
79. Fujioka T., Kashiwada Y., Kilkuskie RE., et al. Anti-aids agents, 11. betulinic acid and platanic acid as anti-HIV principles from *Syzigium claviflorum*, and the anti-HIV activity of structurally related triterpenoids. *J Nat Prod.* 1994;57(2):243–247. Crossref PubMed.
80. Cichewicz RH., Kouzi SA. Chemistry, biological activity, and chemotherapeutic potential of betulinic acid for the prevention and treatment of cancer and HIV infection. *Med Res Rev.* 2004;24(1):90–114. Crossref PubMed.
81. Zhu YM., Shen JK., Wang HK., Cosentino LM., Lee KH. Synthesis and anti-HIV activity of oleanolic acid derivatives. *Bioorg Med Chem Lett.* 2001;11(24):3115–3118. Crossref PubMed.
82. Esimone CO., Eck G., Nworu CS., et al. Dammarenolic acid, a secodammarane triterpenoid from *Aglaia sp.* shows potent anti-retroviral activity in vitro. *Phytomedicine.* 2010;17(7):540–547. Crossref PubMed.
83. Ma C., Nakamura N., Miyashiro H., Hattori M., Shimotohno K. Inhibitory effects of constituents from *Cynomorium songaricum* and related triterpene derivatives on HIV-1 protease. *Chem Pharm Bull.* 1999;47(2):141–145. Crossref PubMed.
84. Creek DJ., Charman WN., Chiu FCK., et al. Relationship between antimalarial activity and heme alkylation for spiro- and dispiro-1,2,4-trioxolane antimalarials. *Antimicrob Agents Chemother.* 2008;52(4):1291–1296. Crossref PubMed.
85. Eckstein-Ludwig U., Webb RJ., Van Goethem IDA., et al. Artemisinins target the SERCA of *Plasmodium falciparum*. *Nature.* 2003;424(6951):957–961. Crossref PubMed.
86. Eichhorn T., Winter D., Büchele B., et al. Molecular interaction of artemisinin with translationally controlled tumor protein (TCTP) of *Plasmodium falciparum*. *Biochem Pharmacol.* 2013;85(1):38–45. Crossref PubMed.
87. Lisewski AM., Quiros JP., Ng CL., et al. Supergenomic network compression and the discovery of EXP1 as a glutathione transferase inhibited by artesunate. *Cell.* 2014;158(4):916–928. Crossref PubMed.
88. Gao S., Liu Z., Li H., Little PJ., Liu P., Xu S. Cardiovascular actions and therapeutic potential of tanshinone IIA. *Atherosclerosis.* 2012;220(1):3–10. Crossref PubMed.
89. Fu J., Huang H., Liu J., Pi R., Chen J., Liu P. Tanshinone IIA protects cardiac myocytes against oxidative stress-triggered damage and apoptosis. *Eur J Pharmacol.* 2007;568(1–3):213–221. Crossref PubMed.
90. Maione F., Cantone V., Chini MG., De Feo V., Mascolo N., Bifulco G. Molecular mechanism of tanshinone IIA and cryptotanshinone in platelet anti-aggregating effects: an integrated study of pharmacology and computational analysis. *Fitoterapia.* 2015;100:174–178. Crossref PubMed.
91. Pan C., Lou L., Huo Y., et al. Salvianolic acid B and tanshinone IIA attenuate myocardial ischemia injury in mice by NO production through multiple pathways. *Ther Adv Cardiovasc Dis.* 2011;5(2):99–111. Crossref PubMed.
92. Wang W., GY S., WQ H., Zhao YQ. Advances in pharmacological effects of ginsenosides on cardiovascular diseases in the past 10 years. *Chin Herb Med.* 2016;47:3736–3741.
93. Jeppesen PB., Dyrskog SE., Agger A., et al. Can stevioside in combination with a soy-based dietary supplement be a new useful treatment of type 2 diabetes? An in vivo study in the diabetic Goto-Kakizaki rat. *Rev Diabet Stud.* 2006;3(4):189–199. Crossref PubMed.
94. Trommer H., Neubert RHH. Overcoming the *Stratum corneum*: the modulation of skin penetration. A review. *Skin Pharmacol Physiol.* 2006;19(2):106–121. Crossref PubMed.
95. Cornwell PA., Barry BW., Bouwstra JA., Gooris GS. Modes of action of terpene penetration enhancers in human skin; differential scanning calorimetry, small-angle X-ray diffraction and enhancer uptake studies. *Int J Pharm.* 1996;127(1):9–26. Crossref.
96. Anjos JLVD., Neto DdeS., Alonso A. Effects of 1,8-cineole on the dynamics of lipids and proteins of *Stratum corneum*. *Int J Pharm.* 2007;345(1–2):81–87. Crossref PubMed.
97. Al-Khalili M., Meidan VM., Michniak BB. Iontophoretic transdermal delivery of buspirone hydrochloride in hairless mouse skin. *AAPS Pharm Sci.* 2003;5(2):61–71. Crossref.
98. Narishetty STK., Panchagnula R. Transdermal delivery of zidovudine: effect of terpenes and their mechanism of action. *J Control Release.* 2004;95(3):367–379. Crossref PubMed.
99. Rhee YS., Choi JG., Park ES., Chi SC. Transdermal delivery of ketoprofen using microemulsions. *Int J Pharm.* 2001;228(1–2):161–170. Crossref PubMed.
100. Lee J., Lee Y., Kim J., Yoon M., Choi YW. Formulation of microemulsion systems for transdermal delivery of aceclofenac. *Arch Pharm Res.* 2005;28(9):1097–1102. Crossref PubMed.
101. Cal K., Kupiec K., Sznitowska M. Effect of physicochemical properties of cyclic terpenes on their ex vivo skin absorption and elimination kinetics. *J Dermatol Sci.* 2006;41(2):137–142. Crossref PubMed.

102. Lan Y., Wang J., Li H., et al. Effect of menthone and related compounds on skin permeation of drugs with different lipophilicity and molecular organization of *Stratum corneum* lipids. *Pharm Dev Technol.* 2016;21(4):389–398. Crossref PubMed.
103. Wang J., Dong C., Song Z., et al. Monocyclic monoterpenes as penetration enhancers of ligustrazine hydrochloride for dermal delivery. *Pharm Dev Technol.* 2017;22(4):571–577. Crossref PubMed.
104. Yang Z., Teng Y., Wang H., Hou H., Huimin H. Enhancement of skin permeation of bufalin by limonene via reservoir type transdermal patch: formulation design and biopharmaceutical evaluation. *Int J Pharm.* 2013;447(1–2):231–240. Crossref PubMed.
105. Ahad A., Aqil M., Kohli K., Sultana Y., Mujeeb M., Ali A. Role of novel terpenes in transcutaneous permeation of valsartan: effectiveness and mechanism of action. *Drug Dev Ind Pharm.* 2011;37(5):583–596. Crossref PubMed.
106. Robledo S., Osorio E., Muñoz D., et al. In vitro and in vivo cytotoxicities and antileishmanial activities of thymol and hemisynthetic derivatives. *Antimicrob Agents Chemother.* 2005;49(4):1652–1655. Crossref PubMed.
107. Kiuchi F., Itano Y., Uchiyama N., et al. Monoterpene hydroperoxides with trypanocidal activity from *Chenopodium ambrosioides. J Nat Prod.* 2002;65(4):509–512. Crossref PubMed.
108. Mafud AC., Silva MPN., Monteiro DC., et al. Structural parameters, molecular properties, and biological evaluation of some terpenes targeting *Schistosoma mansoni* parasite. *Chem Biol Interact.* 2016;244:129–139. Crossref PubMed.
109. López-García S., Castañeda-Sanchez JI., Jiménez-Arellanes A., et al. Macrophage activation by ursolic and oleanolic acids during mycobacterial infection. *Molecules.* 2015;20(8):14348–14364. Crossref PubMed.
110. Ge JF., Li J., Hu CM., et al. Study on the immunomodulatory effects of triterpenic acid of *Lycium barbarum* L. *Chin J Pharm.* 2006;10:1194–1198.
111. Xie J-T., Shao Z-H., Vanden Hoek TL., et al. Antioxidant effects of ginsenoside re in cardiomyocytes. *Eur J Pharmacol.* 2006;532(3):201–207. Crossref PubMed.
112. Wan Q., Ma X., Zhang Z-J., et al. Ginsenoside reduces cognitive impairment during chronic cerebral hypoperfusion through brain-derived neurotrophic factor regulated by epigenetic modulation. *Mol Neurobiol.* 2017;54(4):2889–2900. Crossref PubMed.

17 Food Alkaloids
Advances, Trends, Applications, and Safety

Ingryd Rodrigues Martins,
Fernanda Wariss Figueiredo Bezerra,
Julia Helena da Silva Martins, Andrea Komesu,
Johnatt Allan Rocha de Oliveira,
and Luiza Helena Da Silva Martins

17.1 INTRODUCTION

Plants commonly contain alkaloids, which are nitrogenous compounds. These compounds are bioactive, making them significant in the human diet and medicine. Alkaloids have become increasingly popular in foods due to their potential health benefits. Recent studies have shown that some alkaloids in foods can have therapeutic effects, such as antioxidant, anti-inflammatory, and neuroprotective properties (Chen & Lin, 2019). These compounds can be found in various food sources, including fruits, vegetables, grains, and herbs. Their chemical composition and presence can vary extensively among plant species (Yubin et al., 2014).

Obtaining pure and safe alkaloids from food sources involves extraction, separation, and purification. Effective methods have been developed to achieve this goal, including organic solvent extraction, ultrasound-assisted extraction, and high-performance liquid chromatography (HPLC) (Rostagno & Prado, 2013). To obtain alkaloids of good quality and safety, choosing the correct technique for extraction and purification is crucial and requires careful consideration.

Essential processes must be followed to obtain pure and safe alkaloids for human consumption. These processes involve extracting, separating, and purifying the compounds from food sources. Several techniques have been developed to extract and purify alkaloids effectively. These techniques range from conventional methods, like organic solvent extraction, to more advanced procedures, such as ultrasound-assisted extraction and HPLC (Wang et al., 2023). The correct method for extracting and purifying alkaloids is essential to guarantee their quality and safety (Yubin et al., 2014).

This book chapter provides an overview of dietary alkaloids, highlighting their sources, chemical composition, extraction, purification and analysis techniques, effects on the human body (both beneficial and adverse), and prospects. Studying the alkaloids in foods is essential due to their wide natural distribution and potential impacts on human health. Therefore, this chapter seeks to provide a solid foundation for researchers, health professionals, and the food industry, contributing to significant advances in the understanding and application of food alkaloids for the benefit of society.

17.2 FOOD AS A SOURCE OF BIOACTIVE COMPOUNDS

Foods are rich sources of bioactive compounds and natural substances present in foods of plant origin, which benefit human health. These compounds include polyphenols, flavonoids, and

 DOI: 10.1201/9781003462804-23

carotenoids, which have already been shown to be anticancer, antioxidant, and anti-inflammatory (González-Castejón & Rodriguez-Casado, 2011).

Scientific studies have highlighted the importance of bioactive compounds in preventing chronic diseases, such as cardiovascular diseases, cancer, and diabetes. Polyphenols, present in fruits, vegetables, teas, and whole grains, have been associated with advantageous results on cardiovascular well-being, such as decreased chances of coronary heart disease (Arts & Hollman, 2005).

In addition to polyphenols, carotenoids are another critical class of bioactive compounds present in foods such as vibrantly colored fruits and vegetables. They have antioxidant effects and may help reduce cancer and degenerative diseases (Kaulmann & Bohn, 2014).

It is essential to highlight that the presence and concentration of bioactive compounds in foods can vary according to factors, such as the variety of the plant, the stage of maturation, agricultural practices, and processing methods. Therefore, a diet rich and diverse in natural and minimally processed foods is essential to ensure adequate intake of these bioactive compounds and their potential health benefits (Arts & Hollman, 2005).

Alkaloids are nitrogenous organic compounds found in various plants, many of which are used in human nutrition. For example, caffeine, one of the best-known alkaloids, is present in coffee, tea, and cocoa and stimulates the central nervous system. Another example is capsaicin, an alkaloid found in peppers, which provides a spicy flavor and has analgesic and anti-inflammatory properties (Govindarajan & Sathyanarayana, 1991).

In addition to sensory and therapeutic effects, alkaloids in foods play an essential role in the human diet due to their bioactive properties. Berberine, an alkaloid found in plants such as *Berberis vulgaris* and *Coptis chinensis*, shows potential health benefits, including antioxidant, anti-inflammatory, and hypoglycemic activities, suggesting its use in the prevention and treatment of various diseases (Imanshahidi & Hosseinzadeh, 2008).

17.3 CHEMISTRY OF ALKALOIDS AND ANALYTICAL DETERMINATION

Organic contaminants, such as alkaloids, are secondary metabolites with natural toxicity and are generally produced unintentionally by plants. These nitrogenous substances can also originate from animals and microorganisms, mostly biosynthesized from amino acids (Dey et al., 2020). The classification of the molecular structure of alkaloids is based on the presence of a primary heterocyclic nucleus, and in recent decades, the groups of alkaloids most recurrent in food contamination and of interest to food safety authorities and research institutions are the pyrrolizidines (PAs), tropane (TAs), and opium (OAs), represented in Figure 17.1.

Pyrrolizidine alkaloids (PAs) and their N-oxides are secondary metabolites produced by plants, which have toxic properties in defense against the action of insects and herbivores. Fabaceae (Crotalaria), Asteraceae (Senecioneae and Europatorieae), Orchidaceae, and Apocynaceae are the main plant families that produce PAs, where about 3–5% of flowering plants can synthesize these compounds (EFSA, 2011). The toxic effects are closely related to the chemical structure of PAs, based on necine and side chains esterified with necic acids, plus two pyrrole rings aggregated with a nitrogen molecule in po™sition number 4 (Figure 17.1a). Currently, around 600 molecular structures are already known, which can be saturated or unsaturated, the last being in position number 1 to 2; such unsaturation implies an increase in the toxicity of this type of alkaloid. In contrast, saturated PAs are not toxic compounds, as they are not activated in the body's metabolic function, thus not exerting their toxicity properties (Casado et al., 2023).

Tropane alkaloids are also toxic secondary metabolites of plants in the family Salanaceae, Brassicaceae, and *Erythroxylaceae*. Around 200 compounds have been identified, and tropane's standard central chemical structure characterizes them. The variation of these molecules occurs through the esterification of tropin with different acids, such as propanoic, isovaleric, acetic, isobutyric, tropic, and (+)-α-hydroxy-β-phenylpropionic, among others (EFSA, 2013). The molecular composition generally contains two pyrrolidine and piperidine rings, with a single nitrogen and

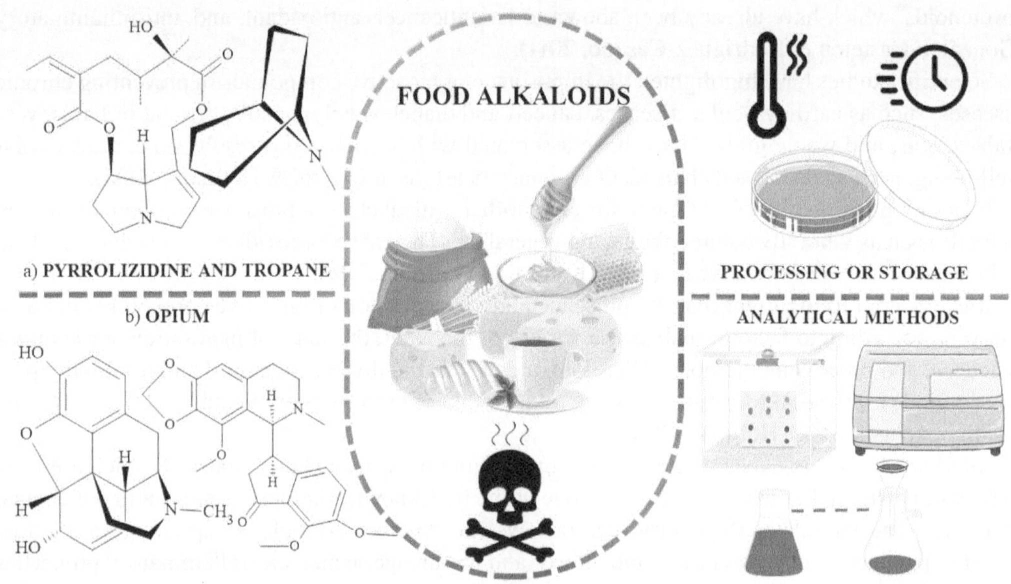

FIGURE 17.1 (a) Standard molecular structure of pyrrolizidine alkaloids and tropane, and (b) two main substances from the phenanthrene (morphine) and benzylisoquinoline (noscapine) families of opium alkaloids, main analytical techniques, toxicity factors, and their food sources.

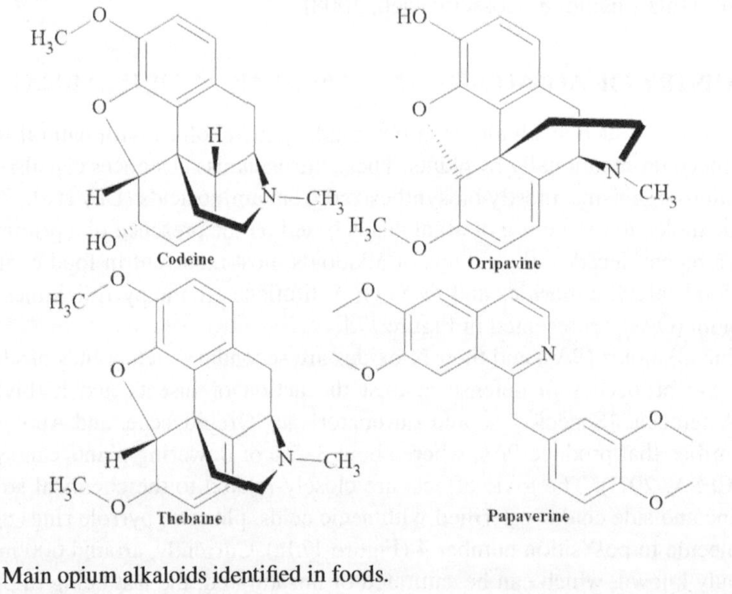

FIGURE 17.2 Main opium alkaloids identified in foods.

two carbon atoms (Eich, 2008). Hyoscine (-)-hyoscyamine and (-)-scopolamine are among the TAs most found in contaminated foods. Atropine and scopolamine are the TAs most explored in studies (González-Gómez et al., 2022); of the two (±)-hyoscyamine, only the (-)-enantiomers, depending on the dosage, may present toxic potential (Kaltner, 2022).

Papaver somniferum L. presents alkaloids, secondary metabolites with around 80 different structures already identified. The best-known OAs are found in two structural chemical groups: phenanthrenes (codeine, grapevine, and thebaine) and benzylisoquinolines (papaverine) (Figure 17.2). Phenanthrenes have an aromatic ring, partially saturated and containing nitrogen, with a bond between carbons 9 and 13. Regarding the biosynthetic route, thebaine is a precursor to oripavine and codeine, which are precursors to morphine, where all molecular structures contain a methylated nitrogen atom and oxygen functional groups (EFSA, 2018). In a previous study, the toxic effects of OAs on humans were highlighted (Martínez & Ballesteros, 2019). However, Kaltner (2022) states that OAs are not presented as an object of study for research with future perspectives.

The most significant occurrence of these natural toxins is in food matrices of plant origin (Figure 17.1), such as cereals, flours, vegetables, teas, spices, and herbs. However, contamination has been observed in foods of animal origin, such as honey, milk, eggs, meat, and derivatives, as well as in processed food products, including dairy products, food supplements, drinks, and snacks. However, the frequency of alkaloid contamination for this last group is relatively low, and in favorable cases, the identified concentration is still low (Casado et al., 2022, 2023). This fact must be related to the contaminant transfer (cross-contamination).

Transformations in molecular structures can occur during food product processing and storage, including undesirable compounds such as alkaloids. Reactions such as demethylation, hydroxylation, hydrolysis, conjugation reactions, or the formation of adducts can pose risks to human health (Bletsou et al., 2015; Kaltner, 2022). It has already been demonstrated that such natural substances contribute to fatal poisonings and are hepatotoxic, pneumotoxic, genotoxic, and carcinogenic (Dusemund et al., 2018).

In contrast, analytical techniques are indispensable for identifying and analyzing new molecules formed from the decomposition of these mycotoxins during food processing. Liquid chromatography coupled with mass spectrometry (LC-MS) is one of the most used and effective methods for identifying specific microorganisms, including alkaloids, as detection is sensitive to toxins of food origin (Ahuja et al., 2023; Kaltner, 2022). Conventional chromatography and immunoassay techniques have limitations. We may encounter difficulties in sample preparation in chromatography, which can be very complex and takes time. At the same time, in immunoassays, the limitation occurs due to the sensitivity of the method and unreliable results on the structure of the compounds already present in the matrix evaluated (Kole et al., 2011; Turner et al., 2009).

Under this assumption, different analytical strategies have been carried out in the last ten years to identify and quantify alkaloids in food matrices. These determinations contribute to significant advancement in food security (Ahuja et al., 2023; Casado et al., 2022). With this advancement, in conjunction with innovation in instrumentation, high-resolution mass spectrometry (HRMS) and tandem mass spectrometry (MS/MS) are highly sensitive and can select toxins in complex mixtures, even at ultralow levels, thus a fundamental tool for the detection and characterization of food alkaloids (Qiu et al., 2022). Table 17.1 briefly summarizes the most common chromatographic methods for detecting different types of alkaloids in foods.

17.4 EXTRACTION AND STABILITY OF FOOD ALKALOIDS METHODS

Alkaloids are chemical compounds that occur naturally in various foods. These compounds can benefit or harm the human health, depending on their concentration and type. Therefore, it is crucial to develop effective extraction methods that enable accurate detection of alkaloids. In addition, the stability of alkaloids is vital for their use in various industries, such as in the manufacture of medicines, cosmetics, and dyes. Ensuring the stability of these compounds is essential to guarantee the final product's quality and effectiveness and prevent any potential adverse effects on consumers' health (Brugnerotto et al., 2021).

Extracting alkaloids from food sources is a crucial part of food chemistry. It allows us to identify and measure these compounds in various sources, such as grains, fruits, vegetables, and herbs. Traditional methods, such as Soxhlet and maceration extraction, have been widely used because of their simplicity and efficiency in recovering alkaloids from food samples. However, these techniques may present some limitations, such as the need for large volumes of solvents and long extraction times, which can lead to the degradation of the target compounds (Silva & Aragão, 2009; Zhang et al., 2018).

TABLE 17.1
Primary Analytical Determination Applied to Characterize Alkaloids Found in Foods

Alkaloid Classification	Alkaloid Identified	Food Matrix	Determination Technique Used	Limit of Detection	Limit of Quantification	Limit of Recoveries (%)	Reference
Pyrrolizidine	Echimidine, echimidine *N*-oxide, intermediate, intermediate *N*-oxide, lycopsamine, retrorsine, retrorsine *N*-oxide, senecionine, and senecionine *N*-oxide	Honey	UHPLC-QTRAP-MS/MS	—	0.03–0.06 μg/kg	63–103	Celano et al. (2019)
	Senecionine, seneciphylline, senkirkin, retrorsine, adonifoline, and chlorine	Milk	DART-IT-MS	0.55–0.85 ng/mL	1.83–2.82 ng/mL	89–112	Chen et al. (2021)
Tropane	Atropine and scopolamine	Maize products	LC-MS/MS	1 μg/kg	1–20 μg/kg	85.3–89.1	Torović et al. (2023)
	Atropine and scopolamine	Leaves of thyme, basil, and coriander	LC-MS/MS	5–200 μg/kg	5–42 μg/kg	70–92	González-Gómez et al. (2020)
	Anisodamine, atropine and scopolamine	Herbal teas	UHPLC-ESI-MA/MS	0.5 μg/kg	5.7–83.9 μg/kg	63–103	Cirlini et al. (2019)
	Apoatropine	Tea and herbal teas	Orbitral LC-MS	2.5–50 μg/L	15–100 μg/kg	28–128	Romera-Torres et al. (2018)
Opium	Morphine, codeine, thebaine, noscapine, papaverine narceine	Poppy seeds, bakery products	UHPLC-MS/MS	0.1 mg/kg	<0.1–106 mg/kg	77–172	López et al. (2018)
	Morphine, codeine, thebaine, papaverine	Additives of hot pot	LC-QqQ_{LIT}-MS/MS	0.05–0.8 μg/kg	0,25–2,5 μg/kg	80.1–115.3	Xu et al. (2019)
	Morphine and codeine	Biscuits and sponge cakes	HPLC-MS/MS	0.1 μg/mL	80 μg/L	30–107	Casado-Hidalgo et al. (2022)

Note: DART-IT-MS: real-time mass spectrometry. LC-MS/MS: liquid chromatography with tandem mass spectrometry. UHPLC-ESI-MA/MS: high-performance liquid chromatography–electrospray ionization tandem mass spectrometry. Orbitral LC-MS: cromatografia líquida acoplada à espectrometria de massa de alta resolução. LC-QqQ_{LIT}-MS/MS: liquid chromatograph coupled to the triple-quadrupole linear ion trap mass spectrometer. HPLC-MS/MS: liquid chromatography coupled to a triple-quadrupole tandem mass detector.

Unconventional techniques, such as ultrasound extraction and extraction with supercritical fluids, have become more popular due to their ability to extract alkaloids more quickly and efficiently. They require smaller amounts of solvents and provide more excellent selectivity in extraction. These techniques contribute significantly to food chemistry research, enabling a more accurate and comprehensive analysis of alkaloids present in food sources.

Therefore, the choice of the extraction technique depends on the sample's specific characteristics and the analysis's objectives (Lopez-Hortas et al., 2022; Zhang et al., 2018).

Combining conventional and unconventional techniques can be useful in obtaining a complete and representative extraction of alkaloids present in foods, providing valuable insights into their distribution, concentration, and potential effects on human health. These advances in the alkaloid chemistry of food sources are essential to ensuring food safety and promoting a healthy, balanced diet (Azmir et al., 2013).

Krska et al. (2008) thoroughly investigated the extraction, purity, and stability of ergot alkaloids. They developed effective extraction methods using various solvents and conditions, such as specific temperature, pressure, and time, and optimized the parameters to ensure maximum alkaloid recovery. Moreover, the scientists evaluated the purity and potency of ergot alkaloid standards to guarantee the accuracy and reliability of analytical results, which forms a solid foundation for future analysis of cereal samples for the presence of these potentially hazardous compounds.

The study identifies several alkaloids, including ergometrine, ergotamine, ergosine, ergocristine, ergocryptine, and their corresponding epimers. To assess their purity, the study employed an exact analytical method, LC-DAD-ESI-TOF-MS, which combines liquid chromatography with diode array detection, electrospray ionization, and time-of-flight mass spectrometry. The rigorous analysis revealed that all standards had purity levels well above 98%, except ergocristinine (94%), ergosine (96%), and ergosinine (95%), underscoring the importance of these compounds in our understanding of ergot alkaloids (Krska et al., 2008).

The authors' research led them to conclude that a combination of acetonitrile and ammonium carbonate buffer was the superior solvent for extracting ergot alkaloids. This choice resulted in significantly higher recoveries than other solvents, demonstrating the effectiveness and precision of their research methods (Krska et al., 2008).

A study by Aguilar-Hernández et al. (2020) explores using ultrasound to extract alkaloids from the pulp and by-products of soursop fruit (*Annona muricata* L.). This research reveals that ultrasound is a highly efficient technique for alkaloid extraction, producing greater yields than traditional methods. Moreover, ultrasound enables faster and more comprehensive extraction, reducing processing time and solvent consumption. The efficiency and sustainability of the alkaloid extraction process from soursop fruit can be significantly improved, which is considered an advantage.

Xiong et al. (2016) investigate the optimization of the extraction of bioactive alkaloids from lotus plume using microwave-assisted response surface methodology. The study highlights that the microwave-assisted extraction method offers several significant advantages compared to conventional methods, including higher efficiency, shorter extraction time, and lower solvent consumption. The use of microwaves provides rapid and uniform heating of the sample, facilitating the release of alkaloids and resulting in higher yields. Furthermore, the response surface methodology allows the optimization of extraction parameters, such as temperature and time, to maximize the yield of alkaloids, thus contributing to a more effective and sustainable approach to obtaining these bioactive compounds from lotus plumule.

Ruan et al. (2017) investigated the extraction of total alkaloids, quinine, and feminine from *Fritillaria thunbergii* Miq flowers (a non-conventional food) using supercritical carbon dioxide (CO_2). This technique was found to have various benefits compared to conventional methods. These advantages include enhanced selectivity, minimal toxicity, the absence of solvent residues, and reduced environmental impact. The study highlights the potential of supercritical CO_2 extraction as a superior alternative to traditional methods. Furthermore, supercritical CO_2 allows the efficient and selective extraction of the desired alkaloids, preserving their bioactive properties and guaranteeing the quality of the extracts obtained. These characteristics make the supercritical CO_2 extraction method a promising and sustainable alternative for obtaining alkaloids from *Fritillaria thunbergii* Miq flowers, with potential application in the pharmaceutical industry and natural product research.

Torres-Vega et al. (2020) investigated the alkaloids extracted from *Peumus boldus* leaves. They used natural deep eutectic solvents (NADES) to extract these compounds sustainably. The study highlights that NADES, composed of a combination of natural compounds, such as organic acids and sugars, are a green and effective alternative to conventional solvents in the extraction of bioactive compounds. Using NADES offers advantages, such as low toxicity, biodegradability, low cost, and abundant renewable raw materials. Furthermore, NADES showed the ability to extract various bioactive compounds from *Peumus boldus* leaves, preserving their antioxidant and phytochemical properties. These results highlight the potential of deep natural eutectic solvents as a promising and sustainable approach to extracting bioactive compounds from natural sources, contributing to producing high-quality natural ingredients in the food and pharmaceutical industry.

Alkaloids were extracted and purified from *Datura stramonium* using a series of steps that involved extracting plant samples and fractionation using chromatographic techniques, such as column chromatography and thin-layer chromatography. The purification method produced highly purified alkaloids, allowing for a better understanding of their chemical composition and properties (Babiker et al., 2017).

Different methods are utilized to purify alkaloids, each with strengths and weaknesses. One of the most used techniques is column chromatography, which separates the mixture of alkaloids into fractions based on their distinct binding affinities with a stationary phase. Another widely used method is thin-layer chromatography, which separates alkaloids on a plate coated with an adsorbent substance. HPLC and gas chromatography (GC) are other techniques often employed to separate and purify alkaloids in complex samples. Advanced techniques, such as affinity chromatography, supercritical fluid extraction, and selective precipitation, have also been developed to achieve excellent selectivity and purity in the purification of alkaloids (Feng et al., 2019).

Several techniques can be applied to the stability of alkaloids, but in this chapter, we will mention encapsulation, which is one of the most promising and convenient techniques. Therefore, we cite the work of Fahmy et al. (2021), who highlighted the encapsulation technique as an effective strategy to increase the stability of *Peganum harmala* alkaloids through self-envelopment; the alkaloids form supramolecular nanocapsules, which demonstrate enhanced antioxidant and cytotoxic activities. This encapsulation approach protects alkaloids from degradation and expands their biological activities, promoting their potential applicability in medicine and other areas.

The choice of the extraction, purification, and stability technique must be made considering the specific objectives of the analysis, the characteristics of the alkaloid compounds, and the quality and safety requirements of food products.

17.5 TRENDS OF FOOD ALKALOIDS APPLICATIONS

The use of food alkaloids is changing as demand for natural and functional goods rises along with advancements in food science. Because of their potential health benefits, alkaloids are becoming an increasingly popular active ingredient in functional products and nutritional supplements. For instance, because caffeine boosts energy and enhances physical performance, it has been commonly used in energy drinks and pre-workout supplements (Koleva et al., 2012, Doughari, 2012).

Another emerging trend is using alkaloids in food and beverages designed to support mental and emotional well-being. Products designed to enhance mood and brain function have been developed because of research into the possible neuroprotective and stimulant effects of alkaloids like theobromine, which is contained in cocoa. These goods include dietary supplements meant to enhance mental health, beverages enhanced with cocoa, and functional chocolates Cova et al. (2019).

Moreover, regular consumption of cocoa, a naturally occurring source of theobromine, has been linked in epidemiological studies to a decreased incidence of mental illnesses, like anxiety and depression (Pase et al., 2013). These results imply that theobromine and other bioactive substances found in cocoa may be crucial in supporting mental wellness and preventing mood disorders. However, additional investigation, including controlled clinical trials, must thoroughly understand the underlying mechanisms and validate theobromine's benefits for mental health.

Newer trends of using capsaicin to improve human health have been studied, and these include its use in the treatment and prevention of neurodegenerative illnesses, including Parkinson's and Alzheimer's (Wang et al., 2020). According to research, capsaicin may have neuroprotective properties, shielding neurons from inflammation and oxidative damage—two significant contributors to the onset of many disorders. Additionally, pre-clinical research has shown that capsaicin might lessen the development of beta-amyloid plaques, a hallmark of Alzheimer's disease, and enhance cognitive performance, suggesting a promising future for therapeutic approaches (Grimm et al., 2020).

The use of capsaicin as an adjuvant in cancer treatment is another new trend, particularly when used in conjunction with more traditional therapies, like radiation and chemotherapy. According to research, capsaicin may have antitumor effects by preventing cancer cells from proliferating and growing some cancer types. Furthermore, research suggests that capsaicin may make cancer cells more sensitive to the effects of anticancer medicines by increasing their sensitivity to apoptosis, or programmed cell death. Capsaicin may represent a new therapeutic approach to combat cancer, although additional research is needed to thoroughly understand the underlying mechanisms and determine clinical efficacy (Adetunji et al., 2022; Díaz-Laviada & Rodríguez-Henche, 2014; Zhang et al., 2020)

Dietary alkaloids are also becoming popular in cosmetics and personal hygiene products. Because of their stimulating and antioxidant qualities, alkaloids like caffeine and capsaicin are being added to creams, lotions, and oral care products. These products can assist in renewing the skin, enhance blood circulation, and promote dental health (Stępniowska et al., 2021). The diverse range of applications highlights the adaptable capabilities of food alkaloids and their increasing significance across multiple industrial domains.

17.6 ALKALOIDS AND THEIR PHARMACOLOGICAL PROPERTIES

Alkaloids, nitrogenous organic compounds found in various plants, are of substantial interest in pharmacological research due to their diverse therapeutic applications and physiological effects on the human system (de Oliveira et al., 2022; Heinrich et al., 2021). They are characterized by their heterocyclic chemical structure and pharmacologically active properties (El-Sayed & Verpoorte, 2007). Among the widely studied food alkaloids, caffeine, theobromine, capsaicin, and morphine stand out (Ferré, 2008).

Another important example of an alkaloid is theobromine, which is present in cocoa-derived products, such as chocolate. It exhibits vasodilatory and mild stimulant properties and contributes to the relaxation of smooth muscles (Al Muqarrabun & Ahmat, 2015). In turn, capsaicin, found in peppers, exerts analgesic and anti-inflammatory effects, reducing the sensation of pain and modulating the immune response (Fattori et al., 2016).

Morphine, an alkaloid extracted from opium, is a potent analgesic used to treat intense pain. It acts as an agonist of opioid receptors in the central nervous system. These examples highlight the diversity of pharmacological effects of food alkaloids, ranging from stimulation to pain relief, providing significant benefits to human health (Pathan & Williams, 2012).

Dietary alkaloids, such as caffeine, theobromine, capsaicin, and morphine, have been investigated in relation to human health. Caffeine has been linked to many other advantages, one of which is to improve cognitive function and increase alertness, continued decreased risk of Alzheimer's and Parkinson's, and even protection from specific cancers, such as liver and colon (Grosso et al., 2017; Liu et al., 2024).

Furthermore, epidemiological research shows an opposite correlation between coffee consumption, an essential source of caffeine, and the risk of developing type 2 diabetes (Ding et al., 2014).

Theobromine found in cocoa products such as chocolate has also demonstrated benefits for cardiovascular health, including lowering blood pressure and improving endothelial function (Desideri et al., 2012; Nehlig, 2013). Similarly, capsaicin's anti-inflammatory, analgesic, and possibly anticancer effects in peppers have been studied (Lee & Clark, 2016; Surh, 2002).

It is essential to consider that the health benefits related to ingesting dietary alkaloids may vary depending on the dose and individual context. There is evidence that moderate use of these substances has positive effects, but excessive use can cause harmful effects, such as insomnia, nervousness, tachycardia, and even dependence (Juliano & Griffiths, 2004).

Recommended levels of dietary alkaloids for therapeutic use vary depending on the compound and clinical condition. For example, the recommended dose of caffeine to improve cognitive performance and alertness ranges between 75 and 300 mg per day, depending on individual sensitivity and tolerance to the compound (Thompson, 2005). Clinical studies use theobromine doses of 250 to 750 mg daily to evaluate its effects on blood pressure and vascular function (Davison et al., 2008).

However, it is essential to remember that these doses are only general recommendations and should be adjusted for each patient according to their age, weight, pre-existing health conditions, and concomitant use of other medications. Additionally, a qualified health-care professional should supervise any therapeutic use of food alkaloids to maximize health benefits and minimize harmful effects.

Food-sourced alkaloids can improve cognitive function, protect the heart, and decrease the risk of certain types of cancer. However, therapy with these substances requires a careful approach, considering the dose, individual sensitivity, and adequate medical supervision. More research is needed to fully understand the effects of these ingredients and determine the ideal doses for each clinical situation (Dhyani et al., 2022).

17.7 THE MECHANISMS OF ACTION OF ALKALOIDS IN THE HUMAN BODY

The mechanisms of action of alkaloids in the human body are varied and often complex. Caffeine was reported in adenosine, blocking receptors in the brain and inhibiting the neurotransmitter responsible for promoting drowsiness. This interaction increases neuronal activity and releases neurotransmitters such as dopamine and noradrenaline, accountable for alertness and well-being (Nehlig et al., 1992).

Capsaicin's analgesic action is mediated by activating TRPV1 (vanilloid type 1) receptors in peripheral sensory neurons. The binding of capsaicin to these receptors triggers the release of neurotransmitter substances that act in the central nervous system, blocking the transmission of pain, providing temporary pain relief (Rossato et al., 2011).

Morphine works by stimulating mu-opioid receptors located in the brain and spinal cord. This stimulation results in the modification of pain perception and the generation of a feeling of well-being and analgesia. This mechanism of action is explained by Ewan and Martin (2011) in their study. Furthermore, morphine also exerts depressant effects on the central nervous system, reducing neuronal activity and inducing a state of relaxation and drowsiness (Lee et al., 2014).

It is possible to report several health benefits associated with moderate consumption of alkaloids from foods. In the case of caffeine, in addition to promoting wakefulness and concentration, it has also been shown to have antioxidant and neuroprotective properties. Neurodegenerative diseases, such as Parkinson's and Alzheimer's, can be prevented with the help of capsaicin. The properties of capsaicin, including vasodilatory and anti-inflammatory effects, have been linked to reducing the risk of cardiovascular diseases and its analgesic effect. According to Liu et al. (2024), capsaicin has been associated with these benefits. Similarly, theobromine, found in cocoa, has been shown to benefit cardiovascular health, helping reduce blood pressure and improve endothelial function (Amirkia & Heinrich, 2014).

It is important to note that consuming too many alkaloids can result in negative effects, like insomnia, nervousness, tachycardia, and dependence (Costantino et al., 2023). It is advisable to consult a medical expert, particularly if you have any pre-existing health conditions or are prone to them (de Oliveira et al., 2022). In summary, food alkaloids represent a valuable source of bioactive compounds with diverse pharmacological applications and health benefits. Its physiological effects and mechanisms of action on the human system offer essential insights for developing new therapies and interventions in health care, highlighting the importance of continued research in this promising field (Pinto et al., 2020).

Koleva et al. (2012) discuss the presence of alkaloids in various foods and their potentially harmful effects on the human body. The study focuses explicitly on pyrrolizidine alkaloids, which can be concerning, because they can transform into reactive alkylating intermediates. Several other alkaloids, such as quinolizidine alkaloids, β-carboline alkaloids, ergot alkaloids, and steroid alkaloids, have neurotoxic properties without needing bioactivation.

The study highlights that many commonly consumed foods contain these alkaloids, such as fruits, vegetables, grains, coffee, and tea. Excessive consumption of alkaloids has been associated with acute and chronic toxicity, gastrointestinal problems, neurobehavioral effects, cardiovascular issues, and even carcinogenicity in some cases (Koleva et al., 2012).

Koleva et al. (2012) emphasize the importance of exploring the natural sources of alkaloids in our diet and the need to understand safe consumption levels to mitigate health risks. Experts recommend conducting more comprehensive research to evaluate the effects of alkaloids on diverse populations and at varying concentrations found in foods. Additionally, they stress the critical importance of regulating and monitoring alkaloid levels in food to ensure food safety and protect public health.

Brugnerotto et al. (2021) conducted a study to examine whether pyrrolizidine alkaloids (PA) are present in beehive products, including honey, beeswax, and propolis. PAs are naturally occurring compounds in various plants, which can harm humans and animals if consumed significantly. The study found that bee products may contain PAs due to nectar and pollen collected from contaminated plants. This research highlights worries about the possible health hazards of consuming bee products that may have been infected. As a result, the article emphasizes the importance of closely monitoring and regulating the existence of PAs in bee products. It also underscores the need for educating producers and consumers about the potentially harmful health outcomes that may result from exposure to these substances.

17.8 SAFETY IN THE USE OF FOOD ALKALOIDS

Food alkaloids may have adverse health effects if ingested insufficiently, so there is increasing concern about using these compounds. In excess, certain alkaloids, such as caffeine found in coffee, tea, and soft drinks, have been linked to adverse consequences, like tachycardia, anxiety, and insomnia (Rodak et al., 2021). Moreover, gastrointestinal issues and drug addiction are two other major health issues that may be linked to long-term high-dose consumption of alkaloids (Juliano & Griffiths, 2004).

It is crucial to emphasize that various dietary alkaloids can positively affect health if taken in moderation and within certain limitations. For instance, capsaicin in peppers has been connected to anti-inflammatory and antioxidant qualities and may help prevent cancer and cardiovascular disease (Srinivasan, 2016). Similarly, theobromine, found in cocoa, has been investigated for possible cardioprotective and neuroprotective benefits. However, high dosages can be toxic (Martínez-Pinilla et al., 2015).

In order to maintain public health and promote responsible usage of these substances, it is essential to have a comprehensive understanding of the potential advantages and disadvantages associated with the consumption of food alkaloids. Along with this understanding, it is also crucial to implement appropriate regulations and practices regarding food safety, as stated by Ekor (2014). Educating consumers about the significance of consuming these foods in moderation and in balance is crucial to encouraging good eating habits and avoiding any health issues associated with alkaloids.

Assessing health risks associated with chemical substances, such as alkaloids, and establishing guidelines and recommendations to ensure food safety are crucial responsibilities of the World Health Organization (WHO) (FAO and WHO, 2023).

To develop international standards and guidelines for food safety, including assessing risks associated with contaminants like alkaloids, the WHO collaborates with other agencies, such as the Food and Agriculture Organization of the United Nations (FAO). These guidelines are usually adopted in national food safety legislation by individual countries.

The WHO also provides technical support and guidance to member countries on safe food production, processing, and preparation practices to minimize exposure to harmful substances, including alkaloids. Although no specific WHO legislation exists on food alkaloids, the organization's comprehensive guidelines and recommendations are essential in promoting global food security and protecting public health.

17.9 FUTURE PERSPECTIVES

Prospects for the future of alkaloids in food indicate a rise in interest in locating and investigating food sources high in these beneficial substances. It is anticipated that new alkaloid sources will be found, and their possible health benefits will be assessed as analytical technology develops. Additionally, a deeper comprehension of how alkaloids function in the human body may pave the way for the creation of novel functional foods enhanced with these substances, providing an additional strategy for disease prevention and health promotion.

New methods for extracting alkaloids from natural sources are anticipated to become more effective, sustainable, and novel. New cost-effective and environmentally friendly extraction techniques may become popular to minimize environmental impact and reduce chemical solvents' usage. The search to enhance the stability and bioavailability of extracted alkaloids could promote the development of innovative encapsulation and formulation technologies. This, in turn, would ensure the secure and effective delivery of these compounds to the human body. By bettering public health and overall well-being, these positive trends suggest a promising future for exploring and utilizing alkaloids in food.

REFERENCES

Adetunji, T. L., Olawale, F., Olisah, C., Adetunji, A. E., & Aremu, A. O. (2022). Capsaicin: A two-decade systematic review of global research output and recent advances against human cancer. *Frontiers in Oncology*, *12*, 908487. https://doi.org/10.3389/fonc.2022.908487

Aguilar-Hernández, G., Zepeda-Vallejo, L. G., García-Magaña, M. D. L., Vivar-Vera, M. D. L. Á., Pérez-Larios, A., Girón-Pérez, M. I., Coria-Tellez, A. V., Rodríguez-Aguayo, C., & Montalvo-González, E. (2020). Extraction of alkaloids using ultrasound from pulp and by-products of soursop fruit (*Annona muricata* L.). *Applied Sciences*, *10*(14), 4869. https://doi.org/10.3390/app10144869

Ahuja, V., Singh, A., Paul, D., Dasgupta, D., Urajová, P., Ghosh, S., . . . Saurav, K. (2023). Recent advances in the detection of food toxins using mass spectrometry. *Chemical Research in Toxicology*, *36*(12), 1834–1863. https://doi.org/10.1021/acs.chemrestox.3c00241

Al Muqarrabun, L. M. R., & Ahmat, N. (2015). Medicinal uses, phytochemistry and pharmacology of family Sterculiaceae: A review. *European Journal of Medicinal Chemistry*, *92*, 514–530. https://doi:10.1016/j.ejmech.2015.01.026

Amirkia, V., & Heinrich, M. (2014). Alkaloids as drug leads—a predictive structural and biodiversity-based analysis. *Phytochemistry Letters*, *10*, xlviii—xlliii. https://doi.org/10.1016/j.phytol.2014.06.015

Arts, I. C., & Hollman, P. C. (2005). Polyphenols and disease risk in epidemiologic studies. *The American Journal of Clinical Nutrition*, *81*(1), 317S–325S. https://doi.org/10.1093/ajcn/81.1.317S

Azmir, J., Zaidul, I. S. M., Rahman, M. M., Sharif, K. M., Mohamed, A., Sahena, F., Jahurul, M. H. A., Ghafoor, K., Norulaini, N. A. N., & Omar, A. K. M. (2013). Techniques for extraction of bioactive compounds from plant materials: A review. *Journal of Food Engineering*, *117*(4), 426–436. https://doi.org/10.1016/j.jfoodeng.2013.01.014

Babiker, F., Jamal, P., Mirghani, M. E. S., & Ansari, A. H. (2017). Characterization, purification, and identification of some Alkaloids in *Datura stramonium. International Food Research Journal*, *24*, S540–S543.

Bletsou, A. A., Jeon, J., Hollender, J., Archontaki, E., & Thomaidis, N. S. (2015). Targeted and non-targeted liquid chromatography-mass spectrometric workflows for identification of transformation products of emerging pollutants in the aquatic environment. *TrAC Trends in Analytical Chemistry*, 66, 32–44. https://doi.org/10.1016/j.trac.2014.11.009

Brugnerotto, P., Seraglio, S. K. T., Schulz, M., Gonzaga, L. V., Fett, R., & Costa, A. C. O. (2021). Pyrrolizidine alkaloids and beehive products: A review. *Food Chemistry*, *342*, 128384. https://doi.org/10.1016/j.foodchem.2020.128384

Casado, N., Casado-Hidalgo, G., González-Gómez, L., Morante-Zarcero, S., & Sierra, I. (2023). Insight into the impact of food processing and culinary preparations on the stability and content of plant alkaloids considered as natural food contaminants. *Applied Sciences*, *13*(3), 1704. https://doi.org/10.1016/j.tifs.2022.01.007

Casado, N., Morante-Zarcero, S., & Sierra, I. (2022). The concerning food safety issue of pyrrolizidine alkaloids: An overview. *Trends in Food Science & Technology*, *120*, 123–139. https://doi.org/10.1016/j.tifs.2022.01.007

Casado-Hidalgo, G., Morante-Zarcero, S., Pérez-Quintanilla, D., & Sierra, I. (2022). Pulsed ultrasound-assisted extraction followed by purification with SBA-15 for the control of opium alkaloids in biscuits and sponge cakes. *Microchemical Journal*, *183*, 108059. https://doi.org/10.1016/j.microc.2022.108059

Celano, R., Piccinelli, A. L., Campone, L., Russo, M., & Rastrelli, L. (2019). Determination of selected pyrrolizidine alkaloids in honey by dispersive liquid–liquid microextraction and ultrahigh-performance liquid chromatography–tandem mass spectrometry. *Journal of Agricultural and Food Chemistry*, *67*(31), 8689–8699. https://doi.org/10.1021/acs.jafc.9b02392

Chen, C., & Lin, L. (2019). Alkaloids in diet. In *Handbook of dietary phytochemicals* (pp. 1–35). Singapore: Springer.

Chen, Y., Li, L., Xiong, F., Xie, Y., Xiong, A., Wang, Z., & Yang, L. (2021). Rapid identification and determination of pyrrolizidine alkaloids in herbal and food samples via direct analysis in real-time mass spectrometry. *Food Chemistry*, *334*, 127472. https://doi.org/10.1016/j.foodchem.2020.127472

Cirlini, M., Cappucci, V., Galaverna, G., Dall'Asta, C., & Bruni, R. (2019). A sensitive UHPLC-ESI-MS/MS method for the determination of tropane alkaloids in herbal teas and extracts. *Food Control*, *105*, 285–291. https://doi.org/10.1016/j.foodcont.2019.05.030

Costantino, A., Maiese, A., Lazzari, J., Casula, C., Turillazzi, E., Frati, P., & Fineschi, V. (2023). The dark side of energy drinks: A comprehensive review of their impact on the human body. *Nutrients*, *15*(18), 3922. https://doi.org/10.3390/nu15183922

Cova, I., Leta, V., Mariani, C., Pantoni, L., & Pomati, S. (2019). Exploring cocoa properties: Is theobromine a cognitive modulator? *Psychopharmacology*, *236*, 561–572. https://doi.org/10.1007/s00213-019-5172-0

Davison, K., Coates, A. M., Buckley, J. D., & Howe, P. R. C. (2008). Effect of cocoa flavanols and exercise on cardiometabolic risk factors in overweight and obese subjects. *International Journal of Obesity*, *32*(8), 1289–1296. https://doi.org/10.1038/ijo.2008.66

de Oliveira, J. A. R., de Paula Menezes Barbosa, P., & Macêdo, G. A. (2022). High concentrate flavonoids extract from citrus pomace using enzymatic and deep eutectic solvents extraction. *Foods*, *11*(20), 3205. https://doi.org/10.3390/foods11203205

Desideri, G., Kwik-Uribe, C., Grassi, D., Necozione, S., Ghiadoni, L., Mastroiacovo, D., . . . Ferri, C. (2012). Benefits in cognitive function, blood pressure, and insulin resistance through cocoa flavanol consumption in elderly subjects with mild cognitive impairment: The Cocoa, Cognition, and Aging (CoCoA) study. *Hypertension*, *60*(3), 794–801. https://doi.org/10.1161/HYPERTENSIONAHA.112.193060

Dey, P., Kundu, A., Kumar, A., Gupta, M., Lee, B. M., Bhakta, T., Dash, S., & Kim, H. S. (2020). Analysis of alkaloids (indole alkaloids, isoquinoline alkaloids, tropane alkaloids). In *Recent advances in natural products analysis* (pp. 505–567). Elsevier. https://doi.org/10.1016/B978-0-12-816455-6.00015-9

Dhyani, P., Quispe, C., Sharma, E., Bahukhandi, A., Sati, P., Attri, D. C., . . . Cho, W. C. (2022). Anticancer potential of alkaloids: a key emphasis to colchicine, vinblastine, vincristine, vindesine, vinorelbine and vincamine. *Cancer Cell International*, *22*(1), 206. https://doi.org/10.1186/s12935-022-02624-9

Díaz-Laviada, I., & Rodríguez-Henche, N. (2014). The potential antitumor effects of capsaicin. *Capsaicin as a Therapeutic Molecule*, 181–208. https://doi.org/10.1007/978-3-0348-0828-6_8

Ding, M., Bhupathiraju, S. N., Chen, M., Van Dam, R. M., & Hu, F. B. (2014). Caffeinated and decaffeinated coffee consumption and risk of type 2 diabetes: A systematic review and a dose-response meta-analysis. *Diabetes Care*, *37*(2), 569–586. https://doi.org/10.2337/dc13-1203

Doughari, J. H. (2012). *Phytochemicals: extraction methods, basic structures and mode of action as potential chemotherapeutic agents* (pp. 1–33). Rijeka, Croatia: INTECH Open Access Publisher. https://doi.org/10.5772/26052

Dusemund, B., Nowak, N., Sommerfeld, C., Lindtner, O., Schäfer, B., & Lampen, A. (2018). Risk assessment of pyrrolizidine alkaloids in food of plant and animal origin. *Food and Chemical Toxicology*, *115*, 63–72. https://doi.org/10.1016/j.fct.2018.03.005

Eich, E. (2008). *Solanaceae and convolvulaceae: Secondary metabolites: Biosynthesis, chemotaxonomy, biological and economic significance (a handbook)*. Springer Science & Business Media. https://doi.org/10.1007/978-3-540-74541-9

Ekor, M. (2014). The growing use of herbal medicines: issues relating to adverse reactions and challenges in monitoring safety. *Frontiers in Pharmacology*, *4*, 66193. https://doi.org/110.3389/fphar.2013.00177

El-Sayed, M., & Verpoorte, R. (2007). Catharanthus terpenoid indole alkaloids: biosynthesis and regulation. *Phytochemistry Reviews*, *6*, 277–305. https://doi.org/10.1007/s11101-006-9047-8

European Food Safety Authority (EFSA). Panel on Contaminants in the Food Chain (CONTAM). (2011). Scientific opinion on pyrrolizidine alkaloids in food and feed. *EFSA Journal*, *9*(11), 2406. https://doi.org/10.2903/j.efsa.2011.2406

European Food Safety Authority (EFSA). Panel on Contaminants in the Food Chain (CONTAM). (2013). Scientific opinion on tropane alkaloids in food and feed. EFSA Panel on Contaminants in the Food Chain. *EFSA Journal*, *11*(10), 3386. https://doi.org/10.2903/j.efsa.2013.3386

European Food Safety Authority (EFSA). Panel on Contaminants in the Food Chain (CONTAM). (2018). Update of the scientific opinion on opium alkaloids in poppy seeds. *EFSA Journal*, *16*(5), e05243. https://doi.org/10.2903/j.efsa.2018.5243

Ewan, E. E., & Martin, T. J. (2011). Opioid facilitation of rewarding electrical brain stimulation is suppressed in rats with neuropathic pain. *The Journal of the American Society of Anesthesiologists*, *114*(3), 624–632. https://doi.org/10.1097/ALN.0b013e31820a4edb

Fahmy, S. A., Issa, M. Y., Saleh, B. M., Meselhy, M. R., & Azzazy, H. M. E. S. (2021). Peganum harmala alkaloids self-assembled supramolecular nanocapsules with enhanced antioxidant and cytotoxic activities. *ACS Omega*, *6*(18), 11954–11963. https://doi.org/10.1021/acsomega.1c00455

Fattori, V., Hohmann, M. S., Rossaneis, A. C., Pinho-Ribeiro, F. A., & Verri Jr, W. A. (2016). Capsaicin: Current understanding of its mechanisms and therapy of pain and other pre-clinical and clinical uses. *Molecules*, *21*(7), 844. https://doi.org/10.3390/molecules21070844

Feng, W., Li, M., Hao, Z., & Zhang, J. (2019). Analytical methods of isolation and identification. In *Phytochemicals in human health*. IntechOpen. https://doi.org/10.5772/intechopen.88122

Ferré, S. (2008). An update on the mechanisms of the psychostimulant effects of caffeine. *Journal of Neurochemistry*, *105*(4), 1067–1079. https://doi.org/10.1111/j.1471-4159.2007.05196.x

Food and Agriculture Organization of the United Nations (FAO) & World Health Organization (WHO). (2023, March 31). *FAO/WHO food safety experts provide guidance on tropane alkaloids*. Retrieved from https://www.fao.org/food-safety/news/news-details/en/c/1363024/

González-Castejón, M., & Rodriguez-Casado, A. (2011). Dietary phytochemicals and their potential effects on obesity: a review. *Pharmacological Research*, *64*(5), 438–455. https://doi.org/10.1016/j.phrs.2011.07.004

González-Gómez, L., Gañán, J., Morante-Zarcero, S., Pérez-Quintanilla, D., & Sierra, I. (2020). Sulfonic acid-functionalized SBA-15 as strong cation-exchange sorbent for solid-phase extraction of atropine and scopolamine in gluten-free grains and flours. *Foods*, *9*(12), 1854. https://doi.org/10.3390/foods9121854

González-Gómez, L., Morante-Zarcero, S., Pérez-Quintanilla, D., & Sierra, I. (2022). Occurrence and chemistry of tropane alkaloids in foods, with a focus on sample analysis methods: A review on recent trends and technological advances. *Foods*, *11*(3), 407. https://doi.org/ 10.3390/foods11030407

Govindarajan, V. S., & Sathyanarayana, M. N. (1991). Capsicum—production, technology, chemistry, and quality. Part V. Impact on physiology, pharmacology, nutrition, and metabolism; structure, pungency, pain, and desensitization sequences. *Critical Reviews in Food Science & Nutrition*, *29*(6), 435–474. https://doi.org/10.1080/10408399109527536

Grimm, M. O., Blümel, T., Lauer, A. A., Janitschke, D., Stahlmann, C., Mett, J., . . . Hartmann, T. (2020). The impact of capsaicinoids on APP processing in Alzheimer's disease in SH-SY5Y cells. *Scientific Reports*, *10*(1), 9164. https://doi.org/10.1038/s41598-020-66009-6

Grosso, G., Godos, J., Galvano, F., & Giovannucci, E. L. (2017). Coffee, caffeine, and health outcomes: An umbrella review. *Annual Review of Nutrition*, *37*, 131–156. https://doi.org/10.1146/annurev-nutr-071816-064941

Heinrich, M., Mah, J., & Amirkia, V. (2021). Alkaloids used as medicines: Structural phytochemistry meets biodiversity—an update and forward look. *Molecules*, *26*(7), 1836. https://doi.org/10.3390/molecules26071836

Imanshahidi, M., & Hosseinzadeh, H. (2008). Pharmacological and therapeutic effects of *Berberis vulgaris* and its active constituent, berberine. *Phytotherapy Research*, *22*(8), 999–1012. https://doi.org/10.1002/ptr.2399

Juliano, L. M., & Griffiths, R. R. (2004). A critical review of caffeine withdrawal: Empirical validation of symptoms and signs, incidence, severity, and associated features. *Psychopharmacology*, *176*, 1–29. https://doi.org/10.1007/s00213-004-2000-x

Kaltner, F. (2022). Fate of food-relevant toxic plant alkaloids during food processing or storing and analytical strategies to unveil potential transformation products. *Journal of Agricultural and Food Chemistry*, *70*(20), 5975–5981. https://doi.org/10.1021/acs.jafc.2c01489

Kaulmann, A., & Bohn, T. (2014). Carotenoids, inflammation, and oxidative stress—implications of cellular signaling pathways and relation to chronic disease prevention. *Nutrition Research*, *34*(11), 907–929. https://doi.org/10.1016/j.nutres.2014.07.010

Kole, P. L., Venkatesh, G., Kotecha, J., & Sheshala, R. (2011). Recent advances in sample preparation techniques for effective bioanalytical methods. *Biomedical Chromatography*, *25*(1–2), 199–217. https://doi.org/10.1002/bmc.1560

Koleva, I. I., van Beek, T. A., Soffers, A. E., Dusemund, B., & Rietjens, I. M. (2012). Alkaloids in the human food chain–natural occurrence and possible adverse effects. *Molecular Nutrition & Food Research*, *56*(1), 30–52. https://doi.org/10.1002/mnfr.201100165

Krska, R., Berthiller, F., Schuhmacher, R., Nielsen, K. F., & Crews, C. (2008). Determination of ergot alkaloids: purity and stability assessment of standards and optimization of extraction conditions for cereal samples. *Journal of AOAC International*, *91*(6), 1363–1371. https://doi.org/10.1093/jaoac/91.6.1363

Lee, M. C., Wanigasekera, V., & Tracey, I. (2014). Imaging opioid analgesia in the humam brain and its potential relevance for understanding opioid use in chronic pain. *Neuropharmacology*, *84*, 123–130. https://doi.org/10.1016/j.neuropharm.2013.06.035

Lee, S. H., & Clark, R. (2016). Anti-tumorigenic effects of capsaicin in colon cancer. *Journal Food Chemistry Nanotechnology*, 2(4), 162–167. https://ar.iiarjournals.org/content/36/3/837

Liu, C., Wang, L., & Lu, W. (2024). Caffeine intake and anxiety: a meta-analysis. *Frontiers in Psychology*, *15*, 1270246. https://doi.org/10.3389/fpsyg.2024.1270246

López, P., Pereboom-de Fauw, D. P., Mulder, P. P., Spanjer, M., de Stoppelaar, J., Mol, H. G., & de Nijs, M. (2018). Straightforward analytical method to determine opium alkaloids in poppy seeds and bakery products. *Food Chemistry*, *242*, 443–450. https://doi.org/10.1016/j.foodchem.2017.08.045

Lopez-Hortas, L., Rodriguez, P., Diaz-Reinoso, B., Gaspar, M. C., de Sousa, H. C., Braga, M. E., & Dominguez, H. (2022). Supercritical fluid extraction as a suitable technology to recover bioactive compounds from flowers. *The Journal of Supercritical Fluids*, *188*, 105652. https://doi.org/10.1016/j.supflu.2022.105652

Martínez, M. A., & Ballesteros, S. (2019). Opium poisoning in modern times. An overview. *Forensic Science International*, *302*, 109848.

Martínez-Pinilla, E., Oñatibia-Astibia, A., & Franco, R. (2015). The relevance of theobromine for the beneficial effects of cocoa consumption. *Frontiers in Pharmacology*, *6*, 126866. https://doi.org/10.3389/fphar.2015.00030

Nehlig, A. (2013). The neuroprotective effects of cocoa flavanol and its influence on cognitive performance. *British Journal of Clinical Pharmacology*, *75*(3), 716–727. https://doi.org/10.1111/j.1365-2125.2012.04378.x

Nehlig, A., Daval, J. L., & Debry, G. (1992). Caffeine and the central nervous system: mechanisms of action, biochemical, metabolic and psychostimulant effects. *Brain Research Reviews*, *17*(2), 139–170. https://doi.org/10.1016/0165-0173(92)90012-B

Pase, M. P., Scholey, A. B., Pipingas, A., Kras, M., Nolidin, K., Gibbs, A., . . . Stough, C. (2013). Cocoa polyphenols enhance positive mood states but not cognitive performance: a randomized, placebo-controlled trial. *Journal of Psychopharmacology*, *27*(5), 451–458. https://doi.org/10.1177/0269881112473791

Pathan, H., & Williams, J. (2012). Basic opioid pharmacology: an update. *British Journal of Pain*, *6*(1), 11–16. https://doi.org/10.1177/2049463712438493

Pinto, D., Delerue-Matos, C., & Rodrigues, F. (2020). Bioactivity, phytochemical profile and pro-healthy properties of *Actinidia arguta*: A review. *Food Research International*, 109449. https://doi.org/10.1016/j.foodres.2020.109449

Qiu, N., Sun, D., Zhou, S., Li, J., Zhao, Y., & Wu, Y. (2022). Rapid and sensitive UHPLC-MS/MS methods for dietary sample analysis of 43 mycotoxins in China total diet study. *Journal of Advanced Research*, *39*, 15–47. https://doi.org/10.1016/j.jare.2021.10.008

Rodak, K., Kokot, I., & Kratz, E. M. (2021). Caffeine as a factor influencing the functioning of the human body—friend or foe? *Nutrients*, *13*(9), 3088. https://doi.org/10.3390/nu13093088

Romera-Torres, A., Romero-González, R., Martínez Vidal, J. L., & Garrido Frenich, A. (2018). Simultaneous analysis of tropane alkaloids in teas and herbal teas by liquid chromatography coupled to high-resolution mass spectrometry (Orbitrap). *Journal of Separation Science*, *41*(9), 1938–1946. https://doi.org/10.1002/jssc.201701485

Rossato, M. F., Trevisan, G., Walker, C. I. B., Klafke, J. Z., de Oliveira, A. P., Villarinho, J. G., . . . Ferreira, J. (2011). Eriodictyol: a flavonoid antagonist of the TRPV1 receptor with antioxidant activity. *Biochemical Pharmacology*, *81*(4), 544–551. https://doi.org/10.1016/j.bcp.2010.11.004

Rostagno, M. A., & Prado, J. M. (Eds.). (2013). *Natural product extraction: principles and applications* (Vol. 21). Royal Society of Chemistry. https://doi.org/10.1039/9781839165894

Ruan, X., Cui, W. X., Yang, L., Li, Z. H., Liu, B., & Wang, Q. (2017). Extraction of total alkaloids, peimine and peiminine from the flower of *Fritillaria thunbergii* Miq using supercritical carbon dioxide. *Journal of CO2 Utilization*, *18*, 283–293. https://doi.org/10.1016/j.jcou.2017.01.024

Silva, I. D. D., & Aragão, C. F. S. (2009). Extraction parameters from *Cinchona pubescens* Vahl using pharmacopeia and no pharmacopeia methods. *Revista Brasileira de Farmacognosia*, *19*, 776–780. https://doi.org/10.1590/S0102-695X2009000500021

Srinivasan, K. (2016). Biological activities of red pepper (*Capsicum annuum*) and its pungent principle capsaicin: a review. *Critical Reviews in Food Science and Nutrition*, *56*(9), 1488–1500. https://doi.org/10.1080/10408398.2013.772090

Stępniowska, A., Cieplińska, P., Fac, W., & Gorska, J. (2021). Selected alkaloids used in the cosmetics industry. *Journal of Cosmetic Science*, *72*(2).

Surh, Y. J. (2002). Anti-tumor promoting potential of selected spice ingredients with antioxidative and anti-inflammatory activities: a short review. *Food and Chemical Toxicology*, *40*(8), 1091–1097. https://doi.org/10.1016/S0278-6915(02)00037-6

Thompson, R. Q. (2005). Encyclopedia of analytical science (worsfold, paul; tonshend, alan; poole, colin). https://doi.org/10.1021/ed082p1313.2

Torović, L., Bursić, V., & Vuković, G. (2023). Health concerns associated to tropane alkaloids in maize food products in Serbia. *Heliyon*, *9*(9). https://doi.org/10.1016/j.heliyon.2023.e19404

Torres-Vega, J., Gómez-Alonso, S., Pérez-Navarro, J., & Pastene-Navarrete, E. (2020). Green extraction of alkaloids and polyphenols from *Peumus boldus* leaves with natural deep eutectic solvents and profiling by HPLC-PDA-IT-MS/MS and HPLC-QTOF-MS/MS. *Plants*, *9*(2), 242. https://doi.org/110.3390/plants9020242

Turner, N. W., Subrahmanyam, S., & Piletsky, S. A. (2009). Analytical methods for determination of mycotoxins: a review. *Analytica Chimica Acta*, *632*(2), 168–180. https://doi.org/10.1016/j.aca.2008.11.010

Wang, H., Chen, Y., Wang, L., Liu, Q., Yang, S., & Wang, C. (2023). Advancing herbal medicine: enhancing product quality and safety through robust quality control practices. *Frontiers in Pharmacology*, *14*, 1265178. https://doi.org/10.3389/fphar.2023.1265178

Wang, J., Sun, B. L., Xiang, Y., Tian, D. Y., Zhu, C., Li, W. W., . . . Wang, Y. J. (2020). Capsaicin consumption reduces brain amyloid-beta generation and attenuates Alzheimer's disease-type pathology and cognitive deficits in APP/PS1 mice. *Translational Psychiatry*, *10*(1), 230. https://doi.org/10.1038/s41398-020-00918-y

Xiong, W., Chen, X., Lv, G., Hu, D., Zhao, J., & Li, S. (2016). Optimization of microwave-assisted extraction of bioactive alkaloids from lotus plumule using response surface methodology. *Journal of Pharmaceutical Analysis*, *6*(6), 382–388. https://doi.org/10.1016/j.jpha.2016.05.007

Xu, F., Liu, F., Wang, C., & Wei, Y. (2019). Amantadine-functionalized magnetic microspheres and stable isotope labeled internal standards for reducing matrix effect in determination of five opium alkaloids by liquid chromatography-quadrupole linear ion trap mass spectrometry. *Journal of the Chinese Chemical Society*, *66*(5), 484–492. https://doi.org/10.1002/jccs.201800310

Yubin, J. I., Miao, Y., Bing, W., & Yao, Z. (2014). The extraction, separation and purification of alkaloids in the natural medicine. *Journal of Chemical and Pharmaceutical Research*, *6*(1), 338–345.

Zhang, Q. W., Lin, L. G., & Ye, W. C. (2018). Techniques for extraction and isolation of natural products: A comprehensive review. *Chinese Medicine*, *20*(13), 1–26. https://doi.org/10.1186/s13020-018-0177-x

Zhang, S., Wang, D., Huang, J., Hu, Y., & Xu, Y. (2020). Application of capsaicin as a potential new therapeutic drug in human cancers. *Journal of Clinical Pharmacy and Therapeutics*, *45*(1), 16–28. https://doi.org/10.1111/jcpt.13039

18 New Food Flavonoids and Their Pharmacological Properties

Magally Araújo Gonçalves, Ingryd Rodrigues Martins, Jonilson de Melo e Silva, Julia Helena da Silva Martins, Andrea Komesu, Johnatt Allan Rocha de Oliveira, and Luiza Helena Da Silva Martins

18.1 INTRODUCTION

Recent investigations into flavonoids found in food have sparked considerable attention, owing to their possible pharmacological attributes and health advantages. Flavonoids, a heterogeneous group of polyphenolic compounds, are extensively present in numerous food sources, like fruits, vegetables, teas, and herbs. Their occurrence in food items has been linked to favorable health outcomes, including antioxidative, anti-inflammatory, antimicrobial, anticancer, and cardioprotective properties (Fuente-Ballesteros et al., 2023).

Research has revealed an increasing diversity of new flavonoids in foods (Li et al., 2014, 2024; Tan et al., 2016; Yang et al., 2018).

These compounds, held to the broad class of polyphenols, have aroused considerable interest due to their potential bioactive properties and health benefits. Several studies have been dedicated to identifying and characterizing these new flavonoids in various food sources (Yang et al., 2018).

As analytical methods like gas chromatography (GC), infrared spectroscopy (IR), high-performance liquid chromatography (HPLC), mass spectrometry (MS), and nuclear magnetic resonance (NMR), among others, continue to evolve, it is now feasible not only to identify but also to accurately elucidate the chemical structure of these compounds (Al-Owaisi et al., 2014; Gbashi et al., 2017; Markham & Geiger, 2017).

These new flavonoids have been associated with diverse health benefits, ranging from antioxidant and anti-inflammatory activities to neuroprotective and cardioprotective properties. Therefore, understanding these new flavonoids' presence and potential properties in foods is essential to exploring their role in promoting health and preventing diseases (Barreca et al., 2021; Zhao et al., 2021).

Extraction methods play a crucial role in efficiently obtaining flavonoids from foods. The appropriate choice of method can directly influence the yield and quality of extracts, affecting their applicability in research and industry. Furthermore, effective extraction techniques guarantee the preservation of flavonoids' bioactive properties, allowing a more accurate analysis of their physiological and pharmacological effects (Chávez-González et al., 2020).

This chapter explores recent advances in discovering new flavonoids in foods and investigates their pharmacological properties. The growing interest in identifying and characterizing these new compounds arises from the need to expand our knowledge about their potential health benefits. Understanding the pharmacological properties of these flavonoids can provide essential insights

DOI: 10.1201/9781003462804-24

into developing therapies and dietary interventions aimed at preventing and treating various health conditions.

This chapter reviews the most recent studies on novel flavonoids in foods to highlight their structural diversity, dietary sources, and potential therapeutic applications. Furthermore, the proposed mechanisms of action behind the pharmacological activities of these flavonoids will be discussed, providing a more comprehensive understanding of how they may influence human health.

18.1.1 Flavonoids

They are substances widely present in foods such as fruits and vegetables, and their potential to promote health has aroused great interest from researchers and industry worldwide. When consumed, these compounds are absorbed into the circulatory system through the intestinal tract, where they employ their beneficial effects (Figure 18.1). However, despite their advantageous properties, flavonoids face significant challenges when considered for application as dietary supplements due to their instability, low bioavailability during gastrointestinal digestion, and lack of a targeted affinity at a biological level (Li et al., 2023).

Flavonoids are part of the bioactive compounds known as polyphenols, produced as secondary metabolites in plants. They perform several functions, including antioxidant action, signaling, and detoxification, protecting plants against stress caused by external factors, whether abiotic or biotic. Flavonoids found in foods are generally called bi-flavonoids and are recognized for their distinct functional properties (Kaushal et al., 2022).

Among the various properties of flavonoids, their antioxidant capacity stands out, which makes them powerful protective agents against free radicals. These radicals are molecules with an odd number of electrons, which makes them unstable and prone to electron capture reactions from cellular proteins, triggering a harmful chain reaction. Flavonoids can neutralize these radicals, including superoxide radicals, hydroxyl radicals, and peroxynitrite, which can cause damage to biomolecules. Furthermore, flavonoids have antimicrobial, antimutagenic, and anti-inflammatory properties and are inhibitors of enzymes harmful to the body (Russo, 2018).

Regarding their structure, flavonoids can be grouped into different classes and categorized based on the heterocycles present. Six of these classes are considered the main ones, comprising flavanols, flavonols, flavones, isoflavones, flavonones, and anthocyanins (Li et al., 2023).

Understanding the chemical structure of flavonoids is fundamental to understanding their antioxidant activity. A widely documented aspect is the relationship between the degree of hydroxylation of the molecule and its antioxidant activity: The greater the degree of hydroxylation, the greater this activity. However, it is essential to consider that substituting methoxyl groups in positions 3 and 5 can reduce the antioxidant capacity of the molecule (Pietta, 2000).

Flavonols, including quercetin, kaempferol, and myricetin, are renowned for their antioxidative, anti-inflammatory, and anticancer attributes. They have shown efficacy in counteracting free radicals, diminishing inflammation, and impeding cancer cell proliferation in both in vitro experiments and animal trials (Shahbaz et al., 2023). Anthocyanins, found in colorful fruits and vegetables, such as blueberries and purple grapes, have potent antioxidant and anti-inflammatory properties and

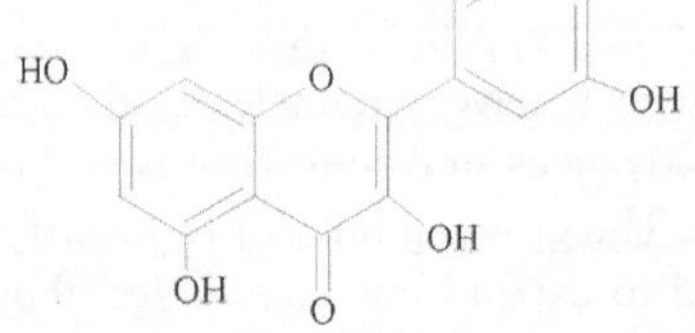

FIGURE 18.1 Molecular structures of known flavonoids frequently found in foods.

contribute to cardiovascular and cognitive health (Mohammadi et al., 2024). Rutin, present in foods such as citrus fruits and red fruits, exhibits antioxidant and anti-inflammatory activity, being associated with vascular protection and reducing the risk of cardiovascular diseases Din et al. (2023). Diosmetin, present in citrus fruits and vegetables, possesses antioxidative and anti-inflammatory characteristics which support cardiovascular health and bolster immune function. These flavonoids are generally acknowledged for their capacity to enhance well-being and deter various chronic ailments, underscoring their significance as essential elements of a well-rounded, nutritious diet (Barreca et al., 2023).

18.1.2 New Flavonoids Found in Foods

New flavonoids food refers to foods that contain recently discovered or identified flavonoids, which may offer additional health benefits due to their properties. These flavonoids can be found in a variety of food sources, such as fruits, vegetables, herbs, and grains, and are the subject of study in nutritional and pharmaceutical research due to their potential to promote health and prevent disease. The term highlights the importance of the continuous discovery of new bioactive compounds in foods and their role in promoting a healthy and balanced diet.

Aoyama et al. (2018) identified two new flavonoids and other constituents in licorice root by examining their relative astringency and radical scavenging effects. The new flavonoids were characterized as glycyrrhizin C and 3',5-di-O-(E)-feruloyl-glycyrrhizin C, demonstrating significant antioxidant properties. Glycyrrhizin C was shown to be more effective in radical scavenging than 3',5-di-O-(E)-feruloyl-glycyrrhizin C, while both exhibited low astringency. These new flavonoids expand knowledge about licorice root constituents and highlight their potential for therapeutic applications due to their antioxidant properties.

Li et al. (2014) unveiled novel flavonoids from *Campylotropis hirtella*, accentuating its immunosuppressive attributes. These recently identified flavonoids were extracted from plant samples with distinct chemical compositions. Research indicates that these newfound flavonoids display notable immunosuppressive effects, suggesting their promise as potential remedies for immune system–related ailments. Among the discoveries are nine fresh flavonoids, encompassing varieties such as (1) one isoflav-3-ene derivative, (2) one coumaronochromone, (3, 4) two isoflavones, (6) one isoflavone derivative, (7) one isoflavone, and (8, 9, 10) three flavonols, along with a previously recognized compound, (5) hydroisoflavone C, pursued in the quest for natural immunosuppressive agents.

Campylotropis hirtella is utilized as a dietary addition in the subtropical areas of China. Through an extensive search for inhibitors of human neutrophil elastase, eight flavonoids were extracted from *C. hirtella*, with three of them (1–3) identified as elastase inhibitors (Tan et al., 2016). Li et al. (2014) discovered new flavonoids in *Campylotropis hirtella*, highlighting its immunosuppressive properties. The newly identified flavonoids are characterized by their unique chemical structure and were isolated from plant extracts. Studies have demonstrated that these new flavonoids exhibit significant immunosuppressive activity, suggesting their potential as therapeutic agents for treating immune system–related diseases. Nine new flavonoids, including (1) one isoflav-3-ene derivative, (2) one coumaronochromone, (3, 4) two isoflavanones, (6) one isoflavone derivative, (7) one isoflavone, (8, 9, 10) three flavonols, as well as one known compound, (5) hydroisoflavone C, were discovered in the pursuit of natural compounds with immunosuppressive activity.

Penthorum chinense Pursh (PCP) is known for its medicinal properties as an edible herbal plant used to prevent the cirrhosis caused by alcohol consumption (ALI) (Li et al., 2024). Three new flavonoids were obtained, namely, (1) pinocembrin-7-O-[3″-O-galloyl]-β-D-glucose, (2) pinocembrin-7-O-[2″-O-galloyl-4″,6″-hexahydroxydiphenoyl]-β-D-glucose, and (3) 2′,6′-dihydroxydihydrochalcone-4′-O-[2″-O-galloyl-4″,6″-hexahydroxydiphenoyl]-β-D-glucopyranoside, in addition to 12 known compounds (4–15) isolated from *Penthorum chinense* Pursh. These new flavonoids exhibited anti-hyperlipidemic activities, as indicated by molecular docking studies (Zhao et al., 2021).

Yang et al. (2018) found new flavonoids in *Portulaca oleracea* L. and their biological activities. The new flavonoids identified are (1) kaempferol-3-O-[β-D-(3″,6″-di-O-acetyl)-galactopyranoside], (2) quercetin-3-O-[β-D-(3 ″,6″-di-O-acetyl)-galactopyranoside], and (3) quercetin-3-O-[β-D-(6″-O-acetyl)-galactopyranoside]. These compounds exhibit potential antioxidant and anti-inflammatory activities, which highlights their potential as candidates for drug development.

Aoyama et al. (2018), for the first time, carried out the isolation of several compounds from *Feijoa sellowiana*, including (4) gossypetin-3-O-β-galactopyranoside, (6) isoquercitrin, (7) reinoutrin, (8) guaijaverin, (12) trifolin, (17) philanthusiin E, (18) naringenin-7-O-β-glucoside, and (19) aromdendrin-7-O-β-glucoside. Furthermore, the possible presence of alkylated derivatives of ellagic acid was observed in the extract of the plant branch. These new flavonoids are potentially interesting for future studies since their chemical structure and biological activities still need to be fully elucidated. Furthermore, the identification of these compounds contributes to the understanding of the chemical composition of *Feijoa sellowiana* and may pave the way for its application in various fields, such as phytotherapy, medicine, and nutrition.

Deng et al. (2007) addressed the discovery of new flavonoids and other constituents in *Lespedeza cuneata*. Cuneatasides A, B, and C stand out among the new flavonoids identified and other substances, such as 3-(4-hydroxyphenyl)propanoic acid and chlorogenic acid. These compounds were isolated and characterized from the plant *Lespedeza cuneata*, contributing to the knowledge of the chemical composition of this plant species.

Olennikov and Kashchenko (2023a) identified nine compounds, including two newly discovered flavonoids, in *Cucumis sativus* L. (Cucurbitaceae) leaves using methods such as UV spectroscopy, NMR, and mass spectrometry. These new flavonoids were characterized as (1) apigenin 6-C-(4-hydroxy-1-ethylbenzene)-8-C-(2"‘-O-β-D-glucopyranosyl)-β-D-glucopyranoside (cucumerin C) and (2) apigenin 6-C-(2‴-O-β-D-glucopyranosyl)-β-D-glucopyranoside-8-C-(4-hydroxy-1-ethylbenzene) (cucumerin D). Both compounds 1 and 2 demonstrated the ability to inhibit pancreatic lipase activity.

Qi et al. (2023) identified a novel flavonoid called (1) (2S,1″R,2″R)-4′-hydroxy-7-methoxy-6-(1,2,3-trihydroxy-3-methyl-butyl)-flavanone, during the isolation of compounds from *Cullen corylifolium* seeds. In addition, seven already-known comments were isolated. Determining the structures, including the absolute configurations, was carried out through a detailed analysis of the spectroscopic data and the work of computational calculation methods.

Yin et al. (2023) discovered a new flavonoid, designated as (1) 2-(phenanthren-9-yl)-4H-chromen-4-one, as well as a derivative of (5) 5-phenylthiophene-2-carbaldehyde, both isolated for the first time from the plant *Rubus rosifolius* S. Vidal (Rosaceae). In addition to these new compounds, five known compounds were identified, including (2–4) three polyketides and (6–7) two sesquiterpenoids, which were isolated from a Chinese medicinal preparation originating from a natural source.

18.2 FLAVONOID EXTRACTION TECHNIQUES

There are several flavonoid extraction techniques using solvents, all to effectively remove these substances from plant matrices. Among the most typical techniques is maceration extraction, where plant material is immersed in a solvent for a set period, allowing the flavonoids to diffuse into the solution. The type of solvent used may vary, including organic solvents, such as acetone, hexane, ethanol, methanol, or mixtures. Choosing the appropriate solvent is crucial, as different flavonoids have different affinities for specific solvents, and some solvents can selectively extract certain types of flavonoids. In addition to maceration, other extraction techniques include Soxhlet extraction, ultrasound extraction, microwave-assisted extraction, and supercritical CO_2, each with its advantages and disadvantages in terms of efficiency and extraction time (Chávez-González et al., 2020).

Before commencing the extraction procedure, it is essential to prepare the matrix thoroughly. This preparatory phase encompasses various methods, including maceration, homogenization, grinding, or drying techniques, like freeze-drying, which safeguard the integrity of flavonoids. Drying serves multiple purposes, including enhancing the yield per unit weight of the raw material, prolonging its shelf life, conserving storage space, and sometimes simplifying subsequent processing stages. These methodologies, such as maceration, grinding, and homogenization, aim to augment the interface area between the solvent and the sample, where flavonoids are nestled within the cellular frameworks of biological specimens. These preparatory measures culminated in the disruption of cellular structures, thereby facilitating a more substantial yield of bioactive compounds. The utilization of vortexing, followed by centrifugation, is commonly practiced within this framework (Montedoro et al., 1992).

Undoubtedly, solvent selection constitutes the primary variable in any extraction procedure. The solvent used must be determined primarily, considering its affinity and dissolution capacity in relation to the matrix of interest. Solvent properties to be evaluated include polarity, pH, viscosity, surface tension, vapor pressure, melting point, boiling point, density, and specific gravity, as well as their impact on the purity and activity of the extracted compound. Meticulous consideration of these factors is essential significantly, as they influence the cavitation threshold, affecting the efficiency of extracting compounds from the matrix (Chaves et al., 2020).

Table 18.1 summarizes some techniques used to extract flavonoids and their advantages and disadvantages.

TABLE 18.1
Techniques Used to Extract Flavonoids, with Their Advantages and Disadvantages

Extraction Technique	Advantage	Disadvantage	Efficiency	Solvents	References
Soxhlet	• High efficiency in total extraction • Relatively fast	• High solvent consumption • Requires specific equipment • Potential for thermal degradation	High	Hexane, ethanol	(Chávez-González et al., 2020)
Green solvents	• Low solvent toxicity • Lower environmental impact • Low toxicity • Potential for reuse and recycling	• Lower efficiency in some cases • Limitations in solubility	Average	Deep eutectic solvents	Chen et al. (2021)
Supercritical CO_2	• High selectivity	• High equipment cost	High	Supercritical CO_2	Vinitha et al. (2022)
Microwave	• Reduction in solvent consumption • Potential for increased selectivity in extraction	• High equipment cost • Requires adequate technical knowledge • Limitation on sample volume	High	Ethanol, water, acetone, methanol, chloroform	Routray and Orsat (2012)
Ultrasound	• Reduction in solvent consumption • Lower risk of thermal degradation	• Limitation on sample volume • High equipment cost	High	Ethanol, water, methanol	Irakli et al. (2018)
Pressurized hot water	• Low cost and availability • Safety, as it is a non-toxic solvent • Lower environmental impact	• Low selectivity • Lower efficiency in extracting lipophilic compounds	Average	Water	Hartonen et al. (2007)

18.3 FLAVONOIDS: PHARMACOLOGICAL PROPERTIES

Current trends in flavonoid research are also directed toward the extraction, identification, and purification of new natural polyphenolic compounds in various food sources, which have new functions and applications to promote health benefits (Liga et al., 2023). A recent review presented a wealth of information on the pharmacological properties of newly isolated flavonoids. Several therapeutic effects demonstrate antioxidant potential and the ability to positively affect health against chronic diseases (Tiwari et al., 2023).

For the first time, the flavonoid hesperetin-5′-O-β-rhamnoglucoside was identified in the study by Wang et al. (2023). The new compound was obtained from an optimized extract of dandelion (*T. mongolicum* Hand.-Mazz), in a solid–liquid ratio of 1:53 (g/mL) and a temperature of 80°C, at which it is solubilized in methanol and acetone. The authors state that the white-colored substance showed more excellent antioxidant activity for the elimination of free radicals compared to other isolated and purified flavonoids (quercetin, hesperetin-7-glucuronide, kaempferol-3-glucoside, baicalein, hyperseroside, and rutin), and encourage new research to explore its pharmacological potential.

Investigated by Soltani et al. (2023) was the profile of secondary metabolites in *Thymelaea tartonraira* leaves. The extract obtained with hexane, dichloromethane, acetone, and methanol from dried powdered leaves presented six compounds, including a new flavonoid called tartonrin (hypolaetin 8-O-β-D-galactopyranoside). The new compound showed the highest antioxidant activity compared to the other five known compounds (daphnoretin, triumbelletin, genkwanin, tiliroside, and yuankanin). Furthermore, the specific inhibitory capacity for α-glucosidase was twice as high as that of acarbose (reference compound). The study showed that *Thymelaea tartonraira* leaves are a promising source of flavonoids with antioxidant potential and specific pharmacological properties for hyperglycemic patients.

Spectroscopic characterization was carried out on *Terminalia arjuna* plant material. Bushra et al. (2023) identified and isolated three flavonoids, one of which was new, called Terminalone A. The three polyphenolic compounds were obtained from methanolic extract, and their inhibitory capacities against microorganisms, *Escherichia coli*, and *Staphylococcus aureus* were subsequently evaluated. The new flavonoid showed more excellent antioxidant activity when compared to known flavonoids (6-hydroxy-2-(4-hydroxyphenyl)-7-methoxy-4H-chromen-4-one and 2-(3,4-dihydroxyphenyl)-5, 7-dihydroxy-4H-chromen-4-one). With the results achieved, the authors recognize the antimicrobial properties of the three isolated compounds, mainly *Terminalone A*, which exhibited more significant antioxidant potential. Given its importance, the authors also highlight its use for research applications in medicinal areas.

Olennikov and Kashchenko (2023b) identified seven phytochemical compounds in cucumber (*Cucumis sativus*) extract, including two new flavonoids. The substances studied were obtained from 60% ethanolic extract at a temperature of 60°C and a ratio of 1:15 (mass/solvent), and their ability to inhibit pancreatic lipase was subsequently evaluated. The study showed that the flavonoid isovitexin-2"-O-(6"‘-O-feruloyl)-glucoside-6"-O-ferulate presented better biological activity as hypolipidemic than the other new flavonoid (isovitexin-2"-O-glucoside-6"-O-p-coumarate). The study demonstrates that the vegetable *Cucumis sativus* is a source of flavonoids that help develop new medicines with therapeutic effects in lipid metabolism disorders.

Chen et al. (2023) isolated eight compounds from the leaves of *Cajanus cajan* (L.) Millsp. (pigeon pea). Among these, five secondary metabolites had never been identified, one of which was a flavonoid designated *Cajanine H.* The extract of pigeon pea leaves was carried out at room temperature using 95% ethanol as a solvent. After the bioactive compounds were isolated, the antimicrobial activity was analyzed against the microorganisms *Staphylococcus aureus*, *Bacillus cereus*, *Escherichia coli*, and anti-SARS-CoV-2 (COVID-19). The new flavonoid exhibited excellent inhibitory activity against *Bacillus cereus* and moderate activity for the other bacteria evaluated, demonstrating a significant effect on COVID-19. The study suggests further research into developing antimicrobial drugs, especially for the microorganisms listed.

Recent research on flavonoids has focused on extraction, isolation, and characterization, with the aim of discovering new compounds and exploring a diversity of properties with positive effects on human health. These discoveries not only contribute to the knowledge of the benefits of flavonoids but also enable the encouragement of studies for the development of new medicines and treatments, which emerge in a range of multi-transdisciplinary areas, such as agricultural sciences and health, but which converge with applicability for therapeutic purposes and improving the quality of life of the population affected by chronic and more complex fatal diseases.

According to research by Chen and Hsieh (2010), the impact of the *Sophora japonica* flower, known as Huaihua, on treating cerebral infarction was investigated. The plant is widely used in traditional Asian medicine to treat various conditions, such as hemorrhoids and bleeding. Furthermore, its composition includes flavonoids such as rutin, quercetin, isorhamnetin, genistein, and kaempferol, which have antioxidant properties, eliminating superoxide anion and free radicals and preventing lipid peroxidation. Quercetin and rutin have demonstrated anti-inflammatory effects, inhibited microglial activation, and reduced capillary permeability of blood vessels.

The effectiveness of *Sophora japonica* flower has been observed in reducing the occurrence of cerebral infarction, attributed to its antioxidant flavonoids that protect brain cells against the ROS during the restoration of blood flow after ischemia. These compounds can also inhibit microglial activation and interleukin-1β release, reducing inflammation and preventing necrosis. Rutin exerts an antiplatelet effect, helping to prevent ischemic strokes. Furthermore, the *Sophora japonica* flower promotes homeostasis, reducing capillary permeability and promoting efficient blood clotting to prevent excessive bleeding. These findings highlight the therapeutic potential of *Sophora japonica* in the context of brain health.

18.4 MECHANISMS OF ACTION OF FLAVONOIDS IN THE HUMAN BODY

Rathee et al. (2009), in their review paper, discuss the mechanisms of action of flavonoids as anti-inflammatory agents, exploring how these compounds exert beneficial effects in combating inflammation and highlighting several mechanisms through which they act. Flavonoids can modulate intracellular signaling pathways, such as the NF-κB and MAP kinase pathways, inhibiting the production mediators of pro-inflammatory (nitric oxide, prostaglandins, and inflammatory cytokines). Moreover, flavonoids also exert antioxidant effects, combating oxidative stress, which is closely linked to inflammation. These collective mechanisms of action enhance the potential of flavonoids to function as anti-inflammatory agents, presenting encouraging avenues for devising therapeutic approaches aimed at addressing inflammatory conditions.

Rendeiro et al. (2015) addressed how flavonoids affect the brain, distinguishing between direct and indirect effects. *Direct effects* refer to the ability of these compounds to cross the blood–brain barrier and interact directly with brain cells, protecting against oxidative damage, reducing inflammation, and promoting synaptic plasticity. *Indirect effects* involve influence on cellular signaling pathways, such as the nitric oxide and neural growth factor pathways, which play a role in regulating mood, cognition, and neuroplasticity. These mechanisms indicate that flavonoids have the potential to provide significant benefits for brain health, which can be explored in the development of strategies for the treatment of these diseases.

According to Safe et al. (2021), understanding the mechanisms of flavonoids' action opens doors to developing new therapies and medicines. Strategies such as the structural modification of flavonoids to improve their bioavailability, the formulation of nanostructures, and the use of encapsulation techniques have been explored to improve the therapeutic efficacy of these compounds. Furthermore, identifying promising flavonoids through high-throughput screening techniques and optimization of formulations represents opportunities for developing new drugs based on these natural compounds.

Flavonoids demonstrate significant efficacy against viruses and microorganisms due to their bioactive properties. They can stop the entry of pathogens into host cells by blocking viral or bacterial binding to cellular receptors. Flavonoids act on the body's immune response, making it stronger, contributing to the increase in cytokines and immune cell activities, such as natural killer (NK) cells and T lymphocytes. These compounds can also interfere with the synthesis of viral or bacterial proteins essential for the replication of the pathogen, inhibiting specific enzymes and interrupting metabolic processes vital for the survival of the infectious agent (Ahmad et al., 2015).

These diverse mechanisms of action of flavonoids give them a broad spectrum of activity against various viruses and microorganisms, making them promising candidates for developing new antiviral and antimicrobial agents. Its ability to act both directly on pathogens and indirectly, stimulating the host's immune response, offers exciting prospects for treating and preventing viral and bacterial infections (Ahmad et al., 2015).

The cancer development process comprises three distinct stages, which involve the disease's initiation, promotion, and progression. Over many years, a normal cell can be transformed into a malignant one, providing opportunities to intervene in the malignantization process, as illustrated in Figure 18.2 (Ahmad et al., 2015). Therefore, a considerable period is required to identify

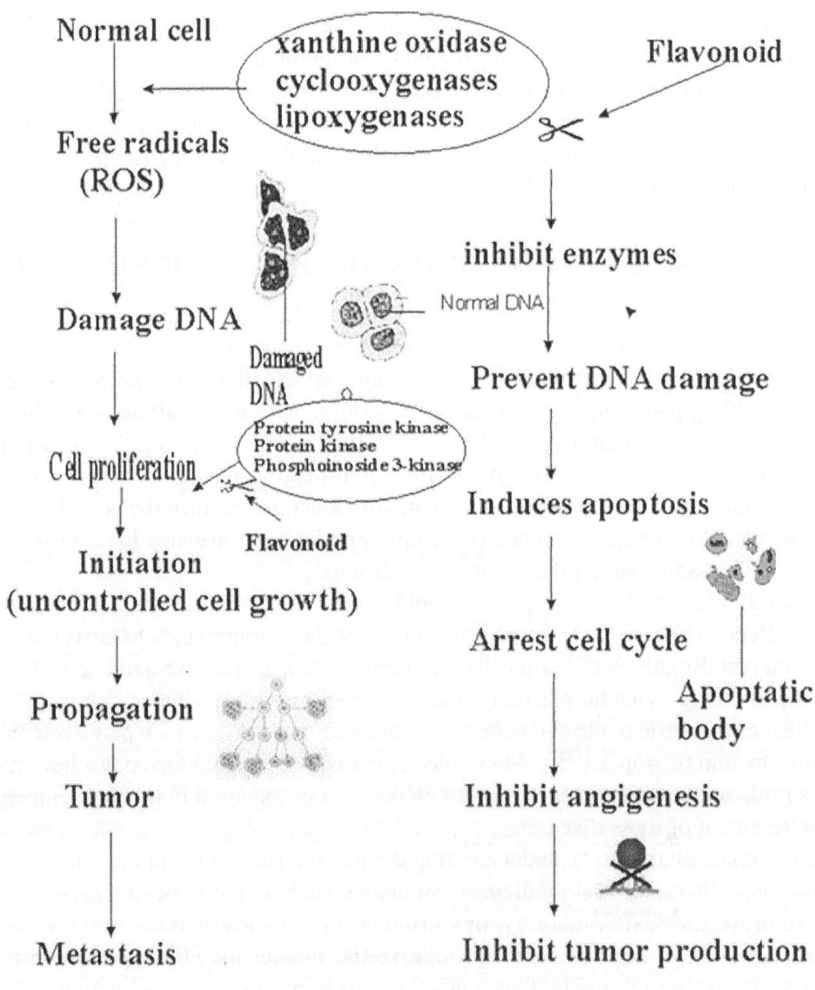

FIGURE 18.2 Stages of cancer formation and the potential effect of flavonoids on the evolution of the disease.

Source: Ahmad et al. (2015). With permission of Food Research International.

compounds with cancer-preventive properties. There is a significant correlation between phenolic compounds' content and green tea's antioxidant capacity (Camellia sinensis). Given that free radicals play a role in cancer development, polyphenol-rich foods such as tea, which exhibit antioxidant activity, may be beneficial in reducing cancer risk (Oliveira et al., 2015).

18.5 IDENTIFICATION AND CHARACTERIZATION OF NEW FLAVONOIDS IN FOODS

Detecting new flavonoids in foods generally involves analytical techniques that aim to identify and characterize these compounds accurately and reliably.

More advanced methods, such as CG coupled to MS (GC-MS), HPLC, and Fourier transform ion cyclotron resonance mass spectrometry (FTICR-MS) with HPLC, offer greater sensitivity and selectivity in the identification of flavonoids (Al-Owaisi et al., 2014). HPLC allows the separation of compounds with high resolution and precision. At the same time, detection is carried out by methods such as UV-vis or diode array detector (DAD), enabling precise quantification of flavonoids. GC-MS, which is especially useful for volatile flavonoids, can be identified based on their unique mass spectra, providing detailed information about the molecular structure of flavonoids (Al-Owaisi et al., 2014).

FTICR-MS with HPLC is a very good tool for identifying flavonoids. This combination offers high sensitivity, resolution, and precision in separating and characterizing compounds, allowing the simultaneous analysis of multiple flavonoids in complex samples. HPLC separates flavonoids based on their retention properties, while FTICR-MS provides detailed information on the molecular mass and composition of the ions generated, enabling precise identification of compounds. This analytical approach has been widely used in metabolomics studies and analysis of plant extracts, providing valuable results on the presence, concentration, and diversity of flavonoids in different biological systems (de Oliveira et al., 2018; Tao et al., 2019).

In addition to these techniques, advanced approaches, such as ultra-performance liquid chromatography (UPLC) coupled with tandem mass spectrometry (MS/MS) (Gbashi et al., 2017) and NMR, are also used in the identification of flavonoids (Markham & Geiger, 2017). UPLC-MS/MS offers greater sensitivity, resolution, and speed of analysis than conventional HPLC, enabling the accurate identification of compounds in complex samples. Meanwhile, NMR provides detailed structural information about flavonoids, allowing the determination of chemical bonds and molecular configurations with high precision. These more advanced methods are essential for thoroughly analyzing flavonoids in complex samples and play a crucial role in advancing research in this field.

The UPLC-tandem mass spectrometry technique is an advanced analytical method combining UPLC with (MS/MS) to identify and quantify compounds in a sample. Gbashi et al. (2017) used this technique to extract and analyze flavonoids from the *Bidens pilosa* plant. The analysis identified several flavonoids in the plant, including quercetin, kaempferol, apigenin, and their glycosylated derivatives. These compounds were characterized and quantified based on their mass spectra and retention times in chromatography. The UPLC-tandem MS allowed a precise and sensitive identification of the flavonoids present in *Bidens pilosa*, contributing to understanding their chemical composition and potential application in pharmacological and nutritional studies.

18.6 FUTURE PERSPECTIVES

The prospects for new flavonoids in foods are up-and-coming, driven by increasing demand for functional foods that offer health benefits. In addition to traditional flavonoid sources such as fruits and vegetables, the search for new compounds now extends to alternative plant sources, including wild plants and food waste. With the development of innovative processing technologies and an increasing focus on sustainability, flavonoid-enriched foods are expected to be produced more effectively, preserving the integrity of these compounds and promoting responsible agricultural practices. Furthermore, personalized health research can lead to the development of products targeted at specific groups based on their health needs and genetic profile, opening up new opportunities in the food industry to promote consumer well-being.

Using high-performance liquid chromatography (HPLC), gas chromatography (GC), ultrafast liquid chromatography (UPLC), mass spectrometry (MS), and nuclear magnetic resonance (NMR), among other techniques, plays a vital role in the identification of new flavonoids in foods. These techniques allow for the precise separation and analysis of complex constituents in food extracts.

REFERENCES

Ahmad, A., Kaleem, M., Ahmed, Z., & Shafiq, H. (2015). Therapeutic potential of flavonoids and their mechanism of action against microbial and viral infections—a review. *Food Research International*, *77*(2), 221–235. https://doi.org/10.1016/j.foodres.2015.08.019

Al-Owaisi, M., Al-Hadiwi, N., & Khan, S. A. (2014). GC-MS analysis, determination of total phenolics, flavonoid content and free radical scavenging activities of various crude extracts of *Moringa peregrina* (Forssk.) Fiori leaves. *Asian Pacific Journal of Tropical Biomedicine*, *4*(12), 964–970. https://doi.org/10.12980/APJTB.4.201414B29

Aoyama, H., Sakagami, H., & Hatano, T. (2018). Three new flavonoids, proanthocyanidin, and accompanying phenolic constituents from *Feijoa sellowiana. Bioscience, Biotechnology, and Biochemistry*, *82*(1), 31–41. https://doi.org/10.1080/09168451.2017.1412246

Barreca, D., Trombetta, D., Smeriglio, A., Mandalari, G., Romeo, O., Felice, M. R., Gattuso, G., & Nabavi, S. M. (2021). Food flavonols: Nutraceuticals with complex health benefits and functionalities. *Trends in Food Science & Technology*, *117*, 194–204. https://doi.org/10.1016/j.tifs.2021.03.030

Barreca, M. M., Alessandro, R., & Corrado, C. (2023). Effects of flavonoids on cancer, cardiovascular and neurodegenerative diseases: Role of NF-κB signaling pathway. *International Journal of Molecular Sciences*, *24*(11), 9236. https://doi.org/10.3390/ijms24119236

Bushra, Rehman, K. U., Khan, D., Almehizia, A. A., Naglah, A. M., Al-Wasidi, A. S., Refat, M. S., El-Sayed, M. Y., Ullah, H., & Khan, S. (2023). Isolation and bioassay of a new terminalone A from *Terminalia arjuna*. *Molecules*, *28*(3), 1015. https://doi.org/10.3390/molecules28031015

Chaves, J. O., De Souza, M. C., Da Silva, L. C., Lachos-Perez, D., Torres-Mayanga, P. C., Machado, A. P. D. F., . . . Rostagno, M. A. (2020). Extraction of flavonoids from natural sources using modern techniques. *Frontiers in Chemistry*, *8*, 507887. https://doi.org/10.3389/fchem.2020.507887

Chávez-González, M. L., Sepúlveda, L., Verma, D. K., Luna-García, H. A., Rodríguez-Durán, L. V., Ilina, A., & Aguilar, C. N. (2020). Conventional and emerging extraction processes of flavonoids. *Processes*, *8*(4), 434. https://doi.org/10.3390/pr8040434

Chen, H. N., & Hsieh, C. L. (2010). Effects of *Sophora japonica* flowers (Huaihua) on cerebral infarction. *Chinese Medicine*, *5*, 1–4. https://doi.org/10.1186/1749-8546-5-34

Chen, J. Y., Lian, X., Fan, Y. W., Ao, Z. Y., Zhang, W. J., Pan, Y. C., Chen, L. P., Yuan, J., & Wu, J. W. (2023). Four new stilbenes and one new flavonoid with potential antibacterial and anti-SARS-CoV-2 activity from *Cajanus cajan*. *Journal of Natural Medicines*, *77*(4), 858–866. https://doi.org/10.1007/s11418-023-01727-5

Chen, X. Q., Li, Z. H., Liu, L. L., Wang, H., Yang, S. H., Zhang, J. S., & Zhang, Y. (2021). Green extraction using deep eutectic solvents and antioxidant activities of flavonoids from two fruits of Rubia species. *LWT*, *148*, 111708. https://doi.org/10.1016/j.lwt.2021.111708

de Oliveira, F. M., Romão, W., & Kuster, R. M. (2018). Identification of phenolic compounds in *Eugenia uniflora* leaves by FTICR MS in association with different ionization sources. *Analytical Methods*, *10*(14), 1647–1655. https://doi.org/10.1039/C8AY00129D

Deng, F. E. I., Chang, J. U. N., & Zhang, J. S. (2007). New flavonoids and other constituents from *Lespedeza cuneata*. *Journal of Asian Natural Products Research*, *9*(7), 655–658. https://doi.org/10.1080/10286020600979894

Din, S. R. U., Saeed, S., Khan, S. U., Kiani, F. A., Alsuhaibani, A. M., & Zhong, M. (2023). Bioactive compounds (BACs): A novel approach to treat and prevent cardiovascular diseases. *Current Problems in Cardiology*, *48*(7), 101664. https://doi.org/10.1016/j.cpcardiol.2023.101664

Fuente-Ballesteros, A., Priovolos, I., Ares, A. M., Samanidou, V., & Bernal, J. (2023). Green sample preparation methods for the analysis of bioactive compounds in bee products: A review. *Advances in Sample Preparation*, *6*, 100060. https://doi.org/10.1016/j.sampre.2023.100060

Gbashi, S., Njobeh, P., Steenkamp, P., & Madala, N. (2017). Pressurized hot water extraction and chemometric fingerprinting of flavonoids from *Bidens pilosa* by UPLC-tandem mass spectrometry. *CyTA-Journal of Food*, *15*(2), 171–180. https://doi.org/10.1080/19476337.2016.1230151

Hartonen, K., Parshintsev, J., Sandberg, K., Bergelin, E., Nisula, L., & Riekkola, M. L. (2007). Isolation of flavonoids from aspen knotwood by pressurized hot water extraction and comparison with other extraction techniques. *Talanta*, *74*(1), 32–38. https://doi.org/10.1016/j.talanta.2007.05.040

Irakli, M., Chatzopoulou, P., & Ekateriniadou, L. (2018). Optimization of ultrasound-assisted extraction of phenolic compounds: Oleuropein, phenolic acids, phenolic alcohols and flavonoids from olive leaves and evaluation of its antioxidant activities. *Industrial Crops and Products*, *124*, 382–388. https://doi.org/10.1016/j.indcrop.2018.07.07

Kaushal, N., Singh, M., & Sangwan, R. S. (2022). Flavonoids: Food associations, therapeutic mechanisms, metabolism and nanoformulations. *Food Research International*, *157*, 111442. https://doi.org/10.1016/j.foodres.2022.111442

Li, R., Wu, D., Hu, J., Ma, Y., Ba, Y., Zou, L., & Hu, Y. (2024). Polyphenol-enriched *Penthorum chinense* Pursh ameliorates alcohol-related liver injury through Ras/Raf/MEK/ERK pathway: Integrating network pharmacology and experiment validation. *Journal of Ethnopharmacology*, *321*, 117513. https://doi.org/10.1016/j.jep.2023.117513

Li, X., Xuan, B., Shou, Q., & Shen, Z. (2014). New flavonoids from *Campylotropis hirtella* with immunosuppressive activity. *Fitoterapia*, *95*, 220–228. https://doi.org/10.1016/j.fitote.2014.03.028

Li, Z., Ren, Z., Zhao, L., Chen, L., Yu, Y., Wang, D., Mao, X., Cao, G., Zhao, Z., & Yang, H. (2023). Unique roles in health promotion of dietary flavonoids through gut microbiota regulation: Current understanding and future perspectives. *Food Chemistry*, *399*, 133959. https://doi.org/10.1016/j.foodchem.2022.133959

Liga, S., Paul, C., & Péter, F. (2023). Flavonoids: Overview of biosynthesis, biological activity, and current extraction techniques. *Plants*, *12*(14), 2732. https://doi.org/10.3390/plants12142732

Markham, K. R., & Geiger, H. (2017). 1 H nuclear magnetic resonance spectroscopy of flavonoids and their glycosides in hexadeuterodimethylsulfoxide. In *The flavonoids advances in research since 1986* (pp. 441–497). Routledge. https://doi.org/10.1201/9780203736692

Mohammadi, N., Farrell, M., O'Sullivan, L., Langan, A., Franchin, M., Azevedo, L., & Granato, D. (2024). Effectiveness of anthocyanin-containing foods and nutraceuticals in mitigating oxidative stress, inflammation, and cardiovascular health-related biomarkers: a systematic review of animal and human interventions. *Food & Function*. https://doi.org/10.1039/d3fo04579j

Montedoro, G., Servili, M., Baldioli, M., & Miniati, E. (1992). Simple and hydrolyzable phenolic compounds in virgin olive oil. 1. Their extraction, separation, and quantitative and semiquantitative evaluation by HPLC. *Journal of Agricultural and Food Chemistry*, *40*(9), 1571–1576. https://doi.org/10.1021/jf00021a019

Olennikov, D. N., & Kashchenko, N. I. (2023a). Acylated flavonoids from *Cucumis sativus* inhibit the activity of human pancreatic lipase. *Applied Biochemistry and Microbiology*, *59*(4), 530–538. https://doi.org/10.1134/S0003683823040099

Olennikov, D. N., & Kashchenko, N. I. (2023b). New flavonoids from *Cucumis sativus*. *Chemistry of Natural Compounds*, *59*(4), 651–654. https://doi.org/10.1007/s10600-023-04079-4

Oliveira, C. C. D., Calado, V. M. D. A., Ares, G., & Granato, D. (2015). Statistical approaches to assess the association between phenolic compounds and the in vitro antioxidant activity of *Camellia sinensis* and *Ilex paraguariensis* teas. *Critical Reviews in Food Science and Nutrition*, *55*(10), 1456–1473. https://doi.org/10.1080/10408398.2012.750233

Pietta, P. (2000). Flavonoids as antioxidants. *Journal of Natural Products*, *63*(7), 1035–1042. https://doi.org/10.1021/np9904509

Qi, S. Z., Li, B. B., Liu, M. S., Zhang, Z., Miao, S., & Gong, K. K. (2023). Chemical constituents from the seeds of *Cullen corylifolium* and their inhibitory activity on diacylglycerol acyltransferase. *Natural Product Research*, *37*(10), 1601–1607. https://doi.org/10.1080/14786419.2022.2103126

Rathee, P., Chaudhary, H., Rathee, S., Rathee, D., Kumar, V., & Kohli, K. (2009). Mechanism of action of flavonoids as anti-inflammatory agents: a review. *Inflammation & Allergy Drug Targets*, *8*(3), 229–235. https://doi.org/10.2174/187152809788681029

Rendeiro, C., Rhodes, J. S., & Spencer, J. P. (2015). The mechanisms of action of flavonoids in the brain: Direct versus indirect effects. *Neurochemistry International*, *89*, 126–139. https://doi.org/10.1016/j.neuint.2015.08.002

Routray, W., & Orsat, V. (2012). Microwave-assisted extraction of flavonoids: a review. *Food and Bioprocess Technology*, *5*, 409–424. https://doi.org/10.1007/s11947-011-0573-z

Russo, D. (2018). Flavonoids and the structure-antioxidant activity relationship. *Journal of Pharmacognosy & Natural Produts*, *4*(1). https://doi.org/10.4172/2472-0992.1000e109

Safe, S., Jayaraman, A., Chapkin, R. S., Howard, M., Mohankumar, K., & Shrestha, R. (2021). Flavonoids: structure–function and mechanisms of action and opportunities for drug development. *Toxicological Research*, *37*, 147–162. https://doi.org/10.1007/s43188-020-00080-z

Shahbaz, M., Imran, M., Alsagaby, S. A., Naeem, H., Al Abdulmonem, W., Hussain, M., Abdelgawad, M. A., El-Ghorab, A. H., Ghoneim, M. M., El-Shaerbiny, M., Atoki, A. V., & Awuchi, C. G. (2023). Anticancer, antioxidant, ameliorative and therapeutic properties of kaempferol. *International Journal of Food Properties*, *26*(1), 1140–1166. https://doi.org/10.1080/10942912.2023.2205040

Soltani, S., Koubaa, I., Dhouib, I., Khemakhem, B., Marchand, P., & Allouche, N. (2023). New specific α-glucosidase inhibitor flavonoid from *Thymelaea tartonraira* leaves: Structure elucidation, biological and molecular docking studies. *Chemistry & Biodiversity*, *20*(3), e202200944. https://doi.org/10.1002/cbdv.202200944

Tan, X. F., Kim, D. W., Song, Y. H., Kim, J. Y., Yuk, H. J., Wang, Y., Curtis-Long, M. J., & Park, K. H. (2016). Human neutrophil elastase inhibitory potential of flavonoids from *Campylotropis hirtella* and their kinetics. *Journal of Enzyme Inhibition and Medicinal Chemistry*, *31*(sup1), 16–22. https://doi.org/10.3109/14756366.2015.1118683

Tao, H., Cui, B., Zhang, H., Bekhit, A. E. D., & Lu, F. (2019). Identification and characterization of flavonoids compounds in cassava leaves (*Manihot esculenta* Crantz) by HPLC/FTICR-MS. *International Journal of Food Properties*, *22*(1), 1134–1145. https://doi.org/10.1080/10942912.2019.1626879

Tiwari, P., Mishra, R., Mazumder, A., Mazumder, R., & Singh, A. (2023). An insight into diverse activities and targets of flavonoids. *Current Drug Targets*, *24*(1), 89–102. https://doi.org/10.2174/1389450123666220915121236

Vinitha, U. G., Sathasivam, R., Muthuraman, M. S., & Park, S. U. (2022). Intensification of supercritical fluid in the extraction of flavonoids: A comprehensive review. *Physiological and Molecular Plant Pathology*, *118*, 101815. https://doi.org/10.1016/j.pmpp.2022.101815

Wang, R., Li, W., Fang, C., Zheng, X., Liu, C., & Huang, Q. (2023). Extraction and identification of new flavonoid compounds in dandelion *Taraxacum mongolicum* Hand.-Mazz. with evaluation of antioxidant activities. *Scientific Reports*, *13*(1), 2166. https://doi.org/10.1038/s41598-023-28775-x

Yang, X., Zhang, W., Ying, X., & Stien, D. (2018). New flavonoids from *Portulaca oleracea* L. and their activities. *Fitoterapia*, *127*, 257–262. https://doi.org/10.1016/j.fitote.2018.02.032

Yin, Y., Huang, Y., Yang, W., Yuan, J., Xie, M., Miao, Y., Yu, J., Wang, J., Zhang, X., & Wang, B. (2023). A novel flavonoid and other constituents from *Rubus rosifolius* S. Vidal (Rosaceae). *Natural Product Research*, 1–9. https://doi.org/10.1080/14786419.2023.2173189

Zhao, W. W., Guo, W. W., Guo, J. F., Wang, X., Chen, X. Q., & Wu, X. (2021). Three new flavonoids from *Penthorum chinense* Pursh and their docking studies. *Natural Product Research*, *35*(1), 49–56. https://doi.org/10.1080/14786419.2019.1613394

19 Açaí—More Than a Traditional Food in the Amazon

A Functional Food and a Promising Raw Material for Pharmaceutical Innovation

Leonardo Oliveira Bittencourt, Felipe Oliveira Nunes, Vinicius Ruan Neves dos Santos, Wallacy Watson Pereira Melo, Kátia Lamarão-Vieira, Hervé Rogez, and Rafael Rodrigues Lima

19.1 INTRODUCTION

As one of the main sources of nutrition and subsistence, açaí (*Euterpe oleracea* Martius) is essential to the way of life in traditional Amazonian communities. Nutrient-dense fruits from this palm are often consumed in juice form. Because of its high antioxidant, fiber, vitamin, and mineral content, its fatty acid profile similar to olive oil's, açaí is highly regarded and plays a large role in the diet of the local community. Açaí is also significant from a cultural and economic standpoint, as it is intricately woven into these communities' customs and eating habits. Açaí extraction and processing are locally driven endeavors that promote social harmony and the passing down of traditional wisdom about the sustainable use of natural resources.

It is impossible to overstate the economic significance of açaí, as it provides a crucial means of subsistence for more than 100,000 Amazon small farmers. The commercialization of açaí has improved riverside villages' financial independence in both domestic and international markets, with the market having tripled in the past 20 years, reaching over 1.5 million tons, 20% of which are external to the Amazon. This increase in demand has resulted in significant production and price increases, culminating in a shortage of açaí for the local market and a re-emergence of food security issues in the past year.

Furthermore, by promoting agricultural methods that do not necessitate deforestation, sustainable açaí production helps preserve the Amazon rainforest. However, the increase in production has primarily been achieved by clearing other species, leading to problems such as increased soil erosion, loss of biodiversity, a rise in the occurrence of Chagas disease through oral transmission via açaí, and scarcity of other traditional foods like shrimp. Thus, while açaí supports environmental conservation and the livelihoods of traditional communities, which is consistent with the values of sustainable development and the preservation of traditional knowledge, the sustainability of the production chain can be questioned.

The sustainable development of the region is presented with strategic potential by the integration of *Euterpe oleracea* Mart. into the Amazonian bioeconomy. Because of its nutritional and pharmacological qualities, açaí has significant potential for use in high-end products like cosmetics, dietary

DOI: 10.1201/9781003462804-25

supplements, and nutraceuticals. The integration of sustainability and biotechnological innovation, known as the bioeconomy, has the potential to revolutionize the açaí supply chain and generate income and employment opportunities for local communities. Moreover, improving the açaí bioeconomy could promote sustainable management techniques and a circular economy applied to its major residue, the kernel.

In order to guarantee *E. oleracea*'s ongoing availability and to protect the Amazonian habitats that it depends on, sustainable utilization is imperative. To ensure the long-term maintenance of açaí palm populations, sustainable management approaches in the açaí supply chain prevent deforestation and encourage biodiversity protection. Furthermore, the organic and sustainable certification of açaí products can boost their market value, benefitting local economies and promoting ethical farming methods. In this context, scientific research is essential because it provides information on the most effective growth, harvesting, and processing techniques that reduce environmental impact and maximize the productivity and quality of goods derived from açaí. In the field of therapeutic strategies and dietary supplementation, açaí has stood out due to its nutritional profile rich in antioxidants, fiber, vitamins, and minerals.

In this chapter, we will cover discussion topics about the botanical characteristics of the *Euterpe oleracea* Martius plant, including mold characteristics of the fruits and their various consumptions from industrial manufacturing, in addition to traditional consumption in the Amazon. Furthermore, considering that the plant species is rich in secondary metabolites with significant pharmacological potential, the main mechanisms already observed that açaí mediates in the face of biochemical and molecular challenges, especially in the face of oxidative stress, will be listed. Finally, we will address how the results evidenced in the literature on the biological processes impacted by intervention with açaí in experimental models can be applied to complex conditions involving the cardiovascular system and central nervous system.

19.2 BOTANICAL ASPECTS AND TRADITIONAL VALUE

Açaí belongs to the Arecaceae family, also known as the palm family, and belongs to the *Euterpe* genus. This genus includes several other species of palm trees found in the tropical Americas, but açaí (*Euterpe oleracea* Martius) is the best-known and widely cultivated species due to its nutritional value and economic importance (ROCHA et al., 2022; BARBOSA et al., 2022; DA SILVA et al., 2023). The three species of palm trees in the genus *Euterpe* are *E. oleracea* (açaí), *E. edulis* (juçara or jussara), and *E. precatoria* (açaí from the Amazonas), which also have some differences. *E. oleracea* is the most known and widely cultivated species due to its nutritional and economic value. It produces smaller, darker fruits, usually dark purple to black in color when ripe. The leaves have a bright green color and are narrower and more elongated compared to those of other species (DA SILVA et al., 2023).

In addition to these morphological and geographic distribution differences, the three species may also present variations in their nutritional profiles and levels of bioactive compounds. While açaí (*E. oleracea*) is the most widely studied in terms of health benefits, the other species also have potential for food and medicinal use, albeit on a smaller scale. For this reason, this chapter will focus on the scientific evidence available about the species *Euterpe oleracea* Martius.

Açaí grows in clusters on palm trees known as açaí trees (as seen in Figure 19.1) and has been an important part of the diet and culture of the Indigenous populations in the Amazon for centuries. It is rich in antioxidants, especially anthocyanins, which are responsible for the fruit's dark purple color. Furthermore, it is a good source of fiber, healthy fats, vitamins (mainly vitamin C and vitamin E), and minerals (such as calcium, magnesium, and potassium) (DA SILVA MENEZES et al., 2008; YUYAMA et al., 2011; DE OLIVEIRA et al., 2021). The plant and its derivatives have gained popularity around the world from 1999, mainly due to their potential to promote health and well-being. It is commonly consumed in the form of frozen pulp, juices, smoothies, ice cream, açaí bowls, and dietary supplements.

FIGURE 19.1 Açaí palm: (a) The *Euterpe oleracea* Martius palm, along with the açaí fruit observed in its canopy during the ripening stage. (b) The black açaí fruit harvested, stage at which it is used in pulp preparation. (c) Açaí pulp prepared for consumption.

The açaí tree is a medium-sized palm tree which can reach heights of up to 25 m. It has a thin and slender trunk, with long, feathery leaves arranged in a fan shape at the top of the plant. The açaí plant produces small white flowers that grow in branched clusters. These flowers give rise to açaí fruits, which are small rounded drupes, dark purple to black in color when ripe. Each fruit contains a single large seed in its center. The açaí tree is a species typically found in the flooded and humid areas of the Amazon rainforest. It thrives in moist soils rich in organic matter and is often found in floodplain areas, close to rivers and streams. The açaí tree can be propagated by seeds or through cuttings. Commercial cultivation generally involves planting seedlings in suitable areas where the plant can receive adequate sunlight and access to water. The cultivation and harvesting of açaí play an important role in the economy and culture of Amazonian communities. However, it is important to ensure that cultivation and harvesting are done in a sustainable manner, to avoid negative impacts on the environment and local populations. In summary, açaí is a highly nutritious tropical fruit rich in antioxidants and essential nutrients. Its consumption has been linked to several health benefits, although more research is needed to fully understand its therapeutic potential (DO NASCIMENTO, 2008; CHAVES et al., 2015; MORAIS, 2019).

Açaí is often promoted for its potential health benefits. Due to its high antioxidant content, it can help fight oxidative stress, reduce inflammation, protect against cellular damage, and improve cardiovascular health (BAPTISTA et al., 2021). Additionally, its fiber content can promote digestive health and aid in weight loss. Although there is limited evidence on the specific health benefits of açaí, early studies suggest that its consumption may be associated with a reduced risk of cardiovascular disease, type 2 diabetes, cancer, and neurodegenerative diseases (DA SILVEIRA et al., 2023). However, more research is necessary to confirm these effects and expand the understanding of the mechanisms through which açaí can benefit health.

19.3 EFFECTS OF AÇAÍ ON BIOLOGICAL PROCESSES

Initially, it is important to highlight that, as well as phytochemical and pharmacological studies of other plant species, studies carried out with açaí have so far provided a wide range of findings that vary according to the part of the plant used (fruits and seeds, leaves and palm hearts), varied extractive processes (phytochemical fractions extracted from solvents with different degrees of polarity and different methods), as well as secondary natural products obtained from the processing of açaí, not necessarily the fresh juice as it is traditionally consumed, that is, there are studies with essential oils, fixed oils, and clarified açaí juice (aqueous extract). This implies the diversity of results found, as well as different applications of açaí. Therefore, whenever we refer to the biological effect triggered by açaí, the way of obtaining and the type of product derived from açaí will be mentioned. It is worth noting that some studies also developed pharmaceutical formulations containing fruit derivatives, which will be discussed later.

Classically, in pre-clinical studies whose objectives are to characterize the possible pharmacological effects of a drug or natural product on cells, tissues, organs, and systems, some common biochemical and molecular parameters are evaluated, such as markers of oxidative stress, inflammatory response, and metabolism. In this sense, initially we will address what the scientific evidence points to so far regarding the use of açaí in these basic cellular biological processes so that, in the next sections, specific questions related to complex pathological processes will be addressed, in animal models or clinical trials, when available.

When referring to fruits with a high content of phenolic compounds, such as anthocyanins, the antioxidant character is always seen as the main biological effect. Thus, antioxidant activity is premised on acting on the oxidative biochemistry of biological systems, with the aim of avoiding or minimizing oxidative and nitrosative stress (HAJAM et al., 2022). Conceptually, oxidative stress is a natural biological process which occurs independently of pathological processes occurring in the living organism, that is, it is part of normal physiology. Traditionally, it is defined as an imbalance between the exacerbated generation of reactive oxygen and nitrogen species and enzymatic and non-enzymatic antioxidant defense, which can trigger direct responses, such as the oxidation of organic macromolecules, such as lipids, proteins, and DNA, but also indirect responses, such as the induction of the expression of genes associated with cell death, expression of inflammatory cytokines, etc. (DAHL et al., 2015; VONA et al., 2021; CALIRI et al., 2021; HAJAM et al., 2022). This is the most studied biological process when it comes to the biological activities of açaí, since due to its high content of phenolic compounds, such as flavonoids, they give it significant antioxidant activity.

Regarding the damage caused by oxidative stress to biochemical macromolecules, in vitro studies have already demonstrated that açaí is capable of minimizing oxidative damage to lipids, genetic material, and proteins. The study by RIBEIRO et al. (2010) pointed out that the administration of pure açaí pulp, commercially acquired, had antigenotoxic effects in a model of genotoxicity induced by doxorubicin in mice. The authors showed that açaí was able to reduce the number of micronuclei found in the bone marrow and peripheral blood, both in the model of simultaneous administration of the drug and açaí and in the prevention model, in which açaí was administered prior to doxorubicin. Regarding the results of the comet assay, which evaluates the integrity of the genetic material, açaí had a significant performance in preventing damage to peripheral blood and kidney cells in the three doses studied (3.33, 10, and 16.67 g of açaí per kilogram of body weight), but only the highest dose had significant effects on liver cells. PINI et al. (2020) designed a clinical study in which human patients had semen collected before and after the oral administration of capsules containing pure, commercially available açaí, with the aim of evaluating the potential benefits on genotoxic parameters, and they found that DNA fragmentation was significantly reduced after the proposed treatment, including the improvement in the sperm motility parameter.

Regarding modifications in proteins markedly associated with oxidative stress, BARBOSA et al. (2021) proposed to test the protective effects of supplementation with pasteurized açaí pulp on the damage caused to the offspring of rats that were fed a diet with high levels of fat and observed a significant reduction in the levels of this marker of protein oxidation. Corroborating these findings, MONTEIRO et al. (2022) showed that in a model of kidney disease induced by a diet rich in adenine, the hydroethanolic extract of açaí seed, administered under a four-week regimen at an approximate dose of 350 mg/kg/day, reduced the levels of carbonyl proteins in kidney tissue.

Evidence indicates that, in vitro, açaí is capable of increasing cellular antioxidant activity by increasing the activity of enzymes. Having said that, the result of greater antioxidant competence is the reduction of the harmful effects of RONS on macromolecules, which has already been evidenced by observing the reduction in lipid peroxidation and damage to the genetic material of eukaryotic cells. However, it is worth highlighting that although in vitro studies are essential for several stages in the process of developing drugs, cosmetics, and other industrialized products, the beneficial effects of açaí have become even more valued based on observations made in in vivo studies, after the oral administration of the product, as traditionally occurs through food. From these studies, it was observed that oxidative stress associated with various conditions, whether infectious or poisoning by xenobiotics, açaí administered orally, even after undergoing biochemical challenges of the digestive tract and metabolism, is still capable of exerting its effects on the modulation of oxidative biochemistry.

19.4 EFFECTS ON BIOLOGICAL SYSTEMS

In this topic, we will address how açaí-derived products have been investigated in pre-clinical and clinical studies with regard to cardiometabolic health. The starting point of this discussion is to understand that the term *cardiometabolic disorders* is a major clinical entity that encompasses cardiac, renal, hepatic, vascular, and endocrinological disorders, such as heart failure, diabetes mellitus, systemic arterial hypertension, dyslipidemia, and obesity. However, it is also important to highlight that one of the biggest challenges of modern medicine is to deal with clinical conditions that sometimes present multiple comorbidities, which can worsen patients' conditions and also make adequate diagnosis and treatment difficult. In this sense, many mechanisms of the pathophysiology of cardiometabolic disorders are common and permeate several of these diseases, such as insulin tolerance, increased lipid markers, and the formation of atheroma plaques.

Cardiometabolic disorders, including hypertension, type 2 diabetes, obesity, dyslipidemia, and insulin resistance, share a common pathophysiological basis. Insulin resistance and chronic inflammation have been identified as central elements in this context. Insulin resistance, characterized by a decreased response of peripheral tissues to insulin, contributes to the development of type 2 diabetes and dyslipidemia, in addition to promoting the accumulation of visceral adipose tissue. This accumulation, in turn, is associated with the release of pro-inflammatory cytokines and the activation of the renin–angiotensin–aldosterone axis, leading to arterial hypertension and endothelial dysfunction (PATEL et al., 2016; ADEVA-ANDANY et al., 2019).

The interconnection between these metabolic and cardiovascular disorders amplifies the risk of adverse cardiovascular events, including coronary artery disease, stroke, and heart failure. Furthermore, the concomitant presence of these conditions increases the complexity of treatment and is associated with worse clinical outcomes. Therefore, an integrated approach that considers not only individual clinical manifestations but also their pathophysiological interrelationships is essential for the successful management of cardiometabolic disorders.

Oxidative stress is a physiological process characterized by the imbalance between the production of reactive oxygen species (ROS) and the body's antioxidant capacity to neutralize them. Oxidative stress contributes to endothelial dysfunction by reducing the bioavailability of nitric oxide (NO), an important mediator of vasodilation (HELMS et al., 2018). The decrease in NO levels favors vasoconstriction, vascular inflammation, and the formation of atherosclerotic plaques, increasing the risk of cardiovascular diseases, such as high blood pressure and atherosclerosis. Furthermore, oxidative stress is closely linked to chronic low-grade inflammation, a phenomenon observed in several cardiometabolic disorders, including obesity and insulin resistance. Considering these potential pharmacological targets, the study by DA COSTA et al. (2012) has already pointed out that açaí seed extract reduces vasodilation induced by acetylcholine, reduces the activity of endothelial nitric oxide synthase (eNOS), and increases the expression of MMP-2, suggesting, therefore, that the extract exerts effects against endothelial dysfunction and structural changes in the vasculature in a model of renal hypertension.

Increased ROS production can activate pro-inflammatory signaling pathways, such as nuclear factor kappa B (NF-κB), leading to exacerbated expression of pro-inflammatory cytokines and adhesion of leukocytes to the endothelium, further promoting the progression of metabolic and cardiovascular diseases (HUSSAIN et al., 2016). Oxidative stress can compromise mitochondrial function, reducing the efficiency of cellular energy production and increasing the formation of ROS within cells. This contributes to the accumulation of intracellular fat, metabolic dysfunction, and insulin resistance, characteristics often seen in conditions such as obesity and type 2 diabetes. In the heart, oxidative stress can cause direct damage to cardiac cells, inducing apoptosis, fibrosis, and contractile dysfunction. These effects may predispose to conditions such as heart failure and ischemic heart disease (GRIENDLING et al., 2003).

Addressing oxidative stress is fundamental for pre-clinical and clinical studies with the aim of investigating secondary outcomes related to the progression of clinical conditions, and in this sense, therapeutic strategies that aim to reduce the production of ROS, increase endogenous antioxidant activity, and improve function of endothelial cells may offer significant benefits in the prevention and treatment of these conditions, thereby improving cardiometabolic health. This can be achieved through lifestyle changes, such as a balanced diet rich in antioxidants, regular physical exercise, and adequate control of underlying medical conditions, in addition to the use of specific pharmacological therapies, when indicated.

The study by DE MORAES ARNOSO et al. (2022) showed that after the development of dyslipidemia in mice due to prolonged consumption of a diet with high levels of fat and administration of *Euterpe oleracea* Mart. seed extract, hyperglycemia and the lipidemic profile in the animals' plasma were reduced compared to the group without açaí. The data even shows that açaí had results comparable to the use of rosuvastatin. Added to this evidence, VILHENA et al. (2021), in a model of renovascular hypertension in experimental animals, which received *E. oleracea* seed extract via intragastric gavage, showed that the levels of plasma markers of renal function, such as urea and creatinine, were reduced after treatment, in addition to the pro-inflammatory markers IL-6, TNF-α, and MCP-1. These two works cited also highlighted a common denominator as the mechanism of action of açaí in the face of cardiometabolic challenges: reduction of lipid peroxidation in the plasma and aortic tissue of the groups treated with açaí.

Açaí also proved to be a promising agent in minimizing adverse reactions caused by medications commonly associated with cardiotoxicity, such as doxorubicin and antineoplastic agents. MATHIAS et al. (2019) supplemented experimental animals with açaí mixed with the rats' daily diet for four weeks and found that there was a significant minimization of cardiotoxic damage triggered by doxorubicin, such as MMP-2 activity, lipid peroxidation levels, increased activity of enzymes of glutathione peroxidase and catalase, and improvements in the levels of molecular markers of mitochondrial energy metabolism in cardiac tissue. Regarding the cardioprotective effects against adverse reactions related to antineoplastic treatment, ALESSANDRA-PERINI et al. (2023) demonstrated that the hydroalcoholic extract of açaí seeds minimized the damage caused by the antineoplastic cocktail containing 5-fluorouracil, cyclophosphamide, docetaxel, and doxorubicin, through the attenuation of morphological changes in cardiac histology (histopathological aspects and increased collagen production), and reduced the levels of creatinine kinase, creatinine kinase MB, troponin I, and CRP, which are important clinical markers used to diagnose cardiac striated muscle damage and inflammation.

The matrix metalloproteinases (MMPs) previously mentioned are enzymes that play a crucial role in the remodeling of the extracellular matrix, a fundamental process in normal physiology and in several pathological conditions, including cardiometabolic disorders. In cardiometabolic disorders, such as atherosclerosis and arterial hypertension, pathological vascular remodeling occurs, characterized by changes in the structure and function of blood vessels. MMPs are involved in the degradation of components of the vascular extracellular matrix, such as collagen, elastin, and proteoglycans, promoting the reorganization of the vascular wall and influencing vascular stiffness, permeability, and vasomotor response. These enzymes contribute to pathological cardiac remodeling by promoting the degradation of interstitial extracellular matrix and facilitating the migration of inflammatory cells to the site of injury (AZEVEDO et al., 2014).

Understanding these mechanisms can provide important insights for the development of therapeutic strategies aimed at preventing and treating these conditions. In cardiometabolic disorders, such as type 2 diabetes, insulin resistance, and metabolic syndrome, glucose and insulin metabolism is often compromised, leading to metabolic and cardiovascular dysfunctions. Insulin resistance is a central feature of cardiometabolic disorders. In this condition, the body's cells have a reduced response to insulin, resulting in high blood glucose levels (DA SILVA et al., 2020). This occurs because cells do not respond properly to insulin signals to transport glucose into cells for use as energy, leading to the accumulation of glucose in the bloodstream. Insulin resistance, together with decreased insulin secretion from the beta cells of the pancreas, contributes to chronic hyperglycemia, characterized by elevated blood glucose levels. Hyperglycemia is a defining feature of type 2 diabetes and is associated with cardiovascular complications, such as coronary artery disease and stroke (KARLSSON et al., 2019)

Hyperinsulinemia and chronic hyperglycemia are associated with the activation of inflammatory pathways and increased oxidative stress, contributing to endothelial dysfunction, which may lead to an increased risk of developing cardiovascular diseases. In cardiometabolic disorders, there is often an accumulation of visceral fat, especially in the abdominal region, that is strongly associated with insulin resistance, chronic inflammation, and increased risk of cardiovascular disease. The accumulation of visceral fat plays a significant role in compromising systemic metabolism, contributing to the pathogenesis of cardiometabolic disorders. This phenomenon is complex and involves a series of interrelated pathophysiological mechanisms (KARLSSON et al., 2019).

In this sense, some pre-clinical evidence indicates that *Euterpe oleracea* plays relevant biological roles in adipose tissue, such as increasing thermogenesis (DA SILVA et al., 2023) and attenuating the pro-inflammatory profile (TNF-alpha and IL-6) in the adipose tissue of mice subjected to a high-calorie diet (SANTOS et al., 2020). Chronic inflammation induced by visceral fat can lead to endothelial dysfunction, compromising the endothelium's ability to regulate vascular tone, vascular permeability, and inflammatory cell adhesion. This contributes to the progression of cardiovascular diseases, such as high blood pressure and atherosclerosis. In addition to its function as an endocrine organ, visceral adipose tissue can influence hormonal regulation through the production and secretion of adipokines. These substances can affect glucose metabolism, vascular function, and energy homeostasis, contributing to the development of cardiometabolic disorders.

Another biological system widely studied in the literature regarding the biological effects of plants of the *Euterpe* genus is the central nervous system, mainly investigating its potential neuroprotective effects. Cell culture studies have investigated the protective effects against oxidant challenges (induced oxidative stress) and exposure to the pollutant methylmercury. Animal study models are diverse and are in advanced stages of exploring the therapeutic potential of plants of the genus and include psychiatric diseases, such as depression and anxiety, and neurological diseases, such as epilepsy, vascular dementia, and Parkinson's disease.

Euterpe oleracea extract reduced damage to lipids and mitochondrial homeostasis induced by oxidative stress in the rotenone exposure model in SH-SY5Y cells (neuron-like cells), in addition to reducing the levels of reactive oxygen species. The same pattern was observed in the primary culture of hippocampal cells, where *Euterpe precatoria* extract reduced the levels of autophagic vacuoles in mouse HT22 neuronal cells and also decreased the expression levels of phospho-mTOR (a marker of cell death signaling) in these cells (TORMA et al., 2017). In astrocytes, which are crucial for tissue function and repair in the central nervous system, *Euterpe oleracea* extract has been shown to inhibit the GABAergic system (ARRIFANO et al., 2018).

Another cell population that has been extensively studied by the scientific community and whose results align with the evidence described previously is the microglial population, which plays a fundamental role in inflammation and tissue repair in the central nervous system, but which also has a deleterious role in microglial activation, which is prolonged and exacerbated. The data so far indicates that *Euterpe oleracea* extract reduces the cellular proliferation of microglia, reduces the levels of ROS (DE SOUZA et al., 2022b), and attenuates the expression of post-inflammatory markers and caspases 1, 3, and 8 (DE SOUZA et al., 2022a).

When evaluating the pre-clinical effects in experimental models, the evidence indicates that the biological effects triggered by extracts from plants of the *Euterpe* genus may be promising in the management of chronic neurodegenerative and psychiatric diseases, since many pathophysiological mechanisms of these diseases are those described previously, such as oxidative stress and inflammation, which trigger other cellular processes associated with tissue death and degeneration. Considering a potential modulator of the GABAergic system, açaí extract (*Euterpe oleracea*) has shown to have effects on minimizing electrophysiological changes in seizure models, in addition to attenuating oxidative damage in the cerebral cortex (DA SILVA et al., 2021; MUTO et al., 2022).

As for Parkinson's disease, in the model induced by injections of 1-methyl-4-phenyl-1,2,3,6-tetrahydropyridine in mice, significant effects were observed, including increased antioxidant parameters, such as catalase, glutathione peroxidase, reduced glutathione, and superoxide dismutase. Additionally, there was a reduction in lipid peroxidation levels, neuronal death, and the expression of pro-inflammatory markers, along with the restoration of dopamine transporter expression levels in the striatum of animals (D'AMICO et al., 2022). In the model of vascular dementia induced by repeated cycles of ischemia and reperfusion, IMPELLIZZERI et al. (2022) demonstrated that açaí reduced neuronal death in the hippocampus of mice by modulating mechanisms involved in autophagy and apoptosis.

19.5 CONCLUSION

It is notable that the scientific literature presents numerous data regarding the therapeutic potentials of products derived from *Euterpe oleracea* Mart., which include not only the classic results on antioxidant activities in various models of oxidative challenges, both in vitro and in vivo, but also some evidence already applied in human studies. However, some paths still need to be explored and better developed for the advancement of its use as a potent therapeutic tool, whether as a dietary supplement or as raw material for the development of new bioactive molecules, such as:

1. Understanding the biological activities triggered by the various secondary metabolites found in açaí extracts, as the diverse benefits reported in the literature can be attributed to the set of biomolecules present.
2. Standardization of the dose of açaí and/or secondary metabolites that are phytochemical markers based on the biological response observed, whether they are markers of antioxidant response or anti-inflammatory response.
3. Clinical trials involving new food and pharmaceutical technologies, such as micro- and nanoencapsulation, aimed at observing the biological responses based on the administered doses.

REFERENCES

Adeva-Andany, María M. et al. Insulin resistance is a cardiovascular risk factor in humans. *Diabetes & Metabolic Syndrome: Clinical Research & Reviews*, v. 13, n. 2, p. 1449–1455, 2019.

Alessandra-Perini, Jéssica et al. *Euterpe oleracea* extract (açaí) exhibits cardioprotective effects after chemotherapy treatment in a breast cancer model. *BMC Complementary Medicine and Therapies*, v. 23, n. 1, p. 301, 2023.

Arrifano, Gabriela P.F. et al. Clarified açaí (Euterpe oleracea) juice as an anticonvulsant agent: In vitro mechanistic study of GABAergic targets. *Oxidative Medicine and Cellular Longevity*, v. 2018, 2018.

Azevedo, Aline et al. Matrix metalloproteinases are involved in cardiovascular diseases. *Basic & Clinical Pharmacology & Toxicology*, v. 115, n. 4, p. 301–314, 2014.

Baptista, Sheyla de L. et al. Biological activities of acai (*Euterpe oleracea* Mart.) and jucara (*Euterpe edulis* Mart.) intake in humans: An integrative review of clinical trials. *Nutrition Reviews*, v. 79, n. 12, p. 1375–1391, 2021.

Barbosa, Jhonatas Rodrigues; De Carvalho Junior, Raul Nunes. Food sustainability trends—how to value the açaí production chain for the development of food inputs from its main bioactive ingredients? *Trends in Food Science & Technology*, v. 124, p. 86–95, 2022.

Barbosa, Priscila O. et al. Açaí (*Euterpe oleracea* Martius) supplementation improves oxidative stress biomarkers in liver tissue of dams fed a high-fat diet and increases antioxidant enzymes' gene expression in offspring. *Biomedicine & Pharmacotherapy*, v. 139, p. 111627, 2021.

Caliri, Andrew W.; Tommasi, Stella; Besaratinia, Ahmad. Relationships among smoking, oxidative stress, inflammation, macromolecular damage, and cancer. *Mutation Research/Reviews in Mutation Research*, v. 787, p. 108365, 2021.

Chaves, Genisson Paes et al. A importância sociocultural do açaí (*Euterpe oleracea* Mart.) na Amazônia brasileira. *Revista Contribuciones a las Ciencias Sociales*, n. 29, 2015.

D'amico, R. et al. Açai berry mitigates Parkinson's disease progression showing dopaminergic neuroprotection via Nrf2-HO1 pathways. *Molecular Neurobiology*, v. 59, n. 10, p. 6519–6533, 2022. https://doi.org/10.1007/s12035-022-02982-5

Da Costa, Cristiane Aguiar et al. Euterpe oleracea Mart.-derived polyphenols prevent endothelial dysfunction and vascular structural changes in renovascular hypertensive rats: role of oxidative stress. *Naunyn-Schmiedeberg's Archives of Pharmacology*, v. 385, p. 1199–1209, 2012.

Da Silva, Alexandre A. et al. Role of hyperinsulinemia and insulin resistance in hypertension: metabolic syndrome revisited. *Canadian Journal of Cardiology*, v. 36, n. 5, p. 671–682, 2020.

Da Silva, Antonio Jorge Barbosa; Sevalho, Elison de Souza; Miranda, Ires Paula de Andrade. Potencial das palmeiras nativas da Amazônia Brasileira para a bioeconomia: análise em rede da produção científica e tecnológica. *Ciência Florestal*, v. 31, p. 1020–1046, 2021.

Da Silva, Ilano Oliveira et al. What we know about Euterpe genus and neuroprotection: A scoping review. *Nutrients*, v. 15, n. 14, p. 3189, 2023.

Da Silva, T.V.N. et al. Dietary *Euterpe oleracea* Mart. attenuates seizures and damage to lipids in the brain of *Colossoma macropomum*. *Fish Physiology and Biochemistry,* v. 47, p. 1851–1864, 2021.

Da Silva Menezes, Ellen Mayra; Torres, Amanda Thiele; Srur, Armando Ubirajara Sabaa. Valor nutricional da polpa de açaí (*Euterpe oleracea* Mart) liofilizada. *Acta Amazonica*, v. 38, n. 2, p. 311–316, 2008.

Da Silveira, Jéssica Teixeira et al. An integrative review of Açaí (*Euterpe oleracea* and *Euterpe precatoria*): Traditional uses, phytochemical composition, market trends, and emerging applications. *Food Research International*, p. 113304, 2023.

Dahl, Jan-Ulrik; Gray, Michael J.; Jakob, Ursula. Protein quality control under oxidative stress conditions. *Journal of Molecular Biology*, v. 427, n. 7, p. 1549–1563, 2015.

De Moraes Arnoso, Bernardo Junqueira et al. Açaí seed extract (ASE) rich in proanthocyanidins improves cardiovascular remodeling by increasing antioxidant response in obese high-fat diet-fed mice. *Chemico-Biological Interactions*, v. 351, p. 109721, 2022.

De Oliveira, Ana Caroline et al. Phenolic composition, antiproliferative and antiulcerogenic activities of a polyphenol-rich purified extract from açai (*Euterpe oleracea*) fruits. *International Journal of Food Science & Technology*, v. 56, n. 12, p. 6626–6634, 2021.

De Souza, Diulie Valente et al. Açaí (*Euterpe oleracea* Mart.) presents anti-neuroinflammatory capacity in LPS-activated microglia cells. *Nutritional Neuroscience*, v. 25, n. 6, p. 1188–1199, 2022a.

De Souza, J. R., et al. Different cell death mechanisms are involved in leprosy pathogenesis. *Microbial Pathogenesis*, v. 166, p. 105511, 2022b.

Do Nascimento, W. M. O. Açaí *Euterpe oleracea* Mart. 2008. ISO 690.

Griendling, Kathy K.; Fitzgerald, Garret A. Oxidative stress and cardiovascular injury: Part I: basic mechanisms and in vivo monitoring of ROS. *Circulation*, v. 108, n. 16, p. 1912–1916, 2003.

Hajam, Younis Ahmad et al. Oxidative stress in human pathology and aging: molecular mechanisms and perspectives. *Cells*, v. 11, n. 3, p. 552, 2022.

Helms, C. C.; Gladwin, M. T.; Kim-Shapiro, D. B. Erythrocytes and vascular function: oxygen and nitric oxide. *Frontiers in Physiology*, v. 9, p. 125, 2018.

Hussain, Tarique et al. Oxidative stress and inflammation: what polyphenols can do for us? *Oxidative Medicine and Cellular Longevity*, v. 2016, 2016.

Impellizzeri, D. et al. Acai berry mitigates vascular dementia-induced neuropathological alterations modulating Nrf-2/beclin1 pathways. *Cells*, v. 11, p. 2616, 2022.

Karlsson, Torgny et al. Contribution of genetics to visceral adiposity and its relation to cardiovascular and metabolic disease. *Nature Medicine*, v. 25, n. 9, p. 1390–1395, 2019.

Mathias, L.M.B.S. et al. *Euterpe oleracea* Mart. (Açai) supplementation attenuates acute doxorubicin-induced cardiotoxicity in rats. *Cellular Physiology and Biochemistry*, v. 53, n. 2, p. 388–399, 2019.

Monteiro, Elisa Bernardes et al. Polyphenol-rich açaí seed extract exhibits reno-protective and anti-fibrotic activities in renal tubular cells and mice with kidney failure. *Scientific Reports*, v. 12, n. 1, p. 20855, 2022.

Morais, Sheldon Christian Nogueira de. Crescimento de mudas de açaizeiro (*Euterpe oleracea* Mart.) Sob efeito de fertilização de liberação lenta. 2019.

Muto, Nilton Akio et al. Extract of *Euterpe oleracea* martius stone presents anticonvulsive activity via the GABAA receptor. *Frontiers in Cellular Neuroscience*, v. 16, p. 872743, 2022.

Patel, Tushar P. et al. Insulin resistance: an additional risk factor in the pathogenesis of cardiovascular disease in type 2 diabetes. *Heart Failure Reviews*, v. 21, p. 11–23, 2016.

Pini, Taylor et al. Mitigating the effects of oxidative sperm DNA damage. *Antioxidants*, v. 9, n. 7, p. 589, 2020.

Ribeiro, Juliana Carvalho et al. Evaluation of the genotoxic and antigenotoxic effects after acute and subacute treatments with açai pulp (*Euterpe oleracea* Mart.) on mice using the erythrocytes micronucleus test and the comet assay. *Mutation Research/Genetic Toxicology and Environmental Mutagenesis*, v. 695, n. 1–2, p. 22–28, 2010.

Rocha, Joaquin Humberto Aquino et al. Circular bioeconomy in the amazon rainforest: evaluation of açaí seed ash as a regional solution for partial cement replacement. *Sustainability*, v. 14, n. 21, p. 14436, 2022.

Santos, Izabelle Barcellos et al. Açaí seed extract prevents the renin-angiotensin system activation, oxidative stress and inflammation in white adipose tissue of high-fat diet–fed mice. *Nutrition Research*, v. 79, p. 35–49, 2020.

Torma, Priscila do Carmo Marchioro Raupp et al. Hydroethanolic extracts from different genotypes of açaí (*Euterpe oleracea*) presented antioxidant potential and protected human neuron-like cells (SH-SY5Y). *Food Chemistry*, v. 222, p. 94–104, 2017.

Vilhena, Juliana Calfa et al. Açaí reverses adverse cardiovascular remodeling in renovascular hypertension: A comparative effect with Enalapril. *Journal of Cardiovascular Pharmacology*, v. 77, n. 5, p. 673–684, 2021.

Vona, Rosa et al. The impact of oxidative stress in human pathology: Focus on gastrointestinal disorders. *Antioxidants*, v. 10, n. 2, p. 201, 2021.

Yuyama, Lucia Kiyoko Ozaki et al. Caracterização físico-química do suco de açaí de Euterpe precatoria Mart. oriundo de diferentes ecossistemas amazônicos. *Acta Amazonica*, v. 41, p. 545–552, 2011.

Section 7

Phenolic Compounds

20 Utilization of Resin by Insects and Its Antibacterial Properties

Showket Ahmad Dar, Ammar Al-Farga, Ayah T. Abdullah Al-Kilani, Kounser Javeed, Wajid Hasan, Yendrembam K. Devi, Naveena Nazim, Dev Karan Bairwa, Kangjam Bumpy, I. Yimjenjang Longkumer, Syed Naseem Zaffar, and Rizwan Rashid

20.1 INTRODUCTION

Bees belonging to the Meliponini family are highly social insects native to tropical and subtropical climates. The stingless bees, which include the Euglossini, Apini, Bombini, and Meliponini families, represent the most diverse and largest group of corbiculate bees, with over 550 species currently known. In addition to exhibiting a bewildering variety of morphologies, behaviors, and life cycles, they account for approximately 70% of all eusocial bee species. It is possible for their colonies to consist of anywhere from a few hundred to many thousands of individuals, and their geographic distribution extends across five continents. The manner in which they construct their nests is extremely diverse, with some species constructing their nests in tree holes, others nesting within active termite or ant nests, and yet others constructing subterranean nests that are as deep as 3 m underground. The practice of stingless beekeeping, also known as meliponiculture, has allowed humans to maintain a relationship with stingless bees for millennia [1]. Stingless beekeepers from Indigenous and rural communities have been generating their ecological information for centuries. Consequently, research on these bees frequently relies on this local ecological expertise, which is in contrast to well-established farms and units. Beekeepers collect resinous compounds from stingless bee nests, which contain a rich chemical composition and information about their several medical applications. The sticky resins that plants release are collected by bees, and these are employed by the bees for a variety of useful reasons. The majority of bee species utilize resin to construct critical nest structures, including brood combs, storage containers for honey and pollen, and a variety of protective structures. Furthermore, resin plays an essential role in nest defense for a multitude of species. For instance, stingless bees utilize resin to construct barriers, capture predators, and eliminate potential intruders. The quantity of resin within the nest cavity affects the physical characteristics of the bees and the microbes that they interact with. The chemical profiles of a multitude of species of stingless bees have been found to be enriched by compounds formed from resin, which also strengthen their nests (Figure 20.1). These compounds play a significant role in the formation of the microbial communities that are connected with stingless bees and their nests.

Despite the essential role of resin in the proper functioning of stingless bee colonies, research on resin and stingless bees is limited. The majority of research on Meliponini species has been conducted by Western scientists, which has contributed to a partial restriction of our understanding of resin utilization. A paucity of studies has concentrated explicitly on the usage of resin, and only

DOI: 10.1201/9781003462804-27

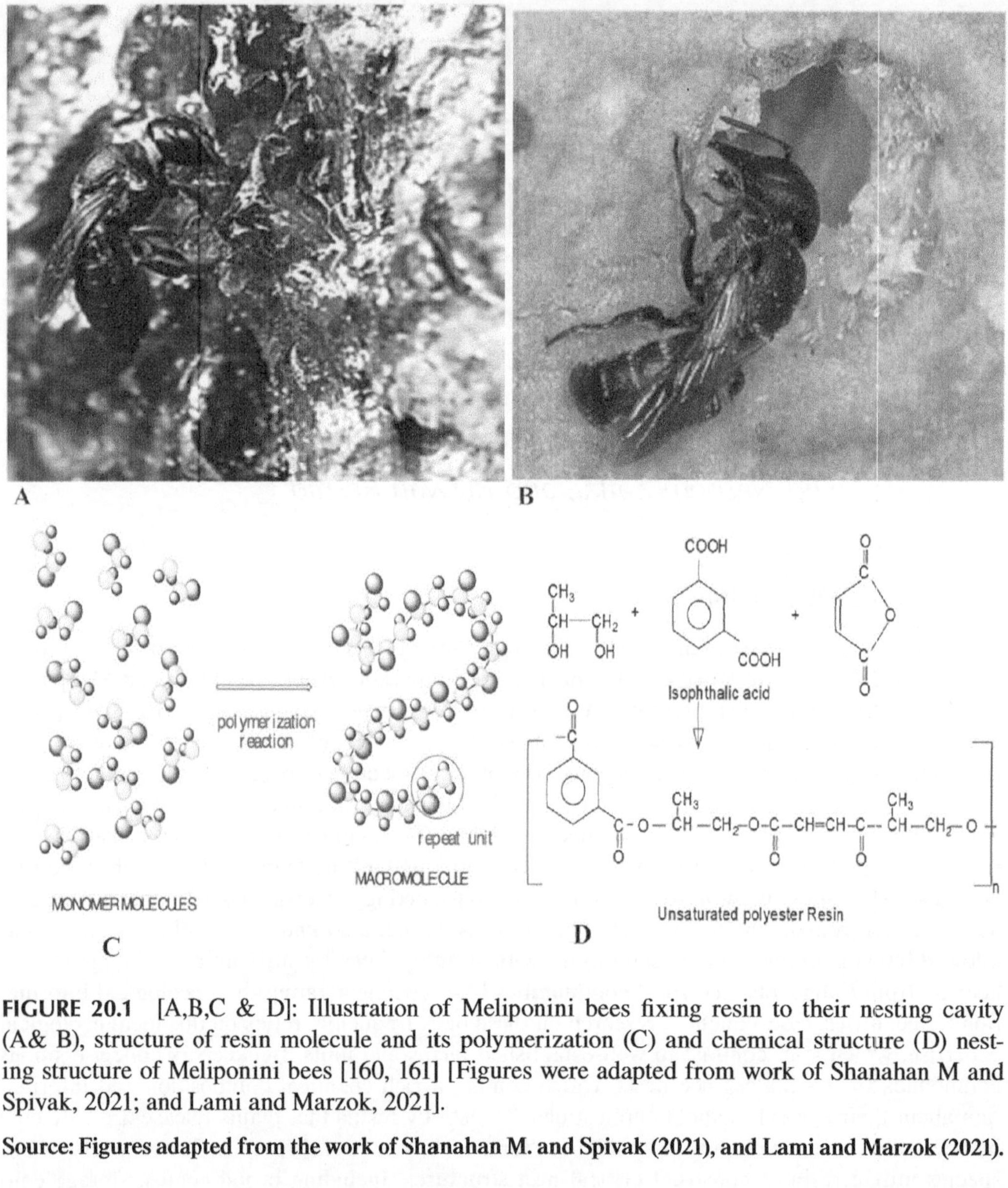

FIGURE 20.1 [A,B,C & D]: Illustration of Meliponini bees fixing resin to their nesting cavity (A& B), structure of resin molecule and its polymerization (C) and chemical structure (D) nesting structure of Meliponini bees [160, 161] [Figures were adapted from work of Shanahan M and Spivak, 2021; and Lami and Marzok, 2021].

Source: Figures adapted from the work of Shanahan M. and Spivak (2021), and Lami and Marzok (2021).

a limited number of these species have been subjected to extensive research. Given the considerable diversity among stingless bees and the high variability in resin consumption across species, it is challenging to obtain comprehensive information on this trait. In contrast, despite numerous challenges, this chapter synthesizes research from diverse geographical regions on a range of disciplines, including natural history, chemical ecology, and microbiology, to examine the utilization of resin by stingless bees. These findings collectively illuminate the pivotal role of resin utilization in the functioning of stingless bee colonies. A review of the existing literature on the role of resin in the construction and defense of stingless bee nests [4], a discussion of resin foraging and resin handling by stingless bees, and a review of the literature on the effects of resin on the cuticular chemical profiles of bees and their microbial companions will be presented in the following sections. In the conclusion, we highlight the knowledge gaps that require further investigation on Meliponini species. This should be carried out in detail, as outlined in the following references [4–9].

20.2 RESIN AND ITS ROLE IN THE INSECT WORLD

Langenheim [10] has provided a comprehensive overview of the chemistry, evolution, ecology, and ethnobotany of plant resins. As a means of self-defense against herbivores and microbes, plants produce resin from their buds, wounds, fruits, flowers, and bark. In select cases, resin secretion is employed to attract pollinators and seed-dispersing birds and disinfect wound sites, thereby aiding in the prevention of the growth of endophytic fungi. Moreover, the resin is employed to ensnare, immobilize, and dissuade predators. The chemical composition of this multifunctional substance is somewhat complex, exhibiting a range of anti-inflammatory, antifungal, antibacterial, and antiviral activities. It is composed of a complex mixture of lipid-soluble volatile and non-volatile phenolic compounds, including flavonoids, aromatic acids, and benzopyrenes, as well as terpenoids, such as mono-, di-, and sesquiterpenes, which exhibit varying chemical compositions depending on the plant species. Even individuals of the same plant species may have different resin compositions.

The resin is a complex and varied mixture of bioactive molecules that makes it difficult for predators and diseases to evolve tolerance to the substance. This resource is utilized by a diverse array of creatures, including humans, coatis, wood ants, and bees, for a multitude of purposes, including medicinal, defensive, and construction applications [11, 12]. All these animals gather resin from plants. A considerable number of bee species, belonging to the genera *Megachilidae* and *Apidae*, utilize plant resins for the construction of nesting structures. A considerable number of corbiculate bees, with the exception of bumblebees (Bombini), are responsible for the collection and utilization of plant resins. A number of scientists have conducted exhaustive research on the utilization of resin by bees. However, the utilization of resin is considerably more prevalent in stingless bees. Additionally, numerous species of stingless bees collect resin in large quantities and employ it to support a variety of function-related features of their colonies, including nest defense and brood protection [13].

20.2.1 Resin in Solitary Bees

A number of challenges must be overcome for solitary bees to successfully rear their offspring. For instance, the substrates chosen for nest construction must be stable and resistant to abiotic changes, accessible in the surrounding environment, and high in plasticity to be utilized in the construction of nests [12]. The internal environment of the nest must provide the appropriate abiotic conditions to prevent food provision spoilage and brood death due to fungal or mold infestations or desiccation. Completed nests must provide protection for developing the young against predators and parasites until they are able to close their eyes. In addition to the utilization of a secreted lining on the inner surface of a cell, which is observed in the Colletidae and Halictidae families, scholars have highlighted the incorporation of external nest material in the creation of nest cells as a significant innovation that distinguishes bee lineages. Indeed, non-floral resources are essential for the survival of more than 30% of the world's bee species. The incorporation of external nest material in the construction of nest cells has been linked to the extensive range expansion and species diversification observed within the Megachilidae family. It is well-documented that representatives of two bee families, namely, the Megachilidae (Anthidiini, Osmiini, Megachilini) and the Apidae (Euglossini, Apini, Meliponini), make regular use of plant-derived resin as a material for the construction of their nests. Resin is but one of the various nest construction materials that are available in the external environment. Plant resin is a versatile material that can be employed in a variety of ways. It can be used to reinforce unstable nest substrates, bind loose structural or camouflaging components, and be easily sculpted when it is still new. Once it has set, the material becomes structurally rigid and waterproof. Additionally, resin has the capacity to regulate moisture levels within the nest. Moreover, numerous resins demonstrate potent antimicrobial properties. Furthermore, resin is waterproof, and it has the potential to offer robust physical defense against some microorganisms and predators, which can preserve brood and prevent the rotting of food supplies. The archives of natural history observations reveal that the gathering of resin by bees and the use of resin by bees to construct nest structures are both extensively documented for a great number of bee species.

A recent study on highly eusocial stingless bees (Apidae: Meliponini) revealed the significance of resin for bees [2].

In contrast, the significance of resin in the ecology of solitary bees has been the subject of only a limited number of studies, in comparison to the significance of traditional floral resources, such as pollen and nectar. This chapter examines the functional utility of resin, with a particular focus on its involvement in the construction of nests, its potential as an adhesive for camouflaging materials and as a chemical camouflage, and in defense, in moisture regulation, for its antimicrobial characteristics, and in microbial communities. The analysis is conducted with a specific emphasis on solitary bees. We describe the limitations that govern the usage of resin based on this knowledge, as well as the ramifications for the ecology of the bees, such as the microbiota that is found in the nest. According to numerous researchers, the existing literature on resin use in bees was initially compiled from references to resin use in a variety of bee genera, subgenera, and species. From in-depth observations of nesting biology and verification of resin sources to straightforward assertions that particular nest structures are made of an unspecified substance—resin (described as, for example, brittle, hard, glass-like, gummy, and strong-smelling)—the literature that was discovered on the use of resin in bees is quite diverse. Depending on whether they are formed by plants in their internal ducts or on their surface glands, resins are classified as combinations of substances that are either terpenoid or phenolic in origin. It is challenging to ascertain whether the substance in question is indeed resin or another plant exudate, such as sap, gum, or latex. We have opted to include studies of this nature because all plant exudates exhibit similar physical characteristics and functions, and they are employed in a comparable manner. Consequently, our research focused on comprehensive observations of the biology of nesting, as well as author perspectives on the function of resin, in order to infer the features and functions. Only 1 of the 100 studies that were initially considered for inclusion provided complete proof of resin function.

20.3 RESIN IN APIDAE: RESIN USE BY MELIPONINI BEES

20.3.1 Nest Construction

The approach that stingless bee species take for the construction of their nests is highly diverse, with various species constructing their nests in a variety of different locations throughout multiple seasons. Some species construct subterranean nests at considerable depths underground, while others construct exposed nests that are affixed to tree branches. A minority of species make use of existing cavities, such as hollow trees, termite nests, or electric light posts. A diverse array of materials is utilized in the construction of nests, including feces, soil, and even residential building structures coated with wet paint and adhesives. It is notable that the majority of stingless bees employ resin in some section of their nest, despite the fact that the construction of nests differs not only between species but also even within species. One of the many reasons resin is an efficient building material is its durability. However, when resins are secreted from plants, they are pliable. Over time, they become more rigid, which allows them to be molded into structures that are long-lasting. Due to its insolubility in water, resin can be employed to create leak-proof nesting areas and watertight storage containers.

Moreover, the antimicrobial properties of resins have the potential to regulate the microbial populations present within stingless bee nests, thereby limiting the deterioration of food materials and the invasion of pathogens. One might posit that the utilization of antibacterial resins in the construction of nests represents a significant factor in the development of sociality in stingless bees. The management of microbial communities is an essential component in the construction and upkeep of a successful nest, as well as in the maintenance of other creatures, including insects. In tropical habitats, where conditions are conducive to the proliferation of microorganisms, and in social insect communities, where the risk of disease transmission is heightened due to the high density of genetically similar individuals living in close proximity to one another, these efforts are of particular significance with regard to the spread of microbial diseases. To prevent the deterioration of food and the transmission of infections to the brood, a considerable number of insects utilize antimicrobial chemicals.

The production of these compounds can occur through two distinct mechanisms: self-production, whereby antimicrobial venom is secreted by insects onto their cuticle and nests, and symbiont production, whereby beneficial microbes secrete antimicrobial compounds that prevent the spoilage of food stores. Alternatively, these compounds can be acquired from the environment, as exemplified by bees, ants, and other insects that bring foreign materials, such as antimicrobial resin, into their nest cavity. It is postulated that this evolutionary adaptation has enabled bees to significantly expand their range and diversify their nesting habits under different climatic conditions [12, 14–22].

The use of foreign materials enables the construction of different types of nests. Resins, in particular, have been postulated to have facilitated the social evolution of stingless bees in tropical ecosystems. Moreover, this process has resulted in the subsequent diversification of stingless bee species in numerous geographic regions [23]. The use of resins helps preserve food stores and allows for the strong construction and defense of nest cavities, which are filled with pollen and have brood populations inside. This hypothesis is supported by the findings of Rasmussen and Camargo [24], who established a molecular phylogeny indicating that resin was likely used by ancestral *Trigona* species in the construction of their nests. Furthermore, resinous materials continue to be an essential component for habituation of different bee species in the construction of nests for stingless bees. The majority of stingless bees combine resin and wax to generate cerumen, which is the material used to construct brood combs, honey and pollen pots, and a variety of other structures within the nest (e.g., *Trigona australis*).

Some scholars have drawn parallels between cerumen and honeybee propolis, citing the presence of both wax and resin in cerumen as a source of its resilience. Moreover, the cerumen is a principal component utilized in the fabrication of the stingless bee hive, exhibiting functional similarities to beeswax. Nevertheless, there are a number of significant differences between the two substances. For instance, the cerumen is continuously processed and recycled within the nest, in contrast to beeswax, which is a part of a permanent comb construction. The physical qualities of cerumen exhibit a considerable range, and the diverse proportions of wax and resin present in the substance are likely responsible for this variability. The composition of the cerumen determines its physical characteristics. It can be soft, flexible, and light brown in color or hard, brittle, and dark brown or black. Roubik [5] observed that the cerumen generated by certain small stingless bees has very little to no resin and is therefore more similar to pure wax. Consequently, there is no evidence to suggest that Schwarzula species make use of any resin at all. Rather, they appear to cultivate scale insects within the nest cavity and combine their wax with waxy materials that they manufacture themselves in order to create a cerumen-like substance [25]. Conversely, the resin content of cerumen in certain species can be as high as 40%. Additionally, there is a lack of understanding regarding the factors that determine the amount of resin that various species of stingless bees incorporate into their cerumen. In addition to using resin for the production of cerumen, stingless bees also incorporate resin into the surroundings of their nests in the form of deposit resins, propolis, and geopropolis, as well as in structures such as the entrance to the nest and batumen (Table 20.1).

However, *propolis* is employed as a catchall term for a resinous material that is found inside the nest, with the exception of cerumen. Moreover, it is crucial to distinguish between these categories, as a single nest may contain a multitude of resin-rich materials and structures constructed on resin, each serving a distinct purpose. Resin is located on the floor or walls of the nest as deposits, which are also known as resin deposits or viscous propolis deposits. In certain species, these caches serve as a temporary storage space for resins, which are collected until they are incorporated into other nest constructions or used for defensive purposes. This is the case for *Trigona pallens* (Trigona) and *Tetragonisca angustula* (Latreille) and *Plebeia* spp. In contrast to the majority of resins, which typically become more rigid when exposed to air, deposit resins maintain their viscosity for an extended period of time. Deposit resins contain a greater proportion of floral resins, which are slow to solidify. This attribute has an important function in nest construction and design. The prolonged viscosity of deposit resins is a result of chemical processing. An investigation of the morphology of head salivary glands and intramandibular glands of bees of varying ages suggests that workers of the species *Plebeia emerina* modify deposit resins by utilizing secretions, which may help maintain the viscosity of the resins. The resins that stingless bees collect and bring back to the nest are mixed

TABLE 20.1
Materials That Are Rich in Resin and Originate from Plants, as Well as Nest Constructions and Construction in Solitary Bee Behavior.

S. No,	Resin Material	Functions in Solitary Bee Nest Construction
1	Cerumen	Bees that do not have stingers employ a combination of wax and resin to construct nest structures such as honey and pollen pots, brood combs, and other things.
2	Deposit resin	Resin deposits, also known as viscous propolis deposits, are caches of resin that are stored by certain species on the floor or walls of their nests. Others call these deposits *resin deposits*.
3	Propolis	The fractures and crevices throughout the nest sealed with resin that had been blended with minor amounts of salivary gland secretions and wax.
4	Geopropolis	Resin that has been combined with particles of soil, silt, and/or sand.
5	Batumen	A construction that resembles a wall and frequently contains resin; many species construct a batumen in order to provide a barrier between the interior environment of the nest and the outside world.
6	Nest entrance	There are several species that have nest entrances that are made up of tubes made of hardened resin that can extend both inside and outside of the nest.

with small amounts of salivary gland secretions and, according to some sources, wax. These materials are referred to as *propolis*, which is composed of multiple types of resins.

A comparative analysis of the chemical diversity of honeybee and stingless bee propolis has revealed that the former exhibits a lower chemical diversity than the latter. Furthermore, several species of stingless bees utilize propolis for the purpose of sealing holes and crevices in their nests, a behavior that is analogous to that of honeybees. It is not uncommon for bees to construct a layer of propolis that is so thick that it requires beekeepers to remove the lid of the hive body in order to gain access to the nest. This is particularly the case with colonies that are kept in box hives. The extent to which stingless bees incorporate wax into their propolis remains unknown, despite the fact that the majority of investigations have found that propolis contains wax. It is evident that the reports on wax content lack precision. Furthermore, it is well-documented that the quantity of wax present in *A. mellifera* propolis is significantly variable [26–28]. A comprehensive investigation was conducted by researchers on the handling of resin from stingless bees. The researchers observed resin deposition in six different species of bees, yet no evidence of resin mixing with wax was found. Although this has not yet been confirmed, it is possible that propolis produced by different species and for different purposes may contain varying amounts of wax. In some instances, it is even possible that it may contain no wax at all.

A few species of stingless bees substitute geopropolis for pure propolis. The combination of plant resins and soil results in the formation of geopropolis. The composition of geopropolis is variable, with particles of soil, silt, and sand comprising up to 90% of the mixture. Its use within the nest is comparable to that of pure propolis, although it is less pliable. Although these two terms are sometimes used interchangeably, it is important to note that they are not synonymous. The geopropolis is the propolis of stingless bees and is distinguished from other propolis compounds by its lack of soil content. In addition to the aforementioned nest constructions, the entrance to the nest and the batumen, two further examples of nest construction, frequently combine propolis and geopropolis. Resin tubes, which can be found both within and outside the nest, are frequently used as material for nest openings. In order to prevent hostile infiltration, these complex labyrinth-like buildings are typically constructed from the inside out. In order to keep the outside world out of their nest, several species of stingless bees construct a wall-like construction called the *batumen*, which is also known as the external involucrum. The most common material used for nest construction is resin, although other materials, such as clay, seeds, wood, and even excrement, can be employed. The construction of batumen varies among species of stingless bees. In certain species, such as *Melipona* spp., the batumen walls can be as thick as 10 cm. In contrast, in other species, such as the *Hypotrigona* and *Trigonisca* spp., batumen is not used.

Batumen, when present, can assume a variety of shapes and sizes. Various researchers have defined various batumen forms and documented their presence or absence for 145 stingless bee species observed in Costa Rica. The study demonstrated that exposed batumen serves as a protective covering for nests that are either entirely or partially exposed to the elements. Bee species such as *Meliponula bocandei*, *Melipona*, and *Cephalotrigona* are capable of adapting the size of their nest cavities to their own needs and those of their fellow bees. *A. mellifera* colonies coat the rough inner surfaces of the hollow tree cavities where they nest with a thin continuous resinous lining, batumen, which is less than 2 mm thick. The construction of laminate batumen is a multilayered structure comprising sheets with passages that bees can use to fly between layers. These channels may also facilitate the flow of air and the evaporation of water. In addition to their protective function, these batumen varieties may also serve to waterproof the nest cavity and facilitate the control of fungal growth.

20.3.2 Defense

Bees are subject to a variety of parasites and predators. The genus *Lestrimelitta* is parasitized by a diverse array of organisms, including lizards, spiders, ants, wasps, beetles, phorid flies, and assassin bugs. Stingless bees employ a diverse array of defensive mechanisms to safeguard their nests, given their inability to sting. Some species employ the tactic of hiding, while others construct cryptic nests, bite, and burrow in hair. These defensive tactics vary among species. A variety of defensive mechanisms are employed by stingless bees to repel, capture, and destroy parasites and predators, including the use of resin [29, 30]. In this study, we categorize resin-based defenses into two groups: *Structural defenses* are those in which bees construct or augment pre-existing structures with resin to ward off invaders. *Direct defenses* are those in which bees smear resin onto intruders' bodies, or theirs, to protect their nests.

20.3.3 Structural Defenses

A number of species of stingless bees, including those belonging to the genera *Lepidotrigona*, *Scaura*, *Tetragona*, *Tetragonula*, and *Trigonisca*, construct a barrier of new resin droplets to safeguard their nest entrances. The adhesive substance functions as a defense mechanism that employs both mechanical and chemical mechanisms. The terpenoid chemicals that are typically contained in resin are effective at deterring a wide range of predators, including ants. Ants attempt to traverse the resin droplets, frequently becoming ensnared in the sticky substance and utilizing the bodies of other ants to traverse the so-called resin "moat." It is anticipated that the adhesive and repellent properties of the resin droplets will diminish over time, necessitating the application of fresh stores of resin droplets to maintain their rigidity [31–35]. The continuous application of fresh resin results in the formation of long and slender entrance tubes for certain species, including *Trigona cilipes*, *Tetragonilla collina*, and other species that are related to these species. The architectural effect of nest entrance resin is striking, with dozens of tubelets branching like coral to create a nest entrance. This is a construction technique used by the stingless bee *Pariotrigona klossi*. This bee is so diminutive that it is said to drink tears. Each tubelet is attached to strings of transparent resin beads which, when viewed as a whole, resemble the "quartz pendants of a chandelier." It is challenging for invading ants to traverse resinous terrain when it has hardened, and it is likely to be inaccessible when it is fresh. Furthermore, this terrain is visually distracting, which further discourages ant attacks. In the event of disturbance at night, such as by the bees *Meliplebeia tanganyilcae medionigra* and *Plebeiella lendliana*, the bees employ particles of resin and cerumen to barricade the entrance to their nests.

It has been observed that certain species of bees, including *Melipona panamica*, *Melipona flavolineata*, and other species of *Melipona*, possess a resin ball that has been specifically designed for this purpose [23]. The bees then roll the hardened resin ball into place and apply new resin to secure it to the entrance of the nest, thus preventing intruders from breaking into the nest. Over time, recycled resin balls tend to gather in close proximity to the entrances of these nests, serving to protect the cavities within. Stingless bees that dwell in active termite or ant nests frequently apply a full protective resin barrier around them. They are capable of adapting to conditions that would otherwise be hostile. Some myrmecophilous stingless bees, such as *Trigona moorei*, commence the construction of a nest by first erecting a preliminary batumen structure, with the objective of creating areas that are free of ants. As they continue to burrow further into the ant colony, they then expand this resinous shield. *Scaura latitarsus* bees exhibit a behavior that is distinct from that of other bees. They construct their nests by excavating a chamber within an active termite nest and then covering that space with a continuous batumen shell. This behavior has been observed in the following references: [36–38].

20.3.4 Direct Defenses

It is a common phenomenon observed in many bee species that they apply resin directly to perceived threats. In addition, they employ resin in nest structures to avoid invasion and brood loss. This practice has been referred to as "resin daubing," and the same strategy was employed by *Austroplebeia australis*, which uses resin to immobilize and mummify invading predators [8]. In order to defend their nests, bees collect resin and cerumen from the deposits that form around the entrance and other sections of the nest. This resin is carried in their mandibles and corbiculae, which they use to defend their nests against parasitic bees that pose a significant threat. Subsequently, nest predators employ two distinct strategies to deter potential threats. One strategy involves the attachment of human hair to the outside of the nest, serving as a visual deterrent. Another strategy involves the capture and mummification of intruders within the nest. Some stingless bee species utilize resin pellets to exclude virgin queens from their nests [39, 40].

In addition to defending their nests from intruders by applying resin, certain species of bees also apply resin to their own bodies, particularly to the cuticle, in order to defend themselves. It has been demonstrated that a number of different species of stingless bees leave the nest with varying amounts of both viscous and solid resin in their corbiculae. This evidence has led to the conclusion that the presence of hardened resin is necessary, despite the fact that soft, sticky resin can be used to ensnare potential invaders. In the species *Melipona subnitida*, resin is present in the corbiculae in approximately 11% of worker bees that were leaving the nest under normal conditions, indicating that they were not disturbed. Approximately half of the bees were observed to be carrying resin that was viscous and malleable, while the remaining half were observed to be carrying resin that was dry, rigid, and strong. As a consequence of the disturbance to the nest, the percentage of worker bees who left the nest carrying resin increased to 90%. The majority of these bees were observed to be carrying resin that had hardened into a substance. In order to investigate the potential role of resin residues on the body in tracking and monitoring colony members, bees that carried resin and those that foraged for nectar were marked with paint, and their activities were tracked. It was hypothesized that the resin carriers were circling the nest rather than transporting materials to a new nest location in response to a predator attack, given the relatively short flight time observed [41–43]. The scientists postulated that the repellent qualities of the resin could be exploited as a form of protection, dissuading potential predators from consuming the resin bearers. Alternatively, the resin bearers may sacrifice themselves for the benefit of the colony. If a predator consumes a resin bearer that is unappetizing, it will be dissuaded from pursuing additional predation. It is evident that the presence of resin on the bodies of stingless bees is not merely a coincidence. The cuticle of *Tetragonula carbonaria* is coated with resin, rendering the entire body sticky. Additionally, a thin layer of resin is observed on the legs, head, and thorax of *Tetragonula angustula* and *Trigona (Tetragonula) melanocephala*, which are crucial nectar foragers. These bees were observed leaving the nest with resin in their corbiculae. Conversely, some researchers have observed bees carrying hardened resin on their bodies as part of an apparent mobilized defense mechanism. These bees then use this resin to construct a nest with the sole purpose of providing individual or collective protection to the nest population. Both visual stimuli and chemical cues have the potential to activate defenses based on resin. An increase in resin transport and the barricading of the nest entrance tube with hardened resin balls mixed with gnathal secretions and mandibular gland extract from *Lestrimelitta limao* (Smith) and *Melipona flavolineata* were observed [44]. The foraging activity of resin by *Tetragonilla collina* was observed to increase when nest entrances were damaged, and it quadrupled following an ant attack. This was attributed to the utilization of resin by worker bees to extend their entrance tubes and fortify them with a barrier of resin droplets. Stingless bees select resins based on their functional properties, which are derived from various plant species. These resins have been found to be effective against a range of pathogenic infections and predators. Furthermore, this indicates that individuals with access to a wide variety of resin sources are essential for the defense of stingless bees [45]. It remains unclear whether, in response to specific dangers, stingless bees alter their resin resource choices. One illustrative example is the bees' response to an attack by a small hive beetle. In this instance, the bees collect resins that are particularly effective in warding off the attacks.

20.4 MICROBIOTA AND RESIN BEES

It is also unclear whether stingless bees utilize resin as a defense mechanism against bacterial or fungal diseases. A propolis envelope has been identified as a means of reducing the severity of numerous hemolymph diseases affecting *A. mellifera*. Additionally, when treated with *Ascosphaera apis*, the organism responsible for larval diseases such as chalkbrood, colonies exhibit an increase in the amount of resin. Furthermore, it has been demonstrated that honeybee colonies may utilize resin as a form of self-medication in response to certain infections. Resin derived from stingless bees has been shown to prevent the growth of a diverse range of microorganisms, and bees have been observed to exhibit behaviors that are comparable to this phenomenon [46, 47]. Despite this, there have been no documented instances of harmful bacteria in stingless bee colonies. The only exception to this is the bacterium *Lysinibacillus sphaericus*, which is responsible for the degeneration of broods. Consequently, the antibacterial activity of resin is typically evaluated against human pathogens. Furthermore, the effect of resin on disease-causing microorganisms in bee colonies has opened a new front of resin research.

20.5 RESIN FORAGING

The role of resin in the development and defense of stingless bee nests is well established; however, the question of how these bees obtain this material remains unanswered. A modest amount of research has been conducted on the collection and foraging of resin by stingless bees. However, studies examining foraging behavior in general have revealed a multitude of behavioral aspects exhibited by these bees. In the case of numerous species, including *Tetragonula minangkabau*, *Heterotrigona itama*, *Trigonella moorei*, *Melipona bicolor bicolor*, *Trigona sapiens*, and *Trigona hockingsi*, resin foragers represent less than 10% of the total foraging force. In contrast, pollen foragers are more prevalent than resin foragers in certain species, including *Melipona bicolor*, *Trigona iridipennis*, and *Melipona fasciculata*. Conversely, resin foragers are more numerous than pollen foragers in *Melipona asilvai*. These species collect vast amounts of resin. For example, resin foragers in *Tetragonula carbonaria* have been observed to account for up to 50% of the foraging force [48]. During the period in which *Tetragonilla collina* constructs its nests, it has been observed that up to 90% of the foragers return with resin. This is in contrast to *A. mellifera*, where resin foragers account for only 1–3% of the entire foraging team. At the level of the colony, the frequency of resin foraging is influenced by a number of environmental circumstances, including temperature, light intensity, humidity, and the availability of resources. Additionally, colony conditions, such as population size and developmental stage, play a role. The impacts of these factors vary depending on the species. For example, resin foraging activity varies seasonally for certain species, including *Trigona iridipennis*, *Melipona asilvai*, *Melipona bicolor schencki*, and *Melipona colimana*. In contrast, resin foraging behaviors remain consistent throughout the year for other species, such as *Melipona fasciculate* [49, 50].

20.6 BIOTIC AND ABIOTIC FACTORS INFLUENCING RESIN COLLECTION

A number of factors may contribute to seasonal fluctuations in resin collection, including the availability of resources, the intensity of pathogen pressure, and the developmental stage of the colony. These factors have not yet been thoroughly investigated. Certain species, such as *Melipona bicolor bicolor*, exhibit a correlation between resin foraging and colony strength and diameter. Resin collection is particularly vigorous in the early phases of colony growth for some individuals, although it may gradually decrease once the structural components of the nest have been created. Increased resin foraging may be a consequence of pathogen pressure for species that employ resin-daubing or resin barriers as a means of defense. Additionally, there is some evidence to suggest that species such as *Plebeia emerina* hoard resin in preparation for periods of resin scarcity. Alternatively, it could be done in preparation for increased pressure from predators or parasites. Worker bees are the primary agents responsible for the collection of resin. The hind tibiae of male stingless bees were observed to contain both resin and pollen loads. Given that male bees were not observed depositing their loads in storage pots or on other nest structures, the scientists concluded that male bees do not make a significant contribution to resin foraging. It is currently unknown whether resin-bearing males contribute to colony function or how they do so. Additionally, there is a lack of clarity on the elements that motivate a forager to prefer resin foraging over nectar or pollen foraging on an individual level. Biesmeijer and Tóth [51] observed that during the course of their foraging career, 50% of the foragers of *Melipona beecheii* specialized in only one resource, while the other 50% rotated between pollen, nectar, resin, and mud. This result is in agreement with the findings of Inoue et al. [47], who investigated the foraging behavior of three species of Sumatran stingless bees. Moreover, research indicated that approximately 50% of the foragers were specialists in a single material.

For instance, the foragers of *A. mellifera* begin collecting resin when they become aware of a requirement for it within the nest. Subsequently, they perform the waggle dance with the objective of recruiting additional resin foragers. However, it remains unclear whether stingless bees respond to stimuli that are analogous to those encountered by other bees, and whether they attract other bees to collect resin. During the process of foraging, stingless bees utilize both visual and olfactory cues to locate and differentiate between resin sources. In particular, species of stingless bees have been observed to focus their attention on specific mixtures of volatile mono- and sesquiterpenes.

This sensory aptitude enables stingless bees to rapidly identify new resin sources and to locate experimentally induced tree wounds. For instance, the seed dispersal mutualism between the eucalyptus tree *Corymbia torelliana* and the stingless bee *Tetragonula carbonaria* is an illustrative example of a situation in which certain resin sources are highly preferred. Even minor experimental modifications to a resin odor, such as changes in single mono- or sesquiterpenes, resulted in a reduction in visitation [44, 45]. This evidence indicates that stingless bees are capable of acquiring complex fragrances and reacting to a multitude of compounds present within the resin bouquet. Moreover, it is conceivable that they exhibit greater selectivity in their choice of resin sources than floral sources. In order to collect resin from plant buds, leaves, flowers, or bark, stingless bees use their mandibles after they have located a source of resin. They then employ tarsi and basitarsi on their front and middle legs in order to load this sticky substance onto their corbiculae. Until they have accumulated a considerable resin burden, which they then transport back to the nest, they continue to repeat this process. Stingless bees bite plant tissues and then collect the resin that oozes from the incision they create. This causes the plant to produce resin, and the foragers are able to milk an active resin source for days or weeks at a time. Furthermore, the practice of resin foraging can potentially cause harm to plant tissues. Consequently, certain stingless bee species, such as *Trigona fuscipennis* and *Trigona nigerrima*, have been regarded as pests for agricultural crops. However, the majority of stingless bee species collect resin in groups and actively guard the resin resources in their nest cavities.

It has been observed that certain species engage in violent conflict over resin, steal cerumen from other nests, or gather resources from nests that have been abandoned. Howard posits that these behaviors suggest that resin is a valuable resource for a great number of stingless bee species. Furthermore, he hypothesizes that the availability of resin resources is probably a factor that restricts the formation of colonies. Some resin-producing plants are evidently preferred by stingless bees, while other plants are ignored by these bees. However, the criteria that determine the resin preferences of stingless bees remain unclear. The antimicrobial or repellent characteristics of certain plant resins may influence the choice of plants by bees. It is possible that morphological features also influence the resources that each species is able to consume. For instance, certain small species, such as *Trigona jatiformis*, search for resin sources that are too small to be seen with the human eye. The resin acquired by these species is only identifiable once it has been deposited in the bees' corbiculae. It is probable that larger bees will ignore small resin sources, but they may be more likely to use their mandibles to cause damage to plants in order to stimulate resin flow. It is not uncommon for smaller bees to utilize the resin sources tapped by larger species. They accomplish this by either collecting resin in conjunction with the larger bees, attempting to displace them, or waiting until the larger bees have vacated the resin source. Thus far, these behaviors have been analyzed from the perspective of competition. However, the interdependence that is latent in these relationships is not only fascinating but also noteworthy. The dependence of certain bee species on others for access to resin resources may have implications for the conservation of stingless bees.

20.7 RESIN HANDLING

The workers who collect resin return to the hive and empty their corbiculae either independently or with the assistance of their legs. They subsequently combine the wax with the brightly colored resin loads, resulting in the formation of cerumen. In other nest constructions, resin loads are mixed over time, resulting in the oxidation of resin terpenoid compounds, which darken in hue and color. It is evident that bees are highly cautious when unloading and processing resin, as they are prone to being caught. The time required for *Plebeia lucii* and *Frieseomelitta varia* foragers to discharge resin back at the hive was found to be between 7 and 13 minutes, as reported in references [52] and [53]. The majority of this time was spent cleaning their tarsi of resin debris. In addition, bees use a sticky material to immobilize and kill their victims. No evidence suggests that stingless bee body parts are anti-adhesive. Scientists compared the mandibles of the stingless bee *Tetragonisca angustula* with the invader ant *Camponotus sericeiventris*'s trombater using electron microscopy and adhesive force experiments. It was found that resin sticks better to the smooth bee mandible than the scaled ant's trochanter. It appears that the secretion of nectar or other substances is a necessary adaptation for stingless bees, as it allows them to reduce the amount of resin on their mandibles. The capacity of bees to synthesize lubricating chemicals and safeguard against adhesive dangers is contingent upon their age and physiological maturation. For instance, in species *T. angustula*, *Plebeia emerina*, and *Trigona (Hypotrigona) grihodoi*, the more mature workers are responsible for resin collection and transportation. However, at a specific developmental stage, *P. emerina* workers produce secretions that maintain propolis viscosity and enable bees to manipulate it without becoming entrapped. This action is carried out by salivary glands and intramandibular glands [54–56].

It is conceivable that only bees that maintain viscous propolis reserves or deposit resins within the hive will develop the head salivary and intramandibular glands in later life, as there are some stingless bees that do not develop these glands throughout their life. Some evidence suggests that *Melipona quadrifasciata* has a genetic basis for "propolis preparation," which is referred to as resin manipulation. The young bees, which were between 1 and 5 days old, were tagged and transferred into three distinct observation hives. Each experimental hive contained bees from all ten source colonies. For a period of 35 days, the bees' actions were monitored. It was observed that bees from particular source colonies were significantly more likely to participate in the manufacture of propolis, despite the fact that resin foraging was common across all members of colonies [57]. The experimental colonies were composed exclusively of workers belonging to a single age cohort. In accordance with the needs of the colony, workers in stingless bee colonies exhibit a certain degree of flexibility, which allows them to alter their activities. The absence of older bees in hives may influence the behavior of younger bees in handling resin. This is because resin handling occurs at a late stage in the lives of several species of bees. In the event that this is the case, the higher incidence of propolis preparation that was identified in bees from specific source colonies may imply higher degrees of plasticity. It is important to note that the correlation does not necessarily reflect a genetic predisposition to resin handling; rather, this option should be investigated further.

20.8 RESIN INFLUENCE ON STINGLESS BEE CUTICULAR CHEMISTRY

Furthermore, the presence of resin within the nest area of certain species of stingless bees has the potential to influence the morphological traits of the bees themselves. The outermost layer of the cuticle of an insect is composed of lipid molecules, which are responsible for a wide variety of functions. The outermost layer of the insect cuticle is composed of lipids. In addition to their role in identifying nestmates and other forms of communication, these chemicals play a part in the defense of insects against potential threats, such as abrasion, desiccation, and predators. The term "cuticular chemical profile" refers to a collection of molecules that include both polar and non-polar molecules from bees and other insects. Some examples of these molecules include n-alkanes, alkenes, and methyl-branched alkanes. In addition to alcohols, esters, ketones, aldehydes, and oxidized terpenes, these substances also comprise a variety of additional compounds. These compounds may be produced by the individual in question or received through environmental causes. Both alternatives are possible. Certain kinds of stingless bees produce unique cuticular chemicals by utilizing resin as a source [58]. Terpenoids, which encompass mono-, sesqui-, and triterpenes, among numerous other terpenoid types, are included in these substances. Not only do stingless bees represent the only known group of social insects that enhance their cuticular chemical profile with compounds derived from resin, but they are also the only group of insects that have been identified as doing so. To the best of our knowledge, no studies have been conducted on these topics in relation to *A. mellifera*. It appears that numerous separate lineages of stingless bees have independently evolved this feature over the course of their evolutionary history. In the majority of cases, this phenomenon is not observed in more fundamental genera. However, there is at least one instance in which it is observed. This feature can be observed in genera that have undergone a more extensive evolutionary development. The consumption of resin-derived compounds results in species-specific terpenoid profiles that are consistent throughout a wide range of geographic regions, despite the fact that different species commonly consume many of the same resin sources [52–54].

It is highly probable that these substances originate from the resin present in the environment of the nest. Furthermore, the chemical profiles of the nest entrance and the cuticular region of a multitude of species of stingless bees are identical, lending credence to this hypothesis. Leonhardt et al. [59] proposed that a filter mechanism of some kind must be able to distinguish between molecules that can be taken in and those that cannot, which is necessary for the mechanism to be considered effective. A comparison of the chemical composition of resin collected by researchers directly from resin wounds with that of resin obtained from the corbiculae of stingless bees revealed that they are chemically identical [60]. Both types of resin are derived from the same source. In order for resin to undergo the requisite modifications, it is essential that this process occur within the hive. It is likely that different species of stingless bees possess distinct enzymes or microbial partners that alter the resins they carry. This would permit the species-specific selective absorption of terpenoid compounds to be achieved. Alternatively, the molecules that bind to the bee could be determined by genetically determined, species-specific alterations in the chemistry of the epidermis. To gain a deeper insight into the influence of resin-derived compounds on the cuticular chemical profile of different species of stingless bees and to ascertain the implications this has for the functioning of the colony, further research is required [61].

A resin-enriched cuticular chemical profile has been identified as a promising candidate for the protection of bees from predators and the reduction of interspecific animosity, which, in turn, facilitates the aggregation of nests. These potential benefits have been demonstrated by studies conducted in recent years. Some species of stingless bees possess a cuticle coated with resin, which confers upon them the ability to resist predation from other insects. In addition to the effects of the genetically determined repellant compounds that the bees produce on their own, these qualities are also present. A study comparing two species observed the repellent properties of cuticular terpenoids: *Tetragonula carbonaria*, a bee that collects a significant amount of resin, of which its cuticular compounds are derived from resin to the extent of 50%, and Austroplebeia australis, a bee that collects only a small amount of resin, of which its cuticular compounds are derived from resin to the extent of 1%. Both species are referred to as "cuticular terpenoids." The results of behavioral experiments indicate that the *T. carbonaria* bees with a high resin content were found to be able to successfully repel predator ants, whereas bees with a low resin content were not successful in doing so. It was demonstrated that the ants exhibited a decreased preference for the bees that had been washed, indicating that the compounds formed from resin and located on the bees' cuticle may be responsible for their repellent properties. It is also possible that terpenes produced from resin and located on the cuticle may play a role in facilitating nest aggression. Although it is possible that these compounds mask the chemical differences between different species of bees, it is also plausible that they contribute to a reduction in the aggression between different species. In a previous study, the researchers examined the aggressive behaviors of bees from different nest aggregations, including those that were not aggregated. The researchers observed that the level of hostility was lower between bees from connected colonies [59, 62].

The idea that the researchers came up with was that the presence of resin-derived terpenoids, and more specifically, sesquiterpenes, is responsible for the significant reduction in aggressive behavior [59, 61, 63, 64]. The chemical profile of *Tetragonula melanocephala*, a stingless bee whose cuticle does not contain sesquiterpenes, was subjected to experimental manipulation, and the researchers discovered that the application of either pure sesquiterpenes or an extract derived from the sesquiterpene-rich cuticle of *Tetragonula collina*, an unusually peaceful bee, led to a reduction in aggression in non-nestmates. This was the result of the researchers' experimentation. Sesquiterpenes may be able to assist nesting aggregations in tropical habitats, according to some hypothesis, which is supported by these findings; nevertheless, additional research is required to identify the precise role that resin-derived compounds play in mediating complex interactions between and within species. It is believed that this profile has an effect on the recognition of nestmates, while the effect of resin on nestmate recognition is not as readily apparent. It is evident that resin-derived chemicals have a role in the cuticular chemical profile of a great number of species of stingless bees, and it is also evident that this profile plays a role in the recognition of nestmates [62].

The results demonstrated that the acceptance rates of *Tetragonisca angustula* workers decreased regardless of the material source when exposed to extracts made from nestmate and non-nestmate resin or wax. This was observed to be the case regardless of the material source and treatment. In an experiment, resin was transferred from the main stores of donor colonies to recipient colonies to assess whether bees use in-hive resin stores as a reference for recognition signals. With regard to the aforementioned study, the results indicated that the acceptability of nestmates from in-donor colonies decreased following the interference, whereas there was no change in the approval of non-nestmates by donors. Conversely, the acceptability of nestmates in donor colonies remained unchanged. Additionally, it was observed that the guards exhibited a general lack of certainty, accompanied by an increase in the acceptance of individuals who were not nestmates in the recipient colonies. The researchers concluded that wax and resin do not play a role in the recognition of nestmates in *T. angustula*, based on the results obtained. Conversely, it is possible that these findings were influenced by the deliberate transfer of molecules derived from resin. Instead of being exposed to raw wax or resin, this would imply that the bees were only exposed to a hexane extract that was enhanced with resin. Consequently, had behavioral trials been conducted relatively rapidly after the relocation of resin stores from donor colonies to recipient colonies, the results would have been confounded by an increase in defensive behavior, which would have obscured the findings regarding nestmate recognition. Further research is required to ascertain whether there is a potential correlation between the chemical compounds utilized in nests and the capacity of stingless bees to recognize their nestmates [65–68].

20.9 MICROBIOTA ASSOCIATED WITH STINGLESS BEES

The sequencing of microbial communities associated with stingless bees and their nests is becoming an increasingly popular topic of study. This interest extends to the investigation of the usefulness of these communities in their respective functions. A significant study has been conducted on the antibacterial properties of stingless bee resin in relation to human health. Moreover, a considerable body of research on stingless bees indicates that resin likely plays a role in shaping the microbiota found within the nest. This is a widely acknowledged phenomenon, although it has been demonstrated that bacteria and fungi play a significant role in the functioning of stingless bee colonies, and resin is assumed to play a significant role in the maintenance of microbial balance. Moreover, the question of whether resin modulates the microbial communities associated with stingless bees and their nest spaces remains unresolved [69]. This is despite the assumption that resin plays a significant role in maintaining microbial balance. Recent research has shed light on the intricate relationships between resin, microbial communities, and bees. Findings from colonies of *A. mellifera* indicate that a propolis envelope plays a stabilizing role in the microbial communities present in bee bellies and on their mouthparts [70]. This was observed by the researchers who conducted a study of the findings of the colonies. It is hypothesized that the propolis envelope stimulates the growth of beneficial bacterial partners, in addition to the observed reduction in the expression of pathogenic or opportunistic microorganisms. In contrast, the role of resin in the establishment of the microbiota associated with stingless bees remains unclear. The microbiomes in question include the cuticular, stomach, whole-bee, and nest microbiomes. A recent study examined the microbiomes of four different species of stingless bees, including *Frieseomelitta varia*, *Melipona quadrifasciata*, *Tetragonisca angustula*, and *Trigona spinipes*. The objective of this study was to examine the bacterial communities associated with the interior nest surfaces of each of these bees. It was found that the diverse materials utilized in the construction of their nests, including clay, resin, wax, and feces, contributed to the observed differences in these bacterial communities. This was the conclusion reached after the investigation was conducted. Furthermore, it is unclear which surfaces were swabbed for germs, and regrettably, this study did not provide a comprehensive examination of the nest architecture for the species under investigation.

Despite these limitations, the findings of this study underscore the importance of comprehending the intricate microbial ecosystems that are present within the nest of the stingless bee, as well as the overall function that the materials utilized to construct the nest, such as resin, play in shaping those ecosystems within nests. Consequently, the significance of this study cannot be overstated. Through the use of sequencing, Leonhardt and Kaltenpoth [71] were able to describe a microbiome that was connected with three species of sympatric Australian stingless bees. In the nests of two of these species, *T. carbonaria* and *T. hockingsii*, substantial quantities of resin are incorporated, whereas the nests of the third species, *Austroplebeia australis*, contain essentially no resin at all. Following the extraction of DNA from six worker bees collected from ten different colonies, a comparative investigation of the microbiomes of complete bees was conducted. This revealed that microbial communities exhibit traits that are distinct to individual species. However, in order to determine whether these alterations may be attributed to the presence or absence of resin, it is necessary to conduct tests on more species. Consequently, comparisons between microbial communities associated with high-resin and low-resin colonies within the same species, as was conducted in research involving *A. mellifera*, could prove to be enlightening. This is due to the fact that numerous additional species-specific elements, including genotype, nutrition, external environment, and nest construction materials, have the ability to alter microbial communities due to their specificity to the species [69].

A second study was conducted with the objective of evaluating whether or not a resin-rich environment imparts antimicrobial characteristics to the stingless bee cuticle. The rate of mold growth on the bodies of some of the same high-resin (*T. carbonaria*) and low-resin (*A. australis*) bees was also assessed. During the course of this research, it was observed that the rate of mold growth remained consistent across the various species, a finding that was deemed significant. Conversely, Roubik [18] postulates that the resin-deficient *A. australis* has developed compensatory physiological traits (such as greater secretion of self-produced antimicrobials) in order to replenish resin resources over time. This hypothesis is supported by the fact that it is probable. If this is indeed the case, then the cuticle of species with high and low resin content should contain antimicrobial substances that inhibit mold growth. The question of whether these compounds are created by the organism itself or received from the surrounding environment is a key differentiating factor between the two species [70, 72, 73].

Additional research is necessary to gain a better understanding of the influence that resin has on the bacteria and fungus that are naturally present on the cuticle of stingless bees and within the nests of stingless bees. Furthermore, it is essential to conduct research to ascertain the relationship between the presence of resin and the self-produced antimicrobial chemicals by the bees. With regard to the significance of microbial companions to the operation of colonies, the mutualism that exists between the stingless bee *Scaptotrigona depilis* and a fungus that belongs to the genus *Zygosaccharomyces* is perhaps the most striking evidence of the importance of microbial associations. A latent form of this fungus is discovered in the cerumen of *S. depilis* colonies with a very low activity level. Upon contact with the liquid larval food contained within the brood cells, the fungus undergoes a phase of growth. During this phase, the fungus extends visible white mycelia from the cell wall of the brood in the direction of the food supply for the larvae. Upon initial discovery, it was postulated that these mycelia were detrimental; however, subsequent studies demonstrated that they actually produce steroid precursors, which are essential for the pupation of *S. depilis* larvae. A substantial body of evidence indicates that the presence of resin plays a role in the expansion of Zygosaccharomyces. This is because resin is a vital component of cerumen and because it prevents the growth of certain microbes, although not all of them [74]. This is because resin is an organic substance. The multitude of stingless bee–microbe interactions that could prove to be essential to the operation of the colony is considerable, and this particular instance of mutualism is but one evident illustration of the many conceivable microbe–bee interactions. This evidence, in conjunction with recent discoveries that demonstrate the role of propolis in shaping the gut and mouthpart microbiomes of *A. mellifera*, suggests that resin may help stabilize and/or support microbial communities that may prove to be essential to the function of stingless bee colonies. Nevertheless, further research is required to substantiate these findings and to elucidate the precise role of propolis in the gut and mouthpart microbiomes of *A. mellifera*.

20.10 RESIN APPLICATIONS IN THE PHARMACEUTICAL INDUSTRY

20.10.1 Antibacterial Properties of Plant Resin

The utilization of plant resins in the pharmaceutical industry preceded the advent of modern antibiotics. Plant resins have been employed in traditional medicine for millennia to treat a multitude of ailments. Research conducted by Bassolé and Juliani [75] revealed that plant resins contain a diverse array of secondary metabolites, the primary function of which is to defend plants from a multitude of predators and microbial pathogens. A variety of chemical substances, including alkaloids, phenols, and terpenes, are responsible for the antimicrobial activity exhibited by plant resins. The World Health Organization (WHO) posits that medicinal plants and the resins that they produce may be the most effective source for acquiring a wide range of medications. The most extensive research has been conducted on propolis and essential oils derived from aromatic plants, which are both natural sources of antimicrobial compounds that are utilized to treat infectious disorders.

Hive bees (*Apis mellifera* L.) collect propolis from tree buds, leaves, and bark. Bees combine resinous plant material with wax and enzymes to produce propolis. A significant proportion of the components in propolis resin, up to 50%, are flavonoids, phenolic acids, and their esters [76–79]. The antimicrobial, antioxidant, and antifungal properties of propolis's phenolic components have been linked to a range of health benefits. A considerable number of studies have identified the active components of plant resins and propolis as having antibacterial properties. Bioautography, a microbiological screening method in conjunction with planar chromatography, is employed to identify bioactive chemicals in crude extracts [80, 81].

The diversity of propolis samples is attributable to a number of factors, including their geographic origin, the plant from which they are derived, the climate in which they are produced, the season during which they are collected, and the species of bee from which they are harvested. A number of studies have established a correlation between propolis samples and their botanical origin, based on the resin collection area and plant sources [82]. A comparison of the chemical composition of propolis with that of the postulated plant source represents the most effective diagnostic tool for determining its botanical origin. Propolis is produced by poplars and conifers. In temperate regions, including Europe, Asia, North America, and the continental section of Australia, propolis is derived from the buds of the *Populus* spp. According to some research, resin has a high concentration of flavonoids, phenolic acids, and derivatives of these compounds [83–85]. A secondary supply of this form of propolis is also provided by multiple plants. The honeybee, *Apis mellifera*, gathers antimicrobial plant resins from its surroundings and stores them in its nests as propolis. This substance has been found to have multiple benefits, including the suppression of bacterial disease symptoms. In order to establish a link between the advantages that bees obtain from propolis and specific resinous plants, several studies identified and determined the botanical source of propolis compounds that are effective against bee pathogens using the bioassay-guided fractionation method. This involved testing the compounds against the bacterium *Paenibacillus larvae*, which is the main causal agent of American foulbrood. A total of 11 dihydroflavonols were extracted from resins, among which pinobanksin-3-octanoate was identified. This compound, along with five additional 3-acyl-dihydroflavonols, exhibited inhibitory efficacy against both *P. larvae* and *Ascosphaera apis*, which is the fungal cause of chalkbrood. A correlation was observed between the size of the acyl group and the antimicrobial action. Longer acyl groups were found to enhance the activity against *P. larvae*, whereas shorter acyl groups were found to enhance the activity against *Apis* [83–85]. Ultimately, it was demonstrated that the resinous material (3-acyl-dihydroflavonols) originated from several social forestry plants. This discovery revealed that these compounds are highly advantageous in the control of several brood and colony diseases [86–89]. Figure 20.2 illustrates the high-performance biomedical applications of resin and an eco-friendly thermosetting polyester resin derived from unsaturated polyesters (UPs) and related cross-linked resins (UPRs). The aforementioned materials utilize bio-based resinous building blocks derived from 2,5-furandicarboxylic acid (FDCA) [90].

As illustrated in Figure 20.3, the constructed organic polymer (polyelectrolyte) resin follows a specific mechanism of action that facilitates the release of the medication from the ion exchange resin, which is based on multiple ionizable functional groups that are highly effective in making organic resin. The ion exchange resin (IER) is a cross-linked synthetic high-molecular-weight, water-insoluble substance that is typically white or yellowish in color. The versatility of IER as drug delivery vehicles has attracted significant attention from scientists working in the pharmaceutical industry. IERs have been demonstrated to be equally suited for drug delivery strategies, such as controlled release, transdermal, nasal, topical, and taste masking. The most significant disadvantage of continuous release over extended release is dose dumping, which increases the risk of toxicity [88, 91–97]. The superior drug-retaining qualities and the ability to prevent dose dumping of ion exchange resins (IERs) have led to their significant role in the development of controls or sustained-release systems. IERs have been utilized in the pharmaceutical and medical industries for the purpose of disguising the taste of drugs or achieving regulated release of the substance. Drug resin

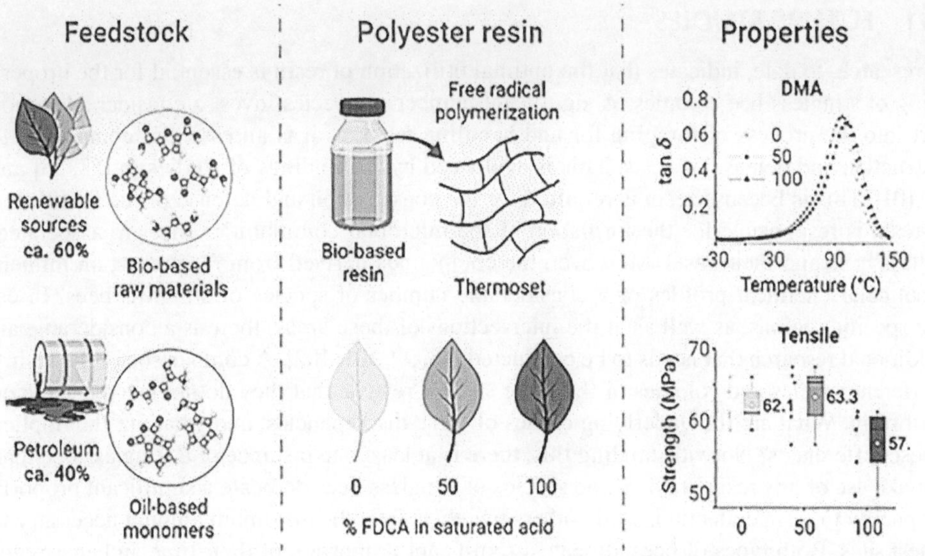

FIGURE 20.2 An eco-friendly thermosetting polyester resin comprised of unsaturated polyesters (UPs) and related cross-linked resins (UPRs) used for high-performance biomedical applications. The bio-based resinous building blocks used in these materials include 2,5-furandicarboxylic acid (FDCA).

Source: Figure adapted from the work of Hofmann et al. (2022) [90].

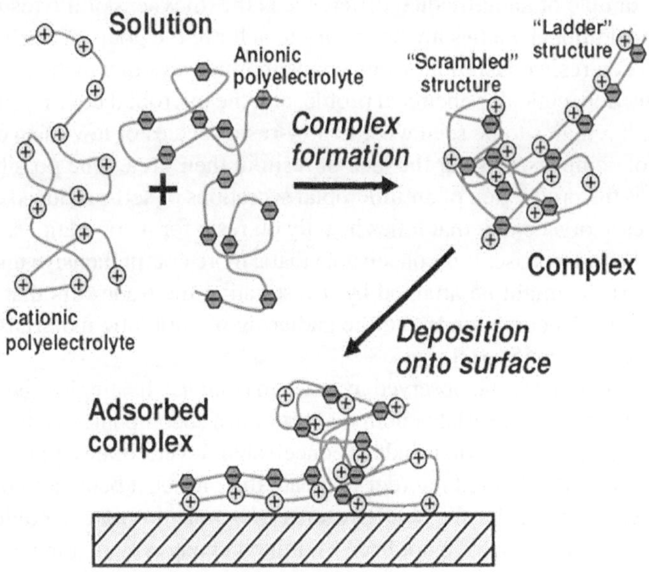

FIGURE 20.3 The mechanism of action that facilitates the release of the medication from the ion exchange resin.

Source: Figure adapted from the work of Sivaneswari et al. [87]

complexation is a process that transforms the drug into an amorphous state, which ultimately results in enhanced drug solubility. This process has been utilized in multiple studies for the purpose of drug delivery at the intended site of action [89, 98, 99]. The pharmaceutical industry makes the most extensive use of sulfonated and carboxylic resins that have a polystyrene backbone [98–101].

20.11 FUTURE STUDIES

The research, to date, indicates that the optimal utilization of resin is essential for the proper functioning of stingless bee colonies. A significant number of species invest a considerable amount of effort into the process of foraging for and handling resin, as it is an essential component for the construction and defense of nests. This is evidenced by the findings of studies in [52, 53] and [98, 100, 101]. This is because resin is required for the construction and defense of the nest. It is likely that resin is responsible for the formation of the microbial communities that are associated with stingless bees and their nests. Moreover, the compounds derived from resin exert an influence on the cuticular chemical profiles of a considerable number of species of stingless bees. In each of these specific regions, as well as at the intersections of these areas, there is a considerable amount of additional research that needs to be conducted [98, 99, 101, 102]. A comparison of the resin usage of different species and colonies of the same species reveals that they do not use the same quantities of resin. What are the underlying causes of these discrepancies, and what are the implications of these differences? Notwithstanding this, there is at least one instance of a stingless bee that does not make use of any resin at all. Some species of stingless bees dedicate a significant proportion of their energy to resin collection, while others merely collect the minimum amount necessary to create nest sites. Both types of bees allocate a significant proportion of their time and energy to resin collection [97, 103–106].

It is possible that variations in resin utilization may occur among species, despite the fact that several of the most significant characteristics, such as species, location, and hive structure, have not altered. Some researchers have identified three potential causes for individual differences in nest building. These include (1) the age of the nest, (2) the genetics of the bees, and (3) the microenvironment, which encompasses factors such as predators, parasites, and symbionts, such as rain, wind, and sun [71]. One example of an individual difference is the thickness of the resinous batumen that surrounds the nest. Additional studies are necessary to achieve the goals of analyzing the effects of each of these criteria on resin consumption and analyzing the ways in which changes in resin usage influence colony function, cuticular chemical profile, and the microbial communities that are linked with stingless bees. It remains to be seen whether low-resin species or low-resin colonies are physiologically capable of compensating for the lack of resin in their area. One possible mechanism for such compensation is the production of antimicrobial secretions or self-produced cuticular chemical compounds. In the case of a colony that relies heavily on resin for its operations, this is an example of a question that is bound to arise. It is conceivable that a more comprehensive understanding of the evolution of social insects might be attained by investigating the trade-offs that exist between the production of antimicrobial compounds and the gathering of antibiotic materials. It is conceivable that this may be the case [52–55, 102].

One potential explanation for the observed increase in resin use in stingless bees is the loss of the stinging apparatus and the antibacterial venom that may have accompanied it. It is therefore reasonable to inquire whether such a scenario is indeed conceivable. Invertebrates typically blend materials that have been secreted and collected in order to create their nests, a behavior commonly observed among stingless bees. Conversely, the selective pressures that promote secretion over gathering are not well understood, and they have not been examined in any way in stingless bees of any kind [23, 24, 26, 27, 107]. In addition to the presence and prevalence of resin in the nest area, there are other aspects of the habitat and function of the colony that are influenced by the presence of resin. It is conceivable that molecules derived from resin might make their way into the food sources of stingless bees, including honey and pollen, and thereby provide them with phytochemicals. Another type of resin present throughout the nest is cerumen, which is found in direct contact with the brood and food storage. These phytochemicals have been identified in honey produced by *A. mellifera*, with propolis being the most probable source. It is possible that some of the phytochemicals found in honey produced by stingless bees may also be contributed by resin. What is the relationship between the quantity of resin incorporated by bees into their nest environment and the quantity of resin-derived compounds found in honey? Additionally, the specific varieties of compounds found in honey remain unclear. The introduction of sugar syrup or honey in *A. mellifera* colonies during periods of scarcity, or the removal of surplus resin stores from a colony for the purpose of colony management, is an example of beekeeping procedure that has the potential to impact the health of bees. It is difficult to imagine how these activities could possibly have an effect on stingless bees, since resin-derived phytochemicals are an essential part of their diet [69–72]. As previously stated, the microbiota associated with stingless bees and their nests is a topic that is both extensive and fascinating. It is conceivable that the expansion of this field will provide answers to questions regarding the chemical ecology of stingless bees, in addition to other aspects of colony function. It has been demonstrated that certain species of stingless bees possess resin-derived chemicals that constitute their cuticular chemical profiles [108]. One might hypothesize that these substances exert an effect on the microbiome that populates the cuticular regions of these bees. What impact does this have on the ability of the cuticular chemical profile to deter prospective predators, reduce aggressive behavior, and other effects similar to these? To what extent does the microbiome of the cuticle influence the chemical profile of the cuticle, and how does this influence manifest itself? It is possible that resin plays a role in the formation of the microbiota associated with stingless bees and the nests they construct. What aspects of the colony's operation are affected by this influence? There are numerous intriguing parallels between honeybees and stingless bees in terms of their utilization of resin. However, there are also notable distinctions between the two types of bees. In each of these study systems, these areas of comparison have the potential to provide information that can be used to inform future research [109, 110].

With regard to chemical profiling in stingless bees, it is possible that the ingestion of resin may serve as a component in the social immunity mechanism. It is unclear whether the application of resin within the confines of the stingless bee nest eliminates the possibility of the development of pathogenic microorganisms. It is possible that the presence of resin may play a role in the modulation of the immune system of stingless bees, just as it does for *A. mellifera*. What are the ways in which the expression of the immune system varies between species that have high levels of resin and those that have low levels of resin? These subjects could be investigated in a future study on stingless bees by drawing on the findings of studies that are currently being conducted on honeybees [111]. In contrast, research on honeybees could benefit from insights gained from studies on stingless bees. This could include investigating whether foragers of *A. mellifera* use olfactory cues to locate resin sources, as is observed in stingless bees, and whether the presence of a propolis envelope influences the cuticular chemical profile of *A. mellifera*. Both of these questions are pertinent to honeybee research. With regard to the natural history of stingless bees, there is a pressing need for additional research, which is the driving force behind all these concepts. A substantial number of the most significant natural history studies in this field have been conducted over a period of more than half a century and have focused on a limited number of species.

It is not possible to consider these studies to be representative, despite the fact that they are helpful. This is due to the fact that stingless bees exhibit such pronounced interspecific differences and that resin consumption is such a variable trait. Further research is necessary to expand the scope of the fundamental studies that play a pivotal role in the existing theories of resin utilization in stingless bees. This allows for a more comprehensive understanding of the topic. In addition to the aforementioned endeavors, there is a significant opportunity to engage with and follow the leadership of stingless bee keepers, Indigenous people, and custodians of local ecological knowledge who have been in relationship with stingless bees for generations. This opportunity is also available in other areas of development. It is evident that this represents a significant opportunity to undertake further extensive explorations of stingless bees, with the objective of improving human living conditions and maintaining a sustainable ecosystem [111–114].

20.12 CONCLUSIONS

The necessity to gain a deeper comprehension of the role that resin plays in the functioning of stingless bee colonies has become more pressing than ever before. This is due to the fact that the changing beekeeping and land use practices that may result in stingless bees having a reduced capacity to integrate resin into their nest environment are becoming increasingly prevalent. In addition to affecting their cuticular chemical profiles and microbial companions, these modifications may restrict the access that these bees have to a wide variety of resin sources, which may hinder their ability to construct and defend their nests. The spread of stingless bee colonies to monocrop fields for the purpose of providing pollination services has increased in recent decades. In addition, beekeeping enterprises have become more prevalent in recent decades due to the recent decline in pollinators. It is possible that these modifications may have a detrimental effect on bees, in addition to other disruptive impacts. Bees are capable of replacing resin in human-made products, such as wet paint, adhesives, and asphalt. This is a well-documented phenomenon, and the potential long-term impact of such substitutes on the functionality of stingless bee colonies in the context of resin scarcity is a cause for concern. While the significance of resin likely extends beyond its adhesive properties, it is nevertheless worthwhile to consider the extent to which such substitutions may impact resin-dependent organisms. It is possible that the conservation of resin-rich non-floral resources, which would otherwise be overlooked, could be encouraged if a more comprehensive understanding of the importance of resin utilization for the functioning of stingless bee colonies were to be made fully available and confirmed. It is possible that the conservation of stingless bees could be enhanced through the implementation of concentrated conservation efforts, thereby improving their ability to defend themselves against viruses, parasites, and predators. Furthermore, the health of their colonies could be enhanced in ways that were not yet discernible. This is due to the fact that bees rely on a wide variety of resin sources in order to carry out a wide range of activities. Resin has a multitude of applications in the biomedical sciences. The synthesis of UPRs has been successfully achieved through the utilization of FDCA and a range of renewable building blocks. Furthermore, nearly 99.8 wt% of the linear UPs were synthesized from renewable sources. A diluent consisting of a mixture of two reactive monomers, styrene and HEMA, is employed to dilute the UPRs. In comparison to conventional UPRs, this combination enables the production of UPRs that are more environmentally and human health–friendly. The resinous medical products exhibit enhanced mechanical and thermomechanical properties across a broad range of viscosities, which are essential for fiber impregnation procedures. These characteristics render UPRs as high-performance materials with distinctive characteristics and environmental safety. The utilization of these materials may or may not be advantageous to the composites sector. However, the task of increasing the number of sustainable UPRs and their diverse categories is a critical one. Further study is required to address these complex issues, which involve identifying the optimal balance between performance, sustainability, and processing capabilities. The development of reactive monomers that can be used as diluents, which will completely replace styrene while maintaining excellent mechanical, thermomechanical, and processing capabilities, and bio-based catalysts (initiators and accelerators) that will enable the production of fully bio-based UPRs should be the focus of future research. In this context, it is of paramount importance to assess and expand the scope of the extensive research conducted on the application of resin in agriculture and medical sciences. Upon attaining this knowledge, we will be able to recognize the significance of resin in the proper functioning of various economic sectors and the overall health of the environment and people, particularly in light of the current climate change and global warming.

REFERENCES

1. Suryanarayanan, S.; Beilin, K. Milpa-Melipona-Maya: Mayan interspecies alliances facing agribiotechnology in Yucatan. *ACME* **2020**, *19*, 469–500.
2. Shanahan, M.; Spivak, M. Resin use by stingless bees: A review. *Insects* **Aug 2021**, *12*(8), https://doi.org/10.3390/insects12080719
3. Al-Lami, H.S.; Al-Marzok, M.A. Regenerating of consuming ion-exchange resin by graft polymerization for steam boiler wastewater treatment. *Desal. Water Treat.* **May 2021**, *222*, 127–136, https://doi.org/10.5004/dwt.2021.27115. www.deswater.com.

4. Duangphakdee, O.; Koeniger, N.; Deowanish, S.; Hepburn, H.R.; Wongsiri, S. Ant repellent resins of honeybees and stingless bees. *Insectes Soc.* **2009**, *56*, 333–339. [CrossRef]
5. Roubik, D.W. Stingless bee nesting biology. *Apidologie* **2006**, *37*, 124–143. [CrossRef]
6. Halcroft, M.; Spooner-Hart, R.; Neumann, P. Behavioral defense strategies of the stingless bee, *Austroplebeia australis*, against the small hive beetle, *Aethina tumida*. *Insectes Soc.* **2011**, *58*, 245–253. [CrossRef]
7. Nunes, T.M.; Von Zuben, L.G.; Costa, L.; Venturieri, G.C. Defensive repertoire of the stingless bee *Melipona flavolineata* Friese (Hymenoptera: Apidae). *Sociobiology* **2014**, *61*, 541–546. [CrossRef]
8. Greco, M.K.; Hoffmann, D.; Dollin, A.; Duncan, M.; Spooner-Hart, R.; Neumann, P. The alternative Pharaoh approach: Stingless bees mummify beetle parasites alive. *Naturwissenschaften* **2010**, *97*, 319–323. [CrossRef]
9. Drescher, N.; Wallace, H.M.; Katouli, M.; Massaro, C.F.; Leonhardt, S.D. Diversity matters: How bees benefit from different resin sources. *Oecologia* **2014**, *176*, 943–953. [CrossRef]
10. Langenheim, J.H. *Plant Resins: Chemistry, Evolution, Ecology, and Ethnobotany*; Timber Press: Portland, OR, 2003.
11. Leonhardt, S.D.; Baumann, A.M.; Wallace, H.M.; Brooks, P.; Schmitt, T. The chemistry of an unusual seed dispersal mutualism: Bees use a complex set of olfactory cues to find their partner. *Anim. Behav.* **2014**, *98*, 41–51. [CrossRef]
12. Wallace, H.M.; Lee, D.J. Resin-foraging by colonies of *Trigona sapiens* and *T. hockingsi* (Hymenoptera: Apidae, Meliponini) and consequent seed dispersal of *Corymbia torelliana* (Myrtaceae). *Apidologie* **2010**, *41*, 428–435. [CrossRef]
13. Gonzalez, V.H.; Amith, J.D.; Stein, T.J. Nesting ecology and the cultural importance of stingless bees to speakers of Yoloxóchitl Mixtec, an endangered language in Guerrero, Mexico. *Apidologie* **2018**, *49*, 625–636. [CrossRef]
14. Armbruster, W.S. The role of resin in angiosperm pollination: Ecological and chemical considerations. *Am. J. Bot.* **1984**, *71*, 1149–1160. [CrossRef]
15. Leonhardt, S.D.; Zeilhofer, S.; Blüthgen, N.; Schmitt, T. Stingless bees use terpenes as olfactory cues to find resin sources. *Chem. Senses* **2010**, *35*, 603–611. [CrossRef]
16. Gompper, M.E.; Hoylman, A.M. Grooming with *Trattinnickia* resin: Possible pharmaceutical plant use by coatis in Panama. *J. Trop. Ecol.* **1993**, *9*, 533–540. [CrossRef]
17. Simone, M.; Evans, J.D.; Spivak, M. Resin collection and social immunity in honey bees. *Evolution* **2009**, *63*, 3016–3022. [CrossRef] [PubMed]
18. Roubik, D.W. *Ecology and Natural History of Tropical Bees*; Cambridge University Press: New York, NY, 1989.
19. Martins, A.C.; Melo, G.A.R.; Renner, S.S. The corbiculate bees arose from New World oil-collecting bees: Implications for the origin of pollen baskets. *Mol. Phylogenet. Evol.* **2014**, *80*, 88–94. [CrossRef] [PubMed]
20. Simone-Finstrom, M.; Spivak, M. Propolis and bee health: The natural history and significance of resin use by honey bees. *Apidologie* **2010**, *41*, 295–311. [CrossRef]
21. Simone-Finstrom, M.; Borba, R.S.; Wilson, M.; Spivak, M. Propolis counteracts some threats to honey bee health. *Insects* **2017**, *8*, 46. [CrossRef]
22. Mountford-McAuley, R.; Prior, J.; Clavijo McCormick, A. Factors affecting propolis production. *J. Apic. Res.* **2021**, *62*(1), 162–170. [CrossRef]
23. Boongird, S.; Michener, C.D. Pollen and propolis collecting by male stingless bees (Hymenoptera: Apidae). *J. Kans. Entomol. Soc.* **2010**, *83*, 47–50. [CrossRef]
24. Rasmussen, C.; Camargo, J.M.F. A molecular phylogeny and the evolution of nest architecture and behavior in *Trigona* s.s. (Hymenoptera: Apidae: Meliponini). *Apidologie* **2008**, *39*, 102–118. [CrossRef]
25. Camargo, J.M.F.; Pedro, S.R.M. Mutualistic association between a tiny Amazonian stingless bee and a wax-producing scale insect. *Biotropica* **2002**, *34*, 446–451. [CrossRef]
26. Stow, A.; Briscoe, D.; Gillings, M.; Holley, M.; Smith, S.; Leys, R.; Silberbauer, T.; Turnbull, C.; Beattie, A. Antimicrobial defences increase with sociality in bees. *Biol. Lett.* **2007**, *3*, 422–424. [CrossRef]
27. Baracchi, D.; Tragust, S. Venom as a component of external immune defense in Hymenoptera. In *Evolution of Venomous Animals and Their Toxins*; Gopalakrishnakone, P.; Malhotra, A., Eds.; Springer Nature: Dordrecht, The Netherlands, 2017; pp. 213–233; ISBN 9789400764576.
28. Baracchi, D.; Francese, S.; Turillazzi, S. Beyond the antipredatory defence: Honey bee venom function as a component of social immunity. *Toxicon* **2011**, *58*, 550–557. [CrossRef]
29. Milborrow, B.V.; Kennedy, J.M.; Dollin, A. Composition of wax made by the Australian stingless bee *Trigona australis*. *Aust. J. Biol. Sci.* **1987**, *40*, 15–26. [CrossRef]
30. Massaro, F.C.; Brooks, P.R.; Wallace, H.M.; Russell, F.D. Cerumen of Australian stingless bees (*Tetragonula carbonaria*): Gas chromatography-mass spectrometry fingerprints and potential anti-inflammatory properties. *Naturwissenschaften* **2011**, *98*, 329–337. [CrossRef]

31. dos Santos, C.G.; Blochtein, B.; Megiolaro, F.L.; Imperatriz-Fonseca, V.L. Age polyethism in *Plebeia emerina* (Friese) (Hymenoptera: Apidae) colonies related to propolis handling. *Neotrop. Entomol.* **2010**, *39*, 691–696. [CrossRef]
32. Massaro, C.F.; Smyth, T.J.; Smyth, W.F.; Heard, T.; Leonhardt, S.D.; Katouli, M.; Wallace, H.M.; Brooks, P. Phloroglucinols from anti-microbial deposit-resins of Australian stingless bees (*Tetragonula carbonaria*). *Phyther. Res.* **2015**, *29*, 48–58. [CrossRef]
33. Roubik, D.W. Nest and colony characteristics of stingless bees from French Guiana (Hymenoptera: Apidae). *J. Kans. Entomol. Soc.* **1979**, *52*, 443–470.
34. Murphy, C.M.; Breed, M.D. Nectar and resin robbing in stingless bees. *Am. Entomol.* **2008**, *54*, 36–44. [CrossRef]
35. Salatino, A.; Salatino, M.L.F. Scientific note: Often quoted, but not factual data about propolis composition. *Apidologie* **2021**, *52*, 312–314. [CrossRef]
36. Lavinas, F.C.; Macedo, E.H.B.C.; Sá, G.B.L.; Amaral, A.C.F.; Silva, J.R.A.; Azevedo, M.M.B.; Vieira, B.A.; Domingos, T.F.S.; Vermelho, A.B.; Carneiro, C.S.; Rodrigues, I.A. Brazilian stingless bee propolis and geopropolis: Promising sources of biologically active compounds. *Rev. Bras. Farmacogn.* **2019**, *29*, 389–399. [CrossRef]
37. Barth, O.M. Melissopalynology in Brazil: A review of pollen analysis of honeys, propolis and pollen loads of bees. *Sci. Agric.* **2004**, *61*, 342–350. [CrossRef]
38. Wittmann, D.; Camargo, J.M.F. Nest architecture and distribution of the primitive stingless bee, *Mourella caerulea* (Hymenoptera Apidae, Meliponinae): Evidence for the origin of *Plebeia* (s. lat.) on the Gondwana continent. *Stud. Neotrop. Fauna Environ.* **1989**, *24*, 213–229. [CrossRef]
39. Seeley, T.D.; Morse, R.A. The nest of the honeybee (*Apis mellifera L.*). *Insectes Soc.* **1976**, *4*, 495–512. [CrossRef]
40. Drumond, P.M.; Bego, L.R.; Melo, G.A.R. Nest architecture of the stingless bee *Plebeia poecilochroa* Moure & Camargo, 1993 and related considerations (Hymenoptera, Apidae, Meliponinae). *Iheringia* **1995**, *79*, 39–45.
41. Wenzel, F. *Smell and repel: Resin based defense mechanisms and interactions between Australian ants and stingless bees.* Master's Thesis, Universität Würzburg, Fakultät für Biologie, Würzburg, Germany, 2011.
42. Leonhardt, S.D.; Wallace, H.M.; Blüthgen, N.; Wenzel, F. Potential role of environmentally derived cuticular compounds in stingless bees. *Chemoecology* **2015**, *25*, 159–167. [CrossRef]
43. Jones, S.M.; van Zweden, J.S.; Grüter, C.; Menezes, C.; Alves, D.A.; Nunes-Silva, P.; Czaczkes, T.; Imperatriz-Fonseca, V.L.; Ratnieks, F.L.W. The role of wax and resin in the nestmate recognition system of a stingless bee, *Tetragonisca angustula. Behav. Ecol. Sociobiol.* **2012**, *66*, 1–12. [CrossRef]
44. Bankova, V.; Popova, M. Propolis of stingless bees: A promising source of biologically active compounds. *Pharmacogn. Rev.* **2007**, *1*, 88–92.
45. de Freitas, P.V.D.X.; da Silva, I.E.; Faquinello, P.; Zanata, R.A.; Arnhold, E.; e Silva-Neto, C.d.M. External activity of the stingless bee *Melipona fasciculata* (Smith) kept in the Brazilian Cerrado. *J. Apic. Res.* **2020**. [CrossRef]
46. Ferreira Junior, N.T.; Blochtein, B.; de Moraes, J.F. Seasonal flight and resource collection patterns of colonies of the stingless bee *Melipona bicolor schencki* Gribodo (Apidae, Meliponini) in an Araucaria forest area in southern Brazil. *Rev. Bras. Entomol.* **2010**, *54*, 630–636. [CrossRef]
47. Inoue, T.; Salmah, S.; Abbas, I.; Yusuf, E. Foraging behavior of individual workers and foraging dynamics of colonies of three sumatran stingless bees. *Res. Popul. Ecol.* **1985**, *27*, 373–392. [CrossRef]
48. Hilário, S.D.; Imperatriz-Fonseca, V.L.; Kleinert, A.d.M.P. Flight activity and colony strength in the stingless bee *Melipona bicolor bicolor* (Apidae, Meliponinae). *Rev. Bras. Biol.* **2000**, *60*, 299–306. [CrossRef]
49. Layek, U.; Karmakar, P. Nesting characteristics, floral resources, and foraging activity of *Trigona iridipennis* Smith in Bankura district of West Bengal, India. *Insectes Soc.* **2018**, *65*, 117–132. [CrossRef]
50. Leonhardt, S.D.; Heard, T.A.; Wallace, H. Differences in the resource intake of two sympatric Australian stingless bee species. *Apidologie* **2014**, *45*, 514–527. [CrossRef]
51. Biesmeijer, J.C.; Tóth, E. Individual foraging, activity level and longevity in the stingless bee *Melipona beecheii* in Costa Rica (Hymenoptera, Apidae, Meliponinae). *Insectes Soc.* **1998**, *45*, 427–443. [CrossRef]
52. Reyes-González, A.; Zamudio, F. Competition interactions among stingless bees (Apidae: Meliponini) for *Croton yucatanensis* Lundell resins. *Int. J. Trop. Insect Sci.* **2020**, *40*, 1099–1104. [CrossRef]
53. López-Guillón, G.; Chamó-Vázquez, E.R.; Ruiz, J.G.; Barrera, J.F. First report of *Trigona fuscipennis* and *T. nigerrima* chewing tissues of *Jatropha curcas* L. in Southern Chiapas, Mexico. *Southwest. Entomol.* **2019**, *44*, 503–507. [CrossRef]

54. Patricio, E.F.L.R.A.; Cruz-López, L.; Maile, R.; Tentschert, J.; Jones, G.R.; Morgan, E.D. The propolis of stingless bees: Terpenes from the tibia of three *Frieseomelitta* species. *J. Insect Physiol.* **2002**, *48*, 249–254. [CrossRef]
55. Gastauer, M.; Campos, L.A.O.; Wittmann, D. Handling sticky resin by stingless bees: Adhesive properties of surface structures. *An. Acad. Bras. Cienc.* **2013**, *85*, 1189–1196. [CrossRef] [PubMed]
56. Ferreira Grosso, A.; Rolani Bego, L. Labor division, average life span, survival curve, and nest architecture of *Tetragonisca angustula angustula* (Hymenoptera, Apinae, Meliponini). *Sociobiology* **2002**, *40*, 615–637.
57. Waldschmidt, A.M.; Campos, L.A.O.; De Marco, P. Genetic variability of behavior in *Melipona quadrifasciata* (Hymenoptera: Meliponinae). *Braz. J. Genet.* **1997**, *20*, 595–599. [CrossRef]
58. Leonhardt, S.D.; Blüthgen, N.; Schmitt, T. Chemical profiles of body surfaces and nests from six Bornean stingless bee species. *J. Chem. Ecol.* **2011**, *37*, 98–104. [CrossRef]
59. Leonhardt, S.D.; Schmitt, T.; Blüthgen, N. Tree resin composition, collection behavior and selective filters shape chemical profiles of tropical bees (Apidae: Meliponini). *PLoS ONE* **2011**, *6*, e23445. [CrossRef]
60. Leonhardt, S.D.; Jung, L.M.; Schmitt, T.; Blüthgen, N. Terpenoids tame aggressors: Role of chemicals in stingless bee communal nesting. *Behav. Ecol. Sociobiol.* **2010**, *64*, 1415–1423. [CrossRef]
61. Leonhardt, S.D.; Wallace, H.M.; Schmitt, T. The cuticular profiles of Australian stingless bees are shaped by resin of the eucalypt tree *Corymbia torelliana. Austral. Ecol.* **2011**, *36*, 537–543. [CrossRef]
62. Leonhardt, S.D.; Form, S.; Blüthgen, N.; Schmitt, T.; Feldhaar, H. Genetic relatedness and chemical profiles in an unusually peaceful eusocial bee. *J. Chem. Ecol.* **2011**, *37*, 1117–1126. [CrossRef] [PubMed]
63. Martin, S.J.; Shemilt, S.; Cândida, C.B.; de Carvalho, C.A.L. Are isomeric alkenes used in species recognition among neo-tropical stingless bees (*Melipona* spp.). *J. Chem. Ecol.* **2017**, *43*, 1066–1072. [CrossRef]
64. Leonhardt, S.D.; Rasmussen, C.; Schmitt, T. Genes versus environment: Geography and phylogenetic relationships shape the chemical profiles of stingless bees on a global scale. *Proc. R. Soc. B Biol. Sci.* **2013**, *280*, 7–9. [CrossRef]
65. Nunes, T.M.; Mateus, S.; Turatti, I.C.; Morgan, E.D.; Zucchi, R. Nestmate recognition in the stingless bee *Frieseomelitta varia* (Hymenoptera, Apidae, Meliponini): Sources of chemical signals. *Anim. Behav.* **2011**, *81*, 463–467. [CrossRef]
66. Jungnickel, H.; Da Costa, A.J.S.; Tentschert, J.; Patricio, E.F.L.R.A.; Imperatriz-Fonseca, V.L.; Drijfhout, F.; Morgan, E.D. Chemical basis for inter-colonial aggression in the stingless bee *Scaptotrigona bipunctata* (Hymenoptera: Apidae). *J. Insect Physiol.* **2004**, *50*, 761–766. [CrossRef] [PubMed]
67. Kwong, W.K.; Medina, L.A.; Koch, H.; Sing, K.W.; Soh, E.J.Y.; Ascher, J.S.; Jaffé, R.; Moran, N.A. Dynamic microbiome evolution in social bees. *Sci. Adv.* **2017**, *3*, 1–17. [CrossRef] [PubMed]
68. Cerqueira, A.E.S.; Hammer, T.J.; Moran, N.A.; Santana, W.C.; Kasuya, M.C.M.; da Silva, C.C. Extinction of anciently associated gut bacterial symbionts in a clade of stingless bees. *ISME J.* **2021**, 2–5. [CrossRef]
69. Ngalimat, M.S.; Abd Rahman, R.N.Z.R.; Yusof, M.T.; Amir Hamzah, A.S.; Zawawi, N.; Sabri, S. A review on the association of bacteria with stingless bees. *Sains Malays.* **2020**, *49*, 1853–1863. [CrossRef]
70. Ngalimat, M.S.; Rahman, R.N.Z.R.A.; Yusof, M.T.; Syahir, A.; Sabri, S. Characterisation of bacteria isolated from the stingless bee, *Heterotrigona itama*, honey, bee bread and propolis. *PeerJ* **2019**, *7*, e7478. [CrossRef]
71. Leonhardt, S.D.; Kaltenpoth, M. Microbial communities of three sympatric Australian stingless bee species. *PLoS ONE* **2014**, *9*, e105718. [CrossRef] [PubMed]
72. de Sousa, L.P. Bacterial communities of indoor surface of stingless bee nests. *PLoS ONE* **2021**, *16*, e0252933. [CrossRef]
73. Campos, J.F.; Das Santos, U.P.; Da Rocha, P.D.S.; Damião, M.J.; Balestieri, J.B.P.; Cardoso, C.A.L.; Paredes-Gamero, E.J.; Estevinho, L.M.; De Picoli Souza, K.; Dos Santos, E.L. Antimicrobial, antioxidant, anti-Inflammatory, and cytotoxic activities of propolis from the stingless bee *Tetragonisca fiebrigi* (Jataí). *Evid. Based Complement. Altern. Med.* **2015**, *2015*, 296186. [CrossRef]
74. Al-Hatamleh, M.A.I.; Boer, J.C.; Wilson, K.L.; Plebanski, M.; Mohamud, R.; Mustafa, M.Z. Antioxidant-based medicinal properties of stingless bee products: Recent progress and future directions. *Biomolecules* **2020**, *10*, 923. [CrossRef] [PubMed]
75. Ivica, D.; Ristivojević, P.; Janakiev, T.; Berić, T.; Trifković, J.; Milojković-Opsenica, D.; Stanković, S. Phenolic profiles and antimicrobial activity of various plant resins as potential botanical sources of *Serbian propolis. Ind. Crops Prod. 94*, 856–871, https://doi.org/10.1016/j.indcrop.2016.09.065.
76. Beneke, C.E.; Viljoen, A.M.; Hamman, J.H. Polymeric plant derived excipients in drug delivery. *Molecules* **2009**, *14*, 2602–2620.

77. Lu, E.X.; Jiang, Z.Q.; Zhang, Q.Z. A water-insoluble drug monolithic osmotic tablet system utilizing gum Arabic as an osmotic, suspending and expanding agent. *J. Control. Release* **2003**, *92*, 375–382.
78. Jain, N.K.; Dixit, V.K. Studies on gums and their derivatives as bind agent. *Ind. J. Pharma. Sci.* **1988**, *50*, 113–114.
79. Owen, S.C. Gum tragacanth. In *Handbook of Pharmaceutical Excipients*; Raymod, C.R.; Paul, J.S.; Paul, J.W., Eds.; The Pharmaceutical Press and the American Pharmaceutical Association, Washington, DC, 2003; pp. 654–656.
80. Munday, D.L.; Philip, J.C. Compressed xanthan and karaya gum matrices: hydration, erosion and drug release mechanism. *Int. J. Pharma.* **2000**, *203*, 179–192.
81. Singh, B.; Sharma, V. Influence of polymer network parameters of tragacanth gum based pH responsive hydrogels on drug delivery. *Carbohydr. Polym.* **2014**, *101*, 928–940.
82. Park, C.R.; Munday, D.L. Evaluation of selected polysaccharide excipients in buccoadhesive tablets for sustained release of nicotine. *Drug Develop. Ind. Pharm.* **2004**, *30*, 609–617.
83. Masoumi, A.; Ghaemy, M. Removal of metal ions from water using nonohydrogel tragacanth gum-g-polyamindoxime: isotherm and kinetic study. *Carbohydr. Polym.* **2014,** *108*, 206–215.
84. Sahraei, R.; Pour, Z.S.; Ghaemy, M. Novel magnetic bio-sorbent hydrogel beads based on modified gum tragancanth/graphene oxide: removal of heavy metals and dyes from water. *J. Clean. Prod.* **2017**, *142*, 2973–2984.
85. Haider, A.; Haider, S.; Kang, I. A comprehensive review summarizing the effect of electrospinning parameters and potential applications of nanofibers in biomedical and biotechnology. *Arab. J. Chem.* **2018,** *11*, 1165–1188.
86. Arora, S.K.; Patel, A.A.; Kumar, N.; Chuhan, O.P. Determination of relationship between sensory viscosity rating instrumental flow behaviour of soluble dietary fibres. *J. Food Sci. Tech.* **2016**, *53*, 2067–2076.
87. Ghayempour, S.; Montazer, M. A modified microemulsion method for fabrication of hydrogel tragacanth nanofibers. *Int. J. Biol. Macromol.* **2018**, *115*, 317–323.
88. Ghayempour, S.; Montazer, M.; Rad, M.M. Encapsulation of *Aloe vera* extract into natural tragacanth gum as a novel green wound healing product. *Int. J. Biol. Macromol.* **2016**, *93*, 344–349.
89. Ghayempour, S.; Montazer, M. Ultrasound irradiation based *in-situ* synthesis of star like tragacanth gum/zinc oxide nonoparticles on cotton fabric. *Ultrason. Sonochem.* **2017**, *34*, 458–465.
90. Hofmann, M.A.; Shahid, A.T.; Garrido, M.; Ferreira, M.J.; Correia, J.R.; Bordado, J.C. *ACS Sustain. Chem. Eng.* **2022**, *10*, 3442–3454, https://doi.org/10.1021/acssuschemeng.1c06969
91. Tlak Gajger, I.; Nejedli, S.; Kozarić, Z. The effectof Nozevit on leucine aminopeptidase and esterase-activity in the midgut of honey bees (*Apis mellifera*). *Vet. Med.* **2013a**, *58*(8), 422–429.31, https://doi.org/10.17221/6982-VETMED
92. Tlak Gajger, I.; Tomljanović, Z.; Stanisavljević, Lj. An environmentally friendly approach tothe control of *Varooa destructor* mite and *Nosema ceranaedisease* in Carnolian honeybee (*Apis melliferacarnica*) colonies. *Arch. Biol. Sci.* **2013b**, *65*(4), 1585–1592.33, https://doi.org/10.2298/ABS1304585G
93. Tlak Gajger, I.; Ribaric, J.; Matak, M.; Svecnjak, L.; Kozaric, Z.; Nejedli, S.; Smodis Skerl, I. M. Zeolite clinoptilolite as a dietary supplement and remedyfor honeybee (Apis mellifera) colonies. *Vet. Med.* **2015**, *60*(12), 696–705.34, https://doi.org/10.17221/8584-VETMED
94. Tlak Gajger, I.; Ribarić, J.; Smodiš Škerl, M. Vlainić, J.; Sikirić, P. Stable gastric pentadecapeptide BPC 157 in honeybee (*Apis mellifera*) therapy, to control *Nosema ceranae* invasions in apiary conditions. *J. Vet. Pharmacol. Ther.* **2018 Aug**, *41*(4), 614–621, https://doi.org/10.1111/jvp.12509. Epub 2018 Apr 23. PMID: 29682749.
95. Tlak Gajger, I.; Vlainić, J.; Šoštarić, P.; Prešern, J.; Bubnič, J.; Smodiš Škerl, M. I. Effects onsome therapeutical, biochemical, and immunologicalparameters of honey bee (*Apis mellifera*) exposed toprobiotic treatments, in field and laboratory conditions. *Insects* **2020**, *11*(9), 638, https://doi.org/10.3390/insects11090638.36.
96. Tlak Gajger, I.; Sušec, P. Efficacy of varroacidalfood additive appliance during summer treatment ofhoneybee colonies (*Apis mellifera*). *Vet. Arh.* **2019**, *89*(1), 87–96.38, https://doi.org/10.24099/vet.arhiv.0441
97. Tlak Gajger, I.; Tomljanović, Z.; Petrinec, Z. Monitoring health status of Croatian honey bee coloniesand possible reasons for winter losses. *J. Apic. Res.* **2010,** *49*(1), 107–108, https://doi.org/10.3896/IBRA.1.49.1.1939.
98. Tlak Gajger, I.; Svečnjak, L.; Bubalo, D.; Žorat, T. Control of *Varroa destructor* mite infestations at experimental apiaries situated in croatia. *Diversity* **2020**, *12*, 12, https://doi.org/10.3390/d12010012
99. Tlak Gajger, I.; Kozarić, Z.; Berta, D.; Nejedli, S.; Petrinec, Z. Effect of the herbal preparation Nozevit on the mid-gut structure of honeybees (Apismellifera) infected with Nosema sp. spores. *Vet. Med.* **2011**, *56*(7), 343–350.30, https://doi.org/10.17221/1587-VETMED

100. Tlak Gajger, I.; Kolodziejek, J.; Bakonyi, T.; Nowotny, N. Prevalence and distribution patterns of seven different honeybee viruses in diseased colonies: a case study from Croatia. *Apidologie* **2014a**, *45*, 701–706, https://doi.org/10.1007/s13592-014-0287-0
101. Tlak Gajger, I.; Bičak, J.; Belužić, R. Theoccurrence of honeybee viruses in apiaries in theKoprivnica-Križevci district in Croatia. *Vet. Arh.* **2014b**, *84*, 421–428.29.
102. Nakamura, J.; Seeley, T.D. The functional organization of resin work in honeybee colonies. *Behav. Ecol. Sociobiol.* **2006**, *60*, 339–349. [CrossRef]
103. da Cunha, M.G.; Franchin, M.; Galvão, L.C.C.; de Ruiz, A.L.T.G.; de Carvalho, J.E.; Ikegaki, M.; de Alencar, S.M.; Koo, H.; Rosalen, P.L. Antimicrobial and antiproliferative activities of stingless bee *Melipona scutellaris* geopropolis. *BMC Complement. Altern. Med.* **2013**, *13*, 23. [CrossRef]
104. Tlak Gajger, I.; Vugrek, O.; Grilec, D.; Petrinec, Z. Prevalence and distribution of Nosemaceranae in Croatian honeybee colonies. *Vet. Med.* **2010b**, *55*(9), 457–462.40, https://doi.org/10.17221/2983-VETMED.
105. Ullah, A.; Tlak Gajger, I.; Majoros, A.; Dar, S.A.; Khan, S.; Kalimullah; Haleem Shah, A.; Nasir Khabir, M.; Hussain, R.; Khan, H.U.; Hameed, M.; Anjum, S.I. Viral impacts on honey bee populations: A review. *Saudi J. Biol. Sci.* **2021 Jan**, *28*(1), 523–530, https://doi.org/10.1016/j.sjbs.2020.10.037. Epub 2020 Oct 28. PMID: 33424335; PMCID: PMC7783639.
106. Tlak Gajger, I.; Sakač, M.; Gregorc, A. Impact of thiamethoxam on honey bee queen (*Apis mellifera Carnica*) reproductive morphology and physiology. *Bull. Environ. Contamin. Toxicol.* **2017**, *99*, 297–302, https://doi.org/10.1007/s00128-017-2144-0
107. Berenbaum, M.R.; Calla, B. Honey as a functional food for *Apis mellifera*. *Annu. Rev. Entomol.* **2021**, *66*, 185–208. [CrossRef]
108. Franchin, M.; Da Cunha, M.G.; Denny, C.; Napimoga, M.H.; Cunha, T.M.; Koo, H.; De Alencar, S.M.; Ikegaki, M.; Rosalen, P.L. Geopropolis from *Melipona scutellaris* decreases the mechanical inflammatory hypernociception by inhibiting the production of IL-1β and TNF-α. *J. Ethnopharmacol.* **2012**, *143*, 709–715. [CrossRef]
109. Aparecida Sanches, M.; Pereira, A.M.S.; Serrão, J.E. Pharmalogical actions of extracts of propolis of stingless bees (Meliponini). *J. Apic.* **2017**, *56*, 50–57. [CrossRef]
110. Machado, J.O. Symbiosis among Brazilian social bees (Meliponinae, Apidae) and a species of bacteria. *Cienc. Cult.* **1971**, *23*, 625–633.
111. Hall, M.A.; Brettell, L.E.; Liu, H.; Nacko, S.; Spooner-Hart, R.; Riegler, M.; Cook, J.M. Temporal changes in the microbiome of stingless bee foragers following colony relocation. *FEMS Microbiol. Ecol.* **2021**, *97*, 1–12. [CrossRef]
112. Bahrndorff, S.; Alemu, T.; Alemneh, T.; Lund Nielsen, J. The microbiome of animals: Implications for conservation biology. *Int. J. Genom.* **2016**, *2016*, 5304028. [CrossRef] [PubMed]
113. Kämper, W.; Kaluza, B.F.; Wallace, H.; Schmitt, T.; Leonhardt, S.D. Habitats shape the cuticular chemical profiles of stingless bees. *Chemoecology* **2019**, *29*, 125–133. [CrossRef]
114. Alqarni, A.S.; Rushdi, A.I.; Owayss, A.A.; Raweh, H.S.; El-Mubarak, A.H.; Simoneit, B.R.T. Organic tracers from asphalt in propolis produced by urban honey bees, *Apis mellifera* Linn. *PLoS ONE* **2015**, *10*, e0128311. [CrossRef]

21 Role of Phenolic Compounds in Plants and Their Pharmacological Properties

Showket Ahmad Dar, Ebtesam A. Gashash, Ahmed M. El-Taher, Ayah T. Abdullah Al-Kilani, Kounser Javeed, Wajid Hasan, Yendrembam K. Devi, Naveena Nazim, Ammar Al-Farga, Kangjam Bumpy, I. Yimjenjang Longkumer, Dev Karan Bairwa, Syed Naseem Zaffar, and Rizwan Rashid

21.1 INTRODUCTION

Phenols are composed of one or more hydroxyl groups that are directly connected to an aromatic ring. A hydroxyl group is connected to a chain of carbon atoms and is comparable to an alcohol with an aliphatic structure. Nevertheless, the presence of an aromatic benzene ring has an effect on the hydrogen of the phenolic hydroxyl, and it is the fact that phenols are often considered to be weak acids. Phenolic compounds are defined by their structural characteristics, which include an aromatic ring attached to one or more hydroxyl substituents. These compounds can range from simple single phenolic molecules to highly polymerized compounds. The classification of phenolic compounds is based on the number of carbon atoms present in their molecular structure. This categorization is presented in Table 21.1.

Scientists have discovered that phenolic chemicals are a universal presence in plants, which are secondary metabolites. These compounds are the consequence of the pentose phosphate, shikimate, and phenylpropanoid pathways [11, 12]. The seeds, leaves, bark, and flowers of plants contain substantial quantities of these compounds. To date, scientists have identified over 8,000 distinct phenolic structures. Lignins, melanins, lignans, and tannins are the most notable examples of compounds that are extensively polymerized, ranging from simple molecules with a low molecular weight, such as phenolic acids, flavonoids, and phenylpropanoids, to compounds with the highest degree of polymerization. Flavonoids represent the most prevalent and extensively dispersed subgroup of flavones, isoflavones, and 2,3-dihydro derivatives of flavone, with a total of at least 4,000 molecules having been identified to date [13]. Among the various categories of phenylpropanoids, lignans are of particular interest and possess unique characteristics [14]. Figure 21.1 depicts the representation of shikimate pathway precursors and the movement of aromatic phenolic and amino acids (AAAs) and intermediates across the membranes of plant plastids, as described in reference [15].

Flavonoids represent another intriguing chemical group with a multitude of biological activities. These include antibacterial, antifungal, antiviral, anti-inflammatory, anticancer, and antioxidant actions, as evidenced by studies [16, 17]. Moreover, it is well-documented that they exhibit a diverse array of biological activities. Moreover, phenolic chemicals are generally associated with the beneficial effects on human health that result from consuming a diet rich in fruits and vegetables [18, 19]. Researchers have determined that the antioxidant activity of phenolics

 DOI: 10.1201/9781003462804-28

TABLE 21.1
Classification of Phenolic Compounds into Simple Compounds, with Respective Number of Aromatic Benzene Rings and Relevant Examples

No. Carbon Atoms	Structure	No. Rings	Class	Discovered/Isolated by Scientist(s), and Year	Examples
6	C_6	One	Simple phenols, benzoquinones	Runge, 1934 [1]	Catechol, hydroquinone, 2,6-dimethoxybenzoquinone
7	C_6–C_1	One	Phenolic acids, phenolic aldehydes	Witting G., 1954 [2]	Gallic, salicylic acids
8	C_6–C_2	One	Acetophenones, tyrosine derivatives, phenylacetic acids	Charles, 1857 [3]	3-acetyl-6-methoxybenzaldehyde, tyrosol, p-hydroxyphenylacetic acid, homogentisic acid
9	C_6–C_3	One	Hydroxycinnamic acids, phenylpropenes, coumarins, isocoumarins, chromones	Vogel, 1820 [4]	Caffeic, ferulic acids, myristicin, eugenol, umbelliferone, aesculetin, bergenin, eugenin
10	C_6–C_4	One	Naphthoquinones	Müller and Leistner, 1978 [5]	Juglone, plumbagin
13	C_6–C_1– C_6	Two	Xanthonoids	Micheal, 1830 [6]	Mangiferin
14	C_6–C_2– C_6	Two	Stilbenoids, anthraquinones	Carl Graebe and Carl Theodore Liebermann, 1868 [7]	Resveratrol, emodin
15	C_6–C_3– C_6	Two	Chalconoids, flavonoids, isoflavonoids, neoflavonoids	Albert Szent-Györgyi, 1930 [8]	Quercetin, cyanidin, genistein
16	C_6–C_4– C_6	Two	Halogenated algal phenolic compounds	Dersormes and Clément, 1813 [9]	Kaviol A, colpol
18	$(C_6–C_3)_2$	Two	Lignans, neolignans	—	Pinoresinol, eusiderin
30	$(C_6–C_3–C_6)_2$	Four	Biflavonoids	—	Amentoflavone
Various	$(C_6–C_3)_n$, $(C_6)_n$, $(C_6–C_3–C_6)_n$	$n > 12$	Lignins, catechol melanins, flavolans, polyphenolic proteins, polyphenols	Edgar Hugo Emil Reinsch, 1939 [10]	Raspberry ellagitannin, tannic acid

is responsible for the positive effects that they provide. According to [20], phenolic compounds are a natural source of antioxidants found in the environment. However, the term "polyphenol" should be limited to molecules that include at least two or more phenolic rings in their structure. Furthermore, the term "polyphenol" is frequently used as an alternative to the phenolic compounds. The organic compounds known as polyphenols, or polyhydroxyphenols, are distinguished by the presence of multiple phenols as structural units. These compounds are defined as "phenolic compounds that are water soluble and have molecular weights ranging from 500 to 3,000 Daltons."

The extraction of various alkaloids, gelatin, and other proteins from the solution is a time-consuming and laborious process. The unique characteristics that these compounds possess due to their plant-based nature are of significant structural and physiological importance, as they belong to the

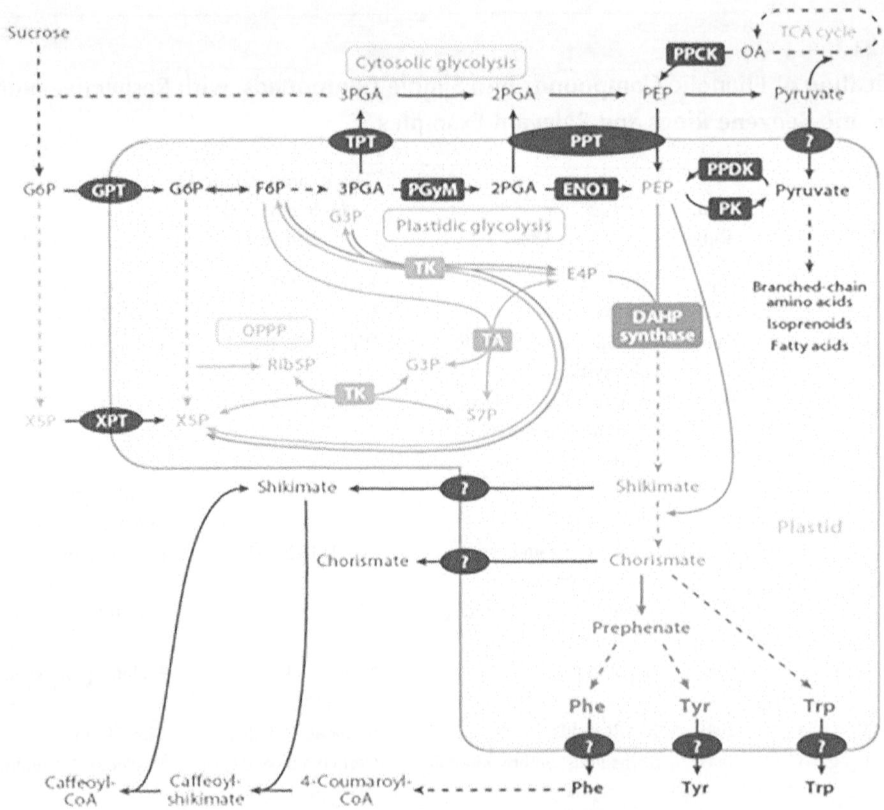

FIGURE 21.1 The provision of the shikimate pathway precursors and the movement of aromatic phenolic and amino acids (AAAs), and pathway intermediates across the membranes of plant plastids.

Source: Adapted from the work of Maeda and Dudareva (2012). [15].

most abundant classes of phytochemicals [21–23]. As stated by Dar [24], phenolics play a significant role in the processes of growth and reproduction, as well as in offering protection against both infections and pests [25, 26]. Moreover, it is involved in the development of color and sensory qualities in fruits and vegetables. The majority of phenolic compounds are found in all parts of plants belonging to numerous families. However, a small number of these compounds are exclusively found in particular plant organs during specific phases of development.

The following figure (Figure 21.2) presents a diagrammatic representation of the various interactions that occur throughout the co-evolutionary process. The development of biochemical systems that effectively protect plants against various biotic and abiotic stresses [27]. Some phenolic compounds are prevalent throughout the plant kingdom, and both functional characteristics and the particular distribution of these compounds are determined by their structural diversity. For example, anthocyanin is the pigment found in the majority of red and blue plant organs. It plays a significant role in attracting pollinators and assisting in pollination while reaching its highest distribution in flowers and fruits as they reach maturity.

The presence of anthocyanin in young leaves is beneficial in maintaining normal growth and protecting them from the harmful effects of photo-degradation. Furthermore, the crimson anthocyanin hue on leaves serves to protect them from the herbivorous effects of insects. Phenolic substances, such as proanthocyanidins (oligomers and polymers of flavan-3-ol) and hydrolyzable tannins (gallotannins and ellagitannins, which are derived from numerous esters of gallic), play a role in the

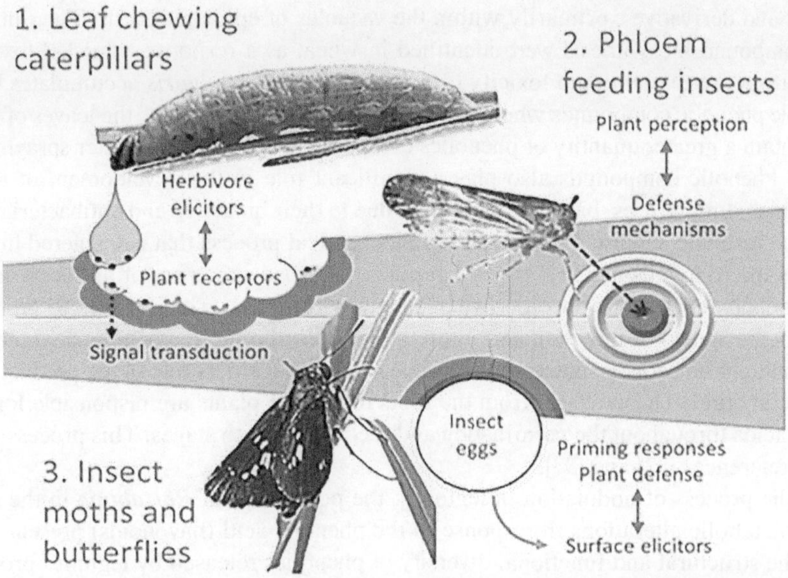

FIGURE 21.2 The diverse interactions during the co-evolution process: an establishment of efficient plant defense by biochemical systems.

Source: Adapted from the work of Ivan and Tomonori (2020). [27].

defense mechanism of plants against herbivores, fungi, and viruses. These substances are most commonly found in the early phases of plant development. There is evidence that flavonoids protect plant tissue from the damaging effects of ultraviolet light. Furthermore, the stress caused by ultraviolet radiation has been imposed on plants as a result of the loss of the ozone layer over the past three decades. Among the most prevalent phenolic derivatives, flavonoids have been demonstrated to be highly effective in reducing the levels of reactive oxidative species produced as a consequence of UVB exposure.

21.2 FUNCTIONS

21.2.1 Functional Properties of Phenolic Compounds

As illustrated in Table 21.2, phenolic compounds exhibit a wide range of functions. In plants, cinnamic acid serves as the primary source of phenolic compounds. These are produced from phenylalanine through the phenylpropanoid pathway, utilizing an enzyme, phenylalanine ammonia–lyase (PAL) [28, 29].

In typical development conditions, 20% of the carbon that plants fix is transported through this system. The existing data lends credence to the significance of this mechanism. Phenolic chemicals are utilized by plants for a variety of purposes. According to numerous researchers, the diverse environmental stress circumstances led to an increase in the accumulation of phenolic compounds and an associated enhancement of the metabolism of phenylpropanoid compounds [30]. The combination of isoflavones with other flavonoids is triggered in plants when they are subjected to a variety of stressors, including disease, insect damage, low temperatures, low nutritional conditions, UV radiation, wilting, and water stress.

In order to prevent the entry of ultraviolet and other harmful rays into the deeper tissues of the plant, plants tend to collect flavonoids that absorb these ultraviolet light using various phenolic compounds and derivatives, primarily within the vacuoles of epidermal cells. The initial stages of phenolic compound biosynthesis were identified in wheat as a response to nickel toxicity, and in maize as a response to aluminum toxicity [31]. Bean *Phaseolus vulgaris* accumulates both soluble and insoluble phenolic compounds when subjected to cadmium. Similarly, the leaves of *Phyllanthus tenellus* contain a greater quantity of phenolics compared to control plants after spraying with copper sulfate. Phenolic compounds also play a significant role in the development of resistance to pathogens, including viruses, bacteria, and fungi, due to their bioactive and antibacterial properties. In plants, the instigate defense mechanism is a biochemical process that is triggered in response to an attack by microorganisms. This process produces a broad spectrum of antibacterial reactions. These compounds elicit a site-specific hypersensitivity response, which serves to protect the host plant against the spread of infection and future attacks. Furthermore, phenolic acid plays a pivotal role in the interaction between plants and microbes, in addition to its role in the production of stress response metabolites. The exudates from the roots of legume plants are responsible for the release of phenolic acids throughout the germination and seedling growth stages. This process is described in detail in references [32] and [33].

During the process of nodulation in legumes, the population of *Rhizobium* in the rhizosphere undergoes metabolic alterations in response to the phenolic acid (flavonoids) present in the environment. The structural and functional diversity of phenolics released by legumes provides selective rhizobial strains with a competitive advantage in the process of nodulation. Root exudates include antibacterial chemicals that help plants resist pathogens. The importance of plant root exudates, which are primarily composed of phenolic compounds, as signaling molecules is illustrated in Figure 21.2, particularly in the context of defense responses. In response to *Fusarium* infection, barley root systems secrete T-cinnamic acid, which has antibacterial activity. This phenomenon represents a dynamic plant defense system at the root level, whereby de novo biosynthesized T-cinnamic acid is released within two days [34]. The ecosystems in which plants develop and exist are of a highly complex and diverse nature. Given that plants lack the mobility to flee from the attacks of herbivores or pathogens, they have evolved constitutive and inducible defense mechanisms [35, 36].

These defensive mechanisms are initiated by signaling systems that are spatially and temporally dynamic in nature. The creation of toxins or defense structures and compounds by plants represents a direct threat to pathogens, eliminating them immediately. Root exudates represent a crucial element of the induced plant defense systems that are deployed. In response to the presence of the pathogenic fungus *Pythium ultimum*, the roots of *Ocimum basilicum* have been observed to release rosmarinic acid [37]. The stimulation of iridoid glycosides in root exudates of *Plantago lanceolata* in the presence of nematodes has been demonstrated to function as a nematicide. Allelochemicals, phytoalexin, phytoanticipins, and nematicides are all examples of compounds that are classified as phenolic compounds. These molecules can be either simple or complex. The allelopathic activity of phenolic compounds, such as terpenoids, hydroquinones, hydroxybenzoates, hydroxycinnamates, and hydroxynaphthoquinones, has been demonstrated in competitive plants and weeds [38, 39]. Among the nematicides that are found in plant defense mechanisms are compounds including cajanin, medicarpin, coumestrol, and limonoids. For instance, hydroxycoumarins and hydroxycinnamate conjugates represent examples of phytoalexins that accumulate in plants in response to pathogen infection [40]. The synthesis, release, and accumulation of these phenolics, in particular salicylic acid (SA), are regarded as the most crucial elements of a number of defensive strategies employed by plants in response to pathogenic attack. The underlying mechanism is illustrated in Figure 21.3. As a consequence of an infection or a stress response, de novo synthesis of salicylic acid (SA) occurs in order to acquire systemic resistance, given that the accumulation of SA represents a crucial molecule for resistance. These volatile derivatives, including salicylic acid methyl ester, have the capacity to induce protection in other areas of the plant as well as in plants in close proximity to it.

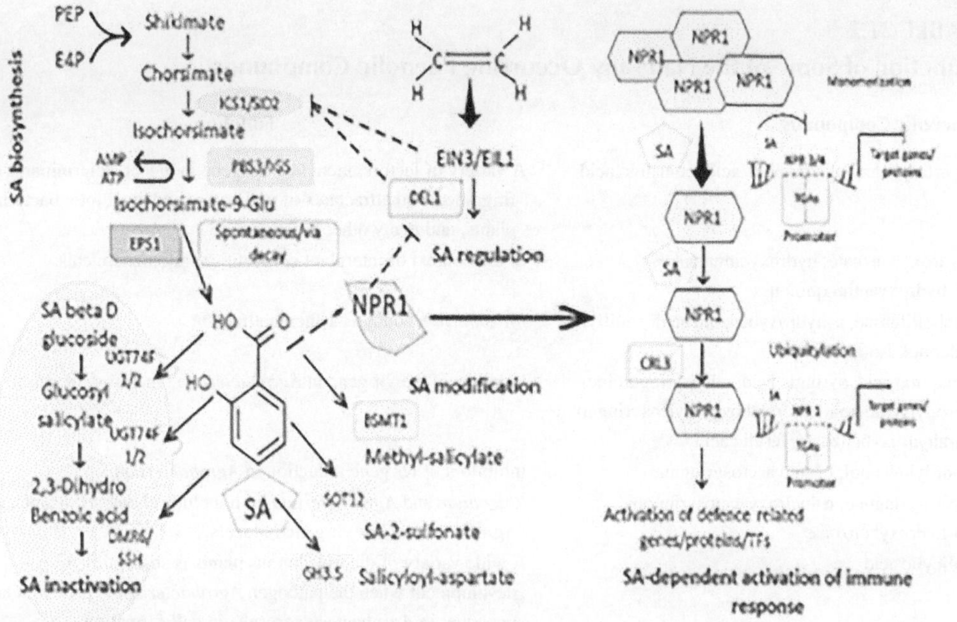

FIGURE 21.3 The mechanism in which salicylic acid (SA) and nitric oxide (NO) interact with microorganisms: the positive and negative effects are shown, which have impact on plant development and surrounding immunity—a mutually beneficial relationship.

Source: Adapted from the work of Rai et al. (2021). [41].

21.3 PHENOLICS AND DERIVATIVES

21.3.1 Flavonoids

Flavonoids are secondary metabolites derived from simple phenols and are the most abundant group. Flavonoids are highly distributed and also the most commonly occurring metabolite. The structural constituents of flavonoids are comprised of a 15-carbon skeleton that is composed of two aromatic rings that are linked together by a three-carbon bridge (C6–C3–C6) [42, 43]. One of the factors that determine the structural and functional distinction of flavonoids is the substituent, especially isoflavonoids, which are formed from 3-phenylchromen-4-one (3-phenyl-1,4-benzopyrone), and neoflavonoids, which are formed from 4-phenylcoumarine (4-phenyl-1,2-benzopyrone). These major classes of flavonoids can be categorized into the following three groups: isoflavonoids, neoflavonoids, and flavones [44]. The structural skeleton of flavonoids is of considerable interest, exhibiting a diverse range of carbon counts. This class of compounds, known as bioflavonoids, represents a novel category within the flavonoid family. The most common flavonoids found in plants include anthocyanidins, anthoxanthins, chalcones, flavones, flavonols, flavandiols, flavans, proanthocyanidins, and tannins, which represent some of the major subgroups derived from flavonoids. Flavonoids play a role in the coloration of flowers, the formation of symbiotic partnerships with prokaryotes, the defense against pathogenic infections, the control of insect pests, the filtration of ultraviolet light, and the provision of antioxidant activity in higher plants. The distinctive colors observed in plants, such as yellow, are attributed to quercetin, while malvidin is responsible for the various hues of blue that plants exhibit [45, 46]. During the stage of rhizobial infection that leads to the formation of symbiotic connections, the flavonoids that are released by the roots of leguminous plants serve as

TABLE 21.2
Function of Some of the Naturally Occurring Phenolic Compounds

Phenolic Compounds	Function
Coniferyl alcohol, sinapinic acid, cinnamic acid	A variety of factors, including vir gene inducers, determinants of fragrance, and attractants of pollinators and symbiotic bacteria in plants, and many other functions.
Hydroxybenzoate, hydroxycinnamates, 5-hydroxyanthraquinones	In the context of interplant competition, allelochemicals.
Umbelliferone, *p*-hydroxybenzoic acid, vanillyl alcohol, isoflavones	*Rhizoborium* contains a chemoattractant.
Sinapinic acid, syringic acid, ethylsyringamide, propylsyringamide, carbethoxyethylensyringamide, parahydroxybenzoate, ferulic acid	Inducers of the vir gene in *Agrobacterium*; vectoring of genes.
Vanillyl alcohol, bromo acetosyringone	Inhibitors of *vir* gene induction in *Agrobacterium*.
Acetosyringone, α-hydroxyacetosyringone, *p*-hydroxybenzoate	*Rhizobium* and *Agrobacterium* both include chemoattractants, while *Agrobacterium* has vir gene inducers.
Salicylic acid	A wide variety of dicotyledonous plants is susceptible to gall development when the pathogen *Agrobacterium* is present. Quorum quencher in *Agrobacterium* results in gall formation.
Hydroquinone	Using allelochemicals to combat plant competition among different types of plants, weeds, and off-type crop varieties, which can have an effect on yield.
Coumarins, xanthones, anthocyanidins	Factors that determine the color of plants and also attract pollinators to certain plants.
Caffeic acid	*Agrobacterium* vir gene inducer: a genetic factor for vectoring.
3,4-dihydroxybenzoic acid	*Agrobacterium* and *Rhizobium* both include a chemoattractant.
Protocatechuic acid, β-resorcylic acid, protocatechuate, *p*-resorcylate, catechol	*Agrobacterium* vir gene inducer: a genetic factor for vectoring.
Chlorogenic acid	Plant protein that serves as a precursor for the production of lignin and suberin.
Lignin, tannins, and suberins	Elements that make up the structure of plant cells.
Catechins	Plant defense against biotic factors.
Flavonoids, *O*-acetyldaidzein, 6-*O*-malonylgenistin, 6-*O*-malonyl daidzin, glycitin, 6-*O*-malonylglycitin	In *Rhizobium*, nod gene inducers are present.
Apigenin, naringenin, luteolin	In *Agrobacterium* and *Rhizobium*, chemoattractants, and in *Rhizobium*, agents that induce the nod gene.
Gallate, gallic acid, pyrogallic acid, syringic acid, kaempferol	*Agrobacterium* vir gene inducer: a genetic factor for vectoring.
Flavanones, quercetin	In *Rhizobium*, nod gene inducers are present.
Isoflavonoids	Inducers of the nod gene and chemoattractants in *Rhizobium*.
Cajanin, medicarpin, glyceoline, rotenone, coumestrol, phaseolin, phaseolin, limonoids, tannins, flavonoids	The involvement of phytoalexins, phytoanticipins, and nematicides in the defense of plants.

a messenger, initiating the process [47]. In order to protect their tissues from damage, plants store flavonoids in the vacuoles of epithelial cells.

During daytime, from May to August, solar radiation exerts a considerable stress on the skin, resulting in a reduction in the body's glucose levels. Flavonoids possess the capacity to absorb radiation within a specific wavelength range (with the highest absorption occurring at 250–270 nm and

335–360 nm) [48, 49]. Another mechanism by which flavonoids protect plants from the harmful effects of metals and their ions is through metal chelation. The superoxide dismutase activity of metal–flavonoid chelates has been confirmed in multiple studies [50].

21.3.2 Coumarins

One further significant by-product of the shikimic acid pathway is coumarin acid [51], which is generated through the hydroxylation of cinnamic acid or the deamination of tyrosine by the enzyme known as tyrosine ammonia-lyase (TAL) [52]. The C9H6O2 is the fundamental unit of its structure, and herbivores are susceptible to the toxicity of coumarins. It is found that coumarin, which has a bitter taste, also prevents animals from consuming plants that contain a significant number of these chemicals.

21.3.3 Tannins

In order to defend plants against herbivory, tannins are a type of polyphenolic molecule that plays an important role. According to some researchers, there are three categories of tannins: (1) hydrolyzable polymers, which are produced by the synthesis of gallic acid; (2) phlorotannins; and (3) condensed tannins and phlobatannins. Tannins are typically found in high concentrations in the bark and leaves of various trees and are also present in leaves. Tannins are extraordinary molecules that can provide plant protection against insects. This is due to their harsh, bitter taste and potential to denature protein.

21.3.4 Salicylic Acid

In the event of a pathogenic infection, the accumulation of pathogenesis-related protein [53] occurs at a distance from the infection site. This process serves to enhance the level of resistance against diseases and insects. As a form of systemic acquired resistance, the accumulation of salicylic acid (SA) and hydrogen peroxides acts on the target of infection and insect damage. This resistance is mediated by a variety of signaling molecules, and it occurs in conjunction with the PR protein [54–56]. The biosynthesis pathway of salicylic acid is depicted in Figure 21.4. Salicylic acid (SA) plays a pivotal role in the initiation of systemic acquired resistance (SAR), a process by which plants develop resistance to pathogens. When exogenously supplied, SA has been demonstrated to induce SAR and provide resistance to infected plants.

21.3.5 Lignin

The polypropanoid pathway is responsible for the synthesis of the polyphenolic molecule known as lignin. Lignin is a structural element found in certain vascular plants and certain algae. Its presence significantly aids the development of cell walls. In response to a pathogenic onslaught, lignin is synthesized both de novo and particularly via a specific biological method [58]. Wheat cultivars that are resistant to the fungal pathogen *Puccinia recondite* f. sp. tritici, which causes leaf rust, exhibit greater lignin accumulation, which provides a higher level of resistance to the disease. A comparable reaction can be observed in wheat when it is exposed to infections caused by *Fusarium graminearum*, which is the root cause of Fusarium head blight. This demonstrates the significance of lignin production as a defense mechanism against pathogenic infections [58–61]. A reduction in stem rust resistance was observed in *Puccinia graminis* Pers f. sp. tritici, in which the enzymes PAL and cinnamyl alcohol dehydrogenase play a pivotal role in the synthesis of lignin. In order to combat pathogenic infections, it is essential to coordinate the expression of lignin biosynthesis.

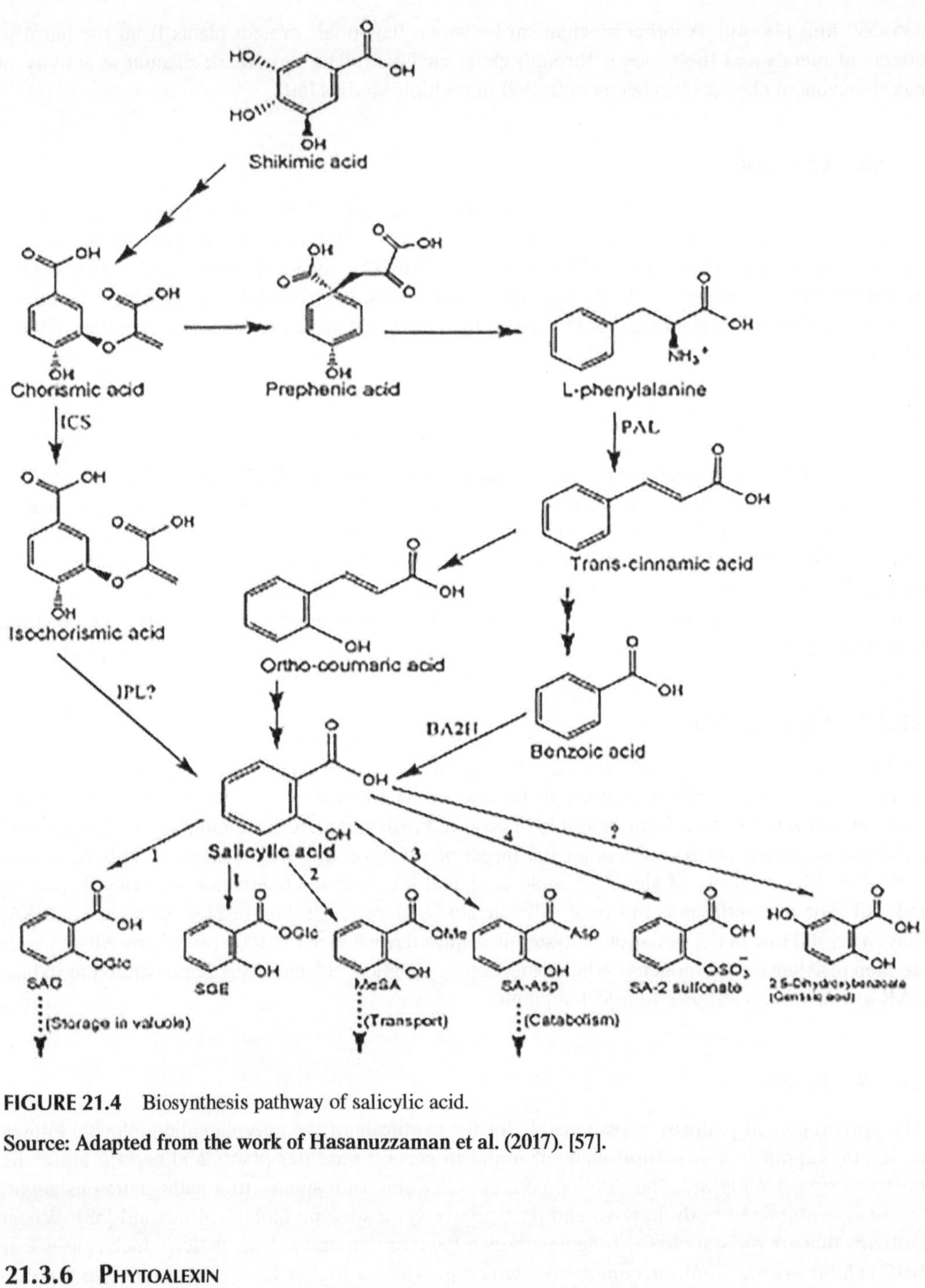

FIGURE 21.4 Biosynthesis pathway of salicylic acid.

Source: Adapted from the work of Hasanuzzaman et al. (2017). [57].

21.3.6 Phytoalexin

In plants, the stress response is responsible for the production of two distinct types of molecules. Both types of responses are in the form of chemical exudates. The first response that a plant has is to produce molecules that are not the intended target and that are effective from a great distance away from the invasion site. In contrast, a different response is observed when a plant produces molecules that have a specific effect at the location of the infection. In general, the term "phytoalexin" refers to the molecules that are created as a result of stress and infections, as well as the metabolites that accompany stress. There is strong evidence that phytoalexin is harmful to pathogens in multiple ways. A significant proportion of phytoalexins belongs to the flavonoids and isoflavonoids group, and they exhibit both antibacterial and antioxidant properties [62]. Phytoalexins possess the capacity to degrade cell walls, interfere with metabolic processes, and inhibit the growth of harmful microorganisms. This significance is evidenced by the observation that different types of phytoalexins demonstrate a greater degree of pathogenic colonization in comparison to wild types, and that the sensitivity of plant tissues rises when phytoalexin synthesis is blocked. The production of phytoalexins represents an essential component of defense mechanisms employed by plants against fungi, bacteria, and other microbes. Trans-resveratrol, a phytoalexin, has been demonstrated to inhibit the growth of *Botrytis cinerea*. The grapevine is capable of producing delta viniferin, which has been demonstrated to be an effective treatment for the infection by *Plasmopara viticola*. Additionally, the phytoalexin known as danielone, which is found in papaya fruit, has been shown to display antifungal effectiveness against the fungus *Colletotrichum gloeosporioides*. It has been demonstrated that chlorogenic acid, which is present in the periderm of potato tubers, is toxic to the bacterium *Streptomyces scabies*, which is responsible for the development of potato scab.

21.3.7 Phenolic Compounds in Abiotic Stress

In numerous instances, the environmental circumstances that surround agricultural fields are not always conducive to crop cultivation, due to a multitude of factors. Annually, abiotic stress, encompassing drought, salinity, and water scarcity, is accountable for substantial economic and yield losses worldwide. Furthermore, the destruction of the land has been exacerbated by climate change and the unsustainable use of land. For a significant proportion of agricultural soils, drought and salinity continue to represent severe abiotic stresses.

21.4 BIOSYNTHESIS

21.4.1 Biosynthesis of Phytoalexin

The biosynthetic pathway of phytoalexins from terpenoid units from monocots using a simplified model of signal transduction is given in Figure 21.5 [63].

It can be observed that there is a correlation between the increasing prevalence of abiotic stresses, such as drought, salt, low/high temperature, and food deprivation, and the formation of reactive oxygen species (ROS). In the process of partial reduction of oxygen species, cells produce reactive oxygen species, including superoxide anion (O_2^-), hydrogen peroxide (H_2O_2), and hydroxyl radical (HO), which are commonly referred to as reactive oxygen species [64]. In order to mitigate the deleterious effects of reactive oxygen species (ROS), plants have evolved a diverse array of antioxidant systems. With regard to antioxidant activity, plants are capable of utilizing both enzymatic and non-enzymatic components. Ascorbate peroxidase, glutathione reductase, and superoxide dismutase are the three enzymes that are considered to be the most significant in terms of enzymatic components. Polyphenolic compounds, including phenolic acids, flavonoids, proanthocyanidins, and anthocyanins, play a pivotal role in the non-enzymatic antioxidant mechanisms responsible for maintaining a normal level of reactive oxygen species (ROS) in the body [65]. There is evidence that suggests the accumulation of antioxidant metabolites in major cellular tissues, including chloroplasts, mitochondria, and mesophyll cells. These tissues are susceptible to significant damage from reactive oxygen species.

21.4.2 Phenolic Compounds in Insect Herbivory

It has been demonstrated that phenolics and insect pests exhibit a complex interaction [66, 67]. According to research, the presence of spotted spider mites (*Tetranychus urticae*) as an insect

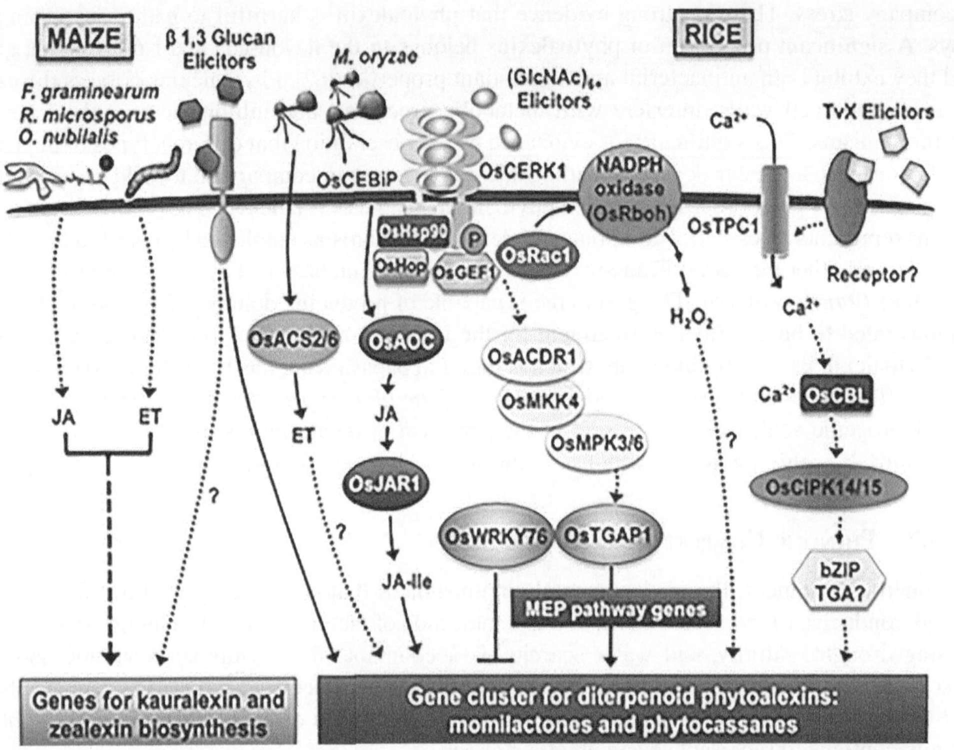

FIGURE 21.5 The biosynthetic pathway of phytoalexins from terpenoid units from monocots using a simplified model of signal transduction. Dotted arrows show areas of ignorance, while solid arrows show well-established connections, whereas the plasma membrane receptor OsCEBiP and related defensome complex proteins (OsHsp90, OsHop, and OsRacGEF1), as well as the small GTPase OsRac1, are bound by chitin fragments [(GlcNAc)6+].

Source: Adapted from the work of Schlelz et al. (2014). [63].

herbivore is positively correlated with the constitutive content of catechol phenolics in strawberry leaves. The cotton phenolic pigment gossypol is toxic to the tobacco bollworm and button worm, respectively. Furthermore, it serves as a deterrent to a multitude of insect pests. It has been demonstrated that the population of mites can be managed by the use of catechol, which binds to the digestive system of mites, rendering them inactive. Tannin is another example of a natural substance with insecticidal properties. According to Feeny's research with *Operophtera brumata* larvae, tannin has the ability to control the larvae of the oak moth. Wheat cultivars that contain phenolics are significantly less appealing to the cereal aphid, *Rhopalosiphum padi*.

21.5 ROLE OF PHENOLS IN BIOMEDICAL SCIENCES AND THE PHARMACOLOGICAL INDUSTRY

21.5.1 Biomedical and Pharmacological Properties of Phenolic Compounds, with Reference to *Acacia confusa* Fine-Leaved Wattle

Acacia confusa is a traditional medicinal plant of Southeast Asia. This section of the chapter collates and synthesizes the phytochemical, antioxidant, and pharmacological research findings on the phenolic components of this plant, with a particular focus on phenolic compounds. This compilation primarily focuses on the existing understanding of phenolic chemicals. The antioxidant and radical scavenging activity, hepatoprotective effect, xanthine oxidase inhibition, semicarbazide-sensitive amine oxidase inhibition, angiotensin-converting enzyme inhibition, antihyperuricemic effect, and anti-inflammatory activity are among the pharmacological activities of this plant that are worthy of note. It is notable that the phenolic compounds, including flavonoids, flavonol glycosides, and phenolic acid derivatives, represent the most significant phytochemical components found in *A. confusa*. The phytochemicals from this species have recently been the focus of pharmacological investigations due to their strong antioxidant activity and other phenolic functions. This section of the chapter will contribute to the growing body of knowledge on the potential uses of this plant and provide compelling evidence that it could have a future in clinical treatment and health functional foods.

21.5.2 Antioxidant and Radical Scavenging Activities

The Folin–Ciocalteu method was employed to assess the plant extracts and antioxidants of *A. confusa*. The results indicated that the root extracts exhibited the highest total phenolic compound (TPC) concentration, with a value of 652.2 mg (GAE)/g. The heartwood extract exhibited the highest content, with a value of 529.7 mg GAE/g. The bark extract followed, with a value of 476.6 mg GAE/g, while the branches, leaves, buds, twigs, and flowers exhibited values of 348.2, 190.2, 173.3, 121.4, and 105.1 mg GAE/g, respectively. In the laboratory, the phenolic groups that form the catechol and pyrogallol groups demonstrate a notable antioxidant and radical scavenging activity. Heartwood and root flavonoids possess 7,8-dihydroxyl structures that enhance antioxidant action. Among the flavonoids, okain and melanoxetin are the most potent antioxidants. It is noteworthy that these compounds exhibit the lowest IC_{50} values for DPPH (2,2-diphenyl-1-picrylhydrazyl) and superoxide radical scavenging, with values of 3.1 and 3.1 μM, respectively, and 2.2 and 2.5 μM, respectively [68].

The highest Trolox equivalent antioxidant capacity (TEAC) was observed in the samples with values of 5.4 and 4.4 mmol TE per mmol. The highest reducing power (TE per mmol) was also observed in these samples, with values of 5.3 and 5.0. Gallic acid has been demonstrated to be an effective scavenger of DPPH and superoxide radicals, with IC_{50} values of 8.2 and 12.4 μM and TEAC of 5.2 mmol/mmol. This makes it the most potent antioxidant among all phenolic acid derivatives [69]. The flavonol glycoside myricetin-3-O-(2"-O-galloyl)-α-rhamnopyranoside, derived from twigs and leaves, exhibits high activity, inhibiting DPPH and superoxide radicals with IC_{50} values of 3.9 and 8.7 μM, respectively, and TEAC of 6.9. In vitro analysis revealed that *A. confusa* plant extracts exhibited high phenolic content and antioxidant activity. Previously, *Caenorhabditis elegans* was utilized to assess the antioxidant efficacy of the extract in *A. confusa* twigs. The results demonstrated that the pretreatment of *C. elegans* with the twig extract resulted in enhanced oxidative stress resistance due to the removal of ROS [70].

21.5.3 Hepatoprotective Effects

CCl4-induced chronic liver injury in rats was prevented by bark extract of ethyl acetate and gallic acid from *A. confusa*. A histological examination of the livers of rats revealed that supplementation with 50 mg/kg gallic acid resulted in a 78% reduction in the development of larger vacuoles, a 78% reduction in the formation of tiny vacuoles, and a 57% reduction in inflammation [71]. The plasma AST and ALT levels, which are indicative of hepatically damaging conditions, exhibited a 94% and 100% reduction, respectively. Gallic acid enzyme activation of SOD, GRD, GPX, and catalase has been demonstrated. Furthermore, gallic acid has been shown to increase erythrocyte SOD, GPX, CAT, and GRD activity, with increases observed for SOD (19 units), GPX (2.3 units), CAT (852 units per milligram), and GRD (2.1 units per gram). In liver tissues, the activities of GPX, CAT, and GRD increased from 0.21, 48, and 0.49 U/g to 0.32, 79, and 0.54 U/g, while the activity of SOD decreased from 6.9 to 5.6 U/g [72]. Gallic acid has been demonstrated to reduce liver and blood lipid peroxidation and oxidative stress. The TBARS method was employed to assess the impact of oxidative stress on plasma and liver tissue samples. The results demonstrated a significant reduction in lipid peroxidation, with plasma samples exhibiting a 2.9-fold decrease in lipid peroxidation from 6.2 μM to 3.3 μM, and liver tissue samples exhibiting a 1.7-fold decrease in lipid peroxidation from 57 nmol/g liver to 39 nmol/g liver. Conversely, oxidized glutathione (GSH) was identified as an oxidative stress factor. In erythrocytes and liver tissues, the ratio of reduced to oxidized glutathione (GSH/GSSG) increased from 27:1 to 107:1 and from 15:1 to 93:1, respectively. In molecular biology studies, gallic acid demonstrated a 45% reduction in hepatic CYP2E1, the major isozyme involved in CCl4 bioactivation and free radical production. Gallic acid protects the liver by supporting antioxidant enzyme activity and reducing lipid peroxidation and CYP2E1 activation.

21.5.4 Xanthine Oxidase (XOD) Inhibitory Activity

A. confusa heartwood extract demonstrated inhibitory activity against XOD in vitro, as evidenced by the latest studies, which indicated an IC_{50} value of 0.076 μM. Both melanoxetin (0.274 μM) and allopurinol also demonstrated the effect. The molecule exhibiting the least degree of suppression was 7,8,3′,4′-tetrahydroxyflavone (26, 10.488 μM) [73]. The competitive substances allopurinol, melanoxetin, and 7,8,3′,4′-tetrahydroxyflavone were compared to the non-competitive okanin, and significant differences were recorded. Moreover, melanoxetin was found to inhibit XOD more effectively than allopurinol for multiple reasons. At equivalent inhibitor concentrations, melanoxetin generated less heat than allopurinol during the exothermic XOD and xanthine reaction. Allopurinol has been observed to bind to melanoxetin, and it is postulated that the presence of carbonyl and multihydroxyl functionalities may increase its XOD binding affinity. Furthermore, luteolin extracted from twigs demonstrated a pronounced inhibitory effect on X-ray oxidation (XOD), with an IC_{50} value of 11.6 μM.

21.5.5 Semicarbazide-Sensitive Amine Oxidase (SSAO) Inhibitory Activity

Five myricetin galloglycosides isolated from the leaves of *A. confusa* were found to inhibit SSAO (EC 1.4.3.6). Myricetin-3-O-(3”-O-galloyl)-α-rhamnopyranoside-7-methyl ether and (2”,3”-di-O-galloyl)-α-rhamnopyranoside demonstrated the highest inhibitory activity, with IC_{50} values of 36.16 μM and 39.35 μM, respectively [74]. The order of inhibition of SSAO and DPPH radical scavenging by the five chemicals was identical. A number of plant, microbial, and mammalian organs contain SSAO, which is responsible for the production of hydrogen peroxide and ammonia by converting primary amines into aldehydes. Atherosclerosis, endothelial damage, and glucose transfer into adipocytes are linked to plasma SSAO, which is elevated in diabetes and heart failure. Consequently, the *A. confusa* leaf extract myricetin galloglycoside has been demonstrated to promote human health [75].

21.5.6 Angiotensin I–Converting Enzyme (ACE) Inhibitory Activity

In addition to myricetin-3-O-(2”,3”-di-O-galloyl), myricetin-3-O-(3”-O-galloyl)-α-rhamnopyranoside is a synthetically produced chemical. Some of the aforementioned chemicals include myricetin-3-O-(2”-O-galloyl) and α-rhamnopyranoside-7-methyl ether [76]. The chemical alpha-rhamnopyranoside has inhibitory actions against ACE (EC 3.4.15.1) and -7-methyl ether, which was produced from the leaf extract of *A. confusa*. Lee et al. posit that the gallic acid groups are responsible for the ACE inhibitory action, and that the exopeptidase ACE, which is responsible for the release of dipeptides, plays a role in the renin–angiotensin system, which is responsible for regulating blood pressure. The antihypertensive drugs that fall into the category of ACE inhibitors are the most popular due to the fact that they have a relatively low incidence of adverse effects, as evidenced by the low number of adverse effects reported in the literature [77].

21.5.7 Antihyperuricemic Effect

A study has demonstrated that the heartwood extract of *A. confusa* has the capacity to successfully lower the levels of uric acid present in the blood of mice stimulated with oxonate [78]. Furthermore, five additional compounds were successfully isolated from the extract. When it was administered at the same dose (100 μmol/kg), the animals treated with melacacidin, 4′-O-methylmelacacidin, melanoxetin, transilitin, okanin, and allopurinol exhibited notable reductions in uric acid levels when compared to the control group. The uric acid levels were reduced by 66%, 72%, 75%, 65%, 69%, and 79%, respectively, as a result of the modifications. Gouty arthritis, gout, stroke, ischemic heart disease, kidney failure, uremia, and urolithiasis are just a few of the serious complications that can result from hyperuricemia. Cinnamaldehyde, the primary constituent of *Cinnamomum osmophloeum* leaves, has been demonstrated to effectively reduce uric acid levels in the blood by 60% when administered at a dosage of 150 mg/kg [79]. In a study conducted by researchers Zhu et al., it was found that the administration of quercetin and rutin at a dosage of 150 mg/kg (496 μmol/kg for quercetin and 245 μmol/kg for rutin) led to a noteworthy reduction in blood uric acid levels by 36% and 32%, respectively. Consequently, the findings of the study conducted by earlier researchers indicated that the heartwood extract of *A. confusa*, in conjunction with its five flavonoids extracted, has the potential to be a feasible therapy for reducing serum uric acid levels and treating gout [80].

21.5.8 Anti-Inflammatory Activity

In LPS-stimulated RAW 264.7 cells, the heartwood extract of *A. confusa* displays slight anti-inflammatory effects, resulting in multiple benefits. The most efficacious chemical, 7-melanoxetin, has the capacity to diminish the synthesis of nitrogen oxides by 6.9 μM (micromolar) and the accumulation of polyglyceride (PGE2) by 60% [81]. Furthermore, the expression of iNOS and COX-2 mRNA was significantly reduced by 76% when melanoxetin was administered at a concentration of 50 μM. In order to avoid the adverse effects associated with existing anti-inflammatory medications, pharmacologists are seeking novel anti-inflammatory agents that can reduce acute inflammation. It is regrettable that chronic inflammation has been linked to a number of health conditions, including cardiovascular disease, Alzheimer's disease, cancer, allergies, autoimmune illnesses, and metabolic problems. The presence of phenolic compounds and flavonoids in the aforementioned substance renders it a potentially valuable source of anti-inflammatory pharmaceuticals and dietary supplements.

21.5.9 Anti–Hepatitis C Virus Activity

Glycosyl lipidoids, which are referred to as ceramides, were discovered in the stems of *A. confusa* and were found to have an inhibitory effect on the hepatitis C virus (HCV) [82]. The human hepatitis virus, also known as HCV, has been linked to chronic hepatitis, cirrhosis, and hepatocellular carcinoma. In addition, it is a single-stranded RNA virus that is enclosed, has a positive sense of orientation, and is a member of the family Flaviviridae. The only treatment that has demonstrated the ability to cure HCV infection is a combination of pegylated interferon-α and ribavirin, which is classified as a nucleoside analogue. Conversely, this treatment is associated with a number of unfavorable side effects, including the development of symptoms that are comparable to those of the flu, hemolytic anemia, and feelings of depression [83].

The activation of nuclear factor-kappa B (NF-κB) results in the development of hepatic diseases that are induced by viruses with the potential to cause persistent inflammation, or the proliferation of hepatoma cells. Human cytomegalovirus (HCMV) is responsible for the production of at least ten distinct proteins. These include six nonstructural proteins (NS2, NS3, NS4A, NS4B, NS5A, and NS5B) and four structural proteins (core, E1, E2, and p7), which strengthen the structure of the cell. A number of these proteins, including core, E2, NS3, and NS5A, have been identified as responsible for the induction of hepatocellular carcinoma [84].

21.5.10 Immunoregulatory Activity

The regulation of immune responses is the responsibility of dendritic cells (DCs), which are a type of antigen-presenting cell that can be found in lymphoid and nonlymphoid organs [85]. It has been demonstrated that the activation of DCs results in the maturation and expression of pro-inflammatory cytokines, including TNF-α, IL-6, and IL-12, by the DCs themselves. The reduction of pro-inflammatory cytokine production in LPS-stimulated dendritic cells (DCs) at a concentration of 12.5 μM was observed with melanoxetin, a compound derived from the heartwood extract of *A. confusa.* This result indicates that melanoxetin has the potential to enhance immunological regulation. Furthermore, LPS is known to synthesize reactive oxygen species (ROS), which have the capacity to activate DCs and induce the development of immature DCs [85]. Treatment with melanoxetin was found to impede the maturation of DCs that were stimulated by LPS. It is postulated that melanoxetin, a potent antioxidant, may enhance immunological modulation by inhibiting the generation of reactive oxygen species (ROS) associated with lipopolysaccharide (LPS). Furthermore, human peripheral blood mononuclear cells (PBMCs), which are also referred to as such, constitute an indispensable element of the immunological function. The enhancement of immunological responses would result from the stimulation and multiplication of peripheral blood mononuclear cells (PBMCs). Myricetin-3-O-(2”-O-galloyl)-α-rhamnopyranoside has been demonstrated to exhibit antiproliferative effects on peripheral blood mononuclear cells (PBMCs), with an IC_{50} value of 11.9 μM [86].

21.5.11 Anti-Osteoclastogenic Effect

The anti-osteoclastogenic properties of 7,3′,4′-trihydroxyflavone and 7,8,3′,4′-tetrahydroxyflavone were demonstrated by the aforementioned researcher, who found it useful in the treatment of bone-lytic illnesses. Mature osteoclasts undergo osteoclastogenesis, as described in reference [87]. Maturation is induced by a number of cytokines, including RANKL and macrophage colony-stimulating factor, which are responsible for this process in 77% of cases. It is evident that conditions such as osteoporosis following menopause, inflammatory arthritis, osteolytic bone metastases, and Paget's bone disease can be caused by an overactive osteoclast. The levels of nuclear factor of activated T cells (NFATc1), a crucial transcription factor in osteoclast formation, and osteoclast marker gene mRNA are decreased by 7,3′,4′-trihydroxyflavone and 7,8,3′,4′-tetrahydroxyflavone [88].

21.5.12 Cancer Cell Cytotoxicity

The in-house brine prawn (*Artemia salina*) lethality assay is a common bioassay for determining the cytotoxicity of plant-based organic compounds, particularly natural products, against cancer cells. The assay is rapid, inexpensive, and suitable for a broad range of substances [89]. Research has demonstrated that the principal components in *A. confusa* leaf extracts were found to contain flavonol galloglycosides, and their ability to prevent brine shrimp from hatching was tested. The IC_{50} values for myricetin-3-O-(2”-O-galloyl)-α-rhamnopyranoside were determined. Myricetin-3-O-(3”-O-galloyl)-α-rhamnopyranoside, myricetin-3-O-(2”-O-galloyl)-α-rhamnopyranoside-7-methyl ether, and myricetin-3-O-(3”-O-galloyl)-α-rhamnopyranoside-7-methyl ether demonstrated moderate brine shrimp lethality [90].

21.5.13 Psychedelic Effects

It is well-documented that imbibing N, N-dimethyltryptamine at high concentrations can result in transient and intermittent visual hallucinations. It has been reported that the psychedelic pharmacological mechanisms of N, N-dimethyltryptamine involve interactions with a variety of receptors (serotonin receptors, trace amine-associated receptors, sigma-1 receptor, and ionotropic and metabotropic glutamate receptors) and neurotransmitters (dopamine, acetylcholine, and glutamate) in the brain [91]. Moreover, N, N-dimethyltryptamine has been identified as a model for psychiatric disorders, including schizophrenia, depression, and anxiety. Moreover, N, N-dimethyltryptamine is regarded as a potent medium for self-discovery and consciousness exploration due to its lower toxicity compared to other synthetic hallucinogenic compounds, such as 25I-NBOMe [92].

21.6 CONCLUSIONS

Phenolic compounds play a pivotal role in the process of plant growth and development, particularly in the processes that are responsible for defense against insects and pathogens. Phenolic compounds are present in plants, and a significant number of these compounds possess potent antioxidant and antimetabolite characteristics, which enable them to combat the effects of oxidative stress. These molecules are found in phenolic compounds. Some of these compounds have the capacity to bind with heavy metal ions, while others lack this ability. Phenolic phytoalexins exhibit antimicrobial and antifungal characteristics. In contrast, phenylpropanoids serve as the initial building blocks for the synthesis of lignin and suberin, which are essential for the formation of cell walls and the imparting of strength to them. Coumarins and tannins are substances that are employed to deter herbivores. Flavonoids are responsible for initiating the symbiotic association between *Rhizobium* and leguminous plants. Flavonoids accomplish this by acting as a chemoattractant.

REFERENCES

1. Liga, S.; Paul, C.; Péter, F. Flavonoids: Overview of biosynthesis, biological activity, and current extraction techniques. *Plants (Basel)* 2023 Jul 23, 12(14), 2732. doi: 10.3390/plants12142732. PMID: 37514347; PMCID: PMC10384615.
2. Tariq, H.; Asif, S.; Andleeb, A.; Hano, C.; Abbasi, B.H. Flavonoid production: Current trends in plant metabolic engineering and de novo microbial production. *Metabolites* 2023, 13, 124. doi: 10.3390/metabo13010124.
3. Okoye, C.O.; Jiang, H.; Wu, Y.; Li, X.; Gao, L.; Wang, Y.; Jiang, J. Bacterial biosynthesis of flavonoids: Overview, current biotechnology applications, challenges, and prospects. *J. Cell. Physiol.* 2023. doi: 10.1002/jcp.31006. Early View.
4. Rehan, M. Biosynthesis of diverse class flavonoids via shikimate and phenylpropanoid pathway. In: Zepka, L.Q.; Nascimento, T.C.D.; Jacob-Lopes, E. (eds) *Bioactive Compounds*. IntechOpen, Rijeka, Croatia, 2021; Ch. 6.
5. Dias, M.C.; Pinto, D.C.G.A.; Silva, A.M.S.; Plant flavonoids: Chemical characteristics and biological activity. *Molecules* 2021, 26, 5377. doi: 10.3390/molecules26175377.
6. Pyrzynska, K. Hesperidin: A review on extraction methods, stability and biological activities. *Nutrients* 2022, 14, 2387. doi: 10.3390/nu14122387.
7. Syahputra, R.A.; Harahap, U.; Dalimunthe, A.; Nasution, M.P.; Satria, D. The role of flavonoids as a cardioprotective strategy against doxorubicin-induced cardiotoxicity: A review. *Molecules* 2022, 27, 1320. doi: 10.3390/molecules27041320.
8. Zhang, Y.; Cai, P.; Cheng, G.; Zhang, Y. A brief review of phenolic compounds identified from plants: Their extraction, analysis, and biological activity. *Nat. Prod. Commun.* 2022, 17(1). doi: 10.1177/1934578X211069721.
9. Cappellini, F.; Marinelli, A.; Toccaceli, M.; Tonelli, C.; Petroni, K. Anthocyanins: From mechanisms of regulation in plants to health benefits in foods. *Front. Plant. Sci.* 2021 Oct 28, 12, 748049. doi: 10.3389/fpls.2021.748049. PMID: 34777426; PMCID: PMC8580863.

10. Kozłowska, A.; Dzierżanowski, T. Targeting inflammation by anthocyanins as the novel therapeutic potential for chronic diseases: An update. *Molecules* 2021, 26, 4380. doi: 10.3390/molecules26144380.
11. LaFountain, A.M.; Yuan, Y.W. Repressors of anthocyanin biosynthesis. *New Phytol.* 2021, 231, 933–949. doi: 10.1111/nph.17397.
12. Gao, P.; Qi, Y.; Li, L.; Yang, S.; Guo, J.; Liu, J.; Wei, H.; Huang, F.; Yu, L. Phenylpropane biosynthesis and alkaloid metabolism pathways involved in resistance of *Amorphophallus spp.* against soft rot disease. *Front. Plant Sci.* 2024, 15, 1334996. doi: 10.3389/fpls.2024.1334996.
13. Zhao, Y.; Zheng, X.; Zhang, X.; Wang, W.; Cai, G.; Bi, G.; et al. PIF3 is phosphorylated by MAPK to modulate plant immunity. *New Phytol.* 2023, 240, 372–381. doi: 10.1111/nph.19139.
14. Zhang, D.; Gao, Z.; Zhang, H.; Yang, Y.; Yang, X.; Zhao, X.; et al. The MAPK-Alfin-like 7 module negatively regulates ROS scavenging genes to promote NLR-mediated immunity. *Proc. Natl. Acad. Sci. U.S.A.* 2023, 120, e2214750120. doi: 10.1073/pnas.2214750120.
15. Sugiyama, A.; Yazaki, K. Root exudates of legume plants and their involvement in interactions with soil microbes. In: Vivanco, J.; Baluška, F. (eds) *Secretions and Exudates in Biological Systems. Signaling and Communication in Plants*, Vol. 12. Springer, Berlin, Heidelberg, 2012. doi: 10.1007/978-3-642-23047-9_2.
16. Hua, L.; Caixing, L.; Guo, Y.; Geoffrey, I.S.; Liheng, L.; Habib, U., Jie, L. Root exudate-driven rhizospheric recruitment of plant growth-promoting rhizobacteria. *Pedosphere* 2024. ISSN 1002-0160. doi: 10.1016/j.pedsph.2024.03.005.
17. Kumar, S.; Abedin, M.M.; Singh, A.K.; Das, S. Role of phenolic compounds in plant-defensive mechanisms. In: Lone, R.; Shuab, R.; Kamili, A. (eds) *Plant Phenolics in Sustainable Agriculture*. Springer, Singapore. 2020. doi: 10.1007/978-981-15-4890-1_22.
18. Rocchetti, G.; Gregorio, R. P.; Lorenzo, J. M.; Barba, F. J.; Oliveira, P. G.; Prieto, M. A.; Simal-Gandara, J.; Mosele, J. I.; Motilva, M.; Tomas, M.; Patrone, V.; Capanoglu, E.; Lucini, L. Functional implications of bound phenolic compounds and phenolics—food interaction: A review. 2022 Mar, 21(2), 811–842.
19. Mohsin, N.; Sun, J.; Shabbir, S.; Khattak, W.A.; Ren, G.; Nie, X.; Bo, Y.; Javed, Q.; Du, D.; Sonne, C. A review of plants strategies to resist biotic and abiotic environmental stressors. *Sci. Total Environ.* 900. doi: 10.1016/j.scitotenv.2023.165832.
20. Dixit, S.; Sivalingam, P.N.; Baskaran, R.K.M.; et al. Plant responses to concurrent abiotic and biotic stress: unravelling physiological and morphological mechanisms. *Plant Physiol. Rep.* 2024, 29, 6–17. doi: 10.1007/s40502-023-00766-0.
21. Ullah, A.; Munir, S.; Badshah, S.L.; Khan, N.; Ghani, L.; Poulson, B.G.; Emwas, A.H.; Jaremko, M. Important flavonoids and their role as a therapeutic agent. *Molecules.* 2020 Nov 11, 25(22), 5243. doi: 10.3390/molecules25225243. PMID: 33187049; PMCID: PMC7697716.
22. Maeda, H.A.; Dudareva, N. The shikimate pathway and aromatic amino acid biosynthesis in plants. *Annu. Rev. Plant Biol.* 2012. doi: 10.1146/annurev-arplant-042811-105439.
23. Ivan, G.; Tomonori, S. Elucidation of plant-insect interactions at the molecular level. In *Plant-Insect Interactions*. Graduate School and Environment and Life Science, 2015.
24. Panche, A.N.; Diwan, A.D.; Chandra, S.R. Flavonoids: an overview. *J. Nut. Sci.* 2016, 5, e47. doi: 10.1017/jns.2016.41.
25. Rai, K.K.; Pandey, N.; Rai, N.; Rai, S.K.; Pandey-Rai, S. Salicylic acid and nitric oxide: Insight into the transcriptional regulation of their metabolism and regulatory functions in plants. *Front. Agron.* 2021, 3, 781027. doi: 10.3389/fagro.2021.781027
26. Silveira Rabelo, A.C.; Borghesi, J.; Noreatto, G.D. The role of dietary polyphenols in osteosarcoma: A possible clue about the molecular mechanisms involved in a process that is just in its infancy. *J. Food Biochem.* 2021, 46(44).
27. Stavenga, D.G.; Leertouwer, H.L.; Dudek, B.; van der Kooi, C.J. Coloration of flowers by flavonoids and consequences of pH dependent absorption. *Front. Plant Sci.* 2021 Jan 8, 11, 600124. doi: 10.3389/fpls.2020.600124. PMID: 33488645; PMCID: PMC7820715.
28. Soni, J.; Sinha, S.; Pandey, R. Understanding bacterial pathogenicity: a closer look at the journey of harmful microbes. *Front. Microbiol.* 2024, 15, 1370818. doi: 10.3389/fmicb.2024.1370818.
29. Wang, J.; Ma, S.; Li, W.; Wang, X.; Huang, D.; Jiang, L.; et al. Salmonella enterica serovar typhi induces host metabolic reprogramming to increase glucose availability for intracellular replication. *Int. J. Mol. Sci.* 2021, 22, 10003. doi: 10.3390/ijms221810003
30. Huang, X.-X.; Wang, Y.; Lin, J.-S.; Chen, L.; Li, Y.-J.; Liu, Q.; Wang, G.-F.; Xu, F.; Liu, L.; Hou, B.-K. The novel pathogen-responsive glycosyltransferase UGT73C7 mediates the redirection of phenylpropanoid metabolism and promotes SNC1-dependent Arabidopsis immunity. *Plant J.* 2021, 107, 149–165.

31. Song, P.; Yu, X.; Yang, W.; Wang, Q. Natural phytoalexin stilbene compound resveratrol and its derivatives as anti-tobacco mosaic virus and anti-phytopathogenic fungus agents. *Sci. Rep.* 2021, 11, 16509.
32. Zhao, D.; Luan, Y.; Shi, W.; Zhang, X.; Meng, J.; Tao, J. A Paeonia ostii caffeoyl-CoA O-methyltransferase confers drought stress tolerance by promoting lignin synthesis and ROS scavenging. *Plant Sci.* 2021, 303, 110765. doi: 10.1016/j.plantsci.2020.110765.
33. Menghani, E.; Khardia, P.; Shilu, A. Metabolic pathway of natural antioxidants, antioxidant enzymes and ROS providence. *Antioxidants* (Basel) 2022 Apr 11, 11(4), 761. doi: 10.3390/antiox11040761. PMID: 35453446; PMCID: PMC9025363.
34. Halliwell, B. Understanding mechanisms of antioxidant action in health and disease. *Nat. Rev. Mol. Cell. Biol.* 2024, 25, 13–33. doi: 10.1038/s41580-023-00645-4
35. Sharma, I.; Thakur, A.; Sharma, A.; Singh, N.; Kumar, R.; Sharma, A. Phytoalexins: Implications in plant defense and human health. In: Sharma, A.K.; Sharma, A. (eds) *Plant Secondary Metabolites*. Singapore: Springer, 2022. doi: 10.1007/978-981-16-4779-6_10
36. Polturak, G.; Misra, R.C.; El-Demerdash, A.; et al. Discovery of isoflavone phytoalexins in wheat reveals an alternative route to isoflavonoid biosynthesis. *Nat. Commun.* 2023, 14, 6977. doi: 10.1038/s41467-023-42464-3
37. Müller, A.T.; Nakamura, Y.; Reichelt, M.; Luck, K.; Cosio, E.; Lackus, N.D.; Gershenzon, J.; Mithöfer, A.; Köllner, T.G. Biosynthesis, herbivore induction, and defensive role of phenylacetaldoxime glucoside. *Plant Physiol.* 2024 Jan, 194(1), 329–346. doi: 10.1093/plphys/kiad448
38. Rabelo, A.C.S.; Borghesi, J.; Noratto, G.D. The role of dietary polyphenols in osteosarcoma: A possible clue about the molecular mechanisms involved in a process that is just in its infancy. *J. Food Biochem.* 2022 Jan, 46(1), e14026. doi: 10.1111/jfbc.14026. Epub 2021 Dec 7. PMID: 34873724.
39. Maeda, H.; Dudareva, N. The shikimate pathway and aromatic amino acid biosynthesis in plants. *Annu. Rev. Plant Biol.* 2012, 63, 73–105. doi: 10.1146/annurev-arplant-042811-105439. PMID: 22554242.
40. Ivan, G.; Tomonori, S. 2020. *Plant Insect Interactions: Understanding of Metabolic Defense Against Herbivores in Plants*. Okayama University: Okayama, Japan. http://www.okayama-u.ac.jp/
41. Hasanuzzaman, M.; Nahar, K.; Bhuiyan, T.F.; Anee, T.I.; Inafuku, M.; Oku, H.; Fujita, M. 2017. Salicylic acid: An all-rounder in regulating abiotic stress responses in plants. Intech Open. In *Phytohormones—Signaling Mechanisms and Crosstalk in Plant Development and Stress Responses*. doi: 10.5772/intechopen.68213.
42. Schmelz, E.A.; Huffaker, A.; Sims, J.W.; Christensen, S.A.; Lu, X.; Okada, K.; Peters, E.J. Biosynthesis, elicitation and roles of monocot terpenoid phytoalexins. *Plant J.* Willey. 2014. https://doi.org/10.1111/tpj.12436.
43. Rudrapal, M.; Khairnar, S.J.; Khan, J.; Dukhyil, A.B.; Ansari, M.A.; Alomary, M.N.; Alshabrmi, F.M.; Palai, S.; Deb, P.K.; Devi, R. Dietary polyphenols and their role in oxidative stress-induced human diseases: Insights into protective effects, antioxidant potentials and mechanism(s) of action. *Front. Pharmacol.* 2022 Feb 14, 13, 806470. doi: 10.3389/fphar.2022.806470. PMID: 35237163; PMCID: PMC8882865.
44. Tlak Gajger, I.; Kolodziejek, J.; Bakonyi, T.; et al. Prevalence and distribution patterns of seven different honeybee viruses in diseased colonies: a case study from Croatia. *Apidologie* 45, 701–706. doi: 10.1007/s13592-014-0287-0
45. Tlak Gajger, I.; Nejedli, S.; Kozarić, Z. The effectof Nozevit on leucine aminopeptidase and esteraseactivity in the midgut of honey bees (*Apis mellifera*). *Vet. Med.* 2013, 58(8), 422–429.31. doi: 10.17221/6982-VETMED.
46. Tlak Gajger, I.; Tomljanović, Z.; Stanisavljević, Lj. An environmentally friendly approach tothe control of *Varooa destructor* mite and *Nosema ceranaedisease* in Carnolian honeybee (*Apis melliferacarnica*) colonies. *Arch. Biol. Sci.* 2013a, 65(4), 1585–1592.33. doi: 10.2298/ABS1304585G.
47. Tlak Gajger, I.; Ribaric, J.; Matak, M.; Svecnjak, L.; Kozaric, Z.; Nejedli, S.; Smodis Skerl, I. M. *Zeolite clinoptilolite* as a dietary supplement and remedyfor honeybee (*Apis mellifera*) colonies. *Vet. Med.* 2015, 60(12), 696–705.34. doi: 10.17221/8584-VETMED.
48. Tlak Gajger, I.; Ribarić, J.; Smodiš Škerl, M.; Vlainić, J.; Sikirić, P. Stable gastric pentadecapeptide BPC 157 in honeybee (*Apis mellifera*) therapy, to control *Nosema ceranae* invasions in apiary conditions. *J. Vet. Pharmacol. Ther.* 2018 Aug, 41(4), 614–621. doi: 10.1111/jvp.12509. Epub 2018 Apr 23. PMID: 29682749.
49. Tlak Gajger, I.; Vlainić, J.; Šoštarić, P.; Prešern, J.; Bubnič, J.; Smodiš Škerl, M. I. Effects onsome therapeutical, biochemical, and immunologicalparameters of honey bee (*Apis mellifera*) exposed toprobiotic treatments, in field and laboratory conditions. *Insects* 2020, 11(9), 638. doi: 10.3390/insects11090638.36.
50. Tlak Gajger, I.; Sušec, P. Efficacy of varroacidalfood additive appliance during summer treatment of honeybee colonies (*Apis mellifera*). *Vet. Arhiv.* 2019, 89(1), 87–96.38. doi: 10.24099/vet.arhiv.0441.

51. Tlak Gajger, I.; Tomljanović, Z.; Petrinec, Z. Monitoring health status of Croatian honey bee coloniesand possible reasons for winter losses. *J. Apicul. Res.* 2010, 49(1), 107–108. doi: 10.3896/IBRA.1.49.1.1939.
52. Tlak Gajger, I.; Vugrek, O.; Grilec, D.; Petrinec, Z. Prevalence and distribution of Nosemaceranae in Croatian honeybee colonies. *Vet. Med.* 2010b, 55(9), 457–462.40. doi: 10.17221/2983-VETMED.
53. Tlak Gajger, I.; Sakač, M.; Gregorc, A. Impact of thiamethoxam on honey bee queen (*Apis mellifera* Carnica) reproductive morphology and physiology. *Bull. Environ. Contam. Toxicol.* 2017, 99, 297–302. doi: 10.1007/s00128-017-2144-0.
54. Marmitt, D.; Shahrajabian, M.H. Plant species used in Brazil and Asia regions with toxic properties. *Phytother. Res.* 2021, 35, 4703–4726.
55. Shahrajabian, M.H.; Sun, W.; Cheng, Q. Clinical aspects and health benefits of ginger (*Zingiber officinale*) in both traditional Chinese medicine and modern industry. *Acta Agric. Scand. B Soil Plant Sci.* 2019, 69, 546–556.
56. Shahrajabian, M.H.; Sun, W.; Shen, H.; Cheng, Q. Chinese herbal medicine for SARS and SARS-CoV-2 treatment and prevention, encouraging using herbal medicine for COVID-19 outbreak. *Acta Agric. Scand. B Soil Plant Sci.* 2020, 70, 437–443.
57. Shahrajabian, M.H.; Sun, W.; Cheng, Q. Product of natural evolution (SARS, MERS, and SARS-CoV-2); deadly diseases, from SARS to SARS-CoV-2. *Hum. Vaccines Immunother.* 2020, 17, 62–83.
58. Sun, W.; Shahrajabian, M.H.; Cheng, Q. Barberry (*Berberis vulgaris*), a medicinal fruit and food with traditional and modern pharmaceutical uses. *Isr. J. Plant Sci.* 2021, 68, 61–71.
59. Marmitt, D.; Shahrajabian, M.H.; Goettert, M.I.; Rempel, C. Clinical trials with plants in diabetes mellitus therapy: A systematic review. *Expert Rev. Clin. Pharmacol.* 2021, 14, 735–747.
60. Sun, W.; Shahrajabian, M.H.; Cheng, Q. Fenugreek cultivation with emphasis on historical aspects and its uses in traditional medicine and modern pharmaceutical sciences. *Mini Rev. Med. Chem.* 2021, 21, 724–730.
61. Sun, W.; Shahrajabian, M.H.; Cheng, Q. Anise (*Pimpinella anisum* L.), a dominant spice and traditional medicinal herb for both food and medicinal purposes. *Cogent. Biol.* 2019, 5, 1673688.
62. Shahrajabian, M.H.; Sun, W.; Cheng, Q. Chemical components and pharmacological benefits of Basil (*Ocimum basilicum*): A review. *Int. J. Food Prop.* 2020, 23, 1961–1970.
63. Supritha, P.; Radha, K.V. Estimation of phenolic compounds present in the plant extracts using high pressure liquid chromatography, antioxidant properties, and its antibacterial activity. *Indian J. Pharm. Educ. Res.* 2018, 52, 321–326.
64. Amessis-Ouchemoukh,N.;Madani,K.;Fale,P.L.V.;Serralheiro,M.L.;Araujo,M.E.M.Antioxidantcapacity and phenolic contents of some Mediterranean medicinal plants and their potential role in the inhibition of cyclooxygenase-1 and acetylcholinesterase activities. *Ind. Crop. Prod.* 2014, 53, 6–15.
65. Subedi, L.; Timalsena, S.; Duwadi, P.; Thapa, R.; Paudel, A.; Parajuli, K. Antioxidant activity and phenol and flavonoid contents of eight medicinal plants from Western Nepal. *J. Tradit. Chin. Med.* 2014, 34, 584–590.
66. Jimenez, N.; Carrillo-Hormaza, L.; Pujol, A.; Alzate, F.; Osorio, E.; Lara-Guzman, O. Antioxidant capacity and phenolic content of commonly used anti-inflammatory medicinal plants in Colombia. *Ind. Crop. Prod.* 2015, 70, 272–279.
67. Cisneros, H.S.; Bertiller, M.B.; Furlong, J.J.P.; Carrera, A.L. Similar structural complexity of phenols in plant morphotypes with contrasting soluble phenol concentration and richness in arid rangelands of Patagonia. *Flora* 2022, 295, 152134.
68. Jaiswal, N.; Kumar, A. HPLC in the discovery of plant phenolics as antifungal molecules against Candida infection related biofilms. *Microchem J.* 2022, 179, 107572.
69. Diaz, P.; Jeong, S.C.; Lee, S.; Khoo, C.; Koyyalamudi, S.R. Antioxidant and anti-inflammatory activities of selected medicinal plants and fungi containing phenolic and flavonoids compounds. *Chin. Med.* 2012, 7, 26.
70. Tukun, A.B.; Shaheen, N.; Banu, C.P.; Mohiduzzaman, M.D.; Islam, S.; Begum, M. Antioxidant capacity and total phenolic contents in hydrophilic extracts of selected Bangladeshi medicinal plants. *Asian Pac. J. Trop. Med.* 2014, 7(Suppl. 1), S568–S573.
71. Manguro, L.O.A.; Lemmen, P. Phenolics of *Moringa oleifera* leaves. *Nat. Prod. Res.* 2007, 21, 56–68. [CrossRef] 35. Sodaeizadeh, H.; Rafieiolhossaini, M.; Havlik, J.; van Damme, P. Allelopathic activity of different plant parts of *Peganum harmala* L. and identification of their growth inhibitors substances. *Plant Growth Regul.* 2009, 59, 227–236.
72. Shah, B.; Cai, Y.Z.; Sun, M.; Corke, H. Antioxidant capacity of 26 spice extracts and characterization of their phenolic constituents. *J. Agric. Food Chem.* 2005, 53, 7749–7759.
73. Vastano, B.C.; Chen, Y.; Zhu, N.Q.; Ho, C.T.; Zhou, Z.Y.; Rosen, R.T. Isolation and identification of stilbenes in two varieties of *Polygonum cuspidatum*. *J. Agric. Food* 2000, 48, 253–256.

74. Ahmad, A.; Misra, L.N. Isolation of herniarin and other constituents from *Matricaria chamomilla* flowers. *Int. J. Pharmacogn.* 1997, 35, 121–125.
75. Morikawa, T.; Luo, F.L.; Manse, Y.; Sugita, H.; Saeki, S.; Chaipech, S.; Pongiriyadacha, Y.; Muraoka, O.; Ninomiya, K. Geranylated coumarins from Thai medicinal plant *Mammea siamensis* with testosterone 5 alpha-reductase inhibitory activity. *Front. Chem.* 2020, 8, 199.
76. Thuong, P.T.; Hung, T.M.; Ngoc, T.M.; Ha, D.T.; Min, B.S.; Kwak, S.J.; Kang, T.S.; Choi, J.S.; Bae, K. Antioxidant activities of coumarins from Korean medicinal plants and their structure-activity relationships. *Phytother. Res.* 2010, 24, 101–106.
77. Singh, M.; Tiwari, N.; Shanker, K.L.; Verma, R.K.; Gupta, A.K.; Gupta, M.M. Two new lignans from *Phyllanthus amarus*. *J. Asian Nat. Prod. Res.* 2009, 11, 562–568.
78. Hu, D.; Yang, Z.Y.; Yao, X.C.; Wang, H.; Han, N.; Liu, Z.H.; Wang, Y.; Yang, J.Y.; Yin, J. Dibenzocyclooctadiene lignans from *Schisanra chinensis* and their inhibitory activity on No production in lipopolysaccharide-activated microglia cells. *Phytochemistry* 2014, 104, 72–78.
79. Baranyai, B.; Backer, C.; Reich, C.; Lindequist, U. The production of 7-methyljuglone, plumbagin, and quercetin in wild and cultivated *Drosera rotundifolia* and *Drosera intermedia*. *Mires Peat.* 2016, 18, 1–8.
80. Tian, K.; Zhang, H.G.; Chen, X.G.; Hu, Z.D. Determination of five anthraquinones in medicinal plants by capillary zone electrophoresis with beta-cyclodextrin addition. *J. Chromatogr. A*, 2006, 1123, 134–137.
81. Ali, I.; Haque, A.; Saleem, K. Separation and identification of curcuminoids in turmeric powder by HPLC using phenyl coloumn. *Anal. Methods* 2014, 6, 2526–2536.
82. Rodrigues, F.; Almeida, I.; Sarmento, B.; Amaral, M.H.; Oliveira, M. Study of the isoflavone content of different extracts of *Medicago* spp. as potential active ingredient. *Ind. Crop. Prod.* 2014, 57, 110–115.
83. Stevenson, P.C.; Kite, G.C.; Lewis, G.P.; Forest, F.; Nyirenda, S.P.; Belmain, S.R.; Sileshi, G.W.; Veitch, N.C. Disntinct chemotypes of *Tephrosia vogelii* and implications for their use in pest control and soil enrichment. *Phytochemistry* 2012, 78, 135–146.
84. Russo, M.; Fanali, C.; Tripodo, G.; Dugo, P.; Muleo, R.; Dugo, L.; de Gara, L.; Mondello, L. Analysis of phenolic compounds in different parts of pomegranate (*Punica granatum*) fruit by HPLC-PDA-ESI/MS and evaluation of their antioxidant activity: Application to different Italian varieties. *Anal. Bioanal. Chem.* 2018, 410, 3507–3520.
85. Bravo, L. Polyphenols: Chemistry, dietary sources, metabolism, and nutritional significance. *Nutr. Rev.* 1998, 56, 317–333. [CrossRef] [PubMed] 53. Balasundram, N.; Sundram, K.; Samman, S. Phenolic compounds in plants and agri-industrial by-products: Antioxidant activity, occurrence, and potential uses. *Food Chem.* 2006, 99, 191–203.
86. De la Rosa, L.A.; Moreno-Escamilla, J.O.; Rodrigo-Garcia, J.; Alvarez-Parrilla, E. Chapter 12—Henolic compounds. In: Yahia, E.M. (ed.) *Postharvest Physiology and Biochemistry of Fruits and Vegetables*. Woodhead Publishing, Sawston, UK, 2019; pp. 253–271.
87. Soleymani, A.; Shahrajabian, M.H.; Naranjani, L. Yield and yield components of berseem clover cultivars in low nitrogen fertilizer input farming. *J. Food Agric. Environ.* 2011, 9, 281–283. 56.
88. Soleymani, A.; Shahrajabian, M.H.; Naranjani, L. Effect of planting dates and different levels of nitrogen on seed yield and yield components of nuts sunflower (*Helianthus annuus* L.). *Afr. J. Agric. Res.* 2013, 8, 5802–5805.
89. Paine, C.S.; van Staden, J.; Finnie, J.F. Elicitation of phenolic compounds of medicinal value in plants. *S. Afr. J. Bot.* 2017, 109, 362. [CrossRef] 58. Salem, M.M.; Davidorf, F.H.; Abdel-Rahman, M.H. In vitro anti-uveal melanoma activity of phenolic compounds from the Egyptian medicinal plant *Acacia nilotica*. *Fitoterapia* 2011, 82, 1279–1284.
90. Carbonell-Capella, J.M.; Buniowska, M.; Barba, F.J.; Esteve, M.J.; Frigola, A. Analytical methods for determining bioavailability and bioaccessibility of bioactive compounds from fruits and vegetables: A review. *Compr. Rev. Food Sci. Food Saf.* 2014, 13, 155–171.
91. Rein, M.J.; Renouf, M.; Cruz-Hernandez, C.; Actis-Goretta, L.; Thakkar, S.K.; da Silva Pinto, M. Bioavailability of bioative food compounds: A challenging journey of bioefficacy. *Br. J. Clin. Pharmacol.* 2013, 75, 588–602.
92. Albergaria, E.T.; Oliveira, A.F.M.; Albuquerque, U.P. The effect of water deficit stress on the composition of phenolic compounds in medicinal plants. *S. Afr. J. Bot.* 2020, 131, 12–17.

22 The Biosynthesis of Lignin and Its Role in Drug/Gene Delivery and Tissue Engineering Applications

Showket Ahmad Dar, Mohmmad Al-Salem, Ebtesam A. Gashash, Ahmed M. El-Taher, Magdy A. Omar, Hany S. Abd El-Raouf, Ayah T. Abdullah Al-Kilani, Kounser Javeed, Wajid Hasan, Yendrembam K. Devi, Naveena Nazim, Ammar Al-Farga, Kangjam Bumpy, I. Yimjenjang Longkumer, Dev Karan Bairwa, Syed Naseem Zaffar, and Rizwan Rashid

22.1 INTRODUCTION

The metabolic process that involves phenylalanine and tyrosine in plant cells is responsible for the production of lignin, which is considered to be one of the most significant secondary metabolites. Given its prevalence in the biosphere, lignin represents a significant proportion of the organic carbon content, accounting for approximately 30% of this material [1]. The production of lignin is discussed in detail in Figure 22.1, which presents a complex network that may be broken down into three distinct processes: (1) the biosynthesis of lignin monomers, (2) transport, and (3) polymerization.

Lignin monomers are created in the cytoplasm and then transported to the apoplast after undergoing a series of biochemical transformations, including deamination, hydroxylation, methylation, and reduction. Finally, peroxidase (POD) and laccase (LAC) in the secondary cell wall are responsible for polymerizing lignin with three primary forms of monolignols: sinapyl alcohol, which has an S-unit; coniferyl alcohol, which has a G-unit; and p-coumaryl alcohol, which has an H-unit [2–5]. Furthermore, it has been demonstrated that a number of additional substances, including hydroxycinnamaldehydes, tricin-flavones, hydroxystilbenes, and xenobiotics, among others, are also considered to be lignin subunits. It is possible for lignin monomers to freely travel in the extracellular space throughout the process of developing protoxylem tracheary elements of *Arabidopsis*. However, the polymerization of lignin in the secondary cell walls is only observed. During the process of plant growth and development, lignin and the associated metabolism play a multitude of critical roles. In addition to facilitating mineral transport through the vascular bundles in plants, lignin, which is a complex phenolic polymer, increases the stiffness of plant cell walls and exhibits hydrophobic qualities [6]. Moreover, lignin serves as an indispensable barrier that safeguards plants against pathogens and pests [7]. Moreover, the resilience of plants to lodging and their capacity to respond to diverse environmental stimuli can also be actively influenced by the metabolism of lignin [8–12]. The capacity of plants to evolve and adjust to their environment is profoundly influenced by lignin, which represents a crucial component of plant cell walls. Moreover, lignin itself can be

 DOI: 10.1201/9781003462804-29

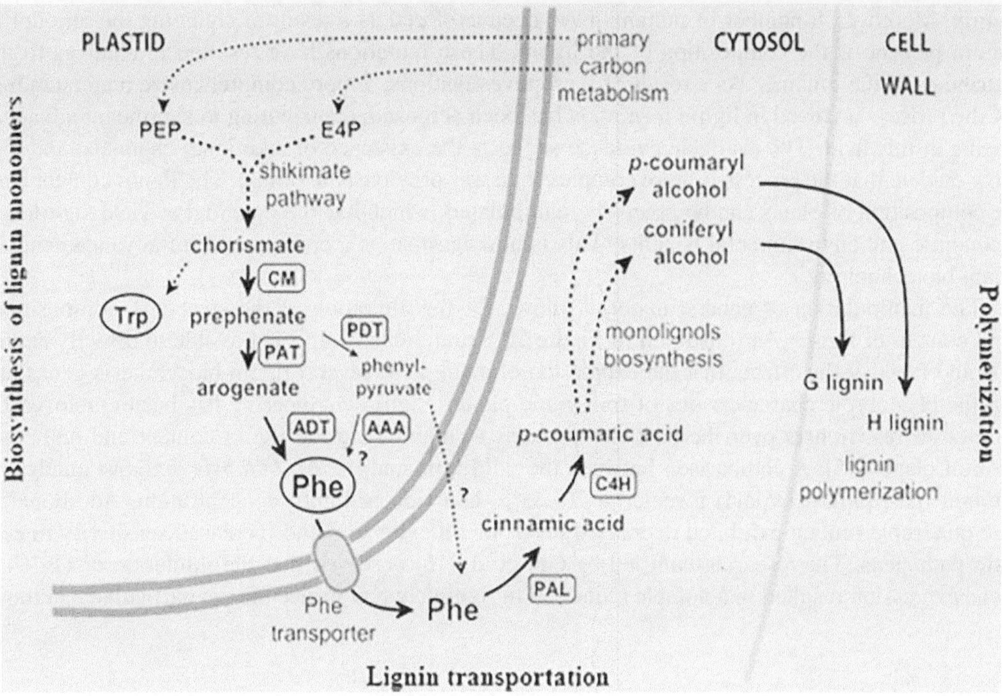

FIGURE 22.1 Biosynthesis of lignin and its pathways.

Source: Adapted from the work of Rao and Barros (2024). [13].

exploited as a resource for the pharmaceutical industry or as a source of energy. It is anticipated that the investigation of lignin biosynthesis and its function will have a substantial influence on the production of industrial goods, including household, furniture, agricultural goods, packaging, and other related activities. It provides a comprehensive overview of lignin biosynthesis, function, and application.

22.2 GENETICS OF LIGNIN SYNTHESIS

22.2.1 Genetic Modification of Lignin Biosynthesis

Lignin, which is defined as a complex heteropolymer of cinnamyl alcohols, is the second most prevalent biopolymer on Earth. While cellulose is one of the most prevalent types of biopolymer, the process of lignification has played a significant role in the adaptation of plants to life on land. Extracellular polymers have also been observed, but lignin is the only one that is responsible for imparting rheological characteristics to plant tissues. Two processes that are likely involved in this process are cell-to-cell communication and the physiology of cells and tissues. It is also likely involved in a broad range of other activities, including determining the quality of wood, influencing the processes of pulp and paper manufacture, and affecting the digestibility of fodder crops. From an economic perspective, lignin is of the greatest significance because it is responsible for determining the quality of wood. Given the aforementioned factors, the route that controls the formation of lignin has been the subject of a considerable amount of research. A significant number of the genes that are responsible for the enzymes involved in the production of lignin have been cloned and characterized at this point in time. Recent studies have reported on the alteration of the expression of these genes through the use of genetic engineering, which results in plants that have modified lignin. Moreover, a number of mutants have been explored as a result of changing the amount of lignin present or the composition of the lignin. These mutations have resulted in changes to the attributes of the mutants. As a result of these investigations, a more comprehensive understanding of the process involved in lignin formation has been achieved, contributing to significant advancements in this field. The available evidence suggests the existence of additional channels, and it is now evident that the process is more complex than was previously assumed. The lignin content and/or composition of plants can be precisely manipulated, which has the potential to yield significant economic and environmental benefits. This is now feasible as a consequence of advancements in plant biotechnology.

The manipulation of genetic material allows for the alteration of the structure, composition, and content of lignin. As illustrated in Figure 22.2 and Table 22.1, it is possible to directly examine and identify the effects that the expression of a single or several lignin biosynthesis genes has on the phenotypic characteristics of transgenic plants. Lignin engineering has been employed by numerous researchers over the past two decades to manipulate the lignin content and composition of plants [15]. A comparison between the wild type and the *Arabidopsis thaliana* quadruple mutant (pal1/pal2/pal3/pal4) revealed a 20–25% reduction in lignin concentration. Additionally, the quadruple mutant exhibited decreased levels of salicylic acid and increased sensitivity to certain pathogens. The research team led by Gui et al. [16] observed that the inhibition of Os4CL3 gene expression resulted in a notable reduction in lignin content in rice plants, particularly in those

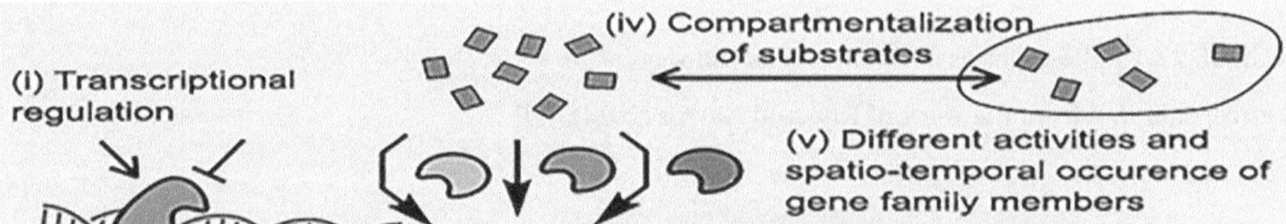

FIGURE 22.2 Schematic of tree lignin pathway modulation at many stages. Full arrowheads indicate metabolic conversions, while hollow arrowheads indicate translocation, protein modification, or signaling. Metabolites are polygons; enzymes are Pac-Man-like spheres.

Source: Adapted from the work of Meester et al. (2022). [4].

TABLE 22.1
The Function of Typical Genes Involved in Lignin Production in the Past Five Years

Genes	Plant Species	Functions	References
4CL	*Oryza sativa* (OsAAEs)	Reduced lignin content and increased response to rice diseases (rice blast)	30
CCR	*Paspalum dilatatum* (pdCCRI)	Improved lignin concentration changed accumulation of lignin	19
CAD	*Populus trichocarpa* (PtrCADI)	Lignin structure and content modified	19
MOMT	*Arabidopsis thaliana* (MOMT4)	Reduced biosynthesis of lignin, and improved biosynthesis and saccharification yields and content	22
COMT	*Sorghum bicolor* SbCOMT	S-lignin biosynthesis facilitated by methylation of tricin precursors	29
CSE	*Populus tremula* × *P. alba* PtxaCSE1/2	Resistant to numerous pathogens and participant in lignin biosynthesis	27
CoAOMT	*Zea maiz* (ZmCCoAMT2)	Adjusted H-lignin conformation	31
HCT	*Populous nigra* (HCT)	G-lignin content diminished, while S units increased	20
5H	*Oryza sativa* (OsCAld5H1)	Contributed to plant growth and development, as well as lignin biosynthesis	20
POD	*Arabidopsis thaliana* AtPrxx72	Lignin biosynthesis and the development of the root, anthers, and vascular system active	23
ABC	*Arabidopsis thaliana* AtABCG29)	Increased salt/osmotic stress tolerance and enhanced lignin deposition	33
MYB	*Betula platyphylla* (BpIMYB46)	Participated in fruit lignification	2
MADS	*Betula platyphylla* BpMADS12	Enhanced lignin accumulation and stress tolerance of *Arabidopsis*	34
HSF	*Eriobotra japonica* EjHSF3	Enhanced lignin accumulation and stress tolerance of *Arabidopsis*	35
SOD	*Potentilla atrosanguica* SOD	Participated in lignin biosynthesis and regulation of brassinosteroid signaling pathway	36
APX	*Rheum australe* (APX)	Enhanced lignin accumulation and stress tolerance of *Arabidopsis*	37

associated with height and straw yield. The suppression of the mRNA level of the 4CL gene in *Pinus radiata* resulted in a dwarfed phenotype of the plant [17], which is largely attributed to the depletion of guaiacyl lignin, which caused the lignin content of transgenic plants to decrease by 36–50%. Cinnamoyl-CoA reductase (CCR) and cinnamyl alcohol dehydrogenase (CAD) are two enzymes that are associated with the monolignols synthesis pathway. The disruption of their function also affects the lignin content, as was demonstrated in *Arabidopsis thaliana* with the ccc triple mutant (cad c cad d ccr1). Once the lignin content was reduced by 50% in comparison to the wild type, accompanied by a male-sterile phenotype [18], it was evident that this mutant was affected. A recent study by Van Acker and colleagues [19] revealed that a deficiency in CAD1 resulted in the formation of distinct metabolic pathways for the synthesis of coniferaldehyde and sinapaldehyde, as well as alterations in the composition and structure of lignin in poplar. The deposition of lignin in the thick-walled parenchyma bundle-sheath cells of C4 forage grasses has been demonstrated to affect the digestibility of the forage by animals. Moreover, the downregulation of CCR genes has been postulated to be involved in the efficient production of low-molecular-weight lignin [19].

The synthesis of sinapyl alcohol (S unit) and coniferyl alcohol (G unit) is catalyzed by enzymes, including p-coumarate 3-hydroxylase (C3H), hydroxycinnamoyl. Additionally, the enzymes CoA transferase (HCT) and caffeate/5-hydroxyconiferaldehyde O-methyltransferase (COMT) are involved. The expression of the CCR1 and COMT1 genes resulted in a notable alteration of ryegrass lignin concentration and composition. This alteration led to an increase in digestibility, without a discernible negative impact on either plant fitness or biomass production. Furthermore, the plants exhibited clear indications of stunting, decreased biomass, and delayed flowering. Shadle et al. [10] and Vanholme [20] demonstrated that the *Populus nigra* hct1 mutant exhibited altered H-lignin distribution. Inhibition of C3H expression in poplars resulted in a reduction in lignin content and a shift in the S/G ratio [21]. The overexpression of the monolignol 4-O-methyltransferase MOMT4 gene resulted in a 28% decrease in the amount of lignin found in *Arabidopsis* cells and an increase in the amount of saccharification produced by transgenic plants, as reported by Zhang et al. [22]. A recent study has demonstrated that the heterologous expression of angiosperm F5H in gymnosperms results in the ectopic production of S-lignin in the cell walls [23]. Conversely, the mutant enzyme ferulate 5-hydroxylase (F5H) is one of the key regulators of the composition of S/G lignin in plants. An increase in the formation of G-lignin was observed in rice when the F5H gene OsCAld5H1 was downregulated, while an increase in the concentration of S units was observed when the gene was overexpressed [24]. A recent study has demonstrated that the enzyme caffeoyl shikimate esterase (CSE) plays a role in the synthesis of lignin in *Arabidopsis thaliana*, *Medicago truncatula*, and hybrid poplar [25–27]. The downregulation of chalcone synthase (CHS) resulted in a significant reduction in the amount of apigenin- and tricin-related flavonoids, which led to a notable decrease in tricin lignin [28]. Additionally, it was demonstrated that tricin reacted with monolignols. The *Sorghum bicolor* (SbCOMT) enzyme was identified by Eudes et al. [29] as having the capacity to methylate the tricin precursors (luteolin and selgin), which are involved in the synthesis of lignin-linked tricin and S-lignin units in sorghum grains.

Following the creation of monomers in the cytoplasm, the subsequent steps involve the transport of these monomers across the cell membrane and the subsequent polymerization of these monomers in the cell wall. The p-coumaryl alcohol transport activity of the AtABCG29 protein resulted in *Arabidopsis* AtABCG29 deletion mutants exhibiting a lower lignin content and greater resistance to p-coumaryl alcohol, as evidenced by studies [2, 33, 37]. The polymerization of monomers is a process that involves the participation of two important enzymes: peroxidases and laccases. POD is overexpressed in plants, which results in an increase in the amount of phenol and lignin. Atprx double mutants (atprx2/atprx25, atprx2/atprx71, and atprx25/atprx71) exhibited a reduced lignin content relative to single mutants. However, these mutants did not exhibit a significant impact on plant growth [38]. The MADS-box transcription factor AGL15 was identified as the regulator of a novel class III peroxidase in *Arabidopsis thaliana*, which is transcribed by PRX17 (At2g22420). This enzyme plays a role in lignin production. In comparison to the genes involved in lignin biosynthesis that were discussed earlier, the role of laccase in the process of lignin biosynthesis has received a comparatively inadequate amount of research over the past several years. A notable increase in the total lignin content of transgenic poplar was observed following the overexpression of the cotton laccase gene [39]. The lignin content of the stems of two *Arabidopsis laccase* double mutants, lac4-1 lac17 and lac4-2 lac17, was observed to decrease by 20% and 40%, respectively [37]. In *Arabidopsis thaliana*, the lac11 lac4 lac17 triple mutant exhibits a striking absence of lignin deposition in roots. This finding is particularly noteworthy [33]. In *Brachypodium distachyon*, a novel lignification-related LACCASE5 was discovered, which resulted in a change in the amount of lignin present and its composition [40].

Research has demonstrated that transcription factors are capable of regulating the metabolism of phenylpropanoids and the synthesis of lignin-specific compounds. The MYB (v-myb avian myeloblastosis viral oncogene homolog) family represents a transcription factor that has been subjected to the most extensive examination in the context of lignin production. Moreover, it was demonstrated that the overexpression of *Arabidopsis thaliana* MYB58 and MYB63 transcription factors resulted in the activation of genes associated with lignin production and the promotion of the lignification of cells. In contrast, the expression of genes associated with lignin production was repressed by other transcription factors, including EgMYB1 from eucalyptus and MusaMYB31 from *Musa*. These transcription factors also negatively regulate lignin buildup [41–43]. A number of members of the NAC transcription factor family in *Arabidopsis* were found to be capable of regulating lignin biosynthesis by regulating the entire set of genes involved in cell wall synthesis [43]. This was in addition to the MYB transcription factor. In addition, WRKY transcription factors have the capacity to function as regulators of genes involved in lignin production. In *Medicago sativa* L., the downregulation of a WRKY transcription factor resulted in an increase in the level of lignin and an increase in the amount of biomass produced [44]. A growing number of genes associated with lignin metabolism have been identified through in vivo experiments in recent years. A number of genes have been identified that are involved in the biosynthesis of lignin. These include the maize Caffeoyl-CoA O-methyl transferase gene (CCoAOMT1), the switchgrass UDP-Arabinomutase gene, the *Betula platyphylla* MADS-box gene (BpMADS12), and the *Eriobotrya japonica* heat shock factors (EjHSF3) [35, 45, 46]. It is imperative that further research be conducted on the growth, development, and environmental adaptability of transgenic plants. Furthermore, genetic engineering technology is employed to alter the lignin concentration and composition.

22.3 LIGNIN AND PLANT DEVELOPMENT

22.3.1 The Importance of Lignin in the Development and Growth of Plants

Lignin, which is one of the primary components of the plant cell wall, is responsible for the provision of functions to a wide variety of cell types that are found in the tissues and organs of plants. As a consequence of interference with lignin production, particularly with regard to the H units, which frequently results in the suppression of plant growth and the formation of deformities [47], lignin metabolism plays a role in the growth and development of living plants. In certain plant species, the accumulation of lignin in cell walls is an essential component for the multiplication of seeds. The accumulation of lignin in the seed coat has the potential to serve as a protective barrier against external substances that may be harmful to the seeds. In comparison to the wild-type *Arabidopsis thaliana*, the mutant seed coat exhibited a significantly lower lignin content, which resulted in a markedly reduced seed germination rate. In addition to exhibiting stunted growth and a decreased number of seeds, the *Arabidopsis thaliana* CCR1 mutant displayed a considerable reduction in the amount of lignin [11, 15]. The growth and accumulation of lignin were both diminished in *Arabidopsis thaliana* C4H mutants, and the mutant also exhibited male sterility with no apical dominance. The simultaneous disruption of CAD and CCR in *Arabidopsis thaliana* resulted in a notable reduction in the quantity of lignin present in the plant, accompanied by a change in the composition of the lignin. This resulted in a notable reduction in plant growth and male sterility, which may be attributed to a lack of lignin in the anther. Figure 22.3 illustrates that *Selaginella* is the sole genus within the family Selaginellaceae, a group of vascular plants that are sometimes referred to as spikemosses or smaller clubmosses.

Arabidopsis thaliana peroxidase AtPrx72 knockout mutants exhibited significantly reduced photosynthetic efficiency, slow growth, fewer branches, and smaller flower and stem sizes compared to wild-type plants [32]. Furthermore, the total lignin and S unit contents were found to be significantly lower in these mutants compared to the wild-type mutants. The precise manner by which lignin affects the growth and development of plants remains poorly understood. The transcriptional process and signaling pathways that respond to cell wall abnormalities may play an essential role in the stunted growth observed in lignin-deficient plants [47]. Lignin, which is one of the primary components of the plant cell wall, is responsible for the provision of functions to a wide variety of cell types that are found in the tissues and organs of plants. In the context of plant growth and

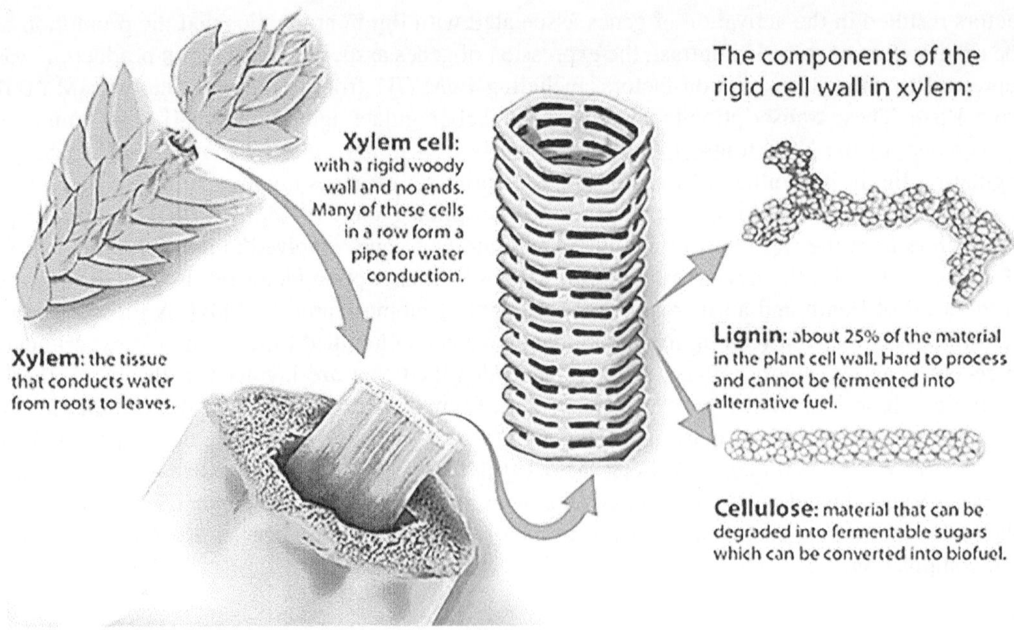

FIGURE 22.3 Among the vascular plants that belong to the family Selaginellaceae, sometimes known as the spikemosses or smaller clubmosses, the only genus that belongs to this family is called **Selaginella**. Location of lignin in a primitive plant: **Selaginella**.

Source: Adapted from the work of the National Science Foundation (2024). [48].

development, lignin metabolism plays a pivotal role, particularly in the interference with lignin biosynthesis, particularly the production of H units, which frequently results in the inhibition [32]. *Arabidopsis thaliana* peroxidase AtPrx72 knockout mutants exhibited significantly reduced photosynthetic efficiency, slow growth, fewer branches, and smaller flowers and stems than wild-type plants [32]. It was observed that the total lignin and S unit contents were significantly reduced in comparison to the wild type. This phenomenon was observed in *Arabidopsis thaliana* cells, where the mechanisms by which lignin influences plant growth and development remain poorly understood. The transcriptional process and signaling pathways that respond to cell wall abnormalities may play an essential role in the stunted growth observed in lignin-deficient plants, as proposed by Bonawitz et al. [47].

22.4 LIGNIN AND PLANT RESISTANCE

22.4.1 The Impact of Lignin on the Resistance of Plants to Lodging

It is one of the most essential characteristics that affects crop growth and grain yield [49]. Lodging resistance can prevent plant stems from bending or breaking simultaneously. The lodging resistance of crops has been demonstrated to be related to a number of factors, including plant height, biomass, stem diameter, and the composition and features of stem cell walls. A multitude of studies have demonstrated this correlation in crops. The mechanical strength of plant stalks is significantly enhanced when they have a considerable buildup of lignin in their cell walls. There are significant implications for the resilience of crops to lodging.

The exogenous application of paclobutrazol to a winter wheat cultivar that exhibited either lodging resistance or susceptibility resulted in a significant reduction in the length of the internode, accompanied by an increase in lateral growth, lignin deposition, the activities of lignin biosynthesis

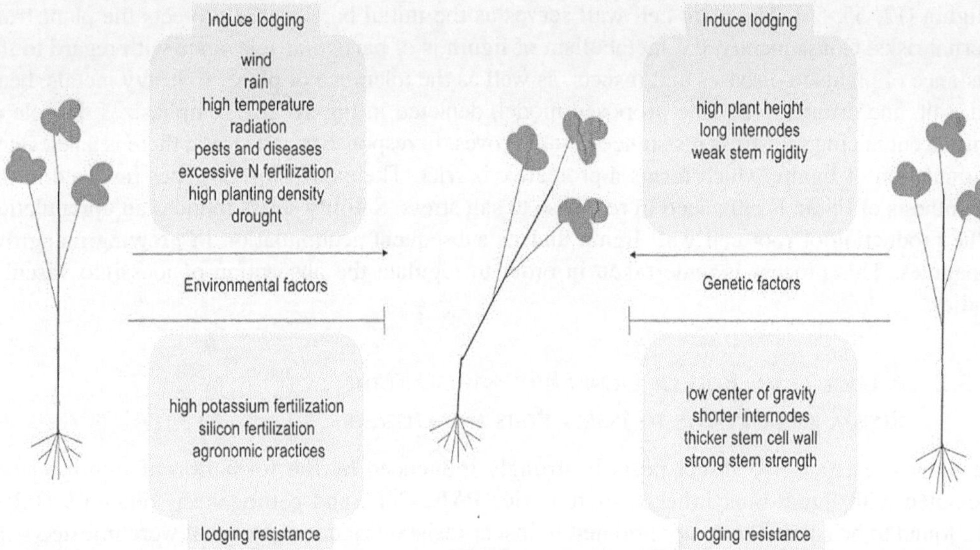

FIGURE 22.4 The factors that influence stem lodging. By "induction factors," we mean the arrow! The lodging resistance factors are represented by the little line.

Source: Adapted from the work of Li et al. (2022). [52].

enzymes, and the thickness of the internode [70]. This resulted in an improvement in the crop's ability to withstand lodging. A further study demonstrated that the density of the crop has the potential to significantly alter the morphological traits and lignin production of the stem, which ultimately results in an increase in the mechanical strength of the stem and a reduction in the risk of lodging [50]. The findings of Hu et al. [51] indicate that analysis of lignin metabolism–related indexes in *Fagopyrum esculentum* Moench varieties with varying degrees of lodging resistance revealed that the lignin content and the activities of lignin biosynthesis enzymes (PAL, 4CL, CAD, and POD) played significant roles in lodging resistance. The factors that influence stem lodging have been identified (2022; Figure 22.4).

The formation of lignin in plants and their resistance to lodging are both significantly influenced by the element silicon, which is present in nutrients. Silicon has the capacity to stimulate the expression of the rice CAD gene, resulting in an augmented accumulation of lignin and enhanced strength of the stalk, which ultimately contributes to an improvement in the stability of the stalk [53]. Buckwheat, rapeseed, and japonica rice were subjected to an excessive amount of nitrogen fertilizer, which resulted in a considerable decrease in the mechanical strength of the stalk and a reduction in resistance to lodging. Kong et al. [54] demonstrated that the detrimental impact of excessive NH4+ on wheat culm strength can be significantly reduced by the addition of K+. This led to the hypothesis that the reduction in lodging resistance observed in response to nitrogen fertilizer may be linked to the suppression of K+ uptake, which, in turn, results in an increase in the deposition of lignin in the vascular bundles.

22.5 LIGNIN AND ADAPTATION TO STRESS

22.5.1 Lignin Production and Plant Adaptation to Stress

One of the fundamental responses of plants subjected to biotic and abiotic challenges is the accumulation of reactive oxygen species (ROS), which is accompanied by an increase in the deposition of lignin [12, 55, 56]. The plant cell wall serves as the initial barrier that protects the plant from external risks. Consequently, the metabolism of lignin is of particular relevance with regard to the resistance of plants to diseases and insects, as well as the tolerance of plants to heavy metals, heat, cold, salt, and drought [12]. The proposed model, depicted in Figure 22.5, emphasizes the role of lignin in enhancing salt stress resistance in mangroves. In response to salt stress, there is an elevated accumulation of lignin, which forms a protective barrier. The expression of genes involved in the biosynthesis of lignin is enhanced in response to salt stress. Salinity stress induces an upregulation in the production of root cell wall lignin and its subsequent accumulation in growing mangrove propagules. This process is undertaken in order to regulate the absorption of ions into vascular bundles.

22.5.2 A Look at the Role of Lignin Biosynthesis in the Resistance of Plants to Insect Pests and Diseases

The plant's resistance to insect pests is strongly influenced by the formation of lignin. Genes associated with lignin biosynthesis, such as rice PAL, C4H, and pathogenesis-related 9 (PR9), were found to be significantly upregulated in insect-resistant rice varieties that were infested with brown plant hoppers. This finding suggests that these genes may participate in lignin biosynthesis in a synergistic manner, thereby regulating the insect resistance of rice. A transcription factor, designated CmMYB19, was identified in *Chrysanthemum* that was induced by aphid penetration. This transcription factor was found to promote the expression of lignin biosynthesis genes and the accumulation of lignin. Consequently, the invasion of aphids was restricted, and the chrysanthemum's tolerance to aphids was enhanced. The ethylene-dependent accumulation of lignin

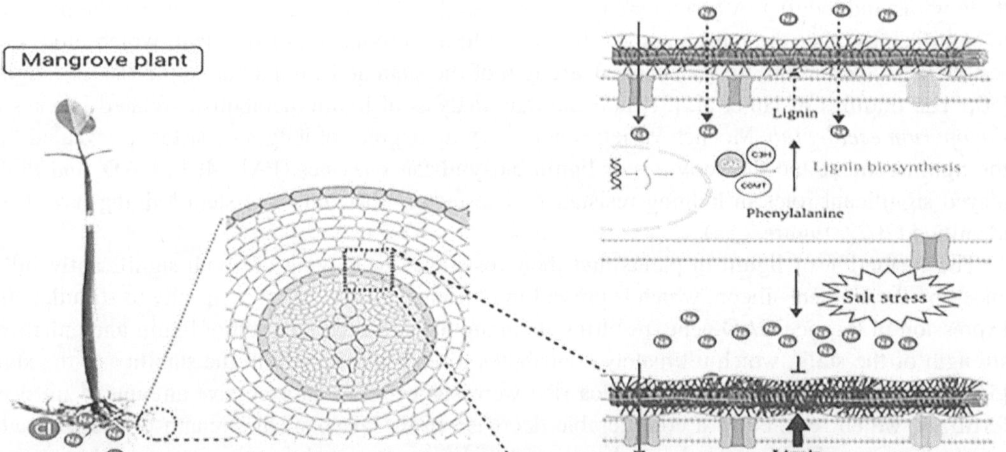

FIGURE 22.5 The lignin-mediated salt stress tolerance in mangroves is the subject of this proposed model. When subjected to salt stress, an increased lignin deposition creates a barrier. In response to salt stress, the expression of essential genes involved in biosynthesis is increased. Under salinity stress, the manufacturing of root cell wall lignin and its accumulation in developing mangrove propagules both increase. This is done in order to regulate the uptake of ions into vascular bundles.

Source: Adapted from the work of Ashifa et al. (2023). [57].

in the roots, as observed by Fujimoto et al. [58], demonstrated that the antimicrobial molecule sclareol increased the resistance of *Arabidopsis thaliana* to the root-knot nematode. Similarly, an insect-specific toxin peptide called LqhIT2, which increased the amount of lignin with leafroller infestation, improved the leafrollers' resistance to rice. The utilization of lignin, either directly as a barrier or indirectly through the related hormone signal system, can result in an increase in insect resistance in plants. Figure 22.6 illustrates the various defense mechanisms employed by plant cells to ward off infectious agents. The thickening of cell walls that occurs as a result of lignification has been found to be an effective method of preventing the development of fungal germ tubes. The production of reactive oxygen species (ROS), phytoalexins, pathogenesis-related protein (PR), and secondary metabolites, such as phenylalanine ammonia lyase, polyphenol oxidase, flavonoids, and other PAMPs/DAMPs, is the biochemical mechanism by which PAMP- and effector-triggered immunity (PTI and ETI) defenses are initiated. PAMPs and DAMPs serve as the initiating triggers for PRR-mediated immunity. A considerable number of secondary metabolites act as signal molecules and intermediates in the process of SAR, which is connected with plant diseases.

Lignin, which contains a greater concentration of H units, accumulates in the cell wall of plants that have been infected with pathogens. This is evidenced by the findings of studies [60, 61]. This results in an increase in the thickness of the cell wall. Lignin-related compounds may result in the inactivation of fungal pathogens and the prevention of their multiplication and dissemination [62, 63, 64]. An augmented accumulation of lignin can diminish the infiltration of fungal enzymes and

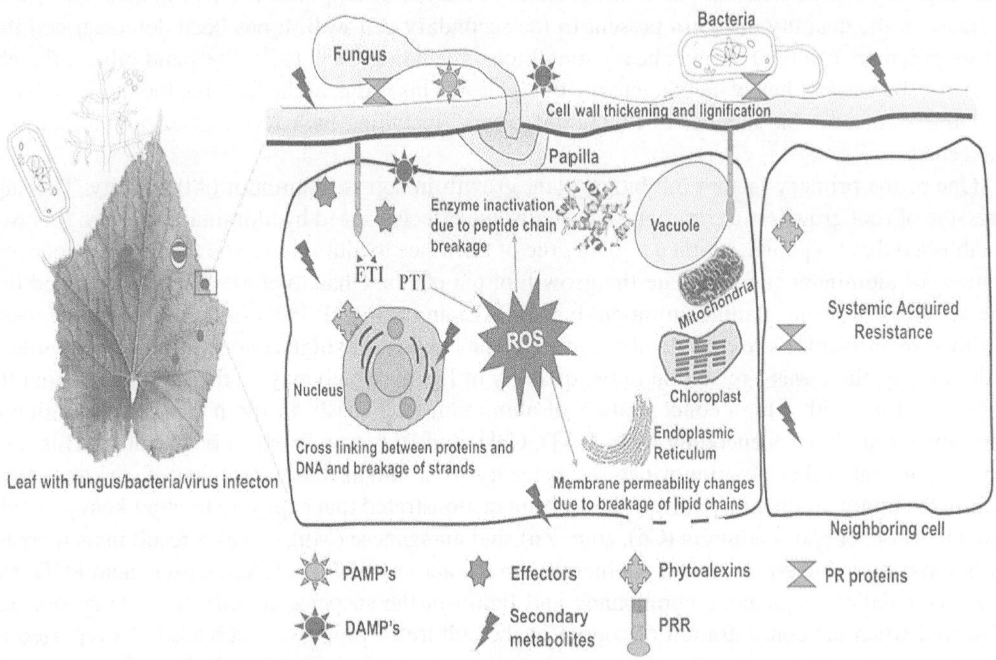

FIGURE 22.6 Plant cell biochemical pathogen defense diagram. Thickening cell walls by lignification and papillas hinder fungal germ tube growth. PAMP- and effector-triggered immunity (PTI and ETI) biochemically defend by creating ROS, phytoalexins, pathogenesis-related (PR) protein, and secondary metabolites, such phenylalanine ammonia lyase, polyphenol oxidase, flavonoids, and others. PAMPs/DAMPs start PRR-mediated immunity. Many secondary metabolites are signal molecules and intermediates for plant pathogen–related SAR.

Source: Adapted from the work of Kaur et al. (2022). [59].

toxins into plant cell walls, thereby providing a fundamental barrier against the dissemination of pathogens and reducing the infiltration of these substances. The research team led by Mandal et al. [65] observed that the quantity of lignin present in tomato cultivars that exhibited resistance to pathogenic bacteria was considerably higher than that observed in cultivars that were susceptible to the bacteria. The expression of the gene responsible for lignin production is directly correlated with the disease resistance exhibited by plants. In *Arabidopsis thaliana*, the expression of *Arabidopsis* CAD5 was demonstrated to be markedly elevated in roots that had undergone substantial lignification and was induced by pathogens that invaded the plant [66]. The overexpression of the GhDIR1 gene resulted in a significant increase in the degree of lignification and enhanced *Verticillium dahliae* resistance in transgenic plants. The cotton dirigent-like gene (DIR) was engaged in the deposition of lignin during the process of genetic modification. The expression of the lignin production pathway gene and the buildup of lignin were both increased as a result of the binding of two HCT genes that were found in maize. These genes were responsible for regulating plant disease resistance. It is conceivable that the observed decline in POD activity and the downregulation of pathogen-related 1a (PR1) are linked to the overexpression of the rice 4CL gene OsAAE3, which resulted in a reduction in lignin accumulation and enhanced sensitivity to rice blast [30]. A correlation was identified between a CCoAOMT gene, designated ZmCCoAOMT2, and resistance to various diseases in maize. This gene may also be involved in the synthesis of lignin and other phenylpropanoid metabolites, as well as the regulation of programmed cell death.

The absorption and transport of heavy metals, as well as the plant reaction to heavy metal stress, are both actively implicated in the cell wall, which serves as the first entry barrier for metal ions [67, 68]. In order to increase the thickness of the cell wall, heavy metal stress can activate the pathway that leads to the production of phenolic secondary metabolic compounds. This can also result in an increase in the quantity of lignin present in the secondary cell wall. It has been demonstrated that lignin polymers can bind various heavy metal ions, including Cu^{2+}, Cd^{2+}, Pb^{2+}, and others, thereby reducing the entry of heavy metals into the cytoplasm. This is due to the fact that the lignin polymer contains a considerable number of functional groups, including hydroxyl, carboxyl, and methoxyl molecules.

One of the primary factors inhibiting plant growth in soil is aluminum (Al) toxicity. The suppression of root growth is a common sign of adverse effects caused by aluminum toxicity. It is well established that tea plants exhibit a high degree of tolerance to aluminum, and that elevated concentrations of aluminum can stimulate the growth of tea plants. Ghanati et al. [69] demonstrated that the activities of phenylalanine ammonia-lyase (PAL) and cell wall POD in the lignin biosynthesis pathway in tea plants were considerably reduced when exposed to high concentrations of aluminum. Additionally, there was a reduction in the quantity of lignin, which may be the rationale behind the capacity of tea with a high concentration of aluminum to flourish. In the presence of aluminum, the expression of genes encoding 4CL, CAD, C3H, and PAL was found to be elevated. This finding demonstrated that aluminum has the capacity to stimulate the expression of enzyme genes within the lignin production pathway. It has been demonstrated that exposure to other heavy metals, including copper (Cu), cadmium (Cd), zinc (Zn), and manganese (Mn), can also result in an increase in lignin content in certain plants. An increase in the activity of PAL, CAD, caffeic acid POD, and the accumulation of phenolic compounds and lignin in the suspension cells of ginseng root was observed when the concentration of copper in the culture medium was increased. As reported by Bhuiyan et al. [70], the growth of soybean roots was found to be inhibited when they were subjected to a Cd treatment. Additionally, the concentration of lignin was elevated, accompanied by an increase in POD and LAC activity. Moreover, the expression of the POD gene, which is associated with lignin production, was also found to be upregulated [71–73].

Furthermore, the absorption, transport, and tolerance of heavy metals in plants are directly correlated with the process of lignin production. When lignin is deposited in the cell wall of the root endoderm, it has the potential to impede the movement of heavy metal elements into the xylem or outward from the vascular bundle. The higher expression level of Zn-induced lignin biosynthesis genes, which were found to be higher in the roots of the Zn/Cd hyperaccumulator of *Thlaspi caerulescens* than in the roots of *Arabidopsis*, was observed. Furthermore, histological analysis of tissue sections revealed that the accumulation of lignin in the endodermis cell wall of *Thlaspi caerulescens* roots was significantly higher than that of *Arabidopsis* roots [74–82]. The findings indicate that the production of lignin plays a pivotal role in the hyperaccumulation of zinc and cadmium in *Thlaspi caerulescens*. This phenomenon may be linked to the process of ion absorption and transport. Our previous research indicated that the deposition of lignin reduced the harmful effects of manganese, which is an essential mechanism of manganese tolerance in *Phytolacca americana*, a manganese hyperaccumulator. Xylem that has undergone lignification in the roots can serve to limit the amount of cadmium transported to the shoots. Our recent findings have demonstrated that the presence of Cu stress can lead to an increase in the accumulation of lignin in the roots of rice by promoting the polymerization of lignin. Hydrogen peroxide (H_2O_2) generated by copper stress was found to regulate the polymerization of lignin monomers and lignin accumulation. This mechanism may influence the transport of copper from the root to the shoot in rice seedlings [34].

22.5.3 The Function of Lignin in the Stress Tolerance of Plants to Salt and Drought

It is a well-established fact that extreme drought and high salt stress typically occur together, leading to an osmotic stress. This stress causes plant cells to lose water or even die, which has a substantial impact on plant growth and development and leads to significant losses in agricultural yield [32]. The maintenance of cell osmotic equilibrium and the integrity of protective membranes can be facilitated by the presence of lignin, which has the capacity to impede the penetration of water and transpiration through plant cell walls. A number of studies have demonstrated that lignin biosynthesis is enhanced in response to drought stress. Both the stem basal zone of *Eucalyptus urograndis* and the stem apical zone of *Eucalyptus globulus* exhibited a notable increase in the quantity of lignin present in the plants following exposure to drought stimulation. The expression levels of lignin biosynthesis–associated genes (CAD and COMT) and the amount of lignin found in the leaves of inbred maize lines were demonstrated to exhibit a strong positive correlation with their observed drought tolerance under drought stress [83–86].

A notable elevation in the expression of CCR1/2 was observed in the root elongation region of maize when the plant was subjected to drought stress. CCR is a crucial gene in the pathway that leads to the production of lignin monomer and the catalysis of lignin production, which plays a pivotal role in the drought stress tolerance of *Leucaena*. Furthermore, it was observed that drought stress significantly elevated lignin content and CCR protein expression in the growing stems of *Leucaena* seedlings. This finding was corroborated by the observation that CCR protein expression was markedly elevated. The expression of CCoAOMT was evidently elevated in the root elongation region of soybeans when they were subjected to drought stress, as evidenced by studies [86–88]. Moreover, the accumulation of lignin in the region was also significantly increased. The increase in lignin content resulted in a reduction in the elongation of the root cell wall, which may have the potential to limit the loss of water from the roots and promote the transport of water into the elongated zone. There is a strong possibility that lignin deposition plays a role in the salt tolerance of plants, as demonstrated by recent research. The overexpression of the IbLEA14 genes in *Ipomoea batatas* resulted in an increase in the accumulation of lignin, which, in turn, led to an improvement in the transgenic callus exhibiting salt and osmotic stress tolerance. In addition to being one of the reactive oxygen species (ROS) that are caused by stress, hydrogen peroxide (H_2O_2) plays a significant regulatory role in the process of lignin production [56, 89]. The overexpression of genes for superoxide dismutase (SOD), ascorbate peroxidase (APX), and *Potentilla atrosanguinea* resulted in an increase in the accumulation of lignin in transgenic *Arabidopsis thaliana* [36]. This was achieved by maintaining the concentration of hydrogen peroxide. The overexpression of *Betula platyphylla* BplMYB46 in transgenic birch resulted in enhanced salt and osmotic stress tolerance and lignin deposition. This indicated that the abiotic stress response and lignin biosynthesis pathways may have a cross talk with each other [34]. MYBs play a pivotal role in the biosynthesis of secondary cell walls and the tolerance to abiotic stress situations. Lignified cell walls, which are rich in lignin and impermeable, make plants more resistant to drought and high osmotic environments [13, 14, 90–94].

22.5.4 The Importance of Lignin in the Adaptability of Plants to Temperature Stress

A rise in temperature can lead to the deterioration of biomacromolecules, such as proteins and nucleic acids, and can also result in the oxidation of membrane lipids, which can disrupt the normal metabolic processes of plants. Climate change exerts an influence on the temperature of the environment in which plants are cultivated. In contrast, low-temperature stress is responsible for the degradation of cell membranes, the reduction of plant photosynthesis and respiration, and the severe inhibition of plant growth and development. C3H may affect the rigidity and permeability of the cell wall by modifying the S/G ratio and participate in the cold resistance of *Rhododendron* [4, 15, –95]. The lignin content of plant tissues exhibited a notable increase during the process of cold acclimation. The expression of the C3H gene was significantly higher in the *Rhododendron* leaves that had been exposed to cold acclimation than in the leaves that had not been exposed to cold acclimation. The expression of an HSF gene, designated EjHSF3, which is involved in fruit lignification through its interaction with the lignin biosynthetic regulator EjAP2-1, was found to be downregulated by heat and low-temperature conditioning (LTC). This further demonstrated that LTC has the ability to reduce low-temperature-induced lignification in loquat fruit, as previously reported in [48, 96].

A positive correlation was established between the temperature of the periods between early September and late October and the amount of lignin deposited in the terminal latewood tracheid of Norway spruce. It was discovered that postharvest high-temperature-induced lignin deposition in Satsuma mandarin pericarp is important in resistance to environmental stress. A notable reduction in lignin content was observed in the *Medicago truncatula* CAD1 mutant relative to the wild type. Moreover, no discernible difference in growth was observed between the wild type and the mutants when the temperature was maintained at a normal level of 22°C. Nevertheless, when the temperature was set at 30°C, the growth of the MtCAD1 mutant was considerably impeded, and the growth phenotype of compensated lines was comparable to that of the natural type [52]. It is possible that the accumulation of reduced lignin may result in vascular injury, which, in turn, may negatively impact transpiration and cause the plant body to overheat.

22.6 LIGNIN IN BIOLOGICAL SYSTEMS

22.6.1 Types and Functions of Lignin in Biological Systems

As illustrated in Figure 22.7, lignin fulfills a multitude of functions within the plant kingdom. The primary applications of plant lignin are as a source of energy and for the development of novel materials, such as lignin-based carbon fibers. This is due to the presence of phenolic hydroxyl groups and aliphatic hydroxyl groups. This is due to the fact that plant lignin is a resource that is both inexpensive and renewable. In general, lignin can be classified into three distinct categories, namely, softwood, hardwood, and grass lignin, depending on the species of plant from which it originates (see references 14, 15, 31, 50, 100–102). Coniferyl alcohol is the sole constituent of softwood lignin, whereas sinapyl alcohol and coniferyl alcohol collectively constitute the predominant components of hardwood lignin. Grass lignin contains three different types of monomers: coniferyl, sinapyl, and p-coumaryl alcohol. Moreover, non-traditional forms, including caffeyl lignin (C-lignin), were identified in the seeds of vanilla orchids and in a number of taxa belonging to the family Cactaceae.

The molecular structure of monomer units has a significant impact on the structural patterns of lignin. This includes the degree of cross-linking with polysaccharides and the branching of polymer units. Consequently, the composition of the monomer exerts a decisive influence on the degradability of lignin and the workability of lignocellulosic biomass [92]. As an illustration, corn and flax

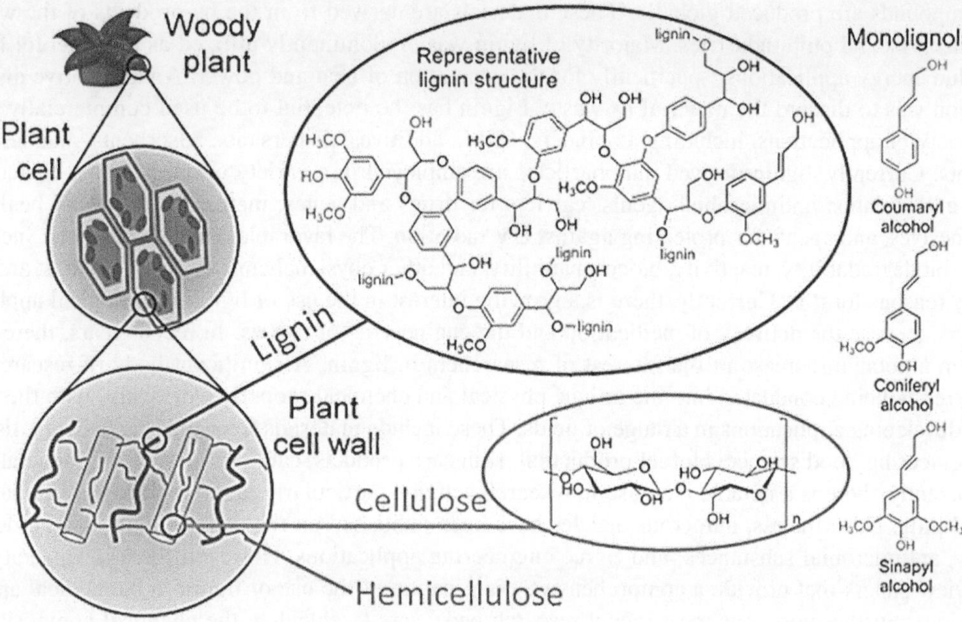

FIGURE 22.7 Diagrammatic representation of the biosynthesis and biological functions of lignin in plants. **Source: Adapted from the work of Liu et al. (2018). [101].**

lignins typically exhibit a high concentration of aliphatic hydroxyl (OH) groups and non-methoxylate phenolic groups, respectively [98–102]. The OH groups are suitable for the production of phenolic resins and polyurethane synthesis, respectively. Conversely, triticale lignin exhibits a balanced OH group content, rendering it suitable for polyester synthesis. Furthermore, low-molecular-weight monomers derived from H and G units exhibit antioxidant properties, indicating that lignin possesses biological activity, including antitumor activity [52, 57, 59, 99–112].

22.7 IMPORTANCE AND APPLICATION OF LIGNIN IN BIOMEDICAL SYSTEMS

22.7.1 Applications of Lignin in Biomedical Sciences

Lignin, derived from the Latin term "lignum," is a vital component of natural lignocellulosic polymers. The composition is comprised of three primary structural units: coniferyl alcohol, p-coumaryl alcohol, and sinapyl alcohol. Lignin is a common biopolymer that is linear, amorphous, and second in abundance only to cellulose. It has the potential to be utilized in pharmaceutical administration. Lignin exhibits a number of advantageous properties, including biodegradability, antimicrobial activity, and greater availability than other polymers. Lignin is a renewable biomacromolecule that exhibits a considerable degree of branching and is second in quantity only to cellulose. Lignin is primarily found in vascular plants, particularly in herbs and grasses. The molar mass of lignin is typically within the range of 1,000 to 2,000 g mol-1. Lignin is distinguished from other biomass constituents, such as carbohydrates, proteins, lipids, vitamins, tastes, and colors, by its complex and amorphous structure. The structure is composed of repeating cross-linked phenylpropanoid units derived from three fundamental monomer units: cumaryl, coniferyl, and sinapyl alcohol. Lignin was obtained from a variety of biomass sources, including wood biomass, trees, wheat straw, prairie cordgrass, switchgrass, and maize stover, through a range of processing methods, including kraft, sulfite, soda, and organosolv. Annually, over 50 million tons of lignin-based compounds are produced globally. These materials are derived from the by-products of the wood hydrolysis and pulp industries. Majority of lignin was predominantly utilized as a biofuel for low-value energy applications, specifically for the generation of heat and power. An alternative disposition was to discard the material as waste. Lignin has the potential to be used commercially in a variety of applications, including as binders, fillers, additives, dispersants, adsorbents, and surfactants. Currently, lignin-derived nanoparticles are employed in a variety of applications, including as antioxidants, antimicrobial agents, carriers for drugs and genes, materials for wound healing, adhesives, and agents for protecting against UV radiation. The favorable qualities of lignin, including biodegradability, reactivity, biocompatibility, and other physicochemical characteristics, are the key reasons for this. Currently, there is a growing interest in the use of lignin in biological applications, such as the delivery of medication and the engineering of tissues. In recent years, there has been a notable increase in the interest of researchers in lignin. A significant body of research is currently being conducted into the unique physical and chemical properties of lignin, with the aim of developing applications in a range of fields. These include materials science, drug delivery, tissue engineering, food science, biofuel production, skin care products, catalysis, and energy generation. Currently, there is a notable increase in research activity, particularly in the biomedical utilization of lignin. Nevertheless, numerous articles have been published on the use of lignin in drug delivery, antimicrobial substances, and tissue engineering applications. However, there is a paucity of review papers that provide a comprehensive examination of the use of lignin in biomedical applications. Furthermore, our most recent research endeavors to elucidate the chemical composition, physical-chemical properties, and extraction techniques of lignin. A comprehensive analysis of the advantages and disadvantages of altering lignin through the use of inorganic nanocomposites/nanomaterials was presented. Furthermore, our research indicates a pressing need to consolidate the information pertaining to the biomedical applications of lignin and the potential challenges that may arise in future advancements. This will facilitate further inquiry by the international scientific community. This survey provides a comprehensive investigation of the diverse applications of lignin in the fields of medication delivery and tissue engineering.

22.7.2 Lignin in Drug Delivery

Figure 22.8(a, b, c) provides a comprehensive overview of drug delivery systems that are based on lignin. Nanoparticle carriers are employed in the safe transport of medicinal formulations to the targeted site in the body, thereby enabling the desired therapeutic effects to be achieved. A multitude of forms of nanoparticles have been subjected to exhaustive investigation. Lignin-based composites are currently in high demand for the purposes outlined in the introduction section. The use of nanoformulation represents a crucial methodology employed in the fabrication of lignin nanoparticles for the purpose of medication delivery systems. Lignin-based nanoparticles are being increasingly utilized in drug delivery and tissue engineering due to their remarkable physicochemical properties, which include antibacterial, antioxidant, and antiparasitic characteristics. Pharmaceutical substances that can be delivered in a systematic manner according to the principles of motion are highly suitable for utilization in drug delivery applications. A substantial body of empirical research has been conducted on the manufacturing of lignin nanoparticles. The synthesis of lignin nanoparticles is achieved through a multitude of diverse methodologies. The most prevalent techniques employed include milling, high-pressure homogenization, solvent shifting, pH shifting, template-based synthesis, ice segregation–induced self-assembly, aerosol processing, electrospinning, supercritical fluid processing, solvent antisolvent precipitation, and acoustic cavitation–assisted nanoparticle preparation. Over the past few years, a multitude of lignin nanoparticles have been utilized in a multitude of ways in accordance with a multitude of approaches to the development process. The therapeutic efficacy of these drugs can also be enhanced by the formation of nanoparticles, which can enhance the bioavailability, absorption, and solubility of a wide range of chemicals. The research community in the domains of biology and medicine has determined that 35–40 lignin is

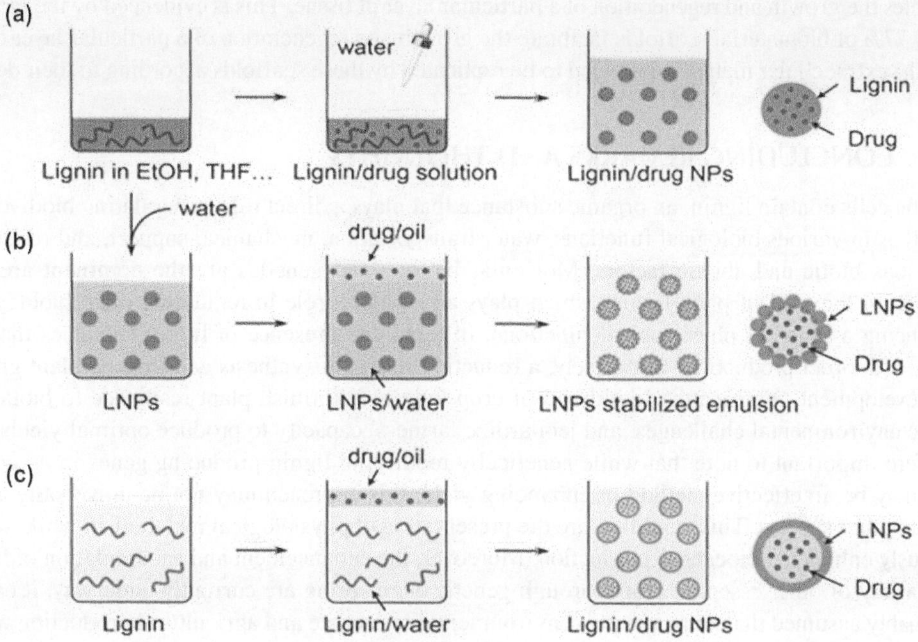

FIGURE 22.8 There are a variety of drug delivery systems based on lignin, and the construction of these systems involves approaches that are commonly used. Nanoparticles can be broken down into three various categories: (a) nanoparticles that are composed of a combination of lignin and drug; (b) drug-loaded lignin nanoparticles that have been stabilized; and (c) lignin itself.

Source: Adapted from the work of Kumar et al. (2021). [112].

the most suitable alternative for the delivery of therapeutic medication due to its potential benefits. Figure 22.8(a, b, c) depicts the formulation of a number of different drug delivery systems based on lignin. It also provides a summary of the strategies typically utilized. The procedure involves the use of ultrasonication at a frequency of 20 kHz and a power of 600 W. This is achieved through the use of ultrasonication. In terms of particle size, the collected particles were spherical and exhibited a size range of 0.01–0.05 μm. The application of acoustic irradiation resulted in the complete degradation of the phenyl coumaran and pinoresinol subunits, while the aryl glycerol aryl ether subunits (β-O-4) exhibited a notable reduction. This finding was corroborated by the experimental results. All these factors were instrumental in the observed reduction in lignin size.

22.7.3 Lignin in Tissue Engineering

The design and development of biomaterials, bioactive chemicals, and cell combinations, as well as the subsequent in vitro and in vivo studies on these subjects, are all included in the discipline of tissue engineering, which is a multidisciplinary field. The ongoing development of tissue engineering technology has made it possible to transplant complete organs or tissues for transplantation purposes. A promising alternative for repairing damaged tissue or pieces, transplantation has become a viable option as a result of this. In order to create regenerated tissue that may be used in medicinal applications, it is necessary to develop natural biomaterials. The following figure illustrates the techniques employed in the creation of the various scaffold materials utilized in tissue engineering. The appropriate substrate is provided by biomaterials, which are key components that facilitate the regeneration and replacement of deformed tissue with functioning tissue. It can be observed that biomaterial scaffolds, when applied in the context of tissue engineering applications, serve as a template that promotes the growth and regeneration of a particular layer of tissue. This is evidenced by the fact that nearly 77% of biomaterial scaffolds facilitate the growth and regeneration of a particular layer of tissue. The extracellular matrix is intended to be replicated by these scaffolds according to their design.

22.8 CONCLUDING REMARKS AND THOUGHTS

Specific cells contain lignin, an organic substance that plays a direct role in regulating biodiversity, as well as in various biological functions, water transportation, mechanical support, and resistance to various biotic and abiotic factors. Molecular biology and genetics are the dominant areas of research in the field of plant lignin, which plays an essential role in regulating plant biology and influencing a range of physiological functions. In trees, the presence of lignin enhances the efficiency of biofuel production. Conversely, a reduction in lignin synthesis can impede plant growth and development, increase the likelihood of crop lodging, diminish plant resilience to biotic and abiotic environmental challenges, and jeopardize farmers' capacity to produce optimal yields. It is therefore important to note that while genetically modifying lignin-producing genes in an energy plant may be an effective method of enhancing yield, this approach may not be universally applicable to other species. This would ensure the preservation of physiological metabolism while simultaneously enhancing bioenergy production. Moreover, the enhancement and accumulation of lignin in a variety of other essential crops through genetic engineering are currently underway. It can be reasonably assumed that an increase in environmental resistance and agricultural production would result in a higher lignin requirement. The manipulation of lignin synthesis genes to alter monomer composition and proportion may also facilitate the application of lignin.

REFERENCES

1. Ralph, J.; Lundquist, K.; Brunow, G.; Lu, F.; Kim, H.; Schatz, P.F.; Marita, J.M.; Hatfield, R.D.; Ralph, S.A.; Christensen, J.H. Lignins: Natural polymers from oxidative coupling of 4-hydroxyphenyl-propanoids. *Phytochem. Rev.* 2004, *3*, 29–60.
2. Alejandro, S.; Lee, Y.; Tohge, T.; Sudre, D.; Osorio, S.; Park, J.; Bovet, L.; Geldner, N.; Fernie, A.R.; Martinoia, E. AtABCG29 is a monolignol transporter involved in lignin biosynthesis. *Curr. Biol.* 2012, *22*, 1207–1212. [CrossRef] [PubMed]
3. Miao, Y.; Liu, C. ATP-binding cassette-like transporters are involved in the transport of lignin precursors across plasma and vacuolar membranes. *Proc. Natl. Acad. Sci. USA* 2010, *107*, 22728–22733. [CrossRef] [PubMed]
4. Bonawitz, N.D.; Chapple, C. The genetics of lignin biosynthesis: Connecting genotype to phenotype. *Annu. Rev. Genet.* 2010, *44*, 337–363. (doi:10.1146/annurev-genet-102209-163508). PMID: 20809799
5. Liu, C.J.; Miao, Y.C.; Zhang, K.W. Sequestration and transport of lignin monomeric precursors. *Molecules* 2011, *16*, 710–727.
6. Schuetz, M.; Benske, A.; Smith, R.A.; Watanabe, Y.; Tobimatsu, Y.; Ralph, J.; Demura, T.; Ellis, B.; Samuels, A.L. Laccases direct lignification in the discrete secondary cell wall domains of protoxylem. *Plant Physiol.* 2014, *166*, 798–807. [CrossRef] [PubMed]
7. Ithal, N.; Recknor, J.; Nettleton, D.; Maier, T.; Baum, T.J.; Mitchum, M.G. Developmental transcript profiling of cyst nematode feeding cells in soybean roots. *Mol. Plant Microbe Interact.* 2007, *20*, 510–525. [CrossRef] [PubMed]
8. Tripathi, S.C.; Sayre, K.D.; Kaul, J.N.; Narang, R.S. Growth and morphology of spring wheat (*Triticum aestivum* L.) culms and their association with lodging: Effects of genotypes, N levels and ethephon. *Field Crop. Res.* 2003, *84*, 271–290. [CrossRef]
9. Rest, B.V.D.; Rochange, S.F. Down-regulation of cinnamoyl-CoA reductase in tomato (*Solanum lycopersicum* L.) induces dramatic changes in soluble phenolic pools. *J. Exp. Bot.* 2006, *57*, 1399–1411. [CrossRef] [PubMed]
10. Shadle, G.; Chen, F.; Srinivasa Reddy, M.S.; Jackson, L.; Nakashima, J.; Dixon, R.A. Down-regulation of hydroxycinnamoyl CoA: Shikimate hydroxycinnamoyl transferase in transgenic alfalfa affects lignification, development and forage quality. *Phytochemistry* 2007, *68*, 1521–1529. [CrossRef] [PubMed]

11. Derikvand, M.M.; Sierra, J.B.; Ruel, K.; Pollet, B.; Do, C.T.; Thévenin, J.; Buffard, D.; Jouanin, L.; Lapierre, C. Redirection of the phenylpropanoid pathway to feruloyl malate in *Arabidopsis* mutants deficient for cinnamoyl-CoA reductase 1. *Planta* 2008, *227*, 943–956. [CrossRef] [PubMed]
12. Moura, J.C.M.S.; Bonine, C.A.V.; Viana, J.D.O.F.; Dornelas, M.C.; Mazzafera, P. Abiotic and biotic stresses and changes in the lignin content and composition in plants. *J. Integr. Plant Biol.* 2010, *52*, 360–376. [CrossRef] [PubMed]
13. Rao, X.; Barros, J. Modeling lignin biosynthesis: a pathway to renewable chemicals. *Trend. Plant Sci.* 2024, *29*(5), 546–559. ISSN 1360-1385.
14. Li, M.; Min, Li.; Wang, L.; Wei, J. Apiaceae medicinal plants: A review of traditional uses, phytochemistry, bolting and flowering, and controlling approaches. 2022. (doi:10.20944/preprints202212.0272.v1)
15. Vanholme, R.; Storme, V.; Vanholme, B.; Sundin, L.; Christensen, J.H.; Goeminne, G.; Halpin, C.; Rohde, A.; Morreel, K.; Boerjan, W. A systems biology view of responses to lignin biosynthesis perturbations in *Arabidopsis*. *Plant Cell* 2012, *24*, 3506–3529. [CrossRef] [PubMed]
16. Gui, J.; Shen, J.; Li, L. Functional characterization of evolutionarily divergent 4-coumarate:coenzyme a ligases in rice. *Plant Physiol.* 2011, *157*, 574–586. [CrossRef] [PubMed]
17. Wagner, A.; Donaldson, L.; Kim, H.; Phillips, L.; Flint, H.; Steward, D.; Torr, K.; Koch, G.; Schmitt, U.; Ralph, J. Suppression of 4-coumarate-CoA ligase in the coniferous gymnosperm *Pinus radiata*. *Plant Physiol.* 2009, *149*, 370–383. [CrossRef] [PubMed]
18. Thevenin, J.; Pollet, B.; Letarnec, B.; Saulnier, L.; Gissot, L.; Maiagrondard, A.; Lapierre, C.; Jouanin, L. The simultaneous repression of CCR and CAD, two enzymes of the lignin biosynthetic pathway, results in sterility and dwarfism in *Arabidopsis thaliana*. *Mol. Plant* 2011, *4*, 70–82. [CrossRef] [PubMed]
19. Van Acker, R.; Déjardin, A.; Desmet, S.; Hoengenaert, L.; Vanholme, R.; Morreel, K.; Laurans, F.; Kim, H.; Santoro, N.; Foster, C.; et al. Different metabolic routes for coniferaldehyde and sinapaldehyde with cinnamyl alcohol dehydrogenase1 deficiency. *Plant Physiol.* 2017, *175*, 1018–1039. [CrossRef] [PubMed]
20. Giordano, A.; Liu, Z.; Panter, S.N.; Dimech, A.M.; Shang, Y.; Wijesinghe, H.; Fulgueras, K.; Ran, Y.; Mouradov, A.; Rochfort, S. Reduced lignin content and altered lignin composition in the warm season forage grass *Paspalum dilatatum* by down-regulation of a Cinnamoyl CoA Reductase Gene. *Transgenic Res.* 2014, *23*, 503–517.
21. Tu, Y.; Rochfort, S.; Liu, Z.; Ran, Y.; Griffith, M.; Badenhorst, P.; Louie, G.V.; Bowman, M.E.; Smith, K.F.; Noel, J.P.; et al. Functional analyses of caffeic acid *O*-methyltransferase and Cinnamoyl-CoA-Reductase genes from perennial ryegrass (*Lolium perenne*). *Plant Cell* 2010, *22*, 3357–3373. [CrossRef] [PubMed]
22. Vanholme, B.; Cesarino, I.; Goeminne, G.; Kim, H.; Marroni, F.; Van Acker, R.; Vanholme, R.; Morreel, K.; Ivens, B.; Pinosio, S.; et al. Breeding with rare defective alleles (BRDA): A natural *Populus nigra HCT* mutant with modified lignin as a case study. *New Phytol.* 2013, *198*, 765–776. [CrossRef] [PubMed]
23. Coleman, H.D.; Park, J.; Nair, R.; Chapple, C.; Mansfield, S.D. RNAi-mediated suppression of p-coumaroyl-CoA 3′-hydroxylase in hybrid poplar impacts lignin deposition and soluble secondary metabolism. *Proc. Natl. Acad. Sci. USA* 2008, *105*, 4501–4506. [CrossRef] [PubMed]
24. Zhang, K.; Bhuiya, M.; Pazo, J.R.; Miao, Y.; Kim, H.; Ralph, J.; Liu, C. An engineered Monolignol 4-*O*-Methyltransferase depresses lignin biosynthesis and confers novel metabolic capability in *Arabidopsis*. *Plant Cell* 2012, *24*, 3135–3152.
25. Wagner, A.; Tobimatsu, Y.; Phillips, L.; Flint, H.; Geddes, B.; Lu, F.; Ralph, J. Syringyl lignin production in conifers: Proof of concept in a Pine tracheary element system. *Proc. Natl. Acad. Sci. USA* 2015, *112*, 6218–6223. [CrossRef] [PubMed]
26. Takeda, Y.; Koshiba, T.; Tobimatsu, Y.; Suzuki, S.; Murakami, S.; Yamamura, M.; Rahman, M.M.; Takano, T.; Hattori, T.; Sakamoto, M.; et al. Regulation of coniferaldehyde 5-hydroxylase expression to modulate cell wall lignin structure in rice. *Planta* 2017, *246*, 1–13. [CrossRef] [PubMed]
27. Vanholme, R.; Cesarino, I.; Rataj, K.; Xiao, Y.; Sundin, L.; Goeminne, G.; Kim, H.; Cross, J.; Morreel, K.; Araujo, P.; et al. Caffeoyl shikimate esterase (CSE) is an enzyme in the lignin biosynthetic pathway in *Arabidopsis*. *Science* 2013, *341*, 1103–1106. [CrossRef] [PubMed]
28. Eloy, N.; Voorend, W.; Lan, W.; Saleme, M.L.; Cesarino, I.; Vanholme, R.; Smith, R.A.; Goeminne, G.; Pallidis, A.; Morreel, K.; et al. Silencing chalcone synthase impedes the incorporation of tricin in lignin and increases lignin content. *Plant Physiol.* 2017, *173*, 998–1016. [CrossRef] [PubMed]
29. Ha, C.M.; Escamilla-Trevino, L.; Yarce, J.C.S.; Kim, H.; Ralph, J.; Chen, F.; Dixon, R.A. An essential role of caffeoyl shikimate esterase in monolignol biosynthesis in *Medicago truncatula*. *Plant J.* 2016, *86*, 363–375. [CrossRef] [PubMed]
30. Saleme, M.; Cesarino, I.; Vargas, L.; Kim, H.; Vanholme, R.; Goeminne, G.; Van, R.A.; Fonseca, F.; Pallidis, A.; Voorend, W.; et al. Silencing caffeoyl shikimate esterase affects lignification and improves saccharification. *Plant Physiol.* 2017, *175*, 1040–1057. [CrossRef] [PubMed]

31. Eudes, A.; Dutta, T.; Deng, K.; Jacquet, N.; Sinha, A.; Benites, V.T.; Eek, B.; Richel, A.; Sattler, S.E.; Northen, T.R.; et al. *SbCOMT* (*Bmr12*) is involved in the biosynthesis of tricin-lignin in sorghum. *PLoS ONE* 2017, *12*, e0178160. [CrossRef] [PubMed]
32. Turlapati, P.V.; Kim, K.W.; Davin, L.B.; Lewis, N.G. The laccase multigene family in *Arabidopsis thaliana*: towards addressing the mystery of their gene function(s). *Planta* 2011, *233*, 439–470. (doi:10.1007/s00425-010-1298-3)
33. Liu, H.; Guo, Z.; Gu, F.; Ke, S.; Sun, D.; Dong, S.; Liu, W.; Huang, M.; Xiao, W.; Yang, G. 4-Coumarate-CoA ligase-like gene *OsAAE3* negatively mediates the rice blast resistance, floret development and lignin biosynthesis. *Front. Plant Sci.* 2016, *7*, 2041. [CrossRef] [PubMed]
34. Yang, Q.; He, Y.; Kabahuma, M.; Chaya, T.; Kelly, A.; Borrego, E.; Bian, Y.; El, K.F.; Yang, L.; Teixeira, P. A gene encoding maize caffeoyl-CoA *O*-methyltransferase confers quantitative resistance to multiple pathogens. *Nat. Genet.* 2017, *49*, 1364–1372.
35. Herrero, J.; Fernándezpérez, F.; Yebra, T.; Novouzal, E.; Pomar, F.; Pedreño, M.Á.; Cuello, J.; Guéra, A.; Estebancarrasco, A.; Zapata, J.M. Bioinformatic and functional characterization of the basic peroxidase 72 from *Arabidopsis thaliana* involved in lignin biosynthesis. *Planta* 2013, *237*, 1599–1612. [CrossRef] [PubMed]
36. Zhao, Q.; Nakashima, J.; Chen, F.; Yin, Y.; Fu, C.; Yun, J.; Shao, H.; Wang, X.; Wang, Z.; Dixon, R.A. Laccase is necessary and nonredundant with peroxidase for lignin polymerization during vascular development in *Arabidopsis*. *Plant Cell* 2013, *25*, 3976–3987. [CrossRef] [PubMed]
37. Li, H.; Yang, Y.; Wang, Z.; Guo, X.; Liu, F.; Jiang, J.; Liu, G. *BpMADS12* gene role in lignin biosynthesis of *Betula platyphylla* Suk by transcriptome analysis. *J. For. Res.* 2016, *27*, 1111–1120. [CrossRef]
38. Shafi, A.; Chauhan, R.; Gill, T.; Swarnkar, M.K.; Sreenivasulu, Y.; Kumar, S.; Kumar, N.; Shankar, R.; Ahuja, P.S.; Singh, A.K. Expression of *SOD* and *APX* genes positively regulates secondary cell wall biosynthesis and promotes plant growth and yield in *Arabidopsis* under salt stress. *Plant Mol. Biol.* 2015, *87*, 615–631. [CrossRef] [PubMed]
39. Berthet, S.; Demontcaulet, N.; Pollet, B.; Bidzinski, P.; Cezard, L.; Bris, P.L.; Borrega, N.; Herve, J.; Blondet, E.; Balzergue, S.; et al. Disruption of LACCASE4 and 17 results in tissue-specific alterations to lignification of *Arabidopsis thaliana* stems. *Plant Cell* 2011, *23*, 1124–1137. [CrossRef] [PubMed]
40. Kim, Y.; Kim, C.Y.; Song, W.; Park, D.; Kwon, S.; Lee, H.; Bang, J.; Kwak, S. Overexpression of sweetpotato swpa4 peroxidase results in increased hydrogen peroxide production and enhances stress tolerance in tobacco. *Planta* 2008, *227*, 867–881. [CrossRef] [PubMed]
41. Shigeto, J.; Itoh, Y.; Hirao, S.; Ohira, K.; Fujita, K.; Tsutsumi, Y. Simultaneously disrupting *AtPrx2*, *AtPrx25* and *AtPrx71* alters lignin content and structure in *Arabidopsis* stem. *J. Integr. Plant Biol.* 2015, *57*, 349–356. [CrossRef] [PubMed]
42. Cosio, C.; Ranocha, P.; Francoz, E.; Burlat, V.; Zheng, Y.; Perry, S.E.; Ripoll, J.J.; Yanofsky, M.; Dunand, C. The class III peroxidase PRX17 is a direct target of the MADS-box transcription factor agamous-like15 (AGL15) and participates in lignified tissue formation. *New Phytol.* 2016, *213*, 250–263. [CrossRef] [PubMed]
43. Wang, J.; Wang, C.; Zhu, M.; Yu, Y.; Zhang, Y.; Wei, Z. Generation and characterization of transgenic poplar plants overexpressing a cotton laccase gene. *Plant Cell Tissue Organ Cult.* 2008, *93*, 303–310. [CrossRef]
44. Wang, Y.; Bouchabke-Coussa, O.; Lebris, P.; Antelme, S.; Soulhat, C.; Gineau, E.; Dalmais, M.; Bendahmane, A.; Morin, H.; Mouille, G.; et al. LACCASE5 is required for lignification of the *Brachypodium distachyon* Culm. *Plant Physiol.* 2015, *168*, 192–204. [CrossRef] [PubMed]
45. Legay, S.; Lacombe, E.; Goicoechea, M.; Briere, C.; Seguin, A.; Mackay, J.; Grimapettenati, J. Molecular characterization of EgMYB1, a putative transcriptional repressor of the lignin biosynthetic pathway. *Plant Sci.* 2007, *173*, 542–549. [CrossRef]
46. Legay, S.; Sivadon, P.; Blervacq, A.; Pavy, N.; Baghdady, A.; Tremblay, L.; Levasseur, C.; Ladouce, N.; Lapierre, C.; Seguin, A.; et al. EgMYB1, an R2R3 MYB transcription factor from *eucalyptus* negatively regulates secondary cell wall formation in *Arabidopsis* and poplar. *New Phytol.* 2010, *188*, 774–786. [CrossRef] [PubMed]
47. Gallego-Giraldo, L.; Shadle, G.; Shen, H.; Barrosrios, J.; Fresquet, C.S.; Wang, H.; Dixon, R.A. Combining enhanced biomass density with reduced lignin level for improved forage quality. *Plant Biotechnol. J.* 2016, *14*, 895–904. [CrossRef] [PubMed]
48. National Science Foundation. https://nsf.gov/news/mmg/mmg_disp.jsp?med_id=62375&f (NSF-2024)
49. Barros, J.; Serk, H.; Granlund, I.; Pesquet, E. The cell biology of lignification in higher plants. *Ann. Bot.* 2015, *115*, 1053–1074. [CrossRef] [PubMed]
50. Berry, P.M.; Sterling, M.; Spink, J.H.; Baker, C.J.; Sylvesterbradley, R.; Mooney, S.J.; Tams, A.R.; Ennos, A.R. Understanding and reducing lodging in cereals. *Adv. Agron.* 2004, *84*, 217–271.

51. Tanaka, K.; Murata, K.; Yamazaki, M.; Onosato, K.; Miyao, A.; Hirochika, H. Three distinct rice cellulose synthase catalytic subunit genes required for cellulose synthesis in the secondary wall. *Plant Physiol.* 2003, *133*, 73–83. [CrossRef] [PubMed]
52. Li, Q.; Fu, C.; Liang, C.; Ni, X.; Zhao, X.; Chen, M.; Ou, L. Crop lodging and the roles of lignin, cellulose, and hemicellulose in lodging resistance. *Agronomy* 2022, *12*, 1795. (doi: 10.3390/agronomy12081795\)
53. Islam, M.S.; Peng, S.; Visperas, R.M.; Ereful, N.; Bhuiya, M.S.U.; Julfiquar, A.W. Lodging-related morphological traits of hybrid rice in a tropical irrigated ecosystem. *Field Crop. Res.* 2007, *101*, 240–248. [CrossRef]
54. Hu, D.; Liu, X.B.; She, H.Z.; Gao, Z.; Ruan, R.W.; Wu, D.Q.; Yi, Z.L. The lignin synthesis related genes and lodging resistance of *Fagopyrum esculentum*. *Biol. Plant.* 2017, *61*, 1–9. [CrossRef]
55. Dorairaj, D.; Ismail, M.R.; Sinniah, U.R.; Tan, K.B. Influence of silicon on growth, yield, and lodging resistance of MR219, a lowland rice of Malaysia. *J. Plant Nutr.* 2017, *40*. [CrossRef]
56. Jie, K.; Sun, Y.; Zhou, M.; Zhang, P.; Zuo, Q.; Wu, J.; Zhou, G. The effect of nitrogen application and planting density on the radiation use efficiency and the stem lignin metabolism in rapeseed (*Brassica napus* L.). *Field Crop. Res.* 2016, *199*, 89–98.
57. Nizam, A.; Thattantavide, A.; Kumar, A. Gene expression pattern, lignin deposition and root cell wall modification of developing mangrove propagules under salinity stress. *J. Plant Growth Regul.* 2024, *43*(9), 3088–3104.
58. Jannoey, P.; Pongprasert, W.; Lumyong, S.; Roytrakul, S.; Nomura, M. Comparative proteomic analysis of two rice cultivars (*Oryza sativa* L.) contrasting in brown planthopper (BPH) stress resistance. *Plant Omics* 2015, *8*, 96–105.
59. Kaur, S.; Samota, M.K.; Choudhary, M.; et al. How do plants defend themselves against pathogens-biochemical mechanisms and genetic interventions. *Physiol. Mol. Biol. Plants* 2022, *28*, 485–504. (doi: 10.1007/s12298-022-01146-y)
60. Duan, C.; Yu, J.; Bai, J.; Zhu, Z.; Wang, X. Induced defense responses in rice plants against small brown planthopper infestation. *Crop J.* 2014, *2*, 55–62. [CrossRef]
61. Jannoey, P.; Channei, D.; Kotcharerk, J.; Pongprasert, W.; Nomura, M. Expression analysis of genes related to rice resistance against brown planthopper, *Nilaparvata lugens*. *Rice Sci.* 2017, *24*, 163–172. [CrossRef]
62. Wang, C.; Hu, D.; Liu, X.; She, H.; Ruan, R.; Yang, H.; Yi, Z.; Wu, D. Effects of uniconazole on the lignin metabolism and lodging resistance of culm in common buckwheat (*Fagopyrum esculentum* M.). *Field Crop. Res.* 2015, *180*, 46–53. [CrossRef]
63. Wang, Y.; Sheng, L.; Zhang, H.; Du, X.; An, C.; Xia, X.; Chen, F.; Jiang, J.; Chen, S. *CmMYB19* overexpression improves aphid tolerance in Chrysanthemum by promoting lignin synthesis. *Int. J. Mol. Sci.* 2017, *18*, 619. [CrossRef] [PubMed]
64. Fujimoto, T.; Mizukubo, T.; Abe, H.; Seo, S. Sclareol induces plant resistance to root-knot nematode partially through ethylene-dependent enhancement of lignin accumulation. *Mol. Plant Microbe Interact.* 2015, *28*, 398–407. [CrossRef] [PubMed]
65. Tianpei, X.; Li, D.; Qiu, P.; Luo, J.; Zhu, Y.; Li, S. Scorpion peptide LqhIT2 activates phenylpropanoid pathways via jasmonate to increase rice resistance to rice leafrollers. *Plant Sci.* 2015, *230*, 1–11. [CrossRef] [PubMed]
66. Guo, H.; Wang, Y.; Wang, L.; Hu, P.; Wang, Y.; Jia, Y.; Zhang, C.; Zhang, Y.; Zhang, Y.; Wang, C. Expression of the MYB transcription factor gene *BplMYB46* affects abiotic stress tolerance and secondary cell wall deposition in *Betula platyphylla*. *Plant Biotechnol. J.* 2017, *15*, 107–121. [CrossRef] [PubMed]
67. Miedes, E.; Vanholme, R.; Boerjan, W.; Molina, A. The role of the secondary cell wall in plant resistance to pathogens. *Front. Plant Sci.* 2014, *5*, 358. [CrossRef] [PubMed]
68. Ma, Q.H.; Zhu, H.H.; Han, J.Q. Wheat ROP proteins modulate defense response through lignin metabolism. *Plant Sci.* 2017, *262*, 32–38. [CrossRef] [PubMed]
69. Guo, X.; Zhang, S.; Shan, X.Q. Adsorption of metal ions on lignin. *J. Hazard. Mater.* 2008, *151*, 134–142. [CrossRef] [PubMed]
70. Wang, G.F.; He, Y.; Strauch, R.; Olukolu, B.A.; Nielsen, D.; Li, X.; Balint-Kurti, P.J. Maize homologs of Hydroxycinnamoyltransferase, a key enzyme in lignin biosynthesis, bind the nucleotide binding leucine-rich repeat Rp1 proteins to modulate the defense response. *Plant Physiol.* 2015, *169*, 2230. [PubMed]
71. Chen, G.; Liu, Y.; Wang, R.; Zhang, J.; Owens, G. Cadmium adsorption by willow root: The role of cell walls and their subfractions. *Environ. Sci. Pollut. Res. Int.* 2013, *20*, 5665–5672. [CrossRef] [PubMed]
72. Dalcorso, G.; Farinati, S.; Furini, A. Regulatory networks of cadmium stress in plants. *Plant Signal. Behav.* 2010, *5*, 663–667. [CrossRef] [PubMed]

73. Demirbas, A. Adsorption of lead and cadmium ions in aqueous solutions onto modified lignin from alkali glycerol delignication. *J. Hazard. Mater.* 2004, *109*, 221–226. [CrossRef] [PubMed]
74. Bhuiyan, N.H.; Liu, W.; Liu, G.; Selvaraj, G.; Wei, Y.; King, J. Transcriptional regulation of genes involved in the pathways of biosynthesis and supply of methyl units in response to powdery mildew attack and abiotic stresses in wheat. *Plant Mol. Biol.* 2007, *64*, 305–318. [CrossRef] [PubMed]
75. Kováčik, J.; Bačkor, M. Phenylalanine ammonia-lyase and phenolic compounds in chamomile tolerance to cadmium and copper excess. *Water Air Soil Pollut.* 2007, *185*, 185–193. [CrossRef]
76. Tahara, K.; Norisada, M.; Hogetsu, T.; Kojima, K. Aluminum tolerance and aluminum-induced deposition of callose and lignin in the root tips of *Melaleuca* and *Eucalyptus* species. *J. For. Res* 2005, *10*, 325–333. [CrossRef]
77. Lima, J.D.; Mazzafera, P.; Moraes, W.D.S.; Silva, R.B.D. Chá: Aspectos relacionados à qualidade e perspectivas tea: Aspects related to the quality and prospects. *Ciênc. Rural* 2009, *39*, 1258–1266. [CrossRef]
78. Ghanati, F.; Morita, A.; Yokota, H. Deposition of suberin in roots of soybean induced by excess boron. *Plant Sci.* 2005, *168*, 397–405. [CrossRef]
79. Zhao, J. Flavonoid transport mechanisms: how to go, and with whom. *Trends Plant Sci.* 2015, *20*, 576–585. (doi:10.1016/j.tplants.2015.06.007)
80. Li, M.; Pu, Y.; Yoo, C.G.; Ragauskas, A.J. The occurrence of tricin and its derivatives in plants. *Green Chem.* 2016, *18*, 1439–1454. (doi:10.1039/C5GC03062E)
81. Schuetz, M.; Benske, A.; Smith, R.A.; Watanabe, Y.; Tobimatsu, Y.; Ralph, J.; Demura, T.; Ellis, B.; Samuels, A.L. Laccases direct lignification in the discrete secondary cell wall domains of protoxylem. *Plant Physiol.* 2014, *166*, 798–807. (doi:10.1104/pp.114.245597).
82. Barros, J.; Serk, H.; Granlund, I; Pesquet, E. The cell biology of lignification in higher plants. *Ann. Bot.* 2015, *115*, 1053–1074. (doi:10.1093/aob/mcv046)
83. Duroux, L.; Welinder, K.G. The peroxidase gene family in plants: a phylogenetic overview. *J. Mol. Evol.* 2003, *57*, 397–407. (doi:10.1007/s00239-003-2489-3)
84. Zhao, Q.; et al. Laccase is necessary and nonredundant with peroxidase for lignin polymerization during vascular development in *Arabidopsis thaliana*. *Plant Cell* 2013, *25*, 3976–3987. (doi:10.1105/tpc.113.117770)
85. Chou, E.Y.; Schuetz, M.; Hoffmann, N.; Watanabe, Y.; Sibout, R.; Samuels, A.L. Distribution, mobility, and anchoring of lignin-related oxidative enzymes in Arabidopsis secondary cell walls. *J. Exp. Bot.* 2018, 69, 1849–1859. (doi:10.1093/jxb/ery067)
86. Gallego-Giraldo, L.; Shadle, G.; Shen, H.; Barros-Rios, J.; Fresquet Corrales, S.; Wang, Dixon R.A. Combining enhanced biomass density with reduced lignin level for improved forage quality. *Plant Biotechnol. J.* 2016, *14*, 895–904. (doi:10.1111/ pbi.12439).
87. Kärkönen, A.; Kuchitsu, K. Reactive oxygen species and cell wall metabolism and development in plants. *Phytochemistry* 2015, *112*, 22–32. (doi:10.1016/j.phytochem.2014.09.016)
88. Lee, Y.; Rubio, M.C.; Alassimone, J.; Geldner, N. A mechanism for localized lignin deposition in the endodermis. *Cell* 2013, *153*, 402–412. (doi:10.1016/j.cell. 2013.02.045)
89. Zhao, Q.; et al. Pinoresinol reductase 1 impacts lignin distribution during secondary cell wall biosynthesis in Arabidopsis. *Phytochemistry* 2014, *112*, 170–178. (doi:10.1016/j.phytochem.2014.07.008)
90. Ralph, J.; Grabber, J.H.; Hatfield, R.D. Ligninferulate cross-links in grasses: active incorporation of ferulate polysaccharide esters into ryegrass lignins. *Carbohydr. Res.* 1995, *275*, 167–178. (doi:10.1016/0008-6215(95)00237-N)
91. Lam, P.Y.; Tobimatsu, Y.; Takeda, Y.; Suzuki, S.; Yamamura, M.; Umezawa, T.; Lo, C. Disrupting flavone synthase II alters lignin and improves biomass digestibility. *Plant Physiol.* 2017, *174*, 972–985. (doi:10.1104/pp.16.01973)
92. Eloy, N.B.; et al. Silencing chalcone synthase in maize impedes the incorporation of tricin into lignin and increases lignin content. *Plant Physiol.* 2017, *173*, 998–1016. (doi:10.1104/pp.16.01108)
93. Wang, P.; Guo, L.; Jaini, R.; Klempen, A.; McCoy, R.; Morgan, J.A.; Dudareva, N.; Chapple, C. A 13C isotope labeling method for the measurement of lignin metabolic flux in Arabidopsis stems. *Plant Methods* 2018, *14*, 51. (doi:10.1186/s13007-018-0318-3)
94. Miao, Y-C.; Liu, C-J. ATP-binding cassette-like transporters are involved in the transport of lignin precursors across plasma and vacuolar membranes. *Proc. Natl Acad. Sci. USA* 2010, *107*, 22 728–22 733. (doi:10.1073/pnas.1007747108)
95. Meester, B.E.; Vanholme, R.; Mota, T.; Boerjan, W. *Plant Communication* 2022, *22*(6), 100465.
96. Dixon, R.A.; Barros, J. Lignin biosynthesis: old roads revisited and new roads explored. *Open Biol.* 2019, *9*, 190215. (doi: 10.1098/rsob.190215)
97. Tlak Gajger, I.; Nejedli, S.; Kozarić, Z. The effectof Nozevit on leucine aminopeptidase and esteraseactivity in the midgut of honey bees (*Apis melli*fera). *Vet. Med.* 2013, *58*(8), 422–429.31. (doi: 10.17221/6982-VETMED)

98. Tlak Gajger, I.; Sušec, P. Efficacy of varroacidalfood additive appliance during summer treatment ofhoneybee colonies (*Apis mellifera*). *Vet. Arh.* 2019, *89*(1), 87–96.38. (doi: 10.24099/vet.arhiv.0441)
99. Tlak Gajger, I.; Tomljanović, Z.; Petrinec, Z. Monitoring health status of Croatian honey bee coloniesand possible reasons for winter losses. *J. Apicul. Res.* 2010, *49*(1), 107–108. (doi: 10.3896/IBRA.1.49.1.1939)
100. Tlak Gajger, I.; Vugrek, O.; Grilec, D.; Petrinec, Z. Prevalence and distribution of Nosemaceranae in Croatian honeybee colonies. *Vet. Med.* 2010b, *55*(9), 457–462.40. (doi: 10.17221/2983-VETMED)
101. Ullah, A.; Tlak Gajger, I.; Majoros, A.; Dar, S.A.; Khan, S.; Kalimullah; Haleem, S.A.; Nasir, K.M.; Hussain, R.; Khan, H.U.; Hameed, M.; Anjum, S.I. Viral impacts on honey bee populations: A review. *Saudi J. Biol. Sci.* 2021 Jan, *28*(1), 523–530. (doi: 10.1016/j.sjbs.2020.10.037). Epub 2020 Oct 28. PMID: 33424335; PMCID: PMC7783639.
102. Gajger, I.T.; Sakač, M.; Gregorc, A. Impact of thiamethoxam on honey bee queen (*Apis mellifera carnica*) reproductive morphology and physiology. *Bull. Environ. Contam. Toxicol.* 2017, *99*, 297–302. (doi: 10.1007/s00128-017-2144-0)
103. Liu, Q.; Luo, L.; Zheng, L. Lignins: Biosynthesis and biological functions in plants. *Int. J. Mol. Sci.* 2018 Jan 24;*19*(2):335. (doi: 10.3390/ijms19020335). PMID: 29364145; PMCID: PMC5855557.
104. Becker, J.; Wittmann. *Biotechnol. Adv.* 2019. (doi: 10.1016/j.biotechadv.2019.02.016)
105. Tlak Gajger, I.; Kolodziejek, J.; Bakonyi, T.; et al. Prevalence and distribution patterns of seven different honeybee viruses in diseased colonies: a case study from Croatia. *Apidologie* 2014, *45*, 701–706. (doi: 10.1007/s13592-014-0287-0)
106. Tlak Gajger, I.; Bičak, J.; Belužić, R. Theoccurrence of honeybee viruses in apiaries in the Koprivnica-Križevci district in Croatia. *Vet. Arhiv* 2014a, 84, 421–428.29.
107. Tlak Gajger, I.; Svečnjak, L.; Bubalo, D.; Žorat, T. Control of *Varroa destructor* mite infestations at experimental apiaries situated in Croatia. *Diversity* 2020, *12*, 12. (doi: 10.3390/d12010012)
108. Tlak Gajger, I.; Kozarić, Z.; Berta, D.; Nejedli, S.; Petrinec, Z. Effect of the herbal preparationNozevit on the mid-gut structure of honeybees (Apismellifera) infected with *Nosema sp.* spores. *Vet. Med.* 2011, *56*(7), 343–350.30. (doi: 10.17221/1587-VETMED)
109. Tlak Gajger, I.; Tomljanović, Z.; Stanisavljević, Lj. An environmentally friendly approach tothe control of *Varooa destructor* mite and *Nosema ceranaedisease* in Carnolian honeybee (*Apis melliferacarnica*) colonies. *Arch. Biol. Sci.* 2013a, *65*(4), 1585–1592.33. (doi: 10.2298/ABS1304585G)
110. Tlak Gajger, I.; Ribaric, J.; Matak, M.; Svecnjak, L.; Kozaric, Z.; Nejedli, S.; SmodisSkerl, I. M. Zeolite clinoptilolite as a dietary supplement and remedyfor honeybee (Apis mellifera) colonies. *Vet. Med.* 2015, *60*(12), 696–705.34. (doi: 10.17221/8584-VETMED)
111. Tlak Gajger, I.; Ribarić-, J.; Smodiš Škerl, M.; Vlainić, J.; Sikirić, P. Stable gastric pentadecapeptide BPC 157 in honeybee (*Apis mellifera*) therapy, to control *Nosema ceranae* invasions in apiary conditions. *J. Vet. Pharmacol. Ther.* 2018 Aug, *41*(4), 614–621. (doi: 10.1111/jvp.12509). Epub 2018 Apr 23. PMID: 29682749.
112. Tlak Gajger, I.; Vlainić, J.; Šoštarić, P.; Prešern, J.; Bubnič, J.; Smodiš Škerl, M. I. Effects onsome therapeutical, biochemical, and immunologicalparameters of honey bee (*Apis mellifera*) exposed toprobiotic treatments, in field and laboratory conditions. *Insects* 2020, *11*(9), 638. (doi: 10.3390/insects11090638.36)

Index